FORMULAS FROM ALGEBRA

Exponents

$$a^m a^n = a^{m+n}$$

$$(a^m)^n = a^{mn}$$

$$\frac{a^m}{a^n} = a^{m-n}$$

$$(ab)^n = a^n b^n$$

$$\left(\frac{a}{b}\right)^n = \frac{a^n}{b^n}$$

Radicals

$$(\sqrt[n]{a})^n = a$$

$$\sqrt[n]{a^n} = a, \text{ if } a \geq 0$$

$$\sqrt[n]{ab} = \sqrt[n]{a}\,\sqrt[n]{b}$$

$$\sqrt[n]{\frac{a}{b}} = \frac{\sqrt[n]{a}}{\sqrt[n]{b}}$$

Logarithms

$$\log_a MN = \log_a M + \log_a N$$

$$\log_a (M/N) = \log_a M - \log_a N$$

$$\log_a (N^p) = p \log_a N$$

Factoring Formulas

$$x^2 - y^2 = (x - y)(x + y)$$

$$x^3 - y^3 = (x - y)(x^2 + xy + y^2)$$

$$x^3 + y^3 = (x + y)(x^2 - xy + y^2)$$

$$x^2 + 2xy + y^2 = (x + y)^2$$

$$x^2 - 2xy + y^2 = (x - y)^2$$

$$x^3 + 3x^2 y + 3xy^2 + y^3 = (x + y)^3$$

Binomial Formula

$$(x + y)^n = {}_nC_0 x^n y^0 + {}_nC_1 x^{n-1} y^1 + \cdots + {}_nC_{n-1} x^1 y^{n-1} + {}_nC_n x^0 y^n$$

Quadratic Formula

The solutions to $ax^2 + bx + c = 0$ are $x = \dfrac{-b \pm \sqrt{b^2 - 4ac}}{2a}$

Complex Numbers

Multiplication: $\quad (a + bi)(c + di) = (ac - bd) + (ad + bc)i$

Polar form: $\quad a + bi = r(\cos \theta + i \sin \theta)$ where $r = \sqrt{a^2 + b^2}$

Powers: $\quad [r(\cos \theta + i \sin \theta)]^n = r^n(\cos n\theta + i \sin n\theta)$

Roots: $\quad \sqrt[n]{r}\left[\cos\left(\dfrac{\theta + k \cdot 360°}{n}\right) + i \sin\left(\dfrac{\theta + k \cdot 360°}{n}\right)\right] \quad k = 0, 1, 2, \cdots, n-1$

GEOMETRY

Triangles

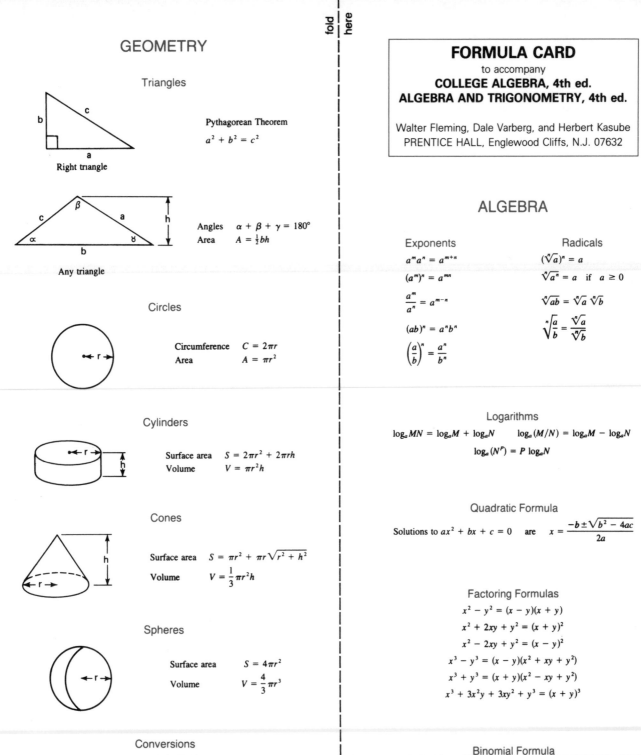

Right triangle

Pythagorean Theorem

$$a^2 + b^2 = c^2$$

Any triangle

Angles $\quad \alpha + \beta + \gamma = 180°$

Area $\quad A = \frac{1}{2}bh$

Circles

Circumference $\quad C = 2\pi r$

Area $\qquad\qquad A = \pi r^2$

Cylinders

Surface area $\quad S = 2\pi r^2 + 2\pi rh$

Volume $\qquad\quad V = \pi r^2 h$

Cones

Surface area $\quad S = \pi r^2 + \pi r\sqrt{r^2 + h^2}$

Volume $\qquad\quad V = \frac{1}{3}\pi r^2 h$

Spheres

Surface area $\quad S = 4\pi r^2$

Volume $\qquad\quad V = \frac{4}{3}\pi r^3$

Conversions

1 inch = 2.54 centimeters
1 liter = 1000 cubic centimeters
1 kilogram = 2.20 pounds
1 kilometer = .62 miles
1 liter = 1.057 quarts
1 pound = 453.6 grams
π radians = 180 degrees

FORMULA CARD
to accompany
COLLEGE ALGEBRA, 4th ed.
ALGEBRA AND TRIGONOMETRY, 4th ed.

Walter Fleming, Dale Varberg, and Herbert Kasube
PRENTICE HALL, Englewood Cliffs, N.J. 07632

ALGEBRA

Exponents

$$a^m a^n = a^{m+n}$$

$$(a^m)^n = a^{mn}$$

$$\frac{a^m}{a^n} = a^{m-n}$$

$$(ab)^n = a^n b^n$$

$$\left(\frac{a}{b}\right)^n = \frac{a^n}{b^n}$$

Radicals

$$(\sqrt[n]{a})^n = a$$

$$\sqrt[n]{a^n} = a \quad \text{if} \quad a \geq 0$$

$$\sqrt[n]{ab} = \sqrt[n]{a}\,\sqrt[n]{b}$$

$$\sqrt[n]{\frac{a}{b}} = \frac{\sqrt[n]{a}}{\sqrt[n]{b}}$$

Logarithms

$$\log_a MN = \log_a M + \log_a N \qquad \log_a(M/N) = \log_a M - \log_a N$$

$$\log_a(N^P) = P\log_a N$$

Quadratic Formula

Solutions to $ax^2 + bx + c = 0 \quad$ are $\quad x = \dfrac{-b \pm \sqrt{b^2 - 4ac}}{2a}$

Factoring Formulas

$$x^2 - y^2 = (x - y)(x + y)$$
$$x^2 + 2xy + y^2 = (x + y)^2$$
$$x^2 - 2xy + y^2 = (x - y)^2$$
$$x^3 - y^3 = (x - y)(x^2 + xy + y^2)$$
$$x^3 + y^3 = (x + y)(x^2 - xy + y^2)$$
$$x^3 + 3x^2 y + 3xy^2 + y^3 = (x + y)^3$$

Binomial Formula

$$(x + y)^n = {}_nC_0 x^n y^0 + {}_nC_1 x^{n-1}y^1 + \cdots + {}_nC_{n-1}x^1 y^{n-1} + {}_nC_n x^0 y^n$$

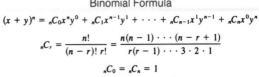

$${}_nC_r = \frac{n!}{(n - r)!\,r!} = \frac{n(n - 1)\cdots(n - r + 1)}{r(r - 1)\cdots 3 \cdot 2 \cdot 1}$$

$${}_nC_0 = {}_nC_n = 1$$

TRIGONOMETRY

Basic Identities

$$\tan t = \frac{\sin t}{\cos t} \qquad \cot t = \frac{\cos t}{\sin t} \qquad \cot t = \frac{1}{\tan t}$$

$$\sec t = \frac{1}{\cos t} \qquad \csc t = \frac{1}{\sin t} \qquad \sin^2 t + \cos^2 t = 1$$

$$1 + \tan^2 t = \sec^2 t \qquad\qquad 1 + \cot^2 t = \csc^2 t$$

Confunction Identities

$$\sin\left(\frac{\pi}{2} - t\right) = \cos t \qquad \cos\left(\frac{\pi}{2} - t\right) = \sin t \qquad \tan\left(\frac{\pi}{2} - t\right) = \cot t$$

Odd-even Identities

$$\sin(-t) = -\sin t \qquad \cos(-t) = \cos t \qquad \tan(-t) = -\tan t$$

Addition Formulas

$$\sin(s + t) = \sin s \cos t + \cos s \sin t \qquad \sin(s - t) = \sin s \cos t - \cos s \sin t$$

$$\cos(s + t) = \cos s \cos t - \sin s \sin t \qquad \cos(s - t) = \cos s \cos t + \sin s \sin t$$

$$\tan(s + t) = \frac{\tan s + \tan t}{1 - \tan s \tan t} \qquad \tan(s - t) = \frac{\tan s - \tan t}{1 + \tan s \tan t}$$

Double Angle Formulas

$$\sin 2t = 2 \sin t \cos t$$

$$\tan 2t = \frac{2 \tan t}{1 - \tan^2 t}$$

$$\cos 2t = \cos^2 t - \sin^2 t = 1 - 2 \sin^2 t = 2 \cos^2 t - 1$$

Half Angle Formulas

$$\sin\frac{t}{2} = \pm\sqrt{\frac{1 - \cos t}{2}} \qquad \cos\frac{t}{2} = \pm\sqrt{\frac{1 + \cos t}{2}} \qquad \tan\frac{t}{2} = \frac{1 - \cos t}{\sin t}$$

Product Formulas

$$2 \sin s \cos t = \sin(s + t) + \sin(s - t)$$
$$2 \cos s \cos t = \cos(s + t) + \cos(s - t)$$
$$2 \cos s \sin t = \sin(s + t) - \sin(s - t)$$
$$2 \sin s \sin t = \cos(s - t) - \cos(s + t)$$

Factoring Formulas

$$\sin s + \sin t = 2 \cos\frac{s - t}{2} \sin\frac{s + t}{2}$$

$$\cos s + \cos t = 2 \cos\frac{s + t}{2} \cos\frac{s - t}{2}$$

$$\sin s - \sin t = 2 \cos\frac{s + t}{2} \sin\frac{s - t}{2}$$

$$\cos s - \cos t = -2 \sin\frac{s + t}{2} \sin\frac{s - t}{2}$$

Laws of Sines and Cosines

$$\frac{\sin \alpha}{a} = \frac{\sin \beta}{b} = \frac{\sin \gamma}{c}$$

$$a^2 = b^2 + c^2 - 2bc \cos \alpha$$

$$\sin t = \sin \theta = y = \frac{b}{r}$$

$$\cos t = \cos \theta = x = \frac{a}{r}$$

$$\tan t = \tan \theta = \frac{y}{x} = \frac{b}{a}$$

$$\cot t = \cot \theta = \frac{x}{y} = \frac{a}{b}$$

Graphs

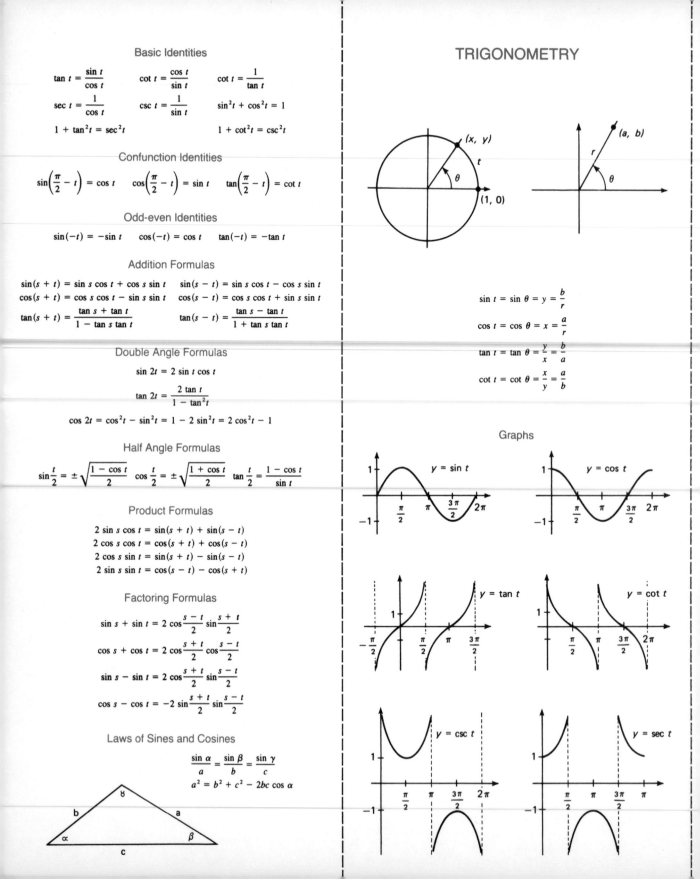

Algebra and Trigonometry

Fourth Edition

Algebra and Trigonometry
A PROBLEM SOLVING APPROACH

Walter Fleming
Hamline University

Dale Varberg
Hamline University

Herbert Kasube
Bradley University

Prentice Hall
Englewood Cliffs, New Jersey 07632

Library of Congress Cataloging-in-Publication Data

Fleming, Walter.
    Algebra and trigonometry / Walter Fleming, Dale Varberg, Herbert
Kasube.—4th ed.
        p.   cm.
    Includes bibliographical references and index.
    ISBN 0-13-028911-6
    1. Algebra.   2. Trigonometry.   I. Varberg, Dale E.   II. Kasube,
Herbert.   III. Title.
QA154.2.F52   1992
512'.13—dc20   91-24122
                                                            CIP

Acquisition Editor: Priscilla McGeehon
Editor-in-Chief: Tim Bozik
Development Editor: Leo Gaffney
Production Editor: Edward Thomas
Marketing Manager: Paul Banks
Designer: Judith A. Matz-Coniglio
Cover Designer: Marjory Dressler
Prepress Buyer: Paula Massenaro
Manufacturing Buyer: Lori Bulwin
Supplements Editor: Susan Black
Editorial Assistant: Marisol L. Torres
Page Layout: Meryl Poweski

 © 1992 by Prentice-Hall, Inc.
A Simon & Schuster Company
Englewood Cliffs, New Jersey 07632

The material in this book has been previously published in portions of Fleming and Varberg's
Algebra and Trigonometry, 3rd Ed., *College Algebra,* 3rd Ed.,
and *Precalculus Mathematics,* 2nd Ed.

Parts of Sections 12-8 and 12-9 are taken from A. Wayne Roberts and Dale E. Varberg:
Faces of Mathematics (1978), Sections 6-1 and 6-2, and used by permission of
Harper & Row, Publishers, Inc.

Credits for quotations used in text appear on page 657.

Printed in the United States of America
10   9   8   7   6   5   4   3   2   1

ISBN 0-13-028911-6

Prentice-Hall International (UK) Limited, *London*
Prentice-Hall of Australia Pty. Limited, *Sydney*
Prentice-Hall Canada Inc., *Toronto*
Prentice-Hall Hispanoamericana, S.A., *Mexico*
Prentice-Hall of India Private Limited, *New Delhi*
Prentice-Hall of Japan, Inc., *Tokyo*
Simon & Schuster Asia Pte. Ltd., *Singapore*
Editora Prentice-Hall do Brasil, Ltda., *Rio de Janeiro*

Contents

4 Coordinates and Curves   129

5 Functions and Their Graphs   172

6 Exponential and Logarithmic Functions   220

7 The Trigonometric Functions 260

8 Trigonometric Identities and Equations 302

9 Applications of Trigonometry 337

10 Theory of Polynomial Equations 386

11 Systems of Equations and Inequalities 417

12 Sequences, Counting Problems, and Probability 473

13 Analytic Geometry 539

Appendix 596

Answers to Odd-Numbered Problems 607

Index of Teaser Problems 659

Index of Names and Subjects 661

George Polya

A great discovery solves a great problem but there is a grain of discovery in the solution of any problem. Your problem may be modest; but if it challenges your curiosities and brings into play your inventive faculties, and if you solve it by your own means, you may experience the tension and enjoy the triumph of discovery. Such experiences at a susceptible age may create a taste for mental work and leave their imprint on the mind and character for a life time.

—How to Solve It (p. v)

Solving a problem is similar to building a house. We must collect the right material, but collecting the material is not enough; a heap of stones is not yet a house. To construct the house or the solution, we must put together the parts and organize them into a purposeful whole.

—Mathematical Discovery (vol. 1, p. 66)

You turn the problem over and over in your mind; try to turn it so it appears simpler. The aspect of the problem you are facing at this moment may not be the most favorable. Is the problem as simply, as clearly, as suggestively expressed as possible? Could you restate the problem?

—Mathematical Discovery (vol. 2, p. 80)

We can scarcely imagine a problem absolutely new, unlike and unrelated to any formerly solved problem; but, if such a problem could exist, it would be insoluable. In fact, when solving a problem, we should always profit from previously solved problems, using their result, or their method, or the experience we acquired solving them. . . . Have you seen it before? Or have you seen the same problem in slightly different form?

—How to Solve It (p. 98)

An insect tries to escape through the windowpane, tries the same hopeless thing again and again, and does not try the next window which is open and through which it came into the room. A mouse may act more intelligently; caught in a trap, he tries to squeeze between two bars, then between the next two bars, then between other bars; he varies his trials, he explores various possibilities. A man is able, or should be able, to vary his trials more intelligently, to explore the various possibilities with more understanding, to learn by his errors and shortcomings. "Try, try again" is popular advice. It is good advice. The insect, the mouse, and the man follow it; but if one follows it with more success than the others it is because he varies his problem more intelligently.

—How to Solve It (p. 209)

These quotations are taken from George Polya, *How to Solve It*, Second Edition (Garden City, NY: Doubleday & Company, Inc., 1957) and George Polya, *Mathematical Discovery*, vols. 1 and 2 (New York: John Wiley & Sons, Inc., 1962).

Preface

This edition of *Algebra and Trigonometry* builds on the strengths of its predecessors.

- **Writing style:** informal but not sloppy
- **Section openers:** an anecdote, quotation, cartoon, or problem
- **Cautions:** warning students of common errors
- **Problem-solving emphasis:** in the spirit of George Polya
- **Carefully graded problem sets:** culminating in a TEASER Problem
- **Chapter reviews:** to help students prepare for tests
- **Attractive design:** featuring open format and use of color
- **Formula card:** a memory aid that can be removed

New to this edition. Following the advice of many reviewers, we have placed all examples in the text proper rather than in the problem sets as in earlier editions. Since this required major changes, we took the opportunity to rewrite every section, adding examples and figures and thereby improving clarity. We have also reworked the section problems sets, which are now divided into two parts. Part A: Skills and Techniques is closely tied to the examples of the section. Part B: Applications and Extensions asks the student to integrate the various techniques, to apply them in real-life situations, and to extend them in novel and challenging ways. Answers to odd-numbered problems can be found at the end of the book. We have also reworked and greatly expanded the chapter review problem sets. Our goal for them is to provide students with a versatile tool to use in preparing for tests. Note that there are answers to all the review problems in the answer key at the back of the book.

The day of the hand-held calculator is here. We expect students to use a scientific calculator freely and no longer bother to identify those problems that require their use. But there is a new development. The day of the graphing calculator (the Casio fx series, the Sharp EL-5200, the HP-28S, and the TI-81) has also arrived. Although a simple scientific calculator (costing under $20) is adequate for this course, the availability of graphing calculators (costing much more but going down in price) allows us to explore many ideas in more depth than in the past. For those who have their own graphing calculator (or for classes where they are required tools), we have written a section (Section 5-3) on how to use a graphing calculator and have thereafter added graphing calculator problems (each identified with a

symbol) at the ends of most sections. Graphing calculators do not diminish the need to understand hand-graphing strategies. Rather, they enrich the subject by allowing the analysis of more complicated functions and the asking of deeper questions. Besides, students will find graphing calculator problems to be great fun. However, we emphasize that they are supplementary material; the course is complete without them.

Flexibility   This book has plenty of material for a two-semester course. It is flexible in that syllabi for many quite different courses can be based on the book. The dependence chart below will help in keeping track of prerequisites. Here are three items worthy of special note.

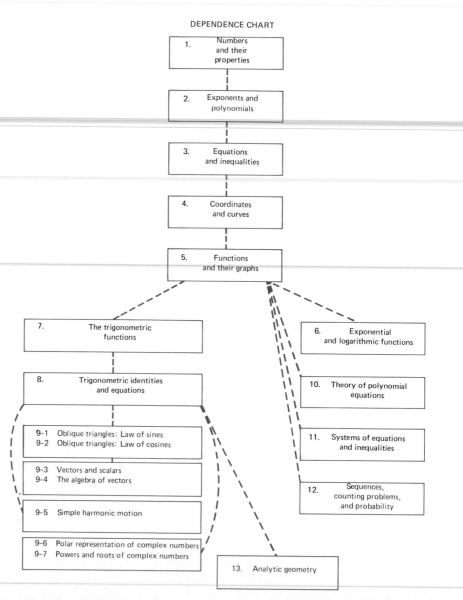

DEPENDENCE CHART

1. Numbers and their properties

2. Exponents and polynomials

3. Equations and inequalities

4. Coordinates and curves

5. Functions and their graphs

7. The trigonometric functions

8. Trigonometric identities and equations

9-1 Oblique triangles: Law of sines
9-2 Oblique triangles: Law of cosines

9-3 Vectors and scalars
9-4 The algebra of vectors

9-5 Simple harmonic motion

9-6 Polar representation of complex numbers
9-7 Powers and roots of complex numbers

6. Exponential and logarithmic functions

10. Theory of polynomial equations

11. Systems of equations and inequalities

12. Sequences, counting problems, and probability

13. Analytic geometry

1. The first three chapters are a review of basic algebra. In some classes, they can be omitted or covered rapidly.
2. Complex numbers are introduced early (Section 1-6). However, this section can be postponed until just before Section 9-6 if desired. To facilitate this, the few problems in the early part of the book that use complex numbers are marked with the symbol $\boxed{i}$ and can be omitted without loss of continuity.
3. An instructor who plans to cover Chapter 13 (Analytic Geometry) may wish to omit the optional section on ellipses and hyperbolas that occurs early in the book (Section 4-5) since this material is handled in much more detail in Chapter 13.

SUPPLEMENTARY MATERIALS

Instructor's Resource Manual was prepared by the authors. It contains the following items:

(a) Teaching Outlines for every section of each chapter.
(b) Complete solutions to all even-numbered problems, including **TEASER** problems.
(c) A Test Item File of more than 1200 problems with answers, designed to be used in conjunction with the computerized Prentice Hall TESTMANAGER.

Prentice Hall TESTMANAGER is a test bank of more than 1200 problems on disk for the IBM PC. This allows the instructor to generate examinations by choosing individual problems, and either editing them or creating completely new problems.

Tutorial Software includes tutorial and drill problems for IBM and MacIntosh.

Student Solutions Manual has worked-out solutions to odd-numbered problems and solutions to all Chapter Review Problem Sets.

ACKNOWLEDGMENTS

This and previous editions have profited from the warm praise and constructive criticism of many reviewers. We offer our thanks to the following people who gave helpful suggestions.

JoAnne B. Brooks, *Blue Mountain College*
Natalie M. Creed, *Belmont Abbey College*
James Daly, *University of Colorado, Boulder*
Milton P. Eisner, *Mount Vernon College*
George T. Fix, *University of Texas, Arlington*

Juan Gatica, *University of Iowa, Iowa City*
Wojciech Komornicki, *Hamline University*
Bruce Lecher, *State University of New York, Binghamton*
Fred Liss, *University of Wisconsin, Rock City*
Carroll Matthews, *Montgomery College, Rockville*
Phil Miles, *University of Wisconsin, Madison*
Michael Montano, *Riverside Community College*
Jim Newsom, *Tidewater Community College, Virginia Beach*
Marvin Papenfuss, *Loras College*
Margot Pullman, *Maryville College*
Cheryl Roberts, *James Madison University*
Richard Semmler, *Northern Virginia Community College*
Cynthia Siegal, *University of Missouri, St. Louis*
John Smashy, *Southwest Baptist University*
Diane Spresser, *James Madison University*
Henry Waldman, *South Dakota School of Mines and Technology*

We thank Ann Phipps of Texas Instruments for granting permission to reproduce a picture of the TI-81 and four diagrams from the TI-81 guidebook, and also for preparing several calculator-generated graphs for inclusion in this text.

The staff at Prentice Hall is to be congratulated on another fine production job. The authors wish to express appreciation especially to Priscilla McGeehon (mathematics editor), Leo Gaffney (development editor), Edward Thomas (production editor), and Judith Matz-Coniglio (designer) for their exceptional contributions.

Walter Fleming
Dale Varberg
Herbert Kasube

Algebra and Trigonometry

CHAPTER **1**

Numbers and Their Properties

■ Numbers are an indispensable tool of civilization, serving to whip its activities into some sort of order . . . The complexity of a civilization is mirrored in the complexity of its numbers.

Philip J. Davis

1-1 WHAT IS ALGEBRA?

"Algebra is a merry science," Uncle Jakob would say. "We go hunting for a little animal whose name we don't know, so we call it x. When we bag our game we pounce on it and give it its right name."

Albert Einstein

Sometimes the simplest questions seem the hardest to answer. One frustrated ninth grader responded, "Algebra is all about x and y, but nobody knows what they are." Albert Einstein was fond of his Uncle Jakob's definition, which is quoted above. A contemporary mathematician, Morris Kline, refers to algebra as generalized arithmetic. There is some truth in all of these statements, but perhaps Kline's statement is closest to the heart of the matter. What does he mean?

In arithmetic we are concerned with numbers and the four operations of addition, subtraction, multiplication, and division. We learn to understand and manipulate expressions like

$$16 - 11 \qquad \frac{3}{24} \qquad (13)(29)$$

In algebra we do the same thing, but we are more likely to write

$$a - b \qquad \frac{x}{y} \qquad mn$$

without specifying precisely what numbers these letters represent. This determination to stay uncommitted (not to know what x and y are) offers some tremendous advantages. Here are two of them.

Generality and Conciseness

All of us know that $3 + 4$ is the same as $4 + 3$ and that $7 + 9$ equals $9 + 7$. We could fill pages and books, even libraries, with the corresponding facts about other numbers. All of them would be correct and all would be important. But we can achieve the same effect much more economically by writing

$$a + b = b + a$$

The simple formula says all there is to be said about adding two numbers in opposite order. It states a general law and does it on one-fourth of a line.

Or take the well-known facts that if I drive 30 miles per hour for 2 hours, I will travel 60 miles, and that if I fly 120 miles per hour for 3 hours, I will cover 360 miles. These and all other similar facts are summarized in the general formula

$$D = RT$$

which is an abbreviation for

$$\text{Distance} = \text{rate} \times \text{time}$$

Formulas

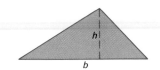

Figure 1

The formula $D = RT$ is just one of many that scientists use almost without thinking. Among these formulas are those of area and volume, which have been known since the time of the Greeks. As a premier example, we mention the formula for the area of a triangle (Figure 1), namely,

$$A = \frac{1}{2}bh$$

a formula that we will have occasion to use innumerable times in this book. Here b denotes the length of the base and h stands for the height (or altitude) of the triangle. Thus a triangle with base $b = 24$ and height $h = 10$ has area

$$A = \frac{1}{2}bh = \frac{1}{2}(24)(10) = 120$$

Of course, we must be careful about units. If the base and height are given in inches, then the area is in square inches.

A more interesting formula is the familiar one

$$A = \pi r^2$$

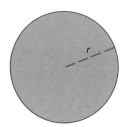

Figure 2

for the area of a circle of radius r (Figure 2). It is interesting because of the appearance of the number π. Perhaps you have learned to approximate π by the fraction 22/7, actually, a rather poor approximation. In this course, we suggest that you use the decimal approximation 3.14159 or the even better approximation that your calculator gives (it should have a π button). Thus a circle of radius 10 centimeters has area

$$A = \pi r^2 = (3.14159)(10)(10) = 314.159$$

The area is about 314 square centimeters.

We are confident that you once learned all the important area and volume formulas but, because your memory may need jogging, we have listed those we will need most often in Figures 3 and 4, which accompany the first problem set.

Problem Solving

Typically, the problems of the real world come to us in words. If we are to use algebra to solve such problems, we must first be able to translate word phrases into algebraic symbols. Here are two simple illustrations.

Example A (Writing Phrases in Algebraic Notation)   Use the symbols x and y to express the following phrases in algebraic notation.
(a) A number divided by the sum of twice that number and another number.
(b) The area A of a triangle whose altitude is 3 feet longer than its base.

Solution

(a) If x is the first number and y is the second, then the given phrase can be expressed by

$$\frac{x}{2x + y}$$

(b) Let x measure the base of the triangle. Then, $x + 3$ measures its altitude. Thus

$$A = \frac{1}{2}x(x + 3) = \frac{1}{2}x^2 + \frac{3}{2}x \quad \blacksquare$$

Uncle Jakob's definition of algebra hinted at something very important in problem solving. Often a problem involves finding a number that is initially unknown but that must satisfy certain conditions. If these conditions can be translated into the symbols of algebra, it may take only a simple manipulation to find the answer or, as Uncle Jakob put it, to bag our game. Examples B and C illustrate this process.

Example B (Stating Sentences as Algebraic Equations)   Express each of the following sentences as an equation using the symbol x. Then solve for x.
(a) The sum of a number and three-fourths of that number is 21.
(b) A rectangular field which is 125 meters longer than it is wide has a perimeter of 650 meters.

Solution

(a) Let x be the number. The sentence can be written as

$$x + \frac{3}{4}x = 21$$

If we multiply both sides by 4, we have

$$4x + 3x = 84$$
$$7x = 84$$
$$x = 12$$

It is always wise to make at least a mental check of the solution. The sum of 12 and three-fourths of 12—that is, the sum of 12 and 9—does equal 21.
(b) Let the width of the field be x meters. Then its length is $x + 125$ meters. Since the perimeter is twice the width plus twice the length, we write

$$2x + 2(x + 125) = 650$$

To solve, we remove parentheses and simplify.

$$2x + 2x + 250 = 650$$

$$4x + 250 = 650$$

$$4x = 400$$

$$x = 100$$

The field is 225 meters long and 100 meters wide. ∎

Example C (Solving a Word Problem)   Roger Longbottom has rented a motor-boat for 5 hours from a river resort. He was told that the boat will travel 6 miles per hour upstream and 12 miles per hour downstream. How far upstream can he go and still return the boat to the resort within the allotted 5-hour period?

Solution   We recognize immediately that this is a distance-rate-time problem: the formula $D = RT$ is certain to be important. Now what is it that we want to know? We want to find a distance, namely, how far upstream Roger dares to go. Let us call that distance x miles. Next we summarize the information that is given, keeping in mind that, since $D = RT$, it is also true that $T = D/R$.

	Going	*Returning*
Distance (miles)	x	x
Rate (miles per hour)	6	12
Time (hours)	$x/6$	$x/12$

There is one piece of information we have not used; it is the key to the whole problem. The total time allowed is 5 hours, which is the sum of the time going and the time returning. Thus

$$\frac{x}{6} + \frac{x}{12} = 5$$

After multiplying both sides by 12, we have

$$2x + x = 60$$

$$3x = 60$$

$$x = 20$$

Roger can travel 20 miles upstream and still return within 5 hours. ∎

We intend to emphasize problem solving in this book. To be able to read a mathematics book with understanding is important. To learn to calculate accurately and to manipulate symbols with ease is a worthy goal. But to be able to solve problems, easy problems and hard ones, practical problems and abstract ones, is a supreme achievement.

It is time for you to try your hand at some problems. If some of them seem difficult, do not become alarmed. All of the ideas of this section will be treated in more detail later. As the title "What Is Algebra?" suggests, we want to give you a preview of what lies ahead.

PROBLEM SET 1-1

A. Skills and Techniques

Express each of the phrases in Problems 1–14 in algebraic notation using the symbols x and y. See Example A.

1. One number plus one-third of another number.
2. The average of two numbers.
3. Twice one number divided by three times another.
4. The sum of a number and its square.
5. Ten percent of a number added to that number.
6. Twenty percent of the amount by which a number exceeds 50.
7. The sum of the squares of two sides of a triangle.
8. One-half the product of the base and the height of a triangle.
9. The distance in miles that a car travels in x hours at y miles per hour.
10. The time in hours it takes to go x miles at y miles per hour.
11. The rate in miles per hour of a boat that traveled y miles in x hours.
12. The total distance a car traveled in 8 hours if its rate was x miles per hour for 3 hours and y miles per hour for 5 hours.
13. The time in hours it took a boat to travel 30 miles upstream and back if its rate upstream was x miles per hour and its rate downstream was y miles per hour.
14. The time in hours it took a boat to travel 30 miles upstream and back if its rate in still water was x miles per hour and the rate of the stream was y miles per hour. Assume that x is greater than y.

In Figures 3 and 4, we have displayed the most important area and volume formulas. Use them to express the quantities of Problems 15–28 in algebraic symbols. In each problem, assume that all dimensions are given in terms of the same unit of length (such as a centimeter).

15. The area of a square of side x.
16. The area of a triangle whose height is $\frac{1}{3}$ the length of its base.
17. The surface area of a cube of side x.

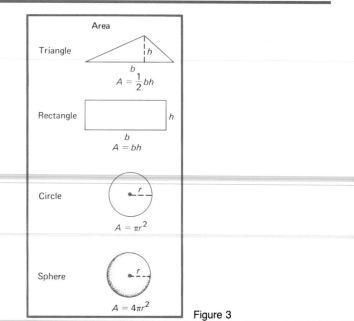

Area

Triangle $A = \frac{1}{2}bh$

Rectangle $A = bh$

Circle $A = \pi r^2$

Sphere $A = 4\pi r^2$

Figure 3

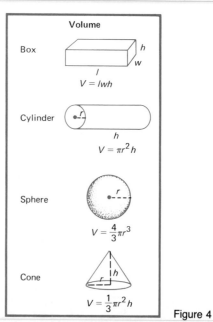

Volume

Box $V = lwh$

Cylinder $V = \pi r^2 h$

Sphere $V = \frac{4}{3}\pi r^3$

Cone $V = \frac{1}{3}\pi r^2 h$

Figure 4

18. The area of the surface of a rectangular box whose dimensions are x, $2x$, and $3x$.

19. The surface area of a sphere whose diameter is x.

20. The area of a Norman window whose shape is that of a square of side x topped with a semicircle.

21. The volume of a box with square base of length x and height 10.

22. The volume of a cylinder whose radius and height are both x.

23. The volume of a sphere of diameter x.

24. The volume of what is left of a cylinder of radius 10 and altitude 12 when a hole of radius x is drilled along the center axis of the cylinder. (Assume that x is less than 10.)

25. The volume of what is left when a round hole of radius 2 is drilled through a cube of side x. (Assume that the hole is drilled perpendicular to a face and that x is greater than 4.)

26. The volume of what is left of a cylinder when a square hole of width x is drilled along the center axis of the cylinder of radius $3x$ and height y.

27. The area and perimeter of a running track in the shape of a square with semicircular ends if the square has side x. (Recall that the circumference of a circle satisfies $C = 2\pi r$, where r is the radius.)

28. The area of what is left of a circle of radius r after removing an isosceles triangle which has a diameter as a base and the opposite vertex on the circumference of the circle.

Express each of the following sentences as an algebraic equation in x and then solve for x. Start by writing down what x represents. Note: Chapter 3 covers this kind of problem in more detail. See Example B.

29. The sum of a number and one-half of that number is 45.

30. Fifteen percent of a number added to that number is 10.35.

31. The sum of two consecutive odd numbers is 168.

32. The sum of three consecutive even numbers is 180.

33. A car going at x miles per hour travels 252 miles in $4\frac{1}{2}$ hours.

34. A circle of radius x centimeters has an area of 64π square centimeters.

B. Applications and Extensions

35. A rectangle has width x meters. The length of the rectangle is 3 meters more than its width and the perimeter is 48 meters. Find x.

36. The area of a rectangle of width x feet is 162 square feet. If the length of the rectangle is twice its width, find x.

37. Two abutting circles of radius 10 feet are inscribed in a rectangle (the rectangle fits tightly around the figure eight formed by the circles). Find the area of the part of the rectangle outside the circles.

38. A goat is tethered to the corner of a square building that is 20 feet on a side by a rope that is 30 feet long. Calculate its grazing area. *Hint:* Draw a picture.

39. If the surface area of a certain sphere is the same as the area of a circle of radius 6 centimeters, find the radius of the sphere.

40. It took exactly the same amount of paint to cover the top and bottom of a cylindrical tank as it did to cover the sides. If the tank is 8 feet high, what is its radius?

41. If a can of oil 5 inches in diameter and 8 inches high sells for $1.80, what is a fair price for a can of oil that is 6 inches in diameter and 10 inches high? Assume that the cost of the cans is negligible.

42. If the radius of a sphere is doubled, what happens to its volume? Its surface area?

43. A can of tennis balls is exactly 6 balls high. If the radius of a ball is r, what is the volume of air around the balls? Assume that the balls fit tightly in the can.

44. A cylindrical can with radius 5 centimeters and height 8 centimeters is full of water. If 30 marbles of radius 1 centimeter are dropped into the can, how much water will remain in the can?

45. Orville Rightman knows that his small plane cruises at 150 miles per hour in still air and that his tank holds enough gas for 6 hours of flying. If he takes off against a constant east wind of 50 miles per hour, how far can he fly in that direction and return safely? (See Example C.)

46. Sylvia Goodspeed drove from Pitstop to Backwater, a distance of 200 miles, in 4 hours. At what speed must she travel on the return trip to average 60 miles per hour for the entire trip?

47. Mabel Savewear rotates the tires on her car so that all 5 tires (spare tire included) have the same wear. After driving 60,000 miles, how many miles were on each tire?

48. Rodney Roller used his spare tire for 2000 miles but managed to get equal mileage on each of the other four tires. After driving 60,000 miles, how many miles were on each of the four tires?

49. Arnold Thinkhard plans to walk around the earth at the equator. Assuming this is possible, how much farther will his nose travel than his feet? To answer, all you need to know is that Arnold holds his nose 6 feet above the ground when he walks.

50. TEASER Arnold Thinkhard walked 1 mile south, 1 mile east, and 1 mile north, thus returning to his starting point. He could have started at the North Pole, but he did not. Where did he start?

1-2 THE INTEGERS AND THE RATIONAL NUMBERS

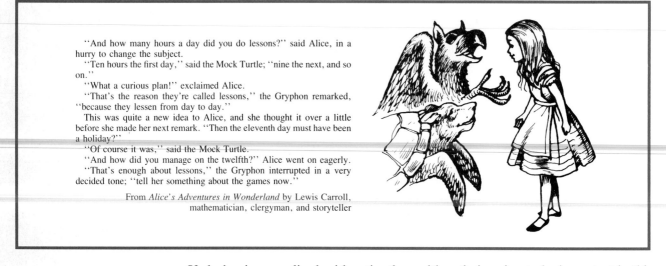

"And how many hours a day did you do lessons?" said Alice, in a hurry to change the subject.

"Ten hours the first day," said the Mock Turtle; "nine the next, and so on."

"What a curious plan!" exclaimed Alice.

"That's the reason they're called lessons," the Gryphon remarked, "because they lessen from day to day."

This was quite a new idea to Alice, and she thought it over a little before she made her next remark. "Then the eleventh day must have been a holiday?"

"Of course it was," said the Mock Turtle.

"And how did you manage on the twelfth?" Alice went on eagerly.

"That's enough about lessons," the Gryphon interrupted in a very decided tone; "tell her something about the games now."

From *Alice's Adventures in Wonderland* by Lewis Carroll, mathematician, clergyman, and storyteller

If algebra is generalized arithmetic, then arithmetic is going to be important in this course. That is why we first review the familiar number systems of mathematics.

The human race learned to count before it learned to write; so did most of us. The **whole numbers** 0, 1, 2, 3, . . . entered our vocabularies before we started school. We used them to count our toys, our friends, and our money. Counting forward to larger and larger numbers was no problem; counting backwards was a different matter: 5, 4, 3, 2, 1, 0. But what comes after 0? It bothered Alice in Wonderland, it bothered mathematicians as late as 1300, and it still bothers some people. However, if we are to talk about debts, cold temperatures, and even moon launch countdowns, we must have an answer. Faced with this problem, mathematicians invented a host of new numbers, −1, −2, −3, . . . , called **negative integers.** Together with the whole numbers, they form the system of integers.

The Integers

The **integers** are the numbers . . . , −5, −4, −3, −2, −1, 0, 1, 2, 3, 4, 5, ... This array is familiar to anyone who has looked at a thermometer lying on its side. In fact, that is an excellent way to think of the integers. There they label points, equally spaced along a line, as in Figure 5.

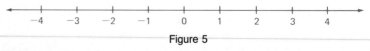

Figure 5

If we are talking about temperatures, -4 means 4 degrees below zero; if we are referring to money, -4 might represent a debt of 4 dollars.

Numbers, used as labels for points, debts, or football players, are significant. But what really makes them useful is our ability to combine them, that is, to add, subtract, and multiply them. All of this is familiar to you, so we shall remind you of only two properties, or rules, that sometimes cause trouble.

(i)
$$(-a)(b) = (a)(-b) = -(ab)$$

(ii)
$$(-a)(-b) = ab$$

These rules actually make perfectly good sense. For example, if I am flat broke today and am losing 4 dollars each day (that is, gaining -4 dollars each day), then 3 days from now I will be worth $3(-4) = -(3 \cdot 4) = -12$ dollars. On the other hand, 3 days ago (that is, on day -3), I must have been worth 12 dollars; that is $(-3)(-4) = 3 \cdot 4 = 12$. The chart in the margin (Figure 6) shows my worth each day starting 4 days ago and continuing to 4 days from now. Can you tell how much I was worth 6 days ago and how much I will be worth 10 days from now?

When a numerical expression involves several operations, we use parentheses and brackets to indicate the order in which these operations are to be performed. Example A reminds us of an important convention.

Example A (Removing Grouping Symbols) Simplify $-3[2 - (6 + x)]$.

Solution Keep these two things in mind. First, always begin work with the innermost parentheses. Second, remember that a minus sign preceeding a set of parentheses (or brackets) means all that is between them must be multiplied by -1, and so the sign of each term must be changed when the parentheses are removed. Thus

$$-3[2 - (6 + x)] = -3[2 - 6 - x] = -3[-4 - x] = 12 + 3x \qquad ∎$$

The problem set will give you practice in working with the integers. In the meantime, we note that with all their beauty and usefulness, the integers are plagued by a serious flaw: You cannot always divide them—that is, if you want an integer as the answer. For, while $\frac{4}{2}$, $\frac{9}{3}$, and $\frac{12}{2}$ are perfectly respectable integers, $\frac{1}{2}$, $\frac{3}{4}$, and $\frac{5}{3}$ are not. To make sense of these symbols requires a further enlargement of the number system.

The Rational Numbers

In a certain sense, mathematicians are master inventors. When new kinds of numbers are needed, they invent them. Faced with the need to divide, mathematicians simply decided that the result of dividing an integer by a nonzero integer should be regarded as a number. That meant

$$\frac{3}{4} \qquad \frac{7}{8} \qquad \frac{-2}{3} \qquad \frac{14}{-16} \qquad \frac{8}{2}$$

and all similar ratios were numbers, numbers with all the rights and privileges of the integers and even a bit more: division, except by zero, was always possible. Naturally, these numbers were called rational numbers (ratio numbers). Stated more ex-

Day	Amount gained each day		My worth
$-4 \cdot$	-4	$=$	$\$16$
$-3 \cdot$	-4	$=$	12
$-2 \cdot$	-4	$=$	8
$-1 \cdot$	-4	$=$	4
$0 \cdot$	-4	$=$	0
$1 \cdot$	-4	$=$	-4
$2 \cdot$	-4	$=$	-8
$3 \cdot$	-4	$=$	-12
$4 \cdot$	-4	$=$	-16

Three days ago → (row -4)
Today → (row 0)
Three days from now → (row 2)

Figure 6

CAUTION

3x − (x + 2) = 2x + 2   (crossed out)

3x − (x + 2) = 2x − 2

The informal use of fractions (ratios) is very old. We know from the Rhind papyrus that the Egyptians were quite proficient with fractions by 1650 B.C. However, they used a different notation and considered only ratios of positive integers.

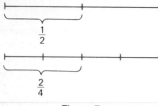

Figure 7

plicitly, a **rational number** is a number that can be expressed as the ratio of two integers p/q, $q \neq 0$.

The rational numbers are admirably suited for certain very practical measurements problems. Take a piece of string of length 1 unit (Figure 7) and divide it into two parts of equal length. We say that each part has length $\frac{1}{2}$. Take the same piece of string and divide it into 4 equal parts. Then each part has length $\frac{1}{4}$ and two of them together have length $\frac{2}{4}$. Thus $\frac{1}{2}$ and $\frac{2}{4}$ must stand for the same number—that is,

$$\frac{1}{2} = \frac{2 \cdot 1}{2 \cdot 2} = \frac{2}{4}$$

Considerations like this suggest an important agreement. We agree that

$$\frac{a}{b} = \frac{k \cdot a}{k \cdot b}$$

for any nonzero number k. Thus $\frac{1}{2}$, $\frac{2}{4}$, $\frac{3}{6}$, $\frac{-4}{-8}$, . . . are all treated as symbols for the same rational number.

We should learn to read equations backwards as well as forwards. Read backwards, the boxed equation tells us that we can divide numerator and denominator (top and bottom) of a quotient by the same nonzero number k. Or, in language that may convey the idea even better, it says that we can cancel a common factor from numerator and denominator.

$$\frac{24}{32} = \frac{\cancel{8} \cdot 3}{\cancel{8} \cdot 4} = \frac{3}{4}$$

Among the many symbols for the same rational number, one is given special honor, the reduced form. If numerator a and denominator b of the rational number a/b have no common integer divisors (factors) greater than 1 and if b is positive, we say a/b is in **reduced form.** Thus $\frac{3}{4}$ is the reduced form of $\frac{24}{32}$, and $-2/3$ is the reduced from of $50/-75$.

CAUTION

$$3\frac{x}{y} = \frac{3x}{3y}$$

$$3\frac{x}{y} = \frac{3x}{y}$$

Example B (Reducing Fractions)   Reduce.

(a) $\dfrac{24}{36}$ (b) $\dfrac{3+9}{3+6}$ (c) $\dfrac{12+3x}{3x}$

Solution   You can cancel common factors (which are multiplied) from numerator and denominator. Do not make the common mistake of trying to cancel common terms (which are added).

(a) $\dfrac{24}{36} = \dfrac{\cancel{12} \cdot 2}{\cancel{12} \cdot 3} = \dfrac{2}{3}$

(b) $\dfrac{3+9}{3+6} = \dfrac{\cancel{3}+9}{\cancel{3}+6} = \dfrac{9}{6}$ Wrong

$\dfrac{3+9}{3+6} = \dfrac{12}{9} = \dfrac{\cancel{3} \cdot 4}{\cancel{3} \cdot 3} = \dfrac{4}{3}$ Right

(c) $\dfrac{12 + 3x}{3x} = \dfrac{12 + \cancel{3x}}{\cancel{3x}} = 12$ Wrong

$\dfrac{12 + 3x}{3x} = \dfrac{\cancel{3}(4 + x)}{\cancel{3} \cdot x} = \dfrac{4 + x}{x}$ Right ■

We call attention to another rather obvious fact. Notice that $\frac{2}{1}$ is technically the reduced form of $\frac{4}{2}, \frac{6}{3}$, and so on. However, we almost never write $\frac{2}{1}$, since the ordinary meaning of division implies that $\frac{2}{1}$ is equal to the integer 2. In fact, for any integer a,

$$\dfrac{a}{1} = a$$

Thus the class of rational numbers contains the integers as a subclass.

Let us go back to that horizontal thermometer, the calibrated line we looked at earlier. Now we can label many more points (Figure 8). In fact, it seems that we can fill the line with labels.

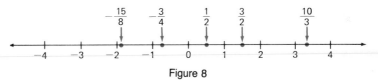

Figure 8

No discussion of the rational numbers is complete without mention of how to add, subtract, multiply, and divide them. You are familiar with these operations but perhaps you need to review them. Examples C, D, and E give several reminders and the problem set will provide plenty of practice.

Example C (Adding and Subtracting Fractions)   Simplify.

(a) $\dfrac{3}{4} + \dfrac{5}{4}$ (b) $\dfrac{3}{5} + \dfrac{5}{4}$ (c) $\dfrac{8}{9} - \dfrac{5}{12}$

Solution   We add fractions with the same denominator by adding numerators. If the fractions have different denominators, we first rewrite them as equivalent fractions with the same denominator and then add. Similar rules apply for subtraction.

(a) $\dfrac{3}{4} + \dfrac{5}{4} = \dfrac{3+5}{4} = \dfrac{8}{4} = 2$ (b) $\dfrac{3}{5} + \dfrac{5}{4} = \dfrac{12}{20} + \dfrac{25}{20} = \dfrac{37}{20}$

(c) $\dfrac{8}{9} - \dfrac{5}{12} = \dfrac{32}{36} - \dfrac{15}{36} = \dfrac{17}{36}$ ■

Example D (Multiplying and Dividing Fractions)   Simplify.

(a) $\dfrac{3}{4} \cdot \dfrac{5}{7}$ (b) $\dfrac{3}{4} \cdot \dfrac{16}{27}$ (c) $\dfrac{\frac{3}{4}}{\frac{9}{16}}$ (d) $4 \div \dfrac{5}{3}$

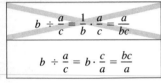
Solution We multiply fractions by multiplying numerators and multiplying denominators. To divide fractions, invert (that is, take the reciprocal of) the divisor and multiply.

(a) $\dfrac{3}{4} \cdot \dfrac{5}{7} = \dfrac{3 \cdot 5}{4 \cdot 7} = \dfrac{15}{28}$

(b) $\dfrac{3}{4} \cdot \dfrac{16}{27} = \dfrac{3 \cdot 16}{4 \cdot 27} = \dfrac{\cancel{3} \cdot \cancel{4} \cdot 4}{\cancel{4} \cdot \cancel{3} \cdot 9} = \dfrac{4}{9}$

(c) $\dfrac{\frac{3}{4}}{\frac{9}{16}} = \dfrac{3}{4} \cdot \dfrac{16}{9} = \dfrac{\cancel{3} \cdot \cancel{4} \cdot 4}{\cancel{4} \cdot \cancel{3} \cdot 3} = \dfrac{4}{3}$

(d) $4 \div \dfrac{5}{3} = \dfrac{4}{1} \cdot \dfrac{3}{5} = \dfrac{4 \cdot 3}{1 \cdot 5} = \dfrac{12}{5}$ ∎

Example E (Complicated Fractions) Simplify.

(a) $\dfrac{\frac{2}{3} + \frac{1}{5}}{\frac{5}{7} - \frac{1}{2}}$ (b) $\dfrac{\frac{5}{6} - \frac{2}{15}}{\frac{11}{30} + \frac{3}{5}}$

Solution We show two methods of attacking four-story expressions.
(a) Start by working with the top portion and bottom portion separately.

$$\dfrac{\dfrac{2}{3} + \dfrac{1}{5}}{\dfrac{5}{7} - \dfrac{1}{2}} = \dfrac{\dfrac{10}{15} + \dfrac{3}{15}}{\dfrac{10}{14} - \dfrac{7}{14}} = \dfrac{\dfrac{13}{15}}{\dfrac{3}{14}} = \dfrac{13}{15} \cdot \dfrac{14}{3} = \dfrac{182}{45}$$

(b) Multiply the top and bottom portions by a common denominator of all the simple fractions.

$$\dfrac{\dfrac{5}{6} - \dfrac{2}{15}}{\dfrac{11}{30} + \dfrac{3}{5}} = \dfrac{30\left(\dfrac{5}{6} - \dfrac{2}{15}\right)}{30\left(\dfrac{11}{30} + \dfrac{3}{5}\right)} = \dfrac{25 - 4}{11 + 18} = \dfrac{21}{29}$$ ∎

PROBLEM SET 1-2

A. Skills and Techniques

Simplify each of the following, as in Example A.

1. $4 - 2(8 - 12)$
2. $-5 + 2(3 - 18)$
3. $-3 + 2[-5 - (12 - 3)]$
4. $-2[1 + 3(6 - 8) + 4] + 3$
5. $-4[3(-6 + 13) - 2(7 - 5)] + 1$
6. $5[-(7 + 12 - 16) + 4] + 2$
7. $-3[4(5 - x) + 2x] + 3x$
8. $-4[x - (3 - x) + 5] + x$
9. $2[-t(3 + 5 - 11) + 4t]$
10. $y - 2[-3(y + 1) + y]$

Reduce each of the following, leaving your answer with a positive denominator. See Example B.

11. $\dfrac{24}{27}$

12. $\dfrac{16}{36}$

13. $\dfrac{45}{-60}$

14. $\dfrac{63}{-81}$

15. $\dfrac{3 - 9x}{6}$

16. $\dfrac{4 + 6x}{4 - 8}$

17. $\dfrac{4x - 6}{4 - 8}$

18. $\dfrac{4x - 12}{8}$

Simplify each of the following. See Example C.

19. $\dfrac{5}{6} + \dfrac{11}{12}$

20. $\dfrac{8}{10} - \dfrac{3}{20}$

21. $\dfrac{4}{5} - \dfrac{3}{20} + \dfrac{3}{10}$

22. $\dfrac{11}{24} + \dfrac{2}{3} - \dfrac{5}{12}$

23. $\dfrac{5}{12} + \dfrac{7}{18} - \dfrac{1}{6}$

24. $\dfrac{11}{24} + \dfrac{3}{4} - \dfrac{5}{6}$

25. $\dfrac{-5}{27} + \dfrac{5}{12} + \dfrac{3}{4}$

26. $\dfrac{23}{30} + \dfrac{2}{25} - \dfrac{3}{5}$

Simplify, as in Example D.

27. $\dfrac{5}{6} \cdot \dfrac{9}{15}$

28. $\dfrac{4}{13} \cdot \dfrac{5}{12}$

29. $\dfrac{3}{4} \cdot \dfrac{6}{15} \cdot \dfrac{5}{2}$

30. $\dfrac{9}{11} \cdot \dfrac{33}{5} \cdot \dfrac{15}{18}$

31. $\dfrac{\frac{5}{6}}{\frac{8}{12}}$

32. $\dfrac{\frac{9}{24}}{\frac{15}{12}}$

33. $\dfrac{\frac{3}{4}}{2}$

34. $\dfrac{6}{7} \div 9$

35. $6 \div \dfrac{7}{9}$

36. $\dfrac{\frac{3}{4}}{\frac{7}{5}} \cdot \dfrac{7}{5}$

Simplify, as in Example E.

37. $\dfrac{\frac{2}{3} + \frac{3}{4}}{\frac{7}{12}}$

38. $\dfrac{\frac{3}{5} - \frac{3}{4}}{\frac{9}{20}}$

39. $\dfrac{\frac{2}{3} + \frac{3}{4}}{\frac{2}{3} - \frac{3}{4}}$

40. $\dfrac{\frac{8}{9} - \frac{2}{27}}{\frac{8}{9} + \frac{2}{27}}$

41. $\dfrac{\frac{5}{6} \div \frac{1}{12}}{\frac{3}{4} + \frac{2}{3}}$

42. $\dfrac{\frac{3}{50} - \frac{1}{2} + \frac{4}{5}}{\frac{4}{25} + \frac{7}{10}}$

B. Applications and Extensions

In Problems 43–58, perform the indicated operations and simplify.

43. $\dfrac{5}{6} + \dfrac{1}{9} - \dfrac{2}{3} + \dfrac{5}{18}$

44. $-2 + \dfrac{3}{4} - \dfrac{2}{3} + \dfrac{7}{6}$

45. $-\dfrac{2}{3}\left(\dfrac{7}{6} - \dfrac{5}{4}\right)$

46. $\dfrac{2}{3}\left[\dfrac{1}{2}\left(\dfrac{2}{3} - \dfrac{3}{4}\right) - \dfrac{5}{6}\right]$

47. $\dfrac{2}{3} - \dfrac{1}{2}\left(\dfrac{2}{3} - \dfrac{1}{2} + \dfrac{1}{12}\right)$

48. $-\dfrac{3}{4}\left[\dfrac{3}{5} - \dfrac{1}{2}\left(\dfrac{1}{3} - \dfrac{1}{5}\right)\right]$

49. $\left(\dfrac{5}{7} + \dfrac{7}{5}\right) \div \dfrac{2}{5}$

50. $\left[2 \div \left(3 + \dfrac{4}{5}\right)\right]\dfrac{2}{5}$

51. $\left[2 - \left(3 \div \dfrac{4}{5}\right)\right]\dfrac{2}{7}$

52. $\dfrac{\frac{11}{25} - \frac{3}{5}}{\frac{11}{25} + \frac{3}{5}} + \dfrac{2}{13}$

53. $\dfrac{\frac{1}{2} + \frac{3}{4} - \frac{7}{8}}{\frac{1}{2} - \frac{3}{4} + \frac{7}{8}} + \dfrac{2}{5}$

54. $3 - \dfrac{\frac{2}{5} + \frac{5}{12}}{\frac{3}{4}}$

55. $\left(3 + \dfrac{2}{3 + \frac{1}{3}}\right) \div \dfrac{6}{11}$

56. $\dfrac{1}{18} + \dfrac{1}{2 - \dfrac{1}{3 - \frac{1}{4}}}$

57. $\left(1 - \dfrac{1}{2}\right)\left(1 - \dfrac{1}{3}\right)\left(1 - \dfrac{1}{4}\right) \cdots \left(1 - \dfrac{1}{19}\right)$

58. $\left(1 + \dfrac{3}{4}\right)\left(1 + \dfrac{3}{5}\right)\left(1 + \dfrac{3}{6}\right) \cdots \left(1 + \dfrac{3}{19}\right)$

59. I got 80, 82, and 98 on my first three algebra tests. What will I have to score on the fourth test to bring my average up to 90?

60. Sylvia Goodspeed drove from A to B at an average speed of 40 miles per hour and returned at an average speed of 60 miles per hour. What was her average speed for the round trip?

61. Professor Umberto Rodriguez gave the same test to both of his algebra classes. The first class of 20 students scored an average of 70; the second class of 30 students scored an average of 60. What was the combined average for the 50 students?

62. A farmer distributed a certain number of acres of land among his three children. First, Julie got 2/3 of the acreage, then Henry got 4/7 of what was left, and finally Rick got the rest, amounting to 18 acres. How many acres of land did Julie get?

63. Ronald Longlegs walked a certain distance, jogged $2\frac{1}{2}$ times as far as he walked, and then sprinted $2\frac{1}{3}$ times as far as he jogged. If he covered 2352 meters in all, how far did he walk?

64. **TEASER** Late at night, Arnold Thinkhard stood outside the haunted house holding an infinite supply of balls numbered 1, 2, 3, ... At 11:30 P.M. (that is, at $12 - \frac{1}{2}$), he tossed balls 1, 2, . . . , 10 in through the window, but somebody (who knows who) tossed ball 1 right back out. At 11:45 P.M. (that is, at $12 - \frac{1}{4}$), Arnold tossed in balls 11, 12, . . . , 20 and out came ball 2. At 11:52$\frac{1}{2}$ P.M. (that is, at $12 - \frac{1}{8}$), Arnold tossed in balls 21, 22, . . . , 30 and out came ball 3. Arnold continued this pattern, tossing in 10 new balls each time: each time the lowest-numbered ball still in the house came flying out.

(a) How many balls were in the house right after $(12 - \frac{1}{16})$ P.M.? Right after $(12 - \frac{1}{32})$ P.M.?

(b) How many balls were in the house right after $(12 - (\frac{1}{2})^n)$ P.M.?

(c) When did ball 9 come flying out?

(d) Would you agree that the number of balls in the house gets larger and larger as time nears midnight?

(e) How many balls were in the house at midnight?

1-3 THE REAL NUMBERS

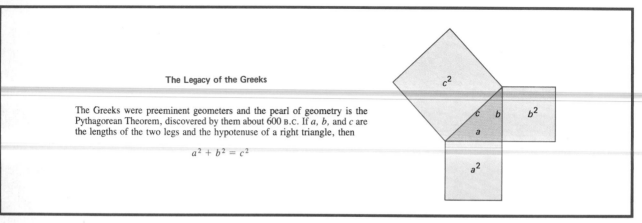

The Legacy of the Greeks

The Greeks were preeminent geometers and the pearl of geometry is the Pythagorean Theorem, discovered by them about 600 B.C. If a, b, and c are the lengths of the two legs and the hypotenuse of a right triangle, then

$$a^2 + b^2 = c^2$$

Figure 9

The Pythagorean Theorem, that beautiful gem of geometry, came to be the great nemesis of Greek mathematics. For hidden in that elegant formula is a consequence that seemed to destroy the Greek conception of numbers. Here is a paraphrased version of an old story reported by Euclid.

"Consider," said Euclid, "a right triangle with legs of unit length and hypotenuse of length C (Figure 9). By the theorem of Pythagoras,

$$C^2 = 1^2 + 1^2 = 2$$

That is, $C = \sqrt{2}$. But then, as I will show by a logical argument, the quantity $\sqrt{2}$, whatever it is, is not a rational number. To put it bluntly, if only rational numbers exist, then the simplest right triangle we know has a hypotenuse whose length cannot be measured."

Euclid's argument (soon to be presented) is flawless; it is also devilishly clever. But to understand it, we need to know some basic facts about prime numbers.

A Digression on Prime Numbers

A **prime number** is a whole number with exactly two whole-number divisors, itself and 1. The first few primes are

$$2, 3, 5, 7, 11, 13, 17, 19, 23, 29$$

Prime numbers are the building blocks of other whole numbers. For example,

$$18 = 2 \cdot 3 \cdot 3 \qquad 40 = 2 \cdot 2 \cdot 2 \cdot 5$$

This type of factorization is possible for all nonprime whole numbers greater than 1.

Fundamental Theorem of Arithmetic

Any nonprime whole number (greater than 1) can be written as the product of a unique set of prime numbers.

To find the prime factorization of large whole numbers, it is wise to use a systematic procedure. We illustrate.

Example A (Prime Factorization)   Find the prime factorizations of 168 and 420.

Solution   Factor out as many 2's as possible, then 3's, then 5's, and so on.

$$168 = 2 \cdot 84 = 2 \cdot 2 \cdot 42 = 2 \cdot 2 \cdot 2 \cdot 21 = 2 \cdot 2 \cdot 2 \cdot 3 \cdot 7$$
$$420 = 2 \cdot 210 = 2 \cdot 2 \cdot 105 = 2 \cdot 2 \cdot 3 \cdot 35 = 2 \cdot 2 \cdot 3 \cdot 5 \cdot 7 \quad \blacksquare$$

The prime factorization is useful in connection with another important idea. The **least common multiple** (lcm) of several whole numbers is the smallest positive whole number that is a multiple of each of them. For example, the lcm of 12 and 18 is 36 and the lcm of 8, 12, and 18 is 72.

Example B (Least Common Multiple)   Find the least common multiple of 168 and 420.

Solution   We found the prime factorizations of these two numbers in Example A. To find the least common multiple of 168 and 420, write down the product of all factors that occur in either number, repeating a factor according to the greatest number of times it occurs in either number.

$$\text{lcm}(168, 420) = 2 \cdot 2 \cdot 2 \cdot 3 \cdot 5 \cdot 7 = 840 \quad \blacksquare$$

The concept of least common multiple plays a role when adding fractions with different denominators.

Example C (Least Common Denominator)   Calculate

$$\frac{5}{168} + \frac{13}{420}$$

Solution   Write both fractions with a common denominator. The best choice of common denominator is the least common multiple of 168 and 420—namely, 840—obtained in Example B. We call it the **least common denominator.**

$$\frac{5}{168} + \frac{13}{420} = \frac{5 \cdot 5}{840} + \frac{13 \cdot 2}{840} = \frac{25 + 26}{840} = \frac{51}{840} = \frac{17}{280} \quad \blacksquare$$

The Fundamental Theorem of Arithmetic is important in many places. Here is another simple consequence. When the square of any whole number is written as a product of primes, each prime occurs as a factor an even number of times. For example,

$$(18)^2 = 18 \cdot 18 = 2 \cdot 3 \cdot 3 \cdot 2 \cdot 3 \cdot 3 = \underbrace{2 \cdot 2}_{\text{two 2's}} \cdot \underbrace{3 \cdot 3 \cdot 3 \cdot 3}_{\cdot \text{ four 3's}}$$

$$(40)^2 = 40 \cdot 40 = 2 \cdot 2 \cdot 2 \cdot 5 \cdot 2 \cdot 2 \cdot 2 \cdot 5 = \underbrace{2 \cdot 2 \cdot 2 \cdot 2 \cdot 2 \cdot 2}_{\text{six 2's}} \cdot \underbrace{5 \cdot 5}_{\text{two 5's}}$$

The Proof that $\sqrt{2}$ Is Irrational

Suppose that $\sqrt{2}$ is a rational number; that is, suppose that $\sqrt{2} = m/n$, where m and n are whole numbers (necessarily greater than 1). Then

$$2 = \frac{m^2}{n^2} \qquad \text{and so} \qquad 2n^2 = m^2$$

> Reductio ad absurdum, which Euclid loved so much, is one of a mathematician's finest weapons. It is a far finer gambit than any chess gambit: a chess player may offer the sacrifice of a pawn or even a piece, but a mathematician offers the game.
>
> G. H. Hardy

Now imagine that both n and m are written as products of primes. As we saw above, both n^2 and m^2 must then have an even number of 2's as factors. Thus in the above equation, the prime 2 appears on the left an odd number of times but on the right an even number of times. This is clearly impossible. What can be wrong? The only thing that can be wrong is our supposition that $\sqrt{2}$ was a rational number.

To let one number, $\sqrt{2}$, through the dike was bad enough. But, as Euclid realized, a host of others came pouring through with it. Exactly the same proof shows that $\sqrt{3}$, $\sqrt{5}$, $\sqrt{7}$, and, in fact, the square roots of all primes are irrational. The Greeks, who steadfastly insisted that all measurements must be based on whole numbers and their ratios, could not find a satisfactory way out of this dilemma. Today we recognize that the only adequate solution is to enlarge the number system.

The Real Numbers

Let us take a bold step and simply declare that every line segment shall have a number that measures its length. The set of all numbers that can measure lengths, together with their negatives and zero, constitute the **real numbers.** Thus the rational numbers are automatically real numbers; the positive rational numbers certainly measure lengths.

Consider again the thermometer on its side, the calibrated line (Figure 10). We may have thought we had labeled every point. Not so; there were many holes corresponding to what we now call $\sqrt{2}$, $\sqrt{5}$, π, and so on. But with the introduction of

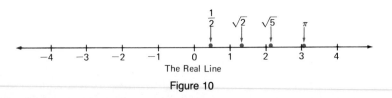

The Real Line

Figure 10

the real numbers, all the holes are filled in; every point has a number label. Because of this, we often call this calibrated line the **real line.**

Decimals

There is another important way to describe the real numbers. It calls for review of a basic idea. Recall that

$$.4 = \frac{4}{10} \qquad .7 = \frac{7}{10}$$

Similarly,

$$.41 = \frac{4}{10} + \frac{1}{100} = \frac{40}{100} + \frac{1}{100} = \frac{41}{100}$$

$$.731 = \frac{7}{10} + \frac{3}{100} + \frac{1}{1000} = \frac{700}{1000} + \frac{30}{1000} + \frac{1}{1000} = \frac{731}{1000}$$

It is a simple matter to locate a decimal on the number line. For example, to locate 1.4, we divide the interval from 1 to 2 into 10 equal parts and pick the fourth point of division (Figure 11).

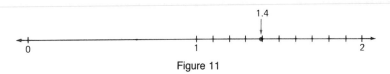

Figure 11

If the interval from 1.4 to 1.5 is divided into 10 equal parts, the second point of division corresponds to 1.42 (Figure 12).

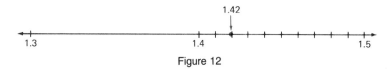

Figure 12

We can find the decimal corresponding to a rational number by long division. For example, the first division in Figure 13 shows that $\frac{7}{8} = .875$. If we try the same procedure on $\frac{1}{3}$, something different happens. The decimal just keeps on going; it is an **unending decimal.** Actually, the terminating decimal .875 can be thought of as unending if we annex zeros. Thus

$$\frac{7}{8} = .875 = .8750000 \ldots$$

$$\frac{1}{3} = .3333333 \ldots$$

$$\frac{1}{6} = .16666 \ldots$$

```
   .875
8)7.000
  64
  ‾‾
   60
   56
   ‾‾
    40
    40
    ‾‾
```

```
    .33333
3)1.00000
  9
  ‾
  10
   9
   ‾
   10
    9
    ‾
    10
     9
     ‾
     10
      9
      ‾
      10
       9
       ‾
       1
```

Figure 13

Let us take an example that is a bit more complicated, $\frac{2}{7}$.

$$
\begin{array}{r}
.285714 \\
7\overline{)2.0000000} \\
1\ 4 \\
\hline
60 \\
56 \\
\hline
40 \\
35 \\
\hline
50 \\
49 \\
\hline
10 \\
7 \\
\hline
30 \\
28 \\
\hline
20 \\
\end{array}
$$

If we continue the division, the pattern must repeat (note the circled 20's). Thus

$$
\frac{2}{7} = .285714285714285714 \ldots
$$

which can also be written

$$
\frac{2}{7} = .\overline{285714}
$$

The bar indicates that the block of digits 285714 repeats indefinitely.

In fact, the decimal expansion of any rational number must inevitably start repeating, because there are only finitely many different possible remainders in the division process (at most as many as the divisor).

Example D (Rational Numbers as Repeating Decimals)   Write $\frac{68}{165}$ as a repeating decimal.

Solution

$$
\begin{array}{r}
.412 \\
165\overline{)68.0000} \\
660 \\
\hline
200 \\
165 \\
\hline
350 \\
330 \\
\hline
200 \\
\end{array}
$$

Therefore,

$$
\frac{68}{165} = .41212 \ldots = .4\overline{12} \quad \blacksquare
$$

Not only is it true that the decimal expansion of any rational number is repeating but, remarkably, any repeating decimal represents a rational number.

Example E (Repeating Decimals as Rational Numbers)   Write $.\overline{24}$ as the ratio of two integers.

Solution   Let $x = .\overline{24} = .242424 \ldots$. Then $100x = 24.242424 \ldots$. Subtract x from $100x$ and simplify.

$$100x = 24.2424 \ldots$$
$$x = .2424 \ldots$$
$$99x = 24$$
$$x = \frac{24}{99} = \frac{8}{33}$$

We multiplied x by 100 because x is a decimal that repeats in a two-digit group. If the decimal had repeated in a three-digit group, we would have multiplied by 1000. ∎

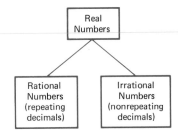

Real Numbers

Rational Numbers (repeating decimals)

Irrational Numbers (nonrepeating decimals)

Figure 14

We conclude that the rational numbers are precisely those numbers that have repeating decimal representations. But what about nonrepeating, unending decimals like

$$.12112111211112 \ldots$$

They represent the **irrational numbers.** And they, together with the rational numbers, constitute the real numbers (Figure 14).

We showed that $\sqrt{2}$ is not rational (that is, it is irrational). It too has a decimal expansion.

$$\sqrt{2} = 1.414213 \ldots$$

Actually, the decimal expansion of $\sqrt{2}$ is known to several thousand places. It does not repeat. It cannot. It is a fact of mathematics.

We have supposed that you are familiar with the symbols $\sqrt{2}, \sqrt{3}, \ldots$ from earlier courses. For completeness, we now make the meaning of these symbols precise.

Square Roots and Cube Roots

Every positive number has two square roots. For example, 3 and -3 are the two square roots of 9: this is so because $3^2 = 9$ and $(-3)^2 = 9$. If a is positive, the symbol $\sqrt{a}$ always denotes the *positive* square root of a. Thus $\sqrt{9} = 3$, and the two square roots of 7 are $\sqrt{7}$ and $-\sqrt{7}$. For the present, $\sqrt{-9}$ is a meaningless symbol since there is no real number whose square is -9. To make sense of the square root of a negative number requires a further extension of the number system, a topic we take up in Section 1-6.

CAUTION

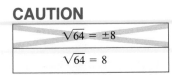

$\sqrt{64} = \pm 8$

$\sqrt{64} = 8$

In contrast, $\sqrt[3]{a}$, called the cube root of a, makes sense for any real number a. It is the unique real number whose cube is a. Thus $\sqrt[3]{8} = 2$ since $2^3 = 8$, and $\sqrt[3]{-64} = -4$ since $(-4)^3 = -64$.

We postpone the discussion of general nth roots until Section 6-1. In the meantime, we shall use square roots and cube roots in examples and problems.

PROBLEM SET 1-3

A. Skills and Techniques

Write the prime factorization of each number. See Example A.

1. 250

2. 504

3. 200

4. 2079

5. 2100

6. 1650

Find each of the following using the information you obtained in Problems 1–6. See Example B.

7. lcm(250, 200)

8. lcm(504, 2079)

9. lcm(250, 2100)

10. lcm(504, 1650)

11. lcm(250, 200, 2100)

12. lcm(504, 2079, 1650)

Calculate each of the following, using the results of Problems 7–12. See Example C.

13. $\dfrac{3}{250} + \dfrac{17}{200}$

14. $\dfrac{5}{504} - \dfrac{1}{2079}$

15. $\dfrac{7}{250} - \dfrac{1}{2100}$

16. $\dfrac{13}{504} + \dfrac{13}{1650}$

17. $\dfrac{3}{250} - \dfrac{17}{200} + \dfrac{11}{2100}$

18. $\dfrac{13}{504} + \dfrac{13}{1650} - \dfrac{17}{2079}$

Find the repeating decimal expansion for each number. Use the bar notation for your answer. See Example D.

19. $\dfrac{2}{3}$

20. $\dfrac{3}{5}$

21. $\dfrac{5}{8}$

22. $\dfrac{13}{11}$

23. $\dfrac{6}{13}$

24. $\dfrac{4}{13}$

Write each of the following as a ratio of two integers, as in Example E.

25. $.\overline{7}$

26. $.\overline{123}$

27. $.2\overline{35}$

28. $.875$

29. $.3\overline{25}$

30. $.5\overline{21}$

31. $.3\overline{21}$

B. Applications and Extensions

32. Write the prime factorizations of 420 and 630.

33. Calculate $\frac{11}{4}20 \cdot \frac{11}{3} - (\frac{13}{4}20 + \frac{11}{6}30)$ and write your answer in reduced form.

34. Find the decimal expansion of $\frac{1}{7} + \frac{1}{9}$.

35. Write $.\overline{27} + .\overline{23}$ as the ratio of two integers.

36. Show that $.2\overline{9}$ and $.3\overline{0}$ are decimal expansions of the same rational number. What numbers have two decimal expansions?

37. Show that $\sqrt{3}$ is irrational by mimicking the proof given in the text for $\sqrt{2}$.

38. Show that $\sqrt{p}$ is irrational if p is a prime.

39. Show that both the sum and product of two rational numbers is rational.

40. Show by example that the sum and product of two irrational numbers can be rational.

41. Show that $\sqrt{2} + \frac{2}{3}$ is irrational. *Hint:* Let $\sqrt{2} + \frac{2}{3} = r$. By adding $-\frac{2}{3}$ to both sides, show that it is impossible for r to be rational.

42. Show that the sum of a rational number and an irrational number is irrational.

43. Show that $\frac{2}{3}\sqrt{2}$ is irrational.

44. Show that the product of a nonzero rational number and an irrational number is irrational.

45. Which of the following numbers are rational?

(a) $\sqrt{5} + .\overline{12}$
(b) $\sqrt{3}(\sqrt{3} + 2)$
(c) $\sqrt[6]{\frac{8}{1}}$
(d) $\sqrt{2}\,\sqrt{8}$
(e) $(1 + \sqrt{2})(1 - \sqrt{2})$
(f) $\sqrt{\frac{27}{7}5}$
(g) $.12\sqrt{2}$
(h) $.\overline{12}\sqrt{2}$
(i) $(.12)(.\overline{12})$

46. Write a positive rational number smaller than $.000001$ and a positive irrational number smaller than $(.000001)\sqrt{2}$. Is there a smallest positive rational number? A smallest positive irrational number?

47. Show that the sum of the squares of the diagonals of a parallelogram is equal to the sum of the squares of the four sides.

48. Consider the rectangular box shown in Figure 15 with sides of length a, b, and c. Find a formula for d, the distance between the diagonally opposite corners A and B.

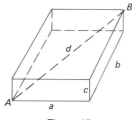

Figure 15

49. For the box of Figure 15, find a formula for d in terms of the lengths e, f, and g of the face diagonals that meet at A.

50. TEASER Consider the box of Figure 15 to be closed. Find (in terms of a, b, and c) the length of the shortest path that a spider could take in crawling from A to B on the surface of the box.

1-4 FUNDAMENTAL PROPERTIES OF THE REAL NUMBERS

Playing by the Rules

Chess has rules; so does mathematics. You can't move a pawn backwards; you must not divide by zero. Chess without rules is no game at all. Numbers without definite properties are worthless curiosities. A good chess player and a good mathematician play by the rules.

Now we face an important question. Are the real numbers adequate to handle all the applications we are likely to encounter? Or will some problems arise that will force us to enlarge our number system again? The answer is that the real numbers are sufficient for most purposes. Except for one type of problem, to be described in Section 1-6, we will do our work within the context of the real numbers. From now on, when we say number with no qualifying adjective, we mean real number. You can count on it.

All of this suggests that we ought to pay some attention to the fundamental properties of the real numbers. By a fundamental property, we mean something so basic that we must understand it, and to understand means more than to memorize. Understanding a property means to see the purpose the property serves, to recognize its implications, and to be able to derive other things from it.

Associative Property

Addition and multiplication are the fundamental operations; subtraction and division are offshoots of them. These operations are binary operations, that is, they work on two numbers at a time. Thus $3 + 4 + 5$ is technically meaningless. We ought to write either $3 + (4 + 5)$ or $(3 + 4) + 5$. But, luckily, it really does not matter; we get the same answer either way. Addition is **associative.**

$$a + (b + c) = (a + b) + c$$

Thus we can write $3 + 4 + 5$ or even $3 + 4 + 5 + 6 + 7$ without ambiguity. The answer will be the same regardless of the order in which the additions are done.

Addition and multiplication are like Siamese twins: What is true for one is quite likely to be true for the other. Thus multiplication, too, is associative.

$$a \cdot (b \cdot c) = (a \cdot b) \cdot c$$

If we wish, we can write $a \cdot b \cdot c$ with no parentheses at all.

Commutative Property

It makes some difference whether you first put on your slippers and then take a bath, or vice versa—the difference between wet and dry slippers. But for addition or multiplication, the order of the two numbers does not matter. Both operations are **commutative.**

$$a + b = b + a$$

$$a \cdot b = b \cdot a$$

Thus

$$3 + 4 = 4 + 3$$

$$3 + 4 + 6 = 4 + 3 + 6 = 6 + 4 + 3$$

and

$$7 \cdot 8 \cdot 9 = 8 \cdot 7 \cdot 9 = 9 \cdot 8 \cdot 7$$

Example A (Associative and Commutative Properties) Calculate each number.

(a) $31.9 + 45 + 68.1 + 43.2 + 155$ (b) $25 \cdot \dfrac{3}{7} \cdot \dfrac{2}{5} \cdot 14$

Solution Write these numbers as follows.

(a) $(31.9 + 68.1) + (45 + 155) + 43.2 = 100 + 200 + 43.2 = 343.2$

(b) $\left(25 \cdot \dfrac{2}{5}\right)\left(\dfrac{3}{7} \cdot 14\right) = 10 \cdot 6 = 60$ ∎

Neutral Elements

To be neutral is to sit on the sidelines and refuse to do battle. A neutral party can be ignored; the outcome is not affected by its presence. Thus we call 0 the **neutral element** for addition; its presence can be ignored in an addition.

$$a + 0 = 0 + a = a$$

Similarly, 1 is the neutral element for multiplication since

$$a \cdot 1 = 1 \cdot a = a$$

The numbers 0 and 1 are also called the **identity elements** for addition and multiplication, respectively.

While 0 is inactive in addition, its effect in multiplication is overwhelming; any number multiplied by 0 is completely wiped out.

$$a \cdot 0 = 0 \cdot a = 0$$

We have not put this property in a box; it is not quite as basic as the others since it can be derived from them (see Problem 37). However, you should still know it.

Inverses

The numbers 3 and -3 are like an acid and a base: When you add them together, they neutralize each other. We refer to them as inverses of each other. In fact, every number a has a unique **additive inverse** $-a$ (also called the negative of a). It satisfies

$$a + (-a) = (-a) + a = 0$$

Similarly, every number a different from 0 has a unique **multiplicative inverse** a^{-1} satisfying

$$a \cdot a^{-1} = a^{-1} \cdot a = 1$$

Thus, $3^{-1} = \frac{1}{3}$ since $3 \cdot \frac{1}{3} = 1$. In fact, for any $a \neq 0$, $a^{-1} = 1/a$, and for this reason we often say the "reciprocal of a" rather than the "multiplicative inverse of a."

Distributive Property

CAUTION

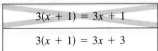

$3(x + 1) = 3x + 1$

$3(x + 1) = 3x + 3$

When we indicate an addition and a multiplication in the same expression, we face a problem. For example, what do we mean by $3 + 4 \cdot 2$? If we mean $(3 + 4) \cdot 2$, the answer is 14; if we mean $3 + (4 \cdot 2)$, the answer is 11. Most of us would not give the first answer. We are so familiar with a convention that we use it without thinking, but now is a good time to emphasize it. In any expression involving additions and multiplications which has no parentheses, we agree to do all the multiplications first. Thus

$$4 \cdot 5 + 3 = 20 + 3 = 23$$

and

$$4 \cdot 5 + 6 \cdot 2 = 20 + 12 = 32$$

We can always overrule this agreement by inserting parentheses. For example,

$$4 \cdot (5 + 3) = 4 \cdot 8 = 32$$

The agreement just described is a matter of convenience; no law forces it upon us. However, the **distributive property,**

$$a \cdot (b + c) = a \cdot b + a \cdot c$$
$$(b + c) \cdot a = b \cdot a + c \cdot a$$

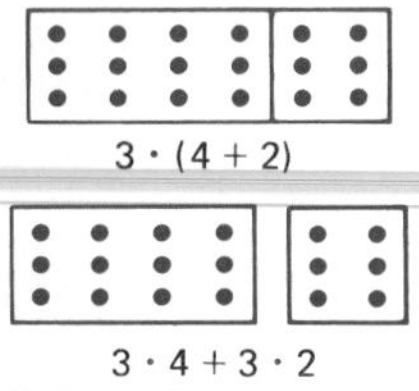

$3 \cdot (4 + 2)$

$3 \cdot 4 + 3 \cdot 2$

Figure 16

is not a matter of choice or convenience. Rather, it is another of the fundamental properties of numbers. That it must hold for the positive integers is almost obvious, as may be seen by examining Figure 16 in the margin. But we assert that it is just as true that

$$\frac{1}{2} \cdot (\sqrt{2} + \pi) = \frac{1}{2} \cdot \sqrt{2} + \frac{1}{2} \cdot \pi$$

```
   34
 × 65
─────
  170
+ 204
─────
 2210
```

Figure 17

Actually, we use the distributive property all the time, often without realizing it. The familiar calculation in Figure 17 is really a shorthand version of

$$65 \cdot 34 = (5 + 60)34 = 5 \cdot 34 + 60 \cdot 34$$
$$= 170 + 2040 = 2210$$

Subtraction and Division

Addition and multiplication are the basic operations; subtraction and division are dependent on them. Subtraction is the addition of an additive inverse and division is multiplication by a reciprocal. Thus

$$a - b = a + (-b)$$

and

$$\frac{a}{b} = a \div b = a \cdot b^{-1} = a \cdot \frac{1}{b}$$

All the familiar manipulations of arithmetic and algebra rest finally on the properties highlighted in boxes and the definitions of subtraction and division. Our next example and Problems 11–20 focus on this.

Example B (Justifying Manipulations) What properties justify each of the following?

(a) $6 + [5 + (-6)] = 5$ (b) $-\dfrac{1}{3}\left(\dfrac{6}{7} - \dfrac{9}{11}\right) = -\dfrac{2}{7} + \dfrac{3}{11}$

Solution
(a) Commutativity of addition; associativity of addition; additive inverse; additive identity:

$$6 + [5 + (-6)] = 6 + [(-6) + 5] = [6 + (-6)] + 5 = 0 + 5 = 5$$

(b) The definition of subtraction and the distributive property:

$$-\frac{1}{3}\left(\frac{6}{7} - \frac{9}{11}\right) = -\frac{1}{3}\left(\frac{6}{7} + \left(-\frac{9}{11}\right)\right)$$

$$= -\frac{1}{3} \cdot \frac{6}{7} + \left(-\frac{1}{3}\right)\left(-\frac{9}{11}\right) = -\frac{2}{7} + \frac{3}{11} \quad\blacksquare$$

Clearly, $3 - 5 \neq 5 - 3$ and $3 \div 5 \neq 5 \div 3$, which shows us that subtraction and division are not commutative operations. Neither are they associative, as Example C and Problem 31 will make clear. That which appears to be true at first glance may be false.

Example C (Apparent Rules Must Be Checked) Which of the following are true for all real numbers a, b, and c?

(a) $(a - b) - c = a - (b - c)$ (b) $(a + b) \div c = (a \div c) + (b \div c)$

Solution
(a) Not true; subtraction is not associative. For example, if $a = 12$, $b = 9$, and $c = 5$, then

$$(a - b) - c = (12 - 9) - 5 = 3 - 5 = -2$$

$$a - (b - c) = 12 - (9 - 5) = 12 - 4 = 8$$

(b) True, provided that $c \neq 0$:

$$(a + b) \div c = (a + b)\left(\frac{1}{c}\right) = a\left(\frac{1}{c}\right) + b\left(\frac{1}{c}\right)$$

$$= (a \div c) + (b \div c) \quad\blacksquare$$

CAUTION

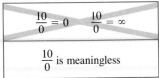

$\dfrac{10}{0} = 0$ $\dfrac{10}{0} = \infty$

$\dfrac{10}{0}$ is meaningless

There is an important restriction with regard to division; we never divide by zero. Why do we exclude $\frac{4}{0}$, $-\frac{6}{0}$, $\frac{10}{0}$, and similar expressions from our consideration? If $\frac{4}{0}$ were a number q, that is, if $\frac{4}{0} = q$, then $4 = 0 \cdot q = 0$, which is nonsense. The symbol $\frac{0}{0}$ is meaningless for a different reason: If $\frac{0}{0} = p$, then $0 = 0 \cdot p$, which is true for any number p. We choose to avoid such ambiguities by excluding division by zero.

PROBLEM SET 1-4

A. Skills and Techniques

Find the following sums or products using the basic properties we have introduced. If you can do it all in your head, that will be fine. See Example A.

1. $420 + 431 + 580$
2. $99{,}985 + 67 + 15$
3. $938 + 400 + 300 + 17$
4. $8.75 + 14 + 36 + 1.25$
5. $\dfrac{11}{13} + 43 + \dfrac{2}{13} + 17$
6. $\dfrac{15}{8} + \dfrac{5}{6} + \dfrac{1}{6} + \dfrac{9}{8}$
7. $6 \cdot \dfrac{3}{4} \cdot \dfrac{1}{6} \cdot 4$
8. $99 + 98 + 97 + 3 + 2 + 1$
9. $5 \cdot \dfrac{1}{3} \cdot \dfrac{2}{5} \cdot 6 \cdot \dfrac{1}{2}$
10. $(.25)(363)(400)\left(\dfrac{1}{3}\right)$

Name the properties that justify each of the following. See Example B.

11. $\dfrac{5}{6}\left(\dfrac{3}{4} \cdot 12\right) = \left(12 \cdot \dfrac{5}{6}\right)\dfrac{3}{4}$
12. $4 + (.52 - 2) = (4 - 2) + .52$
13. $(2 + 3) + 4 = 4 + (2 + 3)$
14. $9(48) = 9(40) + 9(8)$
15. $6 + [-6 + 5] = [6 + (-6)] + 5 = 5$
16. $6\left[\dfrac{1}{6} \cdot 4\right] = \left[6 \cdot \dfrac{1}{6}\right]4 = 4$
17. $-(\sqrt{5} + \sqrt{3} - 5) = -\sqrt{5} - \sqrt{3} + 5$
18. $\left(\dfrac{b}{c}\right)\left(\dfrac{b}{c}\right)^{-1} = 1$
19. $(x + 4)(x + 2) = (x + 4)x + (x + 4)2$
20. $(a + b)a^{-1} = 1 + \dfrac{b}{a}$

Which of the following equalities in Problems 21–32 are true for all choices of a, b, and c? If false, provide a demonstration similar to that in part (b) of Example C. Assume that divisions by 0 are excluded.

21. $a - (b - c) = a - b + c$
22. $a + bc = ac + bc$
23. $a \div (b + c) = (a \div b) + (a \div c)$
24. $(a + b)^{-1} = a^{-1} + b^{-1}$
25. $ab(a^{-1} + b^{-1}) = b + a$
26. $a(a + b + c) = a^2 + ab + ac$
27. $(a + b)(a^{-1} + b^{-1}) = 1$
28. $(a + b)(a + b)^{-1} = 1$
29. $(a + b)(a + b) = a^2 + b^2$
30. $(a + b)(a - b) = a^2 - b^2$
31. $a \div (b \div c) = (a \div b) \div c$
32. $a \div (b \div c) = (a \cdot c) \div b$

B. Applications and Extensions

33. Let # denote the exponentiation operation, that is $a \# b = a^b$.
    - **(a)** Calculate $4 \# 3$, $2(3 \# 2)$, and $2 \# (3 \# 2)$.
    - **(b)** Is # commutative?
    - **(c)** Is # associative?
34. Which of the following operations commute with each other, that is, which give the same answer when performed on a number in either order?
    - **(a)** Doubling and tripling
    - **(b)** Doubling and cubing
    - **(c)** Squaring and cubing
    - **(d)** Squaring and taking the negative
    - **(e)** Squaring and taking the reciprocal
    - **(f)** Adding 3 and taking the negative
    - **(g)** Multiplying by 3 and dividing by 2
35. Show that if a and b are nonzero, then $(ab)^{-1} = b^{-1}a^{-1}$, $(a^{-1})^{-1} = a$, and $(a/b)^{-1} = b/a$.
36. Show that if b, c, and d are nonzero, then $a/b \div c/d = (a \div c)/(b \div d)$.
37. We claimed in the text that we could show that $a \cdot 0 = 0$ for all a using only the properties displayed in boxes. Justify each of the following equalities by using one of these properties.

$$0 = -(a \cdot 0) + a \cdot 0$$
$$= -(a \cdot 0) + a \cdot (0 + 0)$$
$$= -(a \cdot 0) + (a \cdot 0 + a \cdot 0)$$
$$= [-(a \cdot 0) + a \cdot 0] + a \cdot 0$$
$$= 0 + a \cdot 0$$
$$= a \cdot 0$$

38. Demonstrate $(-a) \cdot b = -(a \cdot b)$ using only the properties displayed in boxes and Problem 37.

39. Demonstrate $(-a) \cdot (-b) = a \cdot b$ using only the properties displayed in boxes and Problems 37 and 38.

40. TEASER Let {Ann, Betty, Connie, Debra, and Eva} be five girls arranged according to height so Ann is shorter than Betty, Betty is shorter than Connie, and so on. Define $X + Y$ to be the taller of X and Y and $X \cdot Y$ to be the shorter of X and Y. It is to be understood that $X + X = X$ and $X \cdot X = X$.

(a) Calculate Connie · (Betty · Connie), Eva + (Debra · Betty), and Ann · (Debra + Connie).

(b) Which of the properties displayed in boxes hold?

(c) The distributive law $X + (Y \cdot Z) = (X + Y) \cdot (X + Z)$ of addition over multiplication fails for numbers. Does it hold in the five-girl system?

1-5 ORDER AND ABSOLUTE VALUE

"I wonder, sir, if you would be willing to move to the end?"

We may question the bartender's tact, but not his mathematical taste. He has a deep feeling for an important notion that we call order; it is intimately tied up with the real number system. To describe this notion, we introduce a special symbol $<$; it stands for the phrase "is less than."

We begin by recalling that every real number (except 0) falls into one of two classes. Either it is positive or it is negative. Then, given two real numbers a and b, we say that

$$a < b \text{ if } b - a \text{ is positive}$$

Thus $-3 < -2$ since $-2 - (-3)$ is positive. Similarly, $3 < \pi$ since

$$\pi - 3 = 3.14159 \ldots - 3 = .14159 \ldots$$

which is a positive number.

Another and more intuitive way to think about the symbol $<$ is to relate it to the real number line. To say that $a < b$ means that a is to the left of b on the real line (Figure 18).

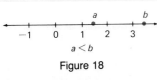

Figure 18

The symbol $<$ has a twin, denoted by $>$, which is read "is greater than." If you know how $<$ behaves, you automatically know how $>$ behaves. Thus there is no need to say much about $>$. It is enough to note that $b > a$ means exactly the same thing as $a < b$. In particular, $b > 0$ and $0 < b$ mean the same thing; both say that b is a positive number. Relations like $a < b$ and $b > a$ are called **inequalities.**

Properties of $<$

Homer   Ichabod   Jehu

Figure 19

If Homer is shorter than Ichabod and Ichabod is shorter than Jehu, then of course Homer is shorter than Jehu (Figure 19). This and other properties of the "less than" relation seem almost obvious. The following is a formal statement of the three most important properties.

Properties of Inequalities

1. (Transitivity) If $a < b$ and $b < c$, then $a < c$.
2. (Addition) If $a < b$, then $a + c < b + c$.
3. (Multiplication) If $a < b$ and $c > 0$, then $a \cdot c < b \cdot c$
   If $a < b$ and $c < 0$, then $a \cdot c > b \cdot c$

Property 2 says that you can add the same number to both sides of an inequality. It also says that you can subtract the same number from both sides, since c can be negative. Property 3 has a catch. Notice that if we multiply both sides by a positive number, we preserve the direction of the inequality: however, if we multiply by a negative number, we reverse the direction of the inequality. Thus

$$2 < 3$$

and

$$2 \cdot 4 < 3 \cdot 4$$

but

$$2(-4) > 3(-4)$$

These facts are shown on the number line in Figure 20.

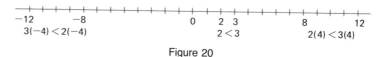

-12 -8 0 2 3 8 12
$3(-4) < 2(-4)$ $2 < 3$ $2(4) < 3(4)$

Figure 20

Division, which is equivalent to multiplication by the reciprocal, also satisfies property 3.

The $\leq$ Relation

In addition to its twin $>$, the symbol $<$ has another relative. It is denoted by $\leq$ and is read "is less than or equal to." We say

$$a \leq b \text{ if } b - a \text{ is either positive or zero}$$

For example, it is correct to say $2 \leq 3$; it is also correct to say $2 \leq 2$. This new relation behaves very much like $<$. In fact, if we put a bar under every $<$ and $>$ in the properties of inequalities displayed above, the resulting statements will be correct. Naturally, $b \geq a$ means the same thing as $a \leq b$.

Example A (Ordering Numbers)   Arrange the numbers π, $\frac{22}{7}$, $3.1415\overline{9}$, $\frac{135}{43}$, and $\frac{88}{28}$ from smallest to largest using the symbols $<$ and $\leq$.

Solution   Since $\frac{88}{28} = \frac{4(22)}{4(7)}$, we see that $\frac{88}{28} = \frac{22}{7}$. Converting to decimals gives $\frac{22}{7} = 3.14285714. \ldots$, $\frac{135}{43} = 3.13953488. \ldots$, and $\pi = 3.14159265. \ldots$ Thus

$$\frac{135}{43} < \pi < 3.1415\overline{9} < \frac{22}{7} \leq \frac{88}{28} \quad \blacksquare$$

Intervals

We can use the order symbols $<$ and $\leq$ to describe intervals on the real line. When we write $-1.5 < x \leq 3.2$, we mean that x is simultaneously greater than -1.5 and less than or equal to 3.2. The set of all such numbers is the interval shown in Figure 21.

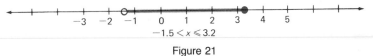

$-1.5 < x \leq 3.2$

Figure 21

The small circle at the left indicates that -1.5 is left out: the heavy dot at the right indicates that 3.2 is included. Figure 22 illustrates other possibilities.

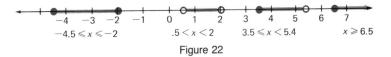

Figure 22

Sometimes we use set notation to describe an interval. For example, to denote the interval at the far left above, we could write $\{x: -4.5 \leq x \leq -2\}$, which is read "the set of all x such that x is greater than or equal to -4.5 and less than or equal to -2."

We should not write nonsense such as

$$3 < x < 2 \quad \text{or} \quad 2 > x < 3$$

The first is simply a contradiction. There is no number both greater than 3 and less than 2. The second says that x is both less than 2 and less than 3. This should be written simply as $x < 2$.

Absolute Value

Often we want to describe the size of a number, not caring whether it is positive or negative. To do this, we introduce the concept of absolute value, symbolized by two vertical bars $| \quad |$. It is defined by

$$|a| = \begin{cases} a & \text{if } a \geq 0 \\ -a & \text{if } a < 0 \end{cases}$$

This two-pronged definition can cause confusion. It says that if a number is positive or zero, its absolute value is itself. But if a number is negative, then its absolute value is its additive inverse (which is a positive number). For example,

$$|7| = 7$$

since $a = 7$ is positive. On the other hand,

$$|-7| = -(-7) = 7$$

since $a = -7$ is negative. Note that $|0| = 0$.

We may also think of $|a|$ geometrically. It represents the distance between a and 0 on the number line. More generally, $|a - b|$ is the distance between a and b (Figure 23)). The number $|b - a|$ represents this same distance.

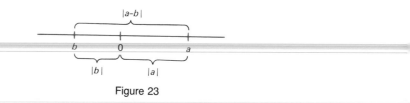

Figure 23

You should satisfy yourself that this is true for arbitrary choices of a and b, for example, that $|7 - (-3)|$ really is the distance between 7 and -3.

Example B (Removing Absolute Values)   Write each of the following without the absolute value symbol. Then show the corresponding interval(s) on the real number line.
(a) $|x| < 3$ (b) $|x - 2| < 3$ (c) $|x - 4| \geq 2$

Solution
(a) Since $|x|$, the distance between x and 0 on the number line, is less than 3, x must be between -3 and 3. We write $-3 < x < 3$; this interval is displayed in Figure 24.

Figure 24

(b) Here $x - 2$, instead of x, must be between -3 and 3, that is,

$$-3 < x - 2 < 3$$

Adding 2 to each quantity gives $-1 < x < 5$, the interval shown in Figure 25.

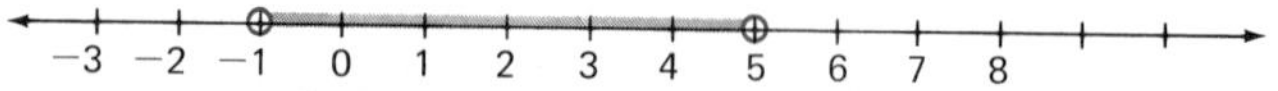

Figure 25

(c) The distance between x and 4 is greater than or equal to 2. This means that $x - 4 \leq -2$ or $x - 4 \geq 2$. That is,

$$x \leq 2 \quad \text{or} \quad x \geq 6$$

The corresponding two-part set is graphed in Figure 26.

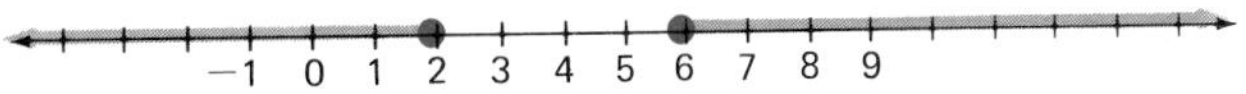

Figure 26

In set notation, we may write the solution set for the inequality in part (c) as $\{x : x \leq 2 \quad \text{or} \quad x \geq 6\}$ or equivalently as $\{x : x \leq 2\} \cup \{x : x \geq 6\}$. ∎

The properties of absolute values are straightforward and easy to remember.

Properties of Absolute Values

1. $|a \cdot b| = |a| \cdot |b|$

2. $\left| \dfrac{a}{b} \right| = \dfrac{|a|}{|b|}$

3. $|a + b| \leq |a| + |b|$

4. $|-a| = |a|$

5. $|a|^2 = a^2$

Actually, Properties 4 and 5 follow immediately from Property 1, since

$$|-a| = |(-1) \cdot a| = |-1| \cdot |a| = 1 \cdot |a| = |a|$$

and

$$|a|^2 = |a| \cdot |a| = |a \cdot a| = |a^2| = a^2$$

There is an important connection between absolute values and square roots, namely,

$$\sqrt{a^2} = |a|$$

For example, $\sqrt{6^2} = |6| = 6$ and $\sqrt{(-6)^2} = |-6| = 6$.

PROBLEM SET 1-5

A. Skills and Techniques

In Problems 1–12, replace the symbol # by the appropriate symbol: $<$, $>$, or $=$.

1. $1.5 \ \# \ -1.6$

2. $-2 \ \# \ -3$

3. $\sqrt{2} \ \# \ 1.4$

4. $\pi \ \# \ 3.15$

5. $\dfrac{1}{5} \ \# \ \dfrac{1}{6}$

6. $-\dfrac{1}{5} \ \# \ -\dfrac{1}{6}$

7. $5 - \sqrt{2} \ \# \ 5 - \sqrt{3}$

8. $\sqrt{2} - 5 \ \# \ \sqrt{3} - 5$

9. $-\dfrac{3}{16}\pi \ \# \ -\dfrac{1}{17}\pi$

10. $\left(\dfrac{16}{17}\right)^2 \ \# \ \left(\dfrac{17}{18}\right)^2$

11. $|-\pi + (-2)| \ \# \ |-\pi| + |-2|$

12. $|\pi - 2| \ \# \ \pi - 2$

13. Order the following numbers from least to greatest. See Example A.

$$\frac{3}{4}, -2, \sqrt{2}, \frac{-\pi}{2}, -\frac{3}{2}\sqrt{2}, \frac{43}{24}$$

14. Order the following numbers from least to greatest.

$$5 - 5, .37, -\sqrt{3}, \frac{14}{33}, -\frac{7}{4}, -\frac{49}{35}, \frac{3}{8}$$

Use a real number line to show the set of numbers that satisfy each given inequality.

15. $x < -4$ **16.** $x < 3$

17. $x \geq -2$ **18.** $x \leq 3$

19. $-1 < x < 3$ **20.** $2 < x < 5$

21. $0 < x \leq 3$ **22.** $-3 \leq x < 2$

23. $-\frac{1}{2} \leq x \leq \frac{3}{2}$ **24.** $-\frac{7}{4} \leq x \leq -\frac{3}{4}$

Write an inequality for each interval.

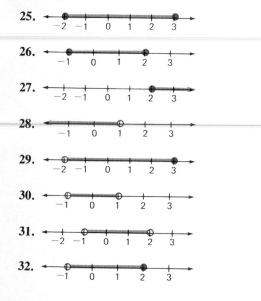

25. (number line marked $-2, -1, 0, 1, 2, 3$)

26. (number line marked $-1, 0, 1, 2, 3$)

27. (number line marked $-2, -1, 0, 1, 2, 3$)

28. (number line marked $-1, 0, 1, 2, 3$)

29. (number line marked $-2, -1, 0, 1, 2, 3$)

30. (number line marked $-1, 0, 1, 2, 3$)

31. (number line marked $-2, -1, 0, 1, 2, 3$)

32. (number line marked $-1, 0, 1, 2, 3$)

Write each of the following without the absolute value symbol. Show the corresponding interval(s) on the real number line. See Example B.

33. $|x| \leq 4$ **34.** $|x| \geq 2$

35. $|x - 3| < 2$ **36.** $|x - 5| < 1$

37. $|x + 1| \leq 3$ **38.** $|x + \frac{3}{2}| \leq \frac{1}{2}$

39. $|x - 5| > 5$ **40.** $|x + 3| < 3$

B. Applications and Extensions

41. Restate each of the following in mathematical symbols.
   (a) x is not greater than 12.
   (b) x is less than 3 and greater than or equal to -11.
   (c) x is not more than 4 units away from y.
   (d) x is at least 3 units away from 7.
   (e) x is closer to 5 than it is to y.

42. Show that if $a < b$, then $a < (a + b)/2 < b$.

43. Given that $\sqrt{2} = 1.414213 \ldots$ arrange the following numbers in increasing order.

$$1.414, \; 1.41\overline{4}, \; 1.4\overline{14}, \; 1.\overline{414}, \; \sqrt{2}, \; \frac{\sqrt{2} + 1.414}{2}$$

44. Show that if a, b, c, and d are positive numbers, then $a/b < c/d$ if and only if $ad < bc$.

45. Indicate which number is the larger.

   (a) $\frac{11}{46}$ or $\frac{6}{25}$ **(b)** $\frac{4}{17}$ or $\frac{7}{29.8}$

   (c) $\frac{17.1}{85}$ or $\frac{33}{165}$ **(d)** $\frac{11}{13}$ or $.\overline{846153}$

46. Given that $0 < a < b$, indicate how each of the following are related and justify your answer.
   (a) a^2 and b^2
   (b) $\sqrt{a}$ and $\sqrt{b}$
   (c) $1/a$ and $1/b$

47. Let a, b, and A denote the altitude, base, and area, respectively, of a triangle. If $10 < b < 12$ and $50 < A < 60$, what can you say about a?

48. If the radius r and height h of a cylinder satisfy $4.9 < r < 5.1$ and $10.2 < h < 10.4$, what can you say about the volume V of the cylinder?

49. If $y = 1/x$ and $|x - 4| < 2$, what can you say about y?

50. Show that $|a| < 1$ and $a \neq 0$ imply $a^2 < |a|$, but $|a| > 1$ implies $a^2 > |a|$.

51. Under what conditions is $|a + b| = |a| + |b|$?

52. Show that if $|a| \leq 1$, then $|2a + 3a^2 + 4a^3| \leq 9$.

53. Show that $\sqrt{a^2 + b^2} \leq |a| + |b|$.

54. **TEASER** Sam Slugger had a better batting average than Wes Weakbat during 1985 and again in 1986. If the hits and at bats are lumped together for the two years, does it follow that Sam has a better combined batting average than Wes? Be sure to justify your conclusion.

1-6 THE COMPLEX NUMBERS

$$i = \sqrt{-1}$$

"The Divine Spirit found a sublime outlet in that wonder of analysis, that portent of the ideal world, that amphibian between being and not-being, which we call the imaginary root of negative unity."

Gottfried Wilhelm Leibniz
1646–1716

As early as 1550, the Italian mathematician Raffael Bombelli had introduced numbers like $\sqrt{-1}$, $\sqrt{-2}$, and so on, to solve certain equations. But mathematicians had a hard time deciding whether they should be considered legitimate numbers. Even Leibniz, who ranks with Newton as the greatest mathematician of the late seventeenth century, called them amphibians between being and not being. He wrote them down, he used them in calculations, but he carefully covered his tracks by calling them imaginary numbers. Unfortunately, that name (which was actually first used by Descartes) has stuck, although these numbers are now well accepted by all mathematicians and have numerous applications in science. Let us see what these new numbers are and why they are needed.

Go back to the whole numbers 0, 1, 2, 3, ... We can easily solve the equation $x + 3 = 7$ within this system ($x = 4$). On the other hand, the equation $x + 7 = 3$ has no whole-number solution. To solve it, we need the negative integer -4. Similarly, we cannot solve $3x = 2$ in the integers. We can say that the solution is $\frac{2}{3}$ only after the rational numbers have been introduced. To their dismay, the Greeks discovered that $x^2 - 2 = 0$ had no rational solution. We conquered that problem by enlarging our family of numbers to the real numbers. But there are still simple equations without solutions. Consider $x^2 + 1 = 0$. Try as you will, you will never solve this equation within the real number system.

By now, our procedure is well established. When we need new numbers, we invent them. This time we invent a number denoted by i (or by $\sqrt{-1}$) which satisfies $i^2 = -1$. However, we cannot get by with just one number. For after we have adjoined it to the real numbers, we still must be able to multiply and add. Thus with i, we also need numbers such as

$$2i \qquad -4i \qquad (\tfrac{3}{2})i$$

which are called pure imaginary numbers. We also need
$$3 + 2i \qquad 11 + (-4i) \qquad \tfrac{3}{4} + \tfrac{3}{2}i$$
and it appears that we need even more complicated things such as
$$(3 + 8i + 2i^2 + 6i^3)(5 + 2i)$$
Actually, this last number can be simplified to $1 + 12i$, as we shall see later. In fact, no matter how many additions and multiplications we do, after the expressions are simplified we shall never have anything more complicated than a number of the form $a + bi$ (a fact that Figure 27 is meant to illustrate). Such numbers, that is, numbers of the form $a + bi$, where a and b are real, are called **complex numbers.** We refer to a as the **real part** and b as the **imaginary part** of $a + bi$. Since we shall agree that $0 \cdot i = 0$, it follows that $a + 0i = a$, and so every real number is automatically a complex number. If $b \neq 0$, then $a + bi$ is nonreal, and in this case $a + bi$ is said to be an imaginary number.

Addition and Multiplication

We cannot say anything sensible about operations for complex numbers until we agree on the meaning of equality. The definition that seems most natural is this:

$$a + bi = c + di \quad \text{means} \quad a = c \text{ and } b = d$$

That is, two complex numbers are equal if and only if their real parts and their imaginary parts are equal. As an example, suppose $x^2 + yi = 4 - 3i$. Then we know that $x = \pm 2$ and $y = -3$.

Now we can consider addition. Actually we have already used the plus sign in $a + bi$. That was like trying to add apples and bananas. The addition can be indicated but no further simplification is possible. We do not get apples and we do not get bananas; we get fruit salad.

When we have two numbers of the form $a + bi$, we can actually perform an addition. We just add the real parts and the imaginary parts separately—that is, we add the apples and we add the bananas. Thus
$$(3 + 2i) + (6 + 5i) = 9 + 7i$$
and more generally,

$$(a + bi) + (c + di) = (a + c) + (b + d)i$$

Example A (Adding Complex Numbers)   Write
$$14i + (3 - 7i) + (-4 - 2i)$$
in the form $a + bi$.

Solution
$$14i + (3 - 7i) + (-4 - 2i) = (3 - 4) + (14 - 7 - 2)i$$
$$= -1 + 5i \quad \blacksquare$$

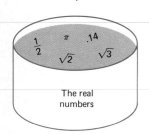

The real numbers

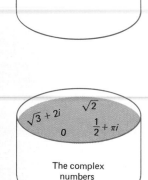

The complex numbers

Figure 27

When we consider multiplication, our desire to maintain the properties of Section 1-4 leads to a definition that looks complicated. Thus let us first look at some examples.

(i) $\qquad$ $2(3 + 4i) = 6 + 8i$ $\qquad$ (distributive property)

(ii) $\qquad$ $(3i)(-4i) = -12i^2 = 12$ $\qquad$ (commutative and associative properties and $i^2 = -1$)

(iii) $\qquad$ $(3 + 2i)(6 + 5i) = (3 + 2i)6 + (3 + 2i)5i$ $\qquad$ (distributive property)

$\qquad$ $= 18 + 12i + 15i + 10i^2$ $\qquad$ (distributive, commutative, and associative properties)

$\qquad$ $= 18 + (12 + 15)i - 10$ $\qquad$ (distributive property)

$\qquad$ $= 8 + 27i$ $\qquad$ (commutative property)

The same kind of reasoning applied to the general case leads to

$$(a + bi)(c + di) = (ac - bd) + (ad + bc)i$$

which we take as the definition of multiplication for complex numbers.

Example B (Multiplying Complex Numbers)   Write the product

$$(-2 + 5i)(3 + 2i) \text{ in the form } a + bi.$$

Solution

$$(-2 + 5i)(3 + 2i) = (-2 \cdot 3 - 5 \cdot 2) + (-2 \cdot 2 + 5 \cdot 3)i = -16 + 11i \qquad \blacksquare$$

Actually there is no need to memorize the formula for multiplication. Just do what comes naturally (that is, use familiar properties) and then replace i^2 by -1 wherever it arises, as in the following example.

$$(2 - 3i)(5 + 4i) = (10 - 12i^2) + (8i - 15i)$$

$$= (10 + 12) + (-7i)$$

$$= 22 - 7i$$

Consider the more complicated expression mentioned earlier. After noting that $i^3 = i^2i = -i$, we have

$$(3 + 8i + 2i^2 + 6i^3)(5 + 2i) = (3 + 8i - 2 - 6i)(5 + 2i)$$

$$= (1 + 2i)(5 + 2i)$$

$$= (5 + 4i^2) + (2i + 10i)$$

$$= 1 + 12i$$

Subtraction and Division

Subtraction is easy; we simply subtract corresponding real and imaginary parts. For example,

$$(3 + 6i) - (5 + 2i) = (3 - 5) + (6i - 2i)$$
$$= -2 + 4i$$

and

$$(5 + 2i) - (3 + 7i) = (5 - 3) + (2i - 7i)$$
$$= 2 + (-5i)$$
$$= 2 - 5i$$

Division is somewhat more difficult. We first note that $a - bi$ is called the **conjugate** of $a + bi$. Thus $2 - 3i$ is the conjugate of $2 + 3i$ and $-2 - 5i$ is the conjugate of $-2 - 5i$. Next, we observe that a complex number times its conjugate is a real number. For example,

$$(3 + 4i)(3 - 4i) = 9 + 16 = 25$$

and in general

$$(a + bi)(a - bi) = a^2 + b^2$$

Thus to simplify the quotient of two complex numbers (that is, to write it in the form $a + bi$), multiply numerator and denominator by the conjugate of the denominator. This replaces the complex denominator by a real one, as we now illustrate.

Example C (Dividing Complex Numbers)   Write the quotient $(2 + 3i)/(3 + 4i)$ in the form $a + bi$.

Solution

$$\frac{2 + 3i}{3 + 4i} = \frac{(2 + 3i)(3 - 4i)}{(3 + 4i)(3 - 4i)} = \frac{18 + i}{9 + 16} = \frac{18}{25} + \frac{1}{25}i \quad \blacksquare$$

Clearly, the procedure just described handles finding the reciprocal of a nonzero complex number, that is, finding the multiplicative inverse of a nonzero complex number.

Example D (Finding Multiplicate Inverses)   Write $(5 + 4i)^{-1}$ in $a + bi$ form.

Solution

$$(5 + 4i)^{-1} = \frac{1}{5 + 4i} = \frac{1(5 - 4i)}{(5 + 4i)(5 - 4i)}$$

$$= \frac{5 - 4i}{25 + 16} = \frac{5 - 4i}{41} = \frac{5}{41} - \frac{4}{41}i \quad \blacksquare$$

Powers of i

Since $i^2 = -1$, $i^3 = -i$, and $i^4 = i^2 i^2 = 1$, we see that all higher powers of i can be expressed as one of the four numbers: i, -1, $-i$, and 1. Thus $i^6 = i^4 i^2 = 1 \cdot i^2 = -1$ and $i^9 = i^4 i^4 i = i$.

Example E (Simplifying Powers of i)   Simplify the following.
(a) i^{51} (b) $(2i)^6 i^{19}$

Solution   We apply the rules of exponents.
(a) $i^{51} = i^{48} i^3 = (i^4)^{12} i^3 = 1^{12}(-i) = -i$
(b) $(2i)^6 (i^{19}) = 2^6 i^6 i^{19} = 64 i^{25} = 64 i^{24} i = 64 i$ ∎

Complex Roots

We say that a is an nth root of b if $a^n = b$. Thus 3 is a 4th root of 81 because $3^4 = 81$. Also, $2i$ is a 6th root of -64 because $(2i)^6 = 2^6 i^6 = 64 i^4 i^2 = -64$.

Example F (Roots of Complex Numbers)   Show that $1 + i$ is a 4th root of -4.

Solution
$$(1 + i)(1 + i)(1 + i)(1 + i) = (1 + i)^2 (1 + i)^2$$
$$= (1 + 2i - 1)(1 + 2i - 1)$$
$$= (2i)(2i)$$
$$= 4i^2$$
$$= -4 \quad ∎$$

A Genuine Extension

We assert that the complex numbers constitute a genuine enlargement of the real numbers. This means first of all that they include the real numbers, since any real number a can be written as $a + 0i$. Second, the complex numbers satisfy all the properties we discussed in Section 1-4 for the real numbers. The order properties of Section 1-5, however, do not apply to the complex numbers.

Figure 28 summarizes our development of the number systems. Is there any need to enlarge the number system again? The answer is no, and for a good reason. With the complex numbers, we can solve any equation that arises in algebra. Right now, we expect you to take this statement on faith.

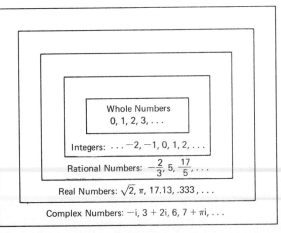

Figure 28

PROBLEM SET 1-6

A. Skills and Techniques

Carry out the indicated operations and write the answer in the form a + bi. See Examples A–C.

1. $(2 + 3i) + (-4 + 5i)$
2. $(3 - 4i) + (-5 - 6i)$
3. $5i - (4 + 6i)$
4. $(-3i + 4) + 8i$
5. $(3i - 6) + (3i + 6)$
6. $(6 + 3i) + (6 - 3i)$
7. $4i^2 + 7i$
8. $i^3 + 2i$
9. $i(4 - 11i)$
10. $(3 + 5i)i$
11. $(3i + 5)(2i + 4)$
12. $(2i + 3)(3i - 7)$
13. $(3i + 5)^2$
14. $(-3i - 5)^2$
15. $(5 + 6i)(5 - 6i)$
16. $(\sqrt{3} + \sqrt{2}i)(\sqrt{3} - \sqrt{2}i)$
17. $\dfrac{5 + 2i}{1 - i}$
18. $\dfrac{4 + 9i}{2 + 3i}$
19. $\dfrac{5 + 2i}{i}$
20. $\dfrac{-4 + 11i}{-i}$
21. $\dfrac{(2 + i)(3 + 2i)}{1 + i}$
22. $\dfrac{(4 - i)(5 - 2i)}{4 + i}$

Write each inverse in a + bi form, as in Example D.

23. $(2 - i)^{-1}$
24. $(3 + 2i)^{-1}$
25. $(\sqrt{3} + i)^{-1}$
26. $(\sqrt{2} - \sqrt{2}i)^{-1}$

Use the results above to perform the following divisions. For example, division by 2 − i is the same as multiplication by $(2 - i)^{-1}$

27. $\dfrac{2 + 3i}{2 - i}$
28. $\dfrac{1 + 2i}{3 + 2i}$
29. $\dfrac{4 - i}{\sqrt{3} + i}$
30. $\dfrac{\sqrt{2} + \sqrt{2}i}{\sqrt{2} - \sqrt{2}i}$

Simplify the numbers in Problems 31–38 as in Example E.

31. i^{94}
32. i^{39}
33. $(-i)^{17}$
34. $(2i)^3 i^{12}$
35. $\dfrac{(3i)^{16}}{(9i)^5}$
36. $\dfrac{2^8 i^{19}}{(-2i)^{11}}$
37. $(1 + i)^3$
38. $(2 - i)^4$

39. Show that i is a 4th root of 1, that is, show that $i^4 = 1$. Can you find three other 4th roots of 1?

40. Show that $-1 - i$ is a 4th root of -4. (See Example F.)
41. Show that $1 - i$ is a 4th root of -4.
42. Show that $-1 + i$ is a 4th root of -4.

B. Applications and Extensions

43. Perform the indicated operations and simplify.
    (a) $2 + 3i - i(4 - 3i)$
    (b) $(i^7 + 2i^2)/i^3$
    (c) $3 + 4i + (3 + 4i)(-1 + 2i)$
    (d) $i^{14} + (5 + 2i)/(5 - 2i)$
    (e) $\dfrac{5 - 2i}{3 + 4i} + \dfrac{5 + 2i}{3 - 4i}$
    (f) $\dfrac{(3 + 2i)(3 - 2i)}{2\sqrt{3} + i}$

44. Simplify.
    (a) $1 + i + i^2 + i^3 + \cdots + i^{16}$
    (b) $1 + \dfrac{1}{i} + \dfrac{1}{i^2} + \dfrac{1}{i^3} + \cdots + \dfrac{1}{i^{16}}$
    (c) $(1 - i)^{16}$

45. Find a and b so that each is true.
    (a) $(2 + i)(2 - i)(a + bi) = 10 - 4i$
    (b) $(2 + i)(a - bi) = 8 - i$

46. Show that -2, $1 + \sqrt{3}i$, and $1 - \sqrt{3}i$ are each cube roots of -8.

47. Let $x = -1/2 + (\sqrt{3}/2)i$. Find and simplify.
    (a) x^3
    (b) $1 + x + x^2$
    (c) $(1 - x)(1 - x^2)$

48. Show that $2 + i$ and $-2 - i$ are both solutions to $x^2 - 3 - 4i = 0$.

49. Find the two square roots of i, writing your answers in the form $a + bi$.

50. Is $1/(a + bi)$ ever equal to $1/a + 1/bi$?

51. A standard notation for the conjugate of the complex number x is $\bar{x}$. Thus, $\overline{a + bi} = a - bi$. Show that each of the following is true in general.
    (a) $\overline{x + y} = \bar{x} + \bar{y}$
    (b) $\overline{xy} = \bar{x}\bar{y}$
    (c) $\overline{x^{-1}} = (\bar{x})^{-1}$
    (d) $\overline{x/y} = \bar{x}/\bar{y}$

52. **TEASER** The absolute value of a complex number x is defined by $|x| = \sqrt{x\bar{x}}$. Note that this is consistent with the meaning of absolute value when x is a real number. Prove the triangle inequality for complex numbers, that is, show $|x + y| \leq |x| + |y|$.

CHAPTER 1 SUMMARY

Algebra is generalized arithmetic. In it, we use letters to stand for numbers and then manipulate these letters according to definite rules.

The number systems used in algebra are the **whole numbers,** the **integers,** the **rational numbers,** the **real numbers,** and the **complex numbers.** Except for the complex numbers, we can visualize all these numbers as labels for points along a line, called the **real line.** Rational numbers can be expressed as ratios of integers. Real numbers (and therefore rational numbers) can be expressed as **decimals.** In fact, the rational numbers can be expressed as **repeating decimals** and the **irrational** (not rational) numbers as **nonrepeating decimals.**

The fundamental properties of numbers include the **commutative, associative,** and **distributive laws.** Because of their special properties, 0 and 1 are called the **neutral elements** for addition and multiplication. Within the rational number system, the real number system, and the complex number system, all numbers have **additive inverses** and all but 0 have **multiplicative inverses.**

The real numbers are **ordered** by the relation $<$ (is less than); its properties should be well understood as should those of its relatives $\leq$, $>$, and $\geq$. The symbol $|\;|$, read **absolute value,** denotes the magnitude of a number regardless of whether it is positive or negative.

CHAPTER 1 REVIEW PROBLEM SET

In Problems 1–10, write True or False in the blank. If false, provide a counterexample. Since all letters are to represent arbitrary real numbers, providing a counterexample to a false statements means giving specific values for the letters which make the statement false.

_____ **1.** $-3(x - 2) = -3x + 6$

_____ **2.** $\dfrac{2x + 1}{2} = x + 1$

_____ **3.** $\sqrt{x^2} = x$

_____ **4.** $.9\overline{9} = 1$

_____ **5.** If $a < b$, then $|a| < |b|$.

_____ **6.** If $a > b > 0$, then $a^2 > b^2$.

_____ **7.** $(a + b)^2 = a^2 + b^2$

_____ **8.** $|-x| = x$

_____ **9.** $(a/b)c = (ac)/b$, provided that $b \neq 0$.

_____ **10.** $(a + b)^{-1} = a^{-1} + b^{-1}$

11. State the associative law of multiplication.

12. State the commutative law of addition.

13. Define the term *rational number*.

14. Perform the indicated operations and simplify.

(a) $\dfrac{11}{24} - \dfrac{7}{12} + \dfrac{4}{3}$ (b) $\dfrac{9}{28} \cdot \dfrac{7}{18} \cdot \dfrac{8}{5}$

(c) $\dfrac{2}{3}\left[\dfrac{1}{4} - \left(\dfrac{5}{6} - \dfrac{2}{3}\right)\right]$

(d) $3\left[\dfrac{5}{4}\left(\dfrac{2}{3} + \dfrac{8}{15} - \dfrac{1}{5}\right) + \dfrac{1}{6}\right]$

15. Perform the indicated operations and simplify.

(a) $\dfrac{\frac{11}{30}}{\frac{33}{25}}$ (b) $\dfrac{\frac{3}{4} - \frac{1}{12} + \frac{3}{8}}{\frac{3}{4} + \frac{5}{12} - \frac{7}{8}}$

(c) $3 + \dfrac{\frac{3}{4} - \frac{7}{8}}{\frac{5}{12}}$ (d) $\dfrac{72}{25}\left(\dfrac{9}{4} - \dfrac{11}{6}\right)^3$

16. Assuming that $\sqrt{3}$ is irrational, show that $2 - \sqrt{3}$ is irrational.

17. The altitude of a triangle is 5 centimeters shorter than its base. Write an expression for the area of the triangle in terms of its base length x.

18. A rectangle has a perimeter of 20 feet. If one side is x feet long, write an expression for the area of the rectangle in terms of x.

19. An open rectangular box (that is, without a top) is to be made from a square piece of cardboard 12 centimeters on a side by cutting identical squares of side length x from each of the four corners. Express the volume of the resulting box in terms of x.

20. Express in terms of x the area of an equilateral triangle of side length x. *Hint:* First determine an expression for the altitude using the Pythagorean Theorem.

21. A closed circular cylinder has base radius x and height

y. Express in terms of x and y the total surface area of the cylinder.

22. The base dimensions of an open rectangular box are x centimeters by y centimeters with $x < y$. The height of the box is half its width. Write a formula for the outer surface area A of the box in terms of x and y.

23. Suppose that an airplane can fly x miles per hour in still air and that the wind is blowing from east at y miles per hour. Express in terms of x and y the time it will take for the plane to fly 100 miles due east and return.

24. A window has the shape of a rectangle x inches wide and y inches high topped by a semicircle. Express the area of the window in terms of x and y.

25. An enclosure has the shape of a rectangle adjoined by an isosceles triangle as shown in Figure 29. Express the area of the enclosure in terms of x, y, and z.

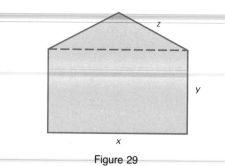

Figure 29

26. Write the prime factorizations of 500 and 360.

27. Determine the least common multiple of 500 and 360.

28. Perform the subtraction $\frac{7}{500} - \frac{7}{360}$ and write your answer in reduced form.

29. Write the least common multiple of $15 \cdot 40 \cdot 18$ and $63 \cdot 72$ in its prime factored form.

30. Write the least common multiple of $15^2 \cdot 40^3 \cdot 18^2$ and $63^4 \cdot 72^2$ in prime factored form.

31. Express $\frac{5}{11}$ and $\frac{5}{13}$ as repeating decimals.

32. Express $.4\overline{68}$ and $3.2\overline{45}$ as ratios of two integers.

33. Which is larger, $\frac{16}{5}$ or $\frac{13}{4}$?

34. Order the following numbers from least to greatest.

$$\frac{29}{20}, \quad 1.4, \quad 1.\overline{4}, \quad \sqrt{2}, \quad \frac{13}{8}$$

35. Show that if $a < b$, then $a < (2a + b)/3$.

36. Write the inequalities $|x - 9| \le 2.5$ and $|x + 3| < 5$ and $|x - 2| > 6$ without using the absolute value symbol. Also show the corresponding intervals on the real line.

37. Write $4 < x < 12$ as a single inequality using the absolute value symbol.

38. Given that a and b are both positive, what condition on a and b implies that $1/a > 1/b$?

39. A firm can make product A at a cost of $5 per unit and product B at a cost of $12 per unit. The firm plans to spend at most $10,000 in making x units of A and y units of B. Write an inequality that describes this situation.

40. Write each of the following in the form $a + bi$.
    (a) $(4 - 5i) + (-8 + 3i) - (6 - 4i)$
    (b) $(5 + 2i)(5 - 2i) - 2i^3 + (2i)^5$
    (c) $3 + 2i + (2 - 5i)/(3 + 2i)$
    (d) $(3 - 3i)^3$ (e) $(2 + i)^{-2}$

41. Simplify
$$1 + 2i + 3i^2 + 4i^3 + 5i^4 - 4i^5 - 3i^6 - 4i^7.$$

42. Show that $3 + 2i$ is a solution to the equation
$$2x^2 - 3x - 1 - 18i = 0$$

43. The sum of three consecutive integers is 225. Find the smallest of these integers.

44. A rectangle with a perimeter of 72 meters is 4 meters longer than it is wide. Find its area.

45. A rectangle, twice as long as wide, has a diagonal of length $\sqrt{125}$ feet. Find the dimensions of the rectangle.

46. If one square is twice as long as another square and if their areas differ by 108 square inches, find the dimensions of the smaller square.

47. What happens to the volume of a circular cylinder if its radius is halved and its height is doubled?

48. Uncle Bill distributed a certain number of rare coins among his four nieces. Mary got one more than Helen, Helen got one more than Eloise, and Jan got 57 coins. If the average number of coins per girl was 30, how many coins did Mary get?

49. In a 4-mile walk, Jenny covered the first 2 miles in 40 minutes. How fast should she walk the rest of the way to average 3.5 miles per hour over all?

50. Show that it is impossible for a car to average 60 miles per hour on a 2-mile stretch of road if it only averaged 30 miles per hour on the first mile.

CHAPTER 2

Exponents and Polynomials

■ Thus under the Descartes-Fermat scheme points became pairs of numbers, and curves became collections of pairs of numbers subsumed in equations. The properties of curves could be deduced by algebraic processes applied to the equations. With this development, the relation between number and geometry had come full circle. The classical Greeks had buried algebra in geometry, but now geometry was eclipsed by algebra. As the mathematicians put it, geometry was arithmetized.

Morris Kline

2-1 INTEGRAL EXPONENTS

Young Franklin has posed a very interesting question. What do you think the answer is? 10 inches? 3 feet? 500 feet? Make a guess and write it in the margin. When we finally work out the solution late in this section, you are likely to be surprised at the answer.

Let us make a start on the problem right away. If the bulletin were c units thick ($c = .01$ inches would be a reasonable value), then after folding once, it would be $2c$ units thick. After two folds, it would measure $2 \cdot 2c$ units thick, and after 40 folds it would have a thickness of

$$2 \cdot 2 \cdot 2 \cdot 2 \cdot 2 \cdot 2 \cdot 2 \cdot 2 \cdot 2 \cdot 2 \cdot 2 \cdot 2 \cdot 2 \cdot 2 \cdot 2 \cdot 2 \cdot 2 \cdot 2 \cdot 2 \cdot 2$$
$$\cdot 2 \cdot c$$

Nobody with a sense of economy and elegance would write a product of forty 2's in this manner. To indicate such a product, most ordinary people and all mathematicians prefer to write 2^{40}. The number 40 is called an **exponent;** it tells you how many 2's to multiply together. The number 2^{40} is called a **power** of 2 and we read it "2 to the 40th power."

In the general case, if b is any number ($\frac{3}{4}$, π, $\sqrt{5}$, i, . . .) and n is a positive integer, then

$$\underbrace{b^n = b \cdot b \cdot b \cdots b}_{n \text{ factors}}$$

Thus

$$b^3 = b \cdot b \cdot b \qquad b^5 = b \cdot b \cdot b \cdot b \cdot b$$

How do we write the product of 1000 b's? (Honey is not the answer we have in mind.) The product of 1000 b's is written as b^{1000}.

Rules for Exponents

The behavior of exponents is excellent, being governed by a few simple rules that are easy to remember. Consider multiplication first. If we multiply 2^5 by 2^8, we have

CAUTION

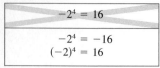

$$-2^4 = -16$$
$$(-2)^4 = 16$$

$$2^5 \cdot 2^8 = \underbrace{(2 \cdot 2 \cdot 2 \cdot 2 \cdot 2)}_{5}\underbrace{(2 \cdot 2 \cdot 2 \cdot 2 \cdot 2 \cdot 2 \cdot 2 \cdot 2)}_{8}$$

$$= \underbrace{2 \cdot 2 \cdot 2 \cdot 2 \cdot 2 \cdot 2 \cdot 2 \cdot 2 \cdot 2 \cdot 2 \cdot 2 \cdot 2 \cdot 2}_{13}$$

$$= 2^{13}$$

$$= 2^{5+8}$$

This suggests that to find the product of powers of 2, you should add the exponents. There is nothing special about 2; it could just as well be 5, $\frac{2}{3}$, or π. We can put the general rule in a nutshell by using the symbols of algebra.

$$b^m \cdot b^n = b^{m+n}$$

Here b can stand for any number, but (for now) think of m and n as positive integers. Be careful with this rule. If you write

$$3^4 \cdot 3^5 = 3^9$$

or

$$\pi^9 \cdot \pi^{12} \cdot \pi^2 = \pi^{23}$$

that is fine. But do not try to use the rule on $2^4 \cdot 3^5$ or $a^2 \cdot b^3$, it just does not apply.

Next consider the problem of raising a power to a power. By definition, $(2^{10})^3$ is $2^{10} \cdot 2^{10} \cdot 2^{10}$, which allows us to apply the rule above. Thus

$$(2^{10})^3 = 2^{10} \cdot 2^{10} \cdot 2^{10} = 2^{10+10+10} = 2^{10 \cdot 3}$$

It appears that to raise a power to a power we should multiply the exponents; in symbols

$$(b^m)^n = b^{m \cdot n}$$

Try to convince yourself that this rule is true for any number b and for any positive integer exponents m and n.

Sometimes we need to simplify quotients like

$$\frac{8^6}{8^6} \qquad \frac{2^9}{2^5} \qquad \frac{10^4}{10^6}$$

The first one is easy enough; it equals 1. Furthermore,

$$\frac{2^9}{2^5} = \frac{2^5 \cdot 2^4}{2^5} = 2^4 = 2^{9-5}$$

and

$$\frac{10^4}{10^6} = \frac{10^4}{10^4 \cdot 10^2} = \frac{1}{10^2} = \frac{1}{10^{6-4}}$$

These illustrate the general rules.

$$\frac{b^m}{b^n} = 1 \qquad \text{if } m = n$$

$$\frac{b^m}{b^n} = b^{m-n} \qquad \text{if } m > n$$

$$\frac{b^m}{b^n} = \frac{1}{b^{n-m}} \qquad \text{if } n > m$$

In each case, we assume $b \neq 0$.

CAUTION

$$\frac{b^{12}}{b^4} = b^3$$

$$\frac{b^{12}}{b^4} = b^8$$

We did not put a box around these rules simply because we are not happy with them. It took three lines to describe what happens when you divide powers of the same number. Surely we can do better than that, but we shall have to extend the notion of exponents to numbers other than positive integers.

Example A (Applying the Basic Rules of Exponents)   Write each of the following as a single power of 5.

(a) $5 \cdot 5^2 \cdot 5^3$ (b) $(5^2 \cdot 5^4)^3$ (c) $\dfrac{5^3(5^2 \cdot 5^3)^2}{5^6}$

Solution

(a) $5 \cdot 5^2 \cdot 5^3 = 5^{1+2+3} = 5^6$

(b) $(5^2 \cdot 5^4)^3 = (5^6)^3 = 5^{18}$

(c) $\dfrac{5^3(5^2 \cdot 5^3)^2}{5^6} = \dfrac{5^3(5^5)^2}{5^6} = \dfrac{5^3 \cdot 5^{10}}{5^6} = \dfrac{5^{13}}{5^6} = 5^7$ ■

Zero and Negative Exponents

So far, symbols like 4^0 and 10^{-3} have not been used. We want to give them meaning and do it in a way that is consistent with what we have already learned. For example, 4^0 must behave so that

$$4^0 \cdot 4^7 = 4^{0+7} = 4^7$$

This can happen only if $4^0 = 1$. More generally, we require that

$$b^0 = 1$$

Here b can be any number except 0 (0^0 will be left undefined).

What about 10^{-3}? If it is to be admitted to the family of powers, it too must abide by the rules. Thus we insist that

$$10^{-3} \cdot 10^3 = 10^{-3+3} = 10^0 = 1$$

This means that 10^{-3} has to be the reciprocal of 10^3. Consequently, we are led to make the definition

$$b^{-n} = \frac{1}{b^n} \qquad b \neq 0$$

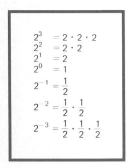

$$
\begin{aligned}
2^3 &= 2 \cdot 2 \cdot 2 \\
2^2 &= 2 \cdot 2 \\
2^1 &= 2 \\
2^0 &= 1 \\
2^{-1} &= \frac{1}{2} \\
2^{-2} &= \frac{1}{2} \cdot \frac{1}{2} \\
2^{-3} &= \frac{1}{2} \cdot \frac{1}{2} \cdot \frac{1}{2}
\end{aligned}
$$

Figure 1

This definition results in the nice pattern illustrated in Figure 1. But what is more significant, it allows us to state the law for division of powers in a very simple form. To lead up to that law, consider the following manipulations.

$$\frac{b^0}{b^{-6}} = \frac{1}{1/b^6} = b^6 = b^{0-(-6)}$$

$$\frac{b^4}{b^9} = \frac{b^4}{b^4 \cdot b^5} = \frac{1}{b^5} = b^{-5} = b^{4-9}$$

$$\frac{b^5}{b^5} = 1 = b^0 = b^{5-5}$$

$$\frac{b^{-3}}{b^{-9}} = \frac{1/b^3}{1/b^9} = \frac{b^9}{b^3} = b^6 = b^{-3-(-9)}$$

In fact, for any choice of integers m and n, we find that

$$\frac{b^m}{b^n} = b^{m-n} \qquad b \neq 0$$

What about the two rules we learned earlier? Are they still valid when m and n are arbitrary (possibly negative) integers? The answer is yes. A few illustrations may help convince you.

$$b^{-3} \cdot b^7 = \frac{1}{b^3} \cdot b^7 = \frac{b^7}{b^3} = b^4 = b^{-3+7}$$

$$(b^{-5})^2 = \left(\frac{1}{b^5}\right)^2 = \frac{1}{b^5} \cdot \frac{1}{b^5} = \frac{1}{b^{10}} = b^{-10} = b^{(-5)(2)}$$

Example B (Removing Negative Exponents)   Rewrite without negative exponents and simplify.

(a) -4^{-2} (b) $(-4)^{-2}$ (c) $\left[\left(\frac{3}{4}\right)^{-1}\right]^2$ (d) $2^5 3^{-2} 2^{-3}$

Solution
(a) The exponent -2 applies just to 4.

$$-4^{-2} = -\frac{1}{4^2} = -\frac{1}{4 \cdot 4} = -\frac{1}{16}$$

(b) The exponent -2 now applies to -4.

$$(-4)^{-2} = \frac{1}{(-4)^2} = \frac{1}{16}$$

(c) First apply the rule for a power of a power.

$$\left[\left(\frac{3}{4}\right)^{-1}\right]^2 = \left(\frac{3}{4}\right)^{-2} = \frac{1}{(\frac{3}{4})^2} = \frac{1}{\frac{9}{16}} = \frac{16}{9}$$

(d) Note that the two powers of 2 can be combined.

$$2^5 \cdot 3^{-2} \cdot 2^{-3} = 2^5 \cdot 2^{-3} \cdot 3^{-2} = 2^2 \cdot \frac{1}{3^2} = \frac{4}{9} \qquad \blacksquare$$

Powers of Products and Quotients

Expressions like $(ab)^n$ and $(a/b)^n$ often arise; we need rules for handling them. Notice that

$$(ab)^n = \underbrace{(ab)(ab) \ldots (ab)}_{n \text{ factors}} = \underbrace{a \cdot a \cdots a}_{n \text{ factors}} \cdot \underbrace{b \cdot b \cdots b}_{n \text{ factors}} = a^n b^n$$

$$\left(\frac{a}{b}\right)^n = \underbrace{\left(\frac{a}{b}\right)\left(\frac{a}{b}\right) \cdots \left(\frac{a}{b}\right)}_{n \text{ factors}} = \frac{a \cdot a \cdots a}{b \cdot b \cdots b} = \frac{a^n}{b^n}$$

Our demonstrations are valid for any positive integer n, but the results are correct even if n is negative or zero. Thus for any integer n,

$$(ab)^n = a^n b^n$$

$$\left(\frac{a}{b}\right)^n = \frac{a^n}{b^n} \qquad b \neq 0$$

Thus

$$(3x^2 y)^4 = 3^4 (x^2)^4 y^4 = 81 x^8 y^4$$

and

$$\left(\frac{2x^{-1}}{y}\right)^3 = \frac{2^3 (x^{-1})^3}{y^3} = \frac{2^3 x^{-3}}{y^3} = \frac{8}{x^3 y^3}$$

Example C (Products and Quotients)   Use the rules above to simplify.

(a) $(2x)^6$ (b) $\left(\dfrac{2x}{3}\right)^4$ (c) $(x^{-1} y^2)^{-3}$

Solution

(a) $(2x)^6 = 2^6 x^6 = 64 x^6$ (b) $\left(\dfrac{2x}{3}\right)^4 = \dfrac{(2x)^4}{3^4} = \dfrac{2^4 x^4}{3^4} = \dfrac{16 x^4}{81}$

(c) $(x^{-1} y^2)^{-3} = (x^{-1})^{-3}(y^2)^{-3} = x^3 y^{-6} = x^3 \cdot \dfrac{1}{y^6} = \dfrac{x^3}{y^6}$ ■

We summarize our discussion of exponents by stating the five main rules together. In using these rules, it is always understood that division by zero is to be avoided.

Rules for Exponents

In each case, a and b are (real or complex) numbers and m and n are any integers.

1. $b^m b^n = b^{m+n}$

2. $(b^m)^n = b^{mn}$

3. $\dfrac{b^m}{b^n} = b^{m-n}$

4. $(ab)^n = a^n b^n$

5. $\left(\dfrac{a}{b}\right)^n = \dfrac{a^n}{b^n}$

Experience suggests that people often misapply these rules or make up new ones that are false. Study the next example; it will help reinforce correct habits.

Example D (Simplifying Complicated Expressions)   Simplify the following.

(a) $\dfrac{4ab^{-2}c^3}{a^{-3}b^3c^{-1}}$ (b) $\left[\dfrac{(2xz^{-2})^3(x^{-2}z)}{2xz^2}\right]^4$ (c) $(a^{-1} + b^{-2})^{-1}$

Solution

(a) $\dfrac{4ab^{-2}c^3}{a^{-3}b^3c^{-1}} = \dfrac{4a(1/b^2) \cdot c^3}{(1/a^3)b^3(1/c)} = \dfrac{4ac^3/b^2}{b^3/(a^3c)} = \dfrac{4ac^3}{b^2} \cdot \dfrac{a^3c}{b^3} = \dfrac{4a^4c^4}{b^5}$

In simplifying expressions like the one above, a *factor can be moved from numerator to denominator, or vice versa, by changing the sign of its exponent.* That is important enough to remember. Let us do part (a) again using this fact.

$$\frac{4ab^{-2}c^3}{a^{-3}b^3c^{-1}} = \frac{4aa^3c^3c}{b^2b^3} = \frac{4a^4c^4}{b^5}$$

(b) $\left[\dfrac{(2xz^{-2})^3(x^{-2}z)}{2xz^2}\right]^4 = \left[\dfrac{8x^3z^{-6}x^{-2}z}{2xz^2}\right]^4 = \left[\dfrac{8xz^{-5}}{2xz^2}\right]^4 = \left[\dfrac{4}{z^2z^5}\right]^4 = \dfrac{256}{z^{28}}$

CAUTION

$(a^{-1} + b^{-1})^{-1} = a + b$ [struck through]

$(a^{-1} + b^{-1})^{-1} = \left(\dfrac{1}{a} + \dfrac{1}{b}\right)^{-1}$

$= \left(\dfrac{b + a}{ab}\right)^{-1} = \dfrac{ab}{b + a}$

(c) $(a^{-1} + b^{-2})^{-1} = \left(\dfrac{1}{a} + \dfrac{1}{b^2}\right)^{-1} = \left(\dfrac{b^2 + a}{ab^2}\right)^{-1} = \dfrac{ab^2}{b^2 + a}$

Note the difference between a product and a sum.

$$(a^{-1} \cdot b^{-2})^{-1} = a \cdot b^2$$

but

$$(a^{-1} + b^{-2})^{-1} \neq a + b^2 \quad \blacksquare$$

The Paper-Folding Problem Again

It is now a simple matter to solve Franklin Figit's paper-folding problem especially if we are satisfied with a reasonable approximation to the answer. To be specific, let us approximate 1 foot by 10 inches, 1 mile by 5000 feet, and 2^{10} (which is really 1024) by 1000. Then a bulletin of thickness .01 inch will make a stack of the following height when folded 40 times ($\approx$ means "is approximately equal to").

$$(.01)2^{40} = (.01) \cdot 2^{10} \cdot 2^{10} \cdot 2^{10} \cdot 2^{10} \text{ inches}$$

$$\approx \frac{1}{10^2} \cdot 10^3 \cdot 10^3 \cdot 10^3 \cdot 10^3 \text{ inches}$$

$$= 10^{10} \text{ inches}$$

$$\approx 10^9 \text{ feet}$$

$$\approx \frac{10 \cdot 10^8}{5 \cdot 10^3} \text{ miles}$$

$$= 2 \cdot 10^5 \text{ miles}$$

$$= 200,000 \text{ miles}$$

That is a stack of paper that would reach almost to the moon.

PROBLEM SET 2-1

A. Skills and Techniques

Use the rules for exponents (as in Example A) to simplify each of the following. Then calculate the result.

1. $\dfrac{3^2 \cdot 3^5}{3^4}$

2. $\dfrac{2^6 \cdot 2^7}{2^{10}}$

3. $\dfrac{(2^2)^4}{2^6}$

4. $\dfrac{(5^4)^3}{5^9}$

5. $\dfrac{(3^2 \cdot 2^3)^3}{6^6}$

6. $6^6\left(\dfrac{2}{3^2}\right)^3$

Write without negative exponents and simplify. See Example B.

7. 5^{-2}

8. $(-5)^{-2}$

9. -5^{-2}

10. 2^{-5}

11. $(-2)^{-5}$

12. $\left(\dfrac{1}{5}\right)^{-2}$

13. $\left(\dfrac{-2}{3}\right)^{-3}$

14. $-\dfrac{2^{-3}}{3}$

15. $\dfrac{2^{-2}}{3^{-3}}$

16. $\left[\left(\dfrac{2}{3}\right)^{-2}\right]^2$

17. $\left[\left(\dfrac{3}{2}\right)^{-2}\right]^{-2}$

18. $\dfrac{4^0 + 0^4}{4^{-1}}$

19. $\dfrac{2^{-2} - 4^{-3}}{(-2)^2 + (-4)^0}$

20. $\dfrac{3^{-1} + 2^{-3}}{(-1)^3 + (-3)^2}$

21. $3^3 \cdot 2^{-3} \cdot 3^{-5}$

22. $4^2 \cdot 4^{-4} \cdot 3^0$

Simplify, writing your answer without negative exponents. See Example C.

23. $(3x)^4$

24. $\left(\dfrac{2}{y}\right)^5$

25. $(xy^2)^6$

26. $\left(\dfrac{y^2}{3z}\right)^4$

27. $\left(\dfrac{2x^2y}{w^3}\right)^4$

28. $\left(\dfrac{\sqrt{2}x}{3}\right)^4$

29. $\left(\dfrac{3x^{-1}y^2}{z^2}\right)^3$

30. $\left(\dfrac{2x^{-2}y}{z^{-1}}\right)^2$

31. $\left(\dfrac{\sqrt{5}i}{x^{-2}}\right)^4$ ⓘ

32. $(i\sqrt{3}x^{-2})^6$ ⓘ

33. $(4y^3)^{-2}$

34. $(x^3z^{-2})^{-1}$

35. $\left(\dfrac{5x^2}{ab^{-2}}\right)^{-1}$

36. $\left(\dfrac{3x^2y^{-2}}{2x^{-1}y^4}\right)^{-3}$

Simplify, leaving your answer free of negative exponents as in Example D.

37. $\dfrac{2x^{-3}y^2z}{x^3y^4z^{-2}}$

38. $\dfrac{3x^{-5}y^{-3}z^4}{9x^2yz^{-1}}$

39. $\left(\dfrac{-2xy}{z^2}\right)^{-1}(x^2y^{-3})^2$

40. $(4ab^2)^3\left(\dfrac{-a^3}{2b}\right)^2$

41. $\dfrac{ab^{-1}}{(ab)^{-1}} \cdot \dfrac{a^2b}{b^{-2}}$

42. $\dfrac{3(b^{-2}d)^4(2bd^3)^2}{(2b^2d^3)(b^{-1}d^2)^5}$

43. $\left[\dfrac{(3b^{-2}d)(2)(bd^3)^2}{12b^3d^{-1}}\right]^5$

44. $\left[\dfrac{(ab^2)^{-1}}{(ba^2)^{-2}}\right]^{-1}$

45. $(a^{-2} + a^{-3})^{-1}$

46. $a^{-2} + a^{-3}$

47. $\dfrac{x^{-1}}{y^{-1}} - \left(\dfrac{x}{y}\right)^{-1}$

48. $(x^{-1} - y^{-1})^{-1}$

B. Applications and Extensions

Simplify the expressions in Problems 49–58, leaving your answer free of negative exponents.

49. $\left(\dfrac{1}{2}x^{-1}y^2\right)^{-3}$

50. $(x + x^{-1})^2$

51. $\dfrac{2^{-2}}{1 + \dfrac{3^{-1}}{1 + 3^{-1}}}$

52. $\left[\left(\dfrac{1}{2} + \dfrac{2}{3}\right)^{-1} + \left(\dfrac{1}{4} + \dfrac{1}{3}\right)^{-1}\right]^{-1}$

53. $\dfrac{(2x^{-1}y^2)^2}{2xy} \cdot \dfrac{x^{-3}}{y^3}$

54. $\left(\dfrac{\sqrt{2}x^2y}{xy^{-2}z^2}\right)^4$

55. $\left[\left(\dfrac{1}{2}x^{-2}\right)^3(4xy^{-1})^2\right]^2$

56. $\left[\dfrac{4y^2z^{-3}}{x^3(2x^{-1}z^2)^3}\right]^{-2}$

57. $(x^{-1} + y^{-1})^{-1}(x + y)$

58. $[1 - (1 + x^{-1})^{-1}]^{-1}$

59. Express each of the following as a single power of 2.

  (a) $\dfrac{1}{2} \cdot \dfrac{1}{4} \cdot \dfrac{1}{8} \cdot \dfrac{1}{16} \cdot \dfrac{1}{32}$

  (b) $\dfrac{1}{2} + \dfrac{1}{4} + \dfrac{1}{8} + \dfrac{1}{16} + \dfrac{1}{32} + \dfrac{1}{32}$

60. Express $8\left(\dfrac{2}{3}\right)^4 - 4\left(\dfrac{2}{3}\right)^5 + 2\left(\dfrac{2}{3}\right)^6 + 6\left(\dfrac{2}{3}\right)^7$ in the form $2^m/3^n$.

61. Which is larger, 2^{1000} or $(10)^{300}$?

62. Consider Franklin Figit's paper-folding problem, with which we began this section. If the pile of paper stands on 1 square inch after 40 folds, about how much area did it cover at the beginning?

63. G. P. Jetty has agreed to pay his new secretary according to the following plan: 1¢ the first day, 2¢ the second day, 4¢ the third day, and so on, doubling each day.

    (a) How much will the secretary make during the first 4 days? 5 days? 6 days?

    (b) From part (a), you should see a pattern. How much will the secretary make during the first n days?

    (c) Assume Jetty is worth $2 billion and that his secretary started work on January 1. About when will Jetty go broke?

64. TEASER   By a^{b^c}, mathematicians mean $a^{(b^c)}$, that is, in a tower of exponents we start at the top and work down. For example $2^{2^{2^2}} = 2^{2^4} = 2^{16} = 65,536$. Arrange the following numbers (all with four 2s) from smallest to largest. You should be able to do it without making use of a calculator.

$$2222, \quad 222^2, \quad 22^{22}, \quad 2^{222}, \quad 22^{2^2}, \quad 2^{22^2}, \quad 2^{2^{22}}$$

2-2   CALCULATORS AND SCIENTIFIC NOTATION

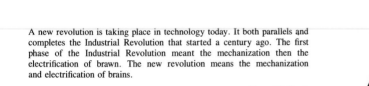

A new revolution is taking place in technology today. It both parallels and completes the Industrial Revolution that started a century ago. The first phase of the Industrial Revolution meant the mechanization then the electrification of brawn. The new revolution means the mechanization and electrification of brains.

Harry M. Davis

Most people can do arithmetic if they have to. But few do it with either enthusiasm or accuracy. Frankly, it is a rather dull subject. The spectacular sales of hand-held electronic calculators demonstrate both our need to do arithmetic and our distaste for it.

Hand-held calculators vary greatly in what they can do. The simplest perform only the four arithmetic operations and may sell for under $10. The most sophisticated can perform hundreds of different operations, have large memories, and are programmable. For most of this course, a standard scientific calculator (with logarithm, exponential, and trigonometric keys, selling for about $25) is ideal. After we have discussed graphing, we will introduce you to the scientific calculator with graphics capability. For one of these you may have to pay $100 or more.

Two kinds of logic are commonly used in hand-held calculators, **reverse Polish** logic and **algebraic** logic. The former avoids the use of parentheses and is highly efficient once you learn its rules. Algebraic logic uses parentheses and mimics the procedures of ordinary algebra. For this reason, we have chosen to illustrate calculator operations using a typical algebraic calculator. However, we warn our readers that not all that we say may be valid for your calculator. We cannot emphasize this too strongly; you must learn the operating rules for your own calculator. If you do, you will find that a calculator will become a powerful tool in helping you solve problems.

Since calculators can display only a fixed number of digits (usually 8 or 10), we face an immediate problem. How shall we handle very large or very small numbers on a calculator? The answer depends on a notational device that was invented long before pocket calculators.

Scientific Notation

It is in science that we are most likely to meet very large or very small numbers. For example, the speed of light is 29,979,000,000 centimeters per second. At the other extreme, the mass of the proton is .00000000000000000000000167 grams. These numbers are unwieldy primarily because of the large number of zeros, zeros which serve only to place the decimal point. The significant digits are 29979 in the first case and 167 in the second. Now note that

$$29,979,000,000 = 2.9979 \times 10^{10}$$

$$.00000000000000000000000167 = 1.67 \times 10^{-24}$$

Both of these numbers have been rewritten in scientific notation.

A positive number N is in **scientific notation** when it is written in the form

$$N = c \times 10^n$$

where n is an integer and c is a real number such that $1 \leq c < 10$. To put N in scientific notation, place the decimal point after the first nonzero digit of N, the so-called standard position. Then count the number of places from there to the original position of the decimal point. This number, taken as positive if counted to the right and negative if counted to the left, is the exponent n to be used as the power of 10. Here is the process illustrated for the number 3,651,000.

$$3651000 = 3.651 \times 10^6$$

6 places

Calculations with large or small numbers are easily accomplished in scientific notation, as we illustrate now.

Example A (Calculating in Scientific Notation)   Calculate

$$p = \frac{(3,200,000,000)(.0000000284)}{.00000000128}$$

Solution   First we write

$$3,200,000,000 = 3.2 \times 10^9$$

$$.0000000284 = 2.84 \times 10^{-8}$$

$$.00000000128 = 1.28 \times 10^{-9}$$

Then

$$p = \frac{(3.2 \times 10^9)(2.84 \times 10^{-8})}{1.28 \times 10^{-9}}$$

$$= \frac{(3.2)(2.84)}{1.28} \times 10^{9-8-(-9)}$$

$$= 7.1 \times 10^{10} \quad \blacksquare$$

The Metric System

Many of the examples and problems in this book will involve metric units; it is time we say a word about the metric system of measurement. It has long been recognized by most countries that the metric system, with its emphasis on 10 and powers of 10, offers an attractive way to measure length, volume and weight. Only in the United States do we hang on to our hodgepodge of inches, feet, miles, pounds, and quarts. Even here, however, it appears that the metric system will gradually win acceptance.

Table 1 summarizes the metric system. It highlights the relationship to powers of 10 within the metric system and gives some of the conversion factors that relate the metric system to our English system. Using the table, you should be able to answer questions such as the following.

1. How many kilometers per hour are equivalent to 65 miles per hour?
2. How many grams are equivalent to 200 pounds?

   The answers to these two questions are

1. 65 miles per hour $\approx 65/.62$, or 105 kilometers per hour;
2. 200 pounds $\approx 200(453.6)$ grams $= 9.07 \times 10^4$ grams.

Table 1 The Metric System

Length	Volume	Weight
kilometer = 10^3 meter	kiloliter = 10^3 liter	kilogram = 10^3 gram
hectometer = 10^2 meter	hectoliter = 10^2 liter	hectogram = 10^2 gram
dekameter = 10 meter	dekaliter = 10 liter	dekagram = 10 gram
meter = 1 meter	liter = 1 liter	gram = 1 gram
decimeter = 10^{-1} meter	deciliter = 10^{-1} liter	decigram = 10^{-1} gram
centimeter = 10^{-2} meter	centiliter = 10^{-2} liter	centigram = 10^{-2} gram
millimeter = 10^{-3} meter	milliliter = 10^{-3} liter	milligram = 10^{-3} gram
1 kilometer $\approx$ .62 miles	1 liter $\approx$ 1.057 quarts	1 kilogram $\approx$ 2.20 pounds
1 inch $\approx$ 2.54 centimeters	1 liter = 10^3 cubic centimeters	1 pound $\approx$ 453.6 grams

Example B (Conversion Between Units)
(a) Convert 2.56×10^4 kilometers to centimeters.
(b) Convert 3.42×10^2 kilograms to ounces.
(c) Convert 43.8 cubic meters to liters.
Give all answers in scientific notation.

Solution

(a) 2.56×10^4 kilometers $= (2.56 \times 10^4)(10^3)$ meters
$$= (2.56 \times 10^4)(10^3)(10^2) \text{ centimeters}$$
$$= 2.56 \times 10^9 \text{ centimeters}$$

(b) 3.42×10^2 kilograms $\approx (3.42 \times 10^2)(2.20)$ pounds
$$= (3.42 \times 10^2)(2.20)(16) \text{ ounces}$$
$$\approx 1.20 \times 10^4 \text{ ounces}$$

(c) 43.8 cubic meters $= (43.8)(100)^3$ cubic centimers
$$= \frac{(43.8)(100)^3}{10^3} \text{ liters}$$
$$= 4.38 \times 10^4 \text{ liters} \quad \blacksquare$$

Entering Data into a Calculator

With scientific notation in hand, we can explain how to get a number into a typical calculator. To enter 238.75, simply press in order the keys 2 3 8 . 7 5.

If the negative of this number is desired, press the same keys and then press the change sign key $\boxed{+/-}$. Numbers larger than 10^8 or smaller than 10^{-8} must be entered in scientific notation. For example, 2.3875×10^{19} would be entered as follows.

$$2.3875 \quad \boxed{EE} \quad 19$$

The key $\boxed{EE}$ (which stands for *enter exponent*) controls the two places at the extreme right of the display. They are reserved for the exponent on 10. After pressing the indicated keys, the display will read

2.3875	19

If you press

$$2.3875 \quad \boxed{+/-} \quad \boxed{EE} \quad 19 \quad \boxed{+/-}$$

the display will read

-2.3875	-19

which stands for the number -2.3875×10^{-19}.

In making a complicated calculation, you may enter some numbers in standard notation and others in scientific notation. The calculator understands either form and makes the proper translations. If any of the entered data is in scientific notation, it will display the answer in this format. Also, if the answer is too large or too small for standard format, the calculator will automatically convert the result of a calculation to scientific notation.

The T1-81 Graphics Calculator

Later in the book, we will use the T1-81 to illustrate the capabilities of graphics calculators. On this calculator, the number

$$2.3875 \times 10^{19}$$

is displayed as

$$\boxed{2.3875E19}$$

Also, the $\boxed{=}$ key is replaced by $\boxed{ENTER}$ and there is no $\boxed{y^x}$ key. The number $2.75^{-.34}$ is calculated by pressing

$$2.75 \; \boxed{\wedge} \; \boxed{(-)} \; .34 \; \boxed{ENTER}$$

Doing Arithmetic

The five keys $\boxed{+}$, $\boxed{-}$, $\boxed{\times}$, $\boxed{\div}$, and $\boxed{=}$ are the workhorses for arithmetic in most calculators using algebraic logic. To perform the calculation

$$175 + 34 - 18$$

simply press the keys indicated below.

The answer 191 will appear in the display.

Or consider (175)(14)/18. Press

and the calculator will display 136.11111.

An expression involving additions (or subtractions) and multiplications (or divisions) may be ambiguous. For example, $2 \times 3 + 4 \times 5$ could have several meanings depending on which operations are performed first.

(i) $\qquad\qquad\qquad\qquad 2 \times (3 + 4) \times 5 = 70$

(ii) $\qquad\qquad\qquad\qquad (2 \times 3) + (4 \times 5) = 26$

(iii) $\qquad\qquad\qquad\qquad [(2 \times 3) + 4] \times 5 = 50$

(iv) $\qquad\qquad\qquad\qquad 2 \times [3 + (4 \times 5)] = 46$

Parentheses are used in mathematics and in calculators to indicate the order in which operations are to be performed. To do calculation (i), press

Similarly, to do calculation (iii), press

Recall that, in arithmetic, we have an agreement that when no parentheses are used, multiplications and divisions are done before additions and subtractions. Thus

$$2 \times 3 + 4 \times 5$$

is interpreted as $(2 \times 3) + (4 \times 5)$. The same convention is used in most scientific calculators. Pressing

will yield the answer 26. Similarly, for calculation (iii), pressing

will yield 50, since within the parentheses, the calculator will do the multiplication first. However, when in doubt use parentheses, since without them it is easy to make errors.

Special Functions

Most scientific calculators have keys for finding powers and roots of a number. On our sample calculator, the $\boxed{y^x}$ key is used to raise a number y to the xth power. For example, to calculate 2.75^{-34}, press

and the correct result, .70896841, will appear in the display.

Finding a root is the inverse of raising to a power. For example, taking a cube root is the inverse operation of cubing. Thus, to calculate $\sqrt[3]{17}$, press 17 $\boxed{\text{INV}}$ $\boxed{y^x}$ 3 $\boxed{=}$ and you will get 2.5712816. In using the $\boxed{y^x}$ key, the calculator insists that y be positive. However, x may be either positive or negative.

Square roots occur so often that there is a special key for them on some calculators. To calculate $\sqrt{17}$, simply press 17 $\boxed{\sqrt{}}$ and you will immediately get 4.1231056.

Keep in mind that most scientific calculators (with algebraic logic) perform operations in the following order.

1. Unary operations, such as taking square roots.
2. Multiplications and divisions from left to right.
3. Additions and subtractions from left to right.

Parentheses are used just as in ordinary algebra. Pressing the $\boxed{=}$ key will cause all pending operations to be performed.

Example C (Hierarchy of Operations in a Calculator)   Calculate

$$\frac{3.12 + (4.15)(5.79)}{5.13 - 3.76}$$

Solution   This can be done in more than one way, and calculators may vary. On the authors' model, either of the following sequences of keys will give the correct result.

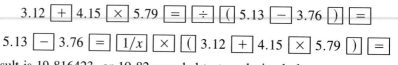

The result is 19.816423, or 19.82 rounded to two decimal places.   ∎

Try the following example on your calculator.

Example D (A Speed of Light Problem)   How long will it take a light ray to reach the earth from the sun? Assume that light travels 2.9979×10^{10} centimeters per second, that 1 mile is equivalent to 1.609 kilometers, and that it is 9.30×10^7 miles from the sun to the earth.

Solution   The speed of light in kilometers per second is 2.9979×10^5, which we shall round to 2.998×10^5. The time required is

$$\frac{(9.30 \times 10^7)(1.609)}{2.998 \times 10^5} = 4.9912 \times 10^2 \approx 499 \text{ seconds.}$$ ∎

Our introduction to hand-held calculators has been very brief. The problem set will give you practice in using your particular model. To clear up any difficulties and to learn other features of your calculator, consult your instruction book.

PROBLEM SET 2-2

A. Skills and Techniques

Write each of the following numbers in scientific notation.

1. 341,000,000
2. 25 billion
3. .0000000513
4. .00000000012
5. .0000000001245
6. .0000000000012578

Calculate, leaving your answers in scientific notation. See Example A.

7. $(1.2 \times 10^5)(7 \times 10^{-9})$
8. $(2.4 \times 10^{-11})(1.2 \times 10^{16})$
9. $\dfrac{(.000021)(240000)}{7000}$
10. $\dfrac{(36,000,000)(.000011)}{.0000033}$
11. $(54)(.00005)(2,000,000)^2$
12. $\dfrac{(3400)^2(400,000)^3}{(.017)^2}$

Express each of the following in scientific notation. See Example B.

13. The number of centimeters in 413.2 meters.
14. The number of millimeters in 1.32×10^4 kilometers.
15. The number of kilometers in 4×10^{15} millimeters.
16. The number of meters in 1.92×10^8 centimeters.
17. The number of millimeters in 1 yard (36 inches).
18. The number of grams in 4.1×10^3 pounds.

Use your pocket calculator as in Example C to perform the calculations in Problems 19–38. We suggest that you begin by making a mental estimate of the answer. For example, the answer to Problem 21 might be estimated as $(3 - 6)(14 \times 50) = -2100$. Similarly, the answer to Problem 35 might be estimated as $(1.5)(10)/(20) = .75$. This will help you catch errors caused by pressing the wrong keys or failing to use parentheses properly.

19. $34.1 - 49.95 + 64.2$
20. $7.465 + 3.12 - .0156$
21. $(3.42 - 6.71)(14.3 \times 51.9)$
22. $(21.34 + 2.37)(74.13 - 26.3)$
23. $\dfrac{514 + 31.9}{52.6 - 50.8}$
24. $\dfrac{547.3 - 832.7}{.0567 - .0416}$

25. $\dfrac{(6.34 \times 10^7)(537.8)}{1.23 \times 10^{-5}}$
26. $\dfrac{(5.23 \times 10^{16})(.0012)}{1.34 \times 10^{11}}$
27. $\dfrac{6.34 \times 10^7}{.00152 + .00341}$
28. $\dfrac{3.134 \times 10^{-8}}{5.123 - 6.1457}$
29. $\dfrac{532 + 1.346}{34.91}(1.75 - 2.61)$
30. $\dfrac{39.95 - 42.34}{15.76 - 16.71}(5.31 \times 10^4)$
31. $(1.214)^3$
32. $(3.617)^{-2}$
33. $\sqrt[3]{1.215}$
34. $\sqrt[3]{1.5789}$
35. $\dfrac{(1.34)(2.345)^3}{\sqrt{364}}$
36. $\dfrac{(14.72)^{12}(59.3)^{11}}{\sqrt{17.1}}$
37. $\dfrac{\sqrt{130} - \sqrt{5}}{15^6 - 4^8}$
38. $\dfrac{\sqrt{143.2} + \sqrt{36.1}}{(234.1)^4 - (11.2)^2}$

Problems 39–44 are related to Example D.

39. How long will it take a light ray to travel from the moon to the earth? (The distance to the moon is 2.39×10^5 miles.)
40. How long would it take a rocket traveling 4500 miles per hour to reach the sun from the earth?
41. A light year is the distance light travels in one year (365.24 days). Our nearest star is 4.300 light years away. How many meters is that?
42. How long would it take a rocket going 4500 miles per hour to get from the earth to our nearest star (see Problem 41)?
43. What is the area in square meters of a rectangular field 2.3 light years by 4.5 light years?
44. What is the volume in cubic meters of a cube 4.3 light years on a side?

B. Applications and Extensions

45. Calculate, writing your answer in scientific notation.

(a) $\dfrac{(3.151 \times 10^2)^4(32,400)}{(21,300)^2}$

(b) $\dfrac{(.433)^3 - (2.31)^{-4} + \sqrt{.0932}}{5.23 \times 10^3}$

46. If, on a flat plane, I walk 24.51 meters due east and then 57.24 meters due north, how far will I be from my starting point?

47. Find the area of the ring (annulus) in Figure 2 if the outer circle has radius 26.25 centimeters and the inner circle has radius 14.42 centimeters.

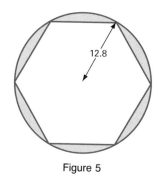

Figure 2

48. Figure 3 shows a trapezoid with unequal sides of lengths 9.2 and 14.3 and altitude 10.4, all in inches. Find its area.

9.2 in.

10.4 in.

14.3 in.

Figure 3

49. Let $x_1, x_2, x_3, \ldots, x_n$ denote n numbers. We define the mean $\bar{x}$ and the standard deviation s by

$$\bar{x} = \frac{1}{n}(x_1 + x_2 + \cdots + x_n)$$

$$s = \sqrt{\frac{1}{n}(x_1^2 + x_2^2 + \cdots + x_n^2) - \bar{x}^2}$$

For the six numbers 121, 132, 155, 161, 133, and 175, calculate $\bar{x}$ and s.

50. A book has 516 pages (that is, 258 sheets) each $7\frac{1}{2}$ inches by $9\frac{1}{4}$ inches. If the paper to make this book could be laid out as a giant square, what would the length of a side be?

51. Mary Cartwright is 86 years old today. If a year has

365.24 days and a heart beats 75 times a minute, how old is Mary measured in heartbeats?

52. The earth has a radius of 3960 miles, the sun a radius of 400,000 miles. In terms of volume, how many times as large as the earth is the sun?

53. Sound waves travel at 1100 feet per second, and radio waves travel at 186,000 miles per second. Hilda sat at the back of an auditorium 200 feet from the speaker. Her husband, Hans, listened on the radio in a city 2000 miles away. Who heard the speaker first and how much sooner?

54. Assuming the universe is a sphere of radius 10^9 light years, what is the volume of the universe in cubic miles?

55. Find the area of the region inside a circle of radius 6.25 centimeters but outside the isosceles triangle with radii as two sides and the third side of length 10.64 centimeters (Figure 4).

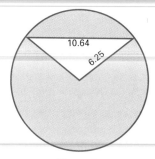

10.64

6.25

Figure 4

56. Find the area of the region inside a circle of radius 12.8 inches but outside an inscribed regular hexagon (Figure 5).

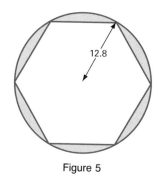

12.8

Figure 5

57. One mole of any substance is an amount equal to its molecular weight in grams. Avogadro's number (6.02×10^{23}) is the number of molecules in a mole.

The atomic weights of hydrogen, carbon, and oxygen are 1, 12, and 16, respectively. Thus, the molecular weights of water (H_2O) and of carbon dioxide (CO_2) are $2 \cdot 1 + 1 \cdot 16 = 18$ and $1 \cdot 12 + 2 \cdot 16 = 44$, respectively.

(a) How many molecules are there in 20 grams of water?

(b) How many molecules are there in 30 grams of carbon dioxide?

58. How much will 10^{24} molecules of cholesterol ($C_{27}H_{46}O$) weigh? (See Problem 57.)

59. Given a seed value x_1, we can generate a sequence of new values by repeated use of a *recursion formula* which relates a new value to an old value. A typical example of a recursion formula stated in words is "new x equals the square of old x." In mathematical notation, this is written as $x_{n+1} = (x_n)^2$. With a seed value of $x_1 = 3$, we see that

$$x_2 = (x_1)^2 = 3^2 = 9$$
$$x_3 = (x_2)^2 = 9^2 = 81$$

and so on.

(a) Find x_8.

(b) If we choose any seed value greater than 1, what happens to x_n as n gets very large?

(c) If we choose any seed value between 0 and 1, what happens to x_n as n gets very large?

60. TEASER This problem will require considerable experimentation with your calculator, using various positive seed values. In each case, describe what happens to x_n as the indicated recursion formula is applied a very large number of times, that is, as n becomes very large. Refer to Problem 59 for the ideas involved.

(a) $x_{n+1} = \sqrt{x_n}$ (b) $x_{n+1} = 2\sqrt{x_n}$

(c) $x_{n+1} = 3\sqrt{x_n}$ (d) $x_{n+1} = k\sqrt{x_n}$

2-3 POLYNOMIALS

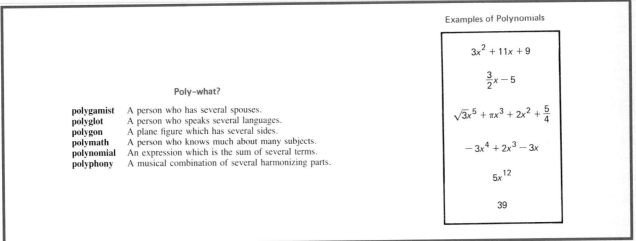

Poly-what?

polygamist	A person who has several spouses.
polyglot	A person who speaks several languages.
polygon	A plane figure which has several sides.
polymath	A person who knows much about many subjects.
polynomial	An expression which is the sum of several terms.
polyphony	A musical combination of several harmonizing parts.

Examples of Polynomials

$$3x^2 + 11x + 9$$

$$\frac{3}{2}x - 5$$

$$\sqrt{3}x^5 + \pi x^3 + 2x^2 + \frac{5}{4}$$

$$-3x^4 + 2x^3 - 3x$$

$$5x^{12}$$

$$39$$

The dictionary definition of polynomial given above is suggestive, but it is not nearly precise enough for mathematicians. The expression $2x^{-2} + \sqrt{x} + 1/x$ is a sum of several terms, but it is not a polynomial. On the other hand, $3x^4$ is just one term; yet it is a perfectly good polynomial. Fundamentally, a **real polynomial** in x is any expression that can be obtained from the real numbers and x using only the operations of addition, subtraction, and multiplication. For example, we can get $3 \cdot x \cdot x \cdot x \cdot x$ or $3x^4$ by multiplication. We can get $2x^3$ by the same process and then have

$$3x^4 + 2x^3$$

by addition. We could never get $2x^{-2} = 2/x^2$ or $\sqrt{x}$; the first involves a division and the second involves taking a root. Thus $2/x^2$ and $\sqrt{x}$ are not polynomials. Try to convince yourself that all of the expressions in the box above are polynomials.

There is another way to define a polynomial in fewer words. A real polynomial in x is any expression of the form

$$a_n x^n + a_{n-1} x^{n-1} + a_{n-2} x^{n-2} + \cdots + a_1 x + a_0$$

where the a's are real numbers and n is a nonnegative integer. We shall not have much to do with complex polynomials in x, but they are defined in exactly the same way, except that the a's are allowed to be complex numbers. In either case, we refer to the a's as **coefficients.** The degree of a polynomial is the largest exponent that occurs in the polynomial in a term with a nonzero coefficient. Here are several more examples.

1. $\frac{3}{2}x - 5$ is a first-degree (or **linear**) polynomial in x.
2. $3y^2 - 2y + 16$ is a second-degree (or **quadratic**) polynomial in y.
3. $\sqrt{3}t^5 - \pi t^2 - 17$ is a fifth-degree polynomial in t.
4. $5x^3$ is a third-degree polynomial in x. It is also called a **monomial,** since it has only one term.
5. 13 is a polynomial of degree zero. Does that seem strange to you? If it helps, think of it as $13x^0$. In general, any nonzero constant polynomial has degree zero. We do not define the degree of the zero polynomial.

The whole subject of polynomials would be pretty dull and almost useless if it stopped with a definition. Fortunately, polynomials—like numbers—can be manipulated. In fact, they behave something like the integers. Just as the sum, difference, and product of two integers are integers, so the sum, difference, and product of two polynomials are polynomials. Notice that we did not mention division; that subject is discussed in Section 2-5 and in Chapter 10.

Addition

Adding two polynomials is a snap. Treat x like a number and use the commutative, associative, and distributive properties freely. When you are all done, you will discover that you have just grouped like terms (that is, terms of the same degree) and added their coefficients.

Example A (Adding Polynomials)   Add $x^3 + 2x^2 + 7x + 5$ and $x^2 - 3x - 4$.

Solution

$(x^3 + 2x^2 + 7x + 5) + (x^2 - 3x - 4)$

$\qquad = x^3 + (2x^2 + x^2) + (7x - 3x) + (5 - 4)$ (associative and commutative properties)

$\qquad = x^3 + (2 + 1)x^2 + (7 - 3)x + 1$ (distributive property)

$\qquad = x^3 + 3x^2 + 4x + 1$ ∎

How important are the parentheses in this example? Actually, they are indispensable only in the third line, where we used the distributive property. Why did we use them in the first and second lines? Because they emphasize what is happening. In the first line, they show which polynomials are added; in the second line, they draw attention to the terms being grouped. To shed additional light, notice that

$$x^3 + 3x^2 + 4x + 1$$

is the correct answer not only for

$$(x^3 + 2x^2 + 7x + 5) + (x^2 - 3x - 4)$$

but also for

$$(x^3 + 2x^2) + (7x + 5 + x^2) + (-3x - 4)$$

and even for

$$(x^3) + (2x^2) + (7x) + (5) + (x^2) + (-3x) + (-4)$$

Subtraction

How do we subtract two polynomials? We replace the subtracted polynomial by its negative and add.

Example B (Subtracting Polynomials)   Subtract $5x^2 + 4x - 4$ from $3x^2 - 5x + 2$.

Solution   We write

$$(3x^2 - 5x + 2) - (5x^2 + 4x - 4)$$

as

$$(3x^2 - 5x + 2) + (-5x^2 - 4x + 4)$$

Then, after grouping like terms, we obtain

$$(3x^2 - 5x^2) + (-5x - 4x) + (2 + 4) = (3 - 5)x^2 + (-5 - 4)x + (2 + 4)$$
$$= -2x^2 - 9x + 6$$

If you can go directly from the original problem to the answer, do so; we do not want to make simple things complicated. But be sure to note that

$$(3x^2 - 5x + 2) - (5x^2 + 4x - 4) \neq 3x^2 - 5x + 2 - 5x^2 + 4x - 4$$

The minus sign in front of $(5x^2 + 4x - 4)$ changes the sign of all three terms. ∎

Multiplication

The distributive property is the basic tool in multiplication. Here is a simple example using $a(b + c) = ab + ac$.

$$(3x^2)(2x^3 + 7) = (3x^2)(2x^3) + (3x^2)(7)$$
$$= 6x^5 + 21x^2$$

Here is a more complicated example, which uses the distributive property $(a + b)c = ac + bc$ at the first step.

Example C (Multiplying Polynomials)   Find the following product.

$$(3x - 4)(2x^3 - 7x + 8)$$

Solution

$$
\begin{aligned}
(3x - 4)(2x^3 - 7x + 8) &= (3x)(2x^3 - 7x + 8) + (-4)(2x^3 - 7x + 8) \\
&= (3x)(2x^3) + (3x)(-7x) + (3x)(8) + (-4)(2x^3) \\
&\quad + (-4)(-7x) + (-4)(8) \\
&= 6x^4 - 21x^2 + 24x - 8x^3 + 28x - 32 \\
&= 6x^4 - 8x^3 - 21x^2 + 52x - 32
\end{aligned}
$$

Notice that each term of $3x - 4$ multiplies each term of $2x^3 - 7x + 8$. If the process just illustrated seems unwieldy, you may find the format below helpful.

$$
\begin{array}{r}
2x^3 - 7x + 8 \\
3x \;\; - 4 \\
\hline
6x^4 \qquad\;\; - 21x^2 + 24x \\
- 8x^3 \qquad\quad + 28x - 32 \\
\hline
6x^4 - 8x^3 - 21x^2 + 52x - 32 \qquad \blacksquare
\end{array}
$$

When both polynomials are linear (that is, of the form $ax + b$), there is a handy shortcut. For example, just one look at $(x + 4)(x + 5)$ convinces us that the product has the form $x^2 + (\quad) + 20$. It is the middle term that may cause a little trouble. Think of it this way.

$$
\begin{aligned}
(x + 4)(x + 5) &= x^2 + (\quad) + 20 \\
&= x^2 + 9x + 20
\end{aligned}
$$

Here are two more illustrations.

Example D (Products of Linear Polynomials)   Find the following products.
(a) $(2x - 3)(x + 5)$ (b) $(3x + 2)(5x - 7)$

Solution
(a)

$$
\begin{aligned}
(2x - 3)(x + 5) &= 2x^2 + (\quad) - 15 \\
&= 2x^2 + 7x - 15
\end{aligned}
$$

The FOIL Method

$(x + 4)(x + 5) =$

F O I L

$x^2 + 5x + 4x + 20$

The four terms are the products of the Firsts, Outers, Inners, and Lasts.

Figure 6

CAUTION

~~$(x + 5)^2 = x^2 + 25$~~

$(x + 5)^2 = x^2 + 10x + 25$

(b) $-21x$ $\quad$ $-21x + 10x$

$10x$

$$\begin{aligned}(3x + 2)(5x - 7) &= 15x^2 + (\quad) - 14\\ &= 15x^2 - 11x - 14 \quad \blacksquare\end{aligned}$$

Soon you should be able to find such simple products in your head. Some people find the FOIL method helpful (Figure 6).

Three Special Products

Some products occur so often that they deserve to be highlighted and remembered.

$$\begin{aligned}(x + a)(x - a) &= x^2 - a^2\\ (x + a)^2 &= x^2 + 2ax + a^2\\ (x - a)^2 &= x^2 - 2ax + a^2\end{aligned}$$

Here are some illustrations.

$$\begin{aligned}(x + 7)(x - 7) &= x^2 - 7^2 = x^2 - 49\\ (x + 3)^2 &= x^2 + 2 \cdot 3x + 3^2 = x^2 + 6x + 9\\ (x - 4)^2 &= x^2 - 2 \cdot 4x + 4^2 = x^2 - 8x + 16\end{aligned}$$

Example E (Using Special Product Formulas) Find the following products.
(a) $(2t + 9)(2t - 9)$ (b) $(2x^2 - 3x)^2$ (c) $[(x^2 + 2) + x][(x^2 + 2) - x]$

Solution We appeal to the boxed results.

(a) $(2t + 9)(2t - 9) = (2t)^2 - 9^2 = 4t^2 - 81$

(b) $(2x^2 - 3x)^2 = (2x^2)^2 - 2(2x^2)(3x) + (3x)^2 = 4x^4 - 12x^3 + 9x^2$

(c) We apply the first formula in the box with $x^2 + 2$ and x playing the roles of x and a, respectively.

$$\begin{aligned}[(x^2 + 2) + x][(x^2 + 2) - x] &= (x^2 + 2)^2 - x^2\\ &= x^4 + 4x^2 + 4 - x^2\\ &= x^4 + 3x^2 + 4 \quad \blacksquare\end{aligned}$$

Cubes of Binomials

The following formulas for cubes will be useful. Note the pattern 1 3 3 1 for the coefficients, with the signs alternating in the second case.

$$\begin{aligned}(x + a)^3 &= x^3 + 3ax^2 + 3a^2x + a^3\\ (x - a)^3 &= x^3 - 3ax^2 + 3a^2x - a^3\end{aligned}$$

Example F (Expanding Cubes)   Expand.

(a) $(x - 5)^3$ (b) $(2x^2 + 3x)^3$

Solution

(a) $(x - 5)^3 = x^3 - 3 \cdot 5x^2 + 3 \cdot 5^2 x - 5^3$

$= x^3 - 15x^2 + 75x - 125$

(b) $(2x^2 + 3x)^3 = (2x^2)^3 + 3(2x^2)^2(3x) + 3(2x^2)(3x)^2 + (3x)^3$

$= 8x^6 + 36x^5 + 54x^4 + 27x^3$ ∎

Polynomials in Several Variables

In the boxed formulas, we assumed that you would think of x as a variable and a as a constant. However, that is not necessary. If we consider both x and a to be variables, then expressions like $x^2 - a^2$ and $x^2 + 2ax + a^2$ are polynomials in two variables. Other examples are

$$x^2 y + 3xy + y \qquad u^3 + 3u^2 v + 3uv^2 + v^3$$

Performing operations on polynomials in two variables is very similar to the one-variable case. Addition and subtraction are primarily a matter of combining like terms. Multiplication is illustrated in our next example.

Example G (Multiplying Polynomials in Several Variables)   Find the following products

(a) $(3x + 2y)(x - 5y)$ (b) $(2xy - 3z)^2$

Solution

(a) $(3x + 2y)(x - 5y) = 3x^2 - 13xy - 10y^2$

(b) $(2xy - 3z)^2 = (2xy)^2 - 2(2xy)(3z) + (3z)^2$

$= 4x^2 y^2 - 12xyz + 9z^2$ ∎

PROBLEM SET 2-3

A. Skills and Techniques

Decide whether the given expression is a polynomial. If it is, give its degree.

1. $3x^2 - x + 2$

2. $4x^5 - x$

3. $\pi s^5 - \sqrt{2}$

4. $3\sqrt{2}t$

5. $16\sqrt{2}$

6. $511/\sqrt{2}$

7. $3t^2 + \sqrt{t} + 1$

8. $t^2 + 3t + 1/t$

9. $3t^{-2} + 2t^{-1} + 5$

10. $5 + 4t + 6t^{10}$

Perform the indicated operations and simplify. Write your answer as a polynomial in descending powers of the variable. See Examples A–D.

11. $(2x - 7) + (-4x + 8)$

12. $(\frac{3}{2}x - \frac{1}{4}) + \frac{5}{6}x$

13. $(2x^2 - 5x + 6) + (2x^2 + 5x - 6)$

14. $(\sqrt{3}t + 5) + (6 - 4 - 2\sqrt{3}t)$

15. $(5 - 11x^2 + 4x) + (x - 4 + 9x^2)$

16. $(x^2 - 5x + 4) + (3x^2 + 8x - 7)$
17. $(2x - 7) - (-4x + 8)$
18. $(\frac{3}{2}x - \frac{1}{4}) - \frac{5}{6}x$
19. $(2x^2 - 5x + 6) - (2x^2 + 5x - 6)$
20. $y^3 - 4y + 6 - (3y^2 + 6y - 3)$
21. $5x(7x - 11) + 19$
22. $-x^2(7x^3 - 5x + 1)$
23. $(t + 5)(t + 11)$
24. $(t - 5)(t + 13)$
25. $(x + 9)(x - 10)$
26. $(x - 13)(x - 7)$
27. $(2t - 1)(t + 7)$
28. $(3t - 5)(4t - 2)$
29. $(4 + y)(y - 2)$
30. $1 + y(y - 2)$
31. $(2x - 5)(3x^2 - 2x + 4)$
32. $(3x + 2)(4x^2 + 3x - 1)$

Use the special product formulas to perform the following multiplications. Write your answer as a polynomial in descending powers of the variable. See Example E.

33. $(x + 10)^2$
34. $(y + 12)^2$
35. $(x + 8)(x - 8)$
36. $(t - 5)(t + 5)$
37. $(2t - 5)^2$
38. $(3s + 11)^2$
39. $(2x^4 + 5x)(2x^4 - 5x)$
40. $(u^3 + 2u^2)(u^3 - 2u^2)$
41. $[(t + 2) + t^3]^2$
42. $[(1 - x) + x^2]^2$
43. $[(t + 2) + t^3][(t + 2) - t^3]$
44. $[(1 - x) + x^2][(1 - x) - x^2]$
45. $(2.3x - 1.4)^2$
46. $(2.43x - 1.79)(2.43x + 1.79)$

Expand the following cubes as in Example F.

47. $(x + 2)^3$
48. $(x - 4)^3$
49. $(2t - 3)^3$
50. $(3u + 1)^3$
51. $(2t + t^2)^3$
52. $(4 - 3t^2)^3$
53. $[(2t + 1) + t^2]^3$
54. $[u^2 - (u + 1)]^3$

Perform the indicated operations and simplify as in Example G.

55. $(x - 3y)^2$
56. $(2x + y)^2$
57. $(3x - 2y)(3x + 2y)$
58. $(u^2 - 2v)(u^2 + 2v)$
59. $(3x - y)(4x + 5y)$
60. $(2x + y)(6x + 5y)$
61. $(2x^2y + z)(x^2y - z)$
62. $(3x + 5y^2z)(x - y^2z)$
63. $(t + 1 + s)(t + 1 - s)$
64. $(3u + 2v + 1)^2$
65. $(2t - 3s)^3$
66. $(u + 4v)^3$

B. Applications and Extensions

In Problems 67–86, perform the indicated operations and simplify.

67. $(2x^2 - 3y)(2x^2 + 3y)$
68. $(3x - y^3)^2$
69. $(2s^3 + 3t)(s^3 - 4t)$
70. $2u(3u - 1)(3u + 1)$
71. $(2u - v^2)^3$
72. $(x^2 + 3y)^3$
73. $2x(3x^2 - 6x + 4) - 3x[2x^2 - 4(x - 1)]$
74. $(y + 3)(2y - 5) - 2(3y - 2)(y + 2)$
75. $(2s + 3)^2 - (2s + 3)(2s - 3)$
76. $(x - \sqrt{2}y)(x + \sqrt{2}y) - (x - \sqrt{2}y)^2$
77. $(x^2 + 2x - 3)(x^2 + 2x + 3)$
78. $(y^2 - 3y + 2)^2$ 79. $(2x^2 + x - 1)(x + 2)$
80. $(y^2 + 2y - 3)(2y + 1)$
81. $(x^2 - 2xy)(x^2 + 2xy)(x + y)$
82. $xy(2x^2 - 3y^2)(3x^2 + 2y^2)$
83. $(x^2 + 2xy + 4y^2)(x - 2y)$
84. $(2x - y)^3 + 12x^2y - 6xy^2$
85. $(x^2 + xy + y^2)(x^2 - xy + y^2)$
86. $(2x + y)^4$
87. Find the coefficient of x^3 in the expansion of $(x^2 + 2x + 3)(x^3 - 3x^2 + 2x + 1)$.

88. Find the coefficient of x^5 in the expansion of $(x^4 + 2x^3 + 3x^2)^2$.

89. A triple (a, b, c) of positive integers is called a **Pythagorean triple** if $a^2 + b^2 = c^2$. For example, $(3, 4, 5)$ and $(5, 12, 13)$ are Pythagorean triples. Show that $(2m, m^2 - 1, m^2 + 1)$ is a Pythagorean triple, provided m is an integer greater than 1.

90. Show that $(m^2 - n^2, 2mn, m^2 + n^2)$ is a Pythagorean triple, provided m and n are positive integers with $m > n$.

91. If $r > 0$ and $(r + r^{-1})^2 = 5$, find $r^3 + r^{-3}$.

92. If $a + b = 1$ and $a^2 + b^2 = 2$, find $a^3 + b^3$.

93. If $x + y = \sqrt{11}$ and $x^2 + y^2 = 16$, find $x^4 + y^4$. *Hint:* It will be helpful to know that $(x + y)^4 = x^4 + 4x^3y + 6x^2y^2 + 4xy^3 + y^4$.

94. **TEASER** Show that 1 plus the product of four consecutive integers is always a perfect square.

2-4 FACTORING POLYNOMIALS

Polynomial to be Factored	Why Johnny Can't Factor Johnny's Answer	Teacher's Comments
1. $x^6 + 2x^2$	$x^2(x^3 + 2)$	Wrong. Have you forgotten that $x^2x^3 = x^5$? Right answer: $x^2(x^4 + 2)$
2. $x^2 + 5x + 6$	$(x + 6)(x + 1)$	Wrong. You didn't check the middle term. Right answer: $(x + 2)(x + 3)$
3. $x^2 - 4y^2$	$(x + 2y)(x - 2y)$	Right.
4. $x^2 + 4y^2$	$(x + 2y)^2$	Wrong. $(x + 2y)^2 = x^2 + 4xy + 4y^2$ Right answer: $x^2 + 4y^2$ doesn't factor using real coefficients.
5. $x^2y^2 + 6xy + 9$	Impossible	Wrong. x^2y^2 and 9 are squares. You should have suspected a perfect square. Right answer: $(xy + 3)^2$

To factor 90 means to write it as a product of smaller numbers; to factor it completely means to write it as a product of primes, that is, numbers that cannot be factored further. Thus we have factored 90 when we write $90 = 9 \cdot 10$, but it is not factored completely until we write

$$90 = 2 \cdot 3 \cdot 3 \cdot 5$$

Similarly, to factor a polynomial means to write it as a product of polynomials of lower degree; to factor a polynomial completely is to write it as a product of polynomials that cannot be factored further. Thus when we write

$$x^3 - 9x = x(x^2 - 9)$$

we have factored $x^3 - 9x$, but not until we write

$$x^3 - 9x = x(x + 3)(x - 3)$$

have we factored $x^3 - 9x$ completely.

Now why can't Johnny factor? He can't factor because he can't multiply. If he doesn't know that

$$(x + 2)(x + 3) = x^2 + 5x + 6$$

he certainly is not going to know how to factor $x^2 + 5x + 6$. That is why we urge you to memorize the special product formulas below. Of course, a product formula is also a factoring formula when read from right to left.

To urge memorization may be a bit old-fashioned, but we suggest that a fact, once memorized, becomes a permanent friend. It is best to memorize in words. For example, read formula 3 as follows: the square of a sum of two terms is the first squared plus twice their product plus the second squared.

Product Formulas ⟶

⟵ **Factoring Formulas**

1. $a(x + y + z) = ax + ay + az$
2. $(x + a)(x + b) = x^2 + (a + b)x + ab$
3. $(x + y)^2 = x^2 + 2xy + y^2$
4. $(x - y)^2 = x^2 - 2xy + y^2$
5. $(x + y)(x - y) = x^2 - y^2$
6. $(x + y)^3 = x^3 + 3x^2y + 3xy^2 + y^3$
7. $(x - y)^3 = x^3 - 3x^2y + 3xy^2 - y^3$
8. $(x + y)(x^2 - xy + y^2) = x^3 + y^3$
9. $(x - y)(x^2 + xy + y^2) = x^3 - y^3$

Taking Out a Common Factor

This, the simplest factoring procedure, is based on formula 1 above. You should always try this process first. Take Johnny's first problem as an example. Both terms of $x^6 + 2x^2$ have x^2 as a factor; so we take it out.

$$x^6 + 2x^2 = x^2(x^4 + 2)$$

Always factor out as much as you can. Taking 2 out of $4xy^2 - 6x^3y^4 + 8x^4y^2$ is not nearly enough, though it is a common factor; taking out $2xy$ is not enough either. You should take out $2xy^2$. Then

$$4xy^2 - 6x^3y^4 + 8x^4y^2 = 2xy^2(2 - 3x^2y^2 + 4x^3)$$

Factoring by Trial and Error

In factoring, as in life, success often results from trying and trying again. What does not work is systematically eliminated; eventually, effort is rewarded. Let us see how this process works on $x^2 - 5x - 14$. We need to find numbers a and b such that

$$x^2 - 5x - 14 = (x + a)(x + b)$$

Since ab must equal -14, two possibilities immediately occur to us: $a = 7$ and $b = -2$ or $a = -7$ and $b = 2$. Try them both to see if one works.

$$(x + 7)(x - 2) = x^2 + 5x - 14$$

$$(x - 7)(x + 2) = x^2 - 5x - 14 \qquad \text{Success!}$$

The brackets help us calculate the middle term, the crucial step in this kind of factoring. A failure of both trials would have led us to try $a = 14$ and $b = -1$ or $a = -14$ and $b = 1$.

Here is a tougher factoring problem.

Example A (Factoring $dx^2 + ex + f$) Factor $2x^2 + 13x - 15$.

Solution   It is a safe bet that if $2x^2 + 13x - 15$ factors at all, then

$$2x^2 + 13x - 15 = (2x + a)(x + b)$$

Since $ab = -15$, we are likely to try combinations of 3 and 5 first.

$$(2x + 5)(x - 3) = 2x^2 - x - 15$$

$$(2x - 5)(x + 3) = 2x^2 + x - 15$$

$$(2x + 3)(x - 5) = 2x^2 - 7x - 15$$

$$(2x - 3)(x + 5) = 2x^2 + 7x - 15$$

Discouraging, isn't it? But that is a poor reason to give up. Maybe we have missed some possibilities. We have, since combinations of 15 and 1 might work.

$$(2x - 15)(x + 1) = 2x^2 - 13x - 15$$

$$(2x + 15)(x - 1) = 2x^2 + 13x - 15 \qquad \text{Success!}$$

When you have had a lot of practice, you will be able to speed up the process. You will simply write

$$2x^2 + 13x - 15 = (2x + ?)(x + ?)$$

and mentally try the various possibilities until you find the right one. Of course, it may happen, as in the case of $2x^2 - 4x + 5$, that you cannot find a factorization. ■

Perfect Squares

Certain second degree (quadratic) polynomials are a breeze to factor.

$$x^2 + 10x + 25 = (x + 5)(x + 5) = (x + 5)^2$$

$$x^2 - 12x + 36 = (x - 6)^2$$

$$4x^2 + 12x + 9 = (2x + 3)^2$$

These are modeled after the special product formulas 3 and 4, which we now write with a and b replacing x and y.

$$a^2 + 2ab + b^2 = (a + b)^2$$

$$a^2 - 2ab + b^2 = (a - b)^2$$

We look for first and last terms that are squares, say of a and b. Then we ask if the middle term is twice their product.

But we need to be very flexible; a and b might be quite complicated. Consider $x^4 + 2x^2y^3 + y^6$. The first term is the square of x^2 and the last is the square of y^3; the middle term is twice their product. Thus,

$$x^4 + 2x^2y^3 + y^6 = (x^2)^2 + 2(x^2)(y^3) + (y^3)^2$$

$$= (x^2 + y^3)^2$$

Example B (Is It a Perfect Square?)   Factor, if possible, $y^2z^2 - 6ayz + 9a^2$ and $a^4b^2 + 6a^2bc + 4c^2$.

Solution   There is hope that both expressions can be factored since in each case the first and last terms are squares. Note that $6ayz = 2(yz)(3a)$, so

$$y^2z^2 - 6ayz + 9a^2 = (yz - 3a)^2$$

However,

$$a^4b^2 + 6a^2bc + 4c^2 \neq (a^2b + 2c)^2$$

since the middle term does not check. In fact, this trinomial cannot be factored.   ■

Difference of Squares

Do you see a common feature in the following polynomials?

$$x^2 - 16 \qquad y^2 - 100 \qquad 4y^2 - 9b^2$$

Each is the difference of two squares. From one of our special product formulas (formula 5), we know that

$$a^2 - b^2 = (a + b)(a - b)$$

Thus

$$x^2 - 16 = (x + 4)(x - 4)$$

$$y^2 - 100 = (y + 10)(y - 10)$$

$$4y^2 - 9b^2 = (2y + 3b)(2y - 3b)$$

Example C (Is It a Difference of Squares?)   Factor $4x^2y^6 - 25z^4$.

Solution $4x^2y^6 - 25z^4$ is of the form $a^2 - b^2$ with $a = 2xy^3$ and $b = 5z^2$. Thus

$$4x^2y^6 - 25z^4 = (2xy^3 + 5z^2)(2xy^3 - 5z^2)$$ ∎

Sum and Difference of Cubes

Now we are ready for some high-class factoring. Consider $8x^3 + 27$ and $x^3z^3 - 1000$. The first is a sum of cubes and the second is a difference of cubes. The secrets to success are the two special product formulas for cubes, restated here.

$$a^3 + b^3 = (a + b)(a^2 - ab + b^2)$$

$$a^3 - b^3 = (a - b)(a^2 + ab + b^2)$$

To factor $8x^3 + 27$, replace a by $2x$ and b by 3 in the first formula.

$$8x^3 + 27 = (2x)^3 + 3^3 = (2x + 3)[(2x)^2 - (2x)(3) + 3^2]$$

$$= (2x + 3)(4x^2 - 6x + 9)$$

CAUTION

$x^3 - 8 = (x - 2)(x^2 + 4)$
$x^3 - 8 = (x - 2)(x^2 - 4x + 4)$

$x^3 - 8 = (x - 2)(x^2 + 2x + 4)$

Similarly, to factor $x^3z^3 - 1000$, let $a = xz$ and $b = 10$ in the second formula.

$$x^3z^3 - 1000 = (xz)^3 - 10^3 = (xz - 10)(x^2z^2 + 10xz + 100)$$

Someone is sure to make a terrible mistake and write

$$x^3 + y^3 = (x + y)^3 \qquad \text{Wrong!}$$

Remember that

$$(x + y)^3 = x^3 + 3x^2y + 3xy^2 + y^3$$

Example D (Is It a Sum or Difference of Cubes?)   Factor $125c^3d^6 + 64m^{12}$.

Solution   Let $a = 5cd^2$ and $b = 4m^4$ and apply the first of the boxed formulas above.

$$125c^3d^6 + 64m^{12} = (5cd^2)^3 + (4m^4)^3$$

$$= (5cd^2 + 4m^4)[(5cd^2)^2 - (5cd^2)(4m^4) + (4m^4)^2]$$

$$= (5cd^2 + 4m^4)(25c^2d^4 - 20cd^2m^4 + 16m^8)$$ ∎

Factoring by Substitution

Examples C and D illustrated in simple cases a technique with much broader application. We refer to the technique of replacing a collection of symbols by a single letter— the method of substitution.

Example E (Will a Substitution Help?) Factor completely.
(a) $3x^4 + 10x^2 - 8$ (b) $(x + 2y)^2 - 3(x + 2y) - 10$

Solution
(a) Replace x^2 by u (or some other favorite letter of yours). Then

$$3x^4 + 10x^2 - 8 = 3u^2 + 10u - 8$$

But we know how to factor the latter:

$$3u^2 + 10u - 8 = (3u - 2)(u + 4)$$

Thus, when we go back to x, we get

$$3x^4 + 10x^2 - 8 = (3x^2 - 2)(x^2 + 4)$$

Neither of these quadratic polynomials factors further (using integer coefficients), so we are done.
(b) Here we could let $u = x + 2y$ and then factor the resulting quadratic polynomial. But this time, let us do that step mentally and write

$$(x + 2y)^2 - 3(x + 2y) - 10 = [(x + 2y) + ?][(x + 2y) - ?]$$
$$= (x + 2y + 2)(x + 2y - 5) \quad \blacksquare$$

Factoring in Stages

Factoring a polynomial completely may involve several steps. A simple example is factoring $3x^3 - 12xy^2$ where we should first take out the common factor $3x$ and then note that the second factor is a difference of squares.

$$3x^3 - 12xy^2 = 3x(x^2 - 4y^2) = 3x(x + 2y)(x - 2y)$$

Example F (Is More Than One Stage Needed?) Factor $x^6 - y^6$ completely.

Solution   First think of this expression as a difference of squares. Then factor again.
$$x^6 - y^6 = (x^3 - y^3)(x^3 + y^3)$$
$$= (x - y)(x^2 + xy + y^2)(x + y)(x^2 - xy + y^2)$$

A harder way to do this example is to begin by factoring $x^6 - y^6$ as a difference of cubes, obtaining

$$x^6 - y^6 = (x^2 - y^2)(x^4 + x^2y^2 + y^4)$$

We could continue to factor each of these expressions, but factoring the second one requires the special trick described in connection with Problems 67–70. $\blacksquare$

Factoring by Grouping

To factor an expression involving more than three terms requires grouping some of the terms together. Thus

$$am - an + bm - bn = (am - an) + (bm - bn)$$
$$= a(m - n) + b(m - n)$$
$$= (a + b)(m - n)$$

At the last step, we took out the common factor $m - n$; that is, we wrote $ax + bx$ as $(a + b)x$ with $x = m - n$.

A different kind of grouping helps in our next example.

Example G (Will Grouping Help?)   Factor $a^2 - 4ab + b^2 - c^2$.

Solution

$$
\begin{aligned}
a^2 - 4ab + 4b^2 - c^2 &= (a^2 - 4ab + 4b^2) - c^2 \\
&= (a - 2b)^2 - c^2 \quad \text{(difference of squares)} \\
&= [(a - 2b) + c][(a - 2b) - c] \\
&= (a - 2b + c)(a - 2b - c) \quad \blacksquare
\end{aligned}
$$

To Factor or Not to Factor

Which of the following can be factored?

(i) $x^2 - 4$

(ii) $x^2 - 6$

(iii) $x^2 + 16$

Did you say only the first one? You are correct if we insist on integer coefficients, or as we say, if we factor over the integers. But if we factor over the real numbers (that is, insist only that the coefficients be real), then the second polynomial can be factored.

$$
x^2 - 6 = (x + \sqrt{6})(x - \sqrt{6})
$$

If we factor over the complex numbers, even the third polynomial can be factored.

$$
x^2 + 16 = (x + 4i)(x - 4i)
$$

For this reason, we should always spell out what kind of coefficients we permit in the answer. We give specific directions in the following problem set. Incidentally, before trying the problems, review the opening panel of this section. It should make good sense now.

PROBLEM SET 2-4

A. Skills and Techniques

Factor completely over the integers (that is, allow only integer coefficients in your answers). Always begin by looking for common factors. See Examples A–D.

1. $x^2 + 5x$
2. $y^3 + 4y^2$
3. $x^2 + 5x - 6$
4. $x^2 + 5x + 4$
5. $y^4 - 6y^3$
6. $t^4 + t^2$
7. $y^2 + 4y - 12$
8. $z^2 - 3z - 40$
9. $y^2 + 8y + 16$
10. $9x^2 + 24x + 16$
11. $4x^2 - 12xy + 9y^2$
12. $9x^2 - 6x + 1$
13. $y^2 - 64$
14. $x^2 - 4y^2$
15. $1 - 25b^2$
16. $9x^2 - 64y^2$
17. $4z^2 - 4z - 3$
18. $7x^2 - 19x - 6$
19. $20x^2 + 3xy - 2y^2$
20. $4x^2 + 13xy - 12y^2$
21. $x^3 + 27$
22. $y^3 - 27$
23. $a^3 - 8b^3$
24. $8a^3 - 27b^3$

25. $x^3 - x^3y^3$

26. $x^6 + x^3y^3$

27. $x^2 - 3$

28. $y^2 - 5$

29. $3x^2 - 4$

Factor completely over the real numbers.

30. $x^2 - 3$

31. $y^2 - 5$

32. $3x^2 - 4$

33. $5z^2 - 4$

34. $t^4 - t^2$

35. $t^4 - 2t^2$

36. $x^2 + 2\sqrt{2}x + 2$

37. $y^2 - 2\sqrt{3}y + 3$

38. $x^2 + 4y^2$

39. $x^2 + 9$

40. $4x^2 + 1$

$\boxed{\text{i}}$ *Factor completely over the complex numbers.*

41. $x^2 + 9$

42. $4x^2 + 1$

Factor completely over the integers. See Example E, but if you can factor without actually making a substitution, that is fine.

43. $x^6 + 9x^3 + 14$

44. $x^4 - x^2 - 6$

45. $4x^4 - 37x^2 + 9$

46. $6y^4 + 13y^2 - 5$

47. $(x + 4y)^2 + 6(x + 4y) + 9$

48. $(m - n)^2 + 5(m - n) + 4$

49. $x^4 - x^2y^2 - 6y^4$

50. $x^4y^4 + 5x^4y^2 + 6x^4$

Factor completely over the integers. See Example F.

51. $x^6 - 64$

52. $x^4 - y^4$

53. $x^8 - x^4y^4$

54. $x^9 - 64x^3$

55. $x^6 + y^6$ (sum of cubes)

56. $a^6 + 64$

Factor completely over the integers as in Example G.

57. $x^3 - 4x^2 + x - 4$

58. $y^3 + 3y^2 - 2y - 6$

59. $4x^2 - 4x + 1 - y^2$

60. $9a^2 - 4b^2 - 12b - 9$

61. $3x + 3y - x^2 - xy$

62. $y^2 - 3y + xy - 3x$

63. $x^2 + 6xy + 9y^2 + 2x + 6y$

64. $y^2 + 4xy + 4x^2 - 3y - 6x$

65. $x^2 + 2xy + y^2 + 3x + 3y + 2$

66. $a^2 - 2ab + b^2 - c^2 + 4cd - 4d^2$

Factor completely over the integers using the trick of adding and subtracting the same thing. For example, $x^4 + 4 = (x^4 + 4x^2 + 4) - 4x^2$ is the difference of squares.

67. $x^4 + 64$

68. $y^8 + 4$

69. $x^4 + x^2 + 1$

70. $x^8 + 3x^4 + 4$

B. Applications and Extensions

In Problems 71–94, factor completely over the integers.

71. $4 - 9m^2$

72. $9m^2 + 6m + 1$

73. $6x^2 - 5x + 1$

74. $6x^2 - 5x - 6$

75. $5x^3 - 20x$

76. $3x^2 - 18x + 27$

77. $6x^3 - 5x^2 + x$

78. $4x^5 - 32x^2$

79. $2u^4 - 7u^2 + 5$

80. $x^3y^2 + x^2y^3 + 2xy^4$

81. $(a + 2b)^2 - 3(a + 2b) - 28$

82. $(2a + b)^3 - 8$

83. $(a + 3b)^4 - 1$

84. $x^2 + 4xy + 4y^2 - x^2y^2$

85. $x^2 - 6xy + 9y^2 + 4x - 12y$

86. $x^2 - 2xy + y^2 + 3x - 3y + 2$

87. $9x^4 - 24x^2y^2 + 16y^4 - y^2$

88. $9x^4 + 15x^2y^2 + 16y^4$

89. $x^4 - 3x^2y^2 + y^4$

90. $8x^2(x - 2) + 8x(x - 2) + 2(x - 2)$

91. $(x + 3)^2(x + 2)^3 - 20(x + 3)(x + 2)^2$

92. $(x - 3y)(x + 5y)^4 - 4(x - 3y)(x + 5y)^2$

93. $x^{2n} + 3x^n + 2$

94. $x^{n+3} + 5x^n + x^3 + 5$

95. Calculate each of the following the easy way.
    (a) $(547)^2 - (453)^2$

    (b) $\dfrac{2^{20} - 2^{17} + 7}{2^{17} + 1}$

    (c) $\left(1 - \dfrac{1}{2^2}\right)\left(1 - \dfrac{1}{3^2}\right)\left(1 - \dfrac{1}{4^2}\right) \cdots \left(1 - \dfrac{1}{29^2}\right)$

96. If n is an integer greater than 1, then $a^n - b^n = (a - b)P$, where P is a polynomial in a and b. Find the general form of P.

97. Note that $2^3 = 3^2 - 1^2$ and $3^3 = 6^2 - 3^2$. Show that the cube of any integer n can be written as the difference of the squares of two integers. *Hint:* One of these integers can be chosen as $n(n + 1)/2$.

98. TEASER If $a + b + c = 1$, $a^2 + b^2 + c^2 = 2$, and $a^3 + b^3 + c^3 = 3$, find $a^4 + b^4 + c^4$. *Hint:* This generalizes Problem 92 of Section 2-3.

2-5 RATIONAL EXPRESSIONS

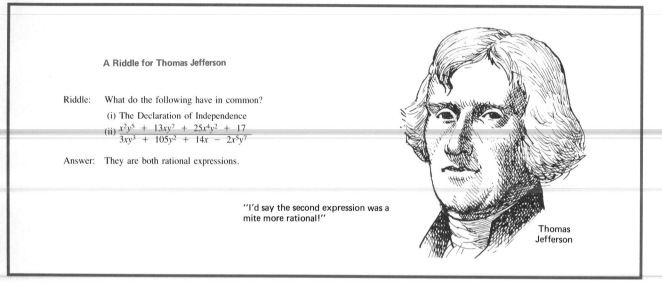

A Riddle for Thomas Jefferson

Riddle: What do the following have in common?

(i) The Declaration of Independence

(ii) $\dfrac{x^2y^5 + 13xy^7 + 25x^4y^2 + 17}{3xy^3 + 105y^2 + 14x - 2x^5y^7}$

Answer: They are both rational expressions.

"I'd say the second expression was a mite more rational!"

Thomas Jefferson

We admit that the riddle above is a bit silly, but it allowed us to insert Jefferson's picture and to make a point that is not well known. Thomas Jefferson was trained in mathematics and wrote that he had been profoundly influenced by his college mathematics teacher.

Recall that the quotient (ratio) of two integers is a rational number. A quotient (ratio) of two polynomials is called a **rational expression.** Here are some examples involving one variable.

$$\frac{x}{x + 1} \qquad \frac{3x^2 + 6}{x^2 + 2x} \qquad \frac{x^{15} + 3x}{6x^2 + 11}$$

The example in the riddle is a rational expression in two variables.

We add, subtract, multiply, and divide rational expressions by the same rules we use for rational numbers (see Section 1-2). The result is always a rational expression. That should not surprise us, since we had the same kind of experience with rational numbers.

Our immediate task is to define the reduced form for a rational expression. This, too, generalizes a notion we had for rational numbers.

Reduced Form

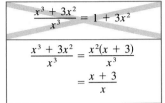
A rational expression is in reduced form if its numerator and denominator have no (nontrivial) common factor. For example, $x/(x + 2)$ is in reduced form, but $x^2/(x^2 + 2x)$ is not, since

$$\frac{x^2}{x^2 + 2x} = \frac{x \cdot x}{x(x + 2)} = \frac{x}{x + 2}$$

To reduce a rational expression, we factor numerator and denominator and divide out, or cancel, common factors.

Example A (Reducing Fractions)   Reduce.

(a) $\dfrac{x^2 + 7x + 10}{x^2 - 25}$ (b) $\dfrac{2x^2 + 5xy - 3y^2}{2x^2 + xy - y^2}$

Solution

(a) $\dfrac{x^2 + 7x + 10}{x^2 - 25} = \dfrac{(x + 5)(x + 2)}{(x + 5)(x - 5)} = \dfrac{x + 2}{x - 5}$

(b) $\dfrac{2x^2 + 5xy - 3y^2}{2x^2 + xy - y^2} = \dfrac{(2x - y)(x + 3y)}{(2x - y)(x + y)} = \dfrac{x + 3y}{x + y}$ ■

Addition and Subtraction

We add (or subtract) rational expressions by rewriting them so that they have the same denominator and then adding (or subtracting) the new numerators. Suppose that we want to add

$$\frac{3}{x} + \frac{2}{x + 1}$$

The appropriate common denominator is $x(x + 1)$. Remember that we can multiply numerator and denominator of a fraction by the same thing. Accordingly,

$$\frac{3}{x} + \frac{2}{x + 1} = \frac{3(x + 1)}{x(x + 1)} + \frac{x \cdot 2}{x(x + 1)} = \frac{3x + 3 + 2x}{x(x + 1)} = \frac{5x + 3}{x(x + 1)}$$

The same procedure is used in subtraction.

$$\frac{2}{x + 1} - \frac{3}{x} = \frac{x \cdot 2}{x(x + 1)} - \frac{3(x + 1)}{x(x + 1)} = \frac{2x - (3x + 3)}{x(x + 1)}$$

$$= \frac{2x - 3x - 3}{x(x + 1)} = \frac{-x - 3}{x(x + 1)}$$

Here is a more complicated example. Study each step carefully.

Example B (Subtracting Fractions)   Subtract and simplify

$$\frac{3x}{x^2 - 1} - \frac{2x + 1}{x^2 - 2x + 1}$$

Solution

Begin by factoring

$$\frac{3x}{x^2 - 1} - \frac{2x + 1}{x^2 - 2x + 1} = \frac{3x}{(x - 1)(x + 1)} - \frac{2x + 1}{(x - 1)^2}$$

$(x - 1)^2(x + 1)$ is the lowest common denominator

$$= \frac{3x(x - 1)}{(x - 1)^2(x + 1)} - \frac{(2x + 1)(x + 1)}{(x - 1)^2(x + 1)}$$

Forgetting the parentheses around $2x^2 + 3x + 1$ would be a serious blunder

$$= \frac{3x^2 - 3x - (2x^2 + 3x + 1)}{(x - 1)^2(x + 1)}$$

$$= \frac{3x^2 - 3x - 2x^2 - 3x - 1}{(x - 1)^2(x + 1)}$$

No cancellation is possible since $x^2 - 6x - 1$ doesn't factor over the integers

$$= \frac{x^2 - 6x - 1}{(x - 1)^2(x + 1)} \quad \blacksquare$$

Sometimes we can simplify subtraction by a manipulation of signs. Recall that a fraction has three sign positions: numerator, denominator, and total fraction. We may change any two of these signs without changing the value of the fraction. Thus

$$\frac{a}{b} = -\frac{a}{-b} = -\frac{-a}{b} = \frac{-a}{-b}$$

Example C (Manipulating Signs)   Simplify

$$\frac{x}{3x - 6} - \frac{2}{2 - x}$$

Solution

$$\frac{x}{3x - 6} - \frac{2}{2 - x} = \frac{x}{3(x - 2)} - \frac{2}{2 - x}$$

Now we make a crucial observation. Notice that

$$-(2 - x) = -2 + x = x - 2$$

That is, $2 - x$ and $x - 2$ are negatives of each other. Thus the expression above may be rewritten as

$$\frac{x}{3(x - 2)} - \frac{2}{2 - x} = \frac{x}{3(x - 2)} - \frac{2}{-(x - 2)}$$

$$= \frac{x}{3(x - 2)} + \frac{2}{x - 2}$$

CAUTION

$$\frac{5}{x - 2} - \frac{x + 2}{x - 2} = \frac{5 - x + 2}{x - 2}$$

$$= \frac{7 - x}{x - 2}$$

$$\frac{5}{x - 2} - \frac{x + 2}{x - 2} = \frac{5 - x - 2}{x - 2}$$

$$= \frac{3 - x}{x - 2}$$

$$= \frac{x}{3(x-2)} + \frac{6}{3(x-2)}$$

$$= \frac{x+6}{3(x-2)}$$

When we replaced $-\dfrac{2}{-(x-2)}$ by $\dfrac{2}{x-2}$, we used the fact that

$$-\frac{a}{-b} = \frac{a}{b} \qquad \blacksquare$$

Multiplication

We multiply rational expressions in the same manner as we do rational numbers; that is, we multiply numerators and multiply denominators. For example,

$$\frac{3}{x+5} \cdot \frac{x+2}{x-4} = \frac{3(x+2)}{(x+5)(x-4)} = \frac{3x+6}{x^2+x-20}$$

Sometimes we need to reduce the product, if we want the simplest possible answer. Here is an illustration.

Example D (Multiplying Fractions)   Simplify

$$\frac{2x-3}{x+5} \cdot \frac{x^2-25}{6xy-9y}$$

Solution

$$\frac{2x-3}{x+5} \cdot \frac{x^2-25}{6xy-9y} = \frac{(2x-3)(x^2-25)}{(x+5)(6xy-9y)}$$

$$= \frac{\cancel{(2x-3)}(x-5)\cancel{(x+5)}}{\cancel{(x+5)}(3y)\cancel{(2x-3)}}$$

$$= \frac{x-5}{3y}$$

This example shows that it is a good idea to do as much factoring as possible at the outset. That is what set up the cancellation. $\blacksquare$

Division

There are no real surprises with division, as we simply invert the divisor and multiply. Here is a division example that involves reduction as a last step.

Example E (Dividing Fractions)   Simplify

$$\frac{x^2-5x+4}{2x+6} \div \frac{2x^2-x-1}{x^2+5x+6}$$

Solution

$$\frac{x^2 - 5x + 4}{2x + 6} \div \frac{2x^2 - x - 1}{x^2 + 5x + 6} = \frac{x^2 - 5x + 4}{2x + 6} \cdot \frac{x^2 + 5x + 6}{2x^2 - x - 1}$$

$$= \frac{(x^2 - 5x + 4)(x^2 + 5x + 6)}{(2x + 6)(2x^2 - x - 1)}$$

$$= \frac{(x - 4)(x - 1)(x + 2)(x + 3)}{2(x + 3)(x - 1)(2x + 1)}$$

$$= \frac{(x - 4)(x + 2)}{2(2x + 1)} \quad \blacksquare$$

Combining Operations

Many fractions that we will face in later work involve combinations of the operations we have discussed. We illustrate with two examples.

Example F (A Quotient Arising in Calculus)   Simplify

$$\frac{\dfrac{2}{x + h} - \dfrac{2}{x}}{h}$$

Solution   This expression may look artificial, but it is one you are apt to find in calculus. It represents the average rate of change in $2/x$ as x changes to $x + h$. We begin by doing the subtraction in the numerator.

$$\frac{\dfrac{2}{x + h} - \dfrac{2}{x}}{h} = \frac{\dfrac{2x - 2(x + h)}{(x + h)x}}{h} = \frac{\dfrac{2x - 2x - 2h}{(x + h)x}}{\dfrac{h}{1}}$$

$$= \frac{-2h}{(x + h)x} \cdot \frac{1}{h} = \frac{-2}{(x + h)x} \quad \blacksquare$$

Example G (Four-Story Fractions)   Simplify

$$\frac{\dfrac{x}{x - 4} - \dfrac{3}{x + 3}}{\dfrac{1}{x} + \dfrac{2}{x - 4}}$$

Solution   *Method 1*   (Simplify the numerator and denominator separately and then divide.)

$$\frac{\dfrac{x}{x - 4} - \dfrac{3}{x + 3}}{\dfrac{1}{x} + \dfrac{2}{x - 4}} = \frac{\dfrac{x(x + 3) - 3(x - 4)}{(x - 4)(x + 3)}}{\dfrac{x - 4 + 2x}{x(x - 4)}}$$

$$= \frac{\dfrac{x^2 + 3x - 3x + 12}{(x - 4)(x + 3)}}{\dfrac{3x - 4}{x(x - 4)}}$$

$$= \frac{x^2 + 12}{(x - 4)(x + 3)} \cdot \frac{x(x - 4)}{3x - 4}$$

$$= \frac{x(x^2 + 12)}{(x + 3)(3x - 4)}$$

Method 2 (Multiply the fractions in the numerator and denominator by a common denominator, in this case, $(x - 4)(x + 3)x$.)

$$\frac{\dfrac{x}{x - 4} - \dfrac{3}{x + 3}}{\dfrac{1}{x} + \dfrac{2}{x - 4}} = \frac{\left(\dfrac{x}{x - 4} - \dfrac{3}{x + 3}\right)(x - 4)(x + 3)x}{\left(\dfrac{1}{x} + \dfrac{2}{x - 4}\right)(x - 4)(x + 3)x}$$

$$= \frac{x^2(x + 3) - 3x(x - 4)}{(x - 4)(x + 3) + 2x(x + 3)}$$

$$= \frac{x^3 + 3x^2 - 3x^2 + 12x}{(x + 3)(x - 4 + 2x)}$$

$$= \frac{x(x^2 + 12)}{(x + 3)(3x - 4)} \quad \blacksquare$$

PROBLEM SET 2-5

A. Skills and Techniques

Reduce each of the following as in Example A.

1. $\dfrac{x + 6}{x^2 - 36}$

2. $\dfrac{x^2 - 1}{4x - 4}$

3. $\dfrac{y^2 + y}{5y + 5}$

4. $\dfrac{x^2 - 7x + 6}{x^2 - 4x - 12}$

5. $\dfrac{(x + 2)^3}{x^2 - 4}$

6. $\dfrac{x^3 + a^3}{(x + a)^2}$

7. $\dfrac{zx^2 + 4xyz + 4y^2z}{x^2 + 3xy + 2y^2}$

8. $\dfrac{x^3 - 27}{3x^2 + 9x + 27}$

Perform the indicated operations and simplify. See Example B and the discussion that precedes it.

9. $\dfrac{5}{x - 2} + \dfrac{4}{x + 2}$

10. $\dfrac{5}{x - 2} - \dfrac{4}{x + 2}$

11. $\dfrac{5x}{x^2 - 4} + \dfrac{3}{x + 2}$

12. $\dfrac{3}{x} - \dfrac{2}{x + 3} + \dfrac{1}{x^2 + 3x}$

13. $\dfrac{2}{xy} + \dfrac{3}{xy^2} - \dfrac{1}{x^2y^2}$

14. $\dfrac{x + y}{xy^3} - \dfrac{x - y}{y^4}$

15. $\dfrac{x + 1}{x^2 - 4x + 4} + \dfrac{4}{x^2 + 3x - 10}$

16. $\dfrac{x^2}{x^2 - x + 1} - \dfrac{x + 1}{x}$

Simplify as in Example C.

17. $\dfrac{4}{2x - 1} + \dfrac{x}{1 - 2x}$

18. $\dfrac{x}{6x - 2} - \dfrac{3}{1 - 3x}$

19. $\dfrac{2}{6y - 2} + \dfrac{y}{9y^2 - 1} - \dfrac{2y + 1}{1 - 3y}$

20. $\dfrac{x}{4x^2 - 1} + \dfrac{2}{4x - 2} - \dfrac{3x + 1}{1 - 2x}$

21. $\dfrac{m^2}{m^2 - 2m + 1} - \dfrac{1}{3 - 3m}$

22. $\dfrac{2x}{x^2 - y^2} + \dfrac{1}{x + y} + \dfrac{1}{y - x}$

In Problems 23–30, multiply and express in simplest form, as illustrated in Example D.

23. $\dfrac{5}{2x - 1} \cdot \dfrac{x}{x + 1}$

24. $\dfrac{3}{x^2 - 2x} \cdot \dfrac{x - 2}{x}$

25. $\dfrac{x + 2}{x^2 - 9} \cdot \dfrac{x + 3}{x^2 - 4}$

26. $\left(1 + \dfrac{1}{x + 2}\right) \dfrac{4}{3x + 9}$

27. $x^2 y^4 \left(\dfrac{x}{y^2} - \dfrac{y}{x^2}\right)$

28. $\dfrac{5x^2}{x^3 + y^3} \left(\dfrac{1}{xy^2} - \dfrac{1}{x^2 y^2} + \dfrac{1}{x^3 y}\right)$

29. $\dfrac{x^2 + 5x}{x^2 - 16} \cdot \dfrac{x^2 - 2x - 24}{x^2 - x - 30}$

30. $\dfrac{x^3 - 125}{2x^3 - 10x^2} \cdot \dfrac{7x}{x^3 + 5x^2 + 25x}$

Express the quotients in Problems 31–37 in simplest form, as illustrated in Example E.

31. $\dfrac{\dfrac{5}{2x - 1}}{\dfrac{x}{x + 1}}$

32. $\dfrac{\dfrac{5}{2x - 1}}{\dfrac{x}{4x^2 - 1}}$

33. $\dfrac{\dfrac{x + 2}{x^2 - 4}}{x}$

34. $\dfrac{\dfrac{x + 2}{x^2 - 3x}}{\dfrac{x^2 - 4}{x}}$

35. $\dfrac{\dfrac{x^2 + a^3}{x^3 - a^3}}{\dfrac{x + 2a}{(x - a)^2}}$

36. $\dfrac{1 + \dfrac{2}{b}}{1 - \dfrac{4}{b^2}}$

37. $\dfrac{\dfrac{y^2 + y - 2}{y^2 + 4y}}{\dfrac{2y^2 - 8}{y^2 + 2y - 8}}$

Simplify as in Example F.

38. $\dfrac{\dfrac{4}{x + h} - \dfrac{4}{x}}{h}$

39. $\dfrac{\dfrac{1}{2x + 2h + 3} - \dfrac{1}{2x + 3}}{h}$

40. $\dfrac{\dfrac{x + h}{x + h + 4} - \dfrac{x}{x + 4}}{h}$

41. $\dfrac{\dfrac{1}{(x + h)^2} - \dfrac{1}{x^2}}{h}$

Simplify, using either method of Example G.

42. $\dfrac{\dfrac{1}{x + 2} - \dfrac{3}{x^2 - 4}}{\dfrac{3}{x - 2}}$

43. $\dfrac{\dfrac{y}{y + 4} - \dfrac{2}{y^2 + 5y + 4}}{\dfrac{4}{y + 1} + \dfrac{3}{y + 4}}$

44. $\dfrac{\dfrac{1}{x} - \dfrac{1}{x - 2} + \dfrac{3}{x^2 - 2x}}{\dfrac{x}{x - 2} + \dfrac{3}{x}}$

45. $\dfrac{\dfrac{a^2}{b^2} - \dfrac{b^2}{a^2}}{\dfrac{a}{b} - \dfrac{b}{a}}$

46. $\dfrac{n - \dfrac{n^2}{n - m}}{1 + \dfrac{m^2}{n^2 - m^2}}$

47. $\dfrac{\dfrac{x^2}{x - y} - x}{\dfrac{y^2}{x - y} + y}$

48. $1 - \dfrac{x - (1/x)}{1 - (1/x)}$

49. $\dfrac{y - \dfrac{1}{1 + (1/y)}}{y + \dfrac{1}{y - (1/y)}}$

B. Applications and Extensions

In Problems 50–59, perform the indicated operations and simplify.

50. $\dfrac{1 - 2x}{x - 4} + \dfrac{6x + 2}{3x - 12}$

51. $x + y + \dfrac{y^2}{x - y}$

52. $x + y + \dfrac{y^2}{x - y} + \dfrac{x^2}{x + y}$

53. $\left(\dfrac{1}{x} + \dfrac{1}{y}\right)\left(x + y - \dfrac{x^2 + y^2}{x + y}\right)$

54. $\dfrac{x}{x^2 - 5x + 6} + \dfrac{3}{x^2 - 7x + 12}$

55. $\dfrac{x}{x^2 + 11x + 30} - \dfrac{5}{x^2 + 9x + 20}$

56. $\dfrac{\dfrac{a - b}{a + b} - \dfrac{a + b}{a - b}}{\dfrac{ab}{a - b}}$

57. $\dfrac{\dfrac{a^2 + 4a + 3}{a} - \dfrac{2a + 2}{a - 1}}{\dfrac{a + 1}{a^2 - a}}$

58. $\dfrac{x^3 - 8y^3}{x^2 - 4y^2} + \dfrac{2xy}{x + 2y}$

59. $\dfrac{18x^2y - 27x^2 - 8y + 12}{6xy - 4y - 9x + 6}$

60. Let x, y, and z be positive numbers. How does $(x + z)/(y + z)$ compare in size with x/y?

61. Arnold Thinkhard simplified

$$\dfrac{x^3 + y^3}{x^3 + (x - y)^3} \quad \text{to} \quad \dfrac{x + y}{x + (x - y)}$$

by canceling all the 3s. Even though his method was entirely wrong, show that he got the right answer.

62. A number of the form

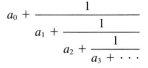

where a_k is an integer with $a_k \geq 0$ if $k \geq 1$, is called a **continued fraction.** Here the dots indicate a pattern that may either continue indefinitely or terminate. Simplify each of the following terminating continued fractions.

(a) $2 + \dfrac{1}{2 + \dfrac{1}{2 + \dfrac{1}{2}}}$ **(b)** $1 + \dfrac{1}{2 + \dfrac{1}{3 + \dfrac{1}{4}}}$

(c) $x + \dfrac{1}{x + \dfrac{1}{x + \dfrac{1}{x}}}$

63. A terminating continued fraction clearly represents a rational number (Why?). Conversely, every rational number can be represented as a terminating continued fraction. For example,

$$\dfrac{7}{4} = 1 + \dfrac{3}{4} = 1 + \dfrac{1}{\dfrac{4}{3}} = 1 + \dfrac{1}{1 + \dfrac{1}{3}}$$

Find the continued fraction expansion of each of the following.

(a) $\dfrac{13}{5}$ **(b)** $\dfrac{29}{11}$ **(c)** $-\dfrac{5}{4}$

64. TEASER Evaluate the nonterminating continued fraction

$$\phi = 1 + \dfrac{1}{1 + \dfrac{1}{1 + \dfrac{1}{1 + \cdots}}}$$

It would spoil the problem to give a real hint but Problem 63 implies that ϕ is irrational and we will tell you that ϕ satisfies a simple quadratic equation. Our choice of the Greek letter ϕ to denote this number is deliberate. The ancient Greeks loved this number and gave it a special name.

CHAPTER 2 SUMMARY

An **exponent** is a numerical superscript placed on a number to indicate that a certain operation is to be performed on that number. In particular

$$b^4 = b \cdot b \cdot b \cdot b \qquad b^0 = 1 \qquad b^{-3} = \frac{1}{b^3} = \frac{1}{b \cdot b \cdot b}$$

Exponents mix together according to five laws called the **rules of exponents.**

A number is in **scientific notation** when it appears in the form $c \times 10^n$ where $1 \le |c| < 10$ and n is an integer. Very small and very large numbers are commonly written this way. Hand-held calculators often use scientific notation in their displays. These electronic marvels are designed to take the drudgery out of arithmetic calculations and are a valuable tool in an algebra-trigonometry course.

An expression of the form

$$a_n x^n + a_{n-1} x^{n-1} + \cdots + a_1 x + a_0$$

is called a **polynomial** in x. The exponent n is its **degree** (provided that $a_n \ne 0$); the a's are its **coeffi-** cients. Polynomials can be added, subtracted, and multiplied, the result in each case being another polynomial.

To **factor** a polynomial is to write it as a product of simpler polynomials; to **factor over the integers** is to write a polynomial as a product of polynomials with integer coefficients. Here are five examples.

Common factor: $x^2 - 2ax = x(x - 2a)$
Difference of squares: $4x^2 - 25 = (2x + 5)(2x - 5)$
Trial and error: $6x^2 + x - 15 = (2x - 3)(3x + 5)$
Perfect square: $x^2 + 14x + 49 = (x + 7)^2$
Sum of cubes: $x^3 + 1000 = (x + 10)(x^2 - 10x + 100)$

A quotient (ratio) of two polynomials is called a **rational expression.** The expression is in **reduced form** if its numerator and denominator have no nontrivial common factors. We add, subtract, multiply, and divide rational expressions in much the same way as we do rational numbers.

CHAPTER 2 REVIEW PROBLEM SET

In Problems 1–10, write True or False in the blank. If false, tell why.

_____ **1.** $2^m 2^n = 2^{mn}$

_____ **2.** $3^n 2^n = 6^n$

_____ **3.** $(-4)^n = 1/4^n$

_____ **4.** 62.345×10^5 is in scientific notation.

_____ **5.** The expression $2^x + 1$ is a polynomial.

_____ **6.** If p is a polynomial of degree 3 and q is a polynomial of degree 4, then pq is a polynomial of degree 12.

_____ **7.** $a^6 - 4b^4 = (a^3 + 2b^2)(a^3 - 2b^2)$

_____ **8.** $x^2 + 13$ can be factored over the complex numbers.

_____ **9.** $x - i$ is a factor of $x^4 + x^2$.

_____ **10.** If $x \ne \dfrac{1}{4}$, then $-\dfrac{2 - 3x}{4x - 1} = \dfrac{3x - 2}{1 - 4x}$

Simplify, leaving your answer free of negative exponents.

11. $\left(\dfrac{4}{3}\right)^{-3}$

12. $\left(\dfrac{4}{7}\right)^4 \left(\dfrac{4}{7}\right)^{-2}$

13. $\left(2 + \dfrac{2}{3} - \dfrac{1}{2}\right)^{-2}$

14. $\dfrac{3a^{-1}b^2}{(2a^{-1})^{-3}b^4}$

15. $\left(\dfrac{4x^{-3}}{y^2}\right)^{-3}$

16. $\dfrac{(x^3 y^{-1})^3 (2x^{-1} y^2)^{-2}}{(x^4 y^{-2})^4}$

After simplifying, if necessary, express in scientific notation.

17. $215,000,000$

18. $.000107$

19. $(402,000)(2)10^{-8}$

20. $\dfrac{(1.44) \cdot 10^4}{(4.8) \cdot 10^{-5}}$

Decide which of the following are polynomials. Give the degree of each polynomial.

21. $5x^3 - 2x + 4\sqrt{x}$

22. $3x^2 + \sqrt{\pi} x - \dfrac{15}{7}$

23. $16x/(2x^2 + 5)$ **24.** $2x^{-3} + 2x^{-1} + 5$

Perform the indicated operations and simplify.

25. $(2x^2 + 7) + (-3x^2 + x - 14)$

26. $4 + x^4 - (2x - 3x^4 + x^2 - 11)$

27. $(2y + 1)(3y - 2)$

28. $(4x + 3)^2 + (4x - 3)^2$

29. $(3xy - 2z^2)^2$

30. $(5a^3 + 2b)(2a^3 - 7b)$

31. $(x^2 - 3x + 2)(x^2 + 3x - 2)$

32. $(x^2 - 3y)(x^4 + 3x^2y + 9y^2)$

33. $(2s^2 - 5s + 3)^2$

34. $(5x^2 - 3yz)(5x^2 + 3yz)$

In Problems 35–46, factor completely over the integers.

35. $x^2 + 3x - 10$

36. $x^2 - 7x - 30$

37. $6x^2 + 13x - 5$

38. $2x^2 + 21x - 11$

39. $3x^6 + 2x^3 - 8$

40. $4x^6 - 9$

41. $8x^3 - 125$

42. $25x^2 - 20xy + 4y^2$

43. $9c^2d^4 - 6bcd^2 + b^2$

44. $x^4 - 16$

45. $4x^2 - y^2 + 6y - 9$

46. $25x^2 - 4y^2 - 15x - 6y$

47. Factor $4x^2 - 29$ over the real numbers.

$\boxed{\text{i}}$ **48.** Factor $x^2 + 2x + 2$ over the complex numbers. *Hint:* $x^2 + 2x + 2 = (x + 1)^2 + 1$.

Reduce each of the following.

49. $\dfrac{3x^2 - 6x}{x^2 + x - 6}$

50. $\dfrac{(x^2 - 4)^3}{(x^3 + 8)^2}$

Perform the indicated operations and simplify.

51. $\dfrac{4}{x + 4} - \dfrac{2}{x - 4} + \dfrac{3x - 1}{x^2 - 16}$

52. $\dfrac{2x^2 + 5x - 3}{x^2 - 4} \cdot \dfrac{x^2 - 5x - 14}{2x^2 - 15x + 7}$

53. $\dfrac{(1 - a^{-1})(1 + a^{-3})}{1 - a^{-2}}$

54. $1 - \left(\dfrac{x^3 - x^{-2}}{1 - x^{-2}}\right)^2$

Equations
and Inequalities

■ As the sun eclipses the stars by its brilliancy, so the man of knowledge will eclipse the fame of others in the assemblies of the people if he proposes algebraic problems, and still more if he solves them.

—Brahmagupta

3-1 EQUATIONS

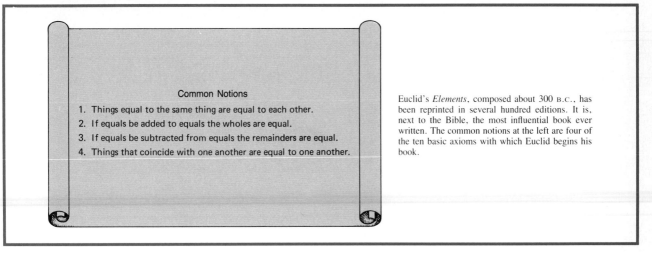

Euclid's *Elements*, composed about 300 B.C., has been reprinted in several hundred editions. It is, next to the Bible, the most influential book ever written. The common notions at the left are four of the ten basic axioms with which Euclid begins his book.

It was part of Euclid's genius to recognize that our usage of the word *equals* is fundamental to all that we do in mathematics. But to describe that usage may not be as simple as Euclid thought. When we write

$$.25 + \frac{3}{4} + \frac{1}{3} - (.3333 \ldots) = 1$$

> **Equality?**
> "We hold these truths to be self-evident, that all men are created equal . . ."

we certainly do not mean that the symbol on the left coincides with the one on the right. Instead, we mean that both symbols, the complicated one and the simple one, stand for (or name) the same number. That is the basic meaning of *equals* as used in this book.

But having said that, we must make another distinction. When we write

$$x^2 - 25 = (x - 5)(x + 5)$$

and

$$x^2 - 25 = 0$$

we have two quite different things in mind. In the first case, we are making an assertion. We claim that no matter what number x represents, the expressions on the left and right of the equality stand for the same number. This certainly cannot be our meaning in the second case. There we are asking a question: What numbers can x symbolize so that both sides of the equality $x^2 - 25 = 0$ stand for the same number?

An equality that is true for all values of the variable is called an **identity.** One that is true only for some values is called a conditional **equation.** And here are the corresponding jobs for us to do. We **prove** identities, but we **solve** (or find the solutions of) equations. Both jobs will be very important in this book; however, it is the latter that interests us most right now.

Solving Equations

Sometimes we can solve an equation by inspection. It takes no mathematical apparatus and little imagination to see that

$$x + 4 = 6$$

has $x = 2$ as a solution. On the other hand, to solve

$$2x^2 + 8x = 8x + 18$$

is quite a different matter. For this kind of equation, we need some machinery. Our general strategy is to modify an equation one step at a time until it is in a form where the solution is obvious. Of course, we must be careful that the modifications we make do not change the solutions. Here, too, Euclid pointed the way.

Rules for Modifying Equations

1. Adding the same quantity to (or subtracting the same quantity from) both sides of an equation does not change its solutions.
2. Multiplying (or dividing) both sides of an equation by the same nonzero quantity does not change its solutions.

Consider $2x^2 + 8x = 8x + 18$ again. One way to solve this equation is to use the following steps.

Given equation:	$2x^2 + 8x = 8x + 18$
Subtract $8x$:	$2x^2 = 18$
Divide by 2:	$x^2 = 9$
Take square roots:	$x = 3$ or -3

Thus the solutions of $2x^2 + 8x = 8x + 18$ are 3 and -3. In Section 3-4, we will solve equations like this one by other methods, but the rules stated above will continue to play a fundamental role.

Linear Equations

The simplest kind of equation to solve is one in which the *variable* (also called the *unknown*) occurs only to the first power. Consider

$$12x - 9 = 5x + 5$$

Our procedure is to use the rules for modifying equations to bring all the terms in x to one side and the constant terms to the other and then to divide by the coefficient of x. The result is that we have x all alone on one side of the equation and a number (the solution) on the other.

Given equation:	$12x - 9 = 5x + 5$
Add 9:	$12x = 5x + 14$
Subtract $5x$:	$7x = 14$
Divide by 7:	$x = 2$

It is always a good idea to check your answer. In the original equation, replace x by the value that you found, to see if a true statement results.

$$12(2) - 9 \overset{?}{=} 5(2) + 5$$

$$15 = 15$$

Here is a similar example but with fractional coefficients.

Example A (Equations Involving Fractions)   Solve

$$\frac{2}{3}x - \frac{3}{4} = \frac{7}{6}x + \frac{1}{2}$$

Solution   When an equation is cluttered up with many fractions, the best first step may be to get rid of them. To do this, multiply both sides by the lowest common denominator (in this case, 12). Then proceed as usual.

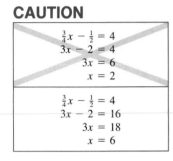

CAUTION

$$\frac{3}{4}x - \frac{1}{2} = 4$$
$$3x - 2 = 4$$
$$3x = 6$$
$$x = 2$$

$$\frac{3}{4}x - \frac{1}{2} = 4$$
$$3x - 2 = 16$$
$$3x = 18$$
$$x = 6$$

$$12\left(\frac{2}{3}x - \frac{3}{4}\right) = 12\left(\frac{7}{6}x + \frac{1}{2}\right)$$

$$12\left(\frac{2}{3}x\right) - 12\left(\frac{3}{4}\right) = 12\left(\frac{7}{6}x\right) + 12\left(\frac{1}{2}\right) \qquad \text{(distributive property)}$$

$$8x - 9 = 14x + 6$$

$$8x = 14x + 15 \qquad \text{(add 9)}$$

$$-6x = 15 \qquad \text{(subtract } 14x)$$

$$x = \frac{15}{-6} = -\frac{5}{2} \qquad \text{(divide by } -6)$$

The solution should now be checked in the original equation. We leave the check to the student.   ■

An equation of the form $ax + b = 0$ ($a \neq 0$) is called a **linear equation.** It has one solution, $x = -b/a$. Many equations not initially in this form can be transformed to it using the rules we have learned.

Equations That Can Be Changed to Linear Form

Consider

$$\frac{2}{x + 1} = \frac{3}{2x - 2}$$

If we agree to exclude $x = -1$ and $x = 1$ from consideration, then $(x + 1)(2x - 2)$ is not zero, and we may multiply both sides by that expression. We get

$$\frac{2}{x + 1}(x + 1)(2x - 2) = \frac{3}{2x - 2}(x + 1)(2x - 2)$$

$$2(2x - 2) = 3(x + 1)$$

$$4x - 4 = 3x + 3$$

$$x = 7$$

As usual, we check our solution in the original equation.

$$\frac{2}{7 + 1} \overset{?}{=} \frac{3}{14 - 2}$$

$$\frac{2}{8} = \frac{3}{12}$$

So $x = 7$ is a solution.

The importance of checking is illustrated by our next example.

Example B (Check Your Solution) Solve

$$\frac{3x}{x - 3} = 1 + \frac{9}{x - 3}$$

Solution   To solve, we multiply both sides by $x - 3$ and then simplify.

$$\frac{3x}{x - 3}(x - 3) = \left(1 + \frac{9}{x - 3}\right)(x - 3)$$

$$3x = x - 3 + 9$$

$$2x = 6$$

$$x = 3$$

When we check in the original equation, we get

$$\frac{3 \cdot 3}{3 - 3} \overset{?}{=} 1 + \frac{9}{3 - 3}$$

This is nonsense, since it involves division by zero. What went wrong? If $x = 3$, then in our very first step we actually multiplied both sides of the equation by zero, a forbidden operation. Thus the given equation has no solution.   ∎

The strategy of multiplying both sides by $x - 3$ in this example was appropriate, even though it initially led us to an incorrect answer. We did not worry, because we knew that in the end we were going to run a check. We should always check answers, especially in any situation in which we have multiplied by an expression involving the unknown. Such a multiplication may introduce an *extraneous* solution (but never results in the loss of a solution).

Here is another example of a similar type of equation.

Example C (Check Your Solution) Solve

$$\frac{x + 4}{(x + 1)(x - 2)} - \frac{3}{x + 1} - \frac{2}{x - 2} = \frac{-8}{(x + 1)(x - 2)}$$

Solution Our first step must be to multiply both sides of the equation by $(x + 1)(x - 2)$. This gives

$$x + 4 - 3(x - 2) - 2(x + 1) = -8$$

$$x + 4 - 3x + 6 - 2x - 2 = -8$$

$$-4x + 8 = -8$$

$$-4x = -16$$

$$x = 4$$

At this point, $x = 4$ is an apparent solution; however, we are not sure until we check it in the original equation.

$$\frac{4 + 4}{(4 + 1)(4 - 2)} - \frac{3}{4 + 1} - \frac{2}{4 - 2} \stackrel{?}{=} \frac{-8}{(4 + 1)(4 - 2)}$$

$$\frac{8}{5 \cdot 2} - \frac{3}{5} - \frac{2}{2} \stackrel{?}{=} \frac{-8}{5 \cdot 2}$$

$$\frac{8}{10} - \frac{6}{10} - \frac{10}{10} = \frac{-8}{10}$$

It works, so $x = 4$ is a solution. ■

Equations with More than One Variable

An equation may contain several variables. It is important that we know how to solve for any one of the variables in terms of others. Consider $xy = 5y + 3xz$. To solve for y, we collect the y terms on one side, factor out y, and then divide by the multiplier of y. Here are the steps.

$$xy = 5y + 3xz$$

$$xy - 5y = 3xz$$

$$(x - 5)y = 3xz$$

$$y = \frac{3xz}{x - 5}$$

Solving for x in a similar fashion gives

$$x = \frac{5y}{y - 3z}$$

Problems of this type occur frequently in science. Here is a typical example.

Example D (Solving for One Variable in Terms of Others) Solve for n in the equation $I = nE/(R + nr)$.

Solution

$$I = \frac{nE}{R + nr} \qquad \text{(original equality)}$$

$$(R + nr)I = nE \qquad \text{(multiply by } R + nr)$$

$$RI + nrI = nE \qquad \text{(distributive property)}$$

$$nrI - nE = -RI \qquad \text{(subtract } nE \text{ and } RI)$$

$$n(rI - E) = -RI \qquad \text{(factor)}$$

$$n = \frac{-RI}{rI - E} \qquad \text{(divide by } rI - E) \qquad \blacksquare$$

PROBLEM SET 3-1

A. Skills and Techniques

Determine which of the following are identities and which are conditional equations.

1. $2(x + 4) = 8$

2. $2(x + 4) = 2x + 8$

3. $3(2x - \frac{2}{3}) = 6x - 2$

4. $2x - 4 - \frac{2}{3}x = \frac{4}{3}x - 4$

5. $\frac{2}{3}x + 4 = \frac{1}{2}x - 1$

6. $3(x - 2) = 2(x - 3) + x$

7. $(x + 2)^2 = x^2 + 4$

8. $x(x + 2) = x^2 + 2x$

9. $x^2 - 9 = (x + 3)(x - 3)$

10. $x^2 - 5x + 6 = (x - 1)(x - 6)$

Solve each of the following equations.

11. $4x - 3 = 3x - 1$

12. $2x + 5 = 5x + 14$

13. $2t + \frac{1}{2} = 4t - \frac{7}{2} + 8t$

14. $y + \frac{1}{3} = 2y - \frac{2}{3} - 6y$

15. $3(x - 2) = 5(x - 3)$

16. $4(x + 1) = 2(x - 3)$

17. $\sqrt{3}z + 4 = -\sqrt{3}z + 8$

18. $\sqrt{2}x + 1 = x + \sqrt{2}$

19. $3.23x - 6.15 = 1.41x + 7.63$

(First obtain $x = \dfrac{7.63 + 6.15}{3.23 - 1.41}$ and then use a calculator.)

20. $42.1x + 11.9 = 1.03x - 4.32$

21. $(6.13 \times 10^{-8})x + (5.34 \times 10^{-6}) = 0$

22. $(5.11 \times 10^{11})x - (6.12 \times 10^{12}) = 0$

Solve by first clearing the fractions as in Example A.

23. $\frac{2}{3}x + 4 = \frac{1}{2}x$

24. $\frac{2}{3}x - 4 = \frac{1}{2}x + 4$

25. $\frac{9}{10}x + \frac{5}{8} = \frac{1}{5}x + \frac{9}{20}$

26. $\frac{1}{3}x + \frac{1}{4} = \frac{1}{5}x + \frac{1}{6}$

27. $\frac{3}{4}(x - 2) = \frac{9}{5}$

28. $x/8 = \frac{2}{3}(2 - x)$

The following equations are nonlinear equations that become linear when cleared of fractions. Solve each equation and check your solutions as some might be extraneous (see Examples B and C).

29. $\dfrac{5}{x + 2} = \dfrac{2}{x - 1}$

30. $\dfrac{10}{2x - 1} = \dfrac{14}{x + 4}$

31. $\dfrac{2}{x - 3} + \dfrac{3}{x - 7} = \dfrac{7}{(x - 3)(x - 7)}$

32. $\dfrac{2}{x - 1} + \dfrac{3}{x + 1} = \dfrac{19}{x^2 - 1}$

33. $\dfrac{x}{x - 2} = 2 + \dfrac{2}{x - 2}$

34. $\dfrac{x}{2x - 4} - \dfrac{2}{3} = \dfrac{7 - 2x}{3x - 6}$

Sometimes an equation that appears to be quadratic is actually equivalent to a linear equation. For example, if we subtract x^2 from both sides of the equation $x^2 + 3x = x^2 + 5$, we see that it is equivalent to $3x = 5$. Use this idea to solve each of the following.

35. $x^2 + 4x = x^2 - 3$

36. $x^2 - 2x = x^2 + 3x + 20$

37. $(x - 4)(x + 5) = (x + 2)(x + 3)$

38. $(2x - 1)(2x + 3) = 4x^2 + 6$

Sometimes an equation involving a radical becomes linear when both sides are raised to the same (appropriate) power. Solve each of the following equations. Check your answers, since raising to powers may introduce extraneous solutions.

39. $\sqrt{5 - 2x} = 5$ **40.** $\sqrt{3x + 7} = 4$

41. $\sqrt[3]{1 - 3x} = 4$ **42.** $\sqrt[3]{4x - 3} = -3$

Solve for the indicated variable in terms of the remaining variables as in Example D.

43. $A = P + Prt$ for P

44. $R = \dfrac{E}{L - 5}$ for L

45. $I = \dfrac{nE}{R + nr}$ for r

46. $mv = Ft + mv_0$ for m

47. $A = 2\pi r^2 + 2\pi rh$ for h

48. $F = \frac{9}{5}C + 32$ for C **49.** $R = \dfrac{R_1 R_2}{R_1 + R_2}$ for R_1

50. $\dfrac{1}{R} = \dfrac{1}{R_1} + \dfrac{1}{R_2} + \dfrac{1}{R_3}$ for R_2

B. Applications and Extensions

In Problems 51–60, solve for x.

51. $\frac{3}{4}x - \frac{4}{3} = \frac{1}{3}x + \frac{5}{6}$

52. $0.3(0.3x - 0.1) = 0.1x - 0.3$

53. $4(3x - \frac{1}{2}) = 5x + \frac{1}{2}$

54. $(x - 2)(x + 5) = (x - 3)(x + 1)$

55. $(x - 2)(3x + 1) = (x - 2)(3x + 5)$

56. $\dfrac{2}{x - 1} = \dfrac{9}{2x - 3}$

57. $\dfrac{2x}{4x + 2} = \dfrac{x + 1}{2x - 1}$

58. $\dfrac{2}{(x + 2)(x - 1)} = \dfrac{1}{x + 2} + \dfrac{3}{x - 1}$

59. $1 + \dfrac{x}{x + 3} = \dfrac{-3}{x + 3}$ **60.** $\dfrac{x^2}{x - 3} = \dfrac{9}{x - 3}$

61. Solve for x in terms of a.

$$1 + \cfrac{1}{1 + \cfrac{1}{a + \cfrac{1}{x}}} = \frac{1}{a}$$

62. Solve for x in terms of a and b.

(a) $\dfrac{ax - b}{bx - a} = \dfrac{a + b}{b}$ **(b)** $\dfrac{x - a}{x + b} = \dfrac{a - ab}{a + b^2}$

63. Celsius and Fahrenheit temperatures are related by the formula $C = \frac{5}{9}(F - 32)$.
   (a) What Fahrenheit temperature corresponds to 30 degrees Celsius?
   (b) How warm is it (in degrees Fahrenheit) when a Celsius thermometer and a Fahrenheit thermometer give the same reading?
   (c) How warm is it (in degrees Celsius) when the Celsius reading is one-half the Fahrenheit reading?

64. When a principal P is invested at the simple interest rate r (written as a decimal), then the accumulated amount A after t years is given by $A = P + Prt$. For example, $2000 invested at 8 percent simple interest for 10 years will accumulate to $2000 + 2000(.08)(10)$, or $3600.
   (a) How long will it take $2000 to grow to $4000 if invested at 9 percent simple interest?
   (b) A principal of $2000 grew to $6000 in 15 years when invested at the simple interest rate r. Find r in percent.

65. What principal invested at 7.5 percent simple interest will accumulate to $3813.75 by the end of 5.5 years? (See Problem 64.)

66. In each of the years 1992 and 1993, June Smythe suffered a 10 percent cut in salary. What percent raise should she get in 1994 to bring her salary up to the 1991 level?

67. As a result of financial difficulties, the ABC Company plans to reduce the salaries of all employees by a certain percent this year but hopes to bring them all back to their original level next year.
   (a) A salary reduction of 10 percent this year will require what percent increase next year?
   (b) A salary reduction of p percent this year will require what percent increase next year?

68. TEASER When Karen inherited a small amount of money, she decided to divide it among her four children. To Alice, she gave $2 plus one-third of the remainder: to Brent, she gave $2 plus one-third of the remainder. Next, to Curtis, she gave $2 plus one-third of the remainder, and, finally, to Debra, she gave $2 plus one-third of the remainder. What was left, she divided equally among the four children. If the girls together got $35 more than the boys together, what was the size of Karen's inheritance and how much did each child get?

3-2 APPLICATIONS USING ONE UNKNOWN

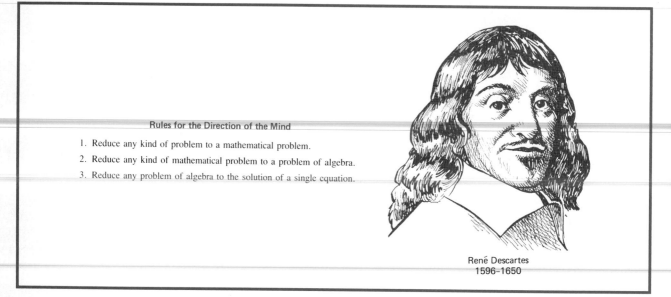

Rules for the Direction of the Mind

1. Reduce any kind of problem to a mathematical problem.
2. Reduce any kind of mathematical problem to a problem of algebra.
3. Reduce any problem of algebra to the solution of a single equation.

René Descartes
1596–1650

Besides being a mathematician, Descartes was a first-rate philosopher. His name appears prominently in every philosophy text. We admit that his rules for the direction of the mind are overstated. Not every problem in life can be solved this way. But they do suggest a style of thinking that has been very fruitful, especially in the sciences. We intend to exploit it.

A Typical Word Problem

Sometimes a little story can illustrate some big ideas.

> John plans to take his wife Helen out to dinner. Concerned about their financial situation, Helen asks him point-blank, "How much money do you have?" Never one to give a simple answer when a complicated one will do, John replies, "If I had $12 more than I have and then doubled that amount, I'd be $60 richer than I am." Helen's response is best left unrecorded.

The problem, of course, is to find out exactly how much money John has. Our task is to take a complicated word description, translate it into mathematical symbols, and then let the machinery of algebra grind out the answer. First we introduce a symbol x. It usually will stand for the principal unknown in the problem. But we need to be very precise. It is not enough to let x be John's money, or even to let x be the amount of John's money, though that is better. The symbol x must represent a number. What we should say is:

Let x be the number of dollars John has.

Our story puts restrictions on x; x must satisfy a specified condition. That condition must be translated into an equation. We shall do it by bits and pieces.

Word Phrase	*Algebraic Translation*
How much John has	x
$12 more than he has	$x + 12$
Double that amount	$2(x + 12)$
$60 richer than he is	$x + 60$

Read John's answer again. It says that the expressions $2(x + 12)$ and $x + 60$ are equal. Thus

$$2(x + 12) = x + 60$$

$$2x + 24 = x + 60$$

$$x = 36$$

John has $36, enough for a pretty good dinner for two—even at today's prices.

Here is another example involving money.

Example A (A Coin Problem)   When Amy opened her coin bank, she discovered that she had 6 more dimes than quarters, twice as many nickels as quarters, and 7 pennies. On counting, she saw that they were worth $4.72 altogether. How many coins of each kind did she have?

Solution   We must be careful to distinguish between the number of coins of a certain kind and their value. Imagine the four kinds of coins to be separated into four piles as in Figure 1.

7 pennies

$x + 6$ dimes

x quarters

$2x$ nickels

Figure 1

Let x be the number of quarters.

Then $x + 6$ and $2x$ denote the corresponding number of dimes and the number of nickels. The values of the four piles, in cents, are $25x$, $10(x + 6)$, $5(2x)$, and 7, respectively. The total value, in cents, is 472. We conclude that

$$25x + 10(x + 6) + 5(2x) + 7 = 472$$

which can be successively written as

$$25x + 10x + 60 + 10x + 7 = 472$$

$$45x + 67 = 472$$

$$45x = 405$$

$$x = \frac{405}{45} = 9$$

Amy has 9 quarters, 15 dimes, 18 nickels, and 7 pennies.   ■

Distance-Rate-Time Problems

Problems involving rates and distances occur very frequently in physics. Usually their solution involves use of the formula $D = RT$, which stands for "distance equals rate multiplied by time."

> At 2:00 P.M., Slowpoke left Kansas City traveling due east at 45 miles per hour. One hour later, Speedy started after him going 60 miles per hour. When will Speedy catch up with Slowpoke?

Most of us can grasp the essential features in a picture more readily than in a mass of words. That is why all good mathematicians make sketches that summarize what is given. One such picture is shown as Figure 2.

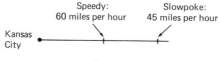

Figure 2

Next assign the unknown. Be precise.

Poor: *Let t be time.*
Better: *Let t be the time when Speedy catches up.*
Good: *Let t be the number of hours after 2:00 P.M. when Speedy catches up with Slowpoke.*

Notice two things:

1. Slowpoke drove t hours: Speedy, starting an hour later, drove only $t - 1$ hours.

2. Both drove the same distance.

Now use the formula $D = RT$ to conclude that Slowpoke drove a distance of $45t$ miles and Speedy drove a distance of $60(t - 1)$ miles. By statement 2, these are equal.

$$45t = 60(t - 1)$$
$$45t = 60t - 60$$
$$60 = 15t$$
$$4 = t$$

Speedy will catch up with Slowpoke 4 hours after 2:00 P.M., or at 6:00 P.M.
    Our next example uses similar ideas.

Example B (Average Speed)   It took the Hogans $10\frac{1}{2}$ hours to drive to their parents' home for Christmas, a distance of 500 miles. Because of fog, they averaged 26 miles per hour less the last 5 hours than they did during the first $5\frac{1}{2}$ hours. What was their average speed for the first $5\frac{1}{2}$ hours?

Solution   *Let s be the average speed in miles per hour during the first $5\frac{1}{2}$ hours.* Then $s - 26$ represents the average speed for the last 5 hours. From the formula $D = RT$, we see that

$$s(5.5) + (s - 26)5 = 500$$
$$5.5s + 5s - 130 = 500$$
$$10.5s = 630$$
$$s = \frac{630}{10.5} = 60$$

The Hogans drove 60 miles per hour during the first part of their trip.   ■

Mixture Problems

Here is a problem from chemistry.

Example C (Mixing Acids)   How many liters of a 60 percent solution of nitric acid should be added to 10 liters of a 30 percent solution to obtain a 50 percent solution?

Solution   We feel sure that Figure 3 will help most students with this problem. We have indicated on the figure what x represents, but let us be specific.

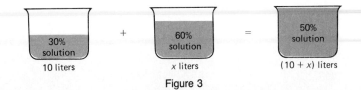

Figure 3

Let x be the number of liters of 60 percent solution to be added.

Now we make a crucial observation, one that will seem obvious once you think about it:

$$\left(\begin{array}{c}\text{amount of}\\ \text{pure acid}\\ \text{we start with}\end{array}\right) + \left(\begin{array}{c}\text{amount of}\\ \text{pure acid}\\ \text{we add}\end{array}\right) = \left(\begin{array}{c}\text{amount of}\\ \text{pure acid}\\ \text{we end with}\end{array}\right)$$

In symbols, this becomes

$$(.30)(10) + (.60)x = (.50)(10 + x)$$

The big job has been accomplished; we have the equation. After multiplying both sides by 10 to clear the equation of decimal fractions, we can easily solve for x.

$$(3)(10) + 6x = 5(10 + x)$$

$$30 + 6x = 50 + 5x$$

$$x = 20$$

We should add 20 liters of 60 percent solution. ∎

A very similar procedure applies to investing money at different rates.

Example D (An Investment Problem) Karla, who has just retired, figures that she needs $10,000 per year to supplement her social security income. She will put some of her savings of $120,000 in a safe fund A at 8 percent simple interest and the rest in a riskier fund B at 12 percent simple interest. How much should she put in each fund to just meet her needs?

Solution *Let x be the number of dollars to be invested in fund A.* Then $120,000 - x$ is the number of dollars to be in fund B. Also $.08x$ and $.12(120,000 - x)$ are the yearly interest payments, which must total 10,000. Thus

$$.08x + .12(120,000 - x) = 10,000$$

$$.08x + 14,400 - .12x = 10,000$$

$$-.04x = -4400$$

$$x = 110,000$$

Karla should put $110,000 in fund A and $10,000 in fund B. ∎

Balancing Weight Problems

We turn to a problem you might meet in a physics course.

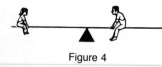

Figure 4

Example E (A Seesaw Problem) Susan, who weighs 80 pounds, wants to ride on a seesaw with her father, who weighs 200 pounds (Figure 4). The plank is 20 feet long and the fulcrum is at the center. If Susan sits at the very end of her side, how far from the fulcrum should her father sit to achieve balance?

Solution   *Let x be the number of feet from the fulcrum to the point where Susan's dad sits.* A law of physics demands that *the weight times the distance from the fulcrum* must be the same for both sides in order to have balance. For Susan, weight times distance is 80 · 10; for her dad, it is 200x. This gives us the equation

$$200x = 800$$

from which we get $x = 4$. Susan's father should sit 4 feet from the fulcrum.   ■

Summary

One hears students say that they have a mental block when it comes to "word problems." Yet most of the problems of the real world are initially stated in words. We want to destroy those mental blocks and give you one of the most satisfying experiences in mathematics. We believe you can learn to do word problems. Here are a few simple suggestions.

1. Read the problem very carefully so that you know exactly what it says and what it asks.
2. Draw a picture or make a chart that summarizes the given information.
3. Identify the unknown quantity and assign a letter to it. Be sure it represents a number (for example, of dollars, of miles, or of liters).
4. Note the condition or restriction that the problem puts on the unknown. Translate it into an equation.
5. Solve the equation. Your result is a specific number. The unknown has become known.
6. State your conclusion in words (for example, Speedy will catch up to Slowpoke at 4 hours after 2:00 P.M., or at 6:00 P.M.).
7. Note whether your answer is reasonable. If, for example, you find that Speedy will not catch up with Slowpoke until 4000 hours after 2:00 P.M., you should suspect that you have made a mistake.

PROBLEM SET 3-2

A. Skills and Techniques

1. The sum of 15 and twice a certain number is 33. Find that number.
2. Tom says to Jerry: I am thinking of a number. When I subtract 5 from that number and then multiply the result by 3, I get 42. What was my number? Jerry figured it out. Can you?
3. Find the number for which twice the number is 12 less than 3 times the number.
4. The result of adding 28 to 4 times a certain number is the same as subtracting 5 from 7 times that number. Find the number.

5. The sum of three consecutive positive integers is 72. Find the smallest one.
6. The perimeter (distance around) of the rectangle shown in Figure 5 is 31 inches. Find the width x.

12 inches   Figure 5

7. A wire 130 centimeters long is bent into the shape of a rectangle which is 3 centimeters longer than it is wide. Find the width of the rectangle.

8. A rancher wants to put 2850 pounds of feed into two empty bins. If she wants the larger bin to contain 750 pounds more of the feed than the smaller one, how much must she put into each?

9. Mary scored 61 on her first math test. What must she score on a second test to bring her average up to 75?

10. Henry has scores of 61, 73, and 82 on his first three tests. How well must he do on the fourth and final test to wind up with an average of 75 for the course?

11. A change box contains 21 dimes. How many quarters must be put in to bring the total amount of change to $3.85? (See Example A.)

12. A change box contains $3.00 in dimes and nothing else. A certain number of dimes are taken out and replaced by an equal number of quarters, with the result that the box now contains $4.20. How many dimes are taken out?

13. A woman has $4.45 in dimes and quarters in her purse. If there are 25 coins in all, how many dimes are there?

14. Young Amy has saved $8.05 in nickels and quarters. She has 29 more nickels than quarters. How many quarters does she have?

15. Read the example (preceding Example B) about Slowpoke and Speedy again. When will Speedy be 100 miles ahead of Slowpoke?

16. Two long-distance runners start out from the same spot on an oval track which is $\frac{1}{2}$ mile around. If one runs at 6 miles per hour and the other at 7 miles an hour, when will the faster runner be one lap ahead of the slower runner?

17. The City of Harmony is 455 miles from the city of Dissension. At 12:00 noon Paul Haymaker leaves Harmony traveling at 60 miles per hour toward Dissension. Simultaneously, Nick Ploughman starts from Dissension heading toward Harmony, managing only 45 miles per hour. When will they meet? (See Example B.)

18. Suppose in Problem 17, Mr. Ploughman starts at 3:00 P.M. At what time will the two drivers meet?

19. Luella can row 1 mile upstream in the same amount of time that it takes her to row two miles downstream. If the rate of the current is 3 miles per hour, how fast can she row in still water?

20. An airplane flew with the wind for 1 hour and returned the same distance against the wind in $1\frac{1}{2}$ hours. If the speed of the plane in still air is 300 miles per hour, find the speed of the wind.

21. A father is three times as old as his son, but 15 years from now he will be only twice as old as his son. How old is his son now?

22. Jim Warmath was in charge of ticket sales at a football game. The price for general admission was $3.50, while reserved seat tickets sold for $5.00. He lost track of the ticket count, but he knew that 110 more general admission tickets had been sold than reserve seat tickets and the total gate receipts were $980. See if you can find out how many general admission tickets were sold.

23. How many cubic centimeters of a 40 percent solution of hydrochloric acid should be added to 2000 cubic centimeters of a 20 percent solution to obtain a 35 percent solution? (See Example C.)

24. In Problem 23, how much of the 40 percent solution would have to be added in order to have a 39 percent solution?

25. A tank contains 1000 liters of 30 percent brine solution. Boiling off water from the solution will increase the percentage of salt. How much water should be boiled off to achieve a 35 percent solution?

26. Sheila Carlson invested $10,000 in a savings and loan association, some at 7 percent (simple interest) per year and the rest at $8\frac{1}{2}$ percent. How much did she invest at 7 percent if the total amount of interest for one year was $796? (See Example D.)

27. At the Style King shop, a man's suit was marked down 15 percent and sold at $123.25. What was the original price?

28. Mr. Titus Canby bickered with a furrier over the price of a fur coat he intended to buy for his wife. The furrier offered to reduce the price by 10 percent. Titus was still not satisfied; he said he would buy the coat if the furrier would come down an additional $200 on the price. The furrier agreed and sold the coat for $1960. What was the original price?

29. The Conkwrights plan to put in a concrete drive from the street to the garage. The drive is 36 feet long and they plan to make it 4 inches thick. Since there is a delivery charge for less than 4 cubic yards of ready mixed concrete, the Conkwrights have decided to make the drive just wide enough to use 4 cubic yards. How wide should they make it?

30. Tom can do a certain job in 3 days, Dick can do it in 4 days, and Harry can do it in 5 days. How long will it take them working together? *Hint:* In one day Tom can do $\frac{1}{3}$ of the job; in x days, he can do $x/3$ of the job.

31. It takes Jack 5 days to hoe his vegetable garden. Jack and Jill together can do it in 3 days. How long would it take Jill to hoe the garden by herself?

Problems 32 and 33 refer to Example E.

32. Where should Susan's father sit if Susan moves 2 feet closer to the fulcrum?

33. If Susan sits at one end with Roscoe, a 12-pound puppy, in her arms, where should her father sit?

Find x in Problems 34–38. Assume in each case that the plank is 20 feet long, that the fulcrum is at the center, and that the weights balance.

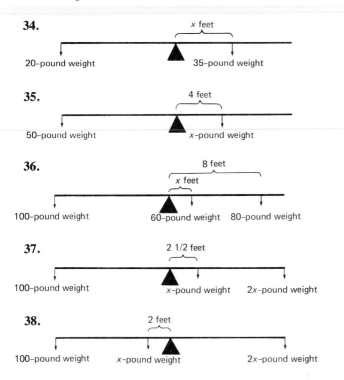

34.

x feet

20-pound weight   35-pound weight

35.

4 feet

50-pound weight   x-pound weight

36.

8 feet

x feet

100-pound weight   60-pound weight   80-pound weight

37.

2 1/2 feet

100-pound weight   x-pound weight   2x-pound weight

38.

2 feet

100-pound weight   x-pound weight   2x-pound weight

B. Applications and Extensions

39. The sidewalk around a square garden is 3 feet wide and has area 249 square feet. What is the width of the garden?

40. What number must be added to both numerator and denominator of $\frac{3}{37}$ to bring this fraction up to $\frac{7}{24}$?

41. Professor Witquick is 4 years older than his wife, Matilda. On their 25th wedding anniversary, Witquick noted that their combined ages were double their combined ages on their wedding day. How old will Matilda be on their 50th anniversary?

42. Sylvia Swindalittle, a lawyer, charges $150 per hour for her own time and $30 per hour for that of her clerical assistant, Donald Duckwork. On a recent case, where Donald worked half again as long as Sylvia, the total charge was $2262. How long did Donald work on the case?

43. In a certain math class, the average weight for the girls was 128 pounds, for the boys, 160 pounds, and for the class as a whole, 146 pounds. How many people are in the class, given that there are 14 girls?

44. Sam Slugger got only 50 hits in his first 200 times at bat of the baseball season. During the rest of the season, he claims he will hit .300 and finish with a season average of .280. How many times at bat does he expect to have during the whole season?

45. When asked to make a contribution to the building fund, Amos Hadwiggle pledged to give $1000 more than the average of all givers. The other 88 givers contributed $44,000. How much did Amos have to give?

46. Opening drain A will empty the swimming pool in 5 hours; opening both A and B will empty it in 3 hours. How long will it take to empty the pool if only drain B is opened?

47. Susan Kahn has invested $7400 at a certain rate of simple interest and $4600 at twice that rate. Find the two rates if she realized an average rate of 7.2625 percent on the total investment.

48. Suppose that 20 liters of 35 percent brine solution is mixed with 30 liters of a 20 percent brine solution. How much water should be added to the resulting solution to obtain a 15 percent solution?

49. Here is an old riddle said to be on the tombstone of Diophantus. Diophantus lived one-sixth of his life as a child, one-twelfth of his life as a youth, and one-seventh of his life as an unmarried adult. A son was born 5 years after Diophantus was married, but this son died 4 years before his father. Diophantus lived twice as long as his son. You figure out how old Diophantus was when he died.

50. Anderson and Benson had been partners for years with Anderson owning three-fifths of the business and Benson the rest. Recently, Christenson offered to pay $100,000 to join the partnership but only on the condition that all three would then be equal partners. Anderson and Benson accepted the offer. How much of the $100,000 did each get, assuming fair division?

51. A column of soldiers 2 miles long is marching at a constant rate of 5 miles per hour. At noon, the general at the rear of the column sent forward two men on horseback traveling at 20 miles per hour.

  (a) The first, A, was told to scout the road ahead of the column and return in 2 hours. After how long should A turn around?

  (b) The second, B, was told to carry a message to the head of the column and return. How long will this take?

52. TEASER Here is a problem attributed to Isaac Newton, coinventor of calculus. Three pastures with grass of identical height, density, and growth rate are of size $3\frac{1}{3}$ acres, 10 acres, and 24 acres, respectively. If 12 oxen can be fed on the first pasture for 4 weeks and 21 oxen can be fed on the second for 9 weeks, how many oxen can be fed on the third pasture for 18 weeks?

3-3 TWO EQUATIONS IN TWO UNKNOWNS

The Farmer's Riddle

I have a collection of hens and rabbits. These animals have 50 heads and 140 feet. How many hens and how many rabbits do I have?

No one has thought more deeply or written more wisely about problem solving than George Polya. In his book *Mathematical Discovery* (Volume 1, Wiley, 1962) Polya uses the farmer's riddle as the starting point for a brilliant essay on setting up equations to solve problems. He suggests three different approaches that we might take to untangle the riddle.

Trial and Error

There are 50 animals altogether. They cannot all be hens as that would give only 100 feet. Nor can all be rabbits; that would give 200 feet. Surely the right answer is somewhere between these extremes. Let us try 25 of each. That gives 50 hen-feet and 100 rabbit-feet, or a total of 150, which is too many. We need more hens and fewer rabbits. Well, try 28 hens and 22 rabbits. It does not work. Try 30 hens and 20 rabbits. There it is! That gives 60 feet plus 80 feet, just what we wanted (Figure 6).

Hens	Rabbits	Feet
50	0	100
0	50	200
25	25	150
28	22	144
30	20	140

Figure 6

Bright Idea

Let us use a little imagination. Imagine that we catch the hens and rabbits engaged in a weird new game. The hens are all standing on one foot and the rabbits are hopping

Figure 7

around on their two hind feet (Figure 7). In this remarkable situation, only half of the feet, that is 70 feet, are in use. We can think of 70 as counting each hen once and each rabbit twice. If we subtract the total number of animals, namely, 50, we will have the number of rabbits. There it is! There have to be $70 - 50 = 20$ rabbits; and that leaves 30 hens.

Algebra

Trial and error is time consuming and inefficient, especially in problems with many possibilities. And we cannot expect a brilliant idea to come along for every problem. We need a systematic method that depends neither on guess work nor on sudden visions. Algebra provides such a method. To use it, we must translate the problem into algebraic symbols and set up equations.

English	*Algebraic Symbols*
The farmer has a certain number of hens and	x
a certain number of rabbits.	y
These animals have 50 heads	$x + y = 50$
and 140 feet.	$2x + 4y = 140$

Now we have two unknowns, x and y, but we also have two equations relating them. We want to find the values for x and y that satisfy both equations at the same time. There are two standard methods.

Method of Addition or Subtraction

We learned two rules for modifying equations in Section 3-1. Here is another rule, especially useful in solving a **system of equations,** that is, a set of several equations in several unknowns.

Rule 3

You may add one equation to another (or subtract one equation from another) without changing the simultaneous solutions of a system of equations.

Here is how this rule is used to solve the farmer's riddle.

Given equations:
$$\begin{cases} x + y = 50 \\ 2x + 4y = 140 \end{cases}$$

Multiply the first equation by (-2): $\qquad -2x - 2y = -100$

Write the second equation: $\qquad\qquad\quad 2x + 4y = 140$

Add the two equations: $\qquad\qquad\qquad\qquad\quad 2y = 40$

Multiply by $\frac{1}{2}$: $\qquad\qquad\qquad\qquad\qquad\boxed{y = 20}$

Substitute $y = 20$ into one of the original equations (in this case, we shall use the first one): $\qquad x + 20 = 50$

Add -20. $\qquad\qquad\qquad\qquad\qquad\qquad\boxed{x = 30}$

The key idea is this: Multiplying the first equation by -2 makes the coefficients of x in the two equations negatives of each other. Addition of the two equations eliminates x, leaving one equation in the single unknown y. The resulting equation can be solved by methods learned earlier (see Section 3-1).

Here is a slightly more complicated example.

Example A (Eliminating y)   Solve the system

$$\begin{cases} 3x - 4y = -3 \\ 4x + 3y = 14.75 \end{cases}$$

Solution   Multiply the first equation by 3 and the second by 4, obtaining an equivalent system in which the coefficients of y are negatives of each other. Add the equations to eliminate y. Then solve for x and substitute to find y.

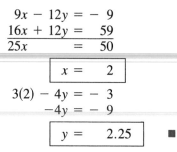

$$\begin{aligned} 9x - 12y &= -9 \\ 16x + 12y &= 59 \\ \hline 25x &= 50 \end{aligned}$$

$$\boxed{x = 2}$$

$$\begin{aligned} 3(2) - 4y &= -3 \\ -4y &= -9 \end{aligned}$$

$$\boxed{y = 2.25} \quad \blacksquare$$

The method we have described extends (with more work) to a system of three equations in three unknowns.

Example B (Three Unknowns)   Find values for x, y, and z that satisfy all three of the following equations.

$$\begin{cases} 2x + y - 3z = -9 \\ x - 2y + 4z = 17 \\ 3x - y - z = 2 \end{cases}$$

Solution   The idea is to eliminate one of the unknowns from two different pairs of equations. This gives us just two equations in two unknowns, which we solve as before. Let us eliminate y from the first two equations. To do this, we multiply the first equation by 2 and add to the second.

$$\begin{aligned} 4x + 2y - 6z &= -18 \\ x - 2y + 4z &= 17 \\ \hline 5x - 2z &= -1 \end{aligned}$$

Next, we eliminate y from the first and third equations by simply adding them.

$$\begin{aligned} 2x + y - 3z &= -9 \\ 3x - y - z &= 2 \\ \hline 5x - 4z &= -7 \end{aligned}$$

Our problem thus reduces to solving the following system of two equations in two unknowns.

$$\begin{cases} 5x - 2z = -1 \\ 5x - 4z = -7 \end{cases}$$

We leave this for you to do. You should get $z = 3$ and $x = 1$. If you substitute these values for x and z in any of the original equations, you will find that $y = -2$. Thus

$$\boxed{x = 1 \qquad y = -2 \qquad z = 3} \quad \blacksquare$$

Method of Substitution

Consider again the system of equations for the hen-rabbit problem.

$$\begin{cases} x + y = 50 \\ 2x + 4y = 140 \end{cases}$$

We may solve the first equation for y in terms of x and substitute the result in the second equation

$$y = 50 - x$$

$$2x + 4(50 - x) = 140$$

$$2x + 200 - 4x = 140$$

$$-2x = -60$$

$$\boxed{x = 30}$$

Then substitute the value obtained for x in the expression for y.

$$\boxed{y = 50 - 30 = 20}$$

Naturally, our results agree with those obtained earlier.

Whichever method we use, it is a good idea to check our answer against the original problem. Thirty chickens and 20 rabbits do have a total of 50 heads and they do have $(30)(2) + (20)(4) = 140$ feet in all.

Let's use the substitution method in another example.

Example C (Substitution in a System) Solve the system

$$3x + 2y = -4$$

$$4x - 5y = 33$$

Solution   Solving the first equation for y gives

$$y = -\frac{3}{2}x - 2$$

When we substitute this expression in the second equation and follow the steps outlined above, we obtain

$$4x - 5(-\frac{3}{2}x - 2) = 33$$

$$4x + \frac{15}{2}x + 10 = 33$$

$$8x + 15x + 20 = 66$$

$$23x = 46$$

$$\boxed{x = 2}$$

$$y = -\frac{3}{2}(2) - 2$$

$$\boxed{y = -5}$$ ∎

Applications

Many applied problems are easily modeled by one equation in one unknown, a process we illustrated in Section 3-2. For other problems, it is more natural to model them as a system of equations in several unknowns. We illustrate with several examples.

Example D (A Digit Problem) A three-digit number has 2 as its second digit. Reversing the digits decreases the number by 198. The sum of the first and last digits is 14. Find the number.

Solution Let x denote the hundreds digit and y the units digit. Thus the number in standard notation is $x2y$, but written out it is $100x + 2(10) + y$. The statement about reversing digits, when translated into an equation, says that

$$100x + 2(10) + y = 100y + 2(10) + x + 198$$

or in simplified form $99x - 99y = 198$. When we combine this with a translation of the last condition, we obtain the system of equations

$$99x - 99y = 198$$
$$x + \quad y = \ 14$$

a system easily solved by the methods illustrated earlier. We obtain $x = 8$ and $y = 6$. The desired number, written in standard notation, is 826. ∎

Example E (A Distance-Rate-Time Problem) An airplane, flying with the help of a strong tail wind, covered 1200 miles in 2 hours. However, the return trip flying into the wind took $2\frac{1}{2}$ hours. How fast would the plane have flown in still air, and what was the speed of the wind, assuming both rates to be constant?

Solution   Let

$$x = \text{speed of the plane in still air in miles per hour}$$

$$y = \text{speed of the wind in miles per hour}$$

Then

$$x + y = \text{speed of the plane with the wind}$$

$$x - y = \text{speed of the plane against the wind}$$

Next, we recall the familiar formula $D = RT$, or distance equals rate multiplied by time. Applying it in the form $TR = D$ to the two trips yields

$$2(x + y) = 1200 \qquad \text{(with wind)}$$

$$\frac{5}{2}(x - y) = 1200 \qquad \text{(against wind)}$$

or equivalently,

$$\begin{cases} 2x + 2y = 1200 \\ 5x - 5y = 2400 \end{cases}$$

Again we have a system easily solved by the methods illustrated earlier. We obtain $x = 540$ and $y = 60$. The plane's speed in still air is 540 miles per hour and the wind speed is 60 miles per hour.

A check against the original statement of the problem shows that our answers are correct. With the wind, the plane will travel at 600 miles per hour and will cover 1200 miles in 2 hours. Against the wind, the plane will fly at 480 miles per hour and take $2\frac{1}{2}$ hours to cover 1200 miles.   ∎

PROBLEM SET 3-3

A. Skills and Techniques

In each of the following, find the values for the two unknowns that satisfy both equations. Use whichever method you prefer. See Examples A and C.

1. $2x + 3y = 13$
        $= 13$

2. $2x - 3y = 7$
        $x = -4$

3. $2u - 5v = 23$
        $2u = 3$

4. $5s + 6t = 2$
        $3t = -4$

5. $7x + 2y = -1$
        $y = 4x + 7$

6. $7x + 2y = -1$
        $x = -5y + 14$

7. $y = -2x + 11$
        $y = 3x - 9$

8. $x = 5y$
        $x = -3y - 24$

9. $x - y = 14$
        $x + y = -2$

10. $2x - 3y = 8$
        $4x + 3y = 16$

11. $2s - 3t = -10$
        $5s + 6t = 29$

12. $2w - 3z = -23$
        $8w + 2z = -22$

13. $5x - 4y = 19$
        $7x + 3y = 18$

14. $4x - 2y = 16$
        $6x + 5y = 24$

Hint: In Problem 13, multiply the top equation by 3 and the bottom one by 4; then add.

15. $7x - 4y = 0$
        $2x + 7y = 57$

16. $2a + 3b = 0$
        $3a - 2b = \frac{13}{2}$

17. $\frac{2}{3}x + 2y = 4$
        $x + 2y = 5$

18. $\frac{3}{4}x - \frac{1}{2}y = 12$
        $x + y = -8$

19. $.125x - .2y = 3$
        $.75x + .3y = 10.5$

20. $.13x - .24y = 1$
        $2.6x + 4y = -2.4$

21. $\dfrac{4}{x} + \dfrac{3}{y} = 17$

      $\dfrac{1}{x} - \dfrac{3}{y} = -7$

22. $\dfrac{4}{x} - \dfrac{2}{y} = 12$

      $\dfrac{5}{x} + \dfrac{1}{y} = 8$

Hint: Let $u = 1/x$ and $v = 1/y$ in Problems 21 and 22. Solve for u and v and then find x and y.

23.
$$\frac{2}{\sqrt{x}} - \frac{1}{\sqrt{y}} = \frac{2}{3}$$
$$\frac{1}{\sqrt{x}} + \frac{2}{\sqrt{y}} = \frac{7}{6}$$

24.
$$\frac{1}{x-2} - \frac{2}{y+3} = 3$$
$$\frac{5}{x-2} + \frac{2}{y+3} = 3$$

Solve each of these systems for x, y, and z as in Example B.

25.
$$\begin{aligned} 4x - y + 2z &= 2 \\ -3x + y - 4z &= -1 \\ x + 5z &= 1 \end{aligned}$$

26.
$$\begin{aligned} x + 4y - 8z &= -10 \\ 3x - y + 5z &= 12 \\ -4x + 2y + z &= -9 \end{aligned}$$

27.
$$\begin{aligned} 2x + 3y + 4z &= -6 \\ -x + 4y - 6z &= 6 \\ 3x - 2y + 2z &= 2 \end{aligned}$$

28.
$$\begin{aligned} 3x + z &= 0 \\ 3x + 2y + z &= 4 \\ 9x + 5y + 10z &= 3 \end{aligned}$$

29.
$$\begin{aligned} 3x - 2y + z &= -2 \\ 4x + 3y - 5z &= 5 \\ 5x - 5y + 3z &= -4 \end{aligned}$$

30.
$$\begin{aligned} 2x - 4y - 2z &= -3 \\ 2x + 6y + 3z &= 7 \\ 3x + 3y + 4z &= 12 \end{aligned}$$

B. Applications and Extensions

31. Solve for r and s if $\frac{1}{3}r + \frac{2}{9}s = 24$ and $\frac{2}{9}r + \frac{1}{3}s = 26$.

32. Solve for x and y if $3/(x-1) + 2/(y-2) = 10$ and $2/(x-1) + 3/(y-2) = 10$.

33. Rodney Roller sold two cars for which he received a total of $13,000. If he received $1400 more for one car than the other, what was the selling price of each car?

34. Janice Stockton's estate is valued at $5000 more than three times her husband's estate. The combined value of their estates is $185,000. Find the value of each estate.

35. Find two numbers whose sum is $\frac{1}{3}$ but whose difference is 3.

36. The sum of the digits of a two-digit number is 12, and reversing the digits increases the value of the number by 54. Find the number. (See Example D.)

37. If the numerator and denominator of a fraction are each increased by 1, the result is $\frac{3}{5}$, but if the numerator and denominator are each decreased by 1, the result is $\frac{5}{9}$. Find the fraction.

38. Ella Goldthorpe needed to borrow $80,000 for a business venture. For part of the needed funds, she was able to get a 10 percent simple interest loan from her credit union; for the rest, she paid 12 percent simple interest at a bank. If her total interest for a year was $9360, how much did she borrow from each source?

39. A wire 39 inches long is cut into two pieces. One is bent to form a square; the other is bent into a rectangle twice as long as wide. If the side of the square is 1 inch longer than the width of the rectangle, at what point is the wire cut?

40. A mixture of some 15 percent nitric acid solution and some 30 percent nitric acid solution was added to 1000 liters of 40 percent solution to form 4000 liters of 28 percent solution (Figure 8). How many liters of 15 percent solution and how many liters of 30 percent solution were used?

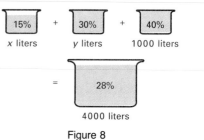

Figure 8

41. The attendance at a professional football game was 45,000 and the gate receipts were $495,000. If each person bought either a $10 ticket or a $15 ticket, how many tickets of each kind were sold?

42. Flying with a wind of 60 miles per hour, it took a plane 2 hours to get from A to B; returning against the wind took 2 hours and 30 minutes. What was the plane's airspeed (speed in still air) and how far is it from A to B? (See Example E.)

43. It took a motorboat 5 hours to go downstream from A to B and 7 hours to return. How long would it have taken Huckleberry Finn to float on his raft from A to B?

44. A certain commuter travels by bus and then by train to work each day. It takes him 45 minutes and costs him $3.05. He estimates that the train averages 50 miles per hour and the bus, only 25 miles per hour. If the bus fare is 12 cents per mile and the train fare is 10 cents per mile, how far does he go by each mode of transportation? Assume that no time is lost in transferring from the bus to the train.

45. A grocer has some coffee worth $3.60 per pound and some worth only $2.90 per pound. How much of each kind should she mix together to get 100 pounds of a blend worth $3.30 per pound?

46. A solution that is 40 percent alcohol is to be mixed with one that is 90 percent alcohol to obtain 50 liters of 50 percent solution. How many liters of each should be used?

47. In stocking her apparel shop, Susan Sharp paid $4800 for some dresses and coats, paying $40 for each dress and $100 for each coat. She was able to sell the dresses at a profit of 20 percent of the selling price and the coats at a profit of 50 percent of the selling price, giving her a total profit of $1800. How many coats and how many dresses did she buy?

48. Workers in a certain factory are classified into two groups depending on the skills required for two jobs. Group 1 workers are paid $12.00 per hour and group 2 workers are paid $7.50 per hour. In negotiations for a new contract, the union is demanding that workers in the second group have their hourly wages brought up to $\frac{2}{3}$ of that for workers in the first group. The factory has 55 workers in group 1 and 40 workers in group 2, all of whom work a 40-hour week. If the management is prepared to increase the weekly payroll by $8640, what hourly wages should it propose for each class of workers?

49. Solve the following system of equations for x and y in terms of z.

$$3x - 4y + 2z = -5$$
$$4x - 5y + 5z = 2$$

50. **TEASER** Here is a slight modification of a problem posed by the famous Swiss mathematician Leonhard Euler. For 100 crowns, a certain farmer bought some horses, goats, and sheep. For a horse, he paid $3\frac{1}{2}$ crowns, for a goat, $1\frac{1}{3}$ crowns, and for a sheep, $\frac{1}{2}$ crown. Assuming he bought at least 7 of each and that he bought 100 animals in all, how many horses, goats, and sheep did he buy? Note that the answers must be integers.

3-4 QUADRATIC EQUATIONS

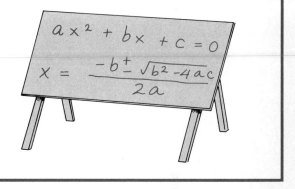

Old Stuff

By 2000 B.C. Babylonian arithmetic had evolved into a well-developed rhetorical algebra. Not only were quadratic equations solved, both by the equivalent of substituting in a general formula and by completing the square, but some cubic (third degree) and biquadratic (fourth degree) equations were discussed.

Howard Eves

A linear (first degree) equation may be put in the form $ax + b = 0$. We have seen that such an equation has exactly one solution, $x = -b/a$. That is simple and straightforward; no one is likely to stumble over it. But even the ancient Babylonians knew that equation solving goes far beyond this simple case. In fact, a good part of mathematical history revolves around attempts to solve more and more complicated equations.

The next case to consider is the second degree, or **quadratic,** equation—that is, an equation of the form

$$ax^2 + bx + c = 0 \qquad (a \neq 0)$$

Here are some examples.

(a) $$x^2 - 4 = 0$$

(b) $$x^2 - x - 6 = 0$$

(c) $$8x^2 - 2x = 1$$

(d) $$x^2 = 6x - 2$$

Although equations (c) and (d) do not quite fit the pattern, we accept them because they readily transform to equations in standard form.

$$\text{(c)} \qquad\qquad 8x^2 - 2x - 1 = 0$$

$$\text{(d)} \qquad\qquad x^2 - 6x + 2 = 0$$

Solution by Factoring

All of us remember that 0 times any number is 0. Just as important but sometimes forgotten is the fact that if the product of two numbers is 0, then one or both of the factors must be 0.

> If $u = 0$ or $v = 0$, then $u \cdot v = 0$
> If $u \cdot v = 0$, then either $u = 0$, or $v = 0$, or both.

CAUTION

$$x^2 - 5x = 0$$
$$x^2 = 5x$$
$$x = 5$$

$$x^2 - 5x = 0$$
$$x(x - 5) = 0$$
$$x = 0 \quad \text{or} \quad x = 5$$

This fact allows us to solve any quadratic equation that has 0 on one side provided we can factor its other side. Simply factor, set each factor equal to 0, and solve the resulting linear equations. We illustrate for the quadratic equations introduced above.

Example A (Quadratics That Can Be Easily Factored)   Solve.
(a) $x^2 - 4 = 0$ (b) $x^2 - x - 6 = 0$ (c) $8x^2 - 2x - 1 = 0$
(d) $x^2 - 6x + 2 = 0$

Solution

(a)
$$x^2 - 4 = 0$$
$$(x - 2)(x + 2) = 0$$
$$x - 2 = 0 \qquad x + 2 = 0$$
$$x = 2 \qquad\qquad x = -2$$

(b)
$$x^2 - x - 6 = 0$$
$$(x - 3)(x + 2) = 0$$
$$x - 3 = 0 \qquad x + 2 = 0$$
$$x = 3 \qquad\qquad x = -2$$

(c)
$$8x^2 - 2x - 1 = 0$$
$$(4x + 1)(2x - 1) = 0$$
$$4x + 1 = 0 \qquad 2x - 1 = 0$$
$$x = -\frac{1}{4} \qquad\quad x = \frac{1}{2}$$

Equation (d) remains unsolved; we do not know how to factor its left side. For this equation, we need a more powerful method. First, however, we need a brief discussion of square roots.   ∎

Square Roots

The number 9 has two square roots, 3 and -3. In fact, every positive number has two square roots, one positive and the other negative. If a is positive, its **positive square root** is denoted by $\sqrt{a}$. Thus $\sqrt{9} = 3$. Do not write $\sqrt{9} = -3$ or $\sqrt{9} = \pm 3$; both are wrong. But you can say that the two square roots of 9 are $\pm\sqrt{9}$ (or ± 3) and that the two square roots of 7 are $\pm\sqrt{7}$.

Here are two important properties of square roots, valid for any positive numbers a and b.

$$\sqrt{ab} = \sqrt{a}\sqrt{b}$$

$$\sqrt{\frac{a}{b}} = \frac{\sqrt{a}}{\sqrt{b}}$$

CAUTION

$$\sqrt{(-4)(-9)} = \sqrt{-4}\sqrt{-9}$$
$$= (2i)(3i) = -6$$

$$\sqrt{(-4)(-9)} = \sqrt{36} = 6$$
The rule $\sqrt{ab} = \sqrt{a}\sqrt{b}$ is not valid if $a < 0$ and $b < 0$.

Thus

$$\sqrt{4 \cdot 16} = \sqrt{4}\sqrt{16} = 2 \cdot 4 = 8$$
$$\sqrt{28} = \sqrt{4 \cdot 7} = \sqrt{4}\sqrt{7} = 2\sqrt{7}$$
$$\sqrt{\frac{4}{9}} = \frac{\sqrt{4}}{\sqrt{9}} = \frac{2}{3}$$

The square roots of a negative number are imaginary. For example, the two square roots of -9 are $3i$ and $-3i$, since

$$(3i)^2 = 3^2 i^2 = 9(-1) = -9$$
$$(-3i)^2 = (-3)^2 i^2 = 9(-1) = -9$$

In fact, if a is positive, the two square roots of $-a$ are $\pm\sqrt{a}i$. And in this case, the symbol $\sqrt{-a}$ will denote $\sqrt{a}i$. Thus $\sqrt{-7} = \sqrt{7}i$.

Example B (Simplifying Square Roots) Simplify.

(a) $\sqrt{54}$ (b) $\dfrac{2 + \sqrt{48}}{4}$ (c) $\dfrac{\sqrt{6}}{\sqrt{150}}$ (d) $\sqrt{-18}$

Solution

(a) $\sqrt{54} = \sqrt{9 \cdot 6} = \sqrt{9}\sqrt{6} = 3\sqrt{6}$

Here we factored out the largest square in 54, namely, 9, and then used the first property of square roots.

(b) $\dfrac{2 + \sqrt{48}}{4} = \dfrac{2 + \sqrt{16 \cdot 3}}{4} = \dfrac{2 + 4\sqrt{3}}{4} = \dfrac{2(1 + 2\sqrt{3})}{4}$

$$= \dfrac{1 + 2\sqrt{3}}{2}$$

If you are tempted to continue as follows,

$$\dfrac{1 + \cancel{2}\sqrt{3}}{\cancel{2}} = 1 + \sqrt{3} \qquad \text{Wrong!}$$

resist the temptation. That cancellation is wrong, because 2 is not a factor of the entire numerator but only of $2\sqrt{3}$.

(c) $\dfrac{\sqrt{6}}{\sqrt{150}} = \sqrt{\dfrac{6}{150}} = \sqrt{\dfrac{1}{25}} = \dfrac{1}{5}$

Sometimes, as in this case, it is best to write a quotient of two square roots as a single square root and then simplify.

(d) $\sqrt{-18} = \sqrt{(9)(2)(-1)} = 3\sqrt{2}i$ ■

Quadratic equations that have the form $u^2 = k$ are easily solved by taking square roots. Keep in mind that every number, except 0, has two square roots.

Example C (Quadratics That Are Already Perfect Squares)   Solve.
(a) $x^2 = 9$ (b) $(x + 3)^2 = 17$ (c) $(2y - 5)^2 = 16$

Solution   The easiest way to solve these equations is to take square roots of both sides.

(a) $x = 3$ or $x = -3$

(b) $x + 3 = \pm\sqrt{17}$

$\qquad x = -3 + \sqrt{17}$ or $x = -3 - \sqrt{17}$

(c) $2y - 5 = \pm 4$

$\qquad 2y = 5 + 4$ or $2y = 5 - 4$

$\qquad y = \frac{9}{2}$ or $y = \frac{1}{2}$ ∎

Completing the Square

Consider equation (d) of Example A again.

$$x^2 - 6x + 2 = 0$$

We may write it as

$$x^2 - 6x = -2$$

Now add 9 to both sides, making the left side a perfect square, and factor.

$$x^2 - 6x + 9 = -2 + 9$$

$$(x - 3)^2 = 7$$

This means that $x - 3$ must be one of the two square roots of 7—that is,

$$x - 3 = \pm\sqrt{7}$$

Hence

$$x = 3 + \sqrt{7} \quad \text{or} \quad x = 3 - \sqrt{7}$$

You may ask how we knew that we should add 9. Any expression of the form $x^2 + px$ becomes a perfect square when $(p/2)^2$ is added, since

$$x^2 + px + \left(\frac{p}{2}\right)^2 = \left(x + \frac{p}{2}\right)^2$$

For example, $x^2 + 10x$ becomes a perfect square when we add $(10/2)^2$ or 25.

$$x^2 + 10x + 25 = (x + 5)^2$$

The rule for completing the square (namely, add $(p/2)^2$) works only when the coefficient of x^2 is 1. However, that fact causes no difficulty for quadratic equations. If the leading coefficient is not 1, we simply divide both sides by this coefficient and then complete the square. We illustrate.

Example D (Complete the Square Method)   Solve $2x^2 - x - 3 = 0$.

Solution   First divide both sides by 2; then add the same number to both sides to complete the square on the left; finally take square roots.

$$x^2 - \frac{1}{2}x \quad - \frac{3}{2} = 0$$

$$x^2 - \frac{1}{2}x \quad\quad = \frac{3}{2}$$

$$x^2 - \frac{1}{2}x + \left(\frac{1}{4}\right)^2 = \frac{3}{2} + \left(\frac{1}{4}\right)^2$$

$$\left(x - \frac{1}{4}\right)^2 \quad\quad = \frac{25}{16}$$

$$x - \frac{1}{4} \quad\quad = \pm\frac{5}{4}$$

$$x = \frac{1}{4} + \frac{5}{4} = \frac{3}{2} \quad \text{or} \quad x = \frac{1}{4} - \frac{5}{4} = -1 \quad \blacksquare$$

The Quadratic Formula

The method of completing the square works on any quadratic equation. But there is a way of doing this process once and for all. Consider the general quadratic equation

$$ax^2 + bx + c = 0$$

with real coefficients $a \neq 0$, b, and c. First add $-c$ to both sides and then divide by a to obtain

$$x^2 + \frac{b}{a}x = -\frac{c}{a}$$

Next complete the square by adding $(b/2a)^2$ to both sides and then simplify.

$$x^2 + \frac{b}{a}x + \left(\frac{b}{2a}\right)^2 = -\frac{c}{a} + \left(\frac{b}{2a}\right)^2$$

$$\left(x + \frac{b}{2a}\right)^2 = -\frac{c}{a} + \frac{b^2}{4a^2}$$

$$\left(x + \frac{b}{2a}\right)^2 = \frac{b^2 - 4ac}{4a^2}$$

Finally take the square root of both sides.

$$x + \frac{b}{2a} = \pm\frac{\sqrt{b^2 - 4ac}}{2a}$$

or

$$x = \frac{-b}{2a} \pm \frac{\sqrt{b^2 - 4ac}}{2a}$$

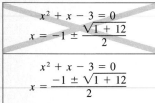
We call this result the **quadratic formula** and normally write it as follows.

$$x = \frac{-b \pm \sqrt{b^2 - 4ac}}{2a}$$

Let us see how it works

Example E (Using the Quadratic Formula)   Solve.
(a) $x^2 - 6x + 2 = 0$ (b) $2x^2 - 4x + \frac{28}{5} = 0$

Solution
(a) Here $a = 1$, $b = -6$, and $c = 2$. Thus

$$x = \frac{-(-6) \pm \sqrt{36 - 4 \cdot 2}}{2} = \frac{6 \pm \sqrt{28}}{2}$$

$$= \frac{6 \pm \sqrt{4 \cdot 7}}{2} = \frac{6 \pm 2\sqrt{7}}{2}$$

$$= \frac{2(3 \pm \sqrt{7})}{2} = 3 \pm \sqrt{7}$$

(b) Now $a = 2$, $b = -4$, and $c = \frac{25}{8}$. The formula gives

$$x = \frac{4 \pm \sqrt{16 - 25}}{4} = \frac{4 \pm \sqrt{-9}}{4} = \frac{4 \pm 3i}{4} \qquad \blacksquare$$

The expression $b^2 - 4ac$ that appears under the square root sign in the quadratic formula is called the **discriminant.** It determines the character of the solutions.

1. If $b^2 - 4ac > 0$, there are two real solutions.
2. If $b^2 - 4ac = 0$, there is one real solution.
3. If $b^2 - 4ac < 0$, there are two nonreal (imaginary) solutions.

Example F (Solving for One Variable in Terms of the Other)   Solve for y in terms of x.
(a) $y^2 - 2xy - x^2 - 2 = 0$ (b) $(y - 3x)^2 + 3(y - 3x) - 4 = 0$

Solution
(a) We use the quadratic formula with $a = 1$, $b = -2x$, and $c = -x^2 - 2$.

$$y = \frac{2x \pm \sqrt{4x^2 + 4(x^2 + 2)}}{2} = \frac{2x \pm \sqrt{8x^2 + 8}}{2}$$

$$= \frac{2x \pm 2\sqrt{2x^2 + 2}}{2}$$

$$= x \pm \sqrt{2x^2 + 2}$$

So $y = x + \sqrt{2x^2 + 2}$ or $y = x - \sqrt{2x^2 + 2}$.

(b) If we substitute z for $y - 3x$, the equation becomes

$$z^2 + 3z - 4 = 0$$

We solve this equation for z.

$$(z + 4)(z - 1) = 0$$

$$z = -4 \quad \text{or} \quad z = 1$$

Thus

$$y - 3x = -4 \quad \text{or} \quad y - 3x = 1$$

$$y = 3x - 4 \quad \text{or} \quad y = 3x + 1 \quad \blacksquare$$

PROBLEM SET 3-4

A. Skills and Techniques

Simplify each of the following as in Example B.

1. $\sqrt{50}$ **2.** $\sqrt{300}$

3. $\sqrt{\frac{1}{4}}$ **4.** $\sqrt{\frac{3}{27}}$

5. $\dfrac{\sqrt{45}}{\sqrt{20}}$ **6.** $\sqrt{.04}$

7. $\sqrt{11^2 \cdot 4}$ **8.** $\dfrac{\sqrt{2^3 \cdot 5}}{\sqrt{2 \cdot 5^3}}$

9. $\dfrac{5 + \sqrt{72}}{5}$ **10.** $\dfrac{4 - \sqrt{12}}{2}$

$\boxed{i}$ **11.** $\dfrac{18 + \sqrt{-9}}{6}$ $\boxed{i}$ **12.** $\dfrac{3 + \sqrt{-8}}{3}$

Solve by taking square roots as in Example C.

13. $x^2 = 25$ **14.** $x^2 = 14$

15. $(x - 3)^2 = 16$ **16.** $(x + 4)^2 = 49$

17. $(2x + 5)^2 = 100$ **18.** $(3y - \frac{1}{3})^2 = 25$

$\boxed{i}$ **19.** $m^2 = -9$ $\boxed{i}$ **20.** $(m - 6)^2 = -36$

Solve by factoring as in Example A.

21. $x^2 = 3x$ **22.** $2x^2 - 5x = 0$

23. $x^2 - 9 = 0$ **24.** $x^2 - \frac{9}{4} = 0$

25. $m^2 - .0144 = 0$ **26.** $x^2 - x - 2 = 0$

27. $x^2 - 3x - 10 = 0$ **28.** $x^2 + 13x + 22 = 0$

29. $3x^2 + 5x - 2 = 0$

30. $3x^2 + x - 2 = 0$

31. $6x^2 - 13x - 28 = 0$

32. $10x^2 + 19x - 15 = 0$

Solve by completing the square (Example D).

33. $x^2 + 8x = 9$ **34.** $x^2 - 12x = 45$

35. $z^2 - z = \frac{3}{4}$ **36.** $x^2 + 5x = 2\frac{3}{4}$

$\boxed{i}$ **37.** $x^2 + 4x = -9$ $\boxed{i}$ **38.** $x^2 - 14x = -65$

Solve by using the quadratic formula (Example E).

39. $x^2 + 8x + 12 = 0$ **40.** $x^2 - 2x - 15 = 0$

41. $x^2 + 5x + 3 = 0$ **42.** $z^2 - 3z - 8 = 0$

43. $3x^2 - 6x - 11 = 0$ **44.** $4t^2 - t - 3 = 0$

45. $x^2 + 5x + 5 = 0$ **46.** $y^2 + 8y + 10 = 0$

47. $2z^2 - 6z + 11 = 0$ $\boxed{i}$ **48.** $x^2 + x + 1 = 0$

Solve using the quadratic formula. Write your answers rounded to four decimal places.

49. $2x^2 - \pi x - 1 = 0$

50. $3x^2 - \sqrt{2}x - 3\pi = 0$

51. $x^2 + .8235x - 1.3728 = 0$

52. $5x^2 - \sqrt{3}x - 4.3213 = 0$

Solve for y in terms of x (Example F).

53. $(y - 2)^2 = 4x^2$

54. $(y + 3x)^2 = 9$

55. $(y + 3x)^2 = 9x^2$

56. $4y^2 + 4xy - 5 + x^2 = 0$

57. $(y + 2x)^2 - 8(y + 2x) + 15 = 0$

58. $(x - 2y + 3)^2 - 3(x - 2y + 3) + 2 = 0$

B. Applications and Extensions

Solve the equations in Problems 59–76.

59. $(2x - 1)^2 = \frac{9}{4}$ **60.** $3x^2 = 1 + 2x$

61. $2x^2 = 4x - 2$

62. $2x^2 + 3x = x^2 + 2x + 12$

63. $y^2 + 2y - 4 = 0$ **64.** $9y^2 - 3y - 1 = 0$

65. $2m^2 + 2m + 1 = 0$

66. $\sqrt{3}m^2 - 2m + 3\sqrt{3} = 0$

67. $x^4 - 5x^2 - 6 = 0$

68. $x - 4\sqrt{x} + 3 = 0$

69. $\left(x - \dfrac{4}{x}\right)^2 - 7\left(x - \dfrac{4}{x}\right) + 12 = 0$

70. $(x^2 + x)^2 - 8(x^2 + x) = -12$

71. $\dfrac{x^2 + 2}{x^2 - 1} = \dfrac{x + 4}{x + 1}$ **72.** $\dfrac{x}{x^2 + 1} = \dfrac{x - 3}{x^2 - 7}$

73. $\dfrac{1}{x} + \dfrac{1}{x - 1} = \dfrac{8}{3}$

74. $\dfrac{1}{x + 2} + \dfrac{1}{x - 2} = 1$

75. $\sqrt{2x + 1} = \sqrt{x} + 1$

76. $\dfrac{7}{x - 1} - \dfrac{2}{\sqrt{x - 1}} + \dfrac{1}{7} = 0$

Solve the systems of equations in Problems 77–80.

77. $xy = 20$ **78.** $x^2 + y^2 = 25$
    $-3 = 2x - y$ $3x + y = 15$

79. $2x^2 - xy + y^2 = 14$ **80.** $x^2 + 4y^2 = 13$
    $x - 2y = 0$ $2x^2 - y^2 = 17$

81. A rectangle has a perimeter of 26 and an area of 30. Find its dimensions.

82. The sum of the squares of three consecutive positive odd integers is 683. Find these integers.

83. A square piece of cardboard was used to construct a tray by cutting 2-inch squares out of each of the four corners and turning up the flaps (Figure 9). Find the size of the original square if the resulting tray has a volume of 128 square inches.

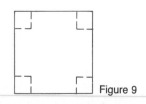
Figure 9

84. To harvest a rectangular field of wheat, which is 720 meters by 960 meters, a farmer cut swaths around the outside, thus forming a steadily growing border of cut wheat and leaving a steadily shrinking rectangle of uncut wheat in the middle. How wide was the border when the farmer was half through?

85. At noon, Tom left point A walking due north; an hour later. Dick left point A walking due east. Both boys walked at 4 miles per hour and both carried walkie-talkies with a range of 8 miles. At what time did they lose contact with each other?

86. The diameter AB of the circle in Figure 10 has length 12, and CD, which is perpendicular to the diameter, has length 5. Find the length of AD.

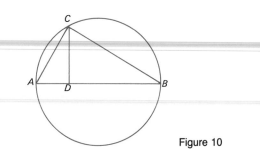
Figure 10

87. It took Samantha 2 hours and 12 minutes to ride her bicycle a distance of 30 miles. She rode 3 miles per hour faster over the first 18 miles than she did the rest of the way. How fast did she ride on the last 12 miles?

88. The hypotenuse of a right triangle measures 24 meters. One of the legs is 3 meters more than twice as long as the other leg. Find the length of the shorter leg.

89. (The golden ratio) As Euclid suggested, divide a line segment into two unequal parts so that the ratio of the whole to the longer part is equal to the ratio of the longer part to the shorter. Find the ratio ϕ of the longer to the shorter.

90. TEASER   Here is a version of a nifty problem posed by the American puzzlist Sam Lloyd. A square formation of soldiers, 50 feet on a side, is marching forward at a constant rate.
   (a) Running at constant speed, the company mascot ran from the left rear to the left front of the formation and back again in the time it took the com-

pany to march forward 50 feet. How many feet did the dog travel?

(b) Running at a faster rate, the mascot ran completely around the square formation in the time the company marched 50 feet forward. How many feet did the dog travel this time?

Suggestion: Because this problem is tricky, we follow Martin Gardner in suggesting a method of attack. Measure distances in units of 50 feet and time in units corresponding to the time for the company to march

50 feet. In these units, let x be the rate of the dog and note that the rate of the company is 1. Now find and simplify the equation x must satisfy. In (a), this leads to a quadratic equation that is easy to solve; in (b), this leads to a fourth-degree equation that is solvable only by advanced methods. However, you can get an approximate answer by experimenting with your calculator. Of course, once you know x in the two parts, you need only multiply by 50 to get the answers to the two questions.

3-5 INEQUALITIES

	Equation	Inequality
Each problem that I solved became a rule which served afterwards to solve other problems. Descartes	$-3x + 7 = 2$ $-3x = -5$ $x = \dfrac{5}{3}$	$-3x + 7 < 2$ $-3x < -5$ $x > \dfrac{5}{3}$

Solving an inequality is very much like solving an equation, as the example above demonstrates. However, there are dangers in proceeding too mechanically. It will be important to think at every step.

Recall the distinction we made between identities and equations in Section 3-1. A similar distinction applies to inequalities. An inequality which is true for all values of the variables is called an **unconditional inequality.** Examples are

$$(x - 3)^2 + 1 > 0$$

and

$$|x| \le |x| + |y|$$

Most inequalities (for example, $-3x + 7 < 2$) are true only for some values of the variables; we call them **conditional inequalities.** Our primary task in this section is to solve conditional inequalities, that is, to find all those numbers which make a conditional inequality true.

Linear Inequalities

To solve the linear inequality $Ax + B < C$, we try to rewrite it in successive steps until the variable x stands by itself on one side of the inequality (see opening display). This depends primarily on the properties stated in Section 1-5 and repeated here.

We illustrate the use of these properties in solving an inequality.

Example A (Solving a Linear Inequality)   Solve the inequality $-2x + 6 < 18 + 4x$.

Solution

$$-2x + 6 < 18 + 4x$$

Add $-4x$: $-6x + 6 < 18$

Add -6: $-6x < 12$

Multiply by $-\frac{1}{6}$: $x > -2$ ∎

By rights, we should check this solution. All we know so far is that any value of x that satisfies the original inequality satisfies $x > -2$. Can we go in the opposite direction? Yes, because every step is reversible. For example, starting with $x > -2$, we can multiply by -6 to get $-6x < 12$. In practice, we do not actually carry out this check as we recognize that Property 2 can be restated:

$$a < b \text{ is equivalent to } a + c < b + c.$$

There are similar restatements of Property 3. Also, we remark that all the inequality properties above are valid with $<$ and $>$ replaced by $\leq$ and $\geq$.

Quadratic Inequalities

To solve

$$x^2 - 2x - 3 > 0$$

we first factor, obtaining

$$(x + 1)(x - 3) > 0$$

Next we ask ourselves when the product of two numbers is positive. There are two cases; either both factors are negative or both factors are positive.

Case 1:   *Both negative.* We want to know when both factors are negative, that is, we seek to solve $x + 1 < 0$ and $x - 3 < 0$ simultaneously. The first gives $x < -1$ and the second gives $x < 3$. Together they give $x < -1$.

Case 2:   *Both positive.* Both factors are positive when $x + 1 > 0$ and $x - 3 > 0$, that is, when $x > -1$ and $x > 3$. These give $x > 3$.

The solution set for the original inequality is the union of the solution sets for the two cases. In set notation, it may be written either as

$$\{x : x < -1 \text{ or } x > 3\}$$

or as

$$\{x : x < -1\} \cup \{x : x > 3\}$$

Figure 11 summarizes what we have learned.

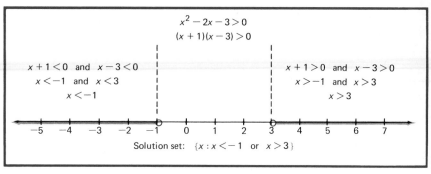

Figure 11

Split-Point Method

The preceding example could have been approached in a slightly different way using the notion of split-points. The solutions of $(x + 1)(x - 3) = 0$, which are -1 and 3, serve as split-points that divide the real line into the three intervals: $x < -1$, $-1 < x < 3$, and $3 < x$. Since $(x + 1)(x - 3)$ can change sign only at a split-point, it must be of one sign (that is, be either always positive or always negative) on each of these intervals. To determine which of them make up the solution set of the inequality $(x + 1)(x - 3) > 0$, all we need to do is pick a single (arbitrary) point from each interval and test it for inclusion in the solution set. If it passes the test, the entire interval from which it was drawn belongs to the solution set.

Here is a further illustration of this method.

Example B (Solving a Quadratic Inequality)   Solve the inequality $2x^2 - 5x - 3 \le 0$.

Solution   Since $2x^2 - 5x - 3 = (2x + 1)(x - 3)$, the split points are $-\frac{1}{2}$ and 3. These points divide the real line into intervals: $x < -\frac{1}{2}$, $-\frac{1}{2} < x < 3$, and $3 < x$. We may test each of these intervals by checking our inequality at one point of the interval. We also check the endpoints of the interval to see if they are included. Our work is summarized in Figure 12.

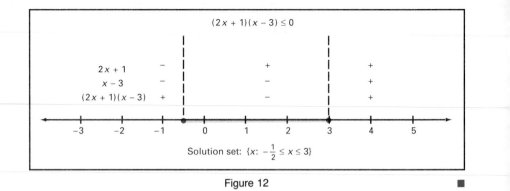

Figure 12 ∎

The method just illustrated works on higher-degree inequalities, too, provided that they can be factored.

Example C (A Cubic Inequality) Solve

$$(x + 2)(x - 1)(x - 4) < 0$$

Solution The solutions of the corresponding equation

$$(x + 2)(x - 1)(x - 4) = 0$$

are -2, 1, and 4. They break the real line into the four intervals $x < -2$, $-2 < x < 1$, $1 < x < 4$, and $4 < x$. Suppose that we pick -3 as the test point for the interval $x < -2$. We see that -3 makes each of the three factors $x + 2$, $x - 1$, and $x - 4$ negative, and so it makes $(x + 2)(x - 1)(x - 4)$ negative. You should pick test points from each of the other three intervals to verify the results shown in Figure 13. ∎

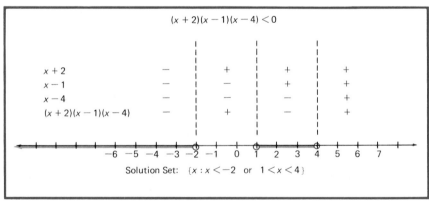

Figure 13

We can even apply the split-point method to inequalities with quotients.

Example D (Inequalities Involving Quotients)   Solve the following inequality.

$$\frac{3}{x-2} > \frac{2}{x}$$

Solution   Our natural inclination might be to multiply both sides by $x(x-2)$ to obtain $3x > 2(x-2)$. But in doing this, we would be assuming that $x(x-2)$ is positive, something that is clearly illegal. Rather, we rewrite the given inequality as follows.

$$\frac{3}{x-2} - \frac{2}{x} > 0 \qquad \text{(add } -\frac{2}{x} \text{ to both sides)}$$

$$\frac{3x - 2(x-2)}{(x-2)x} > 0 \qquad \text{(combine fractions)}$$

$$\frac{x+4}{(x-2)x} > 0 \qquad \text{(simplify numerator)}$$

The factors $x+4$, x, and $x-2$ in the numerator and denominator determine the three split points -4, 0, and 2. The chart in Figure 14 shows the signs of $x+4$, x, $x-2$, and $(x+4)/(x-2)x$ on each of the intervals determined by the split points, as well as the solution set.   ■

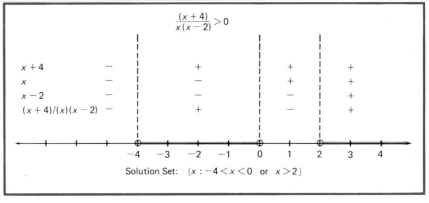

Figure 14

Inequalities with Absolute Values

In solving inequalities involving absolute values, it will be helpful to recall two basic facts we learned in Section 1-5.

Let $a > 0$. Then,

1. $|x| < a$ is equivalent to $-a < x < a$;
2. $|x| > a$ is equivalent to $x < -a$ or $x > a$.

Example E (Rewriting without Absolute Values)   Write the solution to each inequality in a form with no absolute value signs.
(a) $|3x - 2| < 4$ (b) $|2x + 1| \geq 5$

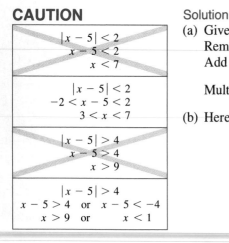
Solution
(a) Given inequality: $|3x - 2| < 4$
    Remove absolute values: $-4 < 3x - 2 < 4$
    Add 2: $-2 < 3x < 6$

    Multiply by $\frac{1}{3}$: $\boxed{-\frac{2}{3} < x < 2}$

(b) Here is the analogous process for the second inequality.

$$|2x + 1| \geq 5$$

$$2x + 1 \leq -5 \quad or \quad 2x + 1 \geq 5$$

$$2x \leq -6 \quad or \quad 2x \geq 4$$

$$\boxed{x \leq -3 \quad or \quad x \geq 2}$$

Be careful not to write this as $2 \leq x \leq -3$ which actually denotes the empty set.   ∎

Example F (Rewriting with Absolute Values)   Write $-2 < x < 8$ as an inequality involving absolute values.

Solution   Look at this interval on the number line (Figure 15). It is 10 units long and its midpoint is at 3. A number x is in this interval provided that it is within a radius of 5 of this midpoint, that is, if

$$|x - 3| < 5$$

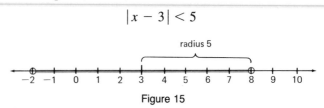

Figure 15

We can check that $|x - 3| < 5$ is equivalent to the original inequality by writing it as

$$-5 < x - 3 < 5$$

and then adding 3 to each member.   ∎

An Application

Here is a word problem whose solution is an inequality.

Example G (Grade Average)   A student wished to get a grade of B in Mathematics 16. On the first four tests, he got 82 percent, 63 percent, 78 percent, and 90 percent, respectively. A grade of B requires an average between 80 percent and 90 percent, inclusive. What grade on the fifth test would qualify this student for a B?

Solution Let x represent the grade (in percent) on the fifth test. The inequality to be satisfied is

$$80 \le \frac{82 + 63 + 78 + 90 + x}{5} \le 90$$

This can be rewritten successively as

$$80 \le \frac{313 + x}{5} \le 90$$

$$400 \le 313 + x \le 450$$

$$87 \le \qquad x \le 137$$

A score greater than 100 is impossible, so the actual solution to this problem is $87 \le x \le 100$. ■

PROBLEM SET 3-5

A. Skills and Techniques

Which of the following inequalities are unconditional and which are conditional?

1. $x \ge 0$

2. $x^2 \ge 0$

3. $x^2 + 1 > 0$

4. $x^2 > 1$

5. $x - 2 < -5$

6. $2x + 3 > -1$

7. $x(x + 4) \le 0$

8. $(x - 1)(x + 2) > 0$

9. $(x + 1)^2 > x^2$

10. $(x - 2)^2 \le x^2$

11. $(x + 1)^2 > x^2 + 2x$

12. $(x - 2)^2 > x(x - 4)$

Solve each of the following inequalities and show the solution set on the real number line. See Examples A–C.

13. $3x + 7 < x - 5$

14. $-2x + 11 > x - 4$

15. $\frac{2}{3}x + 1 > \frac{1}{2}x - 3$

16. $3x - \frac{1}{2} < \frac{1}{2}x + 4$

Hint: First get rid of the fractions.

17. $\frac{3}{4}x - \frac{1}{2} < \frac{1}{6}x + 2$

18. $\frac{2}{7}x + \frac{1}{3} \le -\frac{2}{3}x + \frac{15}{14}$

19. $(x - 2)(x + 5) \le 0$

20. $(x + 1)(x + 4) \ge 0$

21. $(2x - 1)(x + 3) > 0$

22. $(3x + 2)(x - 2) < 0$

23. $x^2 - 5x + 4 \ge 0$

24. $x^2 + 4x + 3 \le 0$

Hint: Factor the left side.

25. $2x^2 - 7x + 3 < 0$

26. $3x^2 - 5x - 2 > 0$

27. $(x + 4)x(x - 3) \ge 0$

28. $(x + 3)x(x - 3) \ge 0$

29. $(x - 2)^2(x - 5) < 0$

30. $(x + 1)^2(x - 1) > 0$

Solve each of the following inequalities as in Example D.

31. $\dfrac{x - 5}{x + 2} \le 0$

32. $\dfrac{x + 3}{x - 2} > 0$

33. $\dfrac{x(x + 2)}{x - 5} > 0$

34. $\dfrac{x - 1}{(x - 3)(x + 3)} \ge 0$

35. $\dfrac{5}{x - 3} > \dfrac{4}{x - 2}$

36. $\dfrac{-3}{x + 1} < \dfrac{2}{x - 4}$

Rewrite each inequality without absolute values. See Example E.

37. $|2x + 3| < 2$

38. $|2x - 4| \le 3$

39. $|-2x - 1| \le 1$

40. $|-2x + 3| > 2$

41. $|5x - 1| \ge 9$

42. $|3x + 2| < 4$

43. $|2x - 3| > 6$

44. $|4x + 3| < 5$

Write each of the following as an inequality involving absolute values. See Example F.

45. $0 < x < 6$

46. $0 < x < 12$

47. $-1 \le x \le 7$

48. $-3 \le x \le 7$

49. $2 < x < 11$

50. $-10 < x < -3$

Solve each of the following inequalities and display the solution set on the number line. Hint: To find split points, you may need the quadratic formula.

51. $x^2 - 7 < 0$

52. $x^2 - 12 > 0$

53. $x^2 - 4x + 2 \ge 0$

54. $x^2 - 4x - 2 \le 0$

55. $x^2 + 6.32x + 3.49 > 0$

56. $x^2 + 4.23x - 2.79 < 0$

Find the least value each of the following expressions can take on. For example, to find the least value of $x^2 - 4x + 9$, write

$$x^2 - 4x + 9 = (x^2 - 4x + 4) + 5 = (x - 2)^2 + 5$$

Since the smallest value $(x - 2)^2$ can take on is zero, the smallest value $(x - 2)^2 + 5$ can assume is 5.

57. $x^2 + 8x + 20$ **58.** $x^2 + 10x + 40$

59. $x^2 - 2x + 101$ **60.** $x^2 - 4x + 104$

B. Applications and Extensions

Solve the inequalities in Problems 61–76.

61. $\frac{1}{2}x + \frac{3}{4} > \frac{2}{3}x - \frac{4}{3}$

62. $3(x - \frac{2}{5}) \le 5x - \frac{1}{3}$

63. $2x^2 + 5x - 3 < 0$

64. $x^2 + x - 1 \le 0$

65. $(x + 1)^2(x - 1)(x - 4)(x - 8) < 0$

66. $(x - 3)(x^3 - x^2 - 6x) \ge 0$

67. $\dfrac{1}{x - 2} + 1 < \dfrac{2}{x + 2}$ **68.** $\dfrac{2}{x - 2} < \dfrac{3}{x - 3}$

69. $|4x - 3| \ge 2$ **70.** $|2x - 1| > 3$

71. $|3 - 4x| < 7$ **72.** $|x| < x + 3$

73. $|x - 2| < |x + 3|$ **74.** $||x + 2| - x| < 5$

75. $|x^2 - 2x - 4| > 4$ **76.** $|x^2 - 14x + 44| < 4$

77. For what values of k will the following equations have real solutions?
(a) $x^2 + 4x + k = 0$ (b) $x^2 - kx + 9 = 0$
(c) $x^2 + kx + k = 0$ (d) $x^2 + kx + k^2 = 0$

78. Amy scored 73, 82, 69, and 94 on four 100-point tests in Mathematics 16. Suppose that a grade of B requires an average between 75 percent and 85 percent.
(a) What score on a 100-point final exam would qualify Amy for a B in the course? (See Example G.)
(b) What score on a 200-point final exam would qualify Amy for a B in the course?

79. The mathematics department at Podunk University has a staff of 6 people with an average salary of $32,000 per year. An additional professor is to be hired. In what range can a salary S be offered if the department average must be between $31,000 and $35,000?

80. Company A will loan out a car for $35 per day plus 10¢ per mile, whereas company B charges $30 per day plus 12¢ per mile. I need a car for 5 days. For what range of mileage will I be ahead financially if I rent a car from company B?

81. A ball thrown upward with a velocity of 64 feet per

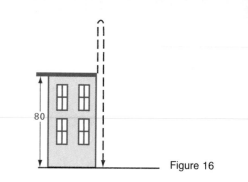

Figure 16

second from the top edge of a building 80 feet high will be at height $(-16t^2 + 64t + 80)$ feet above the ground after t seconds (Figure 16).
(a) What is the greatest height attained by the ball?
(b) During what time period is the ball higher than 96 feet?
(c) At what time did the ball hit the ground, assuming it missed the building on the way down?

82. For what numbers is it true that the sum of the number and its reciprocal is greater than 2?

83. The period T in seconds of a simple pendulum of length l centimeters (Figure 17) is given by $T = 2\pi \sqrt{l/980}$. For what values of l does the period T lie between 2 and 3 seconds?

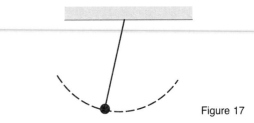

Figure 17

84. For what positive numbers is it true that the sum of a number and its positive square root lies between 8.75 and 15.75?

85. Let a, b, and c denote the lengths of the two legs and hypotenuse of a right triangle so that $a^2 + b^2 = c^2$. Show that if n is an integer greater than 2, then $a^n + b^n < c^n$.

86. TEASER   Sophus Slybones, the proprietor of a specialty coffee shop, weighs coffee for customers by using the two-pan balance shown in Figure 18. Long ago, he dropped the balance on the concrete floor and

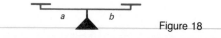

Figure 18

now the left arm is slightly longer than the right one, that is, $a > b$. This poses no problem for Sophus. When a customer orders 2 kilograms of coffee, he first places a 1-kilogram weight in the left pan balancing it with some coffee in the right pan, which he then pours in a sack. Next, he places the 1-kilogram weight in the right pan balancing it with coffee in the left pan, which he also pours in the sack. That makes exactly 2 kilograms says Sophus, a claim you must analyze.

Assume that the balance arms are weightless but that the pans weigh 100 grams each.

(a) Show that Sophus has been cheating himself all these years.

(b) If the amount Sophus gives a customer is actually 2.01 kilograms, determine a/b.

If you want to make the problem a little harder, show that the conclusion in part (a) is valid even if you don't assume the balance arms are weightless.

3-6 MORE APPLICATIONS (OPTIONAL)

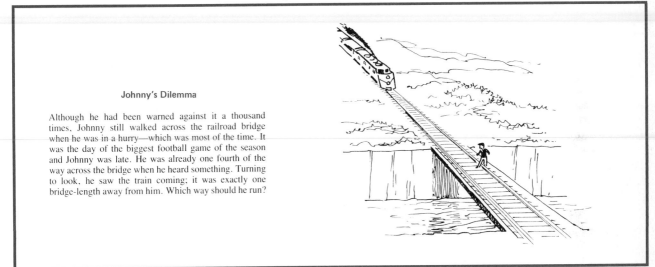

Johnny's Dilemma

Although he had been warned against it a thousand times, Johnny still walked across the railroad bridge when he was in a hurry—which was most of the time. It was the day of the biggest football game of the season and Johnny was late. He was already one fourth of the way across the bridge when he heard something. Turning to look, he saw the train coming; it was exactly one bridge-length away from him. Which way should he run?

We would not suggest that the problem above is a typical application of algebra. Johnny would be well advised to forget about algebra, make an instant decision, and hope for the best. Nevertheless, Johnny's problem is intriguing. Moreover, it can serve to reemphasize the important principles of real-life problem solving that we mentioned briefly at the end of Section 3-2.

First, **clarify the question that is asked.** "Which way should he run?" must mean "Which way should Johnny run to have the best chance of surviving?" That is still too vaguely stated for mathematical analysis. We think the question really means, "Which of the two directions allows Johnny to run at the slowest rate and still avoid the train?" Most questions that come to us from the real world are loosely stated. Our first job is always to pin down precisely (at least in our own minds) what the real question is. It is foolish to try to answer a question that we do not understand. That should be obvious, but it is often overlooked.

Johnny's problem appears to be difficult for another reason; there does not seem to be enough information. We do not know how long the bridge is, we do not know how fast the train is traveling, and we do not know how fast Johnny can run. We could easily despair of making any progress on the problem!

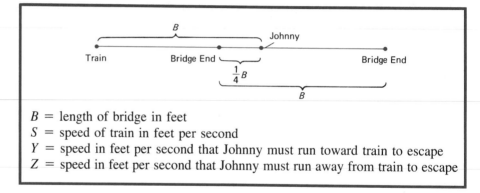

B = length of bridge in feet
S = speed of train in feet per second
Y = speed in feet per second that Johnny must run toward train to escape
Z = speed in feet per second that Johnny must run away from train to escape

Figure 19

This leads us to state our second principle. **Organize the information you have.** The best way to do this is to draw a diagram or picture that somehow captures the essence of the problem. It should be an abstract or idealized picture. We shall represent people and trains by points and train tracks and bridges by line segments. These are only approximations to the real situation, but they are necessary if progress is to be made. After the picture is drawn, **label the key quantities of the problem with letters and write down what they represent.** One way to represent Johnny's situation is in Figure 19.

Our next principle is this. **Write down the algebraic relationships that exist between the symbols you have introduced.** In Johnny's case, we need to remember the formula $D = RT$ (or $T = D/R$), which relates distance, rate, and time. Now, if Johnny runs toward the train just fast enough to escape, then his time to run the distance $B/4$ must equal the time for the train to go $3B/4$. That gives

(i)
$$\frac{B/4}{Y} = \frac{3B/4}{S}$$

On the other hand, if Johnny runs away from the train, he will have to cover $3B/4$ in the same time that the train goes $7B/4$. Thus

(ii)
$$\frac{3B/4}{Z} = \frac{7B/4}{S}$$

Now we have done what is often the hardest part of the problem, translating it into algebraic equations. Our job now is to **manipulate the equations until a solution appears.** In the example we are considering, we must solve the equations for Y and Z in terms of S. When we do this, we get

(i)
$$Y = \frac{1}{3}S$$

(ii)
$$Z = \frac{3}{7}S$$

There remains a crucial step: **Interpret the result in the language of the original problem.** In Johnny's case, this step is easy. If he runs toward the train, he will need to run at $\frac{1}{3}$ the speed of the train to escape. If he runs away from the train, he must run at $\frac{3}{7}$ of the speed of the train. Clearly, he will have a better chance of making it if he runs toward the train.

Perhaps by now you need to be reassured. We do not expect you to go through a long-winded analysis like ours for every problem you meet. But we do think that you will have to apply the principles we have stated if you are to be successful at problem solving. The problem set that follows is designed to let you apply these principles to problems from several areas of applied mathematics.

PROBLEM SET 3-6

The problems below are arranged according to type. Only the first set (rate-time problems) relates directly to the example of this section. However, all of them should require some use of the principles we have enunciated. Do not expect to solve all of the problems. But do accept them as challenges worthy of genuine effort.

Rate-Time Problems

1. Sound travels at 1100 feet per second in air and at 5000 feet per second in water. An explosion on a distant ship was recorded by the underwater recording device at a monitoring station. Thirteen seconds later the operator heard the explosion. How far was the ship from the station?

2. The primary wave and secondary wave of an earthquake travel away from the epicenter at rates of 8 and 4.8 kilometers per second, respectively. If the primary wave arrives at a seismic station 15 seconds ahead of the secondary wave, how far is the station from the epicenter?

3. Ricardo often flies from Clear Lake to Sun City and returns on the same day. On a windless day, he can average 120 miles per hour for the round trip, while on a windy day, he averaged 140 miles per hour one way and 100 miles per hour the other. It took him 15 minutes longer on the windy day. How far apart are Clear Lake and Sun City?

4. A classic puzzle problem goes like this: If a column of men 3 miles long is marching at 5 miles per hour, how long will it take a courier on a horse traveling at 25 miles per hour to deliver a message from the end of the column to the front and then return?

5. How long after 4:00 P.M. will the minute hand on a clock overtake the hour hand?

6. At what time between 4:00 and 5:00 do the hands of a clock form a straight line?

7. An old machine is able to do a certain job in 8 hours. Recently a new machine was installed. Working together, the two machines did the same job in 3 hours. How long would it take the new machine to do the job by itself?

8. Center City uses fire trucks to fill the village swimming pool. It takes one truck 3 hours to do the job and another 2 hours. How long would it take them working together?

9. An airplane takes off from a carrier at sea and flies west for 2 hours at 600 miles per hour. It then returns at 500 miles per hour. In the meantime, the carrier has traveled west at 30 miles per hour. When will the two meet?

10. A passenger train 480 feet long traveling at 75 miles per hour meets a freight train 1856 feet long traveling on parallel tracks at 45 miles per hour. How long does it take the trains to pass each other? *Note:* Sixty miles per hour is equivalent to 88 feet per second.

11. A medical company makes two types of heart valves, standard and deluxe. It takes 5 minutes on the lathe and 10 minutes on the drill press to make a standard valve, but 9 minutes on the lathe and 15 minutes on the drill press for a deluxe valve. On a certain day the lathe will be available for 4 hours and the drill press for 7 hours. How many valves of each kind should the company make that day if both machines are to be fully utilized?

12. Two boats travel at right angles to each other after leaving the dock at 1:00 P.M. At 3:00 P.M., they are 16 miles apart. If the first boat travels 6 miles per hour faster than the second, what are their rates?

13. A car is traveling at an unknown rate. If it traveled 15 miles per hour faster, it would take 90 minutes less to go 450 miles. How fast is the car going?

14. A boy walked along a level road for awhile, then up

a hill. At the top of the hill he turned around and walked back to his starting point. He walked 4 miles per hour on level ground, 3 miles per hour uphill, and 6 miles per hour downhill, with the total trip taking 5 hours. How far did he walk altogether?

15. Jack and Jill live at opposite ends of the same street. Jack wanted to deliver a box at Jill's house and Jill wanted to leave some flowers at Jack's house. They started at the same moment, each walking at a constant speed. They met the first time 300 meters from Jack's house. On their return trip, they met 400 meters from Jill's house. How long is the street? (Assume that neither loitered at the other house nor when they met.)

16. Two cars are traveling toward each other on a straight road at the same constant speed. A plane flying at 350 miles per hour passes over the second car 2 hours after passing over the first car. The plane continues to fly in the same direction and is 2400 miles from the cars when they pass. Find the speed of the cars.

Science Problems

17. The illumination I in foot-candles on a surface d feet from a light source of c candlepower is given by the formula $I = c/d^2$. How far should an 80-candlepower light be placed from a surface to give the same illumination as a 20-candlepower light at 10 feet?

18. By experiment, it has been found that a car traveling v miles per hour will require d feet to stop, where $d = .044v^2 + 1.1v$. Find the velocity of a car if it took 176 feet to stop.

19. A bridge 200 feet long was built in the winter with no provision for expansion. In the summer, the supporting beams expanded in length by 8 inches, forcing the center of the bridge to drop. Assuming, for simplicity, that the bridge took the shape of a V, how far did the center drop? First guess at the answer and then work it out.

20. The distance s (in feet) traveled by an object in t seconds when it has an initial velocity v_0 and a constant acceleration a is given by $s = v_0 t + \frac{1}{2}at^2$. An object was observed to travel 32 feet in 4 seconds and 72 feet in 6 seconds. Find the initial velocity and the acceleration.

21. A chemist has 5 kiloliters of 20 percent sulfuric acid solution. She wishes to increase its strength to 30 percent by draining off some and replacing it with 80 percent solution. How much should she drain off?

22. How many liters each of a 35 percent alcohol solution and a 95 percent alcohol solution must be mixed to obtain 12 liters of an 80 percent alcohol solution?

23. One atom of carbon combines with 2 atoms of oxygen to form one molecule of carbon dioxide. The atomic weights of carbon and oxygen are 12.0 and 16.0, respectively. How many milligrams of oxygen are required to produce 4.52 milligrams of carbon dioxide?

24. Four atoms of iron (atomic weight 55.85) combine with 6 atoms of oxygen (atomic weight 16.00) to form 2 molecules of rust. How many grams of iron would there be in 79.85 grams of rust?

25. A sample weighing .5000 gram contained only sodium chloride and sodium bromide. The chlorine and bromine from this sample were precipitated together as silver chloride and silver bromide. This precipitate weighed 1.100 grams. Sodium chloride is 60.6 percent chlorine, sodium bromide is 77.6 percent bromine, silver chloride is 24.7 percent chlorine, and silver bromide is 42.5 percent bromine. Calculate the weights of sodium chloride and sodium bromide in the sample.

Business Problems

26. Sarah Tyler bought stock in the ABC Company on Monday. The stock went up 10 percent on Tuesday and then dropped 10 percent on Wednesday. If she sold the stock on Wednesday for $1000, what did she pay on Monday?

27. If Jane Witherspoon has $4182 in her savings account and wants to buy stock in the ABC Company at $59 a share, how many shares can she buy and still maintain a balance of at least $2000 in her account?

28. Six men plan to take a charter flight to Bear Lake in Canada for a fishing trip, sharing the cost equally. They discover that if they took three more men along, each share of the original six would be reduced by $150. What is the total cost of the charter?

29. Susan has a job at Jenny's Nut Shop. Jenny asks Susan to prepare 25 pounds of mixed nuts worth $1.74 per pound by using walnuts valued at $1.30 per pound and cashews valued at $2.30 per pound. How many pounds of each kind should Susan use?

30. Alec Brown plans to sell toy gizmos at the state fair. He can buy them at 40¢ apiece and will sell them for 65¢ each. It will cost him $200 to rent a booth for the 7-day fair. How many gizmos must he sell to just break even?

31. The cost of manufacturing a product is the sum of fixed plant costs (real estate taxes, utilities, and so on)

and variable costs (labor, raw materials, and so on) that depend on the number of units produced. The profit P that the company makes in a year is given by

$$P = TR - (FC + VC)$$

where TR is the total revenue (total sales), FC is the total fixed cost, and VC is the total variable cost. A company that makes one product has $32,000 in total fixed costs. If the variable cost of producing one unit is $4 and if units can be sold at $6 each, find out how many units must be produced to give a profit of $15,000.

32. The ABC Company has total fixed costs of $100,000 and total variable costs equal to 80 percent of total sales. What must the total sales be to yield a profit of $40,000? (See Problem 31.)

33. Do Problem 32 assuming that the company pays 30 percent income taxes on its profit and wants a profit of $40,000 after taxes.

34. The XYZ Company has total fixed yearly costs of $120,000 and last year had total variable costs of $350,000 while selling 200,000 gizmos at $2.50 each. Competition will force the manager to reduce the sales price to $2.00 each next year. If total fixed costs and variable costs per unit are expected to remain the same, how many gizmos will the company have to sell to have the same profit as last year?

35. Podunk University wishes to maintain a student-faculty ratio of 1 faculty member for every 15 undergraduates and 1 faculty member for each 6 graduate students. It costs the university $600 for each undergraduate student and $900 for each graduate student over and above what is received in tuition. If the university expects $2,181,600 in gifts (beyond tuition) next year and will have 300 faculty members, how many undergraduate and how many graduate students should it admit?

36. A department store purchased a number of smoke detectors at a total cost of $2000. In unpacking them, the stock boys damaged 8 of them so badly that they could not be sold. The remaining detectors were sold at a profit of $25 each, and a total profit of $400 was realized when all of them were sold. How many smoke detectors were originally purchased?

Geometry Problems

37. A flag that is 6 feet by 8 feet has a blue cross of uniform width on a gray background (Figure 20). The cross extends to all 4 edges of the flag. The area of the

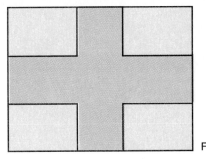

Figure 20

cross and the background are equal. Find the width of the cross.

38. If a right triangle has hypotenuse 2 units longer than one leg and 4 units longer than the other, find the dimensions of the triangle.

39. The area and perimeter of a right triangle are both 30. Find its dimensions.

40. Assume that the earth is a sphere of radius 4000 miles. How far is the horizon from an airplane 5 miles high?

41. Three mutually tangent circles have centers A, B, and C and radii a, b, and c, respectively. The lengths of the segments AB, BC, and CA are 13, 15, and 18, respectively. Find the lengths of the radii. Assume that each circle is outside of the other two.

42. A rectangle is inscribed in a circle of radius 5. Find the dimensions of the rectangle if its area is 40.

43. A ladder is standing against a house with its lower end 10 feet from the house. When the lower end is pulled 2 feet farther from the house, the upper end slides 3 feet down the house. How long is the ladder?

44. A trapezoid (a quadrilateral with two sides parallel) is inscribed in a square 12 inches on a side. One of the parallel sides is the diagonal of the square. If the trapezoid has area 24 square inches, how far apart are its parallel sides?

45. A 40-inch length of wire is cut in two. One of the pieces is bent to form a square and the other is bent to form a rectangle three times as long as wide. If the combined area of the square and the rectangle is $55\frac{3}{4}$ square inches, where was the wire cut?

46. TEASER Figure 21 shows a ring-shaped sidewalk surrounding an irregular pond. We wish to determine the area A of the sidewalk but do not know the radii of the two circles. In fact, we do not even know where their common center is and we have no intention of getting our feet wet. However, we can easily make the measurements a and b shown in the figure. Express A in terms of a and b. Note that this shows that all an-

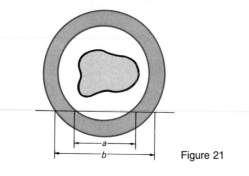

Figure 21

nuli (ring-shaped regions) that intersect a line in the same two line segments have the same area.

47. **TEASER**   H. E. Dudeney, an English puzzlist, gave us this problem. A metal flagpole broke at a certain point with the top part tipping over like a hinge and the tip hitting the ground at a point 20 feet from the base. It was rewelded but again it broke, this time at a point 5 feet lower with the tip hitting the ground at a point 30 feet from the base. How tall was the flagpole?

48. **TEASER**   Audrey, Betty, and Candy, who live in different houses, agree to take a cab to the ball game and share expenses. The cab picked up Audrey first, Betty 3 miles later, and Candy 2 miles after Betty. The ballpark is 1 mile from Candy's house. Their total bill was $19.80. The cab driver wanted to charge Candy $3.30 since she went one-sixth of the way, but she rightly refused. What should each of the women pay?

49. **TEASER**   We do not know the origin of this old teaser, which appears in many mathematical books. Arnold Thinkhard rowed upstream with a uniform (stillwater) speed of v miles per hour against a current of c miles per hour. When he had gone 1 mile, the wind blew his hat into the water just out of reach. It was an old hat, so he decided to let it go. After rowing another hour, he suddenly remembered that he had stuck his train ticket in the hatband of that old hat. Turning around immediately, he rowed back and caught up with his hat at his original starting point. Determine the rate of the current.

50. **TEASER**   This problem is attributed to that great eighteenth-century mathematician Leonhard Euler. A group of heirs divided the estate left by their rich uncle according to the following novel plan. The first heir took c dollars plus an nth of the remainder, the second took $2c$ dollars plus an nth of what still remained, and so on. Each succeeding heir took c dollars more than the previous heir plus one-nth of the new remainder. They were able to follow this scheme and still all heirs got the same amount of money. Express the total inheritance in terms of n and c. Then determine the number of heirs if $n = 16$.

CHAPTER 3 SUMMARY

The equalities $(x + 1)^2 = x^2 + 2x + 1$ and $x^2 = 4$ are quite different in character. The first, called an **identity,** is true for all values of x. The second, called a **conditional equation,** is true only for certain values of x, in fact, only for $x = 2$ and $x = -2$. To **solve** an equation is to find those values of the unknown which make the equality true; it is one of the major tasks of mathematics.

The equation $ax + b = 0$ $(a \neq 0)$ is called a **linear equation** and has exactly one solution, $x = -b/a$. Similarly, $ax^2 + bx + c = 0$ $(a \neq 0)$ is a **quadratic equation** and usually has two solutions. Sometimes they can be found by **factoring** the left side and setting both factors equal to zero. Another method that always works is **completing the square,** but the best general method is simply substituting in the **quadratic formula.**

$$x = \frac{-b \pm \sqrt{b^2 - 4ac}}{2a}$$

Here $b^2 - 4ac$, called the **discriminant,** plays a critical role. The equation has two real solutions, one real solution, or two nonreal solutions according as the discriminant is positive, zero, or negative.

Equations arise naturally in the study of word problems. Such problems may lead to one equation in one unknown but often lead to a **system** of several equations in several unknowns. In the latter case, our task is to find the values of the unknowns that satisfy all the equations of the system simultaneously.

Inequalities look like equations with the equal sign replaced by $<$, $\leq$, $>$, or $\geq$. The methods for solving **conditional inequalities** are very similar to those for conditional equations. One difference is that the direction of an inequality sign is reversed upon multiplication or division by a negative number. Another is that the set of solutions normally consists of one or more intervals of numbers, rather than a finite set. For example, the inequality $3x - 2 < 5$ has the solution set $\{x: x < \frac{7}{3}\}$.

CHAPTER 3 REVIEW PROBLEM SET

In Problems 1–10, write True or False in the blank. If false, tell why.

_____ 1. Multiplying both sides of an equation by a variable quantity gives an equivalent equation.

_____ 2. Dividing both sides of an equation by a nonzero constant gives an equivalent equation.

_____ 3. Every quadratic equation has two solutions.

_____ 4. The system of equations $x + y = 9$ and $2x + 2y = 18$ has (1, 8) as its only solution.

_____ 5. If $(x^2 + 1)(x - 1) = 0$ and x is real, then $x = 1$.

_____ 6. If $b^2 - 4ac < 0$, then $ax^2 + bx + c = 0$ has no solution.

_____ 7. $\sqrt{9} = \pm 3$.

_____ 8. $|x - 3| < 4$ is equivalent to $-1 < x < 7$.

_____ 9. $|x| > 5$ is equivalent to $5 < x < -5$.

_____ 10. $\dfrac{1}{x + 1} < 4$ is equivalent to $1 < 4(x + 1)$.

11. Which of the following are identities, and which are conditional equations?
(a) $5(1 - x) = 5 - 5x$
(b) $(x + 2)^2 - x^2 = 4$
(c) $x^2 - a^2 = (x - a)(x + a)$
(d) $(x + 2)(x - 3) - x^2 + x = -6$

12. Show that each of the following equations has no solution.
(a) $2x + 3 = 2(x - 1)$
(b) $\dfrac{x + 1}{x + 2} = 1$

13. Solve the following equations.
(a) $2(x + \frac{2}{3}) = x + \frac{1}{2}$
(b) $\dfrac{9}{x - 4} = \dfrac{15}{x - 2}$
(c) $(x + 3)(3x - 2) = (3x + 1)(x - 4)$
(d) $\dfrac{x}{2x + 2} - 2 = \dfrac{4 - 9x}{6x + 6}$

14. In the equation $1 + rt = (1 - 2s)/s$, solve for s in terms of the other variables.

15. Fahrenheit and Celsius temperatures are related by the equation $F = \frac{9}{5}C + 32$.
(a) How warm is it in degrees Fahrenheit when the Fahrenheit reading is three times the Celsius reading?
(b) For what Fahrenheit temperatures is the Fahrenheit reading higher than the Celsius reading?

16. Solve the following systems.
(a) $2x + y = 3$
$3x + 2y = 6$
(b) $\frac{1}{2}x - \frac{1}{4}y = 1$
$x + 2y = 5$

17. Simplify.
(a) $\dfrac{6 + \sqrt{18}}{12}$
(b) $\dfrac{\sqrt{28}}{\sqrt{63}}$
(c) $\sqrt{13^2 \cdot 2^4}$
(d) $\dfrac{3 - \sqrt{-9}}{3}$

18. What real values of x satisfy the inequality $x^2 < -1$?

Solve the equations in Problems 19–32.

19. $3x^2 = 192$
20. $(x - 1)^2 = 4$
21. $x^2 + 4x + 4 = 0$
22. $x^2 - 3x = 0$
23. $x^2 - 2x - 35 = 0$
24. $3x^2 - x - 2 = 0$
25. $(x - 2)(x - 3) = 12$
26. $3x^2 + x = 1$
27. $x^2 + 2x - 2 = 0$
28. $\dfrac{1}{x + 1} = \dfrac{x}{x - 1}$
29. $(x - 1)(x - 2) = 1$
30. $x - 2\sqrt{x} - 8 = 0$
31. $(x - 1)^2 + 2(x - 1) - 8 = 0$
32. $\dfrac{6}{(x + 2)^2} + \dfrac{1}{x + 2} - 1 = 0$

In Problems 33 and 34, solve for y in terms of x.

33. $(3x - 2y)^2 = 9$
34. $(y + 2x^2)^2 - 5(y + 2x^2) + 4 = 0$

Solve the inequalities in Problems 35–42.

35. $2x - 3 < 4 - 3x$

36. $5 - 2x \geq 2(x + 4)$

37. $x^2 - 2x - 24 > 0$

38. $x^2 + 4x + 4 > 0$

39. $\dfrac{2x + 1}{x - 3} \geq 0$

40. $1 - \dfrac{4}{x^2} < 0$

41. $x^2 + 2x < 2$

42. $x^2 + 6x + 20 > 0$

Write the inequalities in Problems 43 and 44 without absolute value signs.

43. $|x + 5| < 5$

44. $|3x + 7| \geq 4$

Solve the equations in Problems 45–48.

45. $|2x + 1| = 7$

46. $|5 - x| = 2$

47. $|x| = 3 - x$

48. $|x + 1| = |x - 2|$

49. In solving $x^2 = 4x$, why is it incorrect to divide by x to obtain $x = 4$ as the solution?

50. Suppose that t seconds after a ball is thrown straight up from the ground, its height s (in feet) is given by $s = -16t^2 + 96t$.

(a) When did the ball hit the ground?

(b) When was its height greater than 80 feet?

51. Hideo Ito has divided $25,000 between two funds and remembers that he is to receive $1981 in interest for the year. One fund yields 7 percent simple interest and the other 8.5 percent simple interest. How much did he put in each fund?

52. A rectangle is 20 centimeters longer than it is wide. If the width is decreased by 2 centimeters and the length increased by 4 centimeters, the area decreases by 16 square centimeters. Find the original dimensions of the rectangle.

53. Chicago and St. Paul are 450 miles apart. A fast train left Chicago at 6:00 A.M. traveling at 60 miles per hour toward St. Paul. At 8:00 A.M., a second train left St. Paul traveling at a constant rate toward Chicago. If the two trains passed each other at 11:00 A.M., what was the rate of the second train?

54. The Toadsville Merchants play baseball in a league that has a season of 150 games. They have won 60 percent of their games through the first three-fifths of the season. What percentage of their games must they win during the rest of the season to bring their overall average up to $66.\overline{66}$ percent?

55. Reserved seats were $7.00; general admission seats were $4.50. When the treasurer totaled the receipts, he had taken in $3375. How many seats of each kind were sold, given that a total of 625 tickets were sold?

56. How much water must be evaporated from 100 liters of a 16 percent salt solution to obtain a 25 percent solution?

CHAPTER 4

Coordinates and Curves

■ And so Fermat and Descartes turned to the application of algebra to the study of geometry. The subject they created is called coordinate, or analytic, geometry; its central idea is the association of algebraic equations with curves and surfaces. This creation ranks as one of the richest and most fruitful veins of thought ever struck in mathematics.

—Morris Kline

4-1 THE CARTESIAN COORDINATE SYSTEM

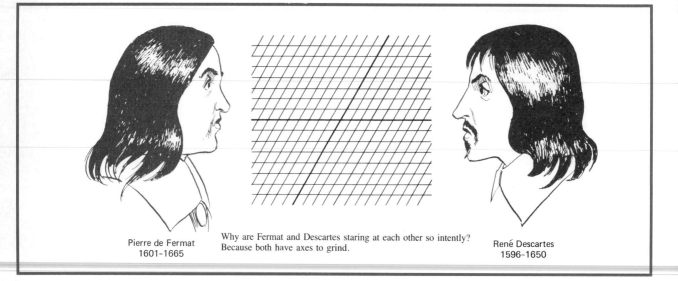

Pierre de Fermat
1601–1665

Why are Fermat and Descartes staring at each other so intently?
Because both have axes to grind.

René Descartes
1596–1650

Two Frenchmen deserve credit for the idea of a coordinate system. Pierre de Fermat was a lawyer who made mathematics his hobby. In 1629, he wrote a paper which makes explicit use of coordinates to describe points and curves. René Descartes was a philosopher who thought mathematics could unlock the secrets of the universe. He published *La Géométrie* in 1637. It is a famous book and though it does emphasize the role of algebra in solving geometric problems, one finds only a hint of coordinates there. By virtue of having the idea first and more explicitly, Fermat ought to get the major credit. History can be a fickle friend; coordinates are known as Cartesian coordinates, named after René Descartes.

No matter who gets the credit, it was an idea whose time had come. It made possible the invention of calculus, one of the greatest inventions of the human mind. That invention was to come in 1665 at the hands of a 23-year-old genius named Isaac Newton. You will probably study calculus later on. There you will use the ideas of this chapter over and over.

Review of the Real Line

Recall the real line (Figure 1), which was introduced in Section 1-3.

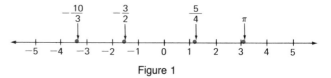

Figure 1

Every point on this line can be given a label, a real number, which specifies exactly where the point is. We call this label the **coordinate** of the point.

Consider now points A and B with coordinates a and b, respectively. We will need a formula for the distance between A and B in terms of the coordinates a and b.

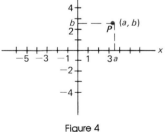

Figure 2

The formula is

$$d(A, B) = |b - a|$$

and it is correct whether A is to the right or to the left of B. Note the two examples in Figure 2. In the first case,

$$d(A, B) = |3 - (-2)| = |5| = 5$$

In the second,

$$d(A, B) = |-1 - 5| = |-6| = 6$$

Example A (Distance on the Real Line) Given that A has coordinate $a = 2.3$, find the coordinate b of two points B on the real line such that $d(A, B) = 5.1$.

Solution The required point to the left of A has coordinate $b = 2.3 - 5.1 = -2.8$; the one to the right has coordinate $b = 2.3 + 5.1 = 7.4$. ■

Cartesian Coordinates

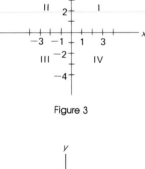

Figure 3

In the plane, produce two copies of the real line, one horizontal and the other vertical, so that they intersect at the zero points of the two lines. The two lines are called **coordinate axes;** their intersection is labeled with O and is called the **origin.** By convention, the horizontal line is called the **x-axis** and the vertical line is called the **y-axis.** The positive half of the x-axis is to the right; the positive half of the y-axis is upward. The coordinate axes divide the plane into four regions called **quadrants,** labeled I, II, III, and IV, as shown in Figure 3.

Each point P in the plane can now be assigned a pair of numbers called its **Cartesian coordinates.** If vertical and horizontal lines through P intersect the x- and y-axes at a and b, respectively, as in Figure 4, then P has coordinates (a, b). We call (a, b) an **ordered pair** of numbers because it makes a difference which number is first. The first number a is the **x-coordinate** (or abscissa); the second number b is the **y-coordinate** (or ordinate).

Figure 4

Conversely, take any ordered pair (a, b) of real numbers. The vertical line through a on the x-axis and the horizontal line through b on the y-axis meet in a point P whose coordinates are (a, b).

Think of it this way: The coordinates of a point are the address of that point. If you have found a house (or a point), you can read its address. Conversely, if you know the address of a house (or a point), you can always locate it. In Figure 5 we have shown the coordinates (addresses) of several points.

The Distance Formula in the Plane

Consider the points P_1 and P_2 with coordinates $(-1, -2)$ and $(3, 1)$, respectively. The segment joining P_1 and P_2 is the hypotenuse of a right triangle with right angle at $P_3(3, -2)$ (see Figure 6). We easily calculate the lengths of the two legs.

$$d(P_1, P_3) = |3 - (-1)| = 4 \qquad d(P_2, P_3) = |1 - (-2)| = 3$$

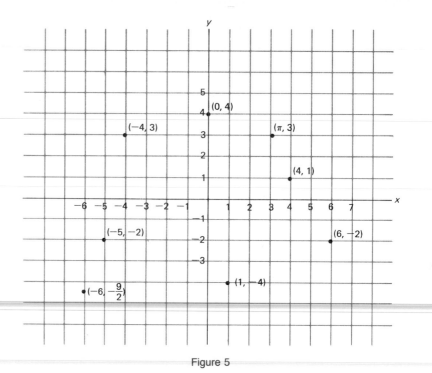

Figure 5

By the Pythagorean Theorem (see Section 1-3).

$$d(P_1, P_2) = \sqrt{4^2 + 3^2} = \sqrt{25} = 5$$

Next consider two arbitrary points $P_1(x_1, y_1)$ and $P_2(x_2, y_2)$ (as in Figure 7) that are not on the same horizontal or vertical line. They determine a right triangle with legs of length $|x_2 - x_1|$ and $|y_2 - y_1|$. By the Pythagorean Theorem,

$$d(P_1, P_2) = \sqrt{(x_2 - x_1)^2 + (y_2 - y_1)^2}$$

This formula is known as the **distance formula.** You should check that it is valid even if P_1 and P_2 lie on the same vertical or horizontal line.

Example B (Using the Distance Formula) Consider the triangle ABC (Figure 8) with vertices $A(-1, 2)$, $B(1, 4)$, and $C(3, -2)$. Find the perimeter of this triangle and show that it is a right triangle.

Solution   We calculate the lengths of the three sides.

$$d(A, B) = \sqrt{(1 + 1)^2 + (4 - 2)^2} = \sqrt{8} = 2\sqrt{2}$$
$$d(A, C) = \sqrt{(3 + 1)^2 + (-2 - 2)^2} = \sqrt{32} = 4\sqrt{2}$$
$$d(B, C) = \sqrt{(3 - 1)^2 + (-2 - 4)^2} = \sqrt{40} = 2\sqrt{10}$$

Thus the perimeter of the triangle is $6\sqrt{2} + 2\sqrt{10} \approx 14.8098$. That the triangle is

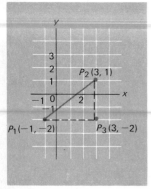

Figure 6

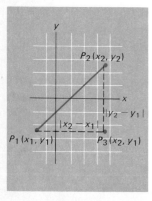

Figure 7

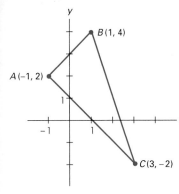

B(1, 4)

A(-1, 2)

C(3, -2)

Figure 8

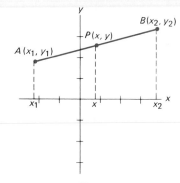

B(x₂, y₂)

P(x, y)

A(x₁, y₁)

Figure 9

a right triangle follows from the converse of the Pythagorean Theorem, which states that

A triangle is a right triangle if its sides satisfy the relationship $a^2 + b^2 = c^2$.

You will be asked to prove this converse in Problem 49. Note in the case of triangle *ABC* that $8 + 32 = 40$. ■

The Midpoint Formula

Consider two points $A(x_1, y_1)$ and $B(x_2, y_2)$ in the plane and let $P(x, y)$ be the midpoint of the segment joining them. Drop perpendiculars from A, P, and B to the x-axis as shown in Figure 9. Then x is midway between x_1 and x_2, so

$$x - x_1 = x_2 - x$$
$$2x = x_1 + x_2$$
$$x = \frac{x_1 + x_2}{2}$$

Similar reasoning applies to y. The result, called the **midpoint formula,** says that the coordinates (x, y) of the midpoint P are given by

$$x = \frac{x_1 + x_2}{2} \qquad y = \frac{y_1 + y_2}{2}$$

For example, the midpoint of the segment joining $A(-3, -2)$ and $B(5, 9)$ has coordinates.

$$\left(\frac{-3 + 5}{2}, \frac{-2 + 9}{2} \right) = \left(1, \frac{7}{2} \right)$$

Example C (Applying the Midpoint Formula) Let $A(1, -1)$ and $B(-5, 6)$ be two points in the plane (Figure 10). Find the midpoint M of segment AB. Also find C so that A is midway between B and C.

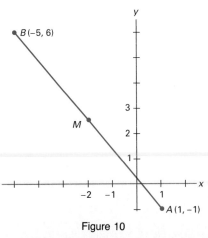

B(-5, 6)

M

A(1, -1)

Figure 10

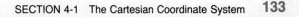

Solution The coordinates of M are $\left(\dfrac{1 + (-5)}{2}, \dfrac{-1 + 6}{2}\right)$ or $(-2, \frac{5}{2})$. If (x, y) denote the coordinates of C, then $\dfrac{x + (-5)}{2} = 1$ and $\dfrac{y + 6}{2} = -1$. Thus C has coordinates $(7, -8)$. ∎

General Point of Division Formula

We can generalize the midpoint formula to a formula for the coordinates of a point that lies a specified fraction of the way from point $A(x_1, y_1)$ to $B(x_2, y_2)$. If t is any number between 0 and 1, then the point $P(x, y)$ on the segment AB satisfying $d(A, P) = t\,d(A, B)$ has coordinates given by

$$x = (1 - t)x_1 + tx_2$$
$$y = (1 - t)y_1 + ty_2$$

Note that the midpoint formula is just the special case corresponding to $t = \frac{1}{2}$.

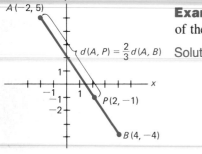

Figure 11

Example D (Finding a Point Between Two Points) Find the point P two-thirds of the way from $A(-2, 5)$ to $B(4, -4)$.

Solution Substitution of $t = \frac{2}{3}$ in the formula gives

$$x = \frac{1}{3}(-2) + \frac{2}{3}(4) = \frac{6}{3} = 2$$

$$y = \frac{1}{3}(5) + \frac{2}{3}(-4) = -\frac{3}{3} = -1$$

The required point has coordinates $(2, -1)$ (see Figure 11). ∎

More Applications of Coordinates

Introducing coordinates can simplify many problems, as we now illustrate.

Example E (Introducing Coordinates to Solve a Problem) Charles City needs to bring a new water line to a factory at A. Point A is exactly 502 meters north and 254 meters east of the water tower T. The city can either run a line directly from the water tower to A or it can run a line from B where it already has water service. Given that B is 419 meters north and 304 meters west of the tower, which option requires the shortest line?

Solution Introduce coordinates with the tower at the origin. Then A has coordinates $(254, 502)$ and B has coordinates $(-304, 419)$ as illustrated in Figure 12. Thus

$$d(T, A) = \sqrt{254^2 + 502^2} \approx 562.6$$
$$d(B, A) = \sqrt{(254 + 304)^2 + (502 - 419)^2} \approx 564.1$$

The line from the tower to A is slightly shorter. ∎

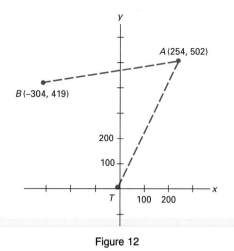

Figure 12

Example F (Introducing Coordinates to Prove a Theorem)   Prove that the diagonals of a parallelogram bisect each other.

Solution   Without loss of generality, we may introduce coordinates for the vertices as shown in Figure 13. Let M_1 and M_2 be the midpoints of the two diagonals. It will be enough to show that M_1 and M_2 have the same coordinates. This is a simple application of the midpoint formula; both have coordinates $\left(\dfrac{a+b}{2}, \dfrac{c}{2}\right)$.

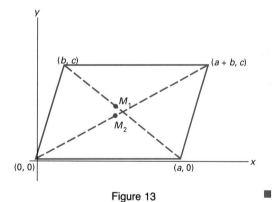

Figure 13   ■

PROBLEM SET 4-1

A. Skills and Techniques

Find $d(A, B)$ where A and B are points on the number line having the given coordinates a and b.

1. $a = -3, b = 2$

2. $a = 5, b = -6$

3. $a = \frac{11}{4}, b = -\frac{5}{4}$

4. $a = \frac{31}{8}, b = \frac{13}{8}$

5. $a = 3.26, b = 4.96$

6. $a = -1.45, b = -5.65$

7. $a = 2 - \pi, b = \pi - 3$

8. $a = 4 + 2\sqrt{3}, b = -6 + \sqrt{3}$

Again, A and B are points on the number line. Given a and $d(A, B)$, what are the two possible values of b? See Example A.

9. $a = 5$, $d(A, B) = 2$ 10. $a = 9$, $d(A, B) = 5$
11. $a = -2$, $d(A, B) = 4$
12. $a = 3$, $d(A, B) = 7$ 13. $a = \frac{5}{2}$, $d(A, B) = \frac{3}{4}$
14. $a = 1.8$, $d(A, B) = 2.4$

Plot A, B, C, and D on a coordinate system. Name the quadrilateral ABCD—that is, is it a square, a rectangle, or what? See Example B.

15. $A(4, 3)$, $B(4, -3)$, $C(-4, 3)$, $D(-4, -3)$
16. $A(-1, 6)$, $B(0, 5)$, $C(3, 2)$, $D(5, 0)$
17. $A(1, 3)$, $B(2, 6)$, $C(4, 7)$, $D(3, 4)$
18. $A(0, 2)$, $B(3, 0)$, $C(2, -1)$, $D(-1, 1)$

In Problems 19–26, find $d(P_1, P_2)$ where P_1 and P_2 have the given coordinates. Also find the coordinates of the midpoint of the segment $P_1 P_2$ (Examples B and C).

19. $(2, -1)$, $(5, 3)$ 20. $(2, 1)$, $(7, 13)$
21. $(4, 2)$, $(2, 4)$ 22. $(-1, 5)$, $(6, 7)$
23. $(\sqrt{3}, 0)$, $(0, \sqrt{6})$ 24. $(\sqrt{2}, 0)$, $(0, -\sqrt{7})$
25. $(1.234, -5.132)$, $(6.714, 8.341)$
26. $(-42.1, 16.3)$, $(12.2, -5.3)$
27. The points A, B, C, and D of Problem 17 form a parallelogram.
    (a) Find $d(A, B)$, $d(B, C)$, $d(C, D)$, and $d(D, A)$.
    (b) Find the coordinates of the midpoints of the diagonals AC and BD of the parallelogram ABCD.
    (c) What facts about a parallelogram do your answers to parts (a) and (b) agree with?
28. The points $(3, -1)$ and $(3, 3)$ are two vertices of a square. Give two pairs of other possible vertices. Can you give a third pair?
29. Let ABCD be a rectangle whose sides are parallel to the coordinate axes. Find the coordinates of B and D if the coordinates of A and C are as given.
    (a) $(-2, 0)$ and $(4, 3)$ (Draw a picture.)
    (b) $(2, -1)$ and $(8, 7)$
30. Show that the triangle whose vertices are $(5, 3)$, $(-2, 4)$ and $(10, 8)$ is isosceles.
31. Use the distance formula to show that the triangle whose vertices are $(2, -4)$, $(4, 0)$, and $(8, -2)$ is a right triangle.
32. (a) Find the point on the y-axis that is equidistant from the points $(3, 1)$ and $(6, 4)$. Hint: Let the unknown point be $(0, y)$.

(b) Find the point on the x-axis that is equidistant from $(3, 1)$ and $(6, 4)$.
33. Use the distance formula to show that the three points are on a line.
    (a) $(0, 0)$, $(3, 4)$, $(-6, -8)$
    (b) $(-4, 1)$, $(-1, 5)$, $(5, 13)$

Use the point-of-division formula to find $P(x, y)$ for the given value of t and given points A and B. Plot the points A, P, and B. See Example D.

34. $A(5, -8)$, $B(11, 4)$, $t = \frac{1}{3}$
35. $A(5, -8)$, $B(11, 4)$, $t = \frac{2}{3}$
36. $A(5, -8)$, $B(11, 4)$, $t = \frac{5}{6}$
37. $A(4, 9)$, $B(104, 209)$, $t = \frac{13}{100}$

B. Applications and Extensions

38. Show that the triangle with vertices $(2, -2)$, $(-2, 2)$, and $(2\sqrt{3}, 2\sqrt{3})$ is equilateral.
39. Cities A, B, and C are located at $(0, 0)$, $(214, 17)$, and $(230, 179)$, respectively, with distances in miles. There are straight roads from A to B and from B to C, but only an air route goes directly from A to C. It costs \$3.71 per mile to ship a certain item by truck and \$4.81 per mile by air. Find the cheaper way to ship this item from A to C and calculate the savings made by choosing this form of shipment.
40. Find x such that the distance between the points $(3, 0)$ and $(x, 3)$ is 5.
41. Find point (x, y) in quadrant 1 that together with $(0, 0)$ and $(-3, 4)$ are vertices of an equilateral triangle.
42. A triangle has vertices $A(0, 0)$, $B(3, 0)$, and $C(0, 4)$. Find the length of the altitude from A to BC.
43. Show that the midpoint of the hypotenuse of a right triangle is equidistant from the three vertices. Hint: Place the triangle in the coordinate system so its vertices are $(0, 0)$, $(a, 0)$, and (a, b) with $a > 0$, $b > 0$.
44. Find the point (x, y) that is equidistant from $(1, 3)$, $(4, 2)$, and $(-3, 1)$.
45. Find the point (x, y) such that $(4, 5)$ is two-thirds of the way from $(2, 1)$ to (x, y) on the segment connecting these points.
46. Find the coordinates of all points P, having equal x- and y-coordinates, that are 5 units from $(5, -2)$.
47. Jane has two married daughters, Susan and Tammy. Susan lives 3 miles east and 5 miles north of Jane. Tammy lives 4 miles west and 7 miles north of Jane. How far does Susan live from Tammy (as the crow flies)? (See Example E.)

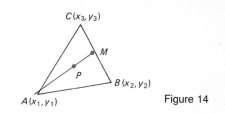

$C(x_3, y_3)$

M

P

$B(x_2, y_2)$

$A(x_1, y_1)$

Figure 14

48. Anne lives two blocks east and one block north of city center. Her sister Joan lives five blocks east and two blocks north of city center. They shop at the same supermarket on Central Avenue (the main north-south avenue through city center), which is equidistant from their homes. Where is the supermarket?

49. Prove the converse of the Pythagorean Theorem. If in a triangle, the sides of lengths a, b, and c satisfy the relation $a^2 + b^2 = c^2$, then the triangle is a right triangle.

50. Show that the midpoints of the sides of a triangle are the vertices of another triangle whose area is one-fourth that of the original triangle. *Hint:* You may suppose the vertices of the original triangle to be $(0, 0)$, $(0, a)$, and (b, c) with a, b, and c positive.

51. Given $A(-4, 3)$ and $B(21, 38)$, find the coordinates of the four points that divide AB into five equal parts.

52. Find the point on the segment joining $(1, 3)$ and $(6, 7)$ that is $\frac{11}{13}$ of the way from the first point to the second.

53. Consider the general triangle shown in Figure 14. The line segment from vertex A to the midpoint M of the opposite side is called a median.
   (a) Express the coordinates of the point P two-thirds of the way from A to M in terms of the coordinates of the three vertices of the triangle.
   (b) What do you conclude from part (a) about the three medians of a triangle?

54. A rhombus is a parallelogram with equal sides. Prove that the diagonals of a rhombus are perpendicular.

55. Let $ABCD$ be the vertices of a rectangle, labeled in cyclic order. Suppose that P is a point in the same plane as the rectangle and at distances a, b, c, and d from A, B, C, and D, respectively. Show that $a^2 + c^2 = b^2 + d^2$. *Hint:* Introduce a coordinate system in an intelligent way.

56. TEASER Arnold Thinkhard has been ordered to attach four guy wires to the top of a tall pole. These four guy wires are anchored to four stakes A, B, C, and D which form the vertices (in cyclic order) of a rectangle on the ground. After attaching wires of length 210, 60, and 180 feet from A, B, and C. Arnold discovers that the fourth wire from D is too short to reach. A new one must be cut, but how long should it be? Even though you don't know the dimensions of the rectangle or the height of the pole, it is your job to solve Arnold's problem. It is not enough to get the answer, you must demonstrate that your answer is correct before poor Arnold climbs back up the pole.

4-2 ALGEBRA AND GEOMETRY UNITED

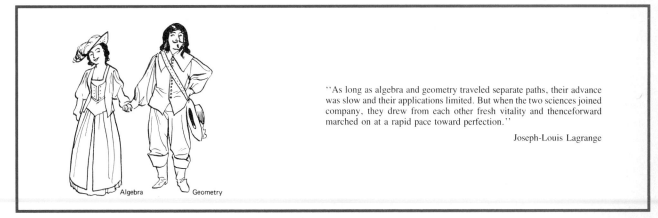

"As long as algebra and geometry traveled separate paths, their advance was slow and their applications limited. But when the two sciences joined company, they drew from each other fresh vitality and thenceforward marched on at a rapid pace toward perfection."

Joseph-Louis Lagrange

Algebra   Geometry

The Greeks were preeminent geometers but poor algebraists. Though they were able to solve a host of geometry problems, their limited algebraic skills kept others beyond their grasp. By 1600, geometry was a mature and eligible bachelor. Algebra was a young woman only recently come of age. Fermat and Descartes were the matchmakers; they brought the two together. The resulting union is called **analytic geometry,** or coordinate geometry.

The Graph of an Equation

An equation is an algebraic object. By means of a coordinate system, it can be transformed into a curve, a geometric object. Here is how it is done.

Consider the equation $y = x^2 - 3$. Its set of solutions is the set of ordered pairs (x, y) that satisfy the equation. These ordered pairs are the coordinates of points in the plane. The set of all such points is called the **graph** of the equation. The graph of an equation in two variables x and y will usually be a curve.

To obtain this graph, we follow a definite procedure.

1. Obtain the coordinates of a few points.
2. Plot those points in the plane.
3. Connect the points with a smooth curve in the order of increasing x values.

The best way to perform step 1 is to make a **table of values.** Assign values to one of the variables, say x, determine the corresponding values of the other, and then list the pairs of values in tabular form. The whole three-step procedure is illustrated in Figure 15 for the previously mentioned equation, $y = x^2 - 3$.

Of course, you need to use common sense and even a little faith. When you connect the points you have plotted with a smooth curve, you are assuming that the

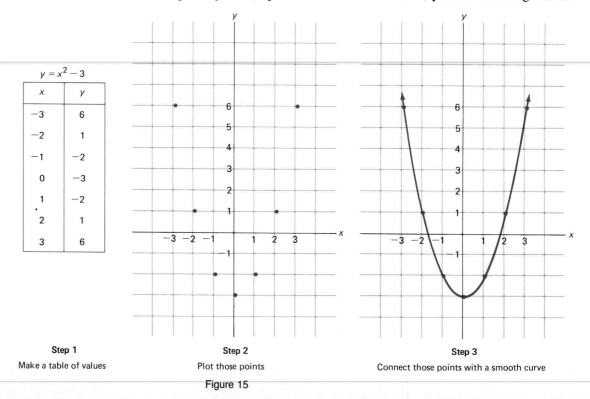

$y = x^2 - 3$

x	y
-3	6
-2	1
-1	-2
0	-3
1	-2
2	1
3	6

Step 1
Make a table of values

Step 2
Plot those points

Step 3
Connect those points with a smooth curve

Figure 15

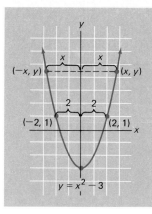

Figure 16

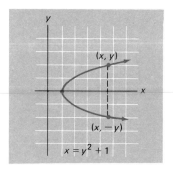

Figure 17

curve behaves nicely between consecutive points; that is faith. This is why you should plot enough points so the outline of the curve seems very clear; the more points you plot, the less faith you will need. Also you should recognize that you can seldom display the whole curve. In our example, the curve has infinitely long arms opening wider and wider. But our graph does show the essential features. That is what we aim to do—show enough of the graph so the essential features are visible.

Symmetry of a Graph

The graph of $y = x^2 - 3$, drawn earlier and again in Figure 16, has a nice property of symmetry. If the coordinate plane were folded along the y-axis the two branches would coincide. For example, $(3, 6)$ would coincide with $(-3, 6)$; $(2, 1)$ would coincide with $(-2, 1)$; and, more generally, (x, y) would coincide with $(-x, y)$. Algebraically, this corresponds to the fact that we may replace x by $-x$ in the equation $y = x^2 - 3$ without changing it. More precisely, $y = x^2 - 3$ and $y = (-x)^2 - 3$ are equivalent equations.

Whenever an equation is unchanged by replacing (x, y) with $(-x, y)$, the graph of the equation is **symmetric with respect to the y-axis.** Likewise, if the equation is unchanged when (x, y) is replaced by $(x, -y)$, its graph is **symmetric with respect to the x-axis.** The equation $x = 1 + y^2$ is of the latter type; its graph is shown in Figure 17.

A third type of symmetry is **symmetry with respect to the origin.** It occurs whenever replacing (x, y) by $(-x, -y)$ produces an equivalent equation. The equation $y = x^3$ is a good example as $-y = (-x)^3$ is equivalent to $y = x^3$. The graph is shown in Figure 18. Note that the dotted line segment from $(-x, -y)$ to (x, y) is bisected by the origin.

$y = x^3$

x	y
-3	-27
-2	-8
-1	-1
0	0
1	1
2	8
3	27

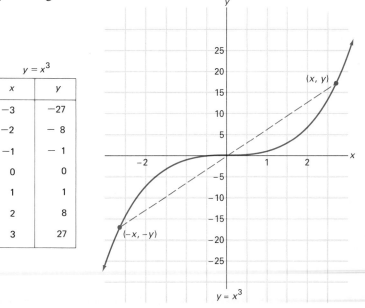

Symmetry with respect to the origin

Figure 18

In graphing $y = x^3$, we used a smaller scale on the y-axis than on the x-axis. This made it possible to show a larger portion of the graph. We suggest that before putting scales on the two axes, you should examine your table of values. Choose scales so that all of your points can be plotted and still keep your graph of reasonable size.

Example A (Graphing $y = \cdots$)   Sketch the graph of $y = x^3 - 4x$.

Solution   To obtain a table of values, we assign numbers to x and calculate the corresponding y. But because $-y = (-x)^3 - 4(-x)$ is equivalent to $y = x^3 - 4x$, it is sufficient to consider only positive values of x. After we have sketched the right half of the graph, we can obtain the left half by reflecting in the origin. The appropriate table and graph are shown in Figure 19.

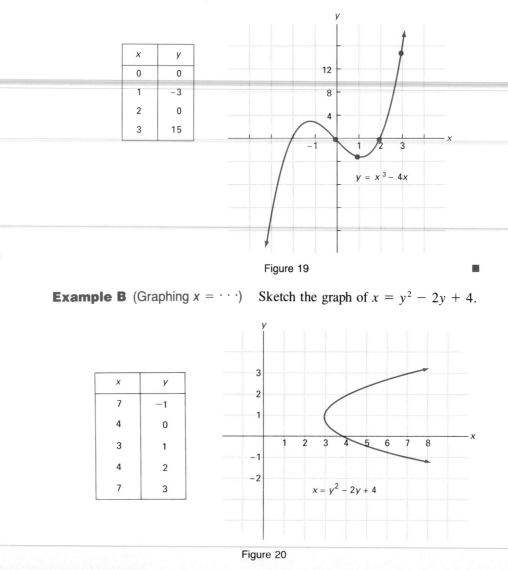

x	y
0	0
1	-3
2	0
3	15

$$y = x^3 - 4x$$

Figure 19

Example B (Graphing $x = \cdots$)   Sketch the graph of $x = y^2 - 2y + 4$.

x	y
7	-1
4	0
3	1
4	2
7	3

$$x = y^2 - 2y + 4$$

Figure 20

Solution This graph does not have any of the symmetries we have mentioned. To get a table of values, we assign values (both positive and negative) to y and calculate x. Such a table and the corresponding graph are shown in Figure 20. ∎

Graphing an equation is an extremely important operation. It gives us a picture to look at. Most of us can absorb qualitative information from a picture much more easily than from symbols. But if we want precise quantitative information, then symbols are better; they are easier to manipulate. That is why we must be able to reverse the process just described, which is our next topic.

The Equation of a Graph

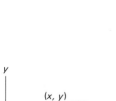

Figure 21

A **graph** is a geometric object, a picture. How can we turn it into an algebraic object, an equation? Sometimes it is easy, but not always. As an example, consider a circle of radius 3. It consists of all points 3 units from a fixed point, called the center. Figure 21 shows this circle with its center at the origin of a coordinate system.

Take *any* point P on the circle and label its coordinates (x, y). It must satisfy the equation

$$d(P, O) = 3$$

From the distance formula of the previous section, we have

$$\sqrt{(x - 0)^2 + (y - 0)^2} = 3$$

or equivalently (after squaring both sides)

$$x^2 + y^2 = 9$$

This is the equation we sought.

We could move to other types of curves, attempting to find their equations. In fact, we will do exactly that in later sections. Right now, we shall consider more general circles, that is, circles with arbitrary radii and arbitrary centers.

The Standard Equation of a Circle

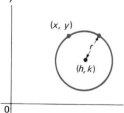

Figure 22

Consider a circle of radius r with center at (h, k), as in Figure 22. To find its equation, take an arbitrary point on the circle with coordinates (x, y). According to the distance formula, it must satisfy the equation

$$\sqrt{(x - h)^2 + (y - k)^2} = r$$

or, equivalently,

$$(x - h)^2 + (y - k)^2 = r^2$$

Every circle has an equation of this form; we call it the **standard equation of a circle.**

Example C (Radius and Center Given) Find the equation of a circle of radius 6 and center $(2, -1)$.

Solution Use the boxed equation to obtain

$$(x - 2)^2 + (y + 1)^2 = 36 \qquad \blacksquare$$

It is just as easy to go in the opposite direction.

Example D (Equation Given in Standard Form) Find the radius and center of the circle with equation $(x + 3)^2 + (y - 4)^2 = 32$.

Solution Comparing the given equation with the boxed equation, we see that the radius is $\sqrt{32} = 4\sqrt{2}$ and the center is $(-3, 4)$. $\blacksquare$

Consider this last equation again. Notice that it could be written as

$$x^2 + 6x + 9 + y^2 - 8y + 16 = 32$$

or equivalently as

$$x^2 + y^2 + 6x - 8y = 7$$

A natural question to ask is whether every equation of the form

$$x^2 + y^2 + Dx + Ey = F$$

is the equation of a circle. The answer is yes, in general.

Example E (Equation Given in Nonstandard Form) Find the center and radius of the circle with equation $x^2 + y^2 - 6y + 16x = 8$.

Solution Recalling a skill we learned in Section 3-4 (completing the square), we may rewrite this as

$$(x^2 + 16x + \quad) + (y^2 - 6y + \quad) = 8$$

or

$$(x^2 + 16x + 64) + (y^2 - 6y + 9) = 8 + 64 + 9$$

or

$$(x + 8)^2 + (y - 3)^2 = 81$$

We recognize this to be the equation of a circle with center at $(-8, 3)$ and radius 9. $\blacksquare$

The complete story is as follows. The equation $x^2 + y^2 + Dx + Ey = F$ can always be transformed to an equation of the form $(x - h)^2 + (y - k)^2 = C$. If $C > 0$, this is the equation of a circle of radius $\sqrt{C}$; if $C = 0$, it is the equation of the point (h, k); if $C < 0$, the equation has an empty graph.

PROBLEM SET 4-2

A. Skills and Techniques

Sketch the graph of each of the following equations, showing enough of the graph to bring out its essential features. Begin by noting any of the three kinds of symmetry discussed in the text. See Example A.

1. $y = 3x - 2$ **2.** $y = 2x + 1$
3. $y = -x^2 + 4$ **4.** $y = -x^2 - 2x$
5. $y = x^2 - 4x$ **6.** $y = x^3 + 2$

7. $y = -x^3$ **8.** $y = \dfrac{12}{x^2 + 4}$

9. $y = \dfrac{4}{x^2 + 1}$ **10.** $y = x^3 + x$

Sketch the graph of each of the following equations (Example B).

11. $x = 2y - 1$ **12.** $x = -3y + 1$
13. $x = -2y^2$ **14.** $x = 2y - y^2$
15. $x = y^3$ **16.** $x = 8 - y^3$

Write the equation of the circle with the given center and radius (Example C).

17. Center $(0, 0)$, radius 6
18. Center $(2, 3)$, radius 3
19. Center $(4, 1)$, radius 5
20. Center $(2, -1)$, radius $\sqrt{7}$
21. Center $(-2, 1)$, radius $\sqrt{3}$
22. Center $(\pi, \frac{3}{4})$, radius $\frac{1}{2}$

Graph the following equations. See Example D.

23. $(x - 2)^2 + y^2 = 16$
24. $(x - 2)^2 + (y + 2)^2 = 25$
25. $(x + 1)^2 + (y - 3)^2 = 64$
26. $(x + 4)^2 + (y + 6)^2 = \frac{49}{4}$

Find the center and radius of each of the following circles (Example E).

27. $x^2 + y^2 + 2x - 10y + 25 = 0$
28. $x^2 + y^2 - 6y = 16$
29. $x^2 + y^2 - 12x + 35 = 0$
30. $x^2 + y^2 - 10x + 10y = 0$

31. $4x^2 + 4y^2 + 4x - 12y + 1 = 0$
32. $3x^2 + 3y^2 - 2x + 4y = \frac{20}{3}$

B. Applications and Extensions

Sketch the graph of the equations in Problems 33–38. Begin by deciding if the graph has any of the three kinds of symmetry discussed in the text.

33. $y = 12/x$ **34.** $y = 3 + 4/x^2$
35. $y = 2(x - 1)^2$ **36.** $x = -y^2 + 8$
37. $x = 4y - y^2$ **38.** $x^2 + y^2 - 2y = 8$

39. Find the equation of the circle with center $(5, -7)$ that is tangent to the x-axis.

40. Find the equation of the circle with center on the line $y = x$ that is tangent to the y-axis at $(0, 5)$.

41. Write the equation of the circle with center $(-3, 2)$ that passes through $(4, 3)$.

42. Find the equation of the circle that has AB as diameter given that $A = (1, 2)$ and $B = (5, 12)$.

43. Identify each of the following as a circle (giving its radius and center), or as a point (giving its coordinates), or as the empty set.
(a) $x^2 + y^2 - 4x + 6y = -13$
(b) $2x^2 + 2y^2 - 2x + 6y = 3$
(c) $4x^2 + 4y^2 - 8x - 4y = -7$
(d) $\sqrt{3}x^2 + \sqrt{3}y^2 - 6y = 2\sqrt{3}$

44. Find the equation of the circle of radius 5 with center in the first quadrant that passes through $(3. 0)$ and $(0, 1)$.

45. Show that the point $(7, \frac{3}{2})$ is outside the circle $x^2 + y^2 - 6x + 4y - 12 = 0$.

46. Find the length of the tangent segment T (Figure 23) from the point $(8, 2)$ to the circle $x^2 + y^2 = 16$.

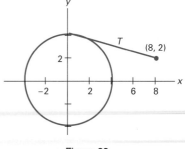

Figure 23

47. The four points $(1, 0)$, $(8, 1)$, $(7, 8)$, and $(0, 7)$ are vertices of a square. Find the equations of (a) the circle circumscribed about this square and (b) the circle inscribed in this square.

48. A belt fits tightly around the two circles (pulleys) with equations $x^2 - 2x + y^2 + 4y = 11$ and $x^2 - 14x + y^2 - 12y = -69$. Find the length of this belt.

49. A belt fits tightly around the two circles with equations $x^2 + y^2 = 1$ and $(x - 6)^2 + y^2 = 16$. Find the length of this belt.

50. TEASER Circles of radius 2 are centered at the three vertices of a triangle with sides of lengths 8, 11, and 12. Find the length of a belt that fits tightly

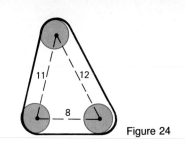
Figure 24

around these three circles as shown in Figure 24. Next, solve the same problem for a quadrilateral with sides of lengths 8, 9, 10, and 12. Finally, discover the very general result of which the examples above are special cases.

4-3 THE STRAIGHT LINE

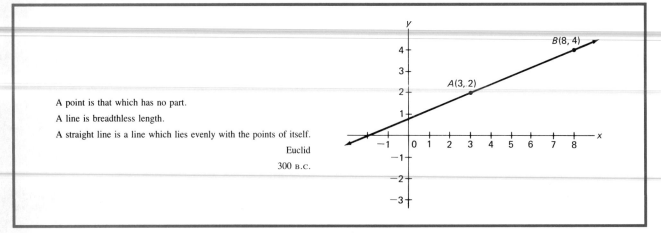

A point is that which has no part.

A line is breadthless length.

A straight line is a line which lies evenly with the points of itself.

Euclid

300 B.C.

Euclid's definition of a straight line is not very helpful, but neither are most of the alternatives we have heard. Fortunately, we all know what we mean by a straight line even if we cannot seem to describe it in terms of more primitive ideas. There is one thing on which we must agree: Given two points (for example, A and B above), there is one and only one straight line that passes through them. And contrary to Euclid, let us agree that the word *line* shall always mean straight line.

A line is a geometric object. When it is placed in a coordinate system, it ought to have an equation just as a circle does. How do we find the equation of a line? To answer this question we will need the notion of slope.

The Slope of a Line

Consider the line in our opening diagram. From point A to point B, there is a **rise** (vertical change) of 2 units and a **run** (horizontal change) of 5 units. We say that the line has a slope of $\frac{2}{5}$. In general (Figure 25), for a line through $A(x_1, y_1)$ and $B(x_2, y_2)$, where $x_1 \neq x_2$, we define the **slope** m of that line by

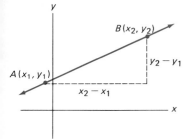

Figure 25

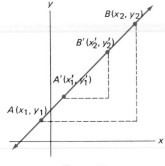

Figure 26

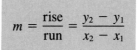

$$m = \frac{\text{rise}}{\text{run}} = \frac{y_2 - y_1}{x_2 - x_1}$$

You should immediately raise a question. A line has many points. Does the value we get for the slope depend on which pair of points we use for A and B? The similar triangles in Figure 26 show us that

$$\frac{y_2' - y_1'}{x_2' - x_1'} = \frac{y_2 - y_1}{x_2 - x_1}$$

Thus points A' and B' would do just as well as A and B. It does not even matter whether A is to the left or right of B since

$$\frac{y_1 - y_2}{x_1 - x_2} = \frac{y_2 - y_1}{x_2 - x_1}$$

All that matters is that we subtract the coordinates in the same order in numerator and denominator.

The slope m is a measure of the steepness of a line, as Figure 27 illustrates. Notice that a horizontal line has zero slope and a line that rises to the right has positive slope. The larger this positive slope is, the more steeply the line rises. A line that falls to the right has negative slope. The concept of slope for a vertical line makes no sense since it would involve division by zero. Therefore the notion of slope for a vertical line is left undefined.

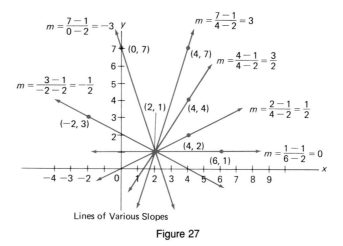

Lines of Various Slopes

Figure 27

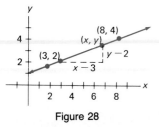

Figure 28

The Point-Slope Form

Consider again the line of our opening diagram: it is reproduced in Figure 28. We know that

1. it passes through $(3, 2)$.
2. it has slope $\frac{2}{5}$.

Take any other point on that line, such as one with coordinates (x, y). If we use this point together with $(3, 2)$ to measure slope, we must get $\frac{2}{5}$; that is,

$$\frac{y - 2}{x - 3} = \frac{2}{5}$$

or, after multiplying by $x - 3$,

$$y - 2 = \frac{2}{5}(x - 3)$$

Notice that this last equation is satisfied by all points on the line, even by $(3, 2)$. Moreover, no points not on the line can satisfy this equation.

What we have just done in an example can be done in general. The line passing through the (fixed) point (x_1, y_1) with slope m has equation

$$y - y_1 = m(x - x_1)$$

We call it the **point-slope** form for the equation of a line.

Consider once more the line of our example. That line passes through $(8, 4)$ as well as $(3, 2)$. If we use $(8, 4)$ as (x_1, y_1), we get the equation

$$y - 4 = \frac{2}{5}(x - 8)$$

which looks quite different from

$$y - 2 = \frac{2}{5}(x - 3)$$

However, both can be simplified to $2x - 5y + 4 = 0$; they are equivalent.

Example A (A Line Through Two Points)   Find the equation of the line through the two points $(5, 2)$ and $(-2, 5)$.

Solution   The slope of the line is $m = (5 - 2)/(-2 - 5) = -\frac{3}{7}$. Thus the equation of the line is $y - 2 = -\frac{3}{7}(x - 5)$. This equation can be written in many other ways. For example, multiplying by 7 and simplifying gives $3x + 7y - 29 = 0$. ∎

The Slope-Intercept Form

The equation of a line can be expressed in various forms. Suppose that we are given the slope m for a line and the y-intercept b [that is, the line intersects the y-axis at $(0, b)$ as in Figure 29]. Choosing $(0, b)$ as (x_1, y_1) and applying the point-slope form, we get

$$y - b = m(x - 0)$$

which we can rewrite as

$$y = mx + b$$

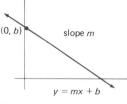

Figure 29

The latter is called the **slope-intercept** form for the equation of a line.

Why get excited about that, you ask? Because any time we see an equation written this way, we recognize it as the equation of a line and can immediately read its slope and y-intercept. For example, consider the equation

$$3x - 2y + 4 = 0$$

If we solve for y, we get

$$y = \frac{3}{2}x + 2$$

It is the equation of a line with slope $\frac{3}{2}$ and y-intercept 2.

Example B (Using the Slope-Intercept Form) Find the equation of the line with y-intercept -3 which has the same slope as the line $3x + 5y = 7$. Draw the graph of this line.

Solution   Solving $3x + 5y = 7$ for y gives $y = -\frac{3}{5}x + \frac{7}{5}$. Thus $m = -\frac{3}{5}$. The required equation is $y = -\frac{3}{5}x - 3$ which can also be written as $5y = -3x - 15$ or as $3x + 5y + 15 = 0$. The graph is shown in Figure 30.

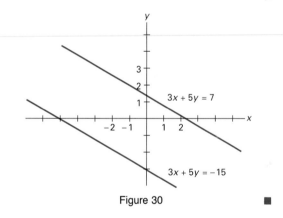

Figure 30

Equations of Vertical and Horizontal Lines

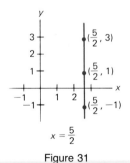

Figure 31

Vertical lines do not fit within the discussion above; their slopes are undefined. But they do have equations, very simple ones. The line in Figure 31 has equation $x = \frac{5}{2}$, since every point on the line satisfies this equation. The equation of any vertical line can be put in the following form.

$$\text{Vertical line:} \qquad x = k$$

where k is a constant. It should be noted that the equation of a horizontal line can be written in the form $y = 0x + b$, that is, in the form $y = k$.

$$\text{Horizontal line:} \qquad y = k$$

The Form $Ax + By + C = 0$

It would be nice to have a form that covered all lines including vertical lines. Consider for example,

(i) $$y - 2 = -4(x + 2)$$

(ii) $$y = 5x - 3$$

(iii) $$x = 5$$

These can be rewritten (by taking everything to the left side) as follows:

(i) $$4x + y + 6 = 0$$

(ii) $$-5x + y + 3 = 0$$

(iii) $$x + 0y - 5 = 0$$

All are of the form

$$Ax + By + C = 0$$

which we call the **general linear equation.** It takes only a moment's thought to see that the equation of any line can be put in this form. Conversely, the graph of $Ax + By + C = 0$ is always a line (if A and B are not both zero; see Problem 54).

Parallel Lines

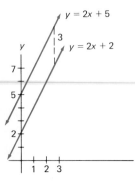

Figure 32

If two lines have the same slope, they are parallel. Thus $y = 2x + 2$ and $y = 2x + 5$ represent parallel lines; both have a slope of 2. The second line is 3 units above the first for every value of x (Figure 32).

Similarly, the lines with equations $-2x + 3y + 12 = 0$ and $4x - 6y = 5$ are parallel. To see this, solve these equations for y (that is, find the slope-intercept form); you get $y = \frac{2}{3}x - 4$ and $y = \frac{2}{3}x - \frac{5}{6}$, respectively. Both have slope $\frac{2}{3}$; they are parallel.

We may summarize by stating that *two nonvertical lines are parallel if and only if they have the same slope.*

Perpendicular Lines

Figure 33

Is there a simple slope condition which characterizes perpendicular lines? Yes; *two nonvertical lines are perpendicular if and only if their slopes are negative reciprocals of each other*. We are not going to prove this, but an example will help explain why it is true. The slopes of the lines $y = \frac{3}{4}x$ and $y = -\frac{4}{3}x$ are negative reciprocals of each other. Both lines pass through the origin. The points $(4, 3)$ and $(3, -4)$ are on the first and second lines, respectively. The two right triangles shown in Figure 33 are congruent with $\angle \alpha = \angle \delta$ and $\angle \beta = \angle \gamma$. But

$$\angle \alpha + \angle \beta = 90°$$

and therefore

$$\angle \alpha + \angle \gamma = 90°$$

That says the two lines are perpendicular to each other.

The lines $2x - 3y = 5$ and $3x + 2y = -4$ are also perpendicular, since —after solving them for y—we see that the first has slope $\frac{2}{3}$ and the second has slope $-\frac{3}{2}$.

Example C (A Line Perpendicular to Another)   Find the equation of the line through $(4, -2)$ and perpendicular to the line $3x - 4y = 2$.

Solution   Solving $3x - 4y = 2$ for y gives $y = \frac{3}{4}x - \frac{1}{2}$, which represents a line with slope $\frac{3}{4}$. The required line has slope $-\frac{4}{3}$. Its equation (point-slope form) is $y + 2 = -\frac{4}{3}(x - 4)$ or, in general linear form, $4x + 3y - 10 = 0$.   ■

Intersecting Lines

If two lines are not parallel, they intersect in a unique point P whose coordinates (x_1, y_1) must satisfy the equations of both lines.

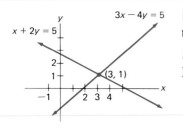

Figure 34

Example D (Intersection of Two Lines)   Find the coordinates of the point of intersection of the lines $3x - 4y = 5$ and $x + 2y = 5$ (Figure 34).

Solution   We simply solve the two equations simultaneously (see Section 3-3). Multiply the second equation by -3 and then add the two equations.

$$\begin{array}{r} 3x - 4y = 5 \\ -3x - 6y = -15 \\ \hline -10y = -10 \end{array}$$

$$y = 1$$

$$x = 3$$

The intersection point has coordinates $(3, 1)$.   ■

Distance from a Point to a Line

The ordinary distance formula (Section 4-1) allows us to calculate the distance from one point to another. Our next formula gives us a simple way of calculating the distance from a point to a line. The distance d from the point (x_1, y_1) to the line $Ax + By + C = 0$ is given by the formula

$$d = \frac{|Ax_1 + By_1 + C|}{\sqrt{A^2 + B^2}}$$

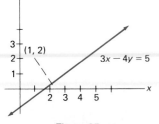

Figure 35

Note that to use this formula, the equation of the line must be in general linear form. The proof of the formula is somewhat lengthy, so we choose to omit it.

Example E (Point-Line Distance)   Find the distance from the point $(1, 2)$ to the line $3x - 4y = 5$ (Figure 35).

Solution   First write the equation as $3x - 4y - 5 = 0$. The formula gives

$$d = \frac{|3 \cdot 1 - 4 \cdot 2 - 5|}{\sqrt{3^2 + (-4)^2}} = \frac{|-10|}{5} = 2 \quad ■$$

PROBLEM SET 4-3

A. Skill and Techniques

Find the slope of the line containing the given two points.

1. (2, 3) and (4, 8)
2. (4, 1) and (8, 2)
3. (−4, 2) and (3, 0)
4. (2, −4) and (0, −6)
5. (3, 0) and (0, 5)
6. (−6, 0) and (0, 6)
7. (−1.732, 5.014) and (4.315, 6.175)
8. $(\pi, \sqrt{3})$ and $(1.642, \sqrt{2})$

Draw the graph and find an equation for each of the following lines. Then write your answer in the form $Ax + By + C = 0$. See Examples A and B.

9. Through (2, 3) with slope 4
10. Through (4, 2) with slope 3
11. Through (3, −4) with slope −2
12. Through (−5, 2) with slope −1
13. With y-intercept 4 and slope −2
14. With y-intercept −3 and slope 1
15. With y-intercept 5 and slope 0
16. With y-intercept 1 and slope −1
17. Through (2, 3) and (4, 8)
18. Through (4, 1) and (8, 2)
19. Through (3, 0) and (0, 5)
20. Through (−6, 0) and (0, 6)
21. Through $(\sqrt{3}, \sqrt{7})$ and $(\sqrt{2}, \pi)$
22. Through $(\pi, \sqrt{3})$ and $(\pi + 1, 2\sqrt{3})$
23. Through (2, −3) and (2, 5)
24. Through (−5, 0) and (−5, 4)

In Problems 25–32, find the slope and y-intercept of each line (Example B).

25. $y = 3x + 5$
26. $y = 6x + 2$
27. $3y = 2x - 4$
28. $2y = 5x + 2$
29. $2x + 3y = 6$
30. $4x + 5y = -20$
31. $y + 2 = -4(x - 1)$
32. $y - 3 = 5(x + 2)$

33. Study Example C and then write the equation of the line through (3, −3)
   (a) parallel to the line $y = 2x + 5$.
   (b) perpendicular to the line $y = 2x + 5$.
   (c) parallel to the line $2x + 3y = 6$.
   (d) perpendicular to the line $2x + 3y = 6$.
   (e) parallel to the line through (−1, 2) and (3, −1).
   (f) parallel to the line $x = 8$.
   (g) perpendicular to the line $x = 8$.

34. Find the value of k for which the line $4x + ky = 5$
   (a) passes through the point (2, 1).
   (b) is parallel to the y-axis.
   (c) is parallel to the line $6x - 9y = 10$.
   (d) has equal x- and y-intercepts.
   (e) is perpendicular to the line $y - 2 = 2(x + 1)$.

35. Write the equation of the line through (0, −4) that is perpendicular to the line $y + 2 = -\frac{1}{2}(x - 1)$.

36. Find the value of k such that the line $kx - 3y = 10$
   (a) is parallel to the line $y = 2x + 4$.
   (b) is perpendicular to the line $y = 2x + 4$.
   (c) is perpendicular to the line $2x + 3y = 6$.

Find the coordinates of the point of intersection in each problem below as in Example D. Then write the equation of the line through that point perpendicular to the line given first.

37. $2x + 3y = 4$
    $-3x + y = 5$

38. $4x - 5y = 8$
    $2x + y = -10$

39. $3x - 4y = 5$
    $2x + 3y = 9$

40. $5x - 2y = 5$
    $2x + 3y = 6$

In each case, find the distance from the given point to the given line (Example E).

41. (−3, 2), $3x + 4y = 6$
42. (4, −1), $2x - 2y + 4 = 0$
43. (−2, −1), $5y = 12x + 1$
44. (3, −1), $y = 2x - 5$

Find the (perpendicular) distance between the given parallel lines. Hint: First find a point on one of the lines.

45. $3x + 4y = 6$, $3x + 4y = 12$
46. $5x + 12y = 2$, $5x + 12y = 7$

B. Applications and Extensions

47. Which pairs of lines are parallel, which are perpendicular, and which are neither?
   (a) $y = 3x - 2$, $6x - 2y = 0$

(b) $x = -2$, $y = 4$
(c) $x = 2(y - 2)$, $y = -\frac{1}{2}(x - 1)$
(d) $2x + 5y = 3$, $10x - 4y = 7$

48. Find the equation of the line through the point $(2, -1)$ that
(a) passes through $(-3, 5)$.
(b) is parallel to $2x - 3y = 5$.
(c) is perpendicular to $x + 2y = 3$.
(d) is perpendicular to the y-axis.

49. Find the equation of the line passing through the intersection of the lines $x + 2y = 1$ and $3x + 2y = 5$ that is parallel to the line $3x - 2y = 4$.

50. A line L is perpendicular to the line $2x + 3y = 6$ and passes through $(-3, 1)$. Where does L cut the y-axis?

51. Find the equation of the perpendicular bisector of the line segment connecting $(3, -2)$ and $(7, 6)$.

52. The center of the circle that circumscribes a triangle is at the intersection of the perpendicular bisectors of the sides (Figure 36). Use this fact to find the center of the circle that goes through the three points $(0, 8)$, $(6, 2)$, and $(12, 14)$.

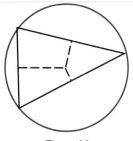

Figure 36

53. Show that the equation of the line with x-intercept a and y-intercept b (both a and b nonzero) can be written in the form

$$\frac{x}{a} + \frac{y}{b} = 1$$

54. Show that the graph of $Ax + Bx + C = 0$ is always a line (provided A and B are not both 0). *Hint:* Consider two cases: (1) $B = 0$ and (2) $B \neq 0$.

55. A line passes through $(3, 2)$ and its nonzero y-intercept is twice its x-intercept. Find the equation of the line.

56. For each k, the equation $2x - y + 4 + k(x + 3y - 6) = 0$ represents a line (why?). One value of k determines a line with slope $\frac{3}{4}$. Where does this line cut the y-axis?

57. The ABC Company makes zeebos, which it sells for $20 apiece. The material and labor to make a zeebo cost $16 and the company has fixed yearly costs (utilities, real estate taxes, and so on) of $8500. Write an expression for the company's profit P in a year in which it makes and sells x zeebos. What is its profit in a year in which it makes only 2000 zeebos?

58. A piece of equipment purchased today for $80,000 will depreciate linearly to a scrap value of $2000 after 20 years. Write a formula for its value V after t years.

59. A house purchased on January 1, 1970 for $60,000 appreciated linearly to a value of $112,000 on January 1, 1990.
(a) Write a formula for V, its value t years after purchase.
(b) When was the house worth $69,100?

60. In 1968, Heidi McGoff invested $5000 in a savings account which yielded 5 percent simple interest for 4 years and 6 percent simple interest after that. Write a formula for A, the accumulated amount after t years if
(a) $0 < t < 4$
(b) $t \geq 4$

61. Find the distance between the parallel lines $12x - 5y = 2$ and $12x - 5y = 7$.

62. Find the formula for the distance between the lines $y = mx + b$ and $y = mx + B$ in terms of m, b, and B.

63. Show that the line through the midpoints of two sides of a triangle is parallel to the third side.

64. Let the three vertices of a triangle lie on a circle with two of them being the ends of a diameter. Show that the triangle is a right triangle. *Hint:* Place the triangle in the coordinate system so two of its vertices are $(-a, 0)$ and $(a, 0)$.

65. Find two points in the plane that are simultaneously equidistant from the lines $y = \pm x$ and the point $(5, 3)$.

66. Using the same axes, sketch the graphs of $x^2 + y^2 = 1$ and $|x| + |y| = 1$ and compute the area of the region between the two graphs.

67. A line through $(4, 4)$ is tangent to the circle $x^2 + y^2 = 4$ at a point P in the fourth quadrant. Find the coordinates of P.

68. TEASER Suppose that 2 million points in the plane are given. Show that there is a line with exactly 1 million of these points on each side of it. Can you always find a circle with exactly 1 million of these points inside it?

4-4 THE PARABOLA

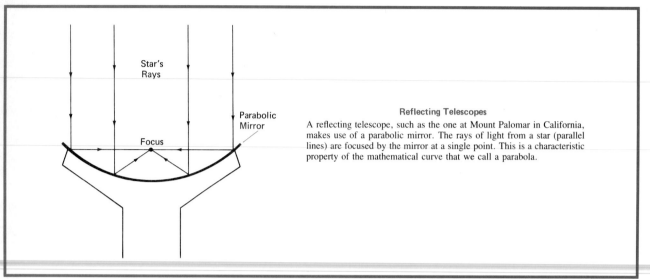

Star's Rays

Parabolic Mirror

Focus

Reflecting Telescopes

A reflecting telescope, such as the one at Mount Palomar in California, makes use of a parabolic mirror. The rays of light from a star (parallel lines) are focused by the mirror at a single point. This is a characteristic property of the mathematical curve that we call a parabola.

We have seen that the graph $y = ax + b$ is always a line. Now we want to study the graph of $y = ax^2 + bx + c$ ($a \neq 0$). As we shall discover, this graph is always a smooth, cup-shaped curve something like the cross section of the mirror shown in the opening diagram. We call it a **parabola.**

The Graph of $y = x^2$

The simplest case of all is $y = x^2$. The graph of this equation is shown in Figure 37. Two important features should be noted.

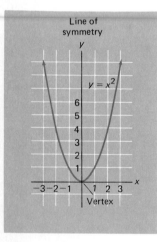

Line of symmetry

$y = x^2$

Vertex

Figure 37

1. The curve is symmetric about the y-axis. This follows from the fact that the equation $y = x^2$ is not changed if we replace (x, y) by $(-x, y)$.
2. The curve reaches its lowest point at $(0, 0)$, the point where the curve intersects the line of symmetry. We call this point the **vertex** of the parabola.

Next we consider how the graph of $y = x^2$ is modified as we look successively at $y = ax^2$, $y = x^2 + k$, $y = (x - h)^2$, and $y = a(x - h)^2 + k$.

The Graph of $y = ax^2$

In Figure 38, we show the graphs of $y = x^2$, $y = 3x^2$, $y = \frac{1}{2}x^2$, and $y = -2x^2$. They suggest the following general facts. The graph of $y = ax^2$, $a \neq 0$, is a parabola with vertex at the origin, opening upward if $a > 0$ and downward if $a < 0$. Increasing $|a|$ makes the graph narrower.

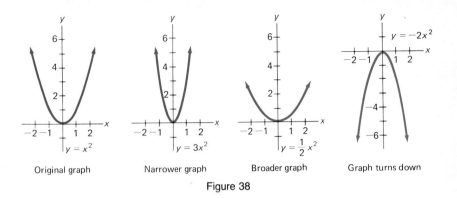

Original graph   Narrower graph   Broader graph   Graph turns down

Figure 38

The Graphs of $y = x^2 + k$ and $y = (x - h)^2$

The graphs of $y = x^2 + 4$, $y = x^2 - 6$, $y = (x - 2)^2$, and $y = (x + 1)^2$ can all be obtained by shifting (translating) the graph of $y = x^2$, while maintaining its shape. They are shown in Figure 39. After studying these graphs carefully, you will understand the general situation we now describe.

The graph of $y = x^2 + k$ is obtained by shifting the graph of $y = x^2$ vertically $|k|$ units, upward if $k > 0$ and downward if $k < 0$. The vertex is at $(0, k)$.

The graph of $y = (x - h)^2$ is obtained by a horizontal shift of $|h|$ units, to the right if $h > 0$ and to the left if $h < 0$. The vertex is at $(h, 0)$.

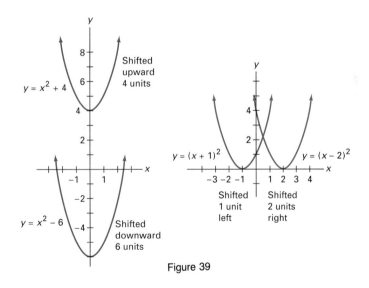

Figure 39

The Graph of $y = a(x - h)^2 + k$ or $y - k = a(x - h)^2$

We can get the graph of $y = 2(x - 3)^2 - 4$ by shifting the graph of $y = 2x^2$ 3 units to the right and 4 units down. This puts the vertex at $(3, -4)$. More gener-

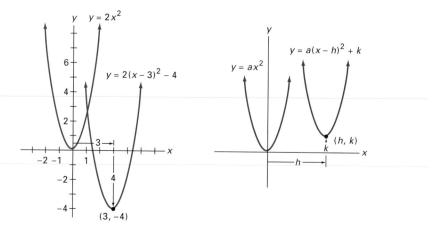

Figure 40

ally, the graph of $y = a(x - h)^2 + k$ is the graph of $y = ax^2$ shifted horizontally $|h|$ units and vertically $|k|$ units, so that the vertex is at (h, k). The graphs in Figure 40 illustrate these facts.

Example A (A Shift of $y = -2x^2$)   Sketch the graph of

$$y = -2(x - 1)^2 + 6$$

Solution   The graph of $y = -2x^2$ was shown in Figure 38. The graph we want has the same shape but is shifted 1 unit right and 6 units up. We show both graphs in Figure 41.   ∎

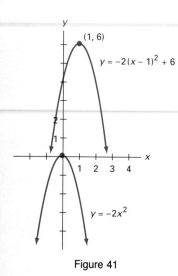

Figure 41

Note the position of the vertex in Figure 41 at $(1, 6)$. Writing the equation $y = -2(x - 1)^2 + 6$ in the equivalent form $y - 6 = -2(x - 1)^2$ allows us to locate the vertex at $(1, 6)$ immediately, as we now explain. Recall that for the circle $x^2 + y^2 = r^2$, replacing x by $x - h$ and y by $y - k$ to obtain $(x - h)^2 + (y - k)^2 = r^2$ had the effect of shifting the center from $(0, 0)$ to (h, k). A similar thing happens to the parabola $y = ax^2$. Replacing x by $x - h$ and y by $y - k$ changes $y = ax^2$ to $y - k = a(x - h)^2$ [which you will note is equivalent to $y = a(x - h)^2 + k$] and correspondingly shifts the vertex from $(0, 0)$ to (h, k).

The Graph of $y = ax^2 + bx + c$

The most general equation considered so far is $y = a(x - h)^2 + k$. If we expand $a(x - h)^2$ and collect terms on the right side, the equation takes the form $y = ax^2 + bx + c$. Conversely, $y = ax^2 + bx + c$ $(a \neq 0)$ always represents a parabola with a vertical line of symmetry, as we shall now show. We use the method of completing the square to rewrite $y = ax^2 + bx + c$ as follows.

$$y = a\left(x^2 + \frac{b}{a}x + \right) + c$$

$$= a\left[x^2 + \frac{b}{a}x + \left(\frac{b}{2a}\right)^2\right] + c - a\left(\frac{b}{2a}\right)^2$$

$$= a\left(x + \frac{b}{2a}\right)^2 + \left(c - \frac{b^2}{4a}\right)$$

This is the equation of a parabola with vertex at $(-b/2a, \; c - b^2/4a)$ and line of symmetry $x = -b/2a$.

What we have just said leads to an important conclusion. The graph of the equation $y = ax^2 + bx + c \; (a \neq 0)$ is always a parabola with vertical axis of symmetry. Moreover, we have the following facts.

The Parabola $y = ax^2 + bx + c$

1. The x-coordinate of the vertex is $-b/2a$; the y-coordinate is found by substitution in the equation.
2. The parabola turns upward if $a > 0$ and downward if $a < 0$. It is a fat or thin parabola, according as $|a|$ is small or large.

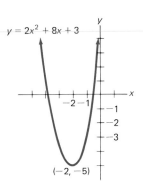

$y = 2x^2 + 8x + 3$

$(-2, -5)$

Figure 42

Example B (Graphing $y = ax^2 + bx + c$)   Sketch the graph of

$$y = 2x^2 + 8x + 3$$

Solution   We know that the graph is a parabola with vertex at $x = -b/2a = -8/4 = -2$. The y-coordinate of the vertex, obtained by substituting $x = -2$ in the equation, is $y = -5$. The parabola turns up and is rather thin (like $y = 2x^2$); it is shown in Figure 42. Note that $x = -2$ is the axis of symmetry.   ■

Horizontal Parabolas

The graph of the equation $x = a(y - k)^2 + h$, which can also be written as $x - h = a(y - k)^2$, is a parabola opening sideways with vertex at (h, k). You should not be surprised at this since we have simply interchanged the roles of x and y in an earlier equation. This parabola will open right if $a > 0$ and left if $a < 0$.

Example C (A Horizontal Parabola and Its Shift)   Sketch the graph.
(a) $x = 2y^2$ (b) $x = 2(y - 3)^2 - 4$

Solution   The vertex in (a) is $(0, 0)$; in (b), it is $(-4, 3)$. The second graph has the same shape as the first but is shifted 4 units left and 3 units up. Both graphs are shown in Figure 43.

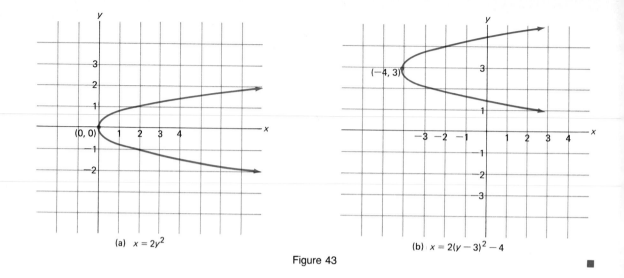

(a) $x = 2y^2$

(b) $x = 2(y-3)^2 - 4$

Figure 43

Intersection of a Line and a Parabola

We have learned how to find the point of intersection of two lines (Example D of Section 4-3). Finding the points of intersection of a line and a parabola is only slightly more difficult.

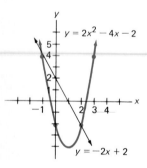

Figure 44

Example D (Finding Points of Intersection.)   Find the points of intersection of the line $y = -2x + 2$ and the parabola $y = 2x^2 - 4x - 2$ (see Figure 44).

Solution   We must solve the two equations simultaneously. This is easy to do by equating the two expressions for y and then solving the resulting equation for x.

$$-2x + 2 = 2x^2 - 4x - 2$$
$$0 = 2x^2 - 2x - 4$$
$$0 = 2(x^2 - x - 2)$$
$$0 = 2(x - 2)(x + 1)$$
$$x = -1 \qquad x = 2$$

By substitution, we find the corresponding values of y to be 4 and -2; the intersection points are therefore $(-1, 4)$ and $(2, -2)$.   ■

Parabolas Determined by Special Conditions

Our approach to parabolas has been algebraic: an equation is given, a graph is drawn, and the resulting curve is discussed. Historically, parabolas were defined and studied in terms of a geometric condition on a curve (see Problems 51 and 52); later it was realized that this condition led to an algebraic equation. We offer next a simple example of using conditions on a parabola to find its algebraic equation.

Example E (Using Side Conditions) Find the equation of the vertical parabola with vertex $(2, -3)$ and passing through the point $(9, -10)$.

Solution   The equation must have the form

$$y + 3 = a(x - 2)^2$$

To find a, we substitute $x = 9$ and $y = -10$.

$$-7 = a(7)^2$$

Thus $a = -\frac{1}{7}$, and the required equation is $y + 3 = -\frac{1}{7}(x - 2)^2$. ■

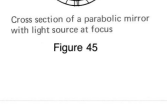

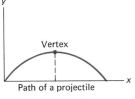

Cross section of a parabolic mirror with light source at focus

Figure 45

Applications

Our opening panel hinted at an important application of the parabola related to its optical properties. When light rays, parallel to the axis of a parabolic mirror, hit the mirror, they are reflected to a single point called the *focus*. Conversely, if a light source is placed at the focus of a parabolic mirror, the reflected rays of light are parallel to the axis, a principle used in flashlights (Figure 45).

The parabola is used also in the design of suspension bridges and load-bearing arches. If equal weights are placed along a line, equally spaced, and suspended from a thin flexible cable, the cable will assume a shape that closely approximates a parabola.

We turn to a very different kind of application. From physics, we learn that the path of a projectile is a parabola. It is known, for example, that a projectile fired at an angle of 45° from the horizontal with an initial speed of $320\sqrt{2}$ feet per second follows a curve with equation

$$y = -\frac{1}{6400}x^2 + x$$

where the coordinate axes are placed as shown in Figure 46. Taking this for granted, we may ask two questions.

1. What is the maximum height attained by the projectile?
2. What is the range (horizontal distance traveled) of the projectile?

To find the maximum height is simply to find the y-coordinate of the vertex. First we find the x-coordinate.

$$x = \frac{-b}{2a} = -\frac{1}{-2/6400} = 3200$$

When we substitute this value in the equation, we get

$$y = -\frac{1}{6400}(3200)^2 + 3200 = -1600 + 3200 = 1600$$

The greatest height is thus 1600 feet.

The range of the projectile is the x-coordinate of the point where it lands. By symmetry, this is simply twice the x-coordinate of the vertex; that is,

$$\text{range} = 2(3200) = 6400 \text{ feet}$$

y

Vertex

Path of a projectile   *x*

Figure 46

This value could also be obtained by solving the quadratic equation

$$-\frac{1}{6400}x^2 + x = 0$$

since the x-coordinate of the landing point is the value of x when $y = 0$.

PROBLEM SET 4-4

A. Skills and Techniques

The equations in Problems 1–10 represent parabolas. Sketch the graph of each parabola, indicating the coordinates of the vertex. See Example A.

1. $y = 3x^2$

2. $y = -\frac{1}{4}x^2$

3. $y = x^2 + 5$

4. $y = 2x^2 - 4$

5. $y = (x - 4)^2$

6. $y = -(x + 3)^2$

7. $y = 2(x - 1)^2 + 5$

8. $y = 3(x + 2)^2 - 4$

9. $y = -4(x - 2)^2 + 1$

10. $y = \frac{1}{2}(x + 3)^2 + 3$

Write each of the following in the form $y = ax^2 + bx + c$.

11. $y = 2(x - 1)^2 + 7$

12. $y = -3(x + 2)^2 + 5$

13. $-2y + 5 = (x - 5)^2$

14. $3y + 6 = (x + 3)^2$

Sketch the graph of each equation. Begin by plotting the vertex and at least one point on each side of the vertex. Recall that the x-coordinate of the vertex for $y = ax^2 + bx + c$ is $x = -b/2a$. See Example B.

15. $y = x^2 + 2x$

16. $y = 3x^2 - 6x$

17. $y = -2x^2 + 8x + 1$

18. $y = -3x^2 + 6x + 4$

In Problems 19–24, sketch the graph of the equation and indicate the coordinates of the vertex. See Example C.

19. $x = -2y^2$

20. $x = -2y^2 + 8$

21. $x = -2(y + 2)^2 + 8$

22. $x = 3(y - 1)^2 + 6$

23. $x = y^2 + 4y + 2$.

  Note: The y-coordinate of the vertex is at $-b/2a$.

24. $x = 4y^2 - 8y + 10$

Find the points of intersection for the given line and parabola as in Example D.

25. $y = -x + 1$
    $y = x^2 + 2x + 1$

26. $y = -x + 4$
    $y = -x^2 + 2x + 4$

27. $y = -2x + 1$
    $y = -x^2 - x + 3$

28. $y = -3x + 15$
    $y = 3x^2 - 3x + 12$

29. $y = 1.5x + 3.2$
    $y = x^2 - 2.9x$

30. $y = 2.1x - 6.4$
    $y = -1.2x^2 + 4.3$

In Problems 31–34, find the equation of the vertical parabola that satisfies the given conditions. See Example E.

31. Vertex $(0, 0)$, passing through $(-6, 3)$

32. Vertex $(0, 0)$, passing through $(3, -6)$

33. Vertex $(-2, 0)$, passing through $(6, -8)$

34. Vertex $(3, -1)$, passing through $(-2, 5)$

35. Find the equation of the parabola passing through $(1, 1)$ and $(2, 7)$ and having the y-axis as the line of symmetry. *Hint:* The equation has the form $y = ax^2 + c$.

36. Find the equation of the parabola passing through $(1, 2)$ and $(-2, -7)$ and having the y-axis as the line of symmetry.

B. Applications and Extensions

37. In each case, find the coordinates of the vertex and then sketch the graph. Use the same coordinate axes for all three graphs.
    (a) $y = 2x^2 - 8x + 4$ **(b)** $y = 2x^2 - 8x + 8$
    (c) $y = 2x^2 - 8x + 11$

38. Sketch the following parabolas using the same coordinate axes.
    (a) $y = -3x^2 + 6x + 9$
    (b) $y = -3x^2 + 6x - 3$
    (c) $y = -3x^2 + 6x - 9$

39. Sketch the graphs of the following parabolas.
    (a) $y = \frac{1}{2}x^2 - 2x$ **(b)** $x = \frac{1}{2}y^2 - 2y$

40. Recall that the discriminant d of $ax^2 + bx + c$ is $d = b^2 - 4ac$ (see page 110). Calculate d for each of the parabolas of Problem 37. Then show that the graph of $y = ax^2 + bx + c$ will cross the x-axis, just touch the x-axis, or not meet the x-axis according as $d > 0$, $d = 0$, or $d < 0$.

41. Find the points of intersection of the following pairs of curves.

  (a) $y = x^2 - 2x + 6$
      $y = -3x + 8$

  (b) $y = x^2 - 4x + 6$
      $y = 2x - 3$

  (c) $y = -x^2 + 2x + 4$
      $y = -2x + 9$

  (d) $y = x^2 - 2x + 7$
      $y = 11 - x^2$

42. For what values of k does the parabola $y = x^2 - kx + 4$ have two x-intercepts?

43. The parabola $y = a(x - 2)(x - 8)$ passes through $(10, 40)$. Find a and the vertex of this parabola.

44. Find the equation of the vertical parabola that passes through $(-1, -2)$, $(0, 3)$, and $(2, 7)$.

45. If the curve shown in Figure 47 is part of a parabola, find the distance $\overline{PQ}$. *Hint:* Begin by finding the equation of the parabola, assuming its vertex is at the origin.

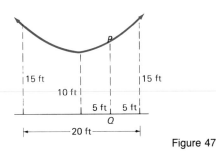

Figure 47

46. The parabolic cable for a suspension bridge is attached to the two towers at points 400 feet apart and 90 feet above the horizontal bridge deck. The cables drop to a point 10 feet above the deck. Find the xy-equation of the cable, assuming it is symmetric about the y-axis with vertex at $(0, 10)$.

47. A retailer has learned from experience that if she charges x dollars apiece for toy trucks, she can sell $300 - 100x$ of them. The trucks cost her \$2 each. Write a formula for her total profit P in terms of x. Then determine what she should charge for each truck to maximize her profit.

48. A company that makes fancy golf carts has fixed overhead costs of \$12,000 per year and direct costs (labor and materials) of \$80 per cart. It sells its carts to a certain retailer at a nominal price of \$120 each. However, the company offers a discount of 1 percent for 100 carts, 2 percent for 200 carts, and, in general, $x/100$ percent for x carts. Assume the retailer will buy as many carts as the company can produce but that its production facilities limit this to a maximum of 1800 units.

  (a) Write a formula for C, the cost of producing x carts.

  (b) Show that its total revenue R in dollars is $R = 120x - .012x^2$.

  (c) What is the smallest number of carts it can produce and still break even?

  (d) What number of carts will produce the maximum profit and what is this profit?

49. Find the area (in terms of a and b) of the triangle shown in Figure 48. The point c is midway between a and b.

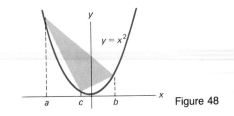

Figure 48

50. Starting at $(0, 0)$, a ball travels along the path $y = ax^2 + bx$. Find a and b if the ball reaches a height of 75 at $x = 50$ and its maximum height at $x = 100$.

51. Let $P(x, y)$ move so that its distance from a fixed point $F(0, 3)$ is always equal to its perpendicular distance from the fixed line $y = -3$. Derive and simplify the equation of the path. *Note:* If you do this correctly, the equation should be that of a parabola with vertex at the origin.

52. Problem 51 gives a special case of the following very important result. Suppose that $P(x, y)$ moves so that its distance from the fixed point $F(0, p)$ is equal to its perpendicular distance from the line $y = -p$. Show that the equation of the path is $x^2 = 4py$, the equation of a parabola. *Note:* The fixed point is called the *focus* and the fixed line, the *directrix*.

53. In Figure 49, FG is parallel to the x-axis and RG is parallel to the y-axis. Show that the length $L = \overline{FR} + \overline{RG}$ is independent of where R is on the parabola by finding a formula for L in terms of p alone.

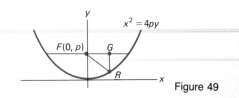

Figure 49

54. TEASER   Let $\{a_1, a_2, \ldots, a_n\}$ and $\{b_1, b_2, \ldots, b_n\}$ be two sets of n numbers each. Show that the inequality

$$(a_1b_1 + a_2b_2 + \cdots + a_nb_n)^2$$
$$\leq (a_1^2 + a_2^2 + \cdots + a_n^2)(b_1^2 + b_2^2 + \cdots + b_n^2)$$

always holds. *Hint*: See if you can use the fact that if $Ax^2 + 2Bx + C \geq 0$ for all x, then $B^2 - AC \leq 0$ (see Problem 40).

4-5   ELLIPSES AND HYPERBOLAS (OPTIONAL)

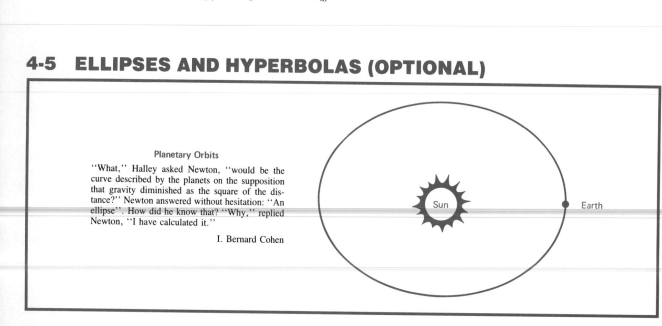

Planetary Orbits

"What," Halley asked Newton, "would be the curve described by the planets on the supposition that gravity diminished as the square of the distance?" Newton answered without hesitation: "An ellipse". How did he know that? "Why," replied Newton, "I have calculated it."

I. Bernard Cohen

Roughly speaking, an ellipse is a flattened circle. In the case of the earth's orbit about the sun, there is very little flattening (less than the opening panel indicates). But all ellipses, be they nearly circular or very flat, have important properties in common. As with parabolas, we shall begin by discussing equations of ellipses.

Equations of Ellipses

Consider the equation

$$\frac{x^2}{25} + \frac{y^2}{16} = 1$$

Because x and y can be replaced by $-x$ and $-y$, respectively, without changing the equation, the graph is symmetric with respect to both axes and the origin. To find the x-intercepts, we let $y = 0$ and solve for x.

$$\frac{x^2}{25} = 1$$

$$x^2 = 25$$

$$x = \pm 5$$

Thus the graph intersects the x-axis at $(\pm 5, 0)$. By a similar procedure (letting

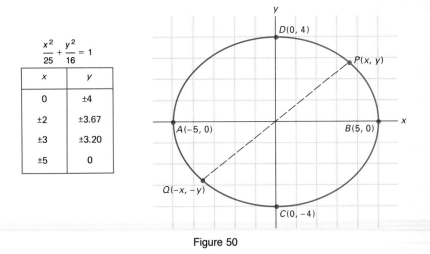

$$\frac{x^2}{25} + \frac{y^2}{16} = 1$$

x	y
0	±4
±2	±3.67
±3	±3.20
±5	0

Figure 50

$x = 0$), we find that the graph intersects the y-axis at $(0, \pm4)$. Plotting these points and a few others leads to the graph in Figure 50.

This curve is an example of an ellipse. The dashed line segment PQ with endpoints on the ellipse and passing through the origin is called a **diameter.** The longest diameter, AB, is the **major diameter** (sometimes called the major axis) and the shortest one, CD, is the **minor diameter.** The origin, which is a bisector of every diameter, is appropriately called the **center** of the ellipse. The endpoints of the major diameter are called the **vertices** of the ellipse.

More generally, if a and b are any positive numbers, the equation

$$\frac{x^2}{a^2} + \frac{y^2}{b^2} = 1$$

represents an ellipse with center at the origin and intersecting the x- and y-axes at $(\pm a, 0)$ and $(0, \pm b)$, respectively. If $a > b$, it is called a *horizontal* ellipse (because the major diameter is horizontal); if $a = b$, it is a *circle* of radius a; and if $a < b$, it is called a *vertical* ellipse (see Figure 51).

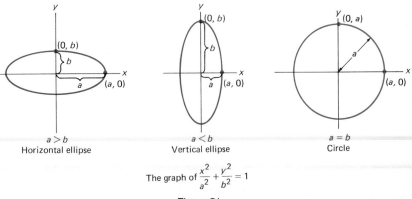

The graph of $\dfrac{x^2}{a^2} + \dfrac{y^2}{b^2} = 1$

Figure 51

Translating (Shifting) the Ellipse

If we move the ellipse $x^2/a^2 + y^2/b^2 = 1$ (without turning it) so that its center is at (h, k) rather than the origin, its equation takes the form

$$\frac{(x - h)^2}{a^2} + \frac{(y - k)^2}{b^2} = 1$$

This is called the **standard form** for the equation of an ellipse. Again, the relative sizes of a and b determine whether it is a horizontal ellipse, a vertical ellipse, or a circle.

Example A (Analyzing the Equation of an Ellipse)   Determine the orientation, the center, and the lengths of the major and minor diameters of the ellipse with equation

$$\frac{(x + 2)^2}{4} + \frac{(y - 4)^2}{9} = 1$$

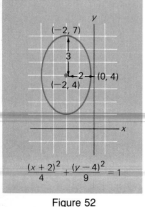

$$\frac{(x + 2)^2}{4} + \frac{(y - 4)^2}{9} = 1$$

Figure 52

Then sketch the graph.

Solution   Because the larger denominator appears in the y-term, we know that the ellipse is vertical. Its center is $(-2, 4)$; the major and minor diameters have length $2 \cdot 3 = 6$ and $2 \cdot 2 = 4$, respectively. All this information allows us to sketch the graph in Figure 52.   ■

Example B (Ellipses Determined by Conditions)   Find the equation of the ellipse with vertices at $(-1, 5)$ and $(9, 5)$ and minor diameter of length 3.

Solution   The ellipse is horizontal with center midway between the two vertices, that is, at $(4, 5)$. Next we note that $a = 5$ (half the length of the horizontal diameter) and $b = 3$. We conclude that the equation of the ellipse is

$$\frac{(x - 4)^2}{25} + \frac{(y - 5)^2}{9} = 1 \quad ■$$

Equations of Hyperbolas

What a difference a change in sign can make! The graphs of

$$\frac{x^2}{25} + \frac{y^2}{16} = 1$$

and

$$\frac{x^2}{25} - \frac{y^2}{16} = 1$$

are as different as night and day. The first is an ellipse; the second is a hyperbola. Let us see what we can find out about this hyperbola.

First note that it has x-intercepts ± 5 but no y-intercepts (setting $x = 0$ in the

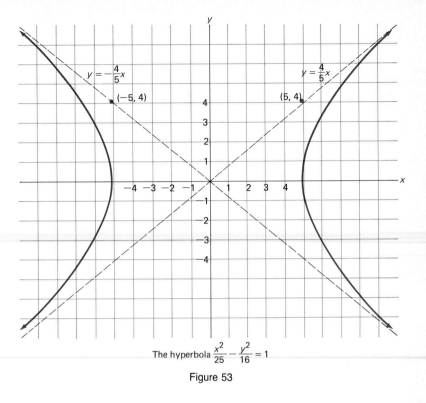

The hyperbola $\dfrac{x^2}{25} - \dfrac{y^2}{16} = 1$

Figure 53

equation yields $y^2 = -16$). Since x can be replaced by $-x$ and y can be replaced by $-y$ without changing the equation, the graph is symmetric with respect to both axes and the origin. This makes it appropriate to call the origin the **center** of the hyperbola. If we solve for y in terms of x, we get

$$y = \pm\tfrac{4}{5}\sqrt{x^2 - 25}$$

This implies first that we must have $|x| \geq 5$; so the graph has no points between $x = -5$ and $x = 5$. Second, since for large $|x|$, $\tfrac{4}{5}\sqrt{x^2 - 25}$ behaves very much like $\tfrac{4}{5}x$, the hyperbola must draw closer and closer to the lines $y = \pm\tfrac{4}{5}x$. These lines are called **asymptotes** of the graph.

When we put all of this information together and plot a few points, we are led to the graph in Figure 53. You can see from it why $(-5, 0)$ and $(5, 0)$ are called **vertices** of the hyperbola.

The example analyzed above suggests the general situation. If a and b are positive numbers, the graphs of

$$\frac{x^2}{a^2} - \frac{y^2}{b^2} = 1 \quad \text{or} \quad \frac{y^2}{b^2} - \frac{x^2}{a^2} = 1$$

are hyperbolas. In the first case, the hyperbola is said to be *horizontal*, since the vertices are at $(\pm a, 0)$, on a horizontal line. In the second case, the hyperbola is *vertical* with vertices $(0, \pm b)$. In both cases, the asymptotes are the lines $y = \pm(b/a)x$. Note in Figure 54 how the numbers a and b determine a rectangle with the asymptotes as diagonals.

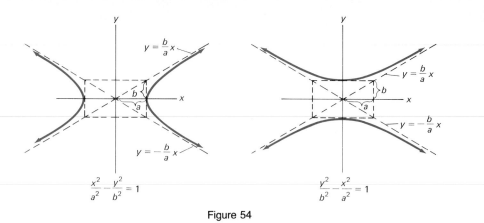

$$\frac{x^2}{a^2} - \frac{y^2}{b^2} = 1$$

$$\frac{y^2}{b^2} - \frac{x^2}{a^2} = 1$$

Figure 54

CAUTION

In contrast to the case of the ellipse, whether a hyperbola is vertical or horizontal is not determined by the relative size of a and b.

Rather, it is determined by whether the minus sign is associated with the x- or the y-term.

As an example of a vertical hyperbola, consider

$$\frac{y^2}{16} - \frac{x^2}{25} = 1$$

The vertices are at $(0, \pm 4)$ and the asymptotes have equations $y = \pm \frac{4}{5}x$.

Translating the Hyperbola

If we translate the general hyperbolas discussed above so that their centers are at (h, k) rather than the origin, their equations take the form

Horizontal Case	Vertical Case
$\dfrac{(x - h)^2}{a^2} - \dfrac{(y - k)^2}{b^2} = 1$	$\dfrac{(y - k)^2}{b^2} - \dfrac{(x - h)^2}{a^2} = 1$

These are called the **standard forms** for the equation of a hyperbola.

Example C (Analyzing the Equation of a Hyperbola) Determine the orientation, the center, and vertices of the hyperbola with equation

$$\frac{(y - 1)^2}{9} - \frac{(x + 2)^2}{4} = 1$$

Then sketch its graph.

Solution   The hyperbola is vertical (the minus is associated with the x-term). The center is $(-2, 1)$ and the vertices are 3 units above and below the center—that is, they are $(-2, 4)$ and $(-2, -2)$. To obtain the graph, we first draw the central dashed rectangle of size $2 \cdot 2$ by $2 \cdot 3$, thus determining the asymptotes. Finally, we sketch the graph (Figure 55). ■

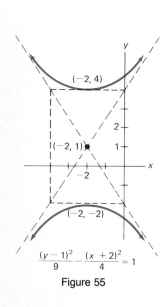

$$\frac{(y - 1)^2}{9} - \frac{(x + 2)^2}{4} = 1$$

Figure 55

Example D (Hyperbolas with Side Conditions)   Find the equation of the hyperbola satisfying the given conditions.

(a) Its vertices are $(4, 0)$ and $(-4, 0)$ and one of its asymptotes has slope $\frac{3}{2}$.

(b) The center is at $(2, 1)$, one vertex is at $(2, -4)$, and the equation of one asymptote is $5x - 7y = 3$.

Solution

(a) This is a horizontal hyperbola with center at $(0, 0)$. Since $\frac{3}{2} = b/a = b/4$, we get $b = 6$. The required equation is

$$\frac{x^2}{16} - \frac{y^2}{36} = 1$$

(b) Since the vertex $(2, -4)$ is 5 units directly below the center, the hyperbola is vertical and $b = 5$. Solving $5x - 7y = 3$ for y in terms of x gives $y = \frac{5}{7}x - \frac{3}{7}$, so the given asymptote has slope $\frac{5}{7}$. Thus $\frac{5}{7} = b/a = 5/a$, so $a = 7$. The required equation is

$$\frac{(y - 1)^2}{25} - \frac{(x - 2)^2}{49} = 1 \quad \blacksquare$$

Equations that are Quadratic in Both x and y

When expanded, the equations for ellipses and hyperbolas take the form $Ax^2 + By^2 + Dx + Ey = F$. What about the converse? Is the graph of such an equation always an ellipse, (or circle) hyperbola and can we tell which is which? The answer is yes, in general, with some fairly obvious exceptions. The key to understanding (just as in the circle case in Section 4-2) is to transform the equation to standard form. We illustrate.

Example E (Changing to Standard Form)   Change each of the following equations to standard form. Decide whether the corresponding curve is an ellipse or a hyperbola and whether it is horizontal or vertical. Find the center and the vertices.

(a) $x^2 + 4y^2 - 8x + 16y = -28$

(b) $4x^2 - 9y^2 + 24x + 36y + 36 = 0$

Solution

(a) We use the familiar process of completing the squares.

$$(x^2 - 8x + \quad) + 4(y^2 + 4y + \quad) = -28$$

$$(x^2 - 8x + 16) + 4(y^2 + 4y + 4) = -28 + 16 + 16$$

$$(x - 4)^2 + 4(y + 2)^2 = 4$$

$$\frac{(x - 4)^2}{4} + \frac{(y + 2)^2}{1} = 1$$

We recognize this to be a horizontal ellipse (the larger denominator is in the x-term). Its center is at $(4, -2)$ and its vertices are at $(2, -2)$ and $(6, -2)$.

(b) Again we complete the squares.

$$4(x^2 + 6x + \quad) - 9(y^2 - 4y + \quad) = -36$$
$$4(x^2 + 6x + 9) - 9(y^2 - 4y + 4) = -36 + 36 - 36$$
$$4(x + 3)^2 - 9(y - 2)^2 = -36$$
$$\frac{(y - 2)^2}{4} - \frac{(x + 3)^2}{9} = 1$$

This is a vertical hyperbola (the y-term is positive). The center is at $(-3, 2)$ and the vertices are at $(-3, 0)$ and $(-3, 4)$, 2 units below and above the center. ■

The Conic Sections

The Greeks called the four curves—circles, parabolas, ellipses, and hyperbolas—*conic sections*. If you take a cone with two nappes and pass planes through it at various angles, the curve of intersection in each instance is one of these four curves (assuming that the intersecting plane does not pass through the apex of the cone). The diagrams in Figure 56 illustrate this important fact.

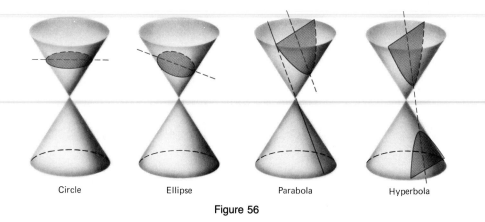

Circle   Ellipse   Parabola   Hyperbola

Figure 56

PROBLEM SET 4-5

A. Skills and Techniques

Each of the following equations represents an ellipse. Find its center and the endpoints of the major and minor diameters and graph the ellipse. See Example A.

1. $\dfrac{x^2}{25} + \dfrac{y^2}{9} = 1$

2. $\dfrac{x^2}{36} + \dfrac{y^2}{16} = 1$

3. $\dfrac{x^2}{9} + \dfrac{y^2}{25} = 1$

4. $\dfrac{x^2}{4} + \dfrac{y^2}{25} = 1$

5. $\dfrac{(x - 2)^2}{25} + \dfrac{(y + 1)^2}{9} = 1$

6. $\dfrac{(x - 3)^2}{36} + \dfrac{(y - 4)^2}{25} = 1$

7. $\dfrac{(x + 3)^2}{9} + \dfrac{y^2}{16} = 1$

8. $\dfrac{x^2}{4} + \dfrac{(y - 4)^2}{49} = 1$

In Problems 9–14, find the center of the ellipse with AB and CD as major and minor diameters, respectively. Then write the equation of the ellipse. See Example B.

9. $A(6, 0)$, $B(-6, 0)$, $C(0, 3)$, $D(0, -3)$
10. $A(5, 0)$, $B(-5, 0)$, $C(0, 1)$, $D(0, -1)$
11. $A(0, 6)$, $B(0, -6)$, $C(4, 0)$, $D(-4, 0)$
12. $A(0, 3)$, $B(0, -3)$, $C(2, 0)$, $D(-2, 0)$
13. $A(-4, 3)$, $B(8, 3)$, $C(2, 1)$, $D(2, 5)$
    Hint: Use the midpoint formula to find the center.
14. $A(3, -4)$, $B(11, -4)$, $C(7, -7)$, $D(7, -1)$

Each of the following equations represents a hyperbola. Find its center and its vertices. Then graph the equation. See Example C.

15. $\dfrac{x^2}{16} - \dfrac{y^2}{9} = 1$

16. $\dfrac{x^2}{36} - \dfrac{y^2}{16} = 1$

17. $\dfrac{y^2}{9} - \dfrac{x^2}{16} = 1$

18. $\dfrac{y^2}{16} - \dfrac{x^2}{36} = 1$

19. $\dfrac{(x-3)^2}{9} - \dfrac{(y+2)^2}{16} = 1$

20. $\dfrac{(x+1)^2}{16} - \dfrac{(y-4)^2}{36} = 1$

21. $\dfrac{(y+3)^2}{4} - \dfrac{x^2}{25} = 1$

22. $\dfrac{y^2}{49} - \dfrac{(x-2)^2}{9} = 1$

In Problems 23–28, find the equation of the hyperbola satisfying the given conditions (Example D).

23. The vertices are $(4, 0)$ and $(-4, 0)$ and one asymptote has slope $\frac{5}{4}$.
24. The vertices are $(7, 0)$ and $(-7, 0)$ and one asymptote has slope $\frac{2}{7}$.
25. The center is $(6, 3)$, one vertex is $(8, 3)$, and one asymptote has slope 1.
26. The center is $(3, -3)$, one vertex is $(-2, -3)$, and one asymptote has slope $\frac{6}{5}$.
27. The vertices are $(4, 3)$ and $(4, 15)$ and the equation of one asymptote is $3x - 2y + 6 = 0$.
28. The vertices are $(5, 0)$ and $(5, 14)$ and the equation of one asymptote is $7x - 5y = 0$.

For each of the following, change the equation to standard form and decide whether the corresponding curve is an ellipse or a hyperbola and whether it is horizontal or vertical. Find the center and the vertices. See Example E.

29. $9x^2 + 16y^2 + 36x - 96y = -36$
30. $x^2 + 4y^2 - 2x + 32y = -61$
31. $4x^2 - 9y^2 - 16x - 18y - 29 = 0$
32. $9x^2 - y^2 + 90x + 8y + 200 = 0$
33. $25x^2 + y^2 - 4y = 96$
34. $25x^2 + 9y^2 - 100x - 125 = 0$
35. $4x^2 - y^2 - 32x - 4y + 69 = 0$
36. $x^2 - y^2 + 8x + 4y + 13 = 0$

B. Applications and Extensions

In Problems 37–44, decide whether the graph of the given equation is a circle, an ellipse, or a hyperbola. Then sketch the graph.

37. $\dfrac{x^2}{64} + \dfrac{y^2}{16} = 1$

38. $\dfrac{x^2}{16} + \dfrac{y^2}{16} = 1$

39. $\dfrac{x^2}{64} - \dfrac{y^2}{16} = 1$

40. $\dfrac{x^2}{4} - \dfrac{y^2}{9} = 1$

41. $\dfrac{(x-2)^2}{25} + \dfrac{(y-1)^2}{4} = 1$

42. $\dfrac{(x-2)^2}{4} - \dfrac{(y+2)^2}{9} = 1$

43. $x^2 - 2y^2 - 6x + 8y = 1$
44. $x^2 + 2y^2 - 6x + 8y = 1$
45. Find the equation of the ellipse whose vertices are $(-4, 2)$ and $(10, 2)$ and whose minor diameter has length 10.
46. Find the equation of the ellipse that passes through the point $(3, 2)$ and has its vertices at $(\pm 5, 0)$.
47. Find the equation of the hyperbola having vertices $(\pm 6, 0)$ and one of its asymptotes passing through $(3, 2)$.
48. Find the equation of the hyperbola with vertices at $(0, \pm 2)$ and passing through the point $(2, 4)$.
49. Find the y-coordinates of the points for which $x = \pm 2.5$ on the ellipse

$$\dfrac{x^2}{24} + \dfrac{y^2}{19} = 1$$

50. Find the y-coordinates of the points for which $x = \pm 5$ on the hyperbola

$$\dfrac{x^2}{17} - \dfrac{y^2}{11} = 1$$

51. The area of the ellipse $x^2/a^2 + y^2/b^2 = 1$ is πab. Find the areas of the following ellipses.

(a) $\dfrac{x^2}{7} + \dfrac{y^2}{11} = 1$ (b) $\dfrac{x^2}{111} + y^2 = 1$

52. Find the radius of the circle which has the same area as the given ellipse.

(a) $\dfrac{x^2}{256} + \dfrac{y^2}{89} = 1$ (b) $\dfrac{x^2}{50} + \dfrac{y^2}{19} = 1$

53. The cross section of the mug in Figure 57 is a hyperbola with center on the base of the mug and vertex 1 centimeter above the base. The mug is 9 centimeters high and has an opening at the top with diameter 8 centimeters. How deep is the mug at a point 2 centimeters from its central axis?

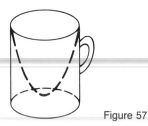

Figure 57

54. A doorway has the shape of an elliptical arch (a semi-ellipse) that is 6 feet wide at the base and 8 feet high at the center (Figure 58). A heavy rectangular box that is 4 feet wide is to be slid through this doorway. How high can the box be?

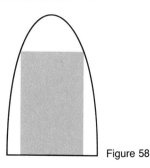

Figure 58

55. An arch has the shape of a horizontal semi-ellipse that is 40 feet wide at the base and 12 feet high at the center. Determine the vertical distance from the base to the arch at a distance 10 feet from the center.

56. Find the points of intersection of the line $x + 2y = 6$ and the ellipse $x^2 + 4y^2 = 20$.

57. Arnold Thinkhard stood high on a ladder 25 feet long that was leaning against a vertical wall (the positive y-axis). When the bottom of the ladder began to slide along the ground (the positive x-axis), Arnold pressed his nose against the ladder and hung on for dear life. If his nose was 4 feet from the top end of the ladder, what was the equation of the path this important organ took in descending to the ground?

58. As the wheel in Figure 59 turns, the point $R(x, y)$ traces out a path in the xy-plane. Find the equation of this path, assuming that R is always on a vertical line through Q and a horizontal line through P.

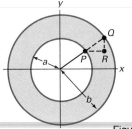

Figure 59

59. Let $F_1(-4, 0)$ and $F_2(4, 0)$ be two given points and let $P(x, y)$ move so that

$$d(P, F_1) + d(P, F_2) = 10$$

By deriving the equation of the path of P, show that this path is an ellipse.

60. Suppose in Problem 59 that the two fixed points F_1 and F_2 are at $(-c, 0)$ and $(c, 0)$ and that $P(x, y)$ moves so that

$$d(P, F_1) + d(P, F_2) = 2a$$

where $a > c > 0$. Show that the equation of the path is $x^2/a^2 + y^2/b^2 = 1$, where $b^2 = a^2 - c^2$. The given distance condition is often taken as the definition of an ellipse and the two fixed points are called *foci* (plural of *focus*).

61. Let $F_1(-5, 0)$ and $F_2(5, 0)$ be two given points and let $P(x, y)$ move so that

$$\left| d(P, F_1) - d(P, F_2) \right| = 6$$

By deriving the equation of the path of P, show that this path is a hyperbola.

62. Suppose in Problem 61 that the two fixed points F_1 and F_2 are at $(-c, 0)$ and $(c, 0)$ and that $P(x, y)$ moves so that

$$\left| d(P, F_1) - d(P, F_2) \right| = 2a$$

where $c > a > 0$. Show that the equation of the path

is $x^2/a^2 - y^2/b^2 = 1$, where $b^2 = c^2 - a^2$. The given distance condition is often taken as the definition of a hyperbola and, as with the ellipse, the two fixed points are called foci.

63. Stakes are driven into the ground at $(\pm 15, 0)$. A loop of rope 80 feet long is thrown over the stakes and attached to Fido's collar. Fido pulls on the rope and, keeping it taut, races around the stakes, thus completing a closed path. Write the xy-equation of Fido's path. *Hint:* See Problem 60.

64. TEASER In Figure 60, A is a fixed point at $(0, 18)$ in the xy-plane. As B painted the elliptical region bounded by $x^2/9 + y^2/16 = 1$, C also painted a region by staying on the line AB and exactly one-third of the way from A to B. Determine the shape of C's region and how much paint C needed if B used exactly 1 gallon.

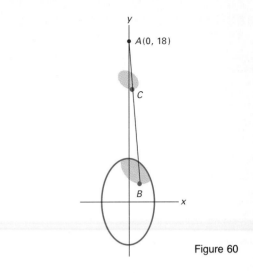

Figure 60

CHAPTER 4 SUMMARY

Like a city planner, we introduce in the plane two main streets, one vertical (the **y-axis**) and the other horizontal (the **x-axis**). Relative to these axes, we can specify any point by giving its address (x, y). The numbers x and y, called **Cartesian coordinates,** measure the directed distances from the vertical and horizontal axes, respectively. And given two points A and B with addresses (x_1, y_1) and (x_2, y_2), we may calculate the distance between them from the **distance formula:**

$$d(A, B) = \sqrt{(x_2 - x_1)^2 + (y_2 - y_1)^2}$$

In **analytic geometry,** we use the notion of coordinates to combine algebra and geometry. Thus we may graph the equation $y = x^2$ (algebra), thereby turning it into a curve (geometry). Conversely, we may take a circle with radius 6 and center $(-1, 4)$ and give it the equation

$$(x + 1)^2 + (y - 4)^2 = 36$$

The simplest of all curves is a **line.** If a line passes through (x_1, y_1) and (x_2, y_2) with $x_1 \neq x_2$, then its **slope** m is given by

$$m = \frac{\text{rise}}{\text{run}} = \frac{y_2 - y_1}{x_2 - x_1}$$

There are two important forms for the equation of a nonvertical line.

Point-slope form: $\qquad y - y_1 = m(x - x_1)$
Slope-intercept form: $\qquad y = mx + b$

Vertical lines do not have slope; their equations take the form $x = k$. All lines (vertical and nonvertical) can be written in the form of the **general linear equation**

$$Ax + By + C = 0$$

The distance d from the point (x_1, y_1) to the line $Ax + By + C = 0$ is

$$d = \frac{|Ax_1 + By_1 + C|}{\sqrt{A^2 + B^2}}$$

Nonvertical lines are parallel if their slopes are equal, and they are perpendicular if their slopes are negative reciprocals.

Somewhat more complicated curves are the *conic sections:* **parabolas, circles, ellipses,** and **hyperbolas.** Here are typical equations for them.

Parabola: $\qquad y - k = a(x - h)^2$
Circle: $\qquad (x - h)^2 + (y - k)^2 = r^2$
Ellipse: $\qquad \dfrac{(x - h)^2}{a^2} + \dfrac{(y - k)^2}{b^2} = 1$
Hyperbola: $\qquad \dfrac{(x - h)^2}{a^2} - \dfrac{(y - k)^2}{b^2} = 1$

CHAPTER 4 REVIEW PROBLEM SET

In Problems 1–10, write True or False in the blank. If false, tell why.

_____ **1.** The equation $3y = 6x - 12$ represents a line with slope 2 and y-intercept -4.

_____ **2.** The line $2x + 5y = 0$ passes through the point $(10, -4)$.

_____ **3.** The distance between the points $(\pi, 3)$ and $(-\pi, 2)$ is $\sqrt{2\pi^2 + 1}$.

_____ **4.** The lines $2x - y = 5$ and $x + 2y = 5$ are perpendicular.

_____ **5.** The radius of the circle $x^2 + y^2 + 4x = 0$ is 4.

_____ **6.** The parabola $y^2 = -3x$ opens down.

_____ **7.** The graph *of* $y = ax^2 + bx + c$ is always a parabola.

_____ **8.** The graph of $x^2 + y^2 + Dx + Ey = F$ is always a circle.

_____ **9.** The graph of $y = ax + b$ is always a line.

_____ **10.** The graph of $2x^2 - y^2 = 4$ is a horizontal hyperbola.

In Problems 11–18, name the graph of the given equation.

11. $(y + 3)^2 = 4(x - 5)$

12. $(x + 4)^2 + (y - 5)^2 = 100$

13. $\dfrac{x^2}{9} + \dfrac{(y - 4)^2}{16} = 1$

14. $3(y - 2) = 4(x + 3)$

15. $x = 3y^2 + y - 1$

16. $\dfrac{(y - 2)^2}{16} - \dfrac{(x - 1)^2}{9} = 1$

17. $4x^2 - 25y^2 = 0$

18. $4(x + 2)^2 + 9(y - 3)^2 = 0$

Sketch the graph of each equation in Problems 19–24.

19. $3x + y = 4$

20. $x^2 + y^2 = 16$

21. $y = x^2 + \dfrac{1}{x}$

22. $y = x^3 - 4x$

23. $y = \sqrt{x} + 2$

24. $\dfrac{x^2}{25} + \dfrac{y^2}{9} = 1$

Problems 25–28 relate to the points $A(3, -1)$, $B(5, 3)$, and $C(-1, 9)$.

25. **(a)** Sketch the triangle ABC.
(b) Find the lengths of its three sides.

26. **(a)** Find the slopes of the lines that contain the three sides of triangle ABC.
(b) Write an equation for each of these lines in the form $Ax + By + C = 0$.

27. **(a)** Find the midpoints of AB and AC.
(b) Verify that the line segment joining these two midpoints is parallel to BC and is one-half the length of BC.

28. **(a)** Find the equation of the line through A parallel to BC.
(b) Find the equation of the line through A perpendicular to BC.

In Problems 29–34, write the equation of the line satisfying the given conditions in the form $Ax + By + C = 0$.

29. It is vertical and passes through $(4, -3)$.

30. It passes through the points $(2, 3)$ and $(5, -1)$.

31. It passes through the origin and is parallel to the line $5x + 7y = 2$.

32. It has y-intercept 6 and is perpendicular to the line $2x + 3y = 7$.

33. It has y-intercept -1 and x-intercept 2.

34. It is tangent to the circle $x^2 + y^2 = 25$ at the point $(3, -4)$.

In Problems 35–38, find the vertex of the parabola and decide which way it opens (up, down, left, or right).

35. $y = 2x^2 - 8x + 1$

36. $x + 2 = \frac{1}{12}(y + 3)^2$

37. $y - 3 = -\frac{2}{3}(x + 1)^2$

38. $x = -2y^2 + 4y$

Write the equation of each curve in Problems 39–46.

39. The vertical parabola with vertex at $(0, 0)$ that goes through $(2, -12)$

40. The horizontal parabola with vertex at $(-2, 3)$ that goes through $(-4, 1)$

41. The ellipse with vertices $(0, \pm 3)$ and 4 as the length of its minor diameter

42. The ellipse with vertices $(3, -3)$ and $(3, 5)$ and 6 as the length of its minor diameter

43. The circle of radius 5 and the same center as the ellipse of Problem 42

44. An ellipse that passes through the points $(-7, 5)$, $(-7, 1)$, $(0, 3)$, and $(-14, 3)$

45. The hyperbola with vertices $(-1, -3)$ and $(5, -3)$ and having an asymptote that goes through $(8, 5)$

46. The circle with diameter having endpoints $(1, -1)$ and $(4, 7)$

Use the technique of completing the square in Problems 47–50 to identify each conic. Then give pertinent information about this conic.

47. $x^2 - 6x + y^2 + 2y = -1$

48. $y^2 + 6y + 4x = 7$

49. $4x^2 - 8x - 9y^2 = 32$

50. $x^2 + 6x + 4y^2 - 40y = -9$

51. Find the points of intersection of the parabola $y = 2x^2 + x - 14$ and the line $x + y = -2$.

52. Find the points of intersection of the ellipse $9x^2 + 4y^2 = 36$ and the hyperbola $y^2 - x^2 = 5/2$.

53. An arch is 9 feet high at its center and 6 feet wide at its base (Figure 61). Find the height of this arch at a point 2 units from its line of symmetry if the arch is

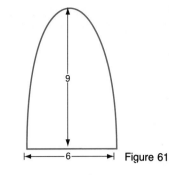

Figure 61

(a) part of a parabola.

(b) a semi-ellipse.

54. A belt fits tightly around the circles $x^2 + y^2 = 9$ and $(x - 6)^2 + (y - 8)^2 = 9$. Find the length of this belt.

55. Find the minimum value of $M = x^2 - 100x + 2486$.

56. A projectile fired from the origin follows the parabolic path $y = -.0032x^2 + x$ where x and y are in feet. Find

(a) the maximum height of the projectile.

(b) the range of the projectile (horizontal distance covered by the projectile).

CHAPTER 5

Functions
and Their Graphs

Mathematicians do not deal in objects, but in relations between objects; thus, they are free to replace some objects by others so long as the relations remain unchanged. Content to them is irrelevant; they are interested in form only.

Henri Poincaré

5-1 FUNCTIONS

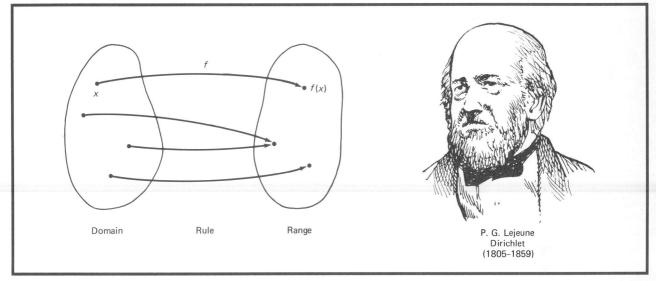

Domain   Rule   Range

P. G. Lejeune
Dirichlet
(1805–1859)

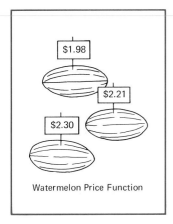

Watermelon Price Function

Figure 1

One of the most important ideas in mathematics is that of a function. For a long time, mathematicians and scientists wanted a precise way to describe the relationships that may exist between two variables. It is somewhat surprising that it took so long for the idea to crystallize into a clear, unambiguous concept. The French mathematician P. G. Lejeune Dirichlet (1805–1859) is credited with the modern definition of function.

> **Definition**
>
> A **function** is a rule which assigns to each element in one set (called the **domain** of the function) exactly one value from another set. The set of all assigned values is called the **range** of the function.

Three examples will help clarify this idea. When a grocer puts a price tag on each of the watermelons for sale (Figure 1), a function is determined. Its domain is the set of watermelons, its range is the set of prices, and the rule is the procedure the grocer uses in assigning prices (perhaps a specified amount per pound.)

When a professor assigns a grade to each student in a class (Figure 2), he or she is determining a function. The domain is the set of students and the range is the set of grades, but who can say what the rule is? It varies from professor to professor; some may even prefer to keep it a secret.

A much more typical function from our point of view is the *squaring* function displayed in Figure 3. It takes a number from the domain $\{-2, -1, 0, 1, 2, 3\}$ and squares it, producing a number in the range $\{0, 1, 4, 9\}$. This function is typical for two reasons: Both the domain and range are sets of numbers and the rule can be specified by giving an algebraic formula. Most functions in this book will be of this type.

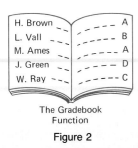

The Gradebook
Function

Figure 2

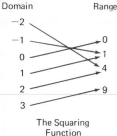

Domain   Range

The Squaring
Function

Figure 3

Not a Function

Figure 4

The definition says that a function assigns exactly one value to each element of the domain. Thus Figure 4 is not a diagram for a function, since two values are assigned to the element 1.

Functional Notation

Long ago, mathematicians introduced a special notation for functions. A single letter like f (or g or h) is used to name the function. Then $f(x)$, read f *of* x or f *at* x, denotes the value that f assigns to x. Thus, if f names the squaring function,

$$f(-2) = 4 \qquad f(2) = 4 \qquad f(-1) = 1$$

and, in general,

$$f(x) = x^2$$

We call this last result the *formula* for the function f. It tells us in a concise algebraic way what f does to any number. Notice that the given formula and

$$f(y) = y^2 \qquad f(z) = z^2$$

all say the same thing; the letter used for the domain variable is a matter of no significance, though it does happen that we shall usually use x. Many functions do not have simple formulas (see Problems 17 and 18), but in this book, most of them do.

For a further example, consider the function that cubes a number and then subtracts 1 from the result. If we name this function g, then

$$g(2) = 2^3 - 1 = 7$$
$$g(-1) = (-1)^3 - 1 = -2$$
$$g(.5) = (.5)^3 - 1 = -.875$$
$$g(\pi) = \pi^3 - 1 \approx 30$$

and, in general,

$$g(x) = x^3 - 1$$

CAUTION

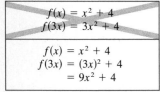

$$f(x) = x^2 + 4$$
$$f(3x) = 3x^2 + 4$$

$$f(x) = x^2 + 4$$
$$f(3x) = (3x)^2 + 4$$
$$= 9x^2 + 4$$

Few students would have trouble using this formula when x is replaced by a specific number. However, it is important to be able to use it when x is replaced by anything whatever, even an algebraic expression. Be sure you understand the following calculations.

$$g(a) = a^3 - 1$$
$$g(y^2) = (y^2)^3 - 1 = y^6 - 1$$
$$g\left(\frac{1}{z}\right) = \left(\frac{1}{z}\right)^3 - 1 = \frac{1}{z^3} - 1$$
$$g(2 + h) = (2 + h)^3 - 1 = 8 + 12h + 6h^2 + h^3 - 1$$
$$= h^3 + 6h^2 + 12h + 7$$
$$g(x + h) = (x + h)^3 - 1 = x^3 + 3x^2h + 3xh^2 + h^3 - 1$$

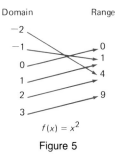

Domain   Range

$f(x) = x^2$

Figure 5

Domain and Range

The rule of correspondence is the heart of a function, but a function is not completely determined until its domain is given. Recall that the **domain** is the set of elements to which the function assigns values. In the case of the squaring function f (reproduced in Figure 5), we gave the domain as the set $\{-2, -1, 0, 1, 2, 3\}$. We could just as well have specified the domain as the set of all real numbers; the formula $f(x) = x^2$ would still make perfectly good sense. In fact, there is a common agreement that if no domain is specified, it is understood to be the largest set of real numbers for which the rule for the function makes sense and gives real number values. We call it the **natural domain** of the function. Thus if no domain is specified for the function with formula $g(x) = x^3 - 1$, it is assumed to be the set of all real numbers. Similarly, for

$$h(x) = \frac{1}{x - 1}$$

we would take the natural domain to consist of all real numbers except 1. Here, the number 1 is excluded to avoid division by zero.

Example A (Finding Natural Domains) Find the natural domain of the following.

(a) $f(x) = 4x/[(x + 2)(x - 3)]$ (b) $g(x) = \sqrt{x^2 - 4}$

Solution

(a) The domain consists of all real numbers except -2 and 3.
(b) We must have $x^2 \geq 4$, which is equivalent to $|x| \geq 2$. Thus the domain is $\{x : |x| \geq 2\}$. Notice that if $|x| < 2$, we would be taking the square root of a negative number, so the result would not be a real number.   ∎

Once the domain is understood and the rule of correspondence is given, the range of the function is determined. It is the set of values of the function. Here are several examples.

Rule	Domain	Range
$F(x) = 4x$	All reals	All reals
$G(x) = \sqrt{x - 3}$	$\{x : x \geq 3)$	Nonnegative reals
$H(x) = \dfrac{1}{(x - 2)^2}$	$\{x : x \neq 2\}$	Positive reals

A function takes a number from its domain, operates on it, and produces a number from its range. This can be illustrated nicely on a handheld calculator, as our next example shows.

Example B (Generating Functions on a Calculator) Show how the function $f(x) = 2\sqrt{x} + 5$ can be generated on an algebraic logic calculator. Then calculate $f(\pi)$.

Calculators vary and you must be aware of differences. Calculators that actually display what is being calculated (such as the TI-81) require that you press the $\boxed{\sqrt{}}$ key before the number x being "square-rooted" since that is the way we write this operation. For example, to calculate $\sqrt{3}$ on the TI-81, you would press $\boxed{\sqrt{}}$ 3 followed by $\boxed{\text{ENTER}}$. Despite these differences, you should be able to follow Example B and work Problems 19–32.

Solution   On a typical algebraic logic calculator,

$$f(x) = 2 \boxed{\times} x \boxed{\sqrt{}} \boxed{+} 5 \boxed{=}$$

In particular,

$$f(\pi) = 2 \boxed{\times} \pi \boxed{\sqrt{}} \boxed{+} 5 \boxed{=}$$

which yields the value 8.544908.   ∎

Independent and Dependent Variables

Scientists like to use the language of variables in talking about functions. Let us illustrate by referring to an object falling under the influence of gravity near the earth's surface. If the object is very dense (so air resistance can be neglected), it will fall according to the formula

$$d = f(t) = 16t^2$$

Here d represents the distance in feet that the object falls during the first t seconds of falling. In this case, t is called the **independent variable** and d, which depends on t, is the **dependent variable.** Also, d is said to be a function of t.

Functions of Two or More Variables

All functions illustrated so far have been functions of one variable. Suppose that a function f associates a value with each ordered pair (x, y). In this case, we say that f is a function of two variables. As an example, let

$$f(x, y) = x^2 + 3y^2$$

Then,

$$f(3, -2) = 3^2 + 3(-2)^2 = 21$$

and

$$f(0, 6) = (0)^2 + 3(6)^2 = 108$$

The natural domain for this function is the set of all ordered pairs (x, y), that is, the whole Cartesian plane. Its range is the set of nonnegative numbers.

For a second example, let $g(x, y, z) = 2x^2 - 4y + z$. We say that g is a function of three variables. Note that

$$g(4, 2, -3) = 2(4)^2 - 4(2) + (-3) = 21$$

and

$$g\left(u^3, 2v, \frac{1}{w}\right) = 2(u^3)^2 - 4(2v) + \frac{1}{w}$$

$$= 2u^6 - 8v + \frac{1}{w}$$

Variation

Sometimes we describe a functional relationship between a dependent variable and one or more independent variables by using the language of variation (or proportion). To say that y **varies directly** as x (or that y is proportional to x) means that $y = kx$ for some constant k. To say that y **varies inversely** as x means that $y = k/x$ for some constant k. Finally, to say that z **varies jointly** as x and y means that $z = kxy$ for some constant k.

In variation problems, we are often given not only the form of the relationship, but also a specific set of values satisfied by the variables involved. This allows us to evaluate the constant k and obtain an explicit formula for the dependent variable in terms of the independent variables. Here is an illustration.

Example C (Direct Variation) If y varies directly as x and $y = 10$ when $x = -2$, find an explicit formula for y in terms of x.

Solution We substitute the given values in the equation $y = kx$ and get $10 = k(-2)$. Thus $k = -5$, and we can write $y = f(x) = -5x$. ∎

The volume V of a right circular cone varies jointly as its altitude h and the square of its radius r. In symbols this means that

$$V = f(h, r) = khr^2$$

Those of you who remember your high school geometry will recall that in this case $k = \pi/3$.

Here is another two-variable illustration.

Example D (Joint Variation) Suppose that w varies jointly as x and the square root of y and that $w = 14$ when $x = 2$ and $y = 4$. Find an explicit formula for $w = f(x, y)$. Then evaluate $f(2, 9)$.

Solution We translate the first statement into mathematical symbols as

$$w = kx\sqrt{y}$$

To evaluate k, we substitute the given values for w, x, and y.

$$14 = k \cdot 2\sqrt{4} = 4k$$

or

$$k = \frac{14}{4} = \frac{7}{2}$$

Thus the explicit formula for w is

$$w = f(x, y) = \frac{7}{2}x\sqrt{y}$$

Finally, $f(2, 9) = \dfrac{7}{2}(2)\sqrt{9} = 21$. ∎

PROBLEM SET 5-1

A. Skills and Techniques

1. If $f(x) = x^2 - 4$, evaluate each expression.
   (a) $f(-2)$ **(b)** $f(0)$
   (c) $f(\frac{1}{2})$ **(d)** $f(.1)$
   (e) $f(\sqrt{2})$ **(f)** $f(a)$
   (g) $f(1/x)$ **(h)** $f(x + 1)$

2. If $f(x) = (x - 4)^2$, evaluate each expression in Problem 1.

3. If $f(x) = 1/(x - 4)$, evaluate each expression.
   (a) $f(8)$ **(b)** $f(2)$
   (c) $f(\frac{9}{2})$ **(d)** $f(\frac{31}{8})$
   (e) $f(4)$ **(f)** $f(4.01)$
   (g) $f(1/x)$ **(h)** $f(x^2)$
   (i) $f(2 + h)$ **(j)** $f(2 - h)$

4. If $f(x) = x^2$ and $g(x) = 2/x$, evaluate each expression.
   (a) $f(-7)$ **(b)** $g(-4)$
   (c) $f(\frac{1}{4})$ **(d)** $1/f(4)$
   (e) $g(\frac{1}{4})$ **(f)** $1/g(4)$
   (g) $g(0)$ **(h)** $g(1)f(1)$
   (i) $f(g(1))$ **(j)** $f(1)/g(1)$

In Problems 5–16, find the natural domain of the given function (Example A).

5. $f(x) = x^2 - 4$ **6.** $f(x) = (x - 4)^2$

7. $g(x) = \dfrac{1}{x^2 - 4}$ **8.** $g(x) = \dfrac{1}{9 - x^2}$

9. $h(x) = \dfrac{2}{x^2 - x - 6}$

10. $h(x) = \dfrac{1}{2x^2 + 3x - 2}$

11. $F(x) = \dfrac{1}{x^2 + 4}$ **12.** $F(x) = \dfrac{1}{9 + x^2}$

13. $G(x) = \sqrt{x - 2}$ **14.** $G(x) = \sqrt{x + 2}$

15. $H(x) = \dfrac{1}{5 - \sqrt{x}}$

16. $H(x) = \dfrac{1}{\sqrt{x + 1} - 2}$

17. Not all functions arising in mathematics have rules given by simple algebraic formulas. Let $f(n)$ be the nth digit in the decimal expansion of

$$\pi = 3.14159265358979323846 \ldots$$

Thus $f(1) = 3$ and $f(3) = 4$. Find (a) $f(6)$; (b) $f(9)$; (c) $f(16)$. What is the natural domain for this function?

18. Let g be the function which assigns to each positive integer the number of factors in its prime factorization. Thus

$$g(2) = 1$$

$$g(4) = g(2 \cdot 2) = 2$$

$$g(36) = g(2 \cdot 2 \cdot 3 \cdot 3) = 4$$

Find (a) $g(24)$; (b) $g(37)$; (c) $g(64)$; (d) $g(162)$. Can you find a formula for this function?

In Problems 19–26, write the sequence of keys that will generate the given function, using x for an arbitrary number. Then use the sequence to calculate f(2.9). See Example B.

19. $f(x) = (x + 2)^2$ **20.** $f(x) = \sqrt{x + 3}$

21. $f(x) = 3(x + 2)^2 - 4$

22. $f(x) = 4\sqrt{x + 3} - 11$

23. $f(x) = \left(3x + \dfrac{2}{\sqrt{x}}\right)^3$ **24.** $f(x) = \left(\dfrac{x}{3} + \dfrac{3}{x}\right)^5$

25. $f(x) = \dfrac{\sqrt{x^5 - 4}}{2 + 1/x}$ **26.** $f(x) = \dfrac{(\sqrt{x} - 1)^3}{x^2 + 4}$

In Problems 27–32, write the algebraic formula for the function that is generated by the given sequence of calculator keys.

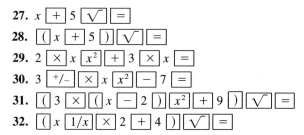

33. $g(2, 5)$ **34.** $g(-1, 3)$

Problems 33–44 deal with functions of two variables. Let $g(x, y) = 3xy - 5x$ and $G(x, y) = (5x + 3y)/(2x - y)$. Find each of the following.

33. $g(2, 5)$ **34.** $g(-1, 3)$
35. $g(5, 2)$ **36.** $g(3, -1)$
37. $G(1, 1)$ **38.** $G(3, 3)$
39. $G(\frac{1}{2}, 1)$ **40.** $G(5, 0)$

41. $g(2x, 3y)$ **42.** $G(2x, 4y)$

43. $g(x, 1/x)$ **44.** $G(x - y, y)$

In Problems 45–50, find an explicit formula for the dependent variable (Examples C and D).

45. y varies directly as x, and $y = 12$ when $x = 3$.

46. y varies directly as x^2, and $y = 4$ when $x = 0.1$.

47. y varies inversely as x (that is, $y = k/x$), and $y = 5$ when $x = \frac{1}{5}$.

48. V varies jointly as r^2 and h, and $V = 75$ when $r = 5$ and $h = 9$.

49. I varies directly as s and inversely as d^2, and $I = 9$ when $s = 4$ and $d = 12$.

50. W varies directly as x and inversely as the square root of yz, and $W = 5$ when $x = 7.5$, $y = 2$, and $z = 18$.

51. The maximum range of a projectile varies as the square of the initial velocity. If the range is 16,000 feet when the initial velocity is 600 feet per second,
  (a) write an explicit formula for R in terms of v, where R is the range in feet and v is the initial velocity in feet per second;
  (b) use this formula to find the range when the initial velocity is 800 feet per second.

52. Suppose that the amount of gasoline used by a car varies jointly as the distance traveled and the square root of the average speed. If a car used 8 gallons on a 100-mile trip going at an average speed of 64 miles per hour, how many gallons would that car use on a 160-mile trip at an average speed of 25 miles per hour?

B. Applications and Extensions

53. If $f(x) = x^2 - (2/x)$, evaluate and simplify each expression.
  (a) $f(2)$ **(b)** $f(-1)$
  (c) $f(\frac{1}{2})$ **(d)** $f(\sqrt{2})$
  (e) $f(2 - \sqrt{2})$ **(f)** $f(.01)$
  (g) $f(1/x)$ **(h)** $f(a^2)$
  (i) $f(a + b)$

54. If $f(x, y) = (x^2 - xy)/(x + y)$, evaluate and simplify each expression.
  (a) $f(2, 1)$ **(b)** $f(3, 0)$
  (c) $f(-2, 1)$ **(d)** $f(3\sqrt{2}, \sqrt{2})$
  (e) $f(1/x, 2y)$ **(f)** $f(a + b, b)$

55. Determine the natural domain of each function.
  (a) $f(t) = \dfrac{t + 2}{t^2(t + 3)}$ **(b)** $g(t) = \sqrt{t^2 - 4}$
  (c) $h(t) = \dfrac{3t + 1}{1 - \sqrt{2t}}$
  (d) $k(s, t) = \dfrac{3st\sqrt{9 - s^2}}{t^2 - 1}$

56. Let $f(n)$ be the nth digit in the decimal expansion of $\frac{5}{13}$. Determine the domain and range of f.

57. A ship leaves port at 10:00 A.M. sailing west at 18 miles per hour. At 12:00 noon, a second ship leaves the same port sailing south at 24 miles per hour. Express the distance $d(t)$ between the ships in terms of t, the number of hours after 12:00 noon.

58. A bus company offers the Seniors Club a special tour at a cost of $100 per person for a group of 40 or less people and will reduce the cost by $.50 per person for every person in excess of 40 until the bus capacity of 60 is reached. Write a (two-part) formula for the total cost $C(x)$ of the tour if x people sign up; then compute $C(53)$.

59. A 2-mile race track has the shape of a rectangle with semicircular ends of radius x (Figure 6). If $A(x)$ is the area of the region inside the track, determine the domain and range of A.

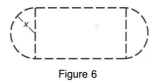

Figure 6

60. Cut a yardstick in two pieces of length x and $3 - x$ and, together with a 1-foot stick, form a triangle of area $A(x)$. Determine the domain and range of A. *Note:* The triangle of maximum area will be isosceles.

61. Write a formula $F(x)$ in each case.
  (a) $F(x)$ is the area of an equilateral triangle of perimeter x.
  (b) $F(x)$ is the area of a regular hexagon inscribed in a circle of radius x.
  (c) $F(x)$ is the volume of water of depth x in a conical tank with vertex downward. The tank is 8 feet high and has diameter 6 feet at the top.
  (d) $F(x)$ is the average cost per unit of producing x refrigerators in a day for a company that has daily

overhead of $1300 and pays direct costs (labor and materials) of $240 to make each refrigerator.

(e) $F(x)$ is the dollar cost of renting a car for 10 days and driving x miles if the rental company charges $18 per day and 22 cents per mile for mileage beyond the first free 100 miles.

62. Some of the following equations determine a function f with formula of the form $y = f(x)$. For those that do, find $f(x)$. *Recall:* A function must associate just one value with each x.

(a) $x^2 + 3y^2 = 1$ **(b)** $xy + 2y = 5 - 2x$
(c) $x = \sqrt{2y - 1}$ **(d)** $2x = (y + 1)/y$

63. The safe load $S(x, y, z)$ of a horizontal beam supported at both ends varies directly as its breadth x and

10 ft   6 in.

2 in.

Load   Figure 7

the square of its depth y and inversely as its length z. If a 2-inch by 6-inch white pine joist 10 feet long safely supports 1000 pounds when placed on edge (as shown in Figure 7), find an explicit formula for $S(x, y, z)$. Then determine its safe load when placed flatwise.

64. Which of the following functions satisfy $f(x + y) = f(x) + f(y)$ for all real numbers x and y?
(a) $f(t) = 2t$ **(b)** $f(t) = t^2$
(c) $f(t) = 2t + 3$ **(d)** $f(t) = -3t$

65. Determine the formula for $f(t)$ if for all real numbers x and y

$$f(x)f(y) - f(xy) = x + y$$

66. TEASER A function f satisfying $f(x + y) = f(x) + f(y)$ for all real numbers x and y is said to be an *additive function*. Prove that if f is additive, then there is a number m such that $f(t) = mt$ for all rational numbers t.

5-2   GRAPHS OF FUNCTIONS

Geometry, however, supplies sustenance and meaning to bare formulas . . . . One can still believe Plato's statement that "geometry draws the soul toward truth."

Morris Kline

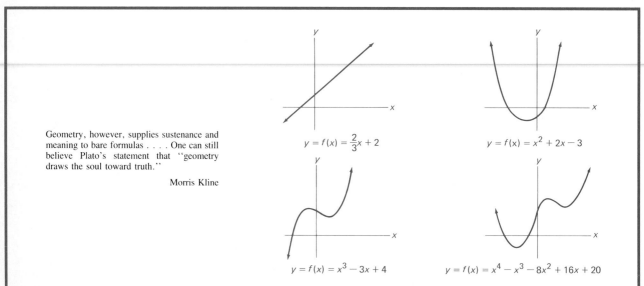

$y = f(x) = \frac{2}{3}x + 2$

$y = f(x) = x^2 + 2x - 3$

$y = f(x) = x^3 - 3x + 4$

$y = f(x) = x^4 - x^3 - 8x^2 + 16x + 20$

We have said that functions are usually specified by giving formulas. Formulas are fine for manipulation and almost essential for exact quantitative information, but to grasp the overall qualitative features of a function, we need a picture. The best picture of a function is its graph. And the graph of a function f is simply the graph of the equation $y = f(x)$. We learned how to graph equations in Chapter 4.

Polynomial Functions

We look first at polynomial functions, that is, functions of the form

$$f(x) = a_n x^n + a_{n-1} x^{n-1} + \cdots + a_1 x + a_0$$

Four typical graphs are shown above. We know from Chapter 4 that the graph of $f(x) = ax + b$ is always a straight line and that, if $a \neq 0$, the graph of $f(x) = ax^2 + bx + c$ is necessarily a parabola.

The graphs of higher degree polynomial functions are harder to describe, but after we have studied two examples, we can offer some general guidelines. Consider first the cubic function

$$f(x) = x^3 - 3x + 4$$

With the help of a table of values, we sketch its graph (Figure 8).

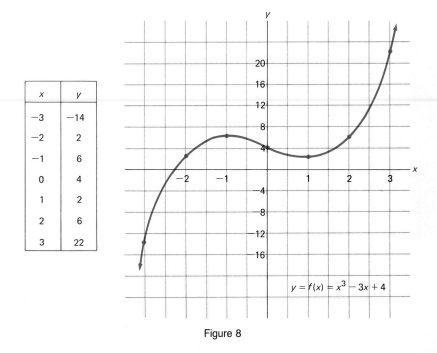

x	y
−3	−14
−2	2
−1	6
0	4
1	2
2	6
3	22

$y = f(x) = x^3 - 3x + 4$

Figure 8

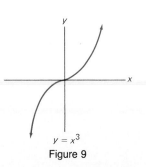

$y = x^3$

Figure 9

Notice that for large positive values of x, the values of y are large and positive; similarly, for large negative values of x, y is large and negative. This is due to the dominance of the leading term x^3 for large $|x|$. This dominance is responsible for a drooping left arm and a right arm held high on the graph. Notice also that the graph has one hill and one valley. This is typical of the graph of a cubic function, though it is possible for it to have no hills or valleys. The graph of $y = x^3$ illustrates this latter behavior (Figure 9).

Next consider a typical fourth-degree polynomial function.

$$f(x) = -x^4 + 4x^3 + 2x^2 - 12x - 3$$

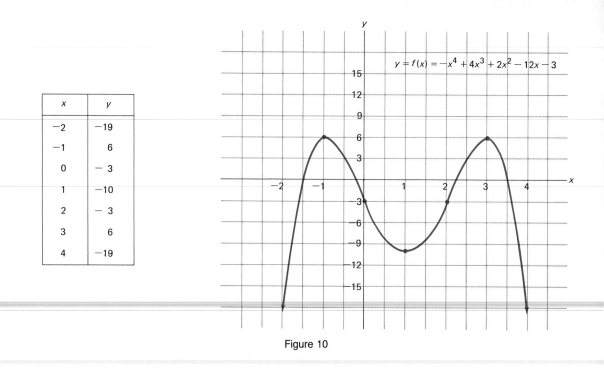

x	y
-2	-19
-1	6
0	-3
1	-10
2	-3
3	6
4	-19

Figure 10

A table of values and the graph are shown in Figure 10.

The leading term $-x^4$, which is negative for all values of x, determines that the graph has two drooping arms. Note that there are two hills and one valley.

In general, we can make the following statements about the graph of

$$f(x) = a_n x^n + a_{n-1} x^{n-1} + \cdots + a_1 x + a_0, \qquad a_n \neq 0$$

1. If n is even and $a_n < 0$, the graph will have two drooping arms; if n is even and $a_n > 0$, it will have both arms raised. This is due to the dominance of $a_n x^n$ for large values of $|x|$.

2. If n is odd, one arm droops and the other is raised. If $a_n > 0$, the right arm is raised; if $a_n < 0$, the right arm droops.

3. The combined number of hills and valleys cannot exceed $n - 1$, although it can be less.

Based on these facts, we expect the graph of a fifth-degree polynomial function with positive leading coefficient to look something like the graph in Figure 11.

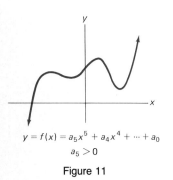

$y = f(x) = a_5 x^5 + a_4 x^4 + \cdots + a_0$
$a_5 > 0$

Figure 11

Symmetry Properties

A function f is called an **even function** if $f(-x) = f(x)$ for all x in its domain. Examples include $f(x) = x^2$ and $f(x) = x^4 - 3x^2$. The graph of an even function is symmetric with respect to the y-axis. A function g is called an **odd function** if $g(-x) = -g(x)$ for all x in its domain. Examples include $g(x) = x^3$ and $g(x) = -1/x^5$. The graph of an odd function is symmetric with respect to the origin. See Section 4-2 for a full discussion of symmetry.

Example A (Determining Symmetry)   Graph the following two functions, making use of their symmetries.
(a) $f(x) = x^4 + x^2 - 3$ (b) $g(x) = x^3 + 2x$

Solution   Notice that

$$f(-x) = (-x)^4 + (-x)^2 - 3 = x^4 + x^2 - 3 = f(x)$$
$$g(-x) = (-x)^3 + 2(-x) = -x^3 - 2x = -g(x)$$

Thus f is even and g is odd. Their graphs are sketched in Figure 12. Note that a polynomial function involving only even powers of x is even, while one involving only odd powers of x is odd. Most functions are neither even nor odd. A typical example is $f(x) = x^3 - 3x + 4$. Its graph (not symmetric with respect to the y-axis or the origin) was shown in Figure 8.

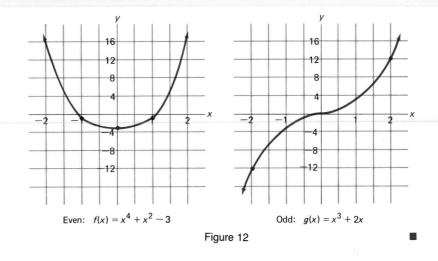

Even: $f(x) = x^4 + x^2 - 3$ Odd: $g(x) = x^3 + 2x$

Figure 12   ∎

Factored Polynomial Functions

The task of graphing can be simplified considerably if our polynomial is factored. The real solutions of $f(x) = 0$ correspond to the x-intercepts of the graph of $y = f(x)$—that is, to the x-coordinates of the points where the graph intersects the x-axis. If the polynomial is factored, these intercepts are easy to find. Consider as an example.

$$y = f(x) = x(x + 3)(x - 1)$$

The solutions of $f(x) = 0$ are 0, -3, and 1; these are the x-intercepts of the graph. Clearly, $f(x)$ cannot change signs between adjacent x-intercepts since only at these points can any of the linear factors change sign. The signs of $f(x)$ on the four intervals determined by $x = -3$, $x = 0$, and $x = 1$ are shown in Figure 13 (to check

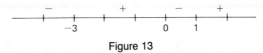

Figure 13

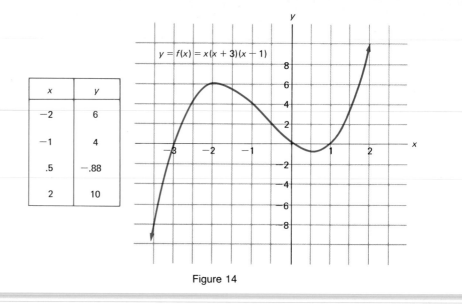

x	y
−2	6
−1	4
.5	−.88
2	10

Figure 14

this, try substituting an x-value from each of these intervals, as in the split-point method of Section 3-5). This information and a few plotted points lead to the graph of Figure 14.

Example B (Graphing Factored Polynomials)   Graph
$$f(x) = (x - 1)^2(x - 3)(x + 2)$$

Solution   The x-intercepts are at 1, 3, and −2. The new feature is that $x - 1$ occurs as a square. The factor $(x - 1)^2$ never changes sign, so the graph does not cross

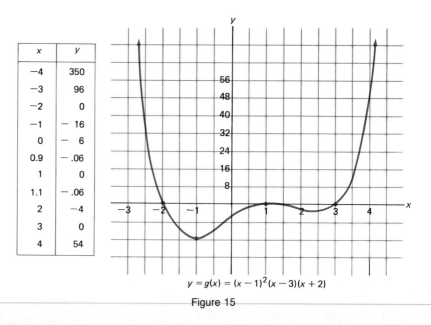

x	y
−4	350
−3	96
−2	0
−1	− 16
0	− 6
0.9	− .06
1	0
1.1	− .06
2	−4
3	0
4	54

$$y = g(x) = (x - 1)^2(x - 3)(x + 2)$$

Figure 15

the x-axis at $x = 1$; it merely touches the axis there (Figure 15). Note the entries in the table of values corresponding to $x = 0.9$ and $x = 1.1$. ■

Piecewise-Defined Functions

Sometimes a function has polynomial components even though it is not a polynomial function. Especially notable is the absolute value function $f(x) = |x|$, which has the two-part rule

$$f(x) = \begin{cases} -x & \text{if } x < 0 \\ x & \text{if } x \geq 0 \end{cases}$$

For $x < 0$, the graph coincides with the line $y = -x$; for $x \geq 0$, it coincides with the line $y = x$. Note the sharp corner at the origin (Figure 16).

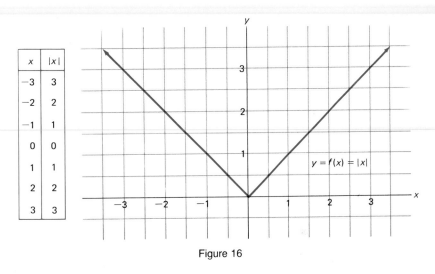

x	$\lvert x \rvert$
-3	3
-2	2
-1	1
0	0
1	1
2	2
3	3

Figure 16

Here is a more complicated example.

Example C (A Function with a Three-Part Rule)   Graph the following function.

$$g(x) = \begin{cases} x + 2 & \text{if } x < 0 \\ x^2 & \text{if } 0 \leq x \leq 2 \\ 4 & \text{if } x > 2 \end{cases}$$

Solution   Though this way of describing a function may seem strange, it is not at all unusual in more advanced courses. The graph of g consists of three pieces (Figure 17).

1. A part of the line $y = x + 2$.
2. A part of the parabola $y = x^2$.
3. A part of the horizontal line $y = 4$.

Note the use of the circle at $(0, 2)$ to indicate that this point is not part of the graph.

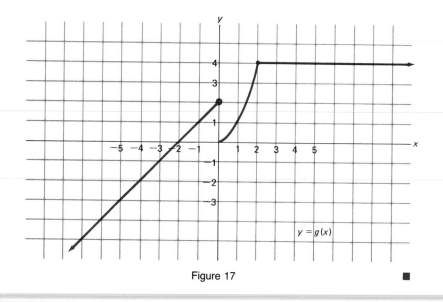

Figure 17

PROBLEM SET 5-2

A. Skills and Techniques

Graph each of the following polynomial functions. The first two are called constant functions.

1. $f(x) = 5$ **2.** $f(x) = -4$

3. $f(x) = -3x + 5$ **4.** $f(x) = 4x - 3$

5. $f(x) = x^2 - 5x + 4$ **6.** $f(x) = x^2 + 2x - 3$

7. $f(x) = x^3 - 9x$ **8.** $f(x) = x^3 - 16x$

9. $f(x) = 2.12x^3 - 4.13x + 2$

10. $f(x) = -1.2x^3 + 2.3x^2 - 1.4x$

Determine which of the following are even functions, which are odd functions, and which are neither. Then sketch the graphs of those that are even or odd, making use of the symmetry properties. See Example A.

11. $f(x) = 2x^2 - 5$ **12.** $f(x) = -3x^2 + 2$

13. $f(x) = x^2 - x + 1$ **14.** $f(x) = -2x^3$

15. $f(x) = 4x^3 - x$ **16.** $f(x) = x^3 + x^2$

17. $f(x) = 2x^4 - 5x^2$ **18.** $f(x) = 3x^4 + x^2$

Sketch the graph of each of the following. See Example B.

19. $f(x) = (x + 1)(x - 1)(x - 3)$

20. $f(x) = x(x - 2)(x - 4)$

21. $f(x) = x^2(x - 4)$ **22.** $f(x) = x(x + 2)^2$

23. $f(x) = (x + 2)^2(x - 2)^2$ **24.** $f(x) = x(x - 1)^3$

Graph each of the following functions. See Example C.

25. $f(x) = 2|x|$ **26.** $f(x) = |x| - 2$

27. $f(x) = |x - 2|$ **28.** $f(x) = |x| + 2$

29. $f(x) = \begin{cases} x & \text{if } x < 0 \\ 2 & \text{if } x \geq 0 \end{cases}$

30. $f(x) = \begin{cases} -1 & \text{if } x \leq 0 \\ 2x & \text{if } x > 0 \end{cases}$

31. $f(x) = \begin{cases} -5 & \text{if } x \leq -3 \\ 4 - x^2 & \text{if } -3 < x \leq 3 \\ -5 & \text{if } x > 3 \end{cases}$

32. $f(x) = \begin{cases} 9 & \text{if } x < 0 \\ 9 - x^2 & \text{if } 0 \leq x \leq 3 \\ x^2 - 9 & \text{if } x > 3 \end{cases}$

B. Applications and Extensions

33. Recall that a function assigns exactly one value to each element in its domain. What must be true about a graph for it to be the graph of a function with rule of the form $y = f(x)$?

34. Which of the graphs in Figure 18 are graphs of functions with rules of the form $y = f(x)$?

In Problems 35–42, sketch the graph of the given function.

35. $f(x) = \sqrt{x}$ **36.** $f(x) = 2\sqrt[3]{x}$

37. $f(x) = (x + 2)(x - 1)^2(x - 3)$

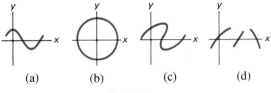

(a)   (b)   (c)   (d)

Figure 18

38. $f(x) = x(x - 2)^3$

39. $f(x) = \begin{cases} 4 - x^2 & \text{if } -2 \le x < 2 \\ x - 1 & \text{if } 2 \le x \le 4 \end{cases}$

40. $f(x) = \begin{cases} |x| & \text{if } -2 \le x \le 1 \\ x^2 & \text{if } 1 < x \le 2 \\ 4 & \text{if } 2 < x \le 4 \end{cases}$

41. $f(x) = x^4 + x^2 + 2, \; -2 \le x \le 2$

42. $f(x) = x^5 + x^3 + x, \; -1 \le x \le 1$

43. The graphs of the three functions $f(x) = x^2$, $g(x) = x^4$, and $h(x) = x^6$ all pass through the points $(-1, 1)$, $(0, 0)$, and $(1, 1)$. Draw sketches of these three functions using the same axes. Be sure to show clearly how they differ for $-1 < x < 1$.

44. Sketch the graph of $f(x) = x^{50}$ for $-1 \le x \le 1$. Be sure to calculate $f(.5)$ and $f(.9)$. What simple figure does the graph resemble?

45. The function $f(x) = [x]$ is called the **greatest integer function.** It assigns to each real number x the largest integer that is less than or equal to x. For example, $[\frac{5}{2}] = 2$, $[5] = 5$, and $[-1.5] = -2$. Graph this function on the interval $-2 \le x \le 6$.

46. Graph each of the following functions on the interval $-2 \le x \le 2$.
   (a) $f(x) = 3[x]$ **(b)** $g(x) = [3x]$

47. Suppose that the cost of shipping a package is 15¢ for anything weighing less than an ounce and 25¢ for anything weighing at least 1 ounce but less than 2 ounces. Beyond that the pattern continues with the cost increased 10¢ for each additional ounce or fraction thereof. Write a formula for the cost $C(x)$ of shipping a package weighing x ounces using the symbol [ ] and graph this function.

48. It costs the XYZ Company $1000 + 10\sqrt{x}$ dollars to make x dolls, which sell for $8 each. Express the total profit $T(x)$ in terms of x and then graph this function.

49. Harold Schwartz plans to fence two identical adjoining rectangular pens, each with 400 square meters of area, as shown in Figure 19.
   (a) Express L, the total length of fence required, in terms of x.

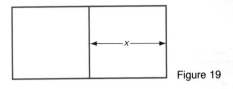

Figure 19

   (b) Calculate the value of L for $x = 5, 10, 15, 20$, and 50.
   (c) Sketch the graph of L on the interval $5 \le x \le 50$.
   (d) Use your graph to estimate the value of x for which L is a minimum.

50. The Schneider Shoe Company wholesales jogging shoes at $50 per pair if 100 or less pairs are ordered but will reduce the price on each pair at the rate of 4¢ times the number beyond 100 that are ordered (up to 1000 pairs).
   (a) Express the amount A that Schneider receives for an order of size n in terms of n, $0 \le n \le 1000$.
   (b) Sketch the graph of a continuous curve that matches the data above for each integer n, $0 \le n \le 1000$.
   (c) Use your graph to estimate the order size that maximizes A.

51. Calculating values for higher-degree polynomials is messy but can be simplified, as we illustrate for $f(x) = 4x^5 - 3x^4 + 2x^3 - x^2 + 7x - 3$. Write this polynomial as

$$f(x) = ((((4x - 3)x + 2)x - 1)x + 7)x - 3$$

Use this to calculate $f(3)$, $f(4.3)$, and $f(-1.6)$. You should be able to make your calculator do these calculations without using any parentheses.

52. An open box is to be made from a piece of 12-inch by 18-inch cardboard by cutting a square of side x inches from each corner and turning up the sides (Figure 20).

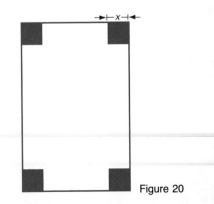

Figure 20

Express the volume $V(x)$ of the box in terms of x and graph the resulting function. What is the domain of V? Use your graph to help you find the value of x that makes $V(x)$ a maximum. What is this maximum value?

53. The function $f(x) = \langle x \rangle$ will denote the **distance to the nearest integer function.** It assigns to each real number x the distance to the integer nearest to x. For example, $\langle 1.2 \rangle = .2$, $\langle 1.7 \rangle = .3$, and $\langle 2 \rangle = 0$. Graph

this function on the interval $0 \le x \le 4$ and then find the area of the region between this graph and the x-axis.

54. **TEASER** Consider the function $f(x) = \langle x \rangle / 10^{[x]}$ on the infinite interval $x \ge 0$. Sketch the graph of this function and calculate the *total* area of the region between the graph and the x-axis. Write this area first as an unending decimal and then as a ratio of two integers.

5-3 GRAPHING CALCULATORS (OPTIONAL)

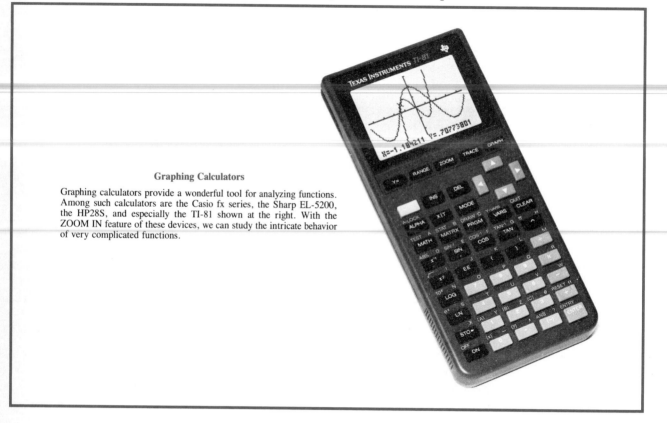

Graphing Calculators

Graphing calculators provide a wonderful tool for analyzing functions. Among such calculators are the Casio fx series, the Sharp EL-5200, the HP28S, and especially the TI-81 shown at the right. With the ZOOM IN feature of these devices, we can study the intricate behavior of very complicated functions.

Graphing complicated functions by hand is a tedious task involving many calculations. Modern technology has greatly simplified the task. First computers with graphics screens appeared. Then came graphing calculators: Casio fx-7000G, Sharp EL-5200, HP 28S, and TI-81. Each of these machines, once mastered, makes graphing a trivial exercise. Perhaps the easiest to use is the TI-81; so this marvelous electronic device will serve as our principal example in what follows. However, we emphasize that any of the other graphing calculators or a computer can be taught to perform all the operations that we will describe.

The Viewing Rectangle

The viewing window on the TI-81 is a rectangle approximately $2\frac{1}{4}$ inches by $1\frac{1}{2}$ inches in size. Its parameters are set from the factory (the so-called *default values*) so that it will display a coordinate system with both the *x*- and *y*-values ranging from -10 to 10 and with one unit between the tick marks (Figure 21). But all of this is easily changed by pressing the $\boxed{\text{RANGE}}$ key, which causes a menu to appear on the screen (Figure 22) from which you may select the parameter values you like: move the cursor, using the arrow keys, to the value you wish to change, press $\boxed{\text{CLEAR}}$, and type in the new value.

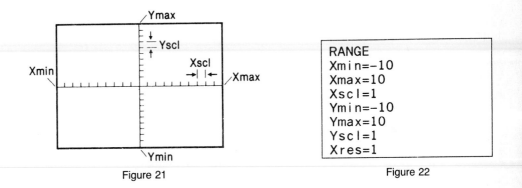

Figure 21

Figure 22

Graphing Functions

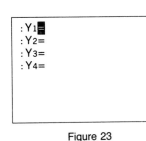

Figure 23

Assuming that the range values are the default values, let us illustrate the graphing of a simple cubic function, namely, $y = f(x) = x^3 - 5x + 1$. Press the $\boxed{\text{Y} =}$ key, thereby displaying the menu shown in Figure 23. With the cursor blinking at $Y_1 =$, type in $X^\wedge 3 - 5X + 1$. Note several things about the TI-81 (remember that calculators vary). It has a special key for the variable X (the $\boxed{\text{X}\,|\,\text{T}}$ key), it uses the $\boxed{\wedge}$ key to denote exponentiation, and it understands juxtaposed symbols to mean multiplication.

With the functions entered, we are ready to graph. Simply press the key $\boxed{\text{GRAPH}}$ and within a few moments, the display shown in Figure 24 will appear.

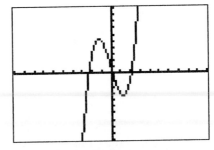

Figure 24

Example A (Changing the Range Values)   Graph $y = \frac{1}{2}x^2 + 11$.

Solution   Proceeding as before, press $\boxed{\text{Y} =}$, and $\boxed{\text{CLEAR}}$, then type in .5X$^\wedge$2 + 11, and press $\boxed{\text{GRAPH}}$. Nothing happens; what is wrong? A little thought should make you realize that the graph is above the viewing window; you will need to change the range values. Press the $\boxed{\text{RANGE}}$ key and change Ymax to 100 and Yscl to 10 (the latter spreads out the tick marks on the y-axis). Finally, press $\boxed{\text{GRAPH}}$ and Figure 25 will appear on the screen.

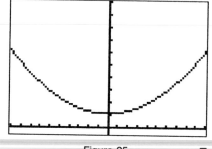

Figure 25   ■

Example B (Graphing Two Functions).   Graph

$$y = f(x) = x^3 - 5x + 3 \qquad \text{and} \qquad y = g(x) = \tfrac{1}{2}x^2$$

in the same coordinate plane.

Using the $\boxed{\text{RANGE}}$ key, return Ymax to 10 and Yscl to 1; then press $\boxed{\text{Y} =}$ and $\boxed{\text{CLEAR}}$. Enter $Y_1 = X^\wedge3 - 5X + 3$, move the cursor down to Y_2 and enter .5X$^\wedge$2, then press $\boxed{\text{GRAPH}}$. In a few moments you should see a display like Figure 26.

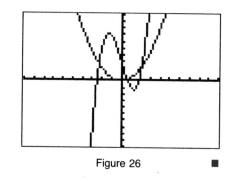

Figure 26   ■

The Zoom Key

Drawing a graph is one thing; analyzing the graph is another. This is where the $\boxed{\text{ZOOM}}$ key plays a perfectly marvelous role. It allows us to magnify any portion of the graph so that we can examine its fine structure. We assume that Figure 26 still appears in the viewing window. Suppose that we are interested in finding the coor-

```
ZOOM
1:Box
2:Zoom In
3:Zoom Out
4:Set Factors
5:Square
6:Standard
7↓Trig
```

Figure 27

dinates of the rightmost of the three intersection points. The cursor, a cross, is at the origin, so you cannot see it initially. Move it (using the arrow keys) as close as you can to the desired intersection point. Press the ⌈ZOOM⌉ key to display a menu (Figure 27), move the cursor to Zoom In, press ⌈ENTER⌉, and the graphs of Figure 26 should reappear with an addition: the coordinates of the cursor are shown at the bottom of the screen. Now press ⌈ENTER⌉ again. With the position of the cursor as center, the graph will be magnified by a factor of 4 (in both directions) from its original size, as shown in Figure 28. Now move the cursor as close as you can to the intersection point (note how the coordinates at the bottom of the screen keep changing) and press ⌈ENTER⌉ again. The graph is magnified by another factor of 4 (Figure 29). Keep this process up until you have a repetition of digits in the coordinates of the intersection point to desired accuracy. We get (2.168, 2.350) to three-decimal-place accuracy.

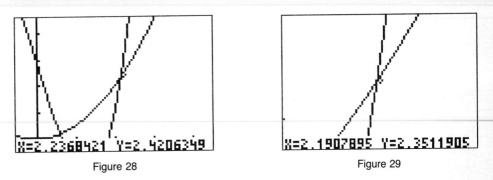

Figure 28   Figure 29

Example C (Intersection Points)   Find the coordinates of the leftmost of the intersection points of the graphs in Example B.

Solution   Press ⌈ZOOM⌉, select Standard, and press ⌈ENTER⌉ to return the graph to its Figure 26 appearance. Move the cursor as close as you can to the desired intersection point. Press ⌈ZOOM⌉ and Zoom In followed by ⌈ENTER⌉ twice to magnify the graph by a factor of 4 with the cursor as center. Move the cursor to the intersection point, magnify again, and so on. When you have desired accuracy, read off the coordinates from the bottom of the screen. We get $(-2.276, 2.590)$. ■

Maxima and Minima

A very important problem is that of finding maximum and minimum points of a graph. First let us distinguish between a local (relative) and a global (absolute) maximum or minimum point. A point (x_1, y_1) on the graph of $y = f(x)$ is a **local maximum point** for f if $y_1 \geq y$ for all points (x, y) of the graph near (x_1, y_1); it is a **global maximum point** if the inequality $y_1 \geq y$ holds for all points (x, y) on the graph of the function. For the definitions of local and global minimum points, simply replace $\geq$ by $\leq$ in the definitions above.

Example D (Finding a Local Maximum Point)   Find the local maximum point for $f(x) = x^3 - 5x + 3$ that is in quadrant II (see Figure 26). Then find the global maximum point, assuming that the domain for f is $-10 \leq x \leq 10$.

Solution   To return the range values to their default values, press ZOOM and then Standard. And to cut down on clutter, press Y = and clear Y_2. Now when you press GRAPH you should have the graph of $y = x^3 - 5x + 3$. Press TRACE and use the arrow keys to move the cursor as close as you can to the local maximum point. (Note that the TRACE key makes the cursor move along the curve.) Press ZOOM and select Zoom In; then press ENTER twice to magnify the graph around the chosen point. Use TRACE to move the cursor as near as you can to the maximum point; then press ZOOM, select Zoom In, and press ENTER twice. Continue until the desired accuracy is achieved and read off the coordinates of the local maximum point at the bottom of the screen. We get $(-1.291, 7.303)$.

One look at the graph of $y = x^3 - 5x + 3$ shows that the global maximum point for the given domain is at the extreme right end of the graph; that is, it is $(10, 953)$.   ■

There is a great deal more that could be said about graphing calculators, but this should be enough to get you started. Read your instruction book for more information.

PROBLEM SET 5-3

A. Skills and Techniques

Draw the graph of each function showing enough of the graph so that the critical features are evident. See Example A.

1. $y = x^3$

2. $y = .1x^3$

3. $y = x^3 + x^2 - 2$

4. $y = .2x^3 - x^2 + 2$

5. $y = -.2x^3 - x^2 + 2$

6. $y = x^3 - 3x + 2$

7. $y = x^4 - x^2 - 3$

8. $y = .1x^4 - x^2 - 3$

9. $y = .1x^4 - x^2 + 2x - 3$

10. $y = x^4 - x^3 - 3x^2 + x - 2$

11. $y = .05x^6 - 15x^2 + 5x - 3$

12. $y = .2x^5 - 3x^3 + 5x - 2$

In Problems 13–18, draw the graphs of both functions using the same axes as in Example B. Then find their intersection points as in Example C. Aim for three-decimal-place accuracy.

13. $y = x^3 - x^2 + 2, y = -2x + 1$

14. $y = x^3 - x^2 + 2, y = -x^2 + 5$

15. $y = x^3 - x^2 + 2, y = \sqrt{x + 8}$

16. $y = x^3 - x^2 + 2, y = \sqrt{x} + 6$

17. $y = x^4 - x^2 - 3, y = \sqrt{x} + 6$

18. $y = x^4 - x^3 - 3x^2 + x - 2, y = 3x - 5$

19. Find the local minimum point of $y = -.2x^3 -$
$x^2 + 2$ in quadrant III. Then find the global minimum point for $-10 \leq x \leq 10$. (See Example D.)

20. Find the local minimum point of $y = .2x^5 - 3x^3 + 5x - 2$ in quadrant IV. Then find the global minimum point for $-10 \leq x \leq 10$.

21. Find the global minimum point of $y = .05x^6 - 15x^2 + 5x - 3$.

22. Find the global minimum point of $y = .1x^4 - x^2 + 2x - 3$.

B. Applications and Extensions

In Problems 23–28, draw the graph of each function showing all essential features.

23. $y = |x| = $ ABS x

24. $y = |x - 2|$

25. $y = 2|x + 1|$

26. $y = 2^{-x}$

27. $y = 8x2^{-x}$

28. $y = 8|x|2^{-x}$

29. Solve the equation $x^3 - 3x - 5 = 0$, that is, find the x-coordinates of all points where the graph of $y = x^3 - 3x - 5$ crosses the x-axis. Your answers should be accurate to three decimal places.

30. Solve $x^4 - x^3 - 3x^2 + x - 2 = 0$.

31. Find the local maximum and minimum points of $y = x^3 - 3x - 5$.

32. Find the global minimum point of $y = x^4 - x^3 - 3x^2 + x - 2$.

Problems 33–36 deal with piecewise defined functions. For the T1-81, an inequality in parentheses like $(x < b)$ has the value 1 if the inequality holds and the value 0 otherwise. (See the instruction book for the TI-81.)

33. Draw the graph of
$$y = -2(x < -1) + 2x(-1 \le x)(x \le 2) + 4(x > 2).$$

34. Draw the graph of
$$y = |x + 2|(x \le 1) + (-2x + 5)(x > 1).$$

35. Draw the graph of the piecewise defined function given by
$$y = \begin{cases} 1 & \text{if } x < -1 \\ x^2 & \text{if } -1 \le x \le 2 \\ 4 & \text{if } x > 2 \end{cases}$$

36. Draw the graph of
$$y = \begin{cases} -7 & \text{if } x < -2 \\ x^3 + 1 & \text{if } -2 \le x \le 1 \\ 2 & \text{if } x > 1 \end{cases}$$

37. Refer to the instruction book and then draw the graph of the greatest integer function $y = [x] = \text{INT } x$.

38. Draw the graph of $y = [.5x] + 2$.

39. Draw the graph of $y = x - [x]$.

40. **TEASER** Figure out a way to graph the distance to the nearest integer function $y = \langle x \rangle$ (see Problem 53 of Section 5-2). Do this by expressing $\langle x \rangle$ in terms of functions appearing previously in this problem set.

5-4 GRAPHING RATIONAL FUNCTIONS

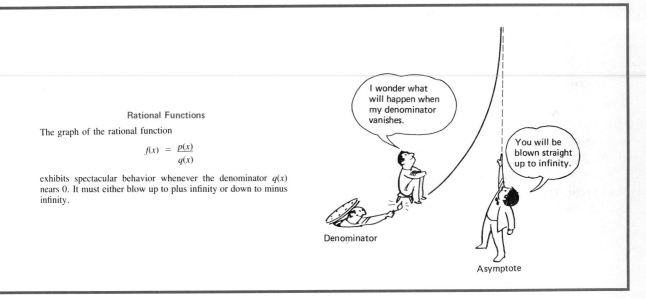

Rational Functions

The graph of the rational function

$$f(x) = \frac{p(x)}{q(x)}$$

exhibits spectacular behavior whenever the denominator $q(x)$ nears 0. It must either blow up to plus infinity or down to minus infinity.

If $f(x)$ is given by

$$f(x) = \frac{p(x)}{q(x)}$$

where $p(x)$ and $q(x)$ are polynomials, then f is called a **rational function.** For simplicity, we shall assume that $f(x)$ is in reduced form, that is, that $p(x)$ and $q(x)$ have no common nontrivial factors. Typical examples of rational functions are

$$f(x) = \frac{x + 1}{x^2 - x + 6} = \frac{x + 1}{(x - 3)(x + 2)}$$

$$g(x) = \frac{(x + 2)(x - 5)}{(x + 3)^3}$$

Graphing a rational function can be tricky, primarily because of the denominator $q(x)$. Whenever it gets close to zero, something dramatic is sure to happen to the graph. That is the point of our opening cartoon.

The Graphs of $1/x$ and $1/x^2$

Let us consider two simple cases.

$$f(x) = \frac{1}{x} \qquad g(x) = \frac{1}{x^2}$$

Notice that f is an odd function $[f(-x) = -f(x)]$, while g is even $[g(-x) = g(x)]$. These facts imply that the graph of f is symmetric with respect to the origin, and that the graph of g is symmetric with respect to the y-axis. Thus we need to use only positive values of x to calculate y-values. Each calculation yields two points on the graph. Observe particularly the behavior of each graph near $x = 0$ (Figure 30).

In both cases, the x- and y-axes play special roles; we call them asymptotes for the graphs. If, as a point moves away from the origin along a curve, the distance between it and a certain line approaches zero, then that line is called an **asymptote** for the curve. Clearly, the line $x = 0$ is a vertical asymptote for both of our curves and the line $y = 0$ is a horizontal asymptote for both of them.

x	$1/x$	$1/x^2$
0	—	—
.01	100	10000
.1	10	100
1	1	1
4	.25	.06

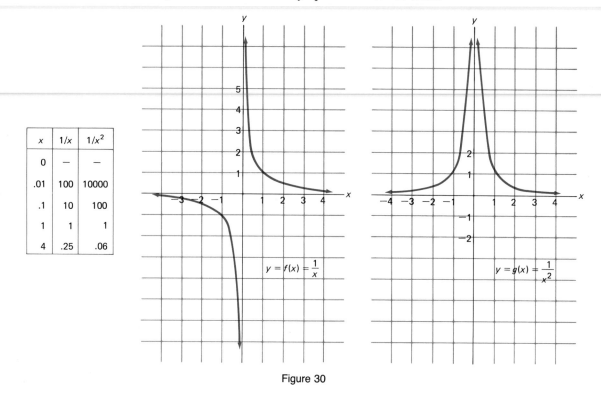

Figure 30

We can describe asymptotes another way by using the symbol $\rightarrow$ for approaches and ∞ for infinity. Thus $x \rightarrow a$ is read "x approaches a." Similarly, $x \rightarrow \infty$ is read "x approaches infinity" but will mean that x grows without bound. Consider now the graph of $y = f(x)$ where f is a rational function. We say that the line $x = a$ is a **vertical asymptote** if $|f(x)| \rightarrow \infty$ as $x \rightarrow a$, and we say that the line $y = b$ is a **horizontal asymptote** if $f(x) \rightarrow b$ as $|x| \rightarrow \infty$. Thus, in the case of $y = 1/x$, $x = 0$ is a vertical asymptote since $|1/x| \rightarrow \infty$ as $x \rightarrow 0$ and $y = 0$ is a horizontal asymptote since $1/x \rightarrow 0$ as $|x| \rightarrow \infty$.

The Graphs of $1/(x - 2)$ and $1/(x - 2)^2$

If we replace x by $x - 2$ in our two functions, we get two new functions.

$$h(x) = \frac{1}{x - 2} \qquad k(x) = \frac{1}{(x - 2)^2}$$

Their graphs are just like those of f and g except that they are shifted two units to the right, as you can see in Figure 31.

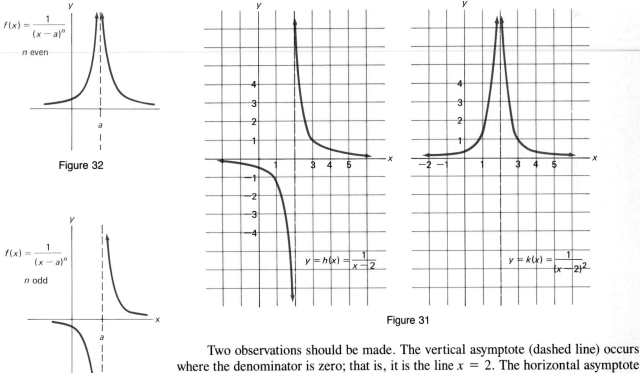

$$f(x) = \frac{1}{(x - a)^n}$$

n even

Figure 32

$$f(x) = \frac{1}{(x - a)^n}$$

n odd

Figure 33

$$y = h(x) = \frac{1}{x - 2}$$

$$y = k(x) = \frac{1}{(x - 2)^2}$$

Figure 31

Two observations should be made. The vertical asymptote (dashed line) occurs where the denominator is zero; that is, it is the line $x = 2$. The horizontal asymptote is again the line $y = 0$.

In general, the graph of $f(x) = 1/(x - a)^n$ has the line $y = 0$ as a horizontal asymptote and the line $x = a$ as a vertical asymptote. The behavior of the graph near $x = a$ for n even and n odd is illustrated in Figures 32 and 33.

More Complicated Examples

Consider next the rational function determined by

$$y = f(x) = \frac{x}{x^2 + x - 6} = \frac{x}{(x - 2)(x + 3)}$$

We expect its graph to have vertical asymptotes at $x = 2$ and $x = -3$. Again, the line $y = 0$ will be a horizontal asymptote since, as $|x|$ gets large, the term x^2 in the denominator will dominate, so that y will behave much like x/x^2 or $1/x$ and will thus approach zero. The graph crosses the x-axis where the numerator is zero, namely, at $x = 0$. Finally, with the help of a table of values, we sketch the graph, shown in Figure 34.

x	y
-6	$-.25$
-4	$-.67$
-2	$.50$
0	0
1	$-.25$
3	$.50$
5	$.21$

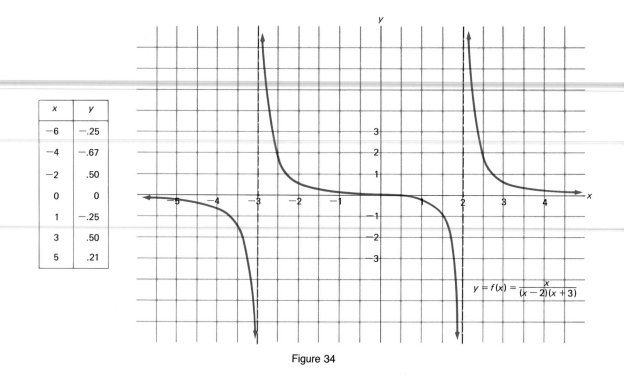

Figure 34

Example A (One Horizontal, One Vertical Asymptote)   Sketch the graph of

$$y = f(x) = \frac{2x^2 + 2x}{x^2 - 4x + 4} = \frac{2x(x + 1)}{(x - 2)^2}$$

Solution   The graph will have one vertical asymptote, at $x = 2$. To check on a horizontal asymptote, we note that for large $|x|$, the numerator behaves like $2x^2$ and the denominator behaves like x^2. It follows that $y = 2$ is a horizontal asymptote. The graph crosses the x-axis where the numerator $2x(x + 1)$ is zero, namely, at $x = 0$ and $x = -1$. Figure 35 exhibits a good approximation to the graph.

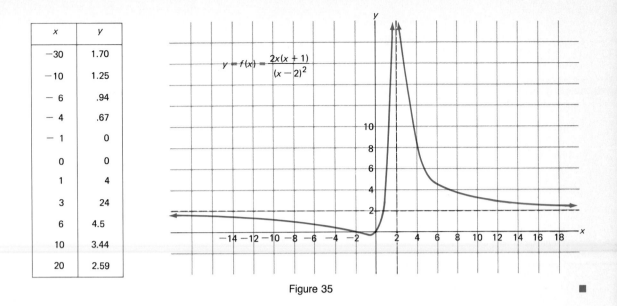

x	y
−30	1.70
−10	1.25
− 6	.94
− 4	.67
− 1	0
0	0
1	4
3	24
6	4.5
10	3.44
20	2.59

Figure 35

It may happen that the denominator of a rational function is nowhere zero. In that case, the graph of the function has no vertical asymptote, as in the next example.

Example B (One Horizontal, No Vertical Asymptote)   Sketch the graph of

$$f(x) = \frac{x^2 - 4}{x^2 + 1} = \frac{(x - 2)(x + 2)}{x^2 + 1}$$

Solution   Note that f is an even function, so the graph will be symmetric with respect to the y-axis. The denominator is not zero for any real x, so there are no vertical asymptotes. The line $y = 1$ is a horizontal asymptote, since for large $|x|$, $f(x)$ behaves like x^2/x^2. The x-intercepts are $x = 2$ and $x = -2$. The graph is shown in Figure 36.

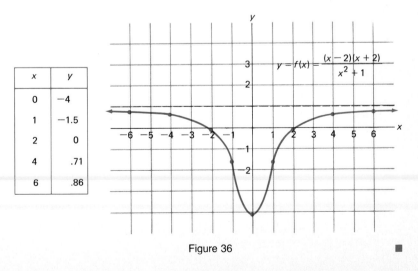

x	y
0	−4
1	−1.5
2	0
4	.71
6	.86

Figure 36

Oblique Asymptotes

It is possible for the graph of a rational function to have an asymptote that is neither vertical nor horizontal. For example, consider

$$f(x) = \frac{x^3 + x + 1}{x^2 + 1} = x + \frac{1}{x^2 + 1}$$

As $|x|$ gets larger and larger, the graph of $y = f(x)$ gets closer and closer to the line $y = x$. This line is an **oblique asymptote,** also called a *slant asymptote*.

Example C (One Vertical, One Oblique Asymptote)   Sketch the graph of

$$f(x) = \frac{x^2}{x + 1}$$

$$\begin{array}{r} x - 1 \\ x + 1 \overline{\smash{)}x^2} \\ \underline{x^2 + x} \\ -x \\ \underline{-x - 1} \\ 1 \end{array}$$

Figure 37

Solution   From our earlier discussion, we expect a vertical asymptote at $x = -1$. There is no horizontal asymptote. However, when we do a long division (as in Figure 37), we find that

$$f(x) = x - 1 + \frac{1}{x + 1}$$

As $|x|$ gets larger and larger, the term $1/(x + 1)$ tends to zero, and so $f(x)$ gets closer and closer to $x - 1$. This means that the line $y = x - 1$ is an oblique asymptote. Its significance is indicated on the graph in Figure 38. In general, we can expect an oblique asymptote for the graph of a rational function whenever the degree of the numerator is exactly one more than that of the denominator.

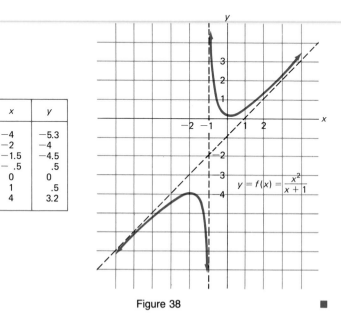

x	y
−4	−5.3
−2	−4
−1.5	−4.5
− .5	.5
0	0
1	.5
4	3.2

$$y = f(x) = \frac{x^2}{x + 1}$$

Figure 38

A General Procedure

Here is an outline of the procedure for graphing a rational function

$$y = f(x) = \frac{p(x)}{q(x)}$$

which is in *reduced form*.

1. Check for symmetry with respect to the y-axis and the origin.
2. Factor the numerator and denominator.
3. Determine the vertical asymptotes (if any) by checking where the denominator is zero. Draw a dashed line for each asymptote. Be sure to examine the behavior of the graph near a vertical asymptote.
4. Determine the horizontal asymptote (if any) by asking what y approaches as $|x|$ becomes large. This is accomplished by examining the quotient of the leading terms from numerator and denominator. Indicate any horizontal asymptote with a dashed line.
5. Determine any oblique asymptote and show it as a dashed line.
6. Determine the x-intercepts (if any). These occur where the numerator is zero.
7. Make a small table of values and plot corresponding points.
8. Sketch the graph.

Rational Functions Not in Reduced Form

So far each of our rational functions has been in reduced form; that is, its numerator and denominator had no nontrivial common factors. Here are some examples of rational functions that are not in reduced form.

$$f(x) = \frac{x(x-1)}{x} \qquad g(x) = \frac{x-2}{x^2-4} \qquad h(x) = \frac{x^3-1}{x^2-1}$$

How do we graph such functions? The next example provides an illustration.

Example D (A Nonreduced Rational Function)   Sketch the graph of

$$f(x) = \frac{x^2 + x - 6}{x - 2}$$

Solution   Notice that

$$f(x) = \frac{(x+3)(x-2)}{x-2}$$

You have the right to expect that we will cancel the factor $x - 2$ from numerator and denominator and graph

$$g(x) = x + 3$$

But note that 2 is in the domain of g but not in the domain of f. Thus f and g and their graphs are exactly alike except at one point, namely, at $x = 2$. Both graphs are shown in Figure 39. You will notice the hole in the graph of $y = f(x)$ at $x = 2$. This technical distinction is occasionally important.

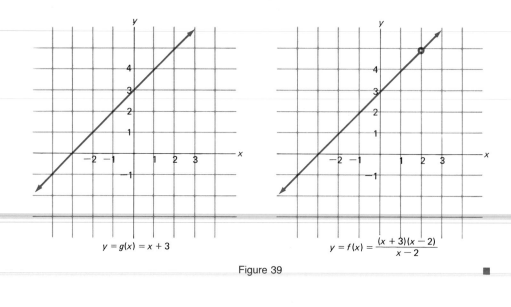

$$y = g(x) = x + 3 \qquad\qquad y = f(x) = \frac{(x + 3)(x - 2)}{x - 2}$$

Figure 39

PROBLEM SET 5-4

A. Skills and Techniques

Sketch the graph of each of the following functions. See Example A.

1. $f(x) = \dfrac{2}{x + 2}$

2. $f(x) = \dfrac{-1}{x + 2}$

3. $f(x) = \dfrac{2}{(x + 2)^2}$

4. $f(x) = \dfrac{1}{(x - 3)^2}$

5. $f(x) = \dfrac{2x}{x + 2}$

6. $f(x) = \dfrac{x + 2}{x - 3}$

7. $f(x) = \dfrac{1}{(x + 2)(x - 1)}$

8. $f(x) = \dfrac{3}{x^2 - 9}$

9. $f(x) = \dfrac{x + 1}{(x + 2)(x - 1)}$

10. $f(x) = \dfrac{3x}{x^2 - 9}$

11. $f(x) = \dfrac{2x^2}{(x + 2)(x - 1)}$

12. $f(x) = \dfrac{x^2 - 4}{x^2 - 9}$

Sketch the graph of each of the following. See Example B.

13. $f(x) = \dfrac{1}{x^2 + 2}$

14. $f(x) = \dfrac{x^2 - 2}{x^2 + 2}$

15. $f(x) = \dfrac{x}{x^2 + 2}$

16. $f(x) = \dfrac{x^3}{x^2 + 2}$

Sketch the graphs of the following rational functions. See Example C.

17. $f(x) = \dfrac{2x^2 + 1}{2x}$

18. $f(x) = \dfrac{x^2 - 2}{x}$

19. $f(x) = \dfrac{x^2}{x - 1}$

20. $f(x) = \dfrac{x^3}{x^2 + 1}$

Sketch the graph of each of the following rational functions, which, you will note, are not in reduced form. See Example D.

21. $f(x) = \dfrac{(x + 2)(x - 4)}{x + 2}$

22. $f(x) = \dfrac{x^2 - 4}{x - 2}$

23. $f(x) = \dfrac{x^3 - x^2 - 12x}{x + 3}$

24. $f(x) = \dfrac{x^3 - 4x}{x^2 - 2x}$

B. Applications and Extensions

Sketch the graphs of the rational functions in Problems 25–30.

25. $f(x) = \dfrac{x}{x + 5}$

26. $f(x) = \dfrac{x - 2}{x + 3}$

27. $f(x) = \dfrac{x^2 - 9}{x^2 - x - 2}$

28. $f(x) = \dfrac{x - 2}{(x + 3)^2}$

29. $f(x) = \dfrac{x^2 - 9}{x^2 - x - 6}$

30. $f(x) = \dfrac{x - 2}{x^2 + 3}$

31. Sketch the graphs of $f(x) = x^n/(x^2 + 1)$ for $n = 1$, 2, and 3, being careful to show all asymptotes.

32. Where does the graph of

$$f(x) = (x^3 + x^2 - 2x + 1)/(x^3 + 2x^2 - 2)$$

cross its horizontal asymptote?

33. Determine all asymptotes (vertical, horizontal, oblique) for the graph of $f(x) = x^4/(x^n - 1)$ in each case.

   (a) $n = 1$ **(b)** $n = 2$
   (c) $n = 3$ **(d)** $n = 4$
   (e) $n = 5$ **(f)** $n = 6$

34. Consider the graph of the general rational function

$$f(x) = \frac{a_n x^n + a_{n-1} x^{n-1} + \cdots + a_1 x + a_0}{b_m x^m + b_{m-1} x^{m-1} + \cdots + b_1 x + b_0},$$

$$a_n \neq 0, \, b_m \neq 0$$

Identify its horizontal asymptote in each case.
   (a) $m > n$ **(b)** $m = n$
   (c) $m < n$

35. A manufacturer of gizmos has overhead of $20,000 per year and direct costs (labor and material) of $50 per gizmo. Write an expression for $U(x)$, the average cost per unit, if the company makes x gizmos per year. Graph the function U and then draw some conclusions from your graph.

36. A cylindrical can is to contain 10π cubic inches. Write a formula for $S(r)$, the total surface area, in terms of the radius r. Graph the function S and use it to estimate the radius of the can that will require the least material to make.

37. A salt solution with a concentration of 50 grams of salt per liter flows into a large tank at the rate of 20 liters per minute (Figure 40). The tank initially contains 400 liters of pure water.

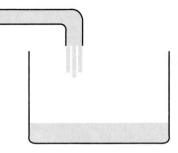

Figure 40

   (a) Write a formula for $A(t)$, the amount of salt in the tank after t minutes.
   (b) Write a formula for $c(t)$, the concentration of salt in the tank after t minutes, and sketch its graph.
   (c) What happens to $c(t)$ as $t \to \infty$?

38. Suppose that three particles with charges -3, $+2$, and -6 are placed on a coordinate line at 1, x, and 5, respectively, as shown in Figure 41. According to Coulomb's law, the force $F(x)$ acting on the particle at x is given by

$$F(x) = \frac{-6k}{(x - 1)^2} + \frac{12k}{(x - 5)^2}$$

for some positive constant k. Sketch the graph of F for $k = \frac{1}{3}$ on the interval $1 < x < 5$ and interpret the behavior near $x = 1$ and $x = 5$.

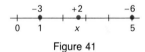

Figure 41

39. Find a formula for $f(x)$ if f is a rational function whose graph goes through (2, 5) and has exactly two asymptotes, namely, $y = 2x + 3$ and $x = 3$.

40. TEASER Sketch the graphs of $f(x) = [1/x]$ and $g(x) = \langle 1/x \rangle$ for $0 < x \leq 1$. The symbols $[\cdot]$ and $\langle \cdot \rangle$ were defined in Problems 45 and 53 of Section 5-2.

41. Graphing calculators with their small screens are not able to give very good global graphs of rational functions. However, they are fine tools for analyzing local behavior such as intersection points. Graph $y = (x + 1)/(x - 2)$ and $y = x^2 + 2$ in the same coordinate plane and find their point of intersection.

42. How many times do the graphs of $y = (x + 1)/(x - 2)$ and $y = x^2 + x + .1$ intersect?

5-5 PUTTING FUNCTIONS TOGETHER

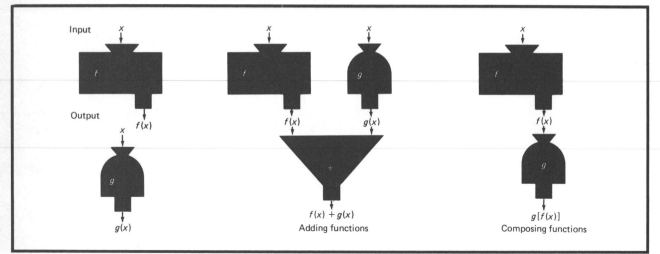

There is still another way to visualize a function. Think of the function named f as a machine. It accepts a number x as input, operates on it, and then presents the number $f(x)$ as output. Machines can be hooked together to make more complicated machines; similarly, functions can be combined to produce more complicated functions. That is the subject of this section.

Sums, Differences, Products, and Quotients

The simplest way to make new functions from old ones is to use the four arithmetic operations. Suppose, for example, that the functions f and g have the formulas

$$f(x) = \frac{x-3}{2} \qquad g(x) = \sqrt{x}$$

We can make a new function $f + g$ by having it assign to x the value $(x - 3)/2 + \sqrt{x}$; that is,

$$(f + g)(x) = f(x) + g(x) = \frac{x-3}{2} + \sqrt{x}$$

Of course, we must be a little careful about domains. Clearly, x must be a number on which both f and g can operate. In other words, the domain of $f + g$ is the intersection (common part) of the domains of f and g.

The functions $f - g$, $f \cdot g$, and f/g are defined in a completely analogous way. Assuming that f and g have their respective natural domains—namely, all reals and the nonnegative reals, respectively—we have the following.

$$(f + g)(x) = f(x) + g(x) = \frac{x-3}{2} + \sqrt{x} \qquad x \geq 0$$

$$(f - g)(x) = f(x) - g(x) = \frac{x-3}{2} - \sqrt{x} \qquad x \geq 0$$

$$(f \cdot g)(x) = f(x) \cdot g(x) = \frac{x-3}{2}\sqrt{x} \qquad x \geq 0$$

$$(f/g)(x) = f(x)/g(x) = \frac{x-3}{2\sqrt{x}} \qquad x > 0$$

To graph the function $f + g$, it is often best to graph f and g separately in the same coordinate plane and then add the y-coordinates together along vertical lines. We illustrate this method (called **addition of ordinates**) in Figure 42.

The graph of $f - g$ can be handled similarly. Simply graph f and g in the same coordinate plane and subtract ordinates. We can even graph $f \cdot g$ and f/g in the same manner, but that is harder.

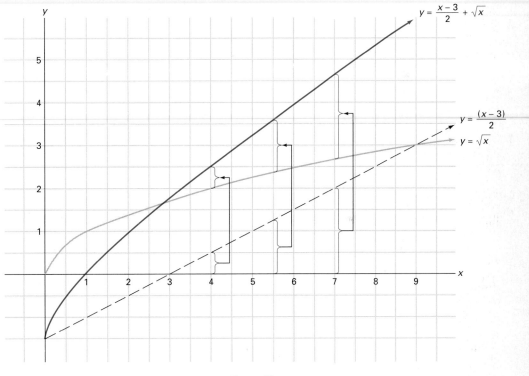

Figure 42

Composition of Functions

To compose functions is to string them together in tandem. Figure 43 shows how this is done. If f operates on x to produce $f(x)$ and then g operates on $f(x)$ to produce $g(f(x))$, we say that we have composed g and f. The resulting function, called the **composite of g with f,** is denoted by $g \circ f$. Thus

$$(g \circ f)(x) = g(f(x))$$

Recall our earlier examples, $f(x) = (x - 3)/2$ and $g(x) = \sqrt{x}$. We may compose them in two ways.

$$(g \circ f)(x) = g(f(x)) = g\left(\frac{x - 3}{2}\right) = \sqrt{\frac{x - 3}{2}}$$

$$(f \circ g)(x) = f(g(x)) = f(\sqrt{x}) = \frac{\sqrt{x} - 3}{2}$$

We note one thing right away: Composition of functions is not commutative; $g \circ f$ and $f \circ g$ are not the same. We must also be careful in describing the domain of a composite function. The domain of $g \circ f$ is that part of the domain of f for which g can accept $f(x)$ as input. In our example, the domain of $g \circ f$ is $x \geq 3$, not all x or $x \geq 0$ as we might have thought at first glance. Figure 44 offers another view of these matters. The shaded portion of the domain of f is not in the domain of $g \circ f$; for x in this portion, $f(x)$ is outside the domain of g.

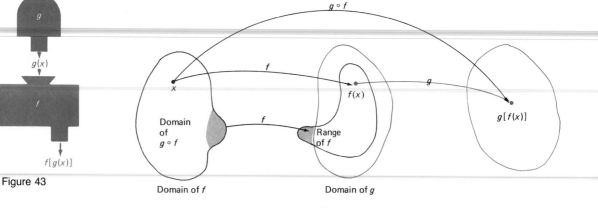

Figure 43

Figure 44

Example A (Composing Functions) Let $f(x) = (x + 1)^2$ and $g(x) = \sqrt{x}$. Write formulas for $(f \circ g)(x)$ and $(g \circ f)(x)$. Also determine the domains of $f \circ g$ and $g \circ f$.

Solution

$$(f \circ g)(x) = f(g(x)) = f(\sqrt{x}) = (\sqrt{x} + 1)^2 = x + 2\sqrt{x} + 1$$
$$(g \circ f)(x) = g(f(x)) = g((x + 1)^2) = \sqrt{(x + 1)^2} = |x + 1|$$

The domain of $f \circ g$ is the set of nonnegative real numbers; for $g \circ f$, it is the set of all real numbers. ■

In calculus, we shall often need to take a given function and decompose it, that is, break it into composite pieces. Usually, this can be done in several ways. Take $p(x) = \sqrt{x^2 + 3}$ for example. We may think of it as

$$p(x) = g(f(x)) \quad \text{where} \quad g(x) = \sqrt{x} \quad \text{and} \quad f(x) = x^2 + 3$$

or as

$$p(x) = g(f(x)) \quad \text{where} \quad g(x) = \sqrt{x + 3} \quad \text{and} \quad f(x) = x^2$$

Example B (Decomposing Functions) In each of the following, H can be thought of as a composite function $g \circ f$. Write formulas for $f(x)$ and $g(x)$.

(a) $H(x) = (2 + 3x)^2$ (b) $H(x) = \dfrac{1}{(x^2 + 4)^3}$

Solution

(a) Think of how you might calculate $H(x)$. You would first calculate $2 + 3x$ and then square the result. That suggests

$$f(x) = 2 + 3x \qquad g(x) = x^2$$

(b) Here there are two obvious ways to proceed. One way would be to let

$$f(x) = x^2 + 4 \qquad g(x) = \frac{1}{x^3}$$

Another selection, which is just as good, is

$$f(x) = \frac{1}{x^2 + 4} \qquad g(x) = x^3$$

We could actually think of H as the composite of four functions. Let

$$f(x) = x^2 \qquad g(x) = x + 4 \qquad h(x) = x^3 \qquad j(x) = \frac{1}{x}$$

Then

$$H = j \circ h \circ g \circ f$$

You should check this result. ∎

Translations

Observing how a function is built up from simpler ones using the operations of this section can be a big aid in graphing. This is especially true of *translations*, which result from the composition $f(x - h)$ and/or the simple addition of a constant k.

Consider, for example, the graphs of

$$y = f(x) \qquad y = f(x - 3) \qquad y = f(x) + 2 \qquad y = f(x - 3) + 2$$

for the case $f(x) = |x|$. The four graphs are shown in Figure 45. Notice that all four graphs have the same shape; the last three are just translations (shifts) of the first. Replacing x by $x - 3$ translates the graph 3 units to the right; adding 2 translates the graph 2 units up. What happened with $f(x) = |x|$ is typical.

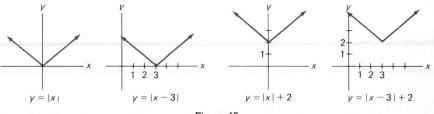

$y = |x|$ $y = |x - 3|$ $y = |x| + 2$ $y = |x - 3| + 2$

Figure 45

Example C (Graphing Using Translations)   Sketch the graph of $y = x^3 + x^2$. Then use translations to sketch the graphs of $y = (x + 1)^3 + (x + 1)^2$, $y = x^3 + x^2 - 2$, and $y = (x + 1)^3 + (x + 1)^2 - 2$.

Solution   The graphs are shown in Figure 46.

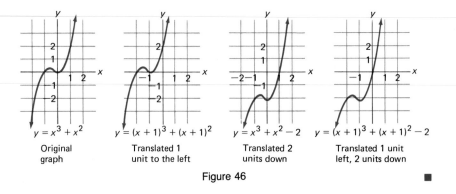

$y = x^3 + x^2$	$y = (x + 1)^3 + (x + 1)^2$	$y = x^3 + x^2 - 2$	$y = (x + 1)^3 + (x + 1)^2 - 2$
Original graph	Translated 1 unit to the left	Translated 2 units down	Translated 1 unit left, 2 units down

Figure 46

Exactly the same principles illustrated in Example C apply in the general situation. Figure 47 shows what happens when h and k are both positive. If $h < 0$, the translation is to the left; if $k < 0$, the translation is downward.

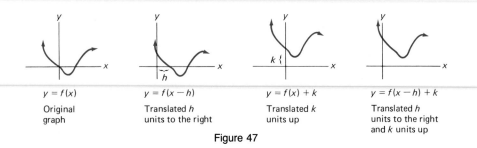

$y = f(x)$	$y = f(x - h)$	$y = f(x) + k$	$y = f(x - h) + k$
Original graph	Translated h units to the right	Translated k units up	Translated h units to the right and k units up

Figure 47

PROBLEM SET 5-5

A. Skills and Techniques

1. Let $f(x) = x^2 - 2x + 2$ and $g(x) = 2/x$. Calculate each of the following.
(a) $(f + g)(2)$ (b) $(f + g)(0)$
(c) $(f - g)(1)$ (d) $(f \cdot g)(-1)$
(e) $(f/g)(2)$ (f) $(g/f)(2)$
(g) $(f \circ g)(-1)$ (h) $(g \circ f)(-1)$
(i) $(g \circ g)(3)$

2. Let $f(x) = 3x + 5$ and $g(x) = |x - 2|$. Perform the calculations in Problem 1 for these functions.

In each of the following, write the formulas for $(f + g)(x)$, $(f - g)(x)$, $(f \cdot g)(x)$, and $(f/g)(x)$ and give the domains of these four functions.

3. $f(x) = x^2$, $g(x) = x - 2$

4. $f(x) = x^3 - 1$, $g(x) = x + 3$

5. $f(x) = x^2$, $g(x) = \sqrt{x}$

6. $f(x) = 2x^2 + 5$, $g(x) = \dfrac{1}{x}$

7. $f(x) = \dfrac{1}{x-2}$, $g(x) = \dfrac{x}{x-3}$

8. $f(x) = \dfrac{1}{x^2}$, $g(x) = \dfrac{1}{5-x}$

Use the method of addition or subtraction of ordinates to graph each of the following. That is, graph $y = f(x)$ and $y = g(x)$ in the same coordinate plane and then obtain the graph of $f + g$ or $f - g$ by adding or subtracting ordinates. See Figure 42.

9. $f + g$, where $f(x) = x^2$ and $g(x) = x - 2$

10. $f + g$, where $f(x) = |x|$ and $g(x) = x$

11. $f - g$, where $f(x) = 1/x$ and $g(x) = x$

12. $f - g$, where $f(x) = x^3$ and $g(x) = -x + 1$

For each of the following, write the formula for $(g \circ f)(x)$ and $(f \circ g)(x)$ and give the domains of these composite functions. See Example A.

13. $f(x) = x^2$, $g(x) = x - 2$

14. $f(x) = x^3 - 1$, $g(x) = x + 3$

15. $f(x) = \dfrac{1}{x}$, $g(x) = x + 3$

16. $f(x) = 2x^2 + 5$, $g(x) = \dfrac{1}{x}$

17. $f(x) = \sqrt{x - 2}$, $g(x) = x^2 - 2$

18. $f(x) = \sqrt{2x}$, $g(x) = x^2 + 1$

19. $f(x) = 2x - 3$, $g(x) = \frac{1}{2}(x + 3)$

20. $f(x) = x^3 + 1$, $g(x) = \sqrt[3]{x - 1}$

In each of the following, write formulas for $g(x)$ and $f(x)$ so that $H = g \circ f$. The answer is not unique. See Example B.

21. $H(x) = (x + 4)^3$

22. $H(x) = (2x + 1)^3$

23. $H(x) = \sqrt{x + 2}$

24. $H(x) = \sqrt[3]{2x + 1}$

25. $H(x) = \dfrac{1}{(2x + 5)^3}$

26. $H(x) = \dfrac{6}{(x + 4)^3}$

27. $H(x) = |x^3 - 4|$

28. $H(x) = |4 - x - x^2|$

In each of the following, graph the function f carefully and then use translations to sketch the graphs of the funtions g, h, and j. See Example C.

29. $f(x) = x^2$, $g(x) = (x - 2)^2$, $h(x) = x^2 - 4$, and $j(x) = (x - 2)^2 + 1$

30. $f(x) = x^3$, $g(x) = (x + 2)^3$, $h(x) = x^3 + 4$, and $j(x) = (x + 2)^3 - 2$

31. $f(x) = \sqrt{x}$, $g(x) = \sqrt{x - 3}$, $h(x) = \sqrt{x} + 2$, and $j(x) = \sqrt{x - 3} - 2$

32. $f(x) = \dfrac{1}{x}$, $g(x) = \dfrac{1}{x-4}$, $h(x) = \dfrac{1}{x} + 3$, and $j(x) = \dfrac{1}{x-4} - 5$

B. Applications and Extensions

33. Let $f(x) = 2x + 3$ and $g(x) = x^3$. Write formulas for each of the following.

(a) $(f + g)(x)$ (b) $(g - f)(x)$

(c) $(f \cdot g)(x)$ (d) $(f/g)(x)$

(e) $(f \circ g)(x)$ (f) $(g \circ f)(x)$

(g) $(f \circ f)(x)$ (h) $(g \circ g \circ g)(x)$

34. If $f(x) = 1/(x - 1)$ and $g(x) = \sqrt{x - 1}$, write formulas for $(f \circ g)(x)$ and $(g \circ f)(x)$ and give the domains of these composite functions.

35. If $f(x) = x^2 - 4$, $g(x) = |x|$, and $h(x) = 1/x$, write a formula for $(h \circ g \circ f)(x)$ and indicate its domain.

36. In general, how many different functions can be obtained by composing three different functions f, g, and h in different orders?

37. In calculus, the *difference quotient*

$$\frac{f(x + h) - f(x)}{h}$$

arises repeatedly. Calculate this expression and simplify it for each of the following.

(a) $f(x) = x^2$ (b) $f(x) = 2x + 3$

(c) $f(x) = 1/x$ (d) $f(x) = 2/(x - 2)$

38. Calculate $[g(x - h) - g(x)]/h$ for each of the following. Simplify your answer.

(a) $g(x) = 4x - 9$ (b) $g(x) = x^2 + 2x$

(c) $g(x) = x + 1/x^3$ (d) $g(x) = x^3$

39. Find $(f \circ g)(x)$ and $(g \circ f)(x)$ in each case.

(a) $f(x) = x^2$, $g(x) = \sqrt{x}$

(b) $f(x) = x^3$, $g(x) = \sqrt[3]{x}$

(c) $f(x) = x^2$, $g(x) = x^3$

(d) $f(x) = x^2$, $g(x) = 1/x^3$

40. Let $f(x) = (x - 3)/(x + 1)$. Show that if $x \neq \pm 1$, then $f(f(f(x))) = x$.

41. Let $f(x) = [(1 - \sqrt{x})/(1 + \sqrt{x})]^2$. Solve for x if $f(f(x)) = x^2 + \frac{1}{4}$ and $0 < x < 1$.

42. Let $f(x) = x^2 + 5x$. Solve for x if $f(f(x)) = f(x)$.

43. At 12:00 noon, Steven left point A walking east at 3.5 miles per hour; at 1:00 P.M., Carole left point A walking north at 4 miles per hour. Suppose that t hours after noon, Steven is at point B, x miles from A, and Carole is at point C, y miles from A (Figure 48). Assume that $t \geq 1$.

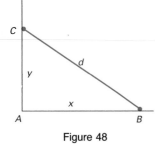

Figure 48

(a) Express x and y as functions of t.
(b) Express the distance d between B and C as a function of x and y.
(c) Express d as a function of t.
(d) Calculate the distance between Steven and Carole at 4:30 P.M.

44. The length x of a rectangular box is twice its width y and three times its height z (Figure 49).

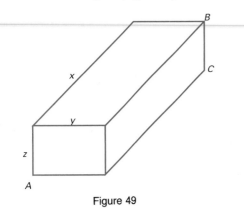

Figure 49

(a) Express the length d of a diagonal (AB in the figure) as a function of x, y, and z.
(b) Express d as a function of x alone.
(c) Calculate the perimeter of triangle ABC when $x = 6$.

45. Sketch the graph of $f(x) = |x + 1| - |x| + |x - 1|$ on the interval $-2 \leq x \leq 2$. Then calculate the area of the region between this graph and the x-axis.

46. The *greatest integer function*, $[\cdot]$, was defined in Problem 45 of Section 5-2. Graph each of the following functions on the interval $-2 \leq x \leq 6$.
(a) $f(x) = 2[x]$ (b) $g(x) = 2 + [x]$
(c) $h(x) = [x - 2]$ (d) $k(x) = x - [x]$

47. Let f be an even function (meaning $f(-x) = f(x)$) and let g be an odd function (meaning $g(-x) = -g(x)$), both functions having the whole real line as their domains. Which of the following are even? Odd? Neither even nor odd?
(a) $f(x)g(x)$ (b) $f(x)/g(x)$
(c) $[g(x)]^2$ (d) $[g(x)]^3$
(e) $f(x) + g(x)$ (f) $g(g(x))$
(g) $f(f(x))$ (h) $3f(x) + [g(x)]^2$
(i) $g(x) + g(-x)$

48. Show that any function f having the whole real line as its domain can be represented as the sum of an even function and an odd function. *Hint:* Consider $f(x) + f(-x)$ and $f(x) - f(-x)$.

49. The *distance to the nearest integer function*, $(\cdot)$, was defined in Problem 53 of Section 5-2.
(a) Sketch the graphs of $f(x) = (x)$, $g(x) = (2x)/2$, $h(x) = (4x)/4$ and $F(x) = f(x) + g(x) + h(x)$ on the interval $0 \leq x \leq 4$.
(b) Find the areas of the regions between the graph of each of these functions and the x-axis on the interval $0 \leq x \leq 4$. Note that the sum of the first three areas is the fourth.

50. TEASER Generalize Problem 49 by considering the graph of

$$F_n(x) = (x) + \frac{1}{2}(2x) + \frac{1}{4}(4x) + \cdots + \frac{1}{2^n}(2^n x)$$

on the interval $0 \leq x \leq 4$. Find a nice formula for the area A_n of the region between this graph and the x-axis. What happens to A_n as n grows without bound? *Note:* The limiting form of F_n plays an important role in advanced mathematics giving an example of a function whose graph is continuous but does not have a tangent line at any point.

51. Draw the graphs of $y = f(x) = \sqrt{|x|}$, $y = g(x) = (x^3 - 4)/(1 + x^2)$, and $y = f(x) + g(x)$ in the same coordinate plane and note how this illustrates the method of addition of ordinates explained in the text. Find the intersection point of the first two graphs.

52. Graph $y = 4x^2/(1 + x^2)$ and $y = 4(x - 2)^2/(1 + (x - 2)^2) + 3.25$ in the same coordinate plane. State how the second curve relates to the first. How many intersection points do the two curves have?

5-6 INVERSE FUNCTIONS

A one-to-one function   has an inverse.

Some processes are reversible; most are not. If I take off my shoes, I may put them back on again. The second operation undoes the first one and brings things back to the original state. But if I throw my shoes in the fire, I will have a hard time undoing the damage I have done.

A function f operates on a number x to produce a number $y = f(x)$. It may be that we can find a function g that will operate on y and give back x. For example, if

$$y = f(x) = 2x + 1$$

then

$$g(x) = \frac{1}{2}(x - 1)$$

is such a function, since

$$g(y) = g(f(x)) = \frac{1}{2}(2x + 1 - 1) = x$$

When we can find such a function g, we call it the *inverse* of f. Not all functions have inverses. Whether they do or not has to do with a concept called one-to-oneness.

One-to-One Functions

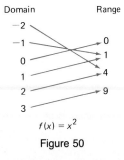

$f(x) = x^2$

Figure 50

In Figure 50 we have reproduced an example we studied earlier, the squaring function with domain $\{-2, -1, 0, 1, 2, 3\}$. It is a perfectly fine function, but it does have one troublesome feature. It may assign the same value to two different x's. In particular, $f(-2) = 4$ and $f(2) = 4$. Such a function cannot possibly have an inverse g. For what would g do with 4? It would not know whether to give back -2 or 2 as the value.

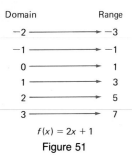

Domain	Range
−2	−3
−1	−1
0	1
1	3
2	5
3	7

$f(x) = 2x + 1$

Figure 51

In contrast, consider $f(x) = 2x + 1$, pictured in Figure 51. Notice that this function never assigns the same value to two different values of x. Therefore there is an unambiguous way of undoing it.

We say that a function f is **one-to-one** if $x_1 \neq x_2$ implies $f(x_1) \neq f(x_2)$, that is, if different values for x always result in different values for $f(x)$. Some functions are one-to-one; some are not. It would be nice to have a graphical criterion for deciding.

Consider the functions $f(x) = x^2$ and $f(x) = 2x + 1$ again, but now let the domains be the set of all real numbers. Their graphs are shown in Figure 52. In the first case, certain horizontal lines (those which are above the x-axis) meet the graph in two points; in the second case, every horizontal line meets the graph in exactly one point. Notice on the first graph that $f(x_1) = f(x_2)$ even though $x_1 \neq x_2$. On the second graph, this cannot happen. Thus we have the important fact that *if every horizontal line meets the graph of a function f in at most one point, then f is one-to-one.*

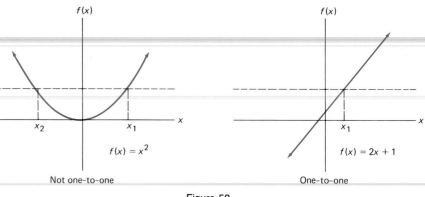

Figure 52

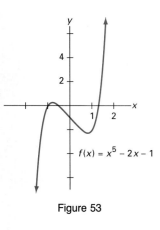

Figure 53

Example A (One-to-One Functions)   Determine whether $f(x) = x^5 - 2x - 1$ (with natural domain) is one-to-one.

Solution   The graph of f is shown in Figure 53. It is clear that some horizontal lines meet the graph in more than one point; f is not one-to-one.   ∎

Inverse Functions

Now we are ready to give a formal definition of the main idea of this section.

Definition

Let f be a one-to-one function with domain X and range Y. Then the function g with domain Y and range X which satisfies

$$g(f(x)) = x$$

for all x in X is called the **inverse of f.**

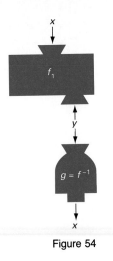

Figure 54

We make several important observations. First, the boxed formula simply says that g undoes what f did (Figure 54). Second, if g undoes f, then f will undo g, that is,

$$f(g(y)) = y$$

for all y in Y. Third, the function g is usually denoted by the symbol f^{-1}. You are cautioned to remember that f^{-1} does *not* mean $1/f$, as you have the right to expect. Mathematicians decided long ago that f^{-1} should stand for the inverse function (the undoing function). Thus

$$(f^{-1} \circ f)(x) = x \quad \text{and} \quad (f \circ f^{-1})(y) = y$$

For example, if $f(x) = 4x$, then $f^{-1}(y) = \frac{1}{4}y$ since

$$(f^{-1} \circ f)(x) = f^{-1}(f(x)) = f^{-1}(4x) = \tfrac{1}{4}(4x) = x$$

and

$$(f \circ f^{-1})(y) = f(f^{-1}(y)) = f(\tfrac{1}{4}y) = 4(\tfrac{1}{4}y) = y$$

The boxed results are illustrated in Figure 55.

CAUTION

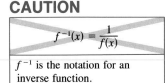

f^{-1} is the notation for an inverse function.

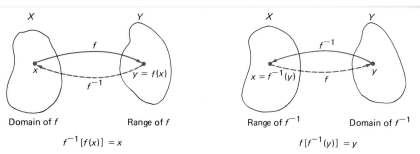

$$f^{-1}[f(x)] = x \qquad\qquad f[f^{-1}(y)] = y$$

Figure 55

Finding a Formula for f^{-1}

If f adds 2, then f^{-1} ought to subtract 2. To say it in symbols, if $f(x) = x + 2$, then we might expect $f^{-1}(y) = y - 2$. And we are right, for

$$f^{-1}[f(x)] = f^{-1}(x + 2) = x + 2 - 2 = x$$

If f divides by 3 and then subtracts 4, we expect f^{-1} to add 4 and multiply by 3. Symbolically, if $f(x) = x/3 - 4$, then we expect $f^{-1}(y) = 3(y + 4)$. Again we are right, for

$$f^{-1}[f(x)] = f^{-1}\left(\frac{x}{3} - 4\right) = 3\left(\frac{x}{3} - 4 + 4\right) = x$$

Note that you must undo things in the reverse order in which you did them (that is, we divided by 3 and then subtracted 4, so to undo this, we first add 4 and then multiply by 3).

When we get to more complicated functions, it is not always easy to find the formula for the inverse function. Here is an important way to look at it.

$$x = f^{-1}(y) \quad \text{if and only if} \quad y = f(x)$$

That means that we can get the formula for f^{-1} by solving the equation $y = f(x)$ for x.

Example B (Finding the Formula for f^{-1})   Let $y = f(x) = 3/(x - 2)$. Find the formula for $f^{-1}(y)$.

Solution   We solve for x in terms of y as follows.

$$y = \frac{3}{x - 2}$$

$$(x - 2)y = 3$$

$$xy - 2y = 3$$

$$xy = 3 + 2y$$

$$x = \frac{3 + 2y}{y}$$

Thus

$$f^{-1}(y) = \frac{3 + 2y}{y} \quad \blacksquare$$

In the formula for f^{-1} just derived, there is no need to use y as the variable. We might use u or t or even x. The formulas

$$f^{-1}(u) = \frac{3 + 2u}{u}$$

$$f^{-1}(t) = \frac{3 + 2t}{t}$$

$$f^{-1}(x) = \frac{3 + 2x}{x}$$

all say the same thing in the sense that they give the same rule. It is conventional to give formulas for functions using x as the variable, and so we would normally write $f^{-1}(x) = (3 + 2x)/x$ as our answer. Let us summarize. To find the formula for $f^{-1}(x)$, use the following steps.

Three-Step Procedure for Finding $f^{-1}(x)$

1. Solve $y = f(x)$ for x in terms of y.
2. Use $f^{-1}(y)$ to name the resulting expression in y.
3. Replace y by x to get the formula for $f^{-1}(x)$.

Example C (Example of the Three-Step Procedure)   Use the three-step procedure to find $f^{-1}(x)$ if $f(x) = 2x^3 - 1$. Check your result by calculating $f(f^{-1}(x))$.

Solution

Step 1 We solve $y = 2x^3 - 1$ for x in terms of y.

$$2x^3 = y + 1$$

$$x^3 = \frac{y + 1}{2}$$

$$x = \sqrt[3]{\frac{y + 1}{2}}$$

Step 2 Call the result $f^{-1}(y)$.

$$f^{-1}(y) = \sqrt[3]{\frac{y + 1}{2}}$$

Step 3 Replace y by x.

$$f^{-1}(x) = \sqrt[3]{\frac{x + 1}{2}}$$

Check: $f(f^{-1}(x)) = 2\left(\sqrt[3]{\frac{x + 1}{2}}\right)^3 - 1 = 2\left(\frac{x + 1}{2}\right) - 1 = x$ ∎

The Graphs of *f* and *f*⁻¹

Since $y = f(x)$ and $x = f^{-1}(y)$ are equivalent, the graphs of these two equations are the same. Suppose that we want to compare the graphs of $y = f(x)$ and $y = f^{-1}(x)$ (where, you will note, we have used x as the domain variable in both cases). To get $y = f^{-1}(x)$ from $x = f^{-1}(y)$, we interchange the roles of x and y. Graphically, this corresponds to folding (reflecting) the graph across the 45° line— that is, across the line $y = x$ (Figure 56). This is the same as saying that if the point (a, b) is on one graph, then (b, a) is on the other.

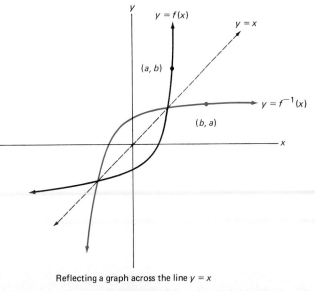

Reflecting a graph across the line $y = x$

Figure 56

Example D (Relating f and f^{-1}) For $f(x) = x^3$, find $f^{-1}(x)$. Then sketch the graphs of $y = f(x)$ and $y = f^{-1}(x)$. Finally, calculate $f(\frac{3}{2})$ and $f^{-1}(\frac{27}{8})$.

Solution The required inverse function has the formula $f^{-1}(x) = \sqrt[3]{x}$. The graphs of $y = x^3$ and $y = \sqrt[3]{x}$ are shown in Figure 57, first separately and then on the same coordinate plane. Note that the second graph is the reflection of the first across the line $y = x$. A simple calculation shows that $f(\frac{3}{2}) = (\frac{3}{2})^3 = \frac{27}{8}$ and $f^{-1}(\frac{27}{8}) = \sqrt[3]{\frac{27}{8}} = \frac{3}{2}$. This illustrates the fact that if $f(a) = b$, then $f^{-1}(b) = a$.

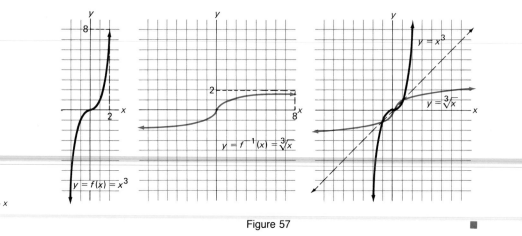

Figure 57

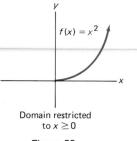

Natural domain

Restricting the Domain

The function $f(x) = x^2$ does not have an inverse if we use its natural domain (all real numbers). However, if we restrict its domain to $x \geq 0$ so that we are considering only the right branch of its graph (Figure 58), then it has the inverse $f^{-1}(x) = \sqrt{x}$. Here is a more complicated example.

Domain restricted to $x \geq 0$

Figure 58

Example E (How to Restrict a Domain) Show that $g(x) = x^2 - 2x - 1$ has an inverse when its domain is appropriately restricted. Then find $g^{-1}(x)$.

Solution The graph of $g(x)$ is shown in Figure 59; it is a parabola with vertex at $x = 1$. Accordingly, we can restrict the domain to $x \geq 1$. To find the formula for $g^{-1}(x)$, we first solve $y = x^2 - 2x - 1$ for x using an old trick, completing the square.

$$y + 1 = x^2 - 2x$$
$$y + 1 + 1 = x^2 - 2x + 1$$
$$y + 2 = (x - 1)^2$$
$$\pm\sqrt{y + 2} = x - 1$$
$$1 \pm \sqrt{y + 2} = x$$

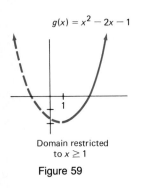

Domain restricted to $x \geq 1$

Figure 59

Notice that there are two expressions for x; they correspond to the two halves of the parabola. We chose to make $x \geq 1$, so $x = 1 + \sqrt{y + 2}$ is the correct expression for $g^{-1}(y)$. Thus

$$g^{-1}(x) = 1 + \sqrt{x + 2}$$

If we had chosen to make $x \leq 1$, the correct answer would have been $g^{-1}(x) = 1 - \sqrt{x + 2}$. ∎

PROBLEM SET 5-6

A. Skills and Techniques

1. Examine the graphs in Figure 60.

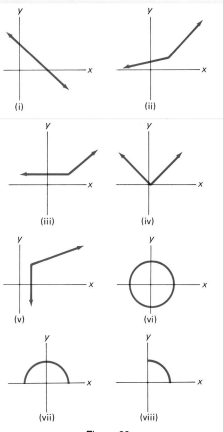

(i)   (ii)

(iii)   (iv)

(v)   (vi)

(vii)   (viii)

Figure 60

(a) Which of these are the graphs of functions with x as domain variable?

(b) Which of these functions are one-to-one?

(c) Which of them have inverses?

2. Let each of the following functions have its natural domain. Which of them are one-to-one? *Hint:* Consider their graphs, as in Example A.

(a) $f(x) = x^4$ **(b)** $f(x) = x^3$

(c) $f(x) = \dfrac{1}{x}$ **(d)** $f(x) = \dfrac{1}{x^2}$

(e) $f(x) = x^2 + 2x + 3$ **(f)** $f(x) = |x|$

(g) $f(x) = \sqrt{x}$ **(h)** $f(x) = -3x + 2$

3. Let $f(x) = 3x - 2$. To find $f^{-1}(2)$, note that $f^{-1}(2) = a$ if $f(a) = 2$, that is, if $3a - 2 = 2$; so $f^{-1}(2) = a = \frac{4}{3}$. Find each of the following.

(a) $f^{-1}(1)$ **(b)** $f^{-1}(-3)$

(c) $f^{-1}(14)$

4. Let $g(x) = 1/(x - 1)$. Find each of the following.

(a) $g^{-1}(1)$ **(b)** $g^{-1}(-1)$

(c) $g^{-1}(14)$

Each of the functions in Problems 5–14 has an inverse (using its natural domain). Find the formula for $f^{-1}(x)$. Then check your result by calculating $f(f^{-1}(x))$. See Examples B and C.

5. $f(x) = 5x$ **6.** $f(x) = -4x$

7. $f(x) = 2x - 7$ **8.** $f(x) = -3x + 2$

9. $f(x) = \sqrt{x} + 2$ **10.** $f(x) = 2\sqrt{x} - 6$

11. $f(x) = \dfrac{x}{x - 3}$ **12.** $f(x) = \dfrac{x - 3}{x}$

13. $f(x) = (x - 2)^3 + 2$ **14.** $f(x) = \frac{1}{3}x^5 - 2$

15. In the same coordinate plane, sketch the graphs of $y = f(x)$ and $y = f^{-1}(x)$ for $f(x) = \sqrt{x} + 2$ (see Problem 9 and Example D).

16. In the same coordinate plane, sketch the graphs of $y = f(x)$ and $y = f^{-1}(x)$ for $f(x) = x/(x - 3)$ (see Problem 11).

17. Sketch the graph of $y = f^{-1}(x)$ if the graph of $y = f(x)$ is as shown in Figure 61.

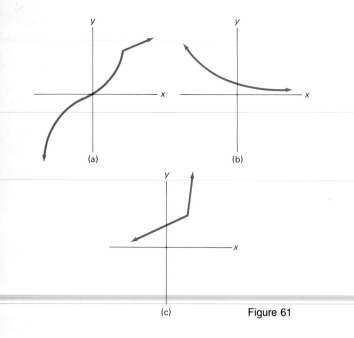

(a) (b)

(c) Figure 61

33. Find the formula for $f^{-1}(x)$ if $f(x) = 1/(x - 1)$ and sketch the graph of $y = f^{-1}(x)$. Compare this graph with the graph of $y = f(x)$ that you sketched in Problem 31.

34. Find the formula for $f^{-1}(x)$ if $f(x) = (x^3 + 2)/(x^3 + 3)$.

35. Sketch the graph of $f(x) = x^2 - 2x - 3$ and observe that it is not one-to-one. Restrict its domain so it is and then find a formula for $f^{-1}(x)$.

36. Let $f(x) = (x - 3)/(x + 1)$ Show that $f^{-1}(x) = f(f(x))$.

37. Let $f(x) = (2x^2 - 4x - 1)/(x - 1)^2$. If we restrict the domain so $x > 1$, then f has an inverse. Find the formula for $f^{-1}(x)$.

38. Suppose that f and g have inverses. Show that in this case $f \circ g$ has an inverse and that $(f \circ g)^{-1} = g^{-1} \circ f^{-1}$.

39. If a ball is dropped from a height of 96 feet above the ground (Figure 62), its height s after t seconds is given by $s = f(t) = -16t^2 + 96, 0 \le t \le \sqrt{6}$.

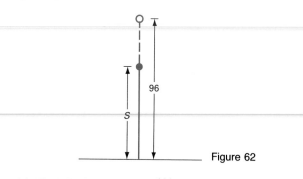

Figure 62

(a) Find the formula for $f^{-1}(s)$.
(b) Use this formula to find the time when the ball is 60 feet above the ground.

40. The length of a rectangular box is 4 centimeters greater than its width, which in turn is twice the height x (Figure 63). Thus the total surface area S of

18. Show that $f(x) = 2x/(x - 1)$ and $g(x) = x/(x - 2)$ are inverses of each other by calculating $f(g(x))$ and $g(f(x))$.

19. Show that $f(x) = 3x/(x + 2)$ and $g(x) = 2x/(3 - x)$ are inverses of each other.

20. Sketch the graph of $f(x) = x^3 + 1$ and note that f is one-to-one. Find a formula for $f^{-1}(x)$.

In each of the following, restrict the domain so that f has an inverse. Describe the restricted domain and find a formula for $f^{-1}(x)$. *Note: Different restrictions of the domain are possible. See Example E.*

21. $f(x) = (x - 1)^2$ 22. $f(x) = (x + 3)^2$
23. $f(x) = (x + 1)^2 - 4$ 24. $f(x) = (x - 2)^2 + 3$
25. $f(x) = x^2 + 6x + 7$ 26. $f(x) = x^2 - 4x + 9$
27. $f(x) = |x + 2|$ 28. $f(x) = 2|x - 3|$
29. $f(x) = \dfrac{(x - 1)^2}{1 + 2x - x^2}$ 30. $f(x) = \dfrac{-1}{x^2 + 4x + 3}$

B. Applications and Extensions

31. Sketch the graph of $f(x) = 1/(x - 1)$. Is f one-to-one? Calculate each of the following.
    (a) $f(3)$ (b) $f^{-1}(\frac{1}{2})$
    (c) $f(0)$ (d) $f^{-1}(-1)$
    (e) $f^{-1}(3)$ (f) $f^{-1}(-2)$

32. If $f(x) = x/(x - 2)$, find the formula for $f^{-1}(x)$.

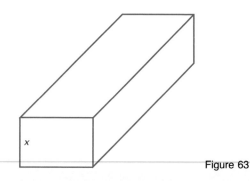

Figure 63

the box can be expressed in terms of x, that is, $S = f(x)$.
(a) Find the formula for $f(x)$.
(b) Find the formula for $f^{-1}(S)$.
(c) Use the latter formula to find the height x if the total surface area is 72 square centimeters.

41. What must be true of the graph of f if f is *self-inverse* (meaning f is its own inverse)? What does this mean for the xy-equation determining f?

42. Let $y = f(x) = x/(x - 1)$. Show that the condition found in Problem 41 is satisfied. Now check that f is self-inverse by showing that $f(f(x)) = x$.

43. Let $f(x) = (ax + b)/(cx + d)$ where $bc - ad \neq 0$.
(a) Find the formula for $f^{-1}(x)$.
(b) Why did we impose the condition $bc - ad \neq 0$?
(c) What relation connecting a and d will make f self-inverse?

44. **TEASER** Let $f_1(x) = x$, $f_2(x) = 1/x$, $f_3(x) = 1 - x$, $f_4(x) = 1/(1 - x)$, $f_5(x) = (x - 1)/x$, and $f_6(x) = x/(x - 1)$. Note that

$$f_4(f_3(x)) = \frac{1}{1 - f_3(x)} = \frac{1}{1 - (1 - x)} = \frac{1}{x} = f_2(x)$$

that is, $f_4 \circ f_3 = f_2$. In fact, if we compose any two of these six functions, we will get one of the six functions. Complete the composition table in Figure 64 and then use it to find each of the following (which will also be one of the six functions).

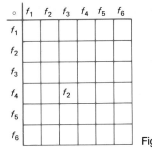

Figure 64

(a) $f_3 \circ f_3 \circ f_3 \circ f_3 \circ f_3$
(b) $f_1 \circ f_2 \circ f_3 \circ f_4 \circ f_5 \circ f_6$
(c) f_6^{-1}
(d) $(f_3 \circ f_6)^{-1}$
(e) F if $f_2 \circ f_5 \circ F = f_5$

45. Draw the graphs of each of the following functions and determine which are one-to-one (which may require use of the ZOOM feature).
(a) $f(x) = x^5 + 2x - 1$
(b) $f(x) = x^5 - .02x - 1$
(c) $f(x) = x^5 + .06x - 1$
(d) $f(x) = .2x^4 - x^3 + 2x - 3, \ x \geq 3.53$

46. Let $f(x) = .01x^6 + x^3 - 5x - 2$. Determine c, as small as possible, so that f has an inverse on the domain $x \geq c$.

CHAPTER 5 SUMMARY

A **function** f is a rule which assigns to each element x in one set (called the **domain**) a value $f(x)$ from another set. The set of all these values is called the **range** of the function. Numerical functions are usually specified by formulas (for example, $g(x) = (x^2 + 1)/(x + 1)$). The **natural domain** for such a function is the largest set of real numbers for which the formula makes sense and gives real values (thus, the natural domain for g consists of all real numbers except $x = -1$). Related to the notion of function is that of **variation.**

The **graph** of a function f is simply the graph of the equation $y = f(x)$. Of special interest are the graphs of **polynomial functions** and **rational functions.** In graphing them, we should show the hills, the valleys, the **x-intercepts,** and, in the case of rational functions, the vertical and horizontal **asymptotes.**

Functions can be combined in many ways. Of these, composition is perhaps the most significant. The **composite** of f with g is defined by $(f \circ g)(x) = f(g(x))$.

Some functions are **one-to-one;** some are not. Those that are one-to-one have undoing functions called inverses. The **inverse** of f, denoted by f^{-1}, satisfies $f^{-1}(f(x)) = x$. Finding a formula for $f^{-1}(x)$ can be tricky; therefore, we described a definite procedure for doing it.

CHAPTER 5 REVIEW PROBLEM SET

In Problems 1–10, write True or False in the blank. If false, tell why.

_____ **1.** If $f(x) = |x + 3|$, then $f(-6.5) = f(.5)$.

_____ **2.** The graphs of $f(x) = (x^2 - 4)/(x - 2)$ and $g(x) = x + 2$ are the same.

_____ **3.** The natural domain of $\sqrt{x^2 - x - 12}$ is the set $\{x : -3 < x < 4\}$.

_____ **4.** The range of the greatest integer function $g(x) = [x]$ is the set of all integers.

_____ **5.** The function $f(x) = (x + 3)^2$, $x \le -3$, is one-to-one.

_____ **6.** If $f(x) = x^3$, then $f(f(x)) = x^6$.

_____ **7.** The graph of $f(x) = (x - 2)/(x^2 + 4)$ has no horizontal asymptotes.

_____ **8.** The graph of $f(x) = (x - 2)(x^2 + 4)/(x + 2)$ crosses the x-axis only once.

_____ **9.** If $g(4)$ is in the domain of f, then 4 is in the domain of $f \circ g$.

_____ **10.** The functions $f(x) = \sqrt[3]{x} - 2$ and $g(x) = x^3 + 2$ are inverse functions.

In Problems 11–14, determine whether or not the given graph is the graph of a function with x as its domain variable.

11. **12.**

13. **14.**

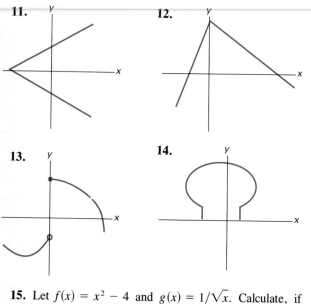

15. Let $f(x) = x^2 - 4$ and $g(x) = 1/\sqrt{x}$. Calculate, if possible.

(a) $g(.81)$ (b) $f(\sqrt{7})$

(c) $f(g(.01))$ (d) $f(2) \cdot g(0)$

(e) $g(g(16))$ (f) $f(0)/g(4)$

16. Determine the natural domain of f if $f(x) = \sqrt{x}/(x - 1)$.

17. If $f(x) = 1/x^2$, find and simplify $[f(x + h) - f(x)]/h$.

18. Suppose that z varies directly as the square of x and inversely as the cube root of y and that $z = -8$ when $x = -2$ and $y = -1$. Find an explicit formula for z in terms of x and y. Then use this formula to evaluate z when $x = 3$ and $y = .001$.

Graph each of the functions in Problems 19–24.

19. $f(x) = (x - 2)^2$ **20.** $f(x) = x^3 + 2x$

21. $f(x) = 2|x - 3| - 3$ **22.** $f(x) = \dfrac{1}{x^2 - x - 2}$

23. $f(x) = [2x]$

24. $f(x) = \begin{cases} 0 & \text{if } x \le 0 \\ x^2 & \text{if } 0 < x < 1 \\ 1 & \text{if } x \ge 1 \end{cases}$

25. Suppose that g is an even function satisfying $g(x) = \sqrt{x}$ for $x \ge 0$. Sketch its graph on $-4 \le x \le 4$.

26. Repeat Problem 25 if g is an odd function.

27. Consider the graphs of $y = f(x)$ and $y = g(x)$ shown

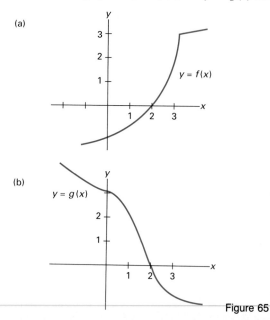

Figure 65

in Figure 65. Using the given axes, sketch the graphs of the following.

(a) $y = f^{-1}(x)$ (b) $y = g^{-1}(x)$

28. For the functions of Problem 27, sketch the graphs of the following.

(a) $y = f(x + 4)$ (b) $y = g(x) - 3$

29. How does the graph of $y = f(x + 3) - 4$ relate to the graph of $y = f(x)$?

30. Sketch the graph of $y = x^2 + 1/x$ on $-4 \leq x \leq 4$, by first graphing $y = x^2$ and $y = 1/x$ and then adding ordinates.

31. If $f(x) = \sqrt[3]{x - 2} + 3$ and $g(x) = (x - 3)^3$, write a formula for each of the following.

(a) $f(x + 2)$ (b) $g(f(x))$
(c) $g^{-1}(x)$ (d) $f^{-1}(x)$

32. If $f(x)$ is obtained by adding 3 to x and then dividing the result by 2, find a formula for $f^{-1}(x)$.

33. If $g(x) = x/(x + 2)$, find $g^{-1}(2)$.

34. If $f(x) = x/(x - 5)$, find a formula for $f^{-1}(x)$.

35. Determine $f(x)$ and $g(x)$ if $h(x) = \sqrt{x^3 - 7}$ is decomposed as $h(x) = (f \circ g)(x)$.

36. Show that 2 is not in the range of $f(x) = (2x + 1)/(x + 2)$.

37. Determine the natural domain of f in Problem 36.

38. Indicate a good way to restrict the domain of $f(x) = (2x - 5)^2$ so that f has an inverse. Then determine a formula for the inverse of this f.

In Problems 39–44, classify each function as to whether it is even, odd, or neither.

39. $f(x) = 2|x|$ **40.** $f(x) = x^2/(x^2 + 1)$

41. $f(x) = 2x/(x^2 + 1)$ **42.** $f(x) = \sqrt[3]{x^3 + 8}$

43. $f(x) = x^5 + x^3 + 1$

44. $f(x) = (x^3 + x)^3(x^2 - 1)$

45. Assume that f is an even function, that g is an odd function, and that both are defined for all real numbers. Show that $f \cdot g$ is an odd function and that $f \circ g$ is an even function.

46. Find the formula for the nowhere positive function g such that $y = g(x)$ and $x^2 + y^2 = 25$.

47. Pedro is on an island at point A, 3 miles from the nearest point B on the straight shoreline of a large lake (Figure 66). He will row toward point C, x miles down the shore from B, at 2.5 miles per hour and then walk to his home D, 12 miles along the shore from B, at 3.5 miles per hour. Write formulas in terms of x for

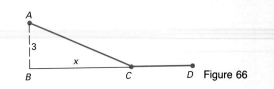

Figure 66

(a) S, the total distance traveled.
(b) T, the time required.

48. A container has the shape of a cylinder topped by a hemisphere (Figure 67). Assuming that the height and radius of the cylindrical part are both x, write formulas for

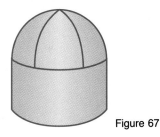

Figure 67

(a) V, the volume of the container.
(b) A, the surface area of the container.

CHAPTER 6

Exponential and Logarithmic Functions

■ The method of logarithms, by reducing to a few days the labor of many months, doubles as it were, the life of the astronomer, besides freeing him from the errors and disgust inseparable from long calculation.

P. S. Laplace

■ The logarithm function is the premier example of a slowly growing function.

Anon.

6-1 RADICALS

Radicals on Top of Radicals

The discovery of the quadratic formula led to a long search for a corresponding formula for the cubic equation. Success had to wait until the sixteenth century, when the Italian school of mathematicians at Bologna found formulas for both the cubic and quartic equations. A typical result is Cardano's solution to $x^3 + ax = b$ which takes the form

$$x = \sqrt[3]{\sqrt{\frac{a^3}{27} + \frac{b^2}{4}} + \frac{b}{2}} - \sqrt[3]{\sqrt{\frac{a^3}{27} + \frac{b^2}{4}} - \frac{b}{2}}$$

Hieronimo Cardano
1501–1576

Historically, interest in radicals has been associated with the desire to solve equations. Even the general cubic equation leads to very complicated radical expressions. Today, powerful iterative methods make results like Cardano's solution historical curiosities. Yet the need for radicals continues; it is important that we know something about them.

Raising a number to the 3rd power (or cubing it) is a process which can be undone. The inverse process—taking the 3rd root—is denoted by $\sqrt[3]{}$. We call $\sqrt[3]{a}$ a *radical* and read it "the cube root of a." Thus $\sqrt[3]{8} = 2$ and $\sqrt[3]{-125} = -5$ since $2^3 = 8$ and $(-5)^3 = -125$.

Our first goal is to give meaning to the symbol $\sqrt[n]{a}$ when n is any positive integer. Naturally, we require that $\sqrt[n]{a}$ be a number which yields a when raised to the nth power; that is

$$(\sqrt[n]{a})^n = a$$

When n is odd, that is all we need to say, since for any real number a, there is exactly one real number whose nth power is a.

When n is even, we face two serious problems, problems that are already apparent when $n = 2$. We have already discussed square roots, using the symbol $\sqrt{}$ rather than $\sqrt[2]{}$ (see Section 3-4). Recall that if $a < 0$, then $\sqrt{a}$ is not a real number (for example, $\sqrt{-4} = 2i$). Even if $a > 0$, we are in trouble since there are always two real numbers with squares equal to a. For example, both -3 and 3 have squares equal to 9. We agree that in this ambiguous case, $\sqrt{a}$ shall always denote the positive square root of a. Thus $\sqrt{9}$ is equal to 3, not -3.

We make a similar agreement about $\sqrt[n]{a}$ for n an even number greater than 2. First, we shall avoid the case $a < 0$. Second, when $a \geq 0$, $\sqrt[n]{a}$ will always denote the nonnegative number whose nth power is a. Thus $\sqrt[4]{81} = 3$, $\sqrt[4]{16} = 2$, and $\sqrt[4]{0} = 0$; however, the symbol $\sqrt[4]{-16}$ will be assigned no meaning in this book. Let us summarize.

> If n is odd, $\sqrt[n]{a}$ is the unique real number satisfying $(\sqrt[n]{a})^n = a$. If n is even and $a \geq 0$, $\sqrt[n]{a}$ is the unique nonnegative real number satisfying $(\sqrt[n]{a})^n = a$.

The symbol $\sqrt[n]{a}$, as we have defined it, is called the **principal nth root of a;** for brevity, we often drop the adjective *principal*.

Rules for Radicals

Radicals, like exponents, obey certain rules. The most important ones are listed below, where it is assumed that all radicals name real numbers.

Rules for Radicals

1. $(\sqrt[n]{a})^n = a$
2. $\sqrt[n]{a^n} = a, \quad n$ odd
3. $\sqrt[n]{a^n} = |a|, \quad n$ even
4. $\sqrt[n]{ab} = \sqrt[n]{a}\sqrt[n]{b}$
5. $\sqrt[n]{\dfrac{a}{b}} = \dfrac{\sqrt[n]{a}}{\sqrt[n]{b}}$

These rules can all be proved, but we believe that the following illustrations will be more helpful to you than proofs.

Example A (Applying the Rules)   Simplify.

(a) $(\sqrt[4]{7})^4$ (b) $\sqrt[7]{(-3)^7}$ (c) $\sqrt[4]{(-3)^4}$ (d) $\sqrt{2}\sqrt{18}$ (e) $\dfrac{\sqrt[3]{750}}{\sqrt[3]{6}}$

Solution

(a) $(\sqrt[4]{7})^4 = 7$ (rule 1)

(b) $\sqrt[7]{(-3)^7} = -3$ (rule 2)

(c) $\sqrt[4]{(-3)^4} = |-3| = 3$ (rule 3)

(d) $\sqrt{2} \cdot \sqrt{18} = \sqrt{36} = 6$ (rule 4)

(e) $\dfrac{\sqrt[3]{750}}{\sqrt[3]{6}} = \sqrt[3]{\dfrac{750}{6}} = \sqrt[3]{125} = 5$ (rule 5)   ∎

Example B (Simplifying Radicals)   Rewrite in simplest form.

(a) $\sqrt[3]{54x^4y^6}$ (b) $\sqrt[4]{x^8 + x^4y^4}$ (c) $\sqrt[3]{-9x^2y^4z^5}\sqrt[3]{9x^4y^2z^4}$

Solution

(a) We start by factoring out the largest third power.

$$\sqrt[3]{54x^4y^6} = \sqrt[3]{(27x^3y^6)(2x)}$$
$$= \sqrt[3]{(3xy^2)^3(2x)}$$
$$= \sqrt[3]{(3xy^2)^3}\sqrt[3]{2x} \qquad \text{(rule 4)}$$
$$= 3xy^2\sqrt[3]{2x} \qquad \text{(rule 2)}$$

(b) Note that $\sqrt[4]{x^8 + x^4y^4} \neq x^2 + xy$ just as $\sqrt[4]{a^4 + b^4} \neq a + b$. Rather,

$$\sqrt[4]{x^8 + x^4y^4} = \sqrt[4]{x^4(x^4 + y^4)}$$
$$= \sqrt[4]{x^4}\sqrt[4]{x^4 + y^4} \qquad \text{(rule 4)}$$
$$= |x|\sqrt[4]{x^4 + y^4} \qquad \text{(rule 3)}$$

We were able to take x^4 out of the radical because it is a 4th power and a factor of $x^8 + x^4y^4$.

(c)
$$\sqrt[3]{-9x^2y^4z^5}\;\sqrt[3]{9x^4y^2z^4} = \sqrt[3]{(-9x^2y^4z^5)(9x^4y^2z^4)} \qquad \text{(rule 4)}$$
$$= \sqrt[3]{3(-27x^6y^6z^9)}$$
$$= \sqrt[3]{3}\;\sqrt[3]{-27x^6y^6z^9} \qquad \text{(rule 4)}$$
$$= -3\sqrt[3]{3}\;x^2y^2z^3 \qquad \text{(rule 2)} \quad \blacksquare$$

Rationalizing Denominators

For some purposes (including hand calculations), fractions with radicals in their denominators are considered to be needlessly complicated. Fortunately, we can usually rewrite a fraction so that its denominator is free of radicals. The process we go through is called **rationalizing the denominator.**

Example C (Rationalizing Denominators) Rewrite each of the following without radical denominators.

(a) $\dfrac{3}{\sqrt{x + 2}}$ (b) $\dfrac{1}{\sqrt[5]{x}}$ (c) $\dfrac{x}{\sqrt{x} + \sqrt{y}}$ (d) $\dfrac{1}{\sqrt[3]{2x^2y^5}}$

Solution

(a) Multiply numerator and denominator by $\sqrt{x + 2}$.

$$\frac{3}{\sqrt{x + 2}} = \frac{3\sqrt{x + 2}}{\sqrt{x + 2}\;\sqrt{x + 2}} = \frac{3\sqrt{x + 2}}{x + 2}$$

(b) Here we multiply numerator and denominator by $\sqrt[5]{x^4}$, which gives the 5th root of a 5th power in the denominator.

$$\frac{1}{\sqrt[5]{x}} = \frac{1 \cdot \sqrt[5]{x^4}}{\sqrt[5]{x} \cdot \sqrt[5]{x^4}} = \frac{\sqrt[5]{x^4}}{\sqrt[5]{x \cdot x^4}} = \frac{\sqrt[5]{x^4}}{\sqrt[5]{x^5}} = \frac{\sqrt[5]{x^4}}{x}$$

(c) This time, we make use of the identity $(a + b)(a - b) = a^2 - b^2$. If we multiply numerator and denominator of the fraction by $\sqrt{x} - \sqrt{y}$, the radicals in the denominator disappear.

$$\frac{x}{\sqrt{x} + \sqrt{y}} = \frac{x(\sqrt{x} - \sqrt{y})}{(\sqrt{x} + \sqrt{y})(\sqrt{x} - \sqrt{y})} = \frac{x\sqrt{x} - x\sqrt{y}}{x - y}$$

We should point out that this manipulation is valid provided $x \neq y$.

(d) In this last case, we multiply numerator and denominator by $\sqrt[3]{4xy}$ to obtain the cube root of a third power in the denominator.

$$\frac{1}{\sqrt[3]{2x^2y^5}} = \frac{\sqrt[3]{4xy}}{\sqrt[3]{2x^2y^5}\ \sqrt[3]{4xy}} = \frac{\sqrt[3]{4xy}}{\sqrt[3]{8x^3y^6}} = \frac{\sqrt[3]{4xy}}{2xy^2} \quad \blacksquare$$

Solving Equations Involving Radicals

The best way to solve an equation involving radicals is to first get rid of the radicals. This is usually a matter of raising both sides of the equation to an appropriate power.

Example D (Equations Involving Radicals) Solve the following equations.

(a) $\sqrt[3]{x - 2} = 3$ (b) $x = \sqrt{2 - x}$ (c) $2\sqrt{x + 5} - \sqrt{x} = 4$

Solution

(a) Raise both sides to the 3rd power and solve for x.

$$(\sqrt[3]{x - 2})^3 = 3^3$$
$$x - 2 = 27$$
$$x = 29$$

(b) Square both sides and solve for x.

$$x^2 = 2 - x$$
$$x^2 + x - 2 = 0$$
$$(x - 1)(x + 2) = 0$$
$$x = 1 \qquad x = -2$$

Let us check our answers in part (b) by substituting them in the original equation. When we substitute these numbers for x in $x = \sqrt{2 - x}$, we find that 1 works but -2 does not.

(c) Separate the two radicals and then square both sides as shown below.

$$2\sqrt{x + 5} - \sqrt{x} = 4$$
$$2\sqrt{x + 5} = \sqrt{x} + 4$$
$$4(x + 5) = x + 8\sqrt{x} + 16$$

We still have a radical but now only one of them, so we have made progress. Isolate that radical on one side and then square again, thus removing the radical.

$$4x + 20 - x - 16 = 8\sqrt{x}$$
$$3x + 4 = 8\sqrt{x}$$
$$9x^2 + 24x + 16 = 64x$$
$$9x^2 - 40x + 16 = 0$$
$$(x - 4)(9x - 4) = 0$$
$$x = 4 \qquad x = \frac{4}{9}$$

When we substitute these numbers for x in the original equation, we find that both answers check. ∎

Combining Fractions Involving Radicals

Sums and differences of fractions involving radicals occur occasionally in calculus. There it will be useful to combine these fractions in the way that we now illustrate.

Example E (Combining Fractions)   Combine in one fraction.

(a) $\dfrac{1}{\sqrt[3]{x+h}} - \dfrac{1}{\sqrt[3]{x}}$ (b) $\dfrac{x}{\sqrt{x^2+4}} - \dfrac{\sqrt{x^2+4}}{x}$

Solution

(a) $\dfrac{1}{\sqrt[3]{x+h}} - \dfrac{1}{\sqrt[3]{x}} = \dfrac{\sqrt[3]{x}}{\sqrt[3]{x}\,\sqrt[3]{x+h}} - \dfrac{\sqrt[3]{x+h}}{\sqrt[3]{x}\,\sqrt[3]{x+h}} = \dfrac{\sqrt[3]{x} - \sqrt[3]{x+h}}{\sqrt[3]{x}\,\sqrt[3]{x+h}}$

CAUTION

$$\begin{array}{|c|}
\hline
\sqrt{a^4 + a^4 b^2} = a^2 + a^2 b \\
\hline
\sqrt{a^4 + a^4 b^2} = \sqrt{a^4(1+b^2)} \\
= a^2\sqrt{1+b^2} \\
\hline
\end{array}$$

(b) $\dfrac{x}{\sqrt{x^2+4}} - \dfrac{\sqrt{x^2+4}}{x} = \dfrac{x^2}{x\sqrt{x^2+4}} - \dfrac{\sqrt{x^2+4}\,\sqrt{x^2+4}}{x\sqrt{x^2+4}}$

$= \dfrac{x^2 - (x^2+4)}{x\sqrt{x^2+4}} = \dfrac{-4}{x\sqrt{x^2+4}}$ ∎

PROBLEM SET 6-1

A. Skills and Techniques

Simplify the following radical expressions. This will involve removing perfect powers from radicals and rationalizing denominators. Assume that all letters represent positive numbers. See Example A–C.

1. $\sqrt{9}$
2. $\sqrt[3]{-8}$
3. $\sqrt[5]{32}$
4. $\sqrt[4]{16}$
5. $(\sqrt[3]{7})^3$
6. $(\sqrt{\pi})^2$
7. $\sqrt[3]{(\frac{3}{2})^3}$
8. $\sqrt[5]{(-2/7)^5}$
9. $(\sqrt{5})^4$
10. $(\sqrt[3]{5})^6$
11. $\sqrt{3}\,\sqrt{27}$
12. $\sqrt{2}\,\sqrt{32}$
13. $\sqrt[3]{16}/\sqrt[3]{2}$
14. $\sqrt[4]{48}/\sqrt[4]{3}$
15. $\sqrt[3]{10^{-6}}$
16. $\sqrt[4]{10^8}$
17. $1/\sqrt{2}$
18. $1/\sqrt{3}$
19. $\sqrt{10}/\sqrt{2}$
20. $\sqrt{6}/\sqrt{3}$
21. $\sqrt[3]{54x^4 y^5}$
22. $\sqrt[3]{-16x^3 y^8}$
23. $\sqrt[4]{(x+2)^4 y^7}$
24. $\sqrt[4]{x^5(y-1)^8}$
25. $\sqrt{x^2 + x^2 y^2}$
26. $\sqrt{25 + 50y^4}$

27. $\sqrt[3]{x^6 - 9x^3 y}$
28. $\sqrt[4]{16x^{12} + 64x^8}$
29. $\sqrt[3]{x^4 y^{-6} z^6}$
30. $\sqrt[3]{32x^{-4} y^9}$
31. $\dfrac{2}{\sqrt{x}+3}$
32. $\dfrac{4}{\sqrt{x}-2}$
33. $\dfrac{2}{\sqrt{x+3}}$
34. $\dfrac{4}{\sqrt{x-2}}$
35. $\dfrac{1}{\sqrt[4]{8x^3}}$
36. $\dfrac{1}{\sqrt[3]{5x^2 y^4}}$
37. $\dfrac{2}{\sqrt{3}-\sqrt{2}}$
38. $\dfrac{1}{\sqrt{5}+\sqrt{3}}$
39. $\sqrt[3]{2x^{-2}y^4}\,\sqrt[3]{4xy^{-1}}$
40. $\sqrt[4]{125x^5 y^3}\,\sqrt[4]{5x^{-9} y^5}$
41. $\sqrt{50} - 2\sqrt{18} + \sqrt{8}$
42. $\sqrt[3]{24} + \sqrt[3]{375}$
43. $\sqrt[3]{192} + \sqrt[3]{-81} + \sqrt[3]{24}$
44. $\sqrt[4]{32} - \sqrt[4]{162} + 3\sqrt[4]{2}$
45. $\dfrac{2 - \sqrt{5}}{2 + \sqrt{5}}$
46. $\dfrac{\sqrt{x} + \sqrt{a}}{\sqrt{x} - \sqrt{a}}$

47. $\dfrac{a}{\sqrt[5]{8a^4b^9}}$

48. $\dfrac{2}{\sqrt[4]{27a^3b^{11}}}$

Solve each of the following equations, as in Example D.

49. $\sqrt{x-1} = 5$

50. $\sqrt{x+2} = 3$

51. $\sqrt[3]{2x-1} = 2$

52. $\sqrt[3]{1-5x} = 6$

53. $\sqrt{\dfrac{x}{x+2}} = 4$

54. $\sqrt[3]{\dfrac{x-2}{x+1}} = -2$

55. $\sqrt{x^2+4} = x+2$

56. $\sqrt{x^2+9} = x-3$

57. $\sqrt{2x+1} = x-1$

58. $\sqrt{x} = 12 - x$

59. $2\sqrt{x+1} = \sqrt{x}+2$

60. $2\sqrt{3x+1} = \sqrt{3x+8}$

Combine the fractions, as in Example E. Do not bother to rationalize denominators.

61. $\dfrac{2}{\sqrt{x+h}} - \dfrac{2}{\sqrt{x}}$

62. $\dfrac{\sqrt{x}}{\sqrt{x+2}} - \dfrac{1}{\sqrt{x}}$

63. $\dfrac{1}{\sqrt{x+6}} + \sqrt{x+6}$

64. $\dfrac{\sqrt{x+1}}{\sqrt{x+3}} - \dfrac{\sqrt{x+3}}{x+1}$

65. $\dfrac{\sqrt[3]{(x+2)^2}}{2} \cdot \dfrac{1}{\sqrt[3]{x+2}}$

66. $\dfrac{\sqrt{x+7}}{\sqrt{x-2}} - \dfrac{\sqrt{x-2}}{x+7}$

67. $\dfrac{1}{\sqrt{x^2+9}} - \dfrac{\sqrt{x^2+9}}{x^2}$

68. $\dfrac{x}{\sqrt{x^2+3}} + \dfrac{\sqrt{x^2+3}}{x}$

B. Applications and Extensions

69. Simplify each expression (including rationalizing denominators). Assume all letters represent positive numbers.

(a) $\sqrt[4]{16a^4b^8}$

(b) $\sqrt{27}\sqrt{3b^3}$

(c) $\sqrt{12} + \sqrt{48} - \sqrt{27}$

(d) $\sqrt{250a^4b^6}$

(e) $\sqrt[3]{\dfrac{-32x^2y^7}{4x^5y}}$

(f) $\left(\sqrt[3]{\dfrac{y}{2x}}\right)^6$

(g) $\sqrt{8a^5} + \sqrt{18a^3}$

(h) $\sqrt[4]{512} - \sqrt{50} + \sqrt[6]{128}$

(i) $\sqrt[4]{a^4 + a^4b^4}$

(j) $\dfrac{1}{\sqrt[3]{7bc^3}}$

(k) $\dfrac{2}{\sqrt{a}-b}$

(l) $\sqrt{a}\left(\sqrt{a} + \dfrac{1}{\sqrt{a^3}}\right)$

70. If a is *any* real number and n is even, then $\sqrt[n]{a^n} = |a|$. Use this to simplify each of the following.

(a) $\sqrt{a^4 + 4a^2}$

(b) $\sqrt[4]{a^4 + a^4b^4}$

(c) $\sqrt{(a-b)^2c^4}$

71. Solve each equation for x.

(a) $\sqrt[3]{1-5x} = -4$

(b) $\sqrt{4x+1} = x+1$

(c) $\sqrt{x+3} = 2 + \sqrt{x-5}$

(d) $\sqrt{12+x} = 4 + \sqrt{4+x}$

(e) $x - \sqrt{x} - 6 = 0$

(f) $\sqrt[3]{x^2} - 2\sqrt[3]{x} - 8 = 0$

72. Most scientific calculators have a key for roots (on some you must use the two keys $\boxed{\text{INV}}\ \boxed{y^x}$). Calculate each of the following.

(a) $\sqrt[3]{31}$

(b) $\sqrt[3]{240}$

(c) $\sqrt[10]{78}$

(d) $\sqrt{282} - \sqrt{280}$

(e) $\sqrt[4]{.012}(\sqrt{30} - \sqrt{29})^2$

(f) $\dfrac{\sqrt[4]{29} + \sqrt[3]{6}}{\sqrt{14}}$

73. We know that $f(x) = x^5$ and $g(x) = \sqrt[5]{x}$ are inverse functions. Sketch their graphs using the same coordinate axes.

74. Rewrite $1/(\sqrt{2} + \sqrt{3} - \sqrt{5})$ with a rational denominator.

75. Figure 1 shows a right triangle. Determine $\overline{AC}$ so that the routes ACB and ADB from A to B have the same length.

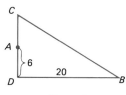

Figure 1

76. In calculus, it is sometimes advantageous to rationalize the numerator. Rewrite each of the following with a rational numerator.

(a) $\dfrac{\sqrt{x} - \sqrt{y}}{\sqrt{x} + \sqrt{y}}$

(b) $\dfrac{\sqrt{x+h} - \sqrt{x}}{h}$

(c) $\dfrac{\sqrt[3]{x} - \sqrt[3]{y}}{x - y}$

77. Determine the length L in Figure 2.

78. Show that $\sqrt{n+1} - \sqrt{n} < 1/2\sqrt{n}$ for all $n > 0$.

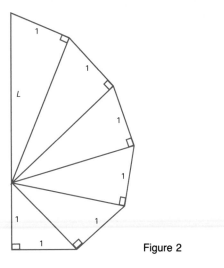

Figure 2

Hint: Begin by rationalizing the numerator in $(\sqrt{n+1} - \sqrt{n})/1$.

79. Show each of the following to be true.

(a) $\dfrac{\sqrt{6} + \sqrt{2}}{2} = \sqrt{2 + \sqrt{3}}$

(b) $\sqrt{2 + \sqrt{3}} + \sqrt{2 - \sqrt{3}} = \sqrt{6}$

(c) $\sqrt[3]{9\sqrt{3} - 11\sqrt{2}} = \sqrt{3} - \sqrt{2}$

80. TEASER Find the exact value of x if $x = \sqrt[3]{9 + 4\sqrt{5}} + \sqrt[3]{9 - 4\sqrt{5}}$. *Hint:* You might guess at the answer by using your calculator but you will not be done until you have given an algebraic demonstration that your guess is correct.

81. Solve $\sqrt{x + 4} + \sqrt{5x} = \sqrt{x - 1} + x^2$.

82. Solve $\sqrt[3]{x + 1} = \sqrt{x^3 - 2}$.

6-2 EXPONENTS AND EXPONENTIAL FUNCTIONS

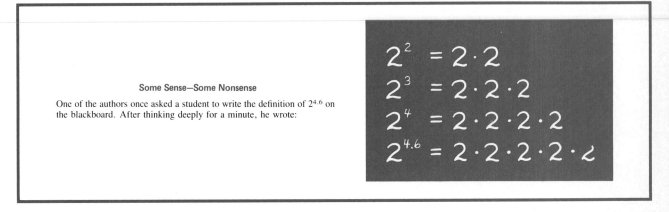

Some Sense—Some Nonsense

One of the authors once asked a student to write the definition of $2^{4.6}$ on the blackboard. After thinking deeply for a minute, he wrote:

$2^2 = 2 \cdot 2$

$2^3 = 2 \cdot 2 \cdot 2$

$2^4 = 2 \cdot 2 \cdot 2 \cdot 2$

$2^{4.6} = 2 \cdot 2 \cdot 2 \cdot 2 \cdot 2$

After you have criticized the student mentioned above, ask yourself how you would define $2^{4.6}$. Of course, integral powers of 2 make perfectly good sense, although 2^{-3} and 2^0 became meaningful only after we had *defined* a^{-n} to be $1/a^n$ and a^0 to be 1 (see Section 2-1). Those were good definitions because they were consistent with the familiar rules of exponents. Now we ask what meaning we can give to powers like $2^{1/2}$, $2^{4.6}$, and even 2^{π} so that these familiar rules still hold.

Rules for Exponents

1. $a^m a^n = a^{m+n}$

2. $\dfrac{a^m}{a^n} = a^{m-n}$

3. $(a^m)^n = a^{mn}$

Rational Exponents

We assume that $a > 0$. If n is any positive integer, we want

$$(a^{1/n})^n = a^{(1/n) \cdot n} = a^1 = a$$

But we know that $(\sqrt[n]{a})^n = a$. Thus we define

$$a^{1/n} = \sqrt[n]{a}$$

For example, $2^{1/2} = \sqrt{2}$, $27^{1/3} = \sqrt[3]{27} = 3$, and $(16)^{1/4} = \sqrt[4]{16} = 2$.
Next, if m and n are positive integers, we want

$$(a^{1/n})^m = a^{m/n} \quad \text{and} \quad (a^m)^{1/n} = a^{m/n}$$

This forces us to define

$$a^{m/n} = (\sqrt[n]{a})^m = \sqrt[n]{a^m}$$

Accordingly,

$$2^{3/2} = (\sqrt{2})^3 = \sqrt{2}\,\sqrt{2}\,\sqrt{2} = 2\sqrt{2}$$

and

$$27^{2/3} = (\sqrt[3]{27})^2 = 3^2 = 9$$

Lastly, we define

$$a^{-m/n} = \frac{1}{a^{m/n}}$$

so that

$$2^{-1/2} = \frac{1}{2^{1/2}} = \frac{1}{\sqrt{2}}$$

and

$$4^{-3/2} = \frac{1}{4^{3/2}} = \frac{1}{(\sqrt{4})^3} = \frac{1}{8}$$

We have just succeeded in defining a^x for all rational numbers x (recall that a rational number is a ratio of two integers). What is more important is that we have done it in such a way that the rules of exponents still hold. Incidentally, we can now answer the question in our opening display.

$$2^{4.6} = 2^4 2^{.6} = 2^4 2^{6/10} = 16(\sqrt[10]{2})^6$$

For simplicity, we have assumed that a is positive in our discussion of $a^{m/n}$. But we should point out that the definition of $a^{m/n}$ given above is also appropriate for the case in which a is negative and n is odd. For example,

$$(-27)^{2/3} = (\sqrt[3]{-27})^2 = (-3)^2 = 9$$

CAUTION

$(-8)^{-1/3} = 8^{1/3} = 2$ [crossed out]

$(-8)^{-1/3} = \dfrac{1}{(-8)^{1/3}} = \dfrac{1}{-2}$

Example A (Changing to Exponential Form)   Write each of the following as a power.

(a) $\sqrt[4]{2}$ (b) $\sqrt[4]{5^3}$ (c) $\dfrac{1}{\sqrt[7]{2^3}}$ (d) $\dfrac{1}{(\sqrt{x+y})^3}$

Solution

(a) $\sqrt[4]{2} = 2^{1/4}$ (b) $\sqrt[4]{5^3} = 5^{3/4}$

(c) $\dfrac{1}{\sqrt[7]{2^3}} = \dfrac{1}{2^{3/7}} = 2^{-3/7}$ (d) $\dfrac{1}{(\sqrt{x+y})^3} = \dfrac{1}{(x+y)^{3/2}} = (x+y)^{-3/2}$ ∎

Example B (Calculating with Rational Exponents)   Calculate.

(a) $8^{1/3}$ (b) $49^{3/2}$ (c) $64^{-5/6}$

Solution

(a) $8^{1/3} = \sqrt[3]{8} = 2$ (b) $49^{3/2} = (\sqrt{49})^3 = 7^3 = 343$

(c) $64^{-5/6} = \dfrac{1}{64^{5/6}} = \dfrac{1}{(\sqrt[6]{64})^5} = \dfrac{1}{2^5} = \dfrac{1}{32}$ ∎

Real Exponents

Irrational powers such as 2^π and $3^{\sqrt{2}}$ are intrinsically more difficult to define than are rational powers. Rather than attempt a technical definition, we ask you to consider what 2^π might mean. The decimal expansion of π is 3.14159 . . . . Thus we could look at the sequence of rational powers

$$2^3,\ 2^{3.1},\ 2^{3.14},\ 2^{3.141},\ 2^{3.1415},\ 2^{3.14159},\ \ldots$$

As you should suspect, when the exponents get closer and closer to π, the corresponding powers of 2 get closer and closer to a definite number. We shall call the number 2^π.

The process of starting with integral exponents and then extending to rational exponents and finally to real exponents can be clarified by means of three graphs (Figure 3). Note the table of values in the margin.

x	2^x
-3	$\dfrac{1}{8}$
-2	$\dfrac{1}{4}$
-1	$\dfrac{1}{2}$
0	1
$\dfrac{1}{2}$	$\sqrt{2} \approx 1.4$
1	2
$\dfrac{3}{2}$	$2\sqrt{2} \approx 2.8$
2	4
3	8

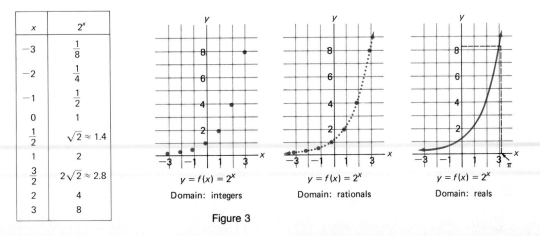

$y = f(x) = 2^x$
Domain: integers

$y = f(x) = 2^x$
Domain: rationals

$y = f(x) = 2^x$
Domain: reals

Figure 3

The first graph suggests a curve rising from left to right. The second graph makes the suggestion stronger. The third graph leaves nothing to the imagination; it is a continuous curve and it shows 2^x for all valus of x, rational and irrational. As x increases in the positive direction, the values of 2^x increase without bound; in the negative direction, the values of 2^x approach 0. Notice that 2^π is a little less than 9; its value correct to seven decimal places is

$$2^\pi = 8.8249778$$

See if your calculator gives this value.

Exponential Functions

The function $f(x) = 2^x$, graphed in Figure 3, is one example of an exponential function. But what has been done with 2 can be done with any positive real number a. In general, the formula

$$f(x) = a^x$$

determines a function called an **exponential function with base a.** Its domain is the set of all real numbers and its range is the set of positive numbers $(a \neq 1)$.

Let us see what effect the size of a has on the graph of $f(x) = a^x$. We choose $a = 2$, $a = 3$, $a = 5$, and $a = \frac{1}{3}$, showing all four graphs in Figure 4.

The graph of $f(x) = 3^x$ looks much like the graph of $f(x) = 2^x$, although it rises more rapidly. The graph of $f(x) = 5^x$ is even steeper. All three of these functions are *increasing functions,* meaning that the values of $f(x)$ increase as x increases; more formally, $x_2 > x_1$ implies $f(x_2) > f(x_1)$. The function $f(x) = (\frac{1}{3})^x$, on the other hand, is a *decreasing function.* In fact, you can get the graph of $f(x) = (\frac{1}{3})^x$ by reflecting the graph of $f(x) = 3^x$ about the y-axis. This is because $(\frac{1}{3})^x = 3^{-x}$.

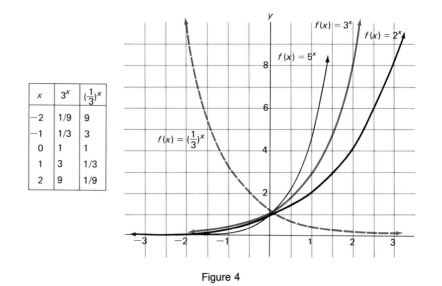

x	3^x	$(\frac{1}{3})^x$
-2	1/9	9
-1	1/3	3
0	1	1
1	3	1/3
2	9	1/9

Figure 4

We can summarize what is suggested by our discussion as follows.

If $a > 1$, $f(x) = a^x$ is an increasing function.
If $0 < a < 1$, $f(x) = a^x$ is a decreasing function.

In both of these cases, the graph of f has the x-axis as an asymptote. The case $a = 1$ is not very interesting since it yields the constant function $f(x) = 1$.

Properties of Exponential Functions

It is easy to describe the main properties of exponential functions, since they obey the rules we learned in Section 2-1. Perhaps it is worth repeating them, since we do want to emphasize that they now hold for all *real* exponents x and y (at least for the case where a and b are both positive).

1. $a^x a^y = a^{x+y}$

2. $\dfrac{a^x}{a^y} = a^{x-y}$

3. $(a^x)^y = a^{xy}$

4. $(ab)^x = a^x b^x$

5. $\left(\dfrac{a}{b}\right)^x = \dfrac{a^x}{b^x}$

Example C (Using the Properties)   Write each expression as a single power.
(a) $3^{1/2} 3^{3/4}$ (b) $\pi^4 / \pi^{5/2}$ (c) $(2^{3/2})^4$ (d) $(5^{-\sqrt{2}} 5^{-2+\sqrt{2}})^{-3}$

Solution
(a) $3^{1/2} 3^{3/4} = 3^{1/2+3/4} = 3^{5/4}$ (b) $\pi^4 / \pi^{5/2} = \pi^{4-5/2} = \pi^{3/2}$
(c) $(2^{3/2})^4 = 2^{4 \cdot 3/2} = 2^6$ (d) $(5^{-\sqrt{2}} 5^{-2+\sqrt{2}})^{-3} = (5^{-2})^{-3} = 5^6$ ∎

Example D (Simplifying Expressions Involving Exponents)   Simplify and write the answer without negative exponents.
(a) $\dfrac{x^{1/3}(8x)^{-2/3}}{x^{-3/4}}$ (b) $\left(\dfrac{2x^{-1/2}}{y}\right)^4 \left(\dfrac{x}{y}\right)^{-1} (3x^{10/3})$

Solution
(a) $\dfrac{x^{1/3}(8x)^{-2/3}}{x^{-3/4}} = x^{1/3} 8^{-2/3} x^{-2/3} x^{3/4} = \dfrac{x^{1/3-2/3+3/4}}{8^{2/3}} = \dfrac{x^{5/12}}{4}$

(b) $\left(\dfrac{2x^{-1/2}}{y}\right)^4 \left(\dfrac{x}{y}\right)^{-1} (3x^{10/3}) = \left(\dfrac{16x^{-2}}{y^4}\right)\left(\dfrac{y}{x}\right)(3x^{10/3})$

$= \dfrac{48x^{-2-1+10/3}}{y^{4-1}} = \dfrac{48x^{1/3}}{y^3}$ ∎

Example E (Combining Fractions)   Perform the following addition.

$$\frac{(x + 1)^{2/3}}{x} + \frac{1}{(x + 1)^{1/3}}$$

Solution

$$\frac{(x + 1)^{2/3}}{x} + \frac{1}{(x + 1)^{1/3}} = \frac{(x + 1)^{2/3}(x + 1)^{1/3}}{x(x + 1)^{1/3}} + \frac{x}{x(x + 1)^{1/3}}$$

$$= \frac{x + 1 + x}{x(x + 1)^{1/3}} = \frac{2x + 1}{x(x + 1)^{1/3}} \quad \blacksquare$$

Mixing Radicals of Different Orders

Square roots and cube roots mix about as well as oil and water, but exponents can serve as a blender. In fact, the use of exponents allows us to write the product of several radicals as a single radical.

Example F (Mixing Radicals)   Express $\sqrt{2}\,\sqrt[3]{5}$ as a single radical.

Solution

$$\sqrt{2}\,\sqrt[3]{5} = 2^{1/2} \cdot 5^{1/3}$$

$$= 2^{3/6} \cdot 5^{2/6}$$

$$= (2^3 \cdot 5^2)^{1/6}$$

$$= \sqrt[6]{200} \quad \blacksquare$$

PROBLEM SET 6-2

A. Skills and Techniques

Write each of the following as a power of 7. See Example A.

1. $\sqrt[3]{7}$

2. $\sqrt[5]{7}$

3. $\sqrt[3]{7^2}$

4. $\sqrt[5]{7^3}$

5. $\dfrac{1}{\sqrt[3]{7}}$

6. $\dfrac{1}{\sqrt[5]{7}}$

7. $\dfrac{1}{\sqrt[3]{7^2}}$

8. $\dfrac{1}{\sqrt[5]{7^3}}$

9. $7\sqrt[3]{7}$

10. $7\sqrt[5]{7}$

Rewrite each of the following using exponents instead of radicals. For example, $\sqrt[5]{x^3} = x^{3/5}$.

11. $\sqrt[3]{x^2}$

12. $\sqrt[4]{x^3}$

13. $x^2\sqrt{x}$

14. $x\sqrt[3]{x}$

15. $\sqrt{(x + y)^3}$

16. $\sqrt[3]{(x + y)^2}$

17. $\sqrt{x^2 + y^2}$

18. $\sqrt[3]{x^3 + 8}$

Rewrite each of the following using radicals instead of fractional exponents. For example, $(xy^2)^{3/7} = \sqrt[7]{x^3y^6}$

19. $4^{2/3}$

20. $10^{3/4}$

21. $8^{-3/2}$

22. $12^{-5/6}$

23. $(x^4 + y^4)^{1/4}$

24. $(x^2 + xy)^{1/2}$

25. $(x^2y^3)^{2/5}$

26. $(3ab^2)^{2/3}$

27. $(x^{1/2} + y^{1/2})^{1/2}$

28. $(x^{1/3} + y^{2/3})^{1/3}$

Simplify each of the following. Give your answer without any exponents (Examples B and C).

29. $25^{1/2}$

30. $27^{1/3}$

31. $8^{2/3}$

32. $16^{3/2}$

33. $9^{-3/2}$

34. $64^{-2/3}$

35. $(-.008)^{2/3}$

36. $(-.027)^{5/3}$

37. $(.0025)^{3/2}$

38. $(1.44)^{3/2}$

39. $5^{2/3}5^{-5/3}$

40. $4^{3/4}4^{-1/4}$

41. $16^{7/6}16^{-5/6}16^{-4/3}$

42. $9^2 9^{2/3} 9^{-7/6}$

43. $(8^2)^{-2/3}$

44. $(4^{-3})^{3/2}$

Simplify, writing your answer without negative exponents, as in Example D.

45. $(3a^{1/2})(-2a^{3/2})$

46. $(2x^{3/4})(5x^{-3/4})$

47. $(2^{1/2}x^{-2/3})^6$

48. $(\sqrt{3}x^{-1/4}y^{3/4})^4$

49. $(xy^{-2/3})^3(x^{1/2}y)^2$

50. $(a^2b^{-1/4})^2(a^{-1/3}b^{1/2})^3$

51. $\dfrac{(2x^{-1}y^{2/3})^2}{x^2 y^{-2/3}}$

52. $\left(\dfrac{a^{1/2}b^{1/3}}{c^{5/6}}\right)^{12}$

53. $\left(\dfrac{x^{-2}y^{3/4}}{x^{1/2}}\right)^{12}$

54. $\dfrac{x^{1/3}y^{-3/4}}{x^{-2/3}y^{1/2}}$

55. $y^{2/3}(2y^{4/3} - y^{-5/3})$

56. $x^{-3/4}\left(-x^{7/4} + \dfrac{2}{\sqrt[4]{x}}\right)$

57. $(x^{1/2} + y^{1/2})^2$

58. $(a^{3/2} + \pi)^2$

Combine the fractions in each of the following, as in Example E.

59. $\dfrac{(x+2)^{4/5}}{3} + \dfrac{2x}{(x+2)^{1/5}}$

60. $\dfrac{(x-3)^{1/3}}{4} - \dfrac{1}{(x-3)^{2/3}}$

61. $(x^2 + 1)^{1/3} - \dfrac{2x^2}{(x^2+1)^{2/3}}$

62. $(x^2 + 2)^{1/4} + \dfrac{x^2}{(x^2+2)^{3/4}}$

Express each of the following in terms of at most one radical in simplest form. See Example F.

63. $\sqrt{2}\sqrt[3]{2}$

64. $\sqrt[3]{2}\sqrt[4]{2}$

65. $\sqrt[4]{2}\sqrt[6]{x}$

66. $\sqrt[3]{5}\sqrt{x}$

67. $\sqrt[3]{x}\sqrt{x}$

68. $\sqrt{x\sqrt[3]{x}}$

Use a calculator to find an approximate value of each of the following.

69. $2^{1.34}$

70. $2^{-.79}$

71. $\pi^{1.34}$

72. π^π

73. $(1.46)^{\sqrt{2}}$

74. $\pi^{\sqrt{2}}$

75. $(.9)^{50.2}$

76. $(1.01)^{50.2}$

Sketch the graph of each of the following functions. See Figure 4.

77. $f(x) = 4^x$

78. $f(x) = 4^{-x}$

79. $f(x) = \left(\tfrac{2}{3}\right)^x$

80. $f(x) = \left(\tfrac{2}{3}\right)^{-x}$

81. $f(x) = \pi^x$

82. $f(x) = (\sqrt{2})^x$

B. Applications and Extensions

83. Rewrite using exponents in place of radicals and simplify.

  (a) $\sqrt[5]{b^3}$

  (b) $\sqrt[8]{x^4}$

  (c) $\sqrt[3]{a^2 + 2ab + b^2}$

84. Simplify.

  (a) $(32)^{-6/5}$

  (b) $(-.008)^{2/3}$

  (c) $(5^{-1/2}\, 5^{3/4}\, 5^{1/8})^{16}$

85. Simplify, writing your answer without either radicals or negative exponents.

  (a) $(27)^{2/3}(.0625)^{-3/4}$

  (b) $\sqrt[3]{4}\sqrt{2} + \sqrt[6]{2}$

  (c) $\sqrt[3]{a^2}\sqrt[4]{a^3}$

  (d) $\sqrt{a\sqrt[3]{a^2}}$

  (e) $[a^{3/2} + a^{-3/2}]^2$

  (f) $[a^{1/4}(a^{-5/4} + a^{3/4})]^{-1}$

  (g) $\left(\dfrac{\sqrt[3]{a^3 b^2}}{\sqrt[4]{a^6 b^3}}\right)^{-1}$

  (h) $\left(\dfrac{a^{-2}b^{2/3}}{b^{-1/2}}\right)^{-4}$

  (i) $\left[\dfrac{(27)^{4/3} - (27)^0}{(3^2 + 4^2)^{1/2}}\right]^{3/4}$

  (j) $(16a^2 b^3)^{3/4} - 4ab^2(a^2 b)^{1/4}$

  (k) $(\sqrt{3})^{3\sqrt{3}} - (3\sqrt{3})^{\sqrt{3}} + (\sqrt{3}^{\sqrt{3}})^{\sqrt{3}}$

  (l) $(a^{1/3} - b^{1/3})(a^{2/3} + a^{1/3}b^{1/3} + b^{2/3})$

86. Combine and simplify, writing your answer without negative exponents.

  (a) $4x^2(x^2 + 2)^{-2/3} - 3(x^2 + 2)^{1/3}$

  (b) $x^3(x^3 - 1)^{-3/4} - (x^3 - 1)^{1/4}$

87. Solve for x.

  (a) $4^{x+1} = (1/2)^{2x}$

  (b) $5^{x^2-x} = 25$

  (c) $2^{4x}4^{x-3} = (64)^{x-1}$

  (d) $(x^2 + x + 4)^{3/4} = 8$

  (e) $x^{2/3} - 3x^{1/3} = -2$

  (f) $2^{2x} - 2^{x+1} - 8 = 0$

Solve the inequalities in Problems 88 and 89. In Problem 89, we suggest that you begin by bringing all terms to one side; then factor.

88. $(2^x)^{x-2} < 1$

89. $6^x > 4 \cdot 3^x - 27 \cdot 2^x + 108$

90. Using the same axes, sketch the graph of each of the following.

  (a) $f(x) = 2^x$

  (b) $g(x) = -2^x$

  (c) $h(x) = 2^{-x}$

  (d) $k(x) = 2^x + 2^{-x}$

  (e) $m(x) = 2^{x-4}$

91. Sketch the graph of $f(x) = 2^{-|x|}$.

92. Using the same axes, sketch the graphs of $f(x) = x^\pi$

and $g(x) = \pi^x$ on the interval $2 \le x \le 3.5$. One solution of $x^{\pi} = \pi^x$ is π. Use your graphs to help you find another one (approximately).

93. Give a simple argument to show that an exponential function $f(x) = a^x$ $(a > 0, a \neq 1)$ is not equivalent to any polynomial function.

94. TEASER If a and b are irrational, does it follow that a^b is irrational? *Hint:* Consider $\sqrt{2}^{\sqrt{2}}$ and $(\sqrt{2}^{\sqrt{2}})^{\sqrt{2}}$.

95. For good practice, reproduce Figure 4. Then redo Problem 92, giving the desired solution accurate to three decimal places.

96. Find the maximum point on the graph of $y = 10x^3 \cdot 2^{-x}$.

6-3 EXPONENTIAL GROWTH AND DECAY

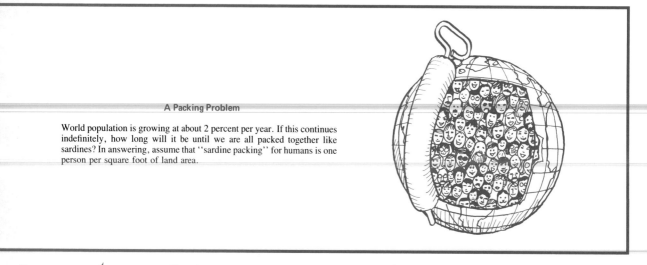

A Packing Problem

World population is growing at about 2 percent per year. If this continues indefinitely, how long will it be until we are all packed together like sardines? In answering, assume that "sardine packing" for humans is one person per square foot of land area.

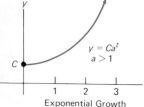

$$y = Ca^t$$
$$a > 1$$

Exponential Growth

Figure 5

The phrase *exponential growth* is used repeatedly by professors, politicians, and pessimists. Population, energy use, mining of ores, pollution, and the number of books about these things are all said to be growing exponentially. Most people probably do not know what exponential growth means, except that they have heard it guarantees alarming consequences. For students of this book, it is easy to explain its meaning. For y to grow exponentially with time t means that it satisfies the relationship

$$y = Ca^t$$

for constants C and a, with $C > 0$ and $a > 1$ (see Figure 5). Why should so many ingredients of modern society behave this way? The basic cause is population growth.

Population Growth

Simple organisms reproduce by cell division. If, for example, there is one cell today, that cell may split so that there are two cells tomorrow. Then each of those cells may divide giving four cells the following day (Figure 6). As this process continues, the numbers of cells on successive days form the sequence

Figure 6

$$1, 2, 4, 8, 16, 32, \ldots$$

t	0	1	2	3	4	5
$f(t)$	100	200	400	800	1600	3200

Figure 7

If we start with 100 cells and let $f(t)$ denote the number present t days from now, we have the results indicated in the table of Figure 7. It seems that

$$f(t) = (100)2^t$$

A perceptive reader will ask if this formula is really valid. Does it give the right answer when $t = 5.7$? Is not population growth a discrete process, occurring in unit amounts at distinct times, rather than a continuous process as the formula implies? The answer is that the exponential growth model provides a very good approximation to the growth of simple organisms, provided the initial population is large.

The mechanism of reproduction is different (and more interesting) for people, but the pattern of population growth is similar. World population is said to be growing at about 2 percent per year. In 1975, there were about 4 billion people. Accordingly, the population in 1976 in billions was $4 + 4(.02) = 4(1.02)$, in 1977 it was $4(1.02)^2$, in 1978 it was $4(1.02)^3$, and so on. If this trend continues, there will be $4(1.02)^{30}$ billion people in the world in 2005, that is, 30 years after 1975. This model says that world population obeys the formula

$$p(t) = 4(1.02)^t$$

where $p(t)$ represents the number of people (in billions) t years after 1975.

In general, if $A(t)$ is the amount at time t of a quantity growing exponentially at the rate of r (written as a decimal), then

$$A(t) = A(0)(1 + r)^t$$

Here $A(0)$ is the initial amount—that is, the amount present at $t = 0$.

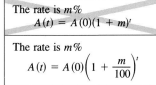

CAUTION

The rate is $m\%$
$$A(t) = A(0)(1 + m)^t$$

The rate is $m\%$
$$A(t) = A(0)\left(1 + \frac{m}{100}\right)^t$$

Example A (A Growing City)   Tooterville, with a present population of 250,000, is growing exponentially at a rate of 1.5 percent per year. What will its population be 18 years from now?

Solution   Appealing to the boxed formula, we may write

$$A(18) = 250,000(1 + .015)^{18} = 250,000(1.015)^{18}$$

$$\approx 326,800$$

Naturally, we used a calculator at the final step.   ∎

Doubling Times

One way to get a feeling for the spectacular nature of exponential growth is via the concept of **doubling time;** this is the length of time required for an exponentially growing quantity to double in size. It is easy to show that if a quantity doubles in an initial time interval of length T, it will double in size in *any* time interval of length

t	$(1.02)^t$
5	1.104
10	1.219
15	1.346
20	1.486
25	1.641
30	1.811
35	2.000
40	2.208
45	2.438
50	2.692
55	2.972
60	3.281
65	3.623
70	4.000
75	4.416
80	4.875
85	5.383
90	5.943

Figure 8

T. Consider the world population problem as an example. By the table of Figure 8, $(1.02)^{35} \approx 2$, so world population doubles in 35 years. Since it was 4 billion in 1975, it should be 8 billion in 2010, 16 billion in 2045, and so on. This alarming information is displayed graphically in Figure 9.

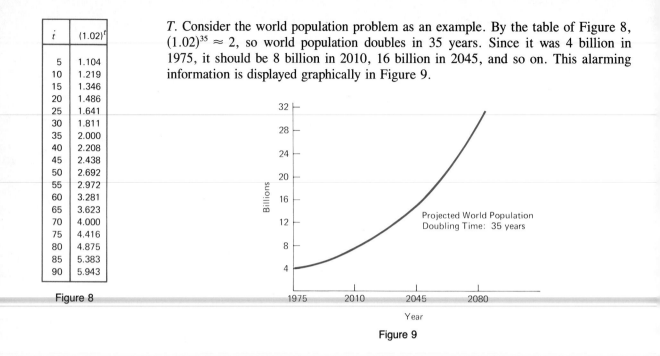

Figure 9

Now we can answer the question about sardine packing in our opening display. There are slightly more than 1,000,000 billion square feet of land area on the surface of the earth. Sardine packing for humans is about 1 square foot per person. Thus we are asking when $4(1.02)^t$ billion will equal 1,000,000 billion. This leads to the equation

$$(1.02)^t = 250,000$$

We call an equation of this type an *exponential equation,* since the unknown is an exponent. In Section 6-5 we will learn an exact method for solving such equations. For now, we will use an approximation method.

$$(1.02)^{35} \approx 2 \qquad 250,000 \approx 2^{18}$$

Our equation can then be rewritten as

$$[(1.02)^{35}]^{t/35} = 2^{18}$$

or

$$2^{t/35} = 2^{18}$$

We conclude that

$$\frac{t}{35} = 18$$

$$t = (18)(35) = 630$$

Thus, after about 600 years, we will be packed together like sardines. If it is any comfort, war, famine, or birth control will alter population growth patterns before then.

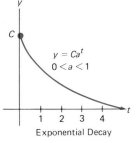

Exponential Decay

Figure 10

Exponential Decay

Not all things grow; many decline or decay. In fact, some quantities decline exponentially. This means that the amount y present at time t satisfies

$$y = Ca^t$$

for some constants C and a with $C > 0$ and $0 < a < 1$ (see Figure 10).

We may describe exponential decay another way. If $A(t)$ is the amount at time t of a quantity declining exponentially at the rate r (written as a decimal), then

$$A(t) = A(0)(1 - r)^t$$

Example B (A Declining City) Abbotsford is losing population at a rate of 2.3 percent per year. It has 34,000 people today. How many people will it have 12 years from now?

Solution According to the formula above,

$$A(12) = 34,000(1 - .023)^{12} = 34,000(.977)^{12} \approx 25,700 \quad \blacksquare$$

Physicists tell us that the radioactive elements decay exponentially. For them, an important notion is that of **half-life,** the time for half of a given amount of a substance to disappear. For example, radium decays with a half-life of 1620 years. Thus if 1000 grams of radium are present now, 1620 years from now 500 grams will be present, $2(1620) = 3240$ years from now only 250 grams will be present, and so on. (Of course, the radium does not evaporate into nothingness; rather, it transforms into other substances.)

Example C (Radioactive Elements) Suppose that a certain element decays with a half-life of 540 years. Starting with 50 grams, how much will be left after 900 years?

Solution Here is a good way to set up this problem. If y is the amount present at time t, then

$$y = 50 \left(\frac{1}{2}\right)^{kt} = 50\left[\left(\frac{1}{2}\right)^k\right]^t$$

with k to be determined. Now a half-life of 540 implies that $y = 25$ when $t = 540$, so

$$25 = 50\left(\frac{1}{2}\right)^{540k}$$

which means that $540k = 1$ or $k = 1/540$. Thus, in general,

$$y = 50\left(\frac{1}{2}\right)^{t/540}$$

and at $t = 900$,

$$y = 50\left(\frac{1}{2}\right)^{900/540} \approx 50\left(\frac{1}{2}\right)^{1.6666667} \approx 15.75 \text{ grams} \quad \blacksquare$$

The phenomenon of radioactive decay is used to date old objects. If an object contains radium and lead (the product to which radium decays) in the ratio 1 to 3, then it is believed that an original amount of pure radium has decayed to one-fourth of its original size. The object must be two half-lives, or 3240 years, old. Two important assumptions have been made: (1) decay of radium is exactly exponential over long periods of time; and (2) no lead was originally present. Recent research raises some question about the correctness of such assumptions.

Compound Interest

One of the best practical illustrations of exponential growth is money earning compound interest. Suppose that Amy puts $1000 in a bank today at 8 percent interest compounded annually. Then at the end of one year the bank adds the interest of $(.08)(1000) = \$80$ to her $1000, giving her a total of $1080. But note that $1080 = 1000(1.08)$. During the second year, $1080 draws interest. At the end of that year, the bank adds $(.08)(1080)$ to the account, bringing the total to

$$1080 + (.08)(1000) = (1080)(1.08)$$

$$= 1000(1.08)(1.08)$$

$$= 1000(1.08)^2$$

Continuing in this way, we see that Amy's account will have grown to $1000(1.08)^3$ by the end of 3 years, $1000(1.08)^4$ by the end of 4 years, and so on. By the end of 15 years, it will have grown to

$$1000(1.08)^{15} \approx 1000(3.172169)$$

$$= \$3172.17$$

Note that this is another application of the formula $A(t) = A(0)(1 + r)^t$.

How long would it take for Amy's money to double—that is, when will

$$1000(1.08)^t = 2000$$

This will occur when $(1.08)^t = 2$. We do not yet have a good algebraic method for solving this exponential equation. (It will come in Section 6-5.) However, experimenting with a calculator leads to the conclusion that $t \approx 9$.

Example D (Compounding Annually)   Roger put $1000 in a money market fund at 15 percent interest compounded annually. How much was it worth after 4 years?

Solution

$$A(4) = 1000(1.15)^4 = \$1749.01 \quad \blacksquare$$

Example E (Compounding at other Periods)   If $1000 is invested at 8 percent compounded quarterly, find the accumulated amount after 15 years.

Solution Interest calculated at 2 percent ($\frac{1}{4}$ of 8 percent) is converted to principal every 3 months. By the end of the first 3-month period, the account has grown to $1000(1.02) = \$1020$; by the end of the second 3-month period, it has grown to $1000(1.02)^2$; and so on. The accumulated amount after 15 years, or 60 conversion periods, is

$$1000(1.02)^{60} \approx 1000(3.28103)$$

$$= \$3281.03$$

Suppose more generally that P dollars is invested at a rate r (written as a decimal), which is compounded m times per year. Then the accumulated amount A after t years is given by

$$A = P\left(1 + \frac{r}{m}\right)^{tm}$$

In our example

$$A = 1000\left(1 + \frac{.08}{4}\right)^{15\cdot4} = 1000(1.02)^{60} \quad \blacksquare$$

PROBLEM SET 6-3

A. Skills and Techniques

Most problems in this set will require a calculator.

1. In each of the following, indicate whether y grows exponentially or decays exponentially with t.
 (a) $y = 128\left(\frac{1}{2}\right)^t$ **(b)** $y = 5\left(\frac{5}{3}\right)^t$
 (c) $y = 4(10)^9(1.03)^t$ **(d)** $y = 1000(.99)^t$

2. Find the values of y corresponding to $t = 0$, $t = 1$, and $t = 2$ for each case in Problem 1.

3. Find each value.
 (a) $(1.08)^{20}$ **(b)** $(1.12)^{25}$
 (c) $1000(1.04)^{40}$ **(d)** $2000(1.02)^{80}$

4. Evaluate each of the following.
 (a) $(1.01)^{100}$ **(b)** $(1.02)^{40}$
 (c) $100(1.12)^{50}$ **(d)** $500(1.04)^{30}$

5. Silver City's present population of 1000 is expected to grow exponentially over the next 10 years at 4 percent per year. How many people will it have at the end of that time? *Hint:* $A(t) = A(0)[1 + r]^t$ (see Example A).

6. The value of houses in Longview is said to be growing exponentially at 12 percent per year. What will a house valued at $100,000 today be worth after 8 years?

7. Under the assumptions concerning world population used in this section, what will be the approximate number of people on earth in each year?
 (a) 2020 (that is, 45 years after 1975)
 (b) 2065

8. As of 1990, the cost of attending a private college was going up at a rate of 7.83 percent per year (from The Independent College 500 Index). What will the annual cost in 2010 be to go to a private college that costs $15,000 per year in 1990? The answer in Figure 11 on the next page is wrong, but it is close.

9. A certain city loses about 2 percent of its population at the beginning of a year by the end of that year. If it has 56,000 people today, how many will it have at the end of 10 years? (See Example B.)

10. An object, worth $1000 today, is losing 5 percent of its value each year (that is, declining exponentially at a rate of 5 percent per year). What will it be worth 12 years from now?

11. A chemical company has set the goal of cutting its use of mercury by 8 percent each year. If it uses 50 kilo-

Figure 11

grams per year now, how much mercury does it plan to use 15 years from now?

12. It is claimed that a certain model car only depreciates by 9 percent of its value from the beginning of a year to the end of that year. What will a model of this car, bought for $21,000, be worth after 10 years?

13. A radioactive element has a half-life of 230 years. Starting with 30 grams, how much will be left after 400 years? (See Example C.)

14. A radioactive element has a half-life of 13 days. What will be left of 12 grams after 50 days?

15. Carbon-14 has a half-life of 5730 years. What fraction of a given amount will be left after 10,000 years?

16. A certain radioactive substance has a half-life of 40 minutes. What fraction of an initial amount will be left after 1 hour and 20 minutes? After 2 hours and 40 minutes?

17. How long does it take a given amount of a radioactive element with a half-life of 210 years to decay to one-seventh of its original size? *Hint:* This leads to the equation $\frac{1}{7} = (\frac{1}{2})^{t/210}$. To solve, try various values of t until you get an approximate answer (or use a graphing calculator).

18. A biological colony is doubling in size every 9 days. To what size will a colony of 800 individuals grow in 36 days?

19. How long will it take the colony of Problem 18 to grow to 2700 individuals? *Hint:* The number y at time t satisfies $y = 800(2)^{t/9}$.

20. A colony of bacteria doubled in size in 30 hours. How long will it take to triple in size?

Problems 21–25 deal with annual compounding. See Example D.

21. If you put $100 in the bank for 8 years, how much will it be worth at the end of that time at
    (a) 8 percent compounded annually?
    (b) 12 percent compounded annually?

22. If you deposit $500 in the bank today, how much will it be worth after 25 years at
    (a) 8 percent compounded annually?
    (b) 4 percent compounded annually?
    (c) 12 percent compounded annually?

23. If you put $3500 in the bank today, how much will it be worth after 40 years at
    (a) 8 percent compounded annually?
    (b) 12 percent compounded annually?

24. Approximately how long will it take for money to accumulate to twice its value if
    (a) it is invested at 8 percent compounded annually?
    (b) it is invested at 12 percent compounded annually?

25. Suppose that you invest P dollars at r percent compounded annually. Write an expression for the amount accumulated after n years.

Find the accumulated amount for the indicated initial principal, compound interest rate, and total time period. See Example E.

26. $2000; 8 percent compounded annually; 15 years
27. $5000; 8 percent compounded semiannually; 5 years
28. $5000; 12 percent compounded monthly; 5 years
29. $3000; 9 percent compounded annually; 10 years
30. $3000; 9 percent compounded semiannually; 10 years
31. $3000; 9 percent compounded quarterly; 10 years
32. $3000; 9 percent compounded monthly; 10 years
33. $1000; 8 percent compounded monthly; 10 years
34. $1000; 8 percent compounded daily; 10 years *Hint:* Assume there are 365 days in a year, so that the interest rate per day is .08/365.

B. Applications and Extensions

35. Let $y = 5400(2/3)^t$. Evaluate y for $t = -1, 0, 1, 2,$ and 3.

36. For what value of t in Problem 35 is $y = 3200/3$?

37. If $(1.023)^T = 2$, find the value of $100(1.023)^{3T}$.

38. If $(.67)^H = \frac{1}{2}$, find the value of $32(.67)^{4H}$.

39. Suppose the population of a certain city follows the formula

$$p(t) = 4600(1.016)^t$$

where $p(t)$ is the population t years after 1980.
  (a) What will the population be in 2020? In 2080?
  (b) Experiment with your calculator to find the doubling time for this population.

40. The number of bacteria in a certain culture is known to triple every hour. Suppose the count at 12:00 noon is 162,000. What was the count at 11:00 A.M.? At 8:00 A.M.?

41. If $100 is invested today, how much will it be worth after 5 years at 8 percent interest if interest is
  (a) compounded annually?
  (b) compounded quarterly?
  (c) compounded monthly?
  (d) compounded daily? (There are 365 days in a year.)

42. About how long does it take money to double if invested at 12 percent compounded monthly?

43. About how long does it take an exponentially growing population to double if its rate of growth is
  (a) 8 percent per year?
  (b) 6.5 percent per year?

44. A manufacturer of radial tires found that the percentage P of tires still usable after being driven m miles was given by

$$P = 100(2.71)^{-.000025m}$$

What percentage of tires are still usable at 80,000 miles?

45. A certain radioactive element has a half-life of 1690 years. Starting with 30 milligrams there will be $q(t)$ milligrams left after t years, where $q(t) = 30(1/2)^{kt}$.
  (a) Determine the constant k.
  (b) How much will be left after 2500 years?

46. One method of depreciation allowed by the IRS is the double-declining-balance method. In this method, the original value C of an item is depreciated each year by $100(2/N)$ percent of its value at the beginning of that year, N being the useful life of the item.
  (a) Write a formula for the value V of the item after n years.

  (b) If an item costs $10,000 and has a useful life of 15 years, calculate its value after 10 years; after 15 years.
  (c) Does the value of an item ever become zero by this depreciation method?

47. (Carbon dating) All living things contain carbon-12, which is a stable element, and carbon-14, which is radioactive. While a plant or animal is alive, the ratio of these two isotopes of carbon remains unchanged, since carbon-14 is constantly renewed; but after death, no more carbon-14 is absorbed. The half-life of carbon-14 is 5730 years. Bones from a human body were found to contain only 76 percent of the carbon-14 in living bones. How long before did the person die?

48. Manhattan Island is said to have been bought from the Indians by Peter Minuit in 1626 for $24. If, instead of making this purchase, Minuit had put the money in a savings account drawing interest at 6 percent compounded annually, what would that account be worth in the year 2000?

49. Hamline University was founded in 1854 with a gift of $25,000 from Bishop Hamline of the Methodist Church. Suppose that Hamline University had wisely put $10,000 of this gift in an endowment drawing 10 percent interest compounded annually, promising not to touch it until 1988 (exactly 134 years later). How much could it then withdraw each year and still maintain this endowment at the 1988 level?

50. TEASER Suppose one water lily growing exponentially at the rate of 8 percent per day is able to cover a certain pond in 50 days. How long would it take 10 of these lilies to cover the pond?

Money invested at compound interest, no matter how small the rate, will eventually grow faster than money invested at simple interest, no matter how large the rate. Problems 51 and 52 illustrate this fact.

51. When will the value of $1 invested at 3 percent compounded annually first surpass the value of $1 invested at 12 percent simple interest? Specifically, find the smallest integer t for which $(1.03)^t > 1 + 0.12t$.

52. After how many years will the value of $1 invested at 1 percent compounded monthly first surpass the value of $1 invested at 10 percent simple interest?

6-4 LOGARITHMS AND LOGARITHMIC FUNCTIONS

An active participant in the political and religious battles of his day, the Scot. John Napier amused himself by studying mathematics and science. He was interested in reducing the work involved in the calculations of spherical trigonometry, especially as they applied to astronomy. In 1614 he published a book containing the idea that made him famous. He gave it the name *logarithm*.

John Napier
(1550-1617)

Napier's approach to logarithms is out of style, but the goal he had in mind is still worth considering. He hoped to replace multiplications by additions. He thought additions were easier to do, and he was right.

Consider the exponential function $f(x) = 2^x$ and recall that

$$2^x \cdot 2^y = 2^{x+y}$$

On the left, we have a multiplication and on the right, an addition. If we are to fulfill Napier's objective, we want logarithms to behave like exponents. That suggests a definition. The logarithm of N to the base 2 is the exponent to which 2 must be raised to yield N. That is,

$$\log_2 N = x \quad \text{if and only if} \quad 2^x = N$$

Thus

$$\log_2 4 = 2 \quad \text{since} \quad 2^2 = 4$$
$$\log_2 8 = 3 \quad \text{since} \quad 2^3 = 8$$
$$\log_2 \sqrt{2} = \tfrac{1}{2} \quad \text{since} \quad 2^{1/2} = \sqrt{2}$$

and, in general,

$$\log_2(2^x) = x \quad \text{since} \quad 2^x = 2^x$$

Has Napier's goal been achieved? Does the logarithm turn a product into a sum? Yes, for note that

$$\log_2(2^x \cdot 2^y) = \log_2(2^{x+y}) \qquad \text{(property of exponents)}$$
$$= x + y \qquad \text{(definition of } \log_2)$$
$$= \log_2(2^x) + \log_2(2^y)$$

Thus

$$\log_2(2^x \cdot 2^y) = \log_2(2^x) + \log_2(2^y)$$

which has the form

$$\log_2(M \cdot N) = \log_2 M + \log_2 N$$

What has been done for 2 can be done for any base $a > 1$. The **logarithm of N to the base a** is the exponent x to which a must be raised to yield N. Thus

$$\log_a N = x \quad \text{if and only if} \quad a^x = N$$

Now we can calculate many kinds of logarithms.

$$\log_4 16 = 2 \quad \text{since} \quad 4^2 = 16$$

$$\log_{10} 1000 = 3 \quad \text{since} \quad 10^3 = 1000$$

$$\log_{10}(.001) = -3 \quad \text{since} \quad 10^{-3} = \frac{1}{1000} = .001$$

What is $\log_{10} 7$? We are not ready to answer that yet, except to say it is a number x satisfying $10^x = 7$.

The Case $0 < a < 1$

The definition of logarithm makes perfectly good sense when $0 < a < 1$. For example,

$$\log_{.5} 8 = -3$$

since

$$(.5)^{-3} = \left(\frac{1}{2}\right)^{-3} = 8$$

However, we will say no more about logarithms to bases between 0 and 1.

Example A (Using the Definition)   Find the value of c.

(a) $\log_5(\frac{1}{25}) = c$ (b) $\log_7(7\sqrt{7}) = c$ (c) $\log_c 16 = 4$ (d) $\log_8 c = \frac{2}{3}$

Solution   In each case, we first change to exponential form.

(a) $5^c = \dfrac{1}{25} = \dfrac{1}{5^2} = 5^{-2}$, so $c = -2$.

(b) $7^c = 7\sqrt{7} = 7 \cdot 7^{1/2} = 7^{3/2}$, so $c = \frac{3}{2}$.

(c) $c^4 = 16 = 2^4$, so $c = 2$.

(d) $c = 8^{2/3} = (\sqrt[3]{8})^2 = 2^2 = 4$.   ■

We point out that negative numbers and zero do not have logarithms. Suppose that -4 and 0 did have logarithms, that is, suppose that

$$\log_a(-4) = m \quad \text{and} \quad \log_a 0 = n$$

Then

$$a^m = -4 \quad \text{and} \quad a^n = 0$$

But that is impossible; we learned earlier that a^x is always positive.

There are three main properties of logarithms.

Properties of Logarithms

1. $\log_a(M \cdot N) = \log_a M + \log_a N$
2. $\log_a(M/N) = \log_a M - \log_a N$
3. $\log_a(M^p) = p \log_a M$

To establish Property 1, let

$$x = \log_a M \quad \text{and} \quad y = \log_a N$$

Then, by definition,

$$M = a^x \quad \text{and} \quad N = a^y$$

so that

$$M \cdot N = a^x \cdot a^y = a^{x+y}$$

Thus $x + y$ is the exponent to which a must be raised to yield $M \cdot N$, that is,

$$\log_a(M \cdot N) = x + y = \log_a M + \log_a N$$

Properties 2 and 3 are demonstrated in a similar fashion.

Example B (Using the Properties)   Given that $\log_a 3 = 1.099$ and $\log_a 5 = 1.609$, calculate each of the following.

(a) $\log_a 15$ (b) $\log_a(\frac{5}{3})$ (c) $\log_a 135$ (d) $\log_a(.12)$

Solution

(a) $\log_a 15 = \log_a(3 \cdot 5) = \log_a 3 + \log_a 5 = 1.099 + 1.609 = 2.708$

(b) $\log_a(\frac{5}{3}) = \log_a 5 - \log_a 3 = 1.609 - 1.099 = .510$

(c) $\log_a 135 = \log_a(27 \cdot 5) = \log_a(3^3) + \log_a 5$

$\qquad = 3 \log_a 3 + \log_a 5 = 3(1.099) + 1.609$

$\qquad = 4.906$

(d) $\log_a(.12) = \log_a(\frac{12}{100}) = \log_a(\frac{3}{25})$

$\qquad = \log_a 3 - \log_a(5^2) = \log_a 3 - 2 \log_a 5$

$\qquad = 1.099 - 2(1.609) = -2.119$ ∎

The function determined by

$$g(x) = \log_a x$$

is called the **logarithmic function with base a.** We can get a feeling for the behavior of this function by drawing its graph for $a = 2$ (Figure 12).

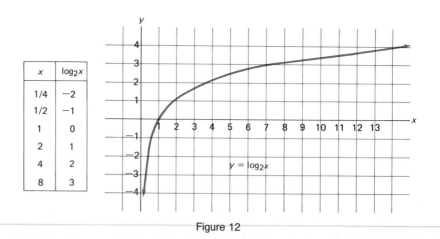

x	$\log_2 x$
1/4	−2
1/2	−1
1	0
2	1
4	2
8	3

$y = \log_2 x$

Figure 12

Several properties of $y = \log_2 x$ are apparent from this graph. The domain consists of all positive real numbers. If $0 < x < 1$, $\log_2 x$ is negative; if $x > 1$, $\log_2 x$ is positive. The y-axis is a vertical asymptote of the graph since very small positive x values yield large negative y values. Although $\log_2 x$ continues to increase as x increases, even this small part of the complete graph indicates how slowly it grows for large x. In fact, by the time x reaches 1,000,000, $\log_2 x$ is still loafing along at about 20. In this sense, it behaves in a manner opposite to the exponential function 2^x, which grows more and more rapidly as x increases. There is a good reason for this opposite behavior; the two functions are inverses of each other.

Inverse Functions

We begin by emphasizing two facts that you must not forget.

<table>
<tr><td>

Equivalents

The statements

(i) $\log_2 8 = 3$

(ii) $2^3 = 8$

(iii) $2^{\log_2 8} = 3$

are equivalent. In words, the third one says

"2 raised to the exponent to which 2 must be raised to yield 8 does yield 8."

</td></tr>
</table>

$$a^{\log_a x} = x$$
$$\log_a(a^x) = x$$

For example, $2^{\log_2 7} = 7$ and $\log_2(2^{-19}) = -19$. Both of these facts are direct consequences of the definition of logarithms; the second is also a special case of Property 3, stated earlier. What these facts tell us is that the logarithmic and exponential functions undo each other.

Example C (The Undoing Properties)   Evaluate.

(a) $\log_5(5^{\pi+3})$ (b) $4^{\log_4 3.2}$ (c) $8^{-2\log_8(2/3)}$

Solution

(a) $\log_5(5^{\pi+3}) = \pi + 3$ (b) $4^{\log_4 3.2} = 3.2$

(c) $8^{-2\log_8(2/3)} = 8^{\log_8[(2/3)^{-2}]} = \left(\frac{2}{3}\right)^{-2} = \left(\frac{3}{2}\right)^2 = \frac{9}{4}$ ∎

Let us express the undoing properties in the language of Section 5-6. If $f(x) = a^x$ and $g(x) = \log_a x$, then

$$f(g(x)) = f(\log_a x) = a^{\log_a x} = x$$

and

$$g(f(x)) = g(a^x) = \log_a(a^x) = x$$

Thus g is really f^{-1}. This fact also tells us something about the graphs of g and f. They are simply reflections of each other about the line $y = x$ (Figure 13).

Note finally that $f(x) = a^x$ has the set of all real numbers as its domain and the positive real numbers as its range. Thus its inverse $f^{-1}(x) = \log_a x$ has domain consisting of the positive real numbers and range consisting of all real numbers. We emphasize again a fact that is important to remember. *Negative numbers and zero do not have logarithms.*

Algebraic Manipulations with Logarithms

First let us note that a complicated-looking expression containing logarithms can sometimes be simplified by using the properties of logarithms.

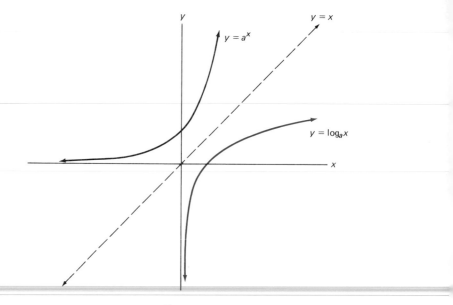

Figure 13

Example D (Combining Logarithms)   Write the following expression as a single logarithm.

$$2 \log_{10} x + 3 \log_{10}(x + 2) - \log_{10}(x^2 + 5)$$

Solution   We use the properties of logarithms to rewrite this as

$$\log_{10} x^2 + \log_{10}(x + 2)^3 - \log_{10}(x^2 + 5) = \log_{10} x^2(x + 2)^3 - \log_{10}(x^2 + 5)$$

$$= \log_{10}\left[\frac{x^2(x + 2)^3}{x^2 + 5}\right] \quad \blacksquare$$

CAUTION

$\log_b 6 + \log_b 4 = \log_b 10$
$\log_b 30 - \log_b 5 = \log_b 25$

$\log_b 6 + \log_b 4 = \log_b 24$
$\log_b 30 - \log_b 5 = \log_b 6$

Equations containing logarithms can often be solved by changing to exponential form. To solve $\log_4(3x + 1) = 2$, rewrite this equation as $3x + 1 = 4^2 = 16$. Then solve the resulting equation to get $x = 5$. Here is a more substantial example.

Example E (Solving Logarithmic Equations)   Solve the equation

$$\log_2 x + \log_2(x + 2) = 3$$

Solution   First we note that we must have $x > 0$ so that both logarithms exist. Next we rewrite the equation using the first property of logarithms and then change it to exponential form.

$$\log_2 x(x + 2) = 3$$

$$x(x + 2) = 2^3$$

$$x^2 + 2x - 8 = 0$$

$$(x + 4)(x - 2) = 0$$

We reject $x = -4$ (because $-4 < 0$) and keep $x = 2$. To make sure that 2 is a solution, we substitute 2 for x in the original equation.

$$\log_2 2 + \log_2(2 + 2) \overset{?}{=} 3$$

$$1 + 2 = 3 \quad \blacksquare$$

Most calculators and most books of tables give only two kinds of logarithms: common logarithms (base 10) and natural logarithms (base e). The latter are discussed in the next section. What do people do when they need logarithms to other bases? They use the following **change-of-base formula.**

$$\log_b x = \frac{\log_a x}{\log_a b}$$

To derive this formula, let $\log_b x = c$ so that $x = b^c$. Take $\log_a$ of both sides of the latter to obtain

$$\log_a x = c \log_a b = \log_b x \log_a b$$

This is equivalent to the change-of-base formula.

Example F (Change of Base)   Calculate $\log_2 13$.

Solution   Use the formula together with the common log key on your calculator (denoted by $\boxed{\log_{10}}$ or simply by $\boxed{\log}$) to write

$$\log_2 13 = \frac{\log_{10} 13}{\log_{10} 2} \approx \frac{1.1139433}{.30103} \approx 3.7004 \quad \blacksquare$$

PROBLEM SET 6-4

A. Skills and Techniques

Write each of the following in logarithmic form. For example, $3^4 = 81$ can be written as $\log_3 81 = 4$.

1. $4^3 = 64$
2. $7^3 = 343$
3. $27^{1/3} = 3$
4. $16^{1/4} = 2$
5. $4^0 = 1$
6. $81^{-1/2} = \frac{1}{9}$
7. $125^{-2/3} = \frac{1}{25}$
8. $2^{9/2} = 16\sqrt{2}$
9. $10^{\sqrt{3}} = a$
10. $5^{\sqrt{2}} = b$
11. $10^a = \sqrt{3}$
12. $b^x = y$

Write each of the following in exponential form. For example, $\log_5 125 = 3$ can be written as $5^3 = 125$.

13. $\log_5 625 = 4$
14. $\log_6 216 = 3$
15. $\log_4 8 = \frac{3}{2}$
16. $\log_{27} 9 = \frac{2}{3}$
17. $\log_{10}(.01) = -2$
18. $\log_3(\frac{1}{27}) = -3$

19. $\log_c c = 1$
20. $\log_b N = x$
21. $\log_c Q = y$

Determine the value of each of the following logarithms.

22. $\log_4 16$
23. $\log_5 25$
24. $\log_7 \frac{1}{7}$
25. $\log_3 \frac{1}{3}$
26. $\log_4 2$
27. $\log_{27} 3$
28. $\log_{10}(10^{-6})$
29. $\log_{10}(.0001)$
30. $\log_8 1$
31. $\log_3 1$
32. $\log_{100} 1000$
33. $\log_8 16$

Find the value of c in each of the following. See Examples A and C.

34. $\log_c 25 = 2$
35. $\log_c 8 = 3$
36. $\log_4 c = -\frac{1}{2}$
37. $\log_9 c = -\frac{3}{2}$
38. $\log_2(2^{5.6}) = c$
39. $\log_3(3^{-2.9}) = c$

40. $8^{\log_8 11} = c$

41. $5^{2\log_5 7} = c$

42. $3^{4\log_3 2} = c$

Given $\log_{10} 2 = .301$ and $\log_{10} 3 = .477$, calculate each of the following without using a calculator, as in Example B.

43. $\log_{10} 6$

44. $\log_{10} \frac{3}{2}$

45. $\log_{10} 16$

46. $\log_{10} 27$

47. $\log_{10} \frac{1}{4}$

48. $\log_{10} \frac{1}{27}$

49. $\log_{10} 24$

50. $\log_{10} 54$

51. $\log_{10} \frac{8}{9}$

52. $\log_{10} \frac{3}{8}$

53. $\log_{10} 5$

54. $\log_{10} \sqrt[3]{3}$

Your scientific calculator has a $\log_{10}$ key (which may be abbreviated log). Use it to find each of the following.

55. $\log_{10} 34$

56. $\log_{10} 1417$

57. $\log_{10}(.0123)$

58. $\log_{10}(.3215)$

59. $\log_{10} 9723$

60. $\log_{10}(\frac{21}{312})$

Write each of the following as a single logarithm, as in Example D.

61. $3\log_{10}(x + 1) + \log_{10}(4x + 7)$

62. $\log_{10}(x^2 + 1) + 5\log_{10} x$

63. $3\log_2(x + 2) + \log_2 8x - 2\log_2(x + 8)$

64. $2\log_5 x - 3\log_5(2x + 1) + \log_5(x - 4)$

65. $\frac{1}{2}\log_6 x + \frac{1}{3}\log_6(x^3 + 3)$

66. $-\frac{2}{3}\log_3 x + \frac{5}{2}\log_3(2x^2 + 3)$

Solve each of the following equations. See Example E.

67. $\log_7(x + 2) = 2$

68. $\log_5(3x + 2) = 1$

69. $\log_2(x + 3) = -2$

70. $\log_4(\frac{1}{64}x + 1) = -3$

71. $\log_2 x - \log_2(x - 2) = 3$

72. $\log_3 x - \log_3(2x + 3) = -2$

73. $\log_2(x - 4) + \log_2(x - 3) = 1$

74. $\log_{10} x + \log_{10}(x - 3) = 1$

Use the method of Example F to find the following.

75. $\log_2 128$

76. $\log_3 128$

77. $\log_3 82$

78. $\log_5 110$

79. $\log_6 39$

80. $\log_2(.26)$

B. Applications and Extensions

81. Find the value of x in each of the following
    (a) $x = \log_6 36$
    (b) $x = \log_4 2$
    (c) $\log_{25} x = \frac{3}{2}$
    (d) $\log_4 x = \frac{5}{2}$
    (e) $\log_x 10\sqrt{10} = \frac{3}{2}$
    (f) $\log_x \frac{1}{8} = -\frac{3}{2}$

82. Write each of the following as a single logarithm.
    (a) $3\log_2 5 - 2\log_2 7$
    (b) $\frac{1}{2}\log_5 64 + \frac{1}{3}\log_5 27 - \log_5(x^2 + 4)$
    (c) $\frac{2}{3}\log_{10}(x + 5) + 4\log_{10} x - 2\log_{10}(x - 3)$

83. Evaluate $\dfrac{(\log_{27} 3)(\log_{27} 9)(3^{2\log_3 2})}{\log_3 27 - \log_3 9 + \log_3 1}$

84. Solve for x.
    (a) $2(\log_4 x)^2 + 3\log_4 x - 2 = 0$
    (b) $(\log_x 8)^2 - \log_x 8 - 6 = 0$
    (c) $(\log_x \sqrt{3} + \log_x 3^5 + \log_x(\frac{1}{27})) = \frac{5}{4}$

85. Solve for x.
    (a) $\log_5(2x - 1) = 2$
    (b) $\log_4\left(\dfrac{x - 2}{2x + 3}\right) = 0$
    (c) $\log_4(x - 2) - \log_4(2x + 3) = 0$
    (d) $\log_{10} x + \log_{10}(x - 15) = 2$
    (e) $\dfrac{\log_2(x + 1)}{\log_2(x - 1)} = 2$
    (f) $\log_8[\log_4(\log_2 x)] = 0$

86. Solve for x.
    (a) $2^{\log_2 x} = 16$ **(b)** $2^{\log_x 2} = 16$
    (c) $x^{\log_2 x} = 16$ **(d)** $\log_2 x^2 = 2$
    (e) $(\log_2 x)^2 = 1$ **(f)** $x = (\log_2 x)^{\log_2 x}$

87. Solve for y in terms of x.
    (a) $\log_a(x + y) = \log_a x + \log_a y$
    (b) $x = \log_a(y + \sqrt{y^2 - 1})$

88. Show that $f(x) = \log_a(x + \sqrt{1 + x^2})$ is an odd function.

89. Figure 14 shows the magnitudes of four famous earthquakes based on the Richter scale. This scale relates the magnitude M of an earthquake to the amplitude A of the largest seismic wave at a distance 100 kilometers from the epicenter by the formula

$$M = \log_{10}\left(\frac{A}{C}\right)$$

where C is a specially defined constant that need not concern us.
    (a) Solve for A in the formula.
    (b) The ratio of the two amplitudes is a measure of

Magnitudes of Some Earthquakes	
1906 San Francisco	8.3
1933 Japan	8.9
1985 Mexico City	8.1
1989 San Francisco	7.1

Figure 14

the relative strengths of two eathquakes. On this basis, how much stronger was the 1933 Japan eathquake than the 1989 San Francisco earthquake?

90. Refer to Problem 89. How much stronger was the 1906 San Francisco earthquake than the 1989 earthquake?

91. Show that $\log_a b = 1/\log_b a$, where a and b are positive.

92. If $\log_b N = 2$, find $\log_{1/b} N$.

93. Graph the equations $y = 3^x$ and $y = \log_3 x$ using the same coordinate axes.

94. Find the solution set for each of the following inequalities.
    (a) $\log_2 x < 0$ (b) $\log_{10} x \geq -1$
    (c) $2 < \log_3 x < 3$
    (d) $-2 \leq \log_{10} x \leq -1$
    (e) $2^x > 10$ (f) $2^x < 3^x$

95. Sketch the graph of each of the following functions using the same coordinate axes.
    (a) $f(x) = \log_2 x$ (b) $g(x) = \log_2(x + 1)$
    (c) $h(x) = 3 + \log_2 x$

96. **TEASER** Let log represent $\log_{10}$. Evaluate.
    (a) $\log \frac{1}{2} + \log \frac{2}{3} + \log \frac{3}{4} + \cdots + \log \frac{98}{99} + \log \frac{99}{100}$
    (b) $\log \frac{3}{4} + \log \frac{8}{9} + \log \frac{15}{16} + \cdots$
    $$+ \log \frac{99^2 - 1}{99^2} + \log \frac{100^2 - 1}{100^2}$$
    (c) $\log_2 3 \cdot \log_3 4 \cdot \log_4 5 \cdot \log_5 6 \cdots \log_{63} 64$

97. Draw the graphs of
    $$f(x) = \frac{10^x - 10^{-x}}{2} \quad \text{and}$$
    $$g(x) = \log_{10}(x + \sqrt{x^2 + 1})$$
    using the same axes.
    (a) Make a conjecture based on your graphs.
    (b) Simplify $f(g(x))$ to throw light on this problem.

98. Draw the graphs of
    $$f(x) = \log_{10}[(x^3 - 3x - 5)^2] \quad \text{and}$$
    $$g(x) = 2 \log_{10}(x^3 - 3x - 5)$$
    (a) Are f and g identical functions? Explain.
    (b) Determine the natural domains for f and g.

6-5 NATURAL LOGARITHMS AND APPLICATIONS

The collected works of this brilliant Swiss mathematician will fill 74 volumes when completed. No other person has written so profusely on mathematical topics. Remarkably, 400 of his research papers were written after he was totally blind. One of his contributions was the introduction of the number $e = 2.71828 \ldots$ as the base for natural logarithms.

Leonhard Euler
1707-1783

Napier invented logarithms to simplify arithmetic calculations. Computers and calculators have reduced that application to minor significance. Here we have in mind deeper applications such as solving exponential equations, defining power functions, and modeling physical phenomena.

Historically, the use of logarithms to simplify arithmetic calculations made logarithms to the base 10 the logarithm of choice. We could, in fact, use such logarithms to do most of what we have in mind for this section. However, we have chosen to introduce you to the **natural logarithm** that is used in all of advanced mathematics including calculus. This logarithm uses as its base the number e (after Euler), which has as its approximate value

$$e \approx 2.71828$$

This number, like the special number π, appears in all kinds of unexpected places (for an example, see Figure 15). It is an irrational number whose decimal expansion is known to hundreds of thousands of places. Your calculator undoubtedly has a $\boxed{\log_e}$ key, but it may be denoted by $\boxed{\ln}$ since ln is the standard shorthand for $\log_e$.

Since ln denotes a genuine logarithm function, we have as in Section 6-4

$$\ln N = x \quad \text{if and only if} \quad e^x = N$$

and consequently

$$\ln e^x = x \quad \text{and} \quad e^{\ln N} = N$$

Moreover, the three properties of logarithms hold.

1. $\ln(MN) = \ln M + \ln N$
2. $\ln(M/N) = \ln M - \ln N$
3. $\ln(N^p) = p \ln N$

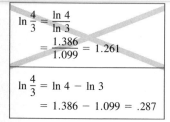

e and Interest

Suppose that one dollar is invested at 100 percent interest compounded n times per year. Then it will be worth $\left(1 + \dfrac{1}{n}\right)^n$ dollars at the end of the year. As n increases without bound (that is, as compounding occurs more and more often)

$$\left(1 + \frac{1}{n}\right)^n \rightarrow e$$

and we have what is called continuous compounding. Under this plan, $1 grows to $2.72 in one year.

Figure 15

CAUTION

$$\ln \frac{4}{3} = \frac{\ln 4}{\ln 3}$$
$$= \frac{1.386}{1.099} = 1.261$$

$$\ln \frac{4}{3} = \ln 4 - \ln 3$$
$$= 1.386 - 1.099 = .287$$

Example A (Using the Properties)   Given that ln 14 = 2.639, calculate each of the following.

(a) $\ln(e^{2.1})$ (b) $\ln(14\, e^2)$ (c) $\ln\left(\dfrac{\sqrt{e}}{14}\right)$ (d) $\ln(14^3)$ (e) $e^{3\ln 4}$

Solution

(a) $\ln(e^{2.1}) = 2.1 \ln e = 2.1$
(b) $\ln(14\, e^2) = \ln 14 + \ln(e^2) = 2.639 + 2 = 4.639$
(c) $\ln(\sqrt{e}/14) = \ln(e^{0.5}) - \ln 14 = .5 - 2.639 = -2.139$
(d) $\ln(14^3) = 3 \ln 14 = 3(2.639) = 7.917$
(e) $e^{3\ln 4} = e^{\ln(4^3)} = 4^3 = 64$ ∎

Solving Exponential Equations

Consider first the simple equation

$$5^x = 1.7$$

We call it an *exponential equation* because the unknown is in the exponent. To solve it, we take natural logarithms of both sides.

$$5^x = 1.7$$

$$\ln(5^x) = \ln 1.7$$

$$x \ln 5 = \ln 1.7$$

$$x = \frac{\ln 1.7}{\ln 5}$$

$$x \approx .3297$$

Example B (Exponential Equations) Solve $4^{3x-2} = 15$ for x.

Solution Take natural logarithms of both sides and then solve for x. Complete the solution using a calculator.

$$\ln 4^{3x-2} = \ln 15$$

$$(3x - 2)\ln 4 = \ln 15$$

$$3x \ln 4 - 2 \ln 4 = \ln 15$$

$$3x \ln 4 = 2 \ln 4 + \ln 15$$

$$x = \frac{2 \ln 4 + \ln 15}{3 \ln 4}$$

$$x \approx 1.3178 \quad \blacksquare$$

Natural Logarithm Tables

Until calculators and computers became widely available in the 1970s, values of logarithms were found by looking in a table. Table A in the Appendix is such a table. We suggest that you study it enough so you will know how to use it when your calculator malfunctions or when its battery dies.

Now we can algebraically solve the doubling time and half-life problems that we had to solve by experiment in Section 6-3.

Example C (Doubling Time) How long will it take money to double in value if it is invested at 9.5 percent interest compounded annually?

Solution For convenience, consider investing $1. From Section 6-3 we know that this dollar will grow to $(1.095)^t$ dollars after t years. Thus we must solve the exponential equation $(1.095)^t = 2$. This we do by the method of Example B, using a calculator. The result is

$$t = \frac{\ln 2}{\ln 1.095} \approx 7.6376 \text{ years}$$

(*Note:* A bank may allow withdrawals with full interest only at the ends of interest periods, in this case, at the ends of years). $\blacksquare$

Example D (Half-lives) Suppose that a substance decays exponentially according to the formula

$$y = 20e^{-.025t}$$

where y is the number of grams present after t years. Determine its half-life.

Solution Since $y = 20$ at $t = 0$, we must solve the exponential equation $20e^{-.025t} = 10$ or after dividing by 20, $e^{-.025t} = .5$. Proceeding as in the examples above, we obtain

$$t = \frac{\ln(.5)}{-.025} \approx 27.726 \text{ years} \quad \blacksquare$$

The Graphs of ln x and e^x

We have already pointed out that $\ln e^x = x$ and $e^{\ln x} = x$. Thus $f(x) = \ln x$ and $g(x) = e^x$ are inverse functions, which means that their graphs are reflections of each other across the line $y = x$. They are shown in Figure 16.

Since $\ln x$ is not defined for $x \le 0$, it is of some interest to consider $\ln |x|$, which is defined for all x except 0. Its graph is shown in Figure 17. Note the symmetry with respect to the y-axis.

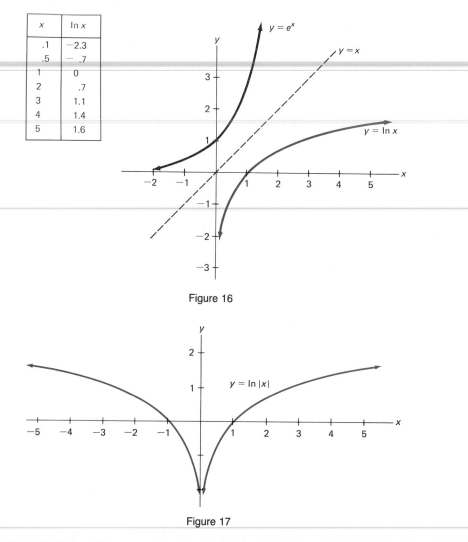

x	$\ln x$
.1	−2.3
.5	− .7
1	0
2	.7
3	1.1
4	1.4
5	1.6

Figure 16

Figure 17

Exponential Functions versus Power Functions

Look closely at the formulas below.

$$f(x) = 2^x \qquad f(x) = x^2$$

They are very different, yet easily confused. The first is an *exponential function,* and the second is called a *power function.* Both grow rapidly for large x, but the exponential function ultimately gets far ahead (see the graphs in Figure 18).

The situation described above is a special instance of two very general classes of functions.

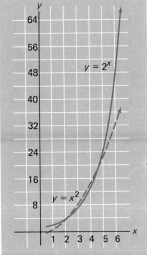

Figure 18

Exponential Functions

$$f(x) = ba^x$$

Power Functions

$$f(x) = bx^a$$

Example E (2^x Versus x^{10})   Show that 2^x ultimately grows faster than x^{10}.

Solution   While this is an example best handled in a calculus course, we can give pretty good evidence by considering the ratio $h(x) = 2^x/x^{10}$ for various values of x. In particular, we discover that

$$h(20) \approx 10^{-7}$$
$$h(40) \approx 10^{-4}$$
$$h(60) \approx 1.9$$
$$h(80) \approx 1.1 \times 10^5$$
$$h(100) \approx 1.3 \times 10^{10} \qquad \blacksquare$$

In fact, it can be shown that a^x, with $a > 1$, ultimately grows faster than any power of x, no matter how large the exponent.

Curve Fitting

A recurring theme in science is to fit a mathematical curve to a set of experimental data. Suppose that a scientist, studying the relationship between two variables x and y, obtained the data plotted in Figure 19. In searching for curves to fit these data, the scientist naturally thought of exponential curves and power curves. How did he or she decide if either was appropriate? The scientist took logarithms. Let us see why.

Figure 19

Model 1	Model 2
$y = ba^x$	$y = bx^a$
$\ln y = \ln b + x \ln a$	$\ln y = \ln b + a \ln x$
$Y = B + Ax$	$Y = B + aX$

Here the scientist made the substitutions $Y = \ln y$, $B = \ln b$, $A = \ln a$, and $X = \ln x$.

In both cases, the final result is a linear equation. But note the difference. In the first case, $\ln y$ is a linear function of x, whereas in the second case, $\ln y$ is a linear function of $\ln x$. These considerations suggest the following procedures. Make two additional plots of the data. In the first, plot $\ln y$ against x, and in the second, plot $\ln y$ against $\ln x$. If the first plotting gives data nearly along a straight line, model 1 is appropriate; if the second does, then model 2 is appropriate. If neither plot approximates a straight line, our scientist should look for a different and perhaps more complicated model.

We have used natural logarithms in the discussion above; we could also have used common logarithms (logarithms to the base 10). In the latter case, special kinds of graph paper are available to simplify the curve fitting process. On semilog paper, the vertical axis has a logarithmic scale; on log-log paper, both axes have logarithmic scales. The xy-data can be plotted *directly* on this paper. If semilog paper gives an (approximately) straight line, model 1 is indicated; if log-log paper does so, then model 2 is appropriate. You will have ample opportunity to use these special kinds of graph paper in your science courses.

t	N
0	100
1	700
2	5000
3	40,000

Figure 20

Example F (Curve Fitting)   The table in Figure 20 shows the number N of bacteria in a certain culture found after t hours. Which is a better description of these data,

$$N = ba^t \quad \text{or} \quad N = bt^a$$

Find a and b.

Solution   In line with our discussion of curve fitting, we begin by plotting $\ln N$ against t. If the resulting points lie along a line, we choose $N = ba^t$ as the appropriate model. If not, we will plot $\ln N$ against $\ln t$ to check on the second model. Since the fit to a line in Figure 21 is quite good, we accept $N = ba^t$ as our model. To find a and b, we write $N = ba^t$ in the form

$$\ln N = \ln b + t \ln a \quad \text{or} \quad Y = \ln b + (\ln a)t$$

Examination of the line shows that it has a Y-intercept of 4.6 and a slope of about 2; so for its equation we write $Y = 4.6 + 2t$. Comparing this with $Y = \ln b + (\ln a)t$ gives

$$\ln b = 4.6 \qquad \ln a = 2$$

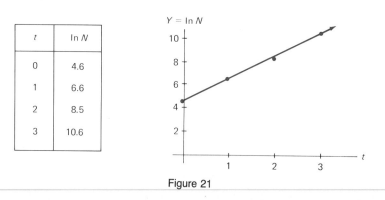

t	$\ln N$
0	4.6
1	6.6
2	8.5
3	10.6

Figure 21

Finally, we use a calculator to find

$$b \approx 100 \qquad a \approx 7.4$$

Thus the original data are described reasonably well by the equation

$$N = 100(7.4)^t \quad \blacksquare$$

Logarithms and Physiology

The human body appears to have a built-in logarithmic calculator. What do we mean by this statement?

In 1834, the German physiologist E. Weber noticed an interesting fact. Two heavy objects must differ in weight by considerably more than two light objects if a person is to perceive a difference between them. Other scientists noted the same phenomenon when human subjects tried to differentiate loudness of sounds, pitches of musical tones, brightness of light, and so on. Experiments suggested that people react to stimuli on a logarithmic scale, a result formulated as the Weber-Fechner law (see Figure 22).

$$S = C \ln\left(\frac{R}{r}\right)$$

Here R is the actual intensity of the stimulus, r is the threshold value (smallest value at which the stimulus is observed), C is a constant depending on the type of stimulus, and S is the perceived intensity of the stimulus. Note that a change in R is not as perceptible for large R as for small R because as R increases, the graph of the logarithmic functions gets steadily flatter.

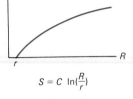

$S = C \ln(\frac{R}{r})$

Figure 22

PROBLEM SET 6-5

A. Skills and Techniques

In Problems 1–10, find the value of each expression. See Example A.

1. $\ln e$

2. $\ln(e^2)$

3. $\ln 1$

4. $\ln\left(\frac{1}{e}\right)$

5. $\ln\sqrt{e}$

6. $\ln(e^{1.1})$

7. $\ln\left(\frac{1}{e^3}\right)$

8. $\ln(e^n)$

9. $e^{2\ln 5}$

10. $e^{-3\ln 2}$

For Problems 11–16, find each value. Assume that $\ln a = 2.5$ and $\ln b = -4$.

11. $\ln(ae)$

12. $\ln\left(\frac{e}{b}\right)$

13. $\ln\sqrt{b}$

14. $\ln(a^2 b^{10})$

15. $\ln\left(\frac{1}{a^3}\right)$

16. $\ln(a^{4/5})$

Use your calculator ($\boxed{\ln}$ or $\boxed{\log_e}$ key) to find each of the following.

17. $\ln 4.31$

18. $\ln 517$

19. $\ln(.127)$

20. $\ln(.00424)$

21. $\ln\left(\frac{6.71}{42.3}\right)$

22. $\ln\sqrt{457}$

23. $\dfrac{\ln 6.71}{\ln 42.3}$

24. $\sqrt{\ln 457}$

25. $\ln(51.4)^3$

26. $\ln(31.2 + 43.1)$

27. $(\ln 51.4)^3$

28. $3 \ln 51.4$

Use your calculator to find N in each of the following. Hint: $\ln N = 5.1$ if and only if $N = e^{5.1}$. On some calculators, there is an $\boxed{e^x}$ key; on others, you use the two keys $\boxed{INV}\ \boxed{\ln}$.

29. $\ln N = 2.12$

30. $\ln N = 5.63$

31. $\ln N = -.125$

32. $\ln N = .00257$

33. $\ln \sqrt{N} = 3.41$

34. $\ln N^3 = .415$

Solve for x using the method of Example B.

35. $3^x = 20$ **36.** $5^x = 40$

37. $2^{x-1} = .3$ **38.** $4^x = 3^{2x-1}$

39. $(1.4)^{x+2} = 19.6$ **40.** $5^x = \frac{1}{2}(4^x)$

Examples C and D relate to Problems 41–46.

41. How long would it take money to double at 12 percent compounded annually?

42. How long would it take money to double at 12 percent compounded monthly? *Hint:* After t months, \$1 is worth $(1.01)^t$ dollars.

43. How long would it take money to double at 15 percent compounded quarterly?

44. A certain substance decays according to the formula $y = 100e^{-.135t}$, where t is in years. Find its half-life.

45. If an element decays so that y grams are present after t years where $y = 50e^{-.0125t}$, find its half-life.

46. What is the half-life of a decaying substance if an amount $y = 8(\frac{1}{3})^{t/100}$ is present after t days?

47. By finding the natural logarithm of the numbers in each pair, determine which is larger.
(a) 10^5, 5^{10} (b) 10^9, 9^{10}
(c) 10^{20}, 20^{10} (d) 10^{1000}, 1000^{10}

48. What do your answers in Problem 47 confirm about the growth of 10^x and x^{10} for large x?

49. On the same coordinate plane, graph $y = 3^x$ and $y = x^3$ for $0 \le x \le 4$.

50. By means of a change of variable(s) (as explained in the text), transform each equation below to a linear equation. Find the slope and Y-intercept of the resulting line.
(a) $y = 3e^{2x}$ (b) $y = 2x^3$
(c) $xy = 12$ (d) $y = x^e$
(e) $y = 5(3^x)$ (f) $y = ex^{1.1}$

For the data sets below, decide whether $y = ba^x$ or $y = bx^a$ is the better model. Then determine a and b. See Example F.

51.

x	1	2	3	4
y	96	145	216	325

52.

x	0	1	2	4
y	243	162	108	48

53.

x	1	2	3	5
y	12	190	975	7490

54.

x	1	4	9
y	16	128	432

B. Applications and Extensions

55. Evaluate without using a calculator.
(a) $\ln(e^{4.2})$ (b) $e^{2\ln 2}$
(c) $\dfrac{\ln 3e}{2 + \ln 9}$

56. Solve for x.
(a) $\ln(5 + x) = 1.2$ (b) $e^{x^2-x} = 2$
(c) $e^{2x} = (.6)8^x$

57. Evaluate without use of a calculator.
(a) $\ln[(e^{3.5})^2]$ (b) $(\ln e^{3.5})^2$
(c) $\ln(1/\sqrt{e})$ (d) $(\ln 1)/(\ln\sqrt{e})$
(e) $e^{3\ln 5}$ (f) $e^{\ln(1/2)+\ln(2/3)}$

58. Since $a^x = e^{\ln a^x} = e^{x\ln a}$, the study of exponential functions can be subsumed under the study of the function e^{kx}. Determine k so that each of the following is true. *Hint:* In (a) rewrite the equation as $3^x = (e^k)^x$ which implies that $3 = e^k$. Now take natural logarithms.
(a) $3^x = e^{kx}$ (b) $\pi^x = e^{kx}$
(c) $(1/3)^x = e^{kx}$

59. Solve for x.
(a) $10^{2x+3} = 200$ (b) $10^{2x} = 8^{x-1}$
(c) $10^{x^2+3x} = 200$ (d) $e^{-.32x} = 1/2$
(e) $x^{\ln x} = 10$ (f) $(\ln x)^{\ln x} = x$
(g) $x^{\ln x} = x$ (h) $\ln x = (\ln x)^{\ln x}$
(i) $(x^2 - 5)^{\ln x} = x$

60. Show that each of the following is an identity. Assume that $x > 0$.
(a) $(\sqrt{3})^{3\sqrt{3}} = (3\sqrt{3})^{\sqrt{3}}$
(b) $2.25^{3.375} = 3.375^{2.25}$
(c) $a^{\ln(x^b)} = (a^{\ln x})^b$
(d) $x^x = e^{x\ln x}$
(e) $(\ln x)^x = e^{x\ln(\ln x)}$
(f) $\dfrac{\ln\left(\dfrac{x+1}{x}\right)^x}{\ln\left(\dfrac{x+1}{x}\right)^{x+1}} = \dfrac{\left(\dfrac{x+1}{x}\right)^x}{\left(\dfrac{x+1}{x}\right)^{x+1}}$

61. A certain substance decays according to the formula $y = 100e^{-3t}$, where y is the amount present after t years. Find its half-life.

62. Suppose that the number of bacteria in a certain culture t hours from now will be $200e^{.468t}$. When will the count reach 10,000?

63. A radioactive substance decays exponentially with a half-life of 240 years. Determine k in the formula $A = A_0e^{-kt}$, where A is the amount present after t years and A_0 is the initial amount.

64. Substances A and B decay exponentially with half-lives 44 and 55 years, respectively. If we have 100 grams of A and 60 grams of B, after how many years will we have equal amounts of the two substances?

65. Sketch the graph of

$$y = \frac{1}{\sqrt{2\pi}} e^{-(1/2)x^2}$$

This is the famous *normal curve,* so important in statistics.

66. In calculus it is shown that for small x

$$e^x \approx 1 + x + \frac{x^2}{2} + \frac{x^3}{6} + \frac{x^4}{24} + \frac{x^5}{120}$$

Use this to approximate e and $e^{-1/2}$.

67. From Problem 66 or from looking at the graph of $y = e^x$, you might guess the true result that $e^x > 1 + x$ for all $x > 0$. Use this and the obvious fact that $(\pi/e) - 1 > 0$ to demonstrate algebraically that $e^\pi > \pi^e$.

68. It is important to have a feeling for how various functions grow for large x. Let $\ll$ symbolize the phrase *grows slower than.* Use a calculator to convince yourself that

$$\ln x \ll \sqrt{x} \ll x \ll x^2 \ll e^x \ll x^x$$

69. In calculus, it is shown that $(1 + r/m)^m$ gets closer and closer to e^r as m gets larger and larger. Now if P dollars is invested at rate r (written as a decimal) compounded m times per year, it will grow to $P(1 + r/m)^{mt}$ dollars at the end of t years (Example E of Section 6-3). If interest is compounded continuously (see the box **e and interest** on page 000), P dollars will grow to Pe^{rt} dollars at the end of t years. Use these facts to calculate the value of $100 after 10 years if interest is at 12 percent (that is, $r = .12$) and is compounded (a) monthly; (b) daily; (c) hourly; (d) continuously.

70. TEASER The *harmonic* sum $S_n = 1 + \frac{1}{2} + \frac{1}{3} + \frac{1}{4} + \cdots + 1/n$ has intrigued both amateur and professional mathematicians since the time of the Greeks. In calculus, it is shown that for $n > 1$

$$\ln n < 1 + \frac{1}{2} + \frac{1}{3} + \frac{1}{4} + \cdots + \frac{1}{n} < 1 + \ln n$$

This shows that S_n grows arbitrarily large but that it grows very very slowly.

(a) About how large would n need to be for $S_n > 100$?

(b) Show how you could stack a pile of identical bricks each of length 1 foot (one brick to a tier as shown in Figure 23) to achieve an overhang of 50 feet. Could you make the overhang 50 million feet? Yes, it does have something to do with part (a).

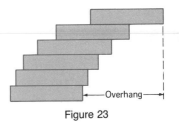

Figure 23

71. The function $f(x) = .01x$ ultimately grows faster than $g(x) = \ln x$. Determine the smallest c such that $f(x) > g(x)$ for all $x > c$.

72. The function $f(x) = e^x$ ultimately grows faster than $g(x) = x^8$. Determine the smallest c such that $f(x) > g(x)$ for all $x > c$.

CHAPTER 6 SUMMARY

The symbol $\sqrt[n]{a}$, the principal nth root of a, denotes one of the numbers whose nth power is a. For odd n, that is all that needs to be said. For n even and $a > 0$, we specify that $\sqrt[n]{a}$ signifies the positive nth root. Thus $\sqrt[3]{-8} = -2$ and $\sqrt{16} = \sqrt[2]{16} = 4$. (It is wrong to write $\sqrt{16} = -4$.) These symbols are also called **radicals.** These radicals obey five carefully prescribed rules (page 000). These rules allow us to simplify complicated radical expressions, in particular, to **rationalize denominators.**

The key to understanding **rational exponents** is the definition $a^{1/n} = \sqrt[n]{a}$, which implies $a^{m/n} = (\sqrt[n]{a})^m$. Thus $16^{5/4} = (\sqrt[4]{16})^5 = 2^5 = 32$. The meaning of **real exponents** is determined by considering rational approximations. For example, 2^π is the number that the sequence $2^3, 3^{3.1}, 2^{3.14}, \ldots$ approaches. The function $f(x) = a^x$ [and more generally, $f(x) = b \cdot a^x$] is called an **exponential function.**

A variable y is **growing exponentially** or **decaying exponentially** according as $a > 1$ or $0 < a < 1$ in the equation $y = b \cdot a^x$. Typical of the former are biological populations; of the latter, radioactive elements. Corresponding key ideas are **doubling times** and **half-lives.**

Logarithms are exponents. In fact, $\log_a N = x$ means $a^x = N$, that is, $a^{\log_a N} = N$. The functions $f(x) = \log_a x$ and $g(x) = a^x$ are **inverses** of each other. Logarithms have three primary properties (page 243). **Natural logarithms** correspond to the choice of base $a = e = 2.71828 \ldots$ and play a fundamental role in advanced courses. **Common logarithms** correspond to base 10 and have historically been used to simplify arithmetic calculations.

CHAPTER 6 REVIEW PROBLEM SET

In Problems 1–10, write True or False in the blank. If false, tell why.

_____ **1.** $(\sqrt[15]{3})^3 = \sqrt[5]{3}$

_____ **2.** $\sqrt[4]{16} = \pm 2$

_____ **3.** $\sqrt{5}\sqrt[3]{5} = \sqrt[6]{5}$

_____ **4.** $(4/9)^{-3/2} = 27/8$

_____ **5.** $\log_5 64 = 3 \log_5 4$

_____ **6.** If a, b, and c are all positive, then $\ln a + 3 \ln b - \frac{1}{2} \ln c = \ln(ab^3/\sqrt{c})$.

_____ **7.** If $c^{17} = 2$, then $c^{68} = 16$.

_____ **8.** The number -2 is not in the domain of $\log_2(x^2 - 2)$.

_____ **9.** If a radioactive substance has a half-life of T days, then one-eighth of the original amount will be left after $3T$ days.

_____ **10.** $f(x) = \ln(x^2)$ and $g(x) = 2 \ln x$ are identical functions.

Write in simplest radical form, being sure to rationalize all denominators.

11. $\sqrt[3]{(-27y^6)/z^{13}}$

12. $\sqrt[5]{64x^{10}y^6z^9}$

13. $\sqrt[3]{5\sqrt{5}}$

14. $5/(2\sqrt{x} - 3)$

15. $\sqrt{81 + 9x^2}$

16. $\sqrt{52} - \sqrt{117} + 39/\sqrt{13}$

Solve for x.

17. $\sqrt{2x - 5} = 9$

18. $3\sqrt{x} = x - 4$

In Problems 19–24, simplify, writing your answer in exponential form but with no negative exponents.

19. $\sqrt[5]{x^3}\, x^{7/5}$

20. $(\sqrt{a}\sqrt{ab})^2$

21. $(x^2y^{-4}z^6)^{-1/2}$

22. $\sqrt[3]{27\sqrt{64x^4}}$

23. $(3a^{1/3}a^{-1/4})^3$

24. $\sqrt[4]{32x^{-2}/x^6y^{-3}}$

25. Sketch the graphs of $y = (\frac{5}{4})^x$ and $y = (\frac{4}{5})^x$ for $-4 \le x \le 4$. How do these graphs relate?

26. The population of a certain city is growing exponentially so that its doubling time is 32 years. If it has 800,000 people now, how many people will it have after 96 years?

27. The half-life of a certain radioactive substance is 1600 years. How long will it take 10 grams to decay to 2.5 grams?

28. To what amount will $100 accumulate in 10 years if interest is
(a) 12 percent simple interest?
(b) 12 percent compounded annually?
(c) 12 percent compounded monthly?

29. To determine the doubling time for money invested at 8.5 percent compounded quarterly, what equation do we have to solve?

30. Write an expression for the value at the end of 2 years of $100 invested at 9 percent compounded daily.

In Problems 31–40, calculate x.

31. $\log_4 x = 3$

32. $\log_{25} x = \frac{-3}{2}$

33. $2^{\log_2 3} = x$

34. $\log_6 6^\pi = x$

35. $x = \log_{10}\sqrt{10,000}$

36. $x = \log_8 32$

37. $\log_x 64 = \frac{3}{2}$

38. $\ln x = 0$

39. $\log_2(x^2 - 1) = 3$

40. $\log_2 x + \log_2(x + 2) = 3$

41. Write $3 \log_4(x^2 + 1) - 2 \log_4 x + 2$ in the form $\log_4(\cdots)$.

42. How does the graph of $y = \ln(x - 1) + 2$ relate to the graph of $y = \ln x$?

43. Sketch the graph of $y = \log_3(x + 2)$ showing the vertical asymptote.

44. If $\log_a b = \frac{2}{3}$, find $\log_b a$.

45. Solve the exponential equation $3^{2x+1} = 20$.

46. How long will it take $100 to accumulate to $160 if it is invested at 8.5 percent compounded quarterly?

47. A radioactive substance decays according to the formula $y = y_0 e^{-.055t}$ where t is measured in years.
(a) Find its half-life.

(b) How long does it take to decay to one-tenth its original size?

48. A radioactive substance initially weighing 40 grams decayed so that $y = 40e^{-kt}$ grams were left after t days. Determine k if it took 20 days to decay to 10 grams.

49. How do the graphs of $y = \log_4 x$ and $y = 4^x$ relate to each other?

50. Solve $x^x = 500$ by experimentation.

CHAPTER **7**

The Trigonometric Functions

■ The great book of Nature lies open before our eyes and true philosophy is written in it. . . . But we cannot read it unless we have first learned the language and characters in which it is written. . . . It is written in mathematical language and the characters are triangles, circles, and other geometrical figures.

Galileo

7-1 RIGHT-TRIANGLE TRIGONOMETRY

What Is Trigonometry?

Sometime before 100 B.C., the Greeks invented trigonometry to solve problems in astronomy, navigation, and geography. The word "trigonometry" comes from Greek and means "triangle measurement." In its most basic form, trigonometry is the study of relationships between the angles and sides of a right triangle.

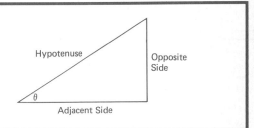

A triangle is called a right triangle if one of its angles is a right angle, that is, a 90° angle. The other two angles are necessarily acute angles (less than 90°) since the sum of all three angles in a triangle is 180°. Let θ (the Greek lower case letter theta) denote one of these acute angles. We may label the three sides relative to θ: adjacent side, opposite side, and hypotenuse, as shown in the diagram above. In terms of these sides, we introduce the three fundamental ratios of trigonometry, sine θ, cosine θ, and tangent θ. Using obvious abbreviations, we give the following definitions.

Your instructor may choose to skip this section, preferring to base all of trigonometry on the unit circle as in Section 7.3. If so, you may still want to peruse this section as it shows how trigonometry was introduced historically.

$$\sin \theta = \frac{\text{opp}}{\text{hyp}}$$

$$\cos \theta = \frac{\text{adj}}{\text{hyp}}$$

$$\tan \theta = \frac{\text{opp}}{\text{adj}}$$

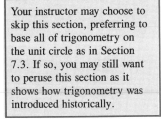

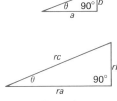

Figure 1

Thus with every acute angle θ, we associate three numbers, $\sin \theta$, $\cos \theta$, and $\tan \theta$. A careful reader might wonder whether these numbers depend only on the size of θ, or if they also depend on the lengths of the sides of the right triangle with which we started. Consider two different right triangles, each with the same angle θ (as in Figure 1). You may think of the lower triangle as a magnification of the upper one. Each of its sides has length r times that of the corresponding side in the upper triangle. If we calculate $\sin \theta$ from the lower triangle, we get

$$\sin \theta = \frac{\text{opp}}{\text{hyp}} = \frac{rb}{rc} = \frac{b}{c}$$

which is the same result we get using the upper triangle. We conclude that for a given θ, $\sin \theta$ has the same value no matter which right triangle is used to compute it. So do $\cos \theta$ and $\tan \theta$.

Special Angles

We can use the Pythagorean Theorem ($a^2 + b^2 = c^2$) to find the values of sine, cosine, and tangent for the special angles 30°, 45°, and 60°. Consider the two right triangles of Figure 2, which involve these angles.

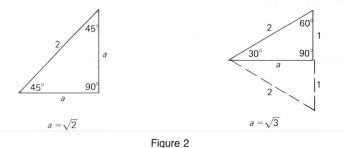

Figure 2

To see that the indicated values of a are correct, note that in the first triangle, $a^2 + a^2 = 2^2$, which gives $a = \sqrt{2}$. In the second, which is half of an equilateral triangle, $a^2 + 1^2 = 2^2$, or $a = \sqrt{3}$.

From these triangles, we obtain the following important facts.

$$\sin 45° = \frac{\sqrt{2}}{2} \qquad \cos 45° = \frac{\sqrt{2}}{2} \qquad \tan 45° = 1$$

$$\sin 30° = \frac{1}{2} \qquad \cos 30° = \frac{\sqrt{3}}{2} \qquad \tan 30° = \frac{1}{\sqrt{3}} = \frac{\sqrt{3}}{3}$$

$$\sin 60° = \frac{\sqrt{3}}{2} \qquad \cos 60° = \frac{1}{2} \qquad \tan 60° = \sqrt{3}$$

Other Angles

When you need the sine, cosine, or tangent of an angle other than the special ones just considered, you may do one of two things. You may push two or three keys on a calculator and have the answer correct to eight or more significant digits. Or you may look up the answer in Table B of the Appendix.

Several facts about Table B should be noted. First, it gives answers usually to four decimal places. Second, angles are measured in degrees and tenths of degrees. By interpolation (see page 597), it is possible to consider angles measured to the nearest hundredth of a degree. Finally, notice that the left column of the table lists angles from 0° to 45°. For angles from 45° to 90°, use the right column; you must then also use the bottom captions. We suggest that you try Examples A and B using both your calculator and Table B to make sure that you can use them correctly.

Example A (General Angles)   Evaluate, correct to four significant digits.
(a) tan 33.1°   (b) sin 26.9°   (c) cos 54.3°   (d) tan 82°

Solution
(a) $\tan 33.1° = .6519$ (b) $\sin 26.9° = .4524$
(c) $\cos 54.3° = .5835$ (d) $\tan 82° = 7.115$ ∎

Example B (Reversing Direction)   Find θ, correct to the nearest hundredth of a degree.
(a) $\sin \theta = .2622$ (b) $\tan \theta = 2.322$ (c) $\cos \theta = .9394$

Solution   On a scientific calculator, this is a simple matter. For part (a), press .2622 $\boxed{\text{INV}}$ $\boxed{\text{sin}}$ or perhaps .2622 $\boxed{\text{sin}^{-1}}$ and similarly for parts (b) and (c). The

Graphing Calculators

On calculators with screen displays, the order of pressing keys is different from that on the simple scientific calculators discussed in the text. On graphing calculators like the TI-81, you press keys in the order in which the symbols are displayed in ordinary writing.

same problems can be done by using Table B backwards. For example, to do part (a), look for .2622 under sin in the body of the table and identify the corresponding angle. In part (c), you will need to interpolate, a matter discussed on page 597. Either way, here are the answers.

(a) $\theta = 15.20°$ (b) $\theta = 66.70°$ (c) $\theta = 20.05°$ ■

Solving a Right Triangle

To solve a triangle means to determine all its unknown parts. We can solve a right triangle if we know one side and one acute angle or if we know two sides.

Example C (Solving a Right Triangle Given an Angle and a Side) Solve the right triangle which has hypotenuse of length 14.6 and an angle measuring 33.2°.

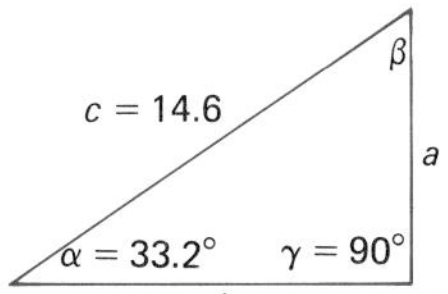

Figure 3

Solution First, we draw the triangle labeling the known parts and assigning letters to the unknown parts. Our convention is to use the first three Greek letters, α, β, and γ (alpha, beta, and gamma) for the angles and a, b, and c for the lengths of the respective sides opposite these angles (see Figure 3). We need to find β, a, and b.

(i) $\beta = 90° - 33.2° = 56.8°$

(ii) $\sin 33.2° = a/14.6$, so

$$a = 14.6 \sin 33.2° = (14.6)(.5476) \approx 7.99$$

(iii) $\cos 33.2° = b/14.6$, so

$$b = 14.6 \cos 33.2° = (14.6)(.8368) \approx 12.2$$

Notice that we gave the answers to three significant digits since the given data have three significant digits. ■

Example D (Solving a Right Triangle Given Two Sides) Solve the right triangle which has legs $a = 42.8$ and $b = 94.1$.

Figure 4

Solution First, we draw the triangle and label its parts (Figure 4). We must find α, β, and c.

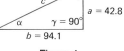

(i) $\tan \alpha = \dfrac{42.8}{94.1} \approx .4548$

Now we can find α, by using Table B backwards, that is, by looking under tangent in the body of the table for .4548 and determining the corresponding angle. Or better, we can use the [INV] [tan] keys on a scientific calculator. On many calculators, press

[(] 42.8 [÷] 94.1 [)] [INV] [tan]

In either case, the result is $\alpha \approx 24.5°$.

(ii) $\beta = 90° - \alpha \approx 90° - 24.5° = 65.5°$

(iii) We could find c by using $c^2 = a^2 + b^2$. Instead, we use $\sin \alpha$.

$$\sin \alpha = \sin 24.5° = \frac{42.8}{c}$$

$$c = \frac{42.8}{\sin 24.5°} \approx 103 \quad ■$$

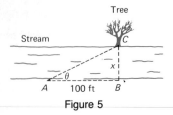

Figure 5

Applications

Suppose that you wish to measure the distance across a stream but do not want to get your feet wet. Here is how you might proceed.

Pick out a tree at C on the opposite shore and set a stone at B directly across from it on your shore (Figure 5). Set another stone at A, 100 feet up the shore from B. With an angle measuring device (for example, a protractor or a transit), measure angle θ between AB and AC. Then x, the length of BC, satisfies the following equation:

$$\tan \theta = \frac{\text{opp}}{\text{adj}} = \frac{x}{100}$$

or

$$x = 100 \tan \theta$$

For example, if θ measures 29°, $x = 100 \tan 29° = 55.43$ feet. Since you used stones and trees for points, this suggests that you should not give your answer with such accuracy. It would be better to say that the distance x is approximately 55 feet.

Example E (The Height of a Steeple)   Find the height of the steeple in Figure 6, given that $\alpha = 35°$ and $\beta = 26°$.

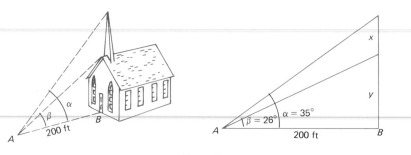

Figure 6

Solution   Let x be the height of the steeple and y be the distance from the ground to the bottom of the steeple. Then

$$\tan 35° = \frac{x + y}{200}$$

$$\tan 26° = \frac{y}{200}$$

If you solve for y in the second equation and substitute the value in the first, you will get the following sequence of equations.

$$\tan 35° = \frac{x + 200 \tan 26°}{200}$$

$$200 \tan 35° = x + 200 \tan 26°$$

$$x = 200 \tan 35° - 200 \tan 26°$$

$$\approx 42.5 \text{ feet} \quad \blacksquare$$

PROBLEM SET 7-1

A. Skills and Techniques

Evaluate each of the following, correct to four significant digits, as in Example A.

1. sin 41.3°
2. tan 54.4°
3. cos 49.2°
4. sin 89.3°
5. tan 72.3°
6. cos 38.7°

Find θ, accurate to the nearest hundredth of a degree, as in Example B.

7. sin θ = .2164
8. tan θ = .3096
9. tan θ = 2.311
10. cos θ = .9354
11. cos θ = .3535
12. sin θ = .7302

Solve for x. See Example C.

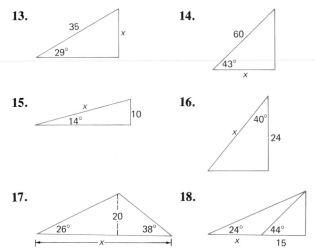

13.
14.
15.
16.
17.
18.

Solve each of the following triangles. First draw the triangle, labeling it as in Example C with $\gamma = 90°$.

19. $\alpha = 42°$, $c = 35$
20. $\beta = 29°$, $c = 50$
21. $\beta = 56.2°$, $c = 91.3$
22. $\alpha = 69.9°$, $c = 10.6$
23. $\alpha = 39.4°$, $a = 120$
24. $\alpha = 40.6°$, $b = 163$

Solve the right triangles satisfying the given information in Problems 25–32, assuming that c is the hypotenuse. See Example D.

25. $a = 9$, $b = 12$
26. $a = 24$, $b = 10$
27. $a = 40$, $c = 50$
28. $c = 41$, $a = 40$
29. $a = 14.6$, $c = 32.5$
30. $a = 243$, $c = 419$
31. $a = 9.52$, $b = 14.7$
32. $a = .123$, $b = .456$

33. A straight path leading up a hill rises 26 feet per 100 horizontal feet. What angle does it make with the horizontal?

34. A 20-foot ladder leans against a wall, making an angle of 76° with the level ground. How high up the wall is the upper end of the ladder?

35. Find the angle of elevation of the sun if a woman 5 feet 9 inches tall casts a shadow 46.8 feet long. (The *angle of elevation* is the upward angle made with the horizontal.)

36. A guy wire to a pole makes an angle of 69° with the level ground and is 14 feet from the pole at the ground. How high above the ground is the wire attached to the pole?

37. Suppose that the woman in Problem 35 is walking with her daughter Sue, who is 3 feet 10 inches tall. How long is Sue's shadow?

38. Find the length of the supporting wire in Problem 36.

B. Applications and Extensions

The following problems can be solved using either a calculator or tables. We recommend using a calculator.

39. Calculate each value.
    (a) tan 14.5°
    (b) 24.6 cos 74.3°
    (c) 15.6 (sin 14°)²/cos 87°

40. Find θ in each case.
    (a) sin θ = .6691
    (b) cos θ = .5519
    (c) tan θ = 5.396

41. From the top of a lighthouse 120 feet above sea level, the *angle of depression* (the downward angle from the horizontal) to a boat adrift on the sea is 9.4° (Figure 7). How far from the foot of the lighthouse is the boat?

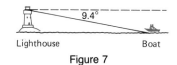

Figure 7

42. Solve the right triangle in which $b = 67.3$ and $c = 82.9$.

43. Find x in Figure 8 on the next page.

44. When the *angle of elevation* (the upward angle from the horizontal) of the sun is 28.4°, the Eiffel Tower in Paris casts a horizontal shadow 1822 feet long. How high is the tower?

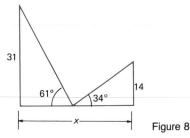

Figure 8

45. With her hands 5 feet above the ground, Sally is pulling on a kite. If the kite is 200 feet above the ground and the kite string makes an angle of 32.4° with the horizontal, how many feet of string are out?

46. A plane is flying directly away from a ground observer at a constant rate, maintaining an elevation of 15,000 feet. At a certain instant, the observer measures the angle of elevation as 44° and 15 seconds later as 31°. How fast is the plane flying in miles per hour?

47. From a window in an office building, I am looking at a television tower that is 600 meters away (horizontally). The angle of elevation of the top of the tower is 19.6° and the angle of depression of the base of the tower is 21.3°. How tall is the tower?

48. The vertical distance from first to second floor of a certain department store is 28 feet. The escalator, which has a horizontal reach of 96 feet, takes 25 seconds to carry a person between floors. How fast does the escalator travel?

49. The Great Pyramid is about 480 feet high and its square base measures 760 feet on a side. Find the angle of elevation of one of its edges, that is, find β in Figure 9.

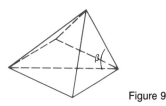

Figure 9

50. Find the angle between a principal diagonal and a face diagonal of a cube.

51. A ship S_1 left the harbor O at 12:00 noon sailing in the direction S45°E at 24 miles per hour. At 1:00 P.M. a second ship S_2 left the same harbor sailing in the direction N39°E at 28 miles per hour (see Figure 10).
   (a) Determine the distance between the ships at 2:30 P.M.
   (b) At 2:30 P.M. the bearing from S_1 to S_2 is NαW. Determine α.

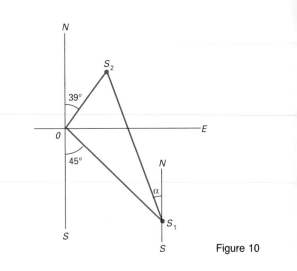

Figure 10

52. Determine the angle α between the chord AC and the diameter AB for the circle in Figure 11.

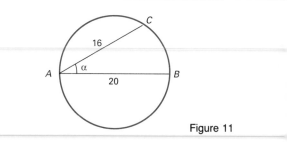

Figure 11

53. A regular hexagon (six equal sides) is inscribed in a circle of radius 4. Find the perimeter P and area A of this hexagon.

54. A regular decagon (10 equal sides) is inscribed in a circle of radius 12. What percent of the area of the circle is the area of the decagon?

55. Find the area of the regular six-pointed Star of David that is inscribed in a circle of radius 1 (Figure 12).

56. TEASER Find the area of the regular five-pointed star (the pentagram) that is inscribed in a circle of radius 1 (Figure 13).

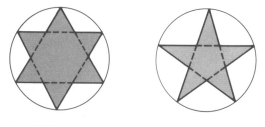

Figure 12   Figure 13

7-2 ANGLES AND ARCS

The Dynamic View of Angles

In geometry, we take a static view of angles. An angle is simply the union of two rays with a common endpoint (the vertex). In trigonometry, angles are thought of in a dynamic way. An angle is determined by rotating a ray about its endpoint from an initial position to a terminal position.

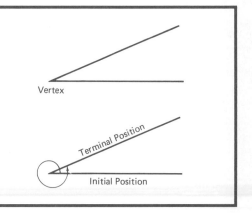

For the solution of right triangles (which involve acute angles), we required only the familiar and simple notion of angle from high-school geometry. But for the broader development of trigonometry, we need the new perspective on angles suggested by our opening display. Not only do we allow arbitrarily large angles, but we also distinguish between positive and negative angles. If an angle is generated by a counterclockwise rotation, it is positive; if generated by a clockwise rotation, it is negative (Figure 14). To know an angle, in trigonometry, is to know how the angle came into being. It is to know the initial side, the terminal side, and the kind of rotation that produced the angle.

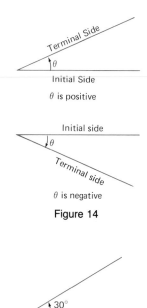

Figure 14

Degree Measurement

Take a circle and divide its circumference into 360 equal parts. The angle with vertex at the center determined by one of these parts has measure one **degree** (written 1°). This way of measuring angles is due to the ancient Babylonians and is so familiar that we used it in Section 7-1 without comment. There is a refinement, however, that we avoid. The Babylonians divided each degree into 60 minutes and each minute into 60 seconds; some people still follow this cumbersome practice. If we need to measure angles to finer accuracy than a degree, we will use decimal parts. Thus we write 40.5° rather than 40°30'.

It is important that we be familiar with measuring both positive and negative angles, as well as angles resulting from large rotations. Three angles are shown in Figure 15. Note that all three have the same initial and terminal sides.

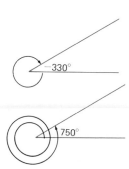

Figure 15

Radian Measurement

The best way to measure angles is in radians. Take a circle of radius r. The familiar formula $C = 2\pi r$ tells us that the circumference has 2π (about 6.28) arcs of length r around it. The angle with vertex at the center of a circle determined by an arc of length equal to its radius measures one **radian** (Figure 16). Thus an angle of size

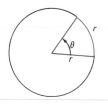

θ measures
one radian
(about 57.3°)

Figure 16

360° measures 2π radians and an angle of size 180° measures π radians. We abbreviate the latter by writing

$$180° = \pi \text{ radians}$$

To convert from degrees to radians, all one needs to remember is the result in the box. By dividing by 2, 3, 4, and 6, respectively, we get the conversions for several special angles.

$$90° = \frac{\pi}{2} \text{ radians} \qquad 45° = \frac{\pi}{4} \text{ radians}$$

$$60° = \frac{\pi}{3} \text{ radians} \qquad 30° = \frac{\pi}{6} \text{ radians}$$

If we divide the boxed formula by 180, we get

$$1° = \frac{\pi}{180} \text{ radians}$$

and if we divide by π, we get

$$\frac{180°}{\pi} = 1 \text{ radian}$$

The following rules thus hold.

To convert from degrees to radians, multiply by $\pi/180$.
To convert from radians to degrees, multiply by $180/\pi$.

Example A (Degrees to Radians) Convert to radians.
(a) 240° (b) −510° (c) 22°

Solution

(a) $240° = 240\left(\frac{\pi}{180}\right) = \frac{4\pi}{3} \text{ radians}$ (b) $-510° = -510\left(\frac{\pi}{180}\right) = -\frac{17\pi}{6} \text{ radians}$

(c) $22° = 22\left(\frac{\pi}{180}\right) = .38397 \text{ radians}$ ∎

Example B (Radians to Degrees) Convert to degrees.

(a) $\frac{3\pi}{2} \text{ radians}$ (b) $\frac{-5\pi}{36} \text{ radians}$ (c) 2.3 radians

Solution

(a) $\frac{3\pi}{2} \text{ radians} = 3(90) = 270°$ (b) $-\frac{5\pi}{36} \text{ radians} = -\left(\frac{5\pi}{36}\right)\left(\frac{180}{\pi}\right) = -25°$

(c) $2.3 \text{ radians} = 2.3\left(\frac{180}{\pi}\right) \approx 131.8°$ ∎

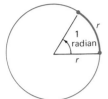

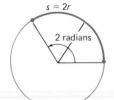

$s = 2r$

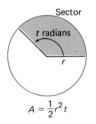

$s = tr$

Figure 17

Sector

$A = \frac{1}{2}r^2t$

Figure 18

Arc Length and Area

Radian measure is almost invariably used in calculus because it is an intrinsic measure. The division of a circle into 360 parts was quite arbitrary; its division into parts of radius length (2π parts) is more natural. Because of this, formulas using radian measure tend to be simple, while those using degree measure are often complicated. As an example, consider arc length. Let t be the radian measure of an angle θ with vertex at the center of a circle of radius r. This angle cuts off an arc of length s which satisfies the simple formula

$$s = rt$$

This follows directly from the fact that an angle of one radian ($t = 1$) cuts off an arc of length r (see Figure 17).

A second nice formula is that for the area of the sector cut off from a circle by a central angle of t radians (Figure 18). Notice that the area A of this sector is to the area of the whole circle as t is to 2π, that is, $A/\pi r^2 = t/2\pi$. Thus

$$A = \frac{1}{2}r^2t$$

Example C (Arc Length)   Find the length of the arc cut off on a circle of radius 5.60 inches by a central angle of $120°$.

Solution $120° = 2\pi/3$ radians. Thus $s = 5.6(2\pi/3) \approx 11.73$ inches.   ■

Example D (Area)   A central angle of $.6\pi$ radians cuts off an arc of length 12.503 centimeters on a circle. Find the area of the corresponding sector.

Solution   First, we determine the radius of the circle.

$$r = \frac{s}{t} = \frac{12.503}{0.6\pi}$$

Then, we apply the area formula.

$$A = \frac{1}{2}r^2t = \frac{1}{2}\left(\frac{12.503}{.6\pi}\right)^2(.6\pi) = \frac{12.503^2}{1.2\,\pi} \approx 41.466 \text{ square centimeters.}   ■$$

The Unit Circle

The formula for arc length takes a particularly simple form when $r = 1$, namely, $s = t$. We emphasize its meaning. *On the unit circle, the length of an arc is the same as the radian measure of the angle it determines.*

Someone is sure to point out a difficulty in what we have just said. What happens when t is greater than 2π or when t is negative? To understand our meaning,

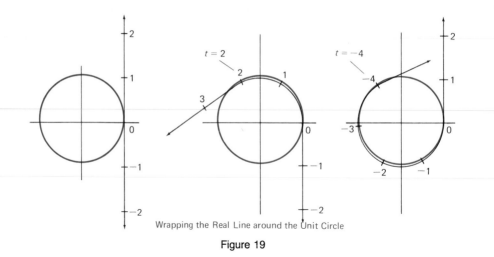

Wrapping the Real Line around the Unit Circle

Figure 19

imagine an infinitely long string on which the real number scale has been marked. Think of wrapping this string around the unit circle as shown in Figure 19.

Now if we think of the directed length (that is, the signed length) of a piece of the string, the formula $s = t$ holds no matter what t is. For example, the length of string corresponding to an angle of 8π radians is 8π. That piece of string wraps counterclockwise around the unit circle exactly 4 times. A piece of string corresponding to an angle of -3π radians would wrap clockwise around the unit circle one and a half times, its directed length being -3π.

In the next section we will relate the coordinates of a point P on the unit circle to the length of the arc from the point $(1, 0)$ to P. As a start in this direction, we consider the simpler problem of locating the quadrant in which P lies.

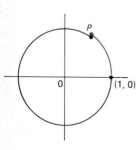

Figure 20

Example E (Locating a Point on the Unit Circle)   Figure 20 shows a unit circle with center at the origin. Suppose that a point P moves in a counterclockwise direction around the circle starting at $(1, 0)$. In which quadrant is P when it has traveled a distance of 4 units? Of 40 units?

Solution   Keep in mind that the distance P travels equals the radian measure of the angle through which OP turns. A distance of 4 units puts P in quadrant III since $\pi < 4 < 3\pi/2$. Once around the circle is $2\pi \approx 6.28$ units. If you divide 40 by 6.28, you get

$$40 = 6(6.28) + 2.32$$

Since 2.32 is between $\pi/2$ and π, traveling 40 units around the unit circle will put P in quadrant II.   ∎

Angular Velocity

A formula closely related to the arc length formula $s = rt$ is the formula

$$v = r\omega$$

which connects the speed (velocity) of a point on the rim of wheel of radius r with the angular velocity ω at which the wheel is turning. Here ω is measured in radians per unit time. Thus, if a wheel of radius 40 centimeters is spinning about its axis at 4.25 radians per second, a point on its rim is traveling at

$$v = r\omega = 40(4.25) = 170 \text{ centimeters per second}$$

Example F (A Bicycle Problem)   Determine the angular velocity of a bicycle wheel of radius 16 inches if the bicycle is being ridden down the road at 30 miles per hour.

Solution   We must use consistent units. You can check that the speed of a point on the rim of the wheel (30 miles per hour) translates to 44 feet per second and that the radius of the wheel is $\frac{4}{3}$ feet. Thus

$$44 = \frac{4}{3}\omega$$

or

$$\omega = \frac{3}{4}(44) = 33 \text{ radians per second} \quad \blacksquare$$

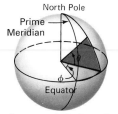

North Pole
Prime Meridian
Equator

θ measures latitude north
ϕ measures longitude east

Figure 21

Geography

Points on the surface of the earth are specified by giving two coordinates, latitude and longitude. *Latitude* specifies the angle north (or south) of the equatorial plane; *longitude* specifies the angle east (or west) of the plane through the prime meridian (Figure 21).

Example G (Distance Between Two Cities)   Cities A and B, both located at longitude 22° east, are at latitude 29° north and 43° south, respectively. How far apart are they along the surface of the earth (assumed to be a sphere of radius 3960 miles)?

Solution   The two cities lie on a circle of radius 3960 miles and determine a central angle of size $(29 + 43) = 72°$. The arc connecting them has length

$$s = rt = 3960\left(\frac{72\pi}{180}\right) \approx 4976 \text{ miles} \quad \blacksquare$$

PROBLEM SET 7-2

A. Skills and Techniques

Convert each of the following to radians. You may leave π in your answer. See Example A.

1. 120°
2. 225°
3. 240°
4. 150°
5. 210°
6. 330°
7. 315°
8. 300°
9. 540°
10. 450°
11. −420°
12. −660°
13. 160°
14. 200°
15. $(20/\pi)°$
16. $(150/\pi)°$

Convert each of the following to degrees. Give your answer correct to the nearest tenth of a degree. See Example B.

17. $\dfrac{4}{3}\pi$ radians

18. $\dfrac{5}{6}\pi$ radians

19. $-\dfrac{2\pi}{3}$ radians

20. $-\dfrac{7\pi}{4}$ radians

21. 3π radians

22. 3 radians

23. 4.52 radians

24. $\dfrac{11}{4}$ radians

25. $\dfrac{1}{\pi}$ radians

26. $\dfrac{4}{3\pi}$ radians

Problems 27–32 relate to Examples C and D.

27. Find the radian measure of the angle at the center of a circle of radius 6 inches which cuts off an arc of length
   (a) 12 inches.   **(b)** 18.84 inches.

28. Find the length of the arc cut off on a circle of radius 3 feet by an angle at the center of
   (a) 2 radians   **(b)** 5.5 radians
   (c) $\dfrac{\pi}{4}$ radians   **(d)** $\dfrac{5\pi}{6}$ radians.

29. Find the radius r for each of the following.
   (a)   **(b)**

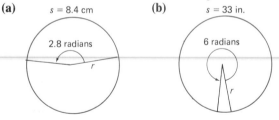

30. Through how many radians does the minute hand of a clock turn in 1 hour? The hour hand in 1 hour? The minute hand in 5 hours?

31. Find the areas of the sectors determined by the specified central angles in parts (a) and (b) of Problem 29.

32. Find the area of the region swept out by the minute hand of a clock between 1:15 and 1:52, assuming the hand is 6 centimeters long.

Find the quadrant in which the point P in Example E lies when it has traveled each of the following distances.

33. 3 units

34. 3.2 units

35. 4.7 units

36. 4.8 units

37. $\left(\dfrac{5\pi}{2}+1\right)$ units

38. $\left(\dfrac{9\pi}{2}-1\right)$ units

39. 100 units

40. 200 units

41. Sally is pedaling her tricycle so the front wheel (radius 8 inches) turns at 4 revolutions per second. How fast is she moving down the sidewalk in feet per second? *Hint:* Four revolutions per second is 8π radians per second. (See Example F).

42. Suppose that the tire on a car has an outer diameter 2.5 feet. How many revolutions per minute does the tire make when the car is traveling 60 miles per hour?

43. A dead fly is stuck to a belt that passes over two pulleys 6 inches and 8 inches in radius, as shown in Figure 22 on the next page. Assuming no slippage, how fast is the fly moving when the larger pulley turns at 20 revolutions per minute?

44. How fast (in revolutions per minute) is the smaller wheel in Problem 43 turning?

B. Applications and Extensions

45. Convert to radians.
   (a) $-1440°$ **(b)** $2\frac{1}{2}$ revolutions
   (c) $(60/\pi)°$

46. Convert to degrees.
   (a) $23\pi/36$ radians   **(b)** -4.63 radians
   (c) $3/(2\pi)$ radians

47. Find the length of arc cut off on a circle of radius 4.25 centimeters by each central angle.
   (a) 6 radians   **(b)** $(18/13\pi)°$
   (c) $17\pi/6$ radians

48. The front wheel of Tony's tricycle has a diameter of 20 inches. How far did he travel in pedaling through 60 revolutions?

49. The pedal sprocket of Maria's bicycle has radius 12 centimeters, the rear wheel sprocket has radius 3 centimeters, and the wheels have radius 40 centimeters. How far did Maria travel if she pedaled continuously for 30 revolutions of the pedal sprocket?

50. A belt traveling at the rate of 60 feet per second drives a pulley (a wheel) at the rate of 900 revolutions per minute. Find the radius of the pulley.

51. Assume that the earth is a sphere of radius 3960 miles. How fast (in miles per hour) is a point on the equator moving as a result of the earth's rotation about its axis?

52. The orbit of the earth about the sun is an ellipse that is nearly circular with radius 93 million miles. Approximately, what is the earth's speed (in miles per hour) in its path around the sun? You will need the fact that a complete orbit takes 365.25 days.

53. If in a certain time, a radius on the smaller wheel in Figure 22 sweeps out an area A, what area does a radius on the larger wheel sweep out?

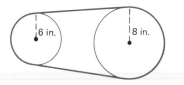

Figure 22

54. Figure 23 shows a circle with diameter AB of length 20. Find
(a) the length of AC.
(b) the area of the shaded region.

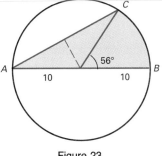

Figure 23

55. The angle subtended by the sun at the earth (93 million miles away) is .0093 radians. Find the diameter of the sun.

56. A nautical mile is the length of 1 minute ($\frac{1}{60}$ of a degree) of arc on the equator of the earth. How many miles are there in a nautical mile?

57. One of the authors (Dale Varberg) lives at exactly 45° latitude north (see Figure 24). How long would it take him to fly to the North Pole at 600 miles per hour (assuming the earth is a sphere of radius 3960 miles)?

58. New York City is located at 40.5° latitude north. How far is it from there to the equator?

59. Oslo, Norway, and Leningrad, Russia, are both located at 60° latitude north. Oslo is at longitude 6° east

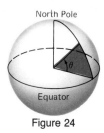

Figure 24

(of the prime meridian) whereas Leningrad is at 30° east. How far apart are these two cities along the 60° parallel?

60. Find the area of the shaded region of the right triangle ABC shown in Figure 25.

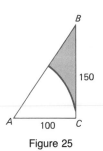

Figure 25

61. The minute hand and hour hand of a clock are both 6 inches long and reach to the edge of the dial. Find the area of the pie-shaped region between the two hands at 5:40.

62. A cone has radius of base R and slant height L. Find the formula for its lateral surface area. *Hint:* Imagine the cone to be made of paper, slit it up the side, and lay it flat in the plane.

63. Find the area of the polar rectangle shown in Figure 26. The two curves are arcs of concentric circles.

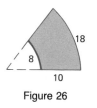

Figure 26

64. TEASER Consider two circles both of radius r and with the center of each lying on the rim of the other. Find the area of the common part of the two circles.

7-3 THE SINE AND COSINE FUNCTIONS

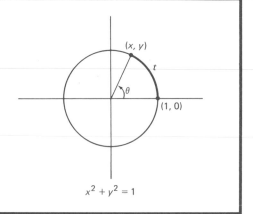

Definitions of Sine and Cosine

Place an angle θ, whose radian measure it t, in **standard position,** that is, put θ in the coordinate plane so that its vertex is at the origin and its initial side is along the positive x-axis. Let (x, y) be the coordinates of the point of intersection of the terminal side with the unit circle. We define both sin θ (sine of θ) and sin t by

$$\sin \theta = \sin t = y.$$

Similarly

$$\cos \theta = \cos t = x.$$

In Section 7-1 we defined the sine and cosine for positive acute angles. The definitions in our opening display are more general and hence more widely applicable. They should be studied carefully. Notice that we have defined the sine and cosine for any angle θ and also for the corresponding number t. Both concepts are important. In geometric situations, angles play a central role; thus we are likely to need sines and cosines of angles. But in most of pure mathematics and in many scientific applications, it is the trigonometric functions of numbers that are important. In this connection, we emphasize that the number t may be positive or negative, large or small. And we may think of it as the radian measure of an angle, as the directed length of an arc on the unit circle, or simply as a number.

If your instructor has chosen to begin the study of trigonometry here using the unit circle approach, you may still wish to review the first two sections of this chapter. They provide a good background for the modern approach to trigonometry.

Consistency with Earlier Definitions

Do the definitions given in Section 7-1 for the sine and cosine of an acute angle harmonize with those given here? Yes. Take a right triangle ABC with an acute angle θ. Place θ in standard position, thus determining a point $B'(x, y)$ on the unit circle and

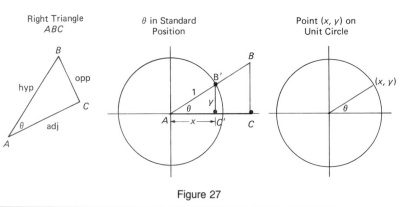

Figure 27

a point $C'(x, 0)$ directly below it on the x-axis (see Figure 27). Notice that triangles ABC and $AB'C'$ are similar. It follows that

$$\frac{\text{opp}}{\text{hyp}} = \frac{BC}{AB} = \frac{B'C'}{AB'} = \frac{y}{1} = y$$

$$\frac{\text{adj}}{\text{hyp}} = \frac{AC}{AB} = \frac{AC'}{AB'} = \frac{x}{1} = x$$

On the left are the old definitions of $\sin \theta$ and $\cos \theta$; on the right are the new ones. They are consistent.

Special Angles

In Section 7-1, we learned that

$$\cos 45° = \frac{\sqrt{2}}{2} \qquad \sin 45° = \frac{\sqrt{2}}{2}$$

$$\cos 30° = \frac{\sqrt{3}}{2} \qquad \sin 30° = \frac{1}{2}$$

$$\cos 60° = \frac{1}{2} \qquad \sin 60° = \frac{\sqrt{3}}{2}$$

Making use of the consistency of the old and new definitions of sine and cosine, we conclude that the point on the unit circle corresponding to $\theta = 45° = \pi/4$ radians must have coordinates $(\sqrt{2}/2, \sqrt{2}/2)$. Similarly, the point corresponding to $\theta = 30° = \pi/6$ radians has coordinates $(\sqrt{3}/2, 1/2)$ and the point corresponding to $\theta = 60° = \pi/3$ radians has coordinates $(1/2, \sqrt{3}/2)$.

Now we can make use of obvious symmetries to find the coordinates of many other points on the unit circle. In the two diagrams of Figure 28, we show a number of these points, noting first the radian measure of the angle and then the coordinates of the corresponding point on the unit circle.

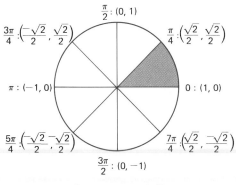

Some multiples of $\pi/4$

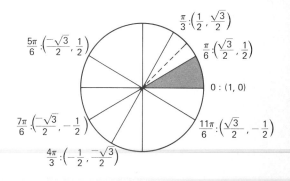

Some multiples of $\pi/6$

Figure 28

Notice, for example, how the coordinates of the points corresponding to $t = 5\pi/6$, $7\pi/6$, and $11\pi/6$ are related to the point corresponding to $t = \pi/6$. You should have no trouble seeing other relationships.

Once we know the coordinates of a point on the unit circle, we can state the sine and cosine of the corresponding angle. In particular, we get the values in the table in Figure 29. They are used so often that you should memorize them.

t	0	$\dfrac{\pi}{6}$	$\dfrac{\pi}{4}$	$\dfrac{\pi}{3}$	$\dfrac{\pi}{2}$	π	$\dfrac{3\pi}{2}$
$\cos t$	1	$\dfrac{\sqrt{3}}{2}$	$\dfrac{\sqrt{2}}{2}$	$\dfrac{1}{2}$	0	-1	0
$\sin t$	0	$\dfrac{1}{2}$	$\dfrac{\sqrt{2}}{2}$	$\dfrac{\sqrt{3}}{2}$	1	0	-1

Figure 29

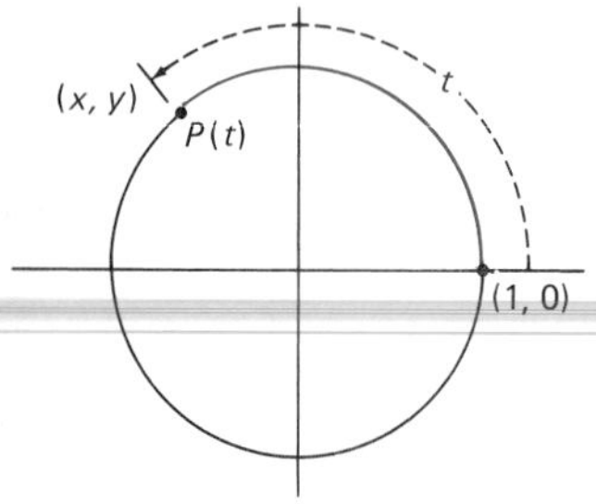

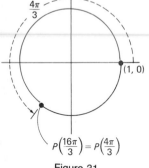

Figure 30

The Trigonometric Point $P(t)$

We have introduced $\cos t$ and $\sin t$ as the x- and y-coordinates of the point on the unit circle whose directed distance from $(1, 0)$ along the unit circle is t. This point is called a **trigonometric point** and will be denoted by $P(t)$ (Figure 30). We may regard $P(t)$ to be a function of t, since for each t there is a unique point $P(t)$. This function, moreover, is **periodic** with period 2π—that is,

$$P(t + 2\pi) = P(t)$$

It follows that

$$P(t + k2\pi) = P(t)$$

for any integer k, a fact that allows us to find the coordinates of $P(t)$ for any t, no matter how large t is.

$$P\left(\frac{16\pi}{3}\right) = P\left(\frac{4\pi}{3}\right)$$

Figure 31

Example A (Mutiples of Special Angles) Find the coordinates of $P(16\pi/3)$.

Solution We first remove the largest possible multiple of 2π by writing

$$\frac{16\pi}{3} = \frac{4\pi}{3} + 4\pi$$

and noting (Figure 31) that $P(16\pi/3) = P(4\pi/3)$. Then we appeal to Figure 28, from which we read the x- and y-coordinates of $P(4\pi/3)$ to be $-1/2$ and $-\sqrt{3}/2$, respectively. Of course, these are also the coordinates of $P(16\pi/3)$. Furthermore, they are the values of $\cos(16\pi/3)$ and $\sin(16\pi/3)$. ■

Example B (Using $P(t)$ to Find Sine and Cosine Values) Find the value.
(a) $\sin(-\pi/6)$ (b) $\cos(29\pi/4)$.

Solution

(a) We locate $P(-\pi/6)$ on the unit circle (Figure 32) and note that its y-coordinate is $-\frac{1}{2}$ because of its position relative to $P(\pi/6)$. Thus $\sin(-\pi/6) = -\frac{1}{2}$.

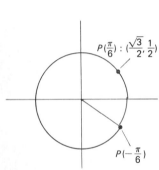

Figure 32

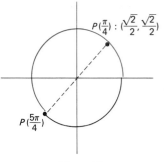

$P(\frac{\pi}{4}) : (\frac{\sqrt{2}}{2}, \frac{\sqrt{2}}{2})$

$P(\frac{5\pi}{4})$

Figure 33

(b) We simplify the problem by removing a large multiple of 2π—that is, by noting that

$$\frac{29\pi}{4} = 6\pi + \frac{5\pi}{4}$$

from which we conclude that $P(29\pi/4) = P(5\pi/4)$. Then we refer to Figure 28, or better yet, we simply observe that $P(5\pi/4)$, being diametrically opposite from $P(\pi/4)$ on the unit circle, has coordinates $(-\sqrt{2}/2, -\sqrt{2}/2)$ (see Figure 33). We conclude that

$$\cos\left(\frac{29\pi}{4}\right) = -\frac{\sqrt{2}}{2} \qquad \blacksquare$$

Properties of Sines and Cosines

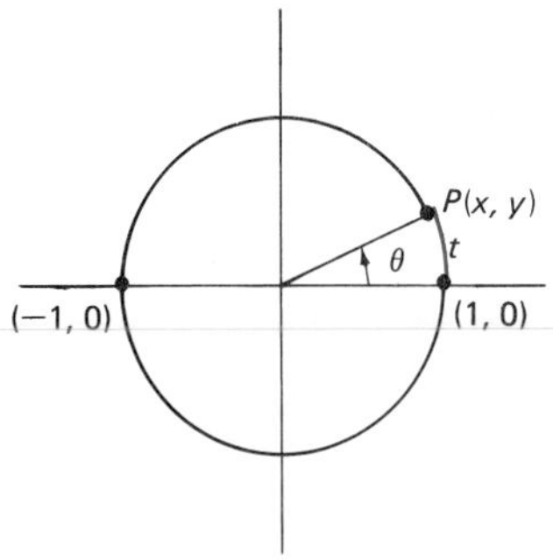

Think of what happens to x and y as t increases from 0 to 2π in Figure 34, that is, as P travels all the way around on the unit circle. For example, x steadily decreases until it reaches its smallest value of -1 at $t = \pi$; then it starts to increase until it is back to 1 at $t = 2\pi$. We have just described the behavior of $\cos t$ (or $\cos \theta$) as t increases from 0 to 2π. You should trace the behavior of $\sin t$ in the same way. Notice that both x and y are always between -1 and 1 (inclusive). It follows that

$$\boxed{\begin{array}{c} -1 \le \sin t \le 1 \\ -1 \le \cos t \le 1 \end{array}}$$

$P(x, y)$

$(-1, 0)$ $(1, 0)$

Figure 34

Since P is on the unit circle, $x^2 + y^2 = 1$, and $x = \cos t$ and $y = \sin t$, it follows that

$$(\sin t)^2 + (\cos t)^2 = 1$$

It is conventional to write $\sin^2 t$ instead of $(\sin t)^2$ and $\cos^2 t$ instead of $(\cos t)^2$. Thus we have

$$\sin^2 t + \cos^2 t = 1$$

This is an identity; it is true for all t. Of course we can just as well write

$$\sin^2 \theta + \cos^2 \theta = 1$$

We have established one basic relationship between the sine and the cosine; here are two others, valid for all t.

$$\sin\left(\frac{\pi}{2} - t\right) = \cos t$$

$$\cos\left(\frac{\pi}{2} - t\right) = \sin t$$

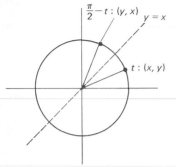

Figure 35

These relationships are easy to see when $0 < t < \pi/2$. Notice that t and $\pi/2 - t$ are measures of complementary angles (two angles with measures totaling 90° or $\pi/2$). That means that t and $\pi/2 - t$ determine points on the unit circle which are reflections of each other about the line $y = x$ (see Figure 35). Thus if one point has coordinates (x, y), the other has coordinates (y, x). The results given above follows from this fact.

Finally, we point out that t, $t \pm 2\pi$, $t \pm 4\pi$, . . . all determine the same point on the unit circle and thus have the same sine and cosine. This repetitive behavior puts the sine and cosine into a special class of functions, for which we give the following definition. A function f is **periodic** if there is a positive number p such that

$$f(t + p) = f(t)$$

for every t in the domain of f. The smallest such p is called the **period** of f. Thus we say that sine and cosine are periodic functions with period 2π and write

$$\sin(t + 2\pi) = \sin t$$
$$\cos(t + 2\pi) = \cos t$$

Example C (Sine and Cosine of $-t$)   Show that for all t

$$\sin(-t) = -\sin t$$
$$\cos(-t) = \cos t$$

that is, sine is an odd function and cosine is an even function.

Solution   The points $P(-t)$ and $P(t)$ are symmetric with respect to the x-axis (Figure 36). Thus if $P(t)$ has coordinates (x, y), $P(-t)$ has coordinates $(x, -y)$ and so

$$\sin(-t) = -y = -\sin t$$
$$\cos(-t) = x = \cos t \quad \blacksquare$$

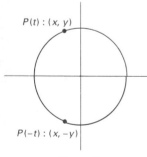

Figure 36

Example D (Finding Cosine, Given Sine)   Suppose that $\pi/2 < t < \pi$ and $\sin t = 3/8$. Find $\cos t$.

Solution   First, we note that cosine is negative for t in the given interval. Then, we use the identity $\cos^2 t + \sin^2 t = 1$ to write

$$\cos t = -\sqrt{1 - \sin^2 t} = -\sqrt{1 - \frac{9}{64}} = -\frac{\sqrt{55}}{8} \quad \blacksquare$$

PROBLEM SET 7-3

A. Skills and Techniques

In Problems 1–8, find the coordinates of the trigonometric point $P(t)$ for the indicated value of t. Hint: Begin by drawing a unit circle and locating $P(t)$ on it; then relate $P(t)$ to the diagrams in Figure 28. See Example A.

1. $t = \dfrac{13\pi}{6}$

2. $t = \dfrac{19\pi}{6}$

3. $t = \dfrac{19\pi}{4}$

4. $t = \dfrac{15\pi}{4}$

5. $t = 24\pi + \dfrac{5\pi}{4}$

6. $t = 16\pi + \dfrac{5\pi}{6}$

7. $t = -\dfrac{7\pi}{6}$

8. $t = \dfrac{13\pi}{4}$

Using the method of Example B. find the value of each of the following.

9. $\sin(-\pi/4)$

10. $\sin(-5\pi/4)$

11. $\sin(9\pi/4)$

12. $\sin(15\pi/4)$

13. $\cos(13\pi/4)$

14. $\cos(-7\pi/4)$

15. $\cos(10\pi/3)$

16. $\cos(25\pi/6)$

17. $\sin(5\pi/2)$

18. $\cos 7\pi$

19. $\sin(-4\pi)$

20. $\cos(7\pi/2)$

21. $\cos(19\pi/6)$

22. $\sin(14\pi/3)$

23. $\cos(-\pi/3)$

24. $\sin(-5\pi/6)$

25. $\cos(125\pi/4)$

26. $\cos(-13\pi/6)$

27. $\sin 510°$

28. $\sin(-390°)$

29. $\cos 840°$

30. $\cos(-720°)$

31. $\cos(-210°)$

32. $\sin 900°$

Refer to Example C for Problems 33–36.

33. If $\sin 1.87 = .95557$ and $\cos 1.87 = -.29476$, find $\sin(-1.87)$ and $\cos(-1.87)$.

34. If $\sin 15.2° = .2622$ and $\cos 15.2° = .9650$, find $\sin(-15.2°)$ and $\cos(-15.2°)$.

35. Given $P(t)$ with coordinates $(1/\sqrt{5}, -2/\sqrt{5})$.
(a) What are the coordinates of $P(-t)$?
(b) What are the values of $\sin(-t)$ and $\cos(-t)$?

36. If t is the radian measure of an angle in quadrant III and $\sin t = -\frac{3}{5}$, evaluate each expression.
(a) $\sin(-t)$

(b) $\cos t$ *Hint:* Use the fact that
$$\sin^2 t + \cos^2 t = 1.$$

(c) $\cos(-t)$

37. Note that $P(t)$ and $P(\pi + t)$ are symmetric with respect to the origin (Figure 37). Use this to show that
(a) $\sin(\pi + t) = -\sin t$. (b) $\cos(\pi + t) = -\cos t$.

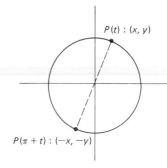

Figure 37

38. Note that $P(t)$ and $P(\pi - t)$ are symmetric with respect to the y-axis. Use this fact to find identities analogous to those in Problem 37 for $\sin(\pi - t)$ and $\cos(\pi - t)$.

39. If $\cos t = \frac{3}{5}$ and $3\pi/2 < t < 2\pi$, find $\sin t$ (see Example D).

40. If $\sin t = -\frac{5}{13}$ and $\pi < t < 3\pi/2$, find $\cos t$.

41. If $\sin t = \frac{12}{13}$ and $\pi/2 < t < \pi$, find $\cos(\pi + t)$ (see Problem 37).

42. If $\cos t = -\frac{4}{5}$ and $\pi < t < 3\pi/2$, find $\sin(\pi + t)$.

B. Applications and Extensions

43. Use the unit circle to determine the sign (plus or minus) of each of the following.
(a) $\cos 2$ (b) $\sin(-3)$
(c) $\cos 428°$ (d) $\sin 21.4$
(e) $\sin(23\pi/32)$ (f) $\sin(-820°)$

44. Find the coordinates of $P(t)$ for the indicated values of t.
(a) $t = \pi$ (b) $t = -3\pi/2$
(c) $t = -3\pi/4$ (d) $t = 5\pi/6$
(e) $t = 44\pi/3$ (f) $t = -93.5\pi$

45. Let $P(t)$ have coordinates $(x, -\frac{1}{2})$.
(a) Find the two possible values of x.
(b) Find the corresponding values of t.

46. With initial point $(0, -1)$, a string of length $4\pi/3$ is wound clockwise around the unit circle. What are the coordinates of the terminal point?

47. Find the straight line distance between $P(\pi/6)$ and $P(-3\pi/4)$ on the unit circle (Figure 38).

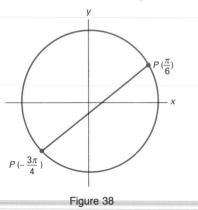

Figure 38

48. Find two values of t between 0 and 2π for which the y-coordinate of $P(t)$ is twice its x-coordinate.

49. For what values of t satisfying $0 \le t \le 2\pi$ are the following true?

(a) $\sin t = \cos t$ **(b)** $\frac{1}{2} < \sin t < \sqrt{3}/2$

(c) $\cos^2 t \ge .25$ **(d)** $\cos^2 t > \sin^2 t$

50. Find the four smallest positive solutions to the following equations.

(a) $\sin t = 1$ **(b)** $|\cos t| = \frac{1}{2}$

(c) $\cos t = -\sqrt{3}/2$ **(d)** $\sin t = -\sqrt{2}/2$

51. In each case, assume that θ is an angle in standard position with terminal side in the fourth quadrant. Use $\sin^2 \theta + \cos^2 \theta = 1$ to determine the indicated value.

(a) $\cos \theta$ if $\sin \theta = -\frac{4}{5}$
(b) $\sin \theta$ if $\cos \theta = \frac{24}{25}$

52. Use the unit circle to find identities for $\sin(2\pi - t)$ and $\cos(2\pi - t)$.

53. If $P(t)$ has coordinates $(\frac{4}{5}, -\frac{3}{5})$, evaluate each of the following. *Hint:* For parts (e) and (f), see Problems 37 and 38.

(a) $\sin(-t)$ **(b)** $\sin(\frac{\pi}{2} - t)$
(c) $\cos(2\pi + t)$ **(d)** $\cos(2\pi - t)$
(e) $\sin(\pi + t)$ **(f)** $\cos(\pi - t)$

54. Fill in all the blanks in the chart of Figure 39.

55. Recall that $[\cdot]$ and $(\!(\cdot)\!)$ denote the *greatest integer in* and the *distance to the nearest integer*, respectively. Determine which of the following functions are periodic and, if so, specify the period.

(a) $f(x) = (\!(x)\!)$ **(b)** $f(x) = (\!(3x)\!)$
(c) $f(x) = [x]$ **(d)** $f(x) = x - [x]$

56. Suppose that $f(x)$ is periodic with period 2 and that $f(x) = 4 - x^2$ for $0 \le x < 2$. Evaluate each of the following.

(a) $f(2)$ **(b)** $f(4.5)$
(c) $f(-.5)$ **(d)** $f(8.8)$

57. Evaluate

$$\sin 1° + \sin 2° + \sin 3° + \cdots +$$
$$\sin 357° + \sin 358° + \sin 359°$$

58. TEASER Evaluate

$$\sin^2 1° + \sin^2 2° + \sin^2 3° + \cdots +$$
$$\sin^2 357° + \sin^2 358° + \sin^2 359°$$

$\sin t$	$\cos t$	$\sin(t + \pi)$	$\cos(t + \pi)$	$\sin(\pi - t)$	$\sin(2\pi - t)$	Least positive value of t
$\sqrt{3}/2$	$-\frac{1}{2}$					
	$\sqrt{2}/2$	$\sqrt{2}/2$				
$-\frac{1}{2}$			$-\sqrt{3}/2$			
-1						
				$\sqrt{3}/2$	$\frac{1}{2}$	
	0				-1	

Figure 39

7-4 FOUR MORE TRIGONOMETRIC FUNCTIONS

"Strange as it may sound, the power of mathematics rests on its evasion of all unnecessary thought and on its wonderful saving of mental operations."

Ernst Mach

New Functions from Old Ones

tangent:	$\tan t = \dfrac{\sin t}{\cos t}$
cotangent:	$\cot t = \dfrac{\cos t}{\sin t}$
secant:	$\sec t = \dfrac{1}{\cos t}$
cosecant:	$\csc t = \dfrac{1}{\sin t}$

Without question, the sine and cosine are the most important of the six trigonometric functions. Not only do they occur most frequently in applications, but the other four functions can be defined in terms of them, as our opening box shows. This means that if you learn all you can about sines and cosines, you will automatically know a great deal about tangents, cotangents, secants, and cosecants. Ernst Mach would say that it is a way to evade unnecessary thought.

Look at the definitions in the opening box again. Naturally, we must rule out any values of t for which a denominator is zero. For example, $\tan t$ is not defined for $t = \pm\pi/2, \pm 3\pi/2, \pm 5\pi/2$, and so on. Similarly, $\csc t$ is not defined for such values as $t = 0, \pm\pi$, and $\pm 2\pi$.

Example A (Using the Definitions)   Find each value.

(a) $\tan\left(\dfrac{\pi}{4}\right)$ (b) $\csc\left(\dfrac{7\pi}{6}\right)$

(c) $\cot 690°$ (d) $\sec\left(\dfrac{5\pi}{2}\right)$

Solution   We make use of the definitions in the opening box and our knowledge of the sines and cosines of special angles.

(a) $\tan\left(\dfrac{\pi}{4}\right) = \dfrac{\sin(\pi/4)}{\cos(\pi/4)} = \dfrac{\sqrt{2}/2}{\sqrt{2}/2} = 1$

(b) $\csc\left(\dfrac{7\pi}{6}\right) = \dfrac{1}{\sin(7\pi/6)} = \dfrac{1}{-\sin(\pi/6)} = \dfrac{1}{-1/2} = -2$

(c) $\cot 690° = \dfrac{\cos 690°}{\sin 690°} = \dfrac{\cos 330°}{\sin 330°} = \dfrac{\cos 30°}{-\sin 30°} = \dfrac{\sqrt{3}/2}{-1/2} = -\sqrt{3}$

(d) $\sec\left(\dfrac{5\pi}{2}\right) = \dfrac{1}{\cos(5\pi/2)} = \dfrac{1}{\cos(\pi/2)}$,   which is undefined.   ∎

Properties of the New Functions

The wisdom of the opening paragraph will now be demonstrated. Recall the identity $\sin^2 t + \cos^2 t = 1$. Out of it come two new identities.

$$1 + \tan^2 t = \sec^2 t$$

$$1 + \cot^2 t = \csc^2 t$$

To show that the first identity is correct, we take its left side, express it in terms of sines and cosines, and do a little algebra.

$$1 + \tan^2 t = 1 + \left(\frac{\sin t}{\cos t}\right)^2 = 1 + \frac{\sin^2 t}{\cos^2 t}$$

$$= \frac{\cos^2 t + \sin^2 t}{\cos^2 t} = \frac{1}{\cos^2 t} = \left(\frac{1}{\cos t}\right)^2 = \sec^2 t$$

The second identity is verified in a similar fashion.

Suppose we wanted to know whether cotangent is an even or an odd function (or neither). We simply recall that $\sin(-t) = -\sin t$ and $\cos(-t) = \cos t$ and write

$$\cot(-t) = \frac{\cos(-t)}{\sin(-t)} = \frac{\cos t}{-\sin t} = -\frac{\cos t}{\sin t} = -\cot t$$

Thus cotangent is an odd function.

In a similar vein, recall the identities

(i)
$$\sin\left(\frac{\pi}{2} - t\right) = \cos t$$

(ii)
$$\cos\left(\frac{\pi}{2} - t\right) = \sin t$$

From them, we obtain

(iii)
$$\tan\left(\frac{\pi}{2} - t\right) = \frac{\sin(\pi/2 - t)}{\cos(\pi/2 - t)} = \frac{\cos t}{\sin t} = \cot t$$

These three identities are examples of what are called **cofunction identities.** Sine and cosine are cofunctions; so are tangent and cotangent; as are secant and cosecant. Notice that identities (i), (ii), and (iii) all have the form

$$\text{function}\left(\frac{\pi}{2} - t\right) = \text{cofunction}(t)$$

With cosecant as the function, we have

$$\csc\left(\frac{\pi}{2} - t\right) = \sec t$$

> **Cofunction Identities**
>
> Angles that sum to 90° (or $\pi/2$ radians) are called *complementary angles*. Thus we may state the cofunction identities in this easy-to-remember form:
>
> "Cofunctions of complementary angles are equal"

Example B (Using the Properties)   Suppose that $\tan t = 2$ and that $\pi < t < 3\pi/2$. Find each value.

(a) $\sec t$ (b) $\cot(-t)$ (c) $\sin\left(\dfrac{\pi}{2} - t\right)$ (d) $\sin t$

Solution

(a) For t in the indicated interval, $\sec t$ is negative, being the reciprocal of $\cos t$. This fact, together with the identity $\sec^2 t = 1 + \tan^2 t$, yield

$$\sec t = -\sqrt{1 + \tan^2 t} = -\sqrt{1 + 4} = -\sqrt{5}$$

(b) $\cot(-t) = -\cot t = -\dfrac{1}{\tan t} = -\dfrac{1}{2}$

(c) Our first calculation shows that $\cos t = -1/\sqrt{5}$. Thus

$$\sin\left(\frac{\pi}{2} - t\right) = \cos t = -\frac{1}{\sqrt{5}}$$

(d) We note first that for the given t, $\sin t$ is negative. The identity $\sin^2 t + \cos^2 t = 1$ implies that

$$\sin t = -\sqrt{1 + \cos^2 t} = -\sqrt{1 - \frac{1}{5}} = -\frac{2}{\sqrt{5}}$$ ∎

Alternative Definitions of the Trigonometric Functions

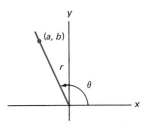

Figure 40

There is another approach to trigonometry favored by some authors. Let θ be an angle in standard position and suppose that (a, b) is any point on its terminal side at a distance r from the origin (Figure 40). Then

$$\sin \theta = \frac{b}{r} \qquad \cos \theta = \frac{a}{r}$$

$$\tan \theta = \frac{b}{a} \qquad \cot \theta = \frac{a}{b}$$

$$\sec \theta = \frac{r}{a} \qquad \csc \theta = \frac{r}{b}$$

To see that these definitions are equivalent to those we gave earlier, consider first an angle θ with terminal side in quadrant I (see Figure 41). By similar triangles,

$$\frac{b}{r} = \frac{y}{1} \quad \text{and} \quad \frac{a}{r} = \frac{x}{1}$$

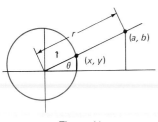

Figure 41

Actually these ratios are equal no matter in which quadrant the terminal side of θ is, since b and y always have the same sign, as do a and x. The first two formulas in the box now follow from our original definitions, which say that

$$\sin \theta = y \quad \text{and} \quad \cos \theta = x$$

The others are a consequence of the fact that the remaining four functions can be expressed in terms of sines and cosines.

Example C (Using the a, b, r Definitions)   Suppose that the point (3, −6) is on the terminal side of an angle in standard position (Figure 42). Find sin θ, tan θ, and sec θ.

Solution   First we find r.

$$r = \sqrt{3^2 + (-6)^2} = \sqrt{45} = 3\sqrt{5}$$

Then

$$\sin \theta = \frac{b}{r} = \frac{-6}{3\sqrt{5}} = -\frac{2}{\sqrt{5}}$$

$$\tan \theta = \frac{b}{a} = \frac{-6}{3} = -2$$

$$\sec \theta = \frac{r}{a} = \frac{3\sqrt{5}}{3} = \sqrt{5} \quad \blacksquare$$

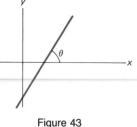

θ

r

(3, −6)

Figure 42

The Tangent Function and Slope

Recall that the slope m of a line is the ratio of rise to run. In particular, if the line goes through the point (a, b) and also the origin, its slope is b/a. But this number b/a is also the tangent of the nonnegative angle θ that the line makes with the positive x-axis (see Figure 43).

In general, the smallest nonnegative angle θ that a line makes with the positive x-axis is called the **angle of inclination** of the line (Figure 43). It follows that for any nonvertical line, the slope m of the line satisfies

$$m = \tan \theta$$

y

θ

x

Figure 43

Example D (Angle of Inclination)   Find the equation of the line that goes through the point (1, 2) with an angle of inclination 120°.

Solution   The slope of the line is $m = \tan 120° = -\sqrt{3}$. The equation of the line is $y - 2 = -\sqrt{3}(x - 1)$.   ■

PROBLEM SET 7-4

A. Skills and Techniques

1. If $\sin t = \frac{4}{5}$ and $\cos t = -\frac{3}{5}$, evaluate each function.
   (a) tan t
   (b) cot t
   (c) sec t
   (d) csc t

2. If $\sin t = -1/\sqrt{5}$ and $\cos t = 2/\sqrt{5}$, evaluate each function.
   (a) tan t
   (b) cot t
   (c) sec t
   (d) csc t

3. Find the values of tan θ and csc θ for the angle θ of Figure 44.

4. Find cot α and sec α for α as shown in Figure 45.

Keeping in mind what you know about the sines and cosines of special angles, find each of the values in Problems 5–22. See Example A.

5. $\tan(\pi/6)$

6. $\cot(\pi/6)$

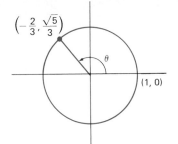

$\left(-\frac{2}{3}, \frac{\sqrt{5}}{3}\right)$

$(1, 0)$

θ

Figure 44

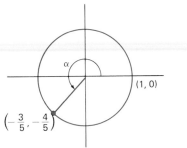

α

$(1, 0)$

$\left(-\frac{3}{5}, -\frac{4}{5}\right)$

Figure 45

7. $\sec(\pi/6)$

8. $\csc(\pi/6)$

9. $\cot(\pi/4)$

10. $\sec(\pi/4)$

11. $\csc(\pi/3)$

12. $\sec(\pi/3)$

13. $\sin(4\pi/3)$

14. $\cos(4\pi/3)$

15. $\tan(4\pi/3)$

16. $\sec(4\pi/3)$

17. $\tan \pi$

18. $\sec \pi$

19. $\tan 330°$

20. $\cot 120°$

21. $\sec 600°$

22. $\csc(-150°)$

23. For what values of t on $0 \le t \le 4\pi$ is each of the following undefined?
(a) $\sec t$ (b) $\tan t$ (c) $\csc t$ (d) $\cot t$

24. For which values of t on $0 \le t \le 4\pi$ is each of the following equal to 1?
(a) $\sec t$ (b) $\tan t$ (c) $\csc t$ (d) $\cot t$

Problems 25–28 relate to Example B.

25. If $\pi/2 < t < \pi$ and $\cot t = -\sqrt{6}$, find
(a) $\csc t$ (b) $\cos t$ (c) $\cot\left(\frac{\pi}{2} - t\right)$.

26. If $3\pi/2 < t < 2\pi$ and $\cos t = \frac{1}{3}$, find
(a) $\sin t$ (b) $\cot t$ (c) $\cot\left(\frac{\pi}{2} - t\right)$.

27. If $180° < \theta < 270°$ and $\sec \theta = -3$, find
(a) $\tan \theta$ (b) $\cot(-\theta)$ (c) $\sin(90° - \theta)$.

28. If $90° < \theta < 180°$ and $\csc \theta = 2$, find
(a) $\cot \theta$ (b) $\tan(-\theta)$ (c) $\tan(\theta - 90°)$.

In Problems 29–32 find $\sin \theta$, $\tan \theta$, and $\sec \theta$, assuming that the given point is on the terminal side of θ. See Example C.

29. $(5, -12)$

30. $(7, 24)$

31. $(-1, -2)$

32. $(-3, 2)$

33. If $\tan \theta = \frac{3}{4}$ and θ is an angle in the first quadrant, find $\sin \theta$ and $\sec \theta$. *Hint:* The point $(4, 3)$ is on the terminal side of θ.

34. If $\tan \theta = \frac{3}{4}$ and θ is an angle in the third quadrant, find $\cos \theta$ and $\csc \theta$. *Hint:* The point $(-4, -3)$ is on the terminal side of θ.

35. If $\sin \theta = \frac{5}{13}$ and θ is an angle in the second quadrant, find $\cos \theta$ and $\cot \theta$. *Hint:* A point with y-coordinate 5 and $r = 13$ is on the terminal side of θ. Thus the x-coordinate must be -12.

36. If $\cos \theta = \frac{4}{5}$ and $\sin \theta < 0$, find $\tan \theta$.

37. Where does the line from the origin to $(5, -12)$ intersect the unit circle?

38. Where does the line from the origin to $(-6, 8)$ intersect the unit circle?

39. Find the equation of the line with angle of inclination $150°$ that passes through $(-3, 4)$ (see Example D).

40. Find the angle of inclination for the line $\sqrt{3}x - 3y = 4$.

41. Find the angle of inclination of the line $5x + 2y = 6$.

42. Find the equation of the line with angle of inclination $75°$ that passes through $(-2, 4)$.

B. Applications and Extensions

43. Evaluate without use of a calculator.
(a) $\sec(7\pi/6)$ (b) $\tan(-2\pi/3)$
(c) $\csc(3\pi/4)$ (d) $\cot(11\pi/4)$
(e) $\csc(570°)$ (f) $\tan(180,045°)$

44. Calculate.
(a) $\tan(\sin 2.4)$ (b) $\cot(\tan 1.49)$
(c) $\csc(\sin 11.8°)$ (d) $\sec^2(\tan 91.2°)$
(e) $\csc(\tan \pi)$ (f) $\tan[\tan(\tan 1.5)]$

45. If $\csc t = 25/24$ and $\cos t < 0$, find each of the following.
(a) $\sin t$ (b) $\cos t$
(c) $\tan t$ (d) $\sec(\frac{\pi}{2} - t)$
(e) $\cot(\frac{\pi}{2} - t)$ (f) $\csc(\frac{\pi}{2} - t)$

46. Show that each of the following are identities.
(a) $\tan(-t) = -\tan t$ (b) $\sec(-t) = \sec t$
(c) $\csc(-t) = -\csc t$

47. Find the two smallest positive values of t that satisfy each of the following.
    (a) $\tan t = -1$ **(b)** $\sec t = \sqrt{2}$
    (c) $|\csc t| = 1$

48. Find the angle of inclination of the line that is perpendicular to the line $4x + 3y = 9$.

49. Write each of the following in terms of sines and cosines and simplify.

    (a) $\dfrac{\sec \theta \csc \theta}{\tan \theta + \cot \theta}$ **(b)** $(\tan \theta)(\cos \theta - \csc \theta)$

    (c) $\dfrac{(1 + \tan \theta)^2}{\sec^2 \theta}$ **(d)** $\dfrac{\sec \theta \cot \theta}{\sec^2 \theta - \tan^2 \theta}$

    (e) $\dfrac{\cot \theta - \tan \theta}{\csc \theta - \sec \theta}$ **(f)** $\tan^4 \theta - \sec^4 \theta$

50. Let θ be a first quadrant angle. Express each of the other five trigonometric functions in terms of $\sin \theta$ alone.

51. Use the identities of Problem 37 in Section 7-3, namely,

$$\sin(t + \pi) = -\sin t \quad \text{and} \quad \cos(t + \pi) = -\cos t$$

to establish each of the following identities.
    (a) $\tan(t + \pi) = \tan t$ **(b)** $\cot(t + \pi) = \cot t$
    (c) $\sec(t + \pi) = -\sec t$ **(d)** $\csc(t + \pi) = -\csc t$

Note: From parts (a) and (b), we conclude that tangent and cotangent are periodic with period π.

52. Show that $|\sec t| \geq 1$ and $|\csc t| \geq 1$ for all t for which these functions are defined.

53. If $\tan \theta = \frac{5}{12}$ and $\sin \theta < 0$, evaluate $\cos^2 \theta - \sin^2 \theta$.

54. A wheel of radius 5, centered at the origin, is rotating counterclockwise at a rate of 1 radian per second. At $t = 0$, a speck of dirt on the rim is at $(5, 0)$. What are the coordinates of the speck at time t?

55. At $t = 2\pi/3$, the speck in Problem 54 came loose and flew off along the tangent line. Where did it hit the x-axis?

56. Find the coordinates of P in Figure 46.

57. Find the length of the shorter arc from $(5, 0)$ to P in Figure 46.

58. Find the length of a side of a regular pentagon which is inscribed in a circle of radius 6.

59. The face of a clock is in the xy-plane with center at the origin and 12 on the positive y-axis. Both hands of the clock are 5 units long.
    (a) Find the slope of the minute hand at 2:24.
    (b) Find the slope of the line through the tips of both hands at 12:50.

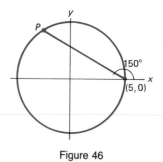

Figure 46

60. From an airplane h miles above the surface of the earth (a sphere of radius 3960 miles), I can just see a bright light on the horizon d miles away. If I measure the angle of depression of the light as 2.1°, help me determine d and h.

61. A wheel of radius 20 centimeters is used to drive a wheel of radius 50 centimeters by means of a belt that fits around the wheels. How long is the belt if the centers of the two wheels are 100 centimeters apart?

62. TEASER Express the length L of the crossed belt that intersects in angle 2α and fits around wheels of radius r and R (Figure 47) in terms of r, R, and α.

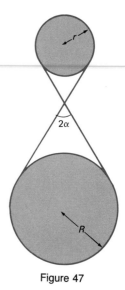

Figure 47

7-5 FINDING VALUES OF THE TRIGONOMETRIC FUNCTIONS

Tables or Calculators

t(rad.)	Sin t	Tan t	Cot t	Cos t
.40	.38942	.42279	2.3652	.92106
.41	.39861	.43463	2.3008	.91712
.42	.40776	.44657	2.2393	.91309
.43	.41687	.45862	2.1804	.90897
.44	.42594	.47078	2.1241	.90475
.45	.43497	.48306	2.0702	.90045
.46	.44395	.49545	2.0184	.89605
.47	.45289	.50797	1.9686	.89157
.48	.46178	.52061	1.9208	.88699
.49	.47063	.53339	1.8748	.88233
.50	.47943	.54630	1.8305	.87758

In order to make significant use of the trigonometric functions, we will have to be able to calculate their values for angles other than the special angles. The simplest procedure is to press the right key on a calculator and read the answer. About the only thing to remember is to make sure the calculator is in the right mode, degree or radian, depending on what we want.

Even though calculators are now standard equipment for most mathematics and science students, we think you should also know how to use tables. This is the subject we take up now. We might call it "what to do when your battery goes dead."

The opening display gives a small portion of a five-place table of values for sin t, tan t, cot t, and cos t. (The complete table appears as Table C at the back of the book.) From it we read the following:

$$\sin .44 = .42594 \qquad \tan .44 = .47078$$

$$\cot .44 = 2.1241 \qquad \cos .44 = .90475$$

These results are not exact; they have been rounded off to five significant digits. Keep in mind that you can think of sin .44 in two ways, as the sine of the number .44 or, if you like, as the sine of an angle of radian measure .44.

Table C appears to have two defects. First, t is given only to 2 decimal places. If we need sin .44736, we have to round or perhaps interpolate (see page 597).

$$\sin .44736 \approx \sin .45 = .43497$$

A more serious defect appears to be the fact that values of t go only to 2.00. This limitation evaporates once we learn about reference angles and reference numbers, our next topic.

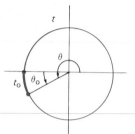

Figure 48

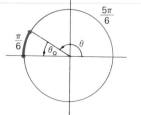

Figure 49

Figure 50

Reference Angles and Reference Numbers

Let θ be any angle in standard position and let t be its radian measure. Associated with θ is an acute angle θ_0, called the **reference angle** and defined to be the smallest positive angle between the terminal side of θ and the x-axis (Figure 48). The radian measure t_0 of θ_0 is called the **reference number** corresponding to t. For example, the reference number for $t = 5\pi/6$ is $t_0 = \pi/6$ (Figure 49).

Example A (Finding Reference Numbers) Find the reference number t_0 for each of the following values of t.

(a) $t = 20.59$ (b) $t = \dfrac{5\pi}{2} - .92$

Solution

(a) To get rid of the irrelevant multiples of 2π, we divide 20.59 by 6.28 ($2\pi \approx 6.28$), obtaining 1.75 as remainder (Figure 50). Since 1.75 is between $\pi/2$ and π, we subtract it from π. Thus

$$t_0 \approx \pi - 1.75 \approx 3.14 - 1.75 = 1.39$$

(b) Since $5\pi/2 - .92 = 2\pi + \pi/2 - .92$, it follows that

$$t_0 = \frac{\pi}{2} - .92 \approx 1.57 - .92 = .65 \quad \blacksquare$$

Once we know t_0, we can find $\sin t$, $\cos t$, and so on, no matter what t is. Here is how we do it. Examine the four diagrams in Figure 51. Each angle θ in B, C, and D has θ_0 as its reference angle and, of course, each t has t_0 as its reference number. Now we make a crucial observation. In each case, the point on the unit circle corresponding to t has the same coordinates, except for sign, as the point corresponding to t_0. It follows from this that

$$\sin t = \pm\sin t_0 \qquad \cos t = \pm\cos t_0$$

with the $+$ or $-$ sign being determined by the quadrant in which the terminal side of the angle falls. For example,

$$\sin \frac{5\pi}{6} = \sin \frac{\pi}{6} \qquad \cos \frac{5\pi}{6} = -\cos \frac{\pi}{6}$$

or, in degree notation,

$$\sin 150° = \sin 30° \qquad \cos 150° = -\cos 30°$$

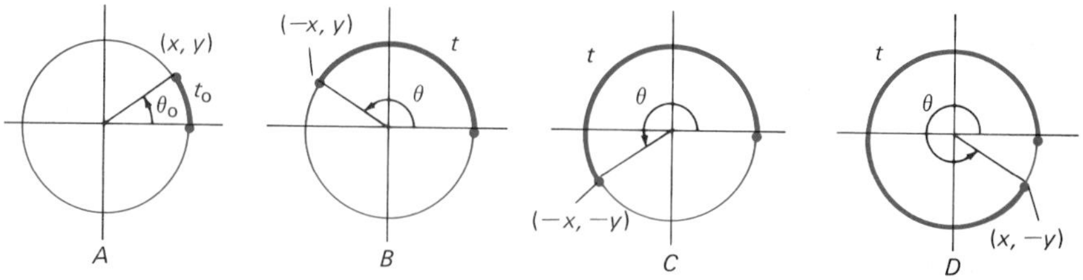

Figure 51

We chose the plus sign for the sine and the minus sign for the cosine because in the second quadrant the sine function is positive, whereas the cosine function is negative.

What we have just said applies to all six trigonometric functions. If T stands for any one of them, then

$$T(t) = \pm T(t_0) \quad \text{and} \quad T(\theta) = \pm T(\theta_0)$$

with the plus or minus sign being determined by the quadrant in which the terminal side of θ lies. Of course $T(t_0)$ itself is always nonnegative since $0 \le t_0 \le \pi/2$.

Using Tables B and C

Example B (Finding Trigonometric Values) Use Table C to evaluate.
(a) cos 2.16   (b) tan 24.95

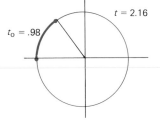

Figure 52

Solution
(a) We must first find the appropriate reference numbers. Approximating π by 3.14, we find that (see Figure 52)

$$t_0 = 3.14 - 2.16 = .98$$

and thus, using Table C,

$$\cos 2.16 = -\cos .98 = -.55702$$

Notice we chose the minus sign because the cosine is negative in quadrant II.

(b) To calculate tan 24.95 is slightly more work. First we remove as large a multiple of 2π as possible from 24.95. Using 6.28 for 2π, we get

$$24.95 = 3(6.28) + 6.11$$

The reference number for 6.11 is (see Figure 53)

$$t_0 = 6.28 - 6.11 = .17$$

Thus

$$\tan 24.95 = \tan 6.11 = -\tan .17 = -.17166$$

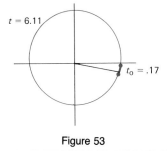

Figure 53

We choose the minus sign because the tangent is negative in quadrant IV.
Now use your handheld calculator to find tan 24.95 the easy way. Be sure you put it in radian mode. You will get $-.18480$ instead of $-.17166$, a rather large discrepancy. Whom should you believe? We suggest that you trust your calculator. The reason we were so far off is that 6.28 is a rather poor approximation for 2π, and multiplying it by 3 made matters worse. Had we used 6.2832 for 2π, we would have obtained $t_0 = .1828$ and tan 24.95 = $-.18486$. ∎

Example C (Finding t When sin t or cos t Is Given) Find two values of t between 0 and 2π for which
(a) sin t = .90863.   (b) cos t = $-.95824$.

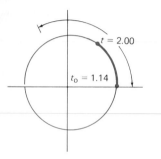

Figure 54

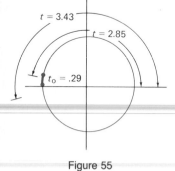

Figure 55

Solution

(a) We get $t = 1.14$ directly from Table C (or using a calculator). Since the sine is also positive in quadrant II, we seek a value of t between $\pi/2$ and π for which 1.14 is the reference number (Figure 54). Only one number fits the bill:

$$\pi - 1.14 \approx 3.14 - 1.14 = 2.00$$

(b) We know that $\cos t_0 = .95824$ and so $t_0 = .29$. Now the cosine is negative in quadrants II and III. Thus we are looking for two numbers between $\pi/2$ and $3\pi/2$ with .29 as reference number (Figure 55). One is $\pi - .29 \approx 3.14 - .29 = 2.85$, and the other $\pi + .29 \approx 3.14 + .29 = 3.43$. ■

In Section 7-1 we used Table B to find sine, cosine, and so on, of positive angles less than 90°. Now we can use this table to evaluate trigonometric functions for angles of arbitrary degree measure.

Example D (Trigonometric Functions for Angles in Degrees)   Find the value of each of the following.
(a) $\cos 214.6°$ (b) $\cot 658°$

Solution

(a) The reference angle is

$$214.6° - 180° = 34.6°$$

$$\cos 214.6° = -\cos 34.6° = -.8231$$

We used the minus sign since cosine is negative in quadrant III.

(b) First we reduce our angle by 360°:

$$658° = 360° + 298°$$

The reference angle for 298° is $360° - 298°$, or 62°. In the column with cot at the bottom and 62° at the right, we find .5317. Therefore, $\cot 658° = -.5317$. ■

CAUTION

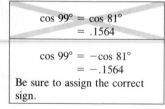

$\cos 99° = \cos 81°$
$= .1564$

$\cos 99° = -\cos 81°$
$= -.1564$
Be sure to assign the correct sign.

PROBLEM SET 7-5

A. Skills and Techniques

Find the value of each of the following using Table B or C. Confirm with a calculator.

1. $\sin 1.38$
2. $\cos .67$
3. $\cos 42.8°$
4. $\tan 18.0°$
5. $\cot .82$
6. $\tan 1.11$
7. $\sin 68.3°$
8. $\cot 49.6°$

Find the reference number t_0 if t has the given value. Use 3.14 for π. See Example A.

9. 1.84
10. 2.14
11. 3.54
12. 3.74
13. 5.18
14. 6.08
15. 10.48
16. 8.38
17. −1.12
18. −1.86
19. −2.64
20. −4.24

Find the reference number for each of the following. You may leave your answer in terms of π.

21. $13\pi/8$
22. $37\pi/36$
23. $40\pi/3$
24. $-11\pi/5$
25. $3\pi + .24$
26. $3\pi/2 + .17$

27. $3\pi - .24$ **28.** $3\pi/2 - .17$

29. $11\pi/2$ **30.** 26π

Find the value of each of the following using Table C and
$\pi = 3.14$ *as in Example B. Calculators will give slightly dif-*
ferent results because of this crude approximation to π.

31. $\cos 1.42$ **32.** $\sin .97$

33. $\tan 1.39$ **34.** $\cot .08$

35. $\sin 2.14$ **36.** $\cos 3.08$

37. $\cot 5.62$ **38.** $\tan 4.11$

39. $\cos(-2.54)$ **40.** $\sin(-4.18)$

Find two values of t between 0 and 2π *for which the given*
equality holds. See Example C.

41. $\sin t = .94898$ **42.** $\cos t = .72484$

43. $\cos t = -.08071$ **44.** $\sin t = -.48818$

45. $\tan t = 4.9131$ **46.** $\cot t = 1.4007$

47. $\tan t = -3.6021$ **48.** $\cot t = -.47175$

Find the reference angle (in degrees) for each of the follow-
ing angles. For example, the reference angle for $\theta = 124.1°$
is $\theta_0 = 180° - 124.1° = 55.9°$.

49. $139.6°$ **50.** $218.1°$

51. $348.7°$ **52.** $375.4°$

53. $-99.8°$ **54.** $-224.4°$

Find the value of each of the following, using Table B. Check
with a calculator.

55. $\sin 156.1°$ **56.** $\cos 138.7°$

57. $\tan 348.9°$ **58.** $\cot 224.9°$

59. $\cos(-66.1°)$ **60.** $\sin 487°$

61. $\cos 441.3°$ **62.** $\sin 180.2°$

63. $\cot(-134°)$ **64.** $\tan 311.6°$

Find two different degree values of θ *between* $0°$ *and* $360°$
for which the given equality holds.

65. $\sin \theta = .3633$ **66.** $\cos \theta = .9907$

67. $\tan \theta = .4942$ **68.** $\cot \theta = 1.2799$

69. $\cos \theta = -.9085$ **70.** $\sin \theta = -.2045$

B. Applications and Extensions

71. Use Table B or C to find each of the following. You
may approximate π by 3.14.

    (a) $\cos 5.63$ **(b)** $\sin 10.34$

(c) $\tan 8.42$ **(d)** $\sin 311.3°$

(e) $\tan(-411°)$ **(f)** $\cos 1989°$

72. Use Tables B and C to calculate. Check with a calculator.

    (a) $\sin(\cos 134°)$ **(b)** $\sin[(\tan 1.5)°]$

    (c) $\tan(-5.4°) + \tan(-5.4)$

73. Calculate.

    (a) $\cos(\sin 2.42°)$ **(b)** $\cos^3(\sin^2 2.42)$

    (c) $\sqrt{\tan 4.21} + \ln(\sin 7.12)$

74. Use Table C and $\pi = 3.14$ to find two values of t between 0 and 2π for which each of the following is
true.

    (a) $\sin t = .62879$ **(b)** $\cos t = -.90045$

    (c) $\tan t = -4.4552$

75. Find two values of t between 0 and 2π for which each
statement is true, giving your answers correct to 6
decimal places.

    (a) $\sin t = .62879$ **(b)** $\cos t = .34176$

    (c) $\tan t = -3.14159$

Note: On many calculators, you would press .62879
$\boxed{\text{INV}}$ $\boxed{\sin}$ to get one answer to (a).

76. If $\pi/2 < t < \pi$, then $t_0 = \pi - t$. In a similar manner, express t_0 in terms of t in each case.

    (a) $3\pi/2 < t < 2\pi$ **(b)** $5\pi < t < 11\pi/2$

    (c) $-2\pi < t < -3\pi/2$

77. If $0° < \phi < 90°$, express the reference angle θ_0 in
terms of ϕ in each case.

    (a) $\theta = 180° + \phi$ **(b)** $\theta = 270° - \phi$

    (c) $\theta = \phi - 90°$

78. Without using tables or a calculator, round to the
nearest degree the smallest positive angle θ satisfying
$\tan \theta = -40,000$.

79. If θ is a fourth quadrant angle whose terminal side coincides with the line $3x + 5y = 0$, find $\sin \theta$.

80. If θ is a third quadrant angle whose terminal side coincides with the line $y = 7x$, find θ.

81. Find θ if θ is a fourth quadrant angle satisfying
$\sec \theta = 3$.

82. In calculus you will learn that

$$\sin t = t - \frac{t^3}{3!} + \frac{t^5}{5!} - \frac{t^7}{7!} + \cdots$$

and

$$\cos t = 1 - \frac{t^2}{2!} + \frac{t^4}{4!} - \frac{t^6}{6!} + \cdots$$

Here, $n! = 1 \cdot 2 \cdot 3 \cdots n$ (for example, $2! = 1 \cdot 2 = 2$ and $3! = 1 \cdot 2 \cdot 3 = 6$). These series are used to construct Tables B and C. If we use just the first three terms in the sine series, we obtain

$$\sin t \approx t - \frac{t^3}{6} + \frac{t^5}{120} = \left[\left(\frac{t^2}{120} - \frac{1}{6}\right)t^2 + 1\right]t$$

Use the first three terms of these series to approximate each of the following and compare with the corresponding value in Table C.

(a) $\sin(.1)$ (b) $\sin(.4)$
(c) $\cos(.2)$

83. Determine ϕ in Figure 56 so that the path ACB has minimum length.

84. TEASER Let α, β, and γ be acute angles such that $\tan \alpha = 1$, $\tan \beta = 2$, and $\tan \gamma = 3$.

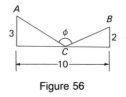

Figure 56

(a) Use your calculator to approximate $\alpha + \beta + \gamma$.
(b) Make a conjecture about the exact value of $\alpha + \beta + \gamma$.
(c) Construct a clever geometric diagram to prove your conjecture.

7-6 GRAPHS OF THE TRIGONOMETRIC FUNCTIONS

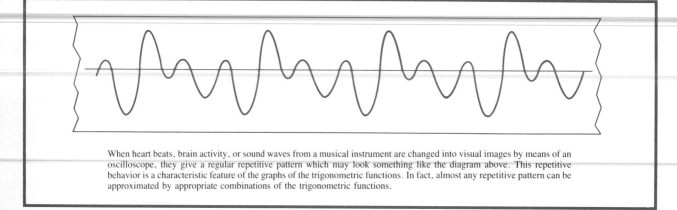

When heart beats, brain activity, or sound waves from a musical instrument are changed into visual images by means of an oscilloscope, they give a regular repetitive pattern which may look something like the diagram above. This repetitive behavior is a characteristic feature of the graphs of the trigonometric functions. In fact, almost any repetitive pattern can be approximated by appropriate combinations of the trigonometric functions.

Recall that to graph $y = f(x)$, we first construct a table of values of ordered pairs (x, y), then plot the corresponding points, and finally connect those points with a smooth curve. Here we want to graph $y = \sin t$, $y = \cos t$, and so on, and we will follow a similar procedure. Notice that we use t rather than x as the independent variable because we used t as the variable (radian measure of an angle) in our definition of the trigonometric functions.

We begin with the graphs of the sine and cosine functions. You should become so well acquainted with these two graphs that you can sketch them quickly whenever you need them. This will aid you in two ways. First, these graphs will help you remember many of the important properties of the sine and cosine functions. Second, knowing them will help you graph other more complicated trigonometric functions.

The Graph of $y = \sin t$

We begin with a table of values (Figure 57). We have listed values of t between 0 and 2π. That is sufficient to graph one period (shown in Figure 58 as a heavy curve).

t	0	$\dfrac{\pi}{6}$	$\dfrac{\pi}{4}$	$\dfrac{\pi}{3}$	$\dfrac{\pi}{2}$	$\dfrac{3\pi}{4}$	π	$\dfrac{5\pi}{4}$	$\dfrac{3\pi}{2}$	$\dfrac{7\pi}{4}$	2π
$y = \sin t$	0	$\dfrac{1}{2}$	$\dfrac{\sqrt{2}}{2}$	$\dfrac{\sqrt{3}}{2}$	1	$\dfrac{\sqrt{2}}{2}$	0	$-\dfrac{\sqrt{2}}{2}$	-1	$-\dfrac{\sqrt{2}}{2}$	0

Figure 57

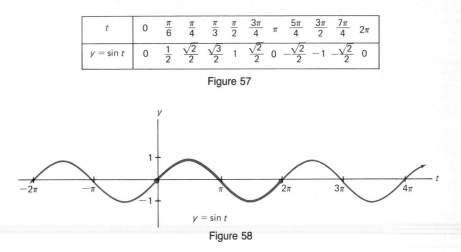

$y = \sin t$

Figure 58

From there on we can continue the curve indefinitely in either direction in a repetitive fashion, for we learned earlier that $\sin(t + 2\pi) = \sin t$.

The Graph of $y = \cos t$

The cosine function is a copycat; its graph is just like that of the sine function but shifted $\pi/2$ units to the left (Figure 59).

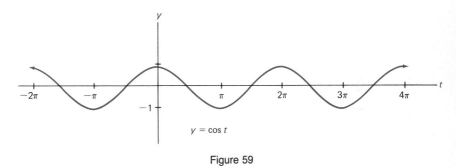

$y = \cos t$

Figure 59

To see that the graph of the cosine function is correct, we might make a table of values and proceed as we did for the sine function. In fact, we ask you to do just that in Problem 1. Alternatively, we can show that

$$\cos t = \sin\left(t + \frac{\pi}{2}\right)$$

This follows directly from identities we have observed earlier.

$$\sin\left(t + \frac{\pi}{2}\right) = \sin\left(\frac{\pi}{2} - (-t)\right)$$

$$= \cos(-t) \qquad \text{(cofuction identity)}$$

$$= \cos t \qquad \text{(cosine is even)}$$

Properties Easily Observed from These Graphs

1. Both sine and cosine are periodic with 2π as period.
2. $-1 \le \sin t \le 1$ and $-1 \le \cos t \le 1$.
3. $\sin t = 0$ if $t = -\pi, 0, \pi, 2\pi$, and so on.
   $\cos t = 0$ if $t = -\pi/2, \pi/2, 3\pi/2$, and so on.
4. $\sin t > 0$ in quadrants I and II.
   $\cos t > 0$ in quadrants I and IV.
5. $\sin(-t) = -\sin t$ and $\cos(-t) = \cos t$.
   The sine is an odd function; its graph is symmetric with respect to the origin. The cosine is an even function; its graph is symmetric with respect to the y-axis.
6. We can see immediately where the sine and cosine functions are increasing and where they are decreasing. For example, the sine function decreases for $\pi/2 \le t \le 3\pi/2$.

The Graph of $y = \tan t$

Since the tangent function is defined by

$$\tan t = \frac{\sin t}{\cos t}$$

we need to beware of values of t for which $\cos t = 0$: $-\pi/2, \pi/2, 3\pi/2$, and so forth. In fact, from Section 5-3, we know that we should expect vertical asymptotes at these places. Notice also that

$$\tan(-t) = \frac{\sin(-t)}{\cos(-t)} = \frac{-\sin t}{\cos t} = -\tan t$$

which means that the graph of the tangent will be symmetric with respect to the origin. Using these two pieces of information, a small table of values, and the fact that the tangent is periodic, we obtain the graph in Figure 60.

To confirm that the graph is correct near $t = \pi/2$, we suggest looking at Table C. Notice that $\tan t$ steadily increases until at $t = 1.57$, we read $\tan t = 1255.8$. But as t takes the short step to 1.58, $\tan t$ takes a tremendous plunge to -108.65. In that short space, t has passed through $\pi/2 \approx 1.5708$ and $\tan t$ has shot up to celestial heights only to fall to a bottomless pit, from which, however, it manages to escape as t moves to the right.

While we knew the tangent would have to repeat itself every 2π units since the sine and cosine do this, we now notice that it actually repeats itself on intervals of length π. Since the word *period* denotes the length of the shortest interval after which a function repeats itself, the tangent function has period π. For an algebraic demonstration, see Problem 51 of Section 7-4.

The Graph of $y = \sec t$

Since $\sec t = 1/\cos t$, one way of getting the graph of the secant is by graphing the cosine and then taking reciprocals of the y-coordinates (Figure 61). Note that since $\cos t = 0$ at $t = -\pi/2, \pi/2, 3\pi/2$, and so on, the graph of $\sec t$ must have vertical asymptotes at these points.

t	0	$\frac{\pi}{4}$	$\frac{\pi}{3}$	$\frac{\pi}{2}$	$\frac{2\pi}{3}$	$\frac{3\pi}{4}$	π	$\frac{5\pi}{4}$	$\frac{3\pi}{2}$	$\frac{7\pi}{4}$	2π
$y = \tan t$	0	1	$\sqrt{3}$	undefined	$-\sqrt{3}$	-1	0	1	undefined	-1	0

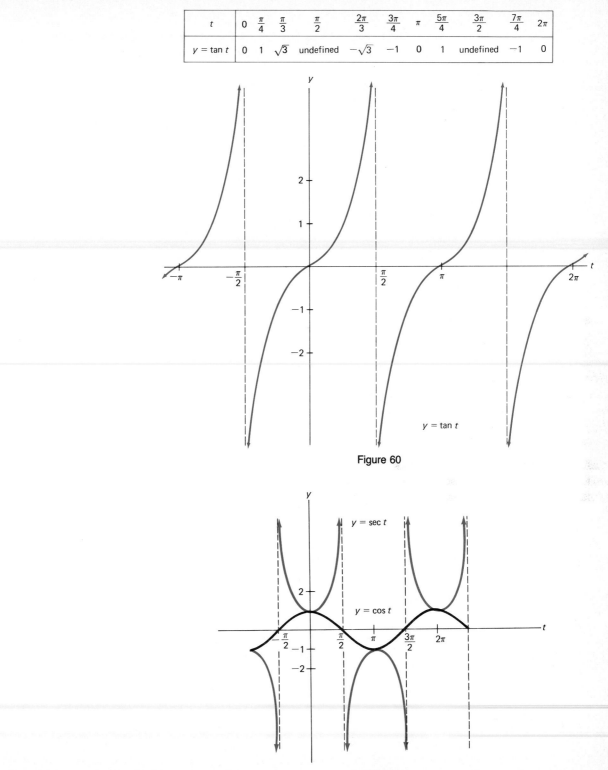

Figure 60

Figure 61

Just like the cosine, the secant is an even function; that is, $\sec(-t) = \sec t$. And, like the cosine, secant has period 2π. However, notice that if $\cos t$ increases or decreases throughout an interval, $\sec t$ does just the opposite. For example, $\cos t$ decreases for $0 < t < \pi/2$, whereas $\sec t$ increases there.

Graphing Modified Sine and Cosine Functions

Two concepts that play a major role in the study of periodic functions are the *period* and the *amplitude*. We have already defined the first of these; the **period** of a periodic function f is the smallest positive number p such that $f(x + p) = f(x)$ for all x in the domain of f. The **amplitude** of a periodic function is the number $(M - m)/2$, where M and m are the largest and smallest values of f, respectively. Both sine and cosine have period 2π and they both have amplitude 1.

Example A (Some Sine-Related Graphs)   Sketch the graph of each of the following for $-2\pi \le t \le 4\pi$.
(a) $y = 2 \sin t$ (b) $y = \sin 2t$

Solution

(a) We could graph $y = 2 \sin t$ from a table of values. It is easier, though, to graph $\sin t$ (dashed graph below) and then multiply the ordinates by 2 (Figure 62). Since the graph bobs up and down between $y = -2$ and $y = 2$, the amplitude is $(2 - (-2))/2 = 2$. The period is 2π, the same as for $\sin t$.

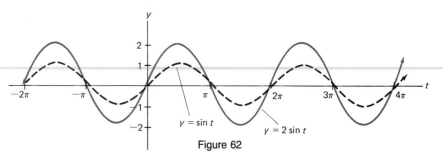

Figure 62

t	$-\pi$	$-\dfrac{3\pi}{4}$	$-\dfrac{\pi}{2}$	$-\dfrac{\pi}{4}$	$-\dfrac{\pi}{12}$	0	$\dfrac{\pi}{12}$	$\dfrac{\pi}{4}$	$\dfrac{\pi}{2}$	$\dfrac{3\pi}{4}$	π
$2t$	-2π	$-\dfrac{3\pi}{2}$	$-\pi$	$-\dfrac{\pi}{2}$	$-\dfrac{\pi}{6}$	0	$\dfrac{\pi}{6}$	$\dfrac{\pi}{2}$	π	$\dfrac{3\pi}{2}$	2π
$\sin 2t$	0	1	0	-1	$-\dfrac{1}{2}$	0	$\dfrac{1}{2}$	1	0	-1	0

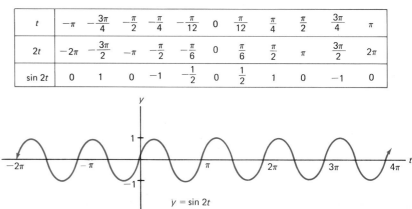

$y = \sin 2t$

Figure 63

(b) Here a table of values is advisable, since this is our first example of this type (Figure 63). This graph goes through a complete cycle as t increases from 0 to π, that is, the period of sin $2t$ is π instead of 2π as it was for sin t. The amplitude is 1, just as for sin t. ∎

Example B (Graphs of A sin Bt and A cos Bt) Sketch the graphs of each of the following for $-2\pi \le t \le 4\pi$.
(a) $y = 3 \sin 4t$ (b) $y = 2 \cos \frac{1}{2}t$

Solution

(a) We can save a lot of work once we recognize how the character of the graph of A sin Bt (and A cos Bt) is determined by the numbers A and B $(B > 0)$. The amplitude (which tells how far the graph rises and falls from its median position) is given by $|A|$. The period is given by $2\pi/B$. Thus for $y = 3 \sin 4t$, the amplitude is 3 and the period is $2\pi/4 = \pi/2$. For a quick sketch, we use these two numbers to determine the high and low points and the t-intercepts, connecting these points with a smooth, wavelike curve (Figure 64).

(b) The amplitude is 2 and the period is $2\pi/\frac{1}{2} = 4\pi$. Thus the graph is just the standard cosine graph stretched by a factor of 2 in both directions (Figure 65).

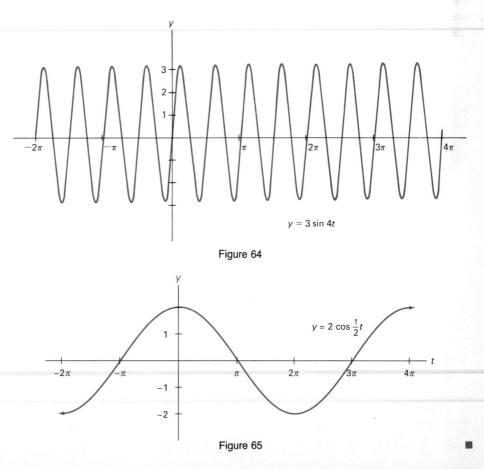

$y = 3 \sin 4t$

Figure 64

$y = 2 \cos \frac{1}{2}t$

Figure 65 ∎

The topic introduced in Example B is extremely important in connection with applications, so much so that we spend a whole section later (Section 9-5) exploring its ramifications.

Example C (Graphing Sums of Trigonometric Functions)   Sketch the graph of the equation $y = 2 \sin t + \cos 2t$.

Solution   We graph $y = 2 \sin t$ and $y = \cos 2t$ on the same coordinate plane (these appear as dashed-line curves in Figure 66) and then add ordinates. Notice that for any t, the ordinates (y-values) of the dashed curves are added to obtain the desired ordinate. The graph of $y = 2 \sin t + \cos 2t$ is quite different from the separate (dashed) graphs but it does repeat itself; it has period 2π.

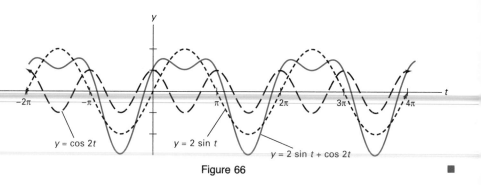

Figure 66

PROBLEM SET 7-6

A. Skills and Techniques

1. Make a table of values and then sketch the graph of $y = \cos t$.

2. What real numbers constitute the domain of the cosine? The range?

3. Sketch the graph of $y = \cot t$ for $-2\pi \le t \le 2\pi$, being sure to show the asymptotes.

4. What real numbers constitute the entire domain of the cotangent? The range?

5. Using the corresponding fact about the cosine, demonstrate algebraically that $\sec(t + 2\pi) = \sec t$.

6. Sketch the graph of $y = \csc t$.

7. What is the domain of the secant? The range?

8. What is the domain of the cosecant? The range?

9. What is the period of the cotangent? The secant?

10. On the interval $-2\pi \le t \le 2\pi$, where is the cotangent increasing?

11. Which is true: $\cot(-t) = \cot t$ or $\cot(-t) = -\cot t$?

12. Which is true: $\csc(-t) = \csc t$ or $\csc(-t) = -\csc t$?

In Problems 13–22, determine the amplitude and the period. Then sketch the graph on the indicated interval. See Examples A and B.

13. $y = 3 \cos t, -\pi \le t \le \pi$
14. $y = \frac{1}{2} \cos t, -\pi \le t \le \pi$
15. $y = -\sin t, -\pi \le t \le \pi$
16. $y = -2 \cos t, -\pi \le t \le \pi$
17. $y = \cos 4t, -\pi \le t \le \pi$
18. $y = \cos 3t, -\pi/2 \le t \le \pi/2$
19. $y = 2 \sin \frac{1}{2}t, -2\pi \le t \le 2\pi$
20. $y = 3 \sin \frac{1}{3}t, -3\pi \le t \le 3\pi$
21. $y = 2 \cos 3t, -\pi \le t \le \pi$
22. $y = 4 \sin 3t, -\pi \le t \le \pi$

Sketch each graph by the method of adding ordinates. Show at least one complete period. See Example C.

23. $y = 2 \sin t + \cos t$
24. $y = \sin t + 2 \cos t$
25. $y = \sin 2t + \cos t$
26. $y = \sin t + \cos 2t$
27. $y = \sin \frac{1}{2}t + \frac{1}{2} \sin t$
28. $y = \cos \frac{1}{2}t + \frac{1}{2} \cos t$

B. Applications and Extensions

In Problems 29–32, sketch each graph on the indicated interval.

29. $y = -\cos t,\ -\pi \le t \le \pi$

30. $y = 3 \sin t,\ -\pi \le t \le \pi$

31. $y = \sin 4t,\ 0 \le t \le \pi$

32. $y = 3 \cos \frac{1}{2}t,\ -2\pi \le t \le 2\pi$

33. What are the amplitudes and periods for the graphs in Problems 29 and 31?

34. What are the amplitudes and periods for the graphs in Problems 30 and 32?

35. Determine the period and sketch the graph of each of the following, showing at least three periods.
(a) $y = \tan 2t$ (b) $y = 3 \tan(t/2)$

36. Follow the directions of Problem 35.
(a) $y = 2 \cot 2t$ (b) $y = \sec 3t$

37. Sketch, using the same axes, the graphs of
(a) $f(t) = \sin t$.   (b) $g(t) = 3 + \sin t$.
(c) $h(t) = \sin(t - \pi/4)$.

38. Sketch, using the same axes, the graphs of
(a) $f(t) = \cos t$.   (b) $g(t) = -2 + \cos t$.
(c) $h(t) = \cos(t + \pi/3)$.

39. Let f be a periodic function with period 3 satisfying $f(x) = x^2$ on the interval $-1 < x \le 2$. Sketch the graph of this function and determine its amplitude.

40. Recall the definitions of the greatest integer function $[\cdot]$ and the distance to the nearest integer function $(\cdot)$ from Section 5-2 (Problems 45 and 51). Sketch the graphs of $f(x) = x - [x]$ and $g(x) = (x)$ and thereby convince yourself that both are periodic. Determine the period and amplitude of each function.

41. Sketch the graph of $y = \cos 3t + 2 \sin t$ for $-\pi \le t \le \pi$. Use the method of adding ordinates.

42. Sketch the graph of $y = t + \sin t$ on $-4\pi \le t \le 4\pi$. Use the method of adding ordinates.

43. Sketch the graph of $y = t - \cos t$ for $0 \le t \le 6$ by actually calculating the y values corresponding to $t = 0, .5, 1, 1.5, 2, 2.5, \ldots, 6$.

44. By sketching the graphs of $y = t$ and $y = 3 \sin t$ on the same coordinate axes, determine approximately all solutions of $t = 3 \sin t$.

45. The strength I of current (in amperes) in a wire of an alternating current circuit might satisfy

$$I = 30 \sin(120\ \pi t)$$

where time t is measured in seconds.
(a) What is the period?
(b) How many cycles (periods) are there in one second?
(c) What is the maximum strength of the current?

46. Sketch the graph of $y = (\sin t)/t$ on $-3\pi \le t \le 3\pi$. Be sure to plot several points for t near 0 (for example, $t = -.5, -.2, -.1, .1, .2, .5$). What value does y seem to approach as t approaches 0?

47. Consider $y = \sin(1/t)$ on the interval $0 < t \le 1$.
(a) Where does its graph cross the t-axis?
(b) Evaluate y for
$t = 2/\pi,\ 2/3\pi,\ 2/5\pi,\ 2/7\pi, \ldots$
(c) Sketch the graph as best you can, using a large unit on the t-axis.

48. **TEASER** How many solutions does each equation have on the indicated interval for t?
(a) $\sin t = t/60,\ t \ge 0$
(b) $\sin(1/t) = t/60,\ t \ge .06$
(c) $\sin(1/t) = t/60,\ t > 0$

49. To help you understand both the power and the limitations of your calculator, redraw each of the figures in this section. (On the T1-81, you may wish to select Trig on the ZOOM menu since this sets the tick marks on the t-axis at multiples of $\pi/2$.) Then draw the graph of $f(x) = 2 \sin 2t - \cos t$ on the interval $-2\pi \le t \le 2\pi$ and find the period and amplitude.

50. Draw the graph of $f(t) = .25t + .5 \sin 2t$ on $-2\pi \le t \le 2\pi$ and find its maximum value there.

51. Find all solutions of the equation $x^2 - 2x = \sin x$, accurate to four decimal places.

52. Determine the number of intersections of the graphs of $y = \sin 2x$ and $y = .25x^2 - x + 1.19$.

CHAPTER 7 SUMMARY

The word **trigonometry** means triangle measurement. In its elementary historical form, it is the study of how to **solve** right triangles when appropriate information is given. The main tools are the three trigonometric ratios $\sin \theta$, $\cos \theta$, and $\tan \theta$, which were first defined only for acute angles θ.

In order to give the subject its modern general form, we first generalized the notion of an angle θ, allowing θ to have arbitrary size and measuring it either in **degrees** or **radians.** Such an angle θ can be placed in **standard position** in a coordinate system, where it will cut off an arc of directed length t (the radian measure of θ) stretching from $(1, 0)$ to (x, y) on the unit circle. This allowed us to make the key definitions

$$\sin \theta = \sin t = y \qquad \cos \theta = \cos t = x$$

on which all of modern trigonometry rests.

From the definitions above, we derived several identities, of which the most important is

$$\sin^2 t + \cos^2 t = 1$$

We also defined four additional functions

$$\tan t = \frac{\sin t}{\cos t} \qquad \cot t = \frac{\cos t}{\sin t}$$

$$\sec t = \frac{1}{\cos t} \qquad \csc t = \frac{1}{\sin t}$$

To evaluate the trigonometric functions, we may use either a scientific calculator or Tables B and C in the Appendix. If the tables are used, the notions of **reference angle** and **reference number** become important. Finally we graphed several of the trigonometric functions, noting especially their **periodic** behavior.

CHAPTER 7 REVIEW PROBLEM SET

In Problems 1–10, write True or False in the blank. If false, tell why.

_____ **1.** $\sin(\pi/3) = \cos(\pi/6)$
_____ **2.** An angle measuring $135°$ measures $3\pi/4$ radians.
_____ **3.** If $\sin \theta = 0$, then $\cos \theta = 1$.
_____ **4.** $\cos(-t) = -\cos t$ for all t.
_____ **5.** If $\beta = \alpha + \pi$, then $\sin \beta = \sin \alpha$.
_____ **6.** If $\cot t = a$, then $|\csc t| = \sqrt{1 + a^2}$.
_____ **7.** The reference angle for $276°$ is $6°$.
_____ **8.** The reference number for $19\pi/4$ is $\pi/4$.
_____ **9.** The amplitude of the periodic function $f(t) = 1 + 2 \sin t$ is 3.
_____ **10.** The period of the periodic function $f(t) = 4 \cos(3t/2)$ is $4\pi/3$.

Solve the right triangles in Problems 11 and 12.

11. $a = 9, c = 15$ **12.** $\alpha = 72.4°, b = 29.6$

Find the exact value of the each of the following without the use of a calculator or tables.

13. $\sin(3\pi/2)$ **14.** $\tan(\pi/6)$
15. $\sec(5\pi/3)$ **16.** $\cot 315°$
17. $\sin(-225°)$ **18.** $\cos 7\pi$
19. $\tan(3\pi/4)$ **20.** $\csc(7\pi/2)$

In Problems 21–24, find the two exact values of t between 0 and 2π for which the following is true.

21. $\cos t = -\frac{1}{2}$ **22.** $\tan t = -1$
23. $\sin t = \sqrt{2}/2$ **24.** $\sec t = 2$
25. If $\sin t = 3/7$ and $\pi/2 < t < \pi$, find $\tan t$.
26. Find θ if $\cos \theta = \frac{1}{2}$ and $-90° < \theta < 90°$.
27. If $\cos \theta = 3/5$, find
    (a) $\cos(-\theta)$.   (b) $\sin(\theta - 90°)$.
    (c) $\cos(\theta + 180°)$.   (d) $|\sin \theta|$.
28. If θ is a second quadrant angle, express each of the following in terms of $\sin \theta$.
    (a) $\cos \theta$ (b) $\csc \theta$ (c) $\tan \theta$
29. If $(-5, 12)$ is on the terminal side of angle θ in standard position, find
    (a) $\tan \theta$.   (b) $\csc \theta$.
30. For what values of t between 0 and 2π is
    (a) $\sin t > 0$?   (b) $\sin 2t > 0$?
31. Determine the range of the function

$$f(t) = 3 \cos t - 2$$

32. Use the fact that the sine function is odd and the cosine function is even to show that the tangent function is odd.
33. Find a and b so that the curve $y = a \sin x + b \cos x$ passes through the points $(\pi/2, 2)$ and $(\pi/3, 4)$.
34. Let $f(t) = 2 + 4 \sin t$ for $0 \le t \le 2\pi$.

(a) Find the maximum and minimum values of $f(t)$.
(b) Where does the graph cross the t-axis?

35. Write in terms of $\cos t$.
   (a) $\cos(t + 4\pi)$ (b) $\cos(t + \pi)$
   (c) $\sin^2 t$ (d) $\sin\left(t - \dfrac{\pi}{2}\right)$

36. If $\tan \theta = 3/4$ and $\sin \theta < 0$, evaluate $\sec^2 \theta - \sin^3 \theta$.

37. Show that the area of a right triangle is $(a^2/2) \cot \alpha$, assuming the standard notation.

38. Sketch the graph of $y = 3 \cos 2t$ for $-\pi \le t \le 2\pi$.

39. Sketch the graph of $y = \sin t + \sin 2t$ for $-\pi \le t \le \pi$.

40. Find two values of t between 0 and 2π at which $\sin 2t$ takes on its maximum value.

41. Angie is pedaling her tricycle so that the front wheel of radius 10 inches is turning at 300° per second.
   (a) Determine her speed in feet per second.
   (b) How far down the road will she travel in 2 minutes?

42. Find the area of the shaded region in Figure 67.

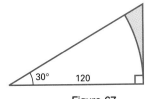

Figure 67

43. A 16-foot ladder leans against a vertical wall. If the ladder makes an angle of 55° with the ground, how high up the wall does it reach?

44. If the bottom of the ladder in Problem 43 is pulled one foot farther from the wall, how far does the top of the ladder slide down the wall?

45. Find x and y in Figure 68.

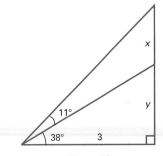

Figure 68

46. The curve on a highway is an arc of a circle of radius 500 meters and subtends an angle of 18° at the center of the circle. Find
   (a) the length of this arc.
   (b) the straight-line distance between the two ends of the arc.

47. Describe in words what happens to the values of $\sin(t/2)$ as t increases from 0 to 2π.

48. At how many points do the graphs of $y = \sin 2t$ and $y = -\cos t$ intersect as t varies from 0 to 2π?

49. A regular octagon (eight sides) is inscribed in a circle of radius 12 centimeters. Determine the perimeter of this octagon.

50. The Pentagon building in Washington, D.C. has the shape of a regular pentagon with five sides each of length 921 feet. Find the area of the region within the perimeter of this building.

CHAPTER 8

Trigonometric Identities and Equations

▪ For just as in nature itself there is no middle ground between truth and falsehood, so in rigorous proofs one must either establish his point beyond doubt, or else beg the question inexcusably. There is no chance of keeping one's feet by invoking limitations, distinctions, verbal distortions, or other mental acrobatics. One must with a few words and at the first assault become Caesar or nothing at all.

Galileo

8-1 IDENTITIES

Complicated combinations of the six trigonometric functions occur often in mathematics. It is important that we, like the professor above, be able to write a complicated trigonometric expression in a simpler or more convenient form. To do this requires two things. We must be good at algebra and we must know the fundamental identities of trigonometry.

The Fundamental Identities

We list eleven fundamental identities, which should be memorized.

1. $\tan t = \dfrac{\sin t}{\cos t}$

2. $\cot t = \dfrac{\cos t}{\sin t} = \dfrac{1}{\tan t}$

3. $\sec t = \dfrac{1}{\cos t}$

4. $\csc t = \dfrac{1}{\sin t}$

5. $\sin^2 t + \cos^2 t = 1$

6. $1 + \tan^2 t = \sec^2 t$

7. $1 + \cot^2 t = \csc^2 t$

8. $\sin\left(\dfrac{\pi}{2} - t\right) = \cos t$

9. $\cos\left(\dfrac{\pi}{2} - t\right) = \sin t$

10. $\sin(-t) = -\sin t$

11. $\cos(-t) = \cos t$

We have seen all these identities before. The first four are actually definitions; the others were established either in the text or the problem sets of Sections 7-3 and 7-4.

Proving New Identities

The professor's work in our opening cartoon can be viewed in two ways. The more likely way of looking at it is that she wanted to simplify the complicated expression

$$(\sec t + \tan t)(1 - \sin t)$$

But it could be that someone had conjectured that

$$(\sec t + \tan t)(1 - \sin t) = \cos t$$

is an identity and that the professor was trying to prove it. It is this second concept we want to discuss now.

Suppose someone claims that a certain equation is an identity—that is, true for all values of the variable for which both sides make sense. How can you check on such a claim? The procedure used by the professor is one we urge you to follow. Start with the more complicated looking side and try to use a chain of equalities to produce the other side.

Example A (Simplify the Complicated Side)   Prove that

$$\sin t + \cos t \cot t = \csc t$$

is an identity.

Solution   We begin with the left side and rewrite it step by step, using algebra and the fundamental identities, until we get the right side.

$$\sin t + \cos t \cot t = \sin t + \cos t\left(\frac{\cos t}{\sin t}\right)$$

$$= \frac{\sin^2 t + \cos^2 t}{\sin t} = \frac{1}{\sin t} = \csc t \quad \blacksquare$$

Example B (Change to Sines and Cosines)   Prove that

$$\csc \theta - \sin \theta = \cot \theta \cos \theta$$

is an identity.

Solution   The left side looks inviting, as $\csc \theta = 1/\sin \theta$. We rewrite it a step at a time.

$$\csc \theta - \sin \theta = \frac{1}{\sin \theta} - \sin \theta$$

$$= \frac{1 - \sin^2 \theta}{\sin \theta}$$

$$= \frac{\cos^2 \theta}{\sin \theta}$$

$$= \frac{\cos \theta}{\sin \theta} \cdot \cos \theta$$

$$= \cot \theta \cos \theta \quad \blacksquare$$

When proving that an equation is an identity, it pays to look before you leap. Changing the more complicated side to sines and cosines, as in the above example, is often the best thing to do. But not always. Sometimes the simpler side gives us a clue as to how we should reshape the other side.

Example C (Note the Character of the Simple Side)   Prove that

$$\tan t = \frac{(\sec t - 1)(\sec t + 1)}{\tan t}$$

is an identity.

Solution   The left side suggests that we try to rewrite the right side in terms of $\tan t$. This can be done by multiplying out the numerator and making use of the fundamental identity $\sec^2 t = 1 + \tan^2 t$.

$$\frac{(\sec t - 1)(\sec t + 1)}{\tan t} = \frac{\sec^2 t - 1}{\tan t} = \frac{\tan^2 t}{\tan t} = \tan t \quad ■$$

Example D [Multiply One Side by $g(t)/g(t)$]   Prove that

$$\frac{\sin t}{1 - \cos t} = \frac{1 + \cos t}{\sin t}$$

is an identity.

Solution   Since both sides are equally complicated, it would seem to make no difference which side we choose to manipulate. We will try to transform the left side into the right side. Seeing $1 + \cos t$ in the numerator of the right side suggests multiplying the left side by $(1 + \cos t)/(1 + \cos t)$.

$$\frac{\sin t}{1 - \cos t} = \frac{\sin t}{1 - \cos t} \cdot \frac{1 + \cos t}{1 + \cos t} = \frac{\sin t (1 + \cos t)}{1 - \cos^2 t}$$

$$= \frac{\sin t (1 + \cos t)}{\sin^2 t} = \frac{1 + \cos t}{\sin t} \quad ■$$

Proving an identity is something like a game in that it requires a strategy. If one strategy does not work, try another, and still another, until you succeed.

Example E (Factor One Side)   Prove that

$$\csc^4 \theta - \cot^4 \theta = \frac{1 + \cos^2 \theta}{1 - \cos^2 \theta}$$

is an identity.

Solution   We are attracted to the left side because we notice that it can be factored. After factoring, we apply fundamental identities in a straightforward way.

$$\csc^4 \theta - \cot^4 \theta = (\csc^2 \theta - \cot^2 \theta)(\csc^2 \theta + \cot^2 \theta)$$

$$= (1 + \cot^2 \theta - \cot^2 \theta)(\csc^2 \theta + \cot^2 \theta)$$

$$= (1)\left(\frac{1}{\sin^2 \theta} + \frac{\cos^2 \theta}{\sin^2 \theta}\right) = \frac{1 + \cos^2 \theta}{\sin^2 \theta} = \frac{1 + \cos^2 \theta}{1 - \cos^2 \theta} \quad ■$$

Expressing All Trigonometric Functions in Terms of One

Clearly we can express all six trigonometric functions in terms of sines and cosines. Can we express all six in terms of just the cosine function? The answer is yes, except for an ambiguity of sign. In fact, if we know the quadrant for the domain angle, we can unambiguously express all six trigonometric functions in terms of any one of them. To do this, all we need are the first seven fundamental identities and a little algebra.

Example F (Expressing All Trigonometric Functions in Terms of Sine) If $\pi/2 < t < \pi$, express $\cos t$, $\tan t$, $\cot t$, $\sec t$, and $\csc t$ in terms of $\sin t$.

Solution Since $\cos^2 t = 1 - \sin^2 t$ and cosine is negative in quadrant II,

$$\cos t = -\sqrt{1 - \sin^2 t}$$

Also

$$\tan t = \frac{\sin t}{\cos t} = -\frac{\sin t}{\sqrt{1 - \sin^2 t}}$$

$$\cot t = \frac{1}{\tan t} = -\frac{\sqrt{1 - \sin^2 t}}{\sin t}$$

$$\sec t = \frac{1}{\cos t} = -\frac{1}{\sqrt{1 - \sin^2 t}}$$

$$\csc t = \frac{1}{\sin t} \quad \blacksquare$$

A Point of Logic

Why all the fuss about working with just one side of a conjectured identity? First of all, it offers good practice in manipulating trigonometric expressions. But there is also a point of logic. If you operate on both sides simultaneously, you are in effect assuming that you already have an identity. That is bad logic and it can be corrected only by carefully checking that each step is reversible. To make this point clear, consider the equation

$$1 - x = x - 1$$

which is certainly not an identity. Yet when we square both sides we get

$$1 - 2x + x^2 = x^2 - 2x + 1$$

which is an identity. The trouble here is that squaring both sides is not a reversible operation.

The situation contrasts sharply with our procedure for solving conditional equations, in which we often perform an operation on both sides. For example, in the case of the equation

$$\sqrt{2x + 1} = 1 - x$$

we even square both sides. We are protected from error here by checking our solutions in the original equation.

PROBLEM SET 8-1

A. Skills and Techniques

Make use of the fundamental identities in Problems 1–4.

1. Express entirely in terms of sin t.
   - (a) $\cos^2 t$
   - (b) $\tan t \cos t$
   - (c) $\dfrac{3}{\csc^2 t} + 2 \cos^2 t - 2$
   - (d) $\cot^2 t$

2. Express entirely in terms of cos t.
   - (a) $\sin^2 t$
   - (b) $\tan^2 t$
   - (c) $\csc^2 t$
   - (d) $(1 + \sin t)^2 - 2 \sin t$

3. Express entirely in terms of tan t.
   - (a) $\cot^2 t$
   - (b) $\sec^2 t$
   - (c) $\sin t \sec t$
   - (d) $2 \sec^2 t - 2 \tan^2 t + 1$

4. Express entirely in terms of sec t.
   - (a) $\cos^4 t$
   - (b) $\tan^2 t$
   - (c) $\tan t \csc t$
   - (d) $\tan^2 t - 2 \sec^2 t + 5$

Prove that each of the following is an identity. See Examples A–E.

5. $\cos t \sec t = 1$

6. $\sin t \csc t = 1$

7. $\tan x \cot x = 1$

8. $\sin x \sec x = \tan x$

9. $\cos y \csc y = \cot y$

10. $\tan y \cos y = \sin y$

11. $\cot \theta \sin \theta = \cos \theta$

12. $\dfrac{\sec \theta}{\csc \theta} = \tan \theta$

13. $\dfrac{\tan u}{\sin u} = \dfrac{1}{\cos u}$

14. $\dfrac{\sin u}{\csc u} + \dfrac{\cos u}{\sec u} = 1$

15. $(1 + \sin z)(1 - \sin z) = \dfrac{1}{\sec^2 z}$

16. $(\sec z - 1)(\sec z + 1) = \tan^2 z$

17. $(1 - \sin^2 x)(1 + \tan^2 x) = 1$

18. $(1 - \cos^2 x)(1 + \cot^2 x) = 1$

19. $\sec t - \sin t \tan t = \cos t$

20. $\sin t (\csc t - \sin t) = \cos^2 t$

21. $\dfrac{\sec^2 t - 1}{\sec^2 t} = \sin^2 t$

22. $\dfrac{1 - \csc^2 t}{\csc^2 t} = \dfrac{-1}{\sec^2 t}$

23. $\cos t (\tan t + \cot t) = \csc t$

24. $\dfrac{1}{\sin t \cos t} - \dfrac{\cos t}{\sin t} = \tan t$

25. $\dfrac{\sin^2 \theta}{\cos \theta} = \sec \theta - \cos \theta$

26. $\tan^2 \theta (1 + \cot^2 \theta) = \dfrac{1}{1 - \sin^2 \theta}$

27. $\dfrac{\tan \theta - \cot \theta}{\sin \theta \cos \theta} = \sec^2 \theta - \csc^2 \theta$

28. $\tan \theta (1 - \cot^2 \theta) = -\cot \theta (1 - \tan^2 \theta)$

29. $\dfrac{\sec t - 1}{\tan t} = \dfrac{\tan t}{\sec t + 1}$

30. $\dfrac{1 - \tan \theta}{1 + \tan \theta} = \dfrac{\cot \theta - 1}{\cot \theta + 1}$

31. $\dfrac{\tan^2 x}{\sec x + 1} = \dfrac{1 - \cos x}{\cos x}$

32. $\dfrac{\cot x}{\csc x + 1} = \dfrac{\csc x - 1}{\cot x}$

33. $\dfrac{\sin t + \cos t}{\tan^2 t - 1} = \dfrac{\cos^2 t}{\sin t - \cos t}$

34. $\dfrac{\sec t - \cos t}{1 + \cos t} = \sec t - 1$

35. $\cot \theta + \tan \theta = \sec \theta \csc \theta$

36. $\dfrac{1 + \tan \theta}{1 - \tan \theta} = \dfrac{\sec^2 \theta + 2 \tan \theta}{2 - \sec^2 \theta}$

Problems 37–40 relate to Example F.

37. If $\pi/2 < t < \pi$, express sin t, tan t, cot t, sec t, and csc t in terms of cos t.

38. If $\pi < t < 3\pi/2$, express sin t, cos t, cot t, sec t, and csc t in terms of tan t.

39. If $\pi/2 < t < \pi$ and $\sin t = \frac{4}{5}$, find the values of the other five functions for the same value of t. *Hint:* Use the results of Example F.

40. If $\pi < t < 3\pi/2$ and $\tan t = 2$, find sin t, cos t, cot t, sec t, and csc t.

B. Applications and Extensions

41. Express $[(\sin x + \cos x)^2 - 1]\sec x \csc^3 x$ as follows.
    - (a) Entirely in terms of sin x.
    - (b) Entirely in terms of tan x.

42. If sec $t = 8$, find the values of (a) cos t; (b) $\cot^2 t$; (c) $\csc^2 t$.

In Problems 43–64, prove that each equation is an identity. Do this by taking one side and showing by a chain of equalities that it is equal to the other side.

43. $(1 + \tan^2 t)(\cos t + \sin t) = (1 + \tan t)\sec t$

44. $1 - (\cos t + \sin t)(\cos t - \sin t) = 2 \sin^2 t$

45. $2 \sec^2 y - 1 = \dfrac{1 + \sin^2 y}{\cos^2 y}$

46. $(\sin x + \cos x)(\sec x + \csc x) = 2 + \tan x + \cot x$

47. $\dfrac{\cos z}{1 + \cos z} = \dfrac{\sin z}{\sin z + \tan z}$

48. $2 \sin^2 t + 3 \cos^2 t + \sec^2 t = (\sec t + \cos t)^2$

49. $(\csc t + \cot t)^2 = \dfrac{1 + \cos t}{1 - \cos t}$

50. $\sec^4 y - \tan^4 y = \dfrac{1 + \sin^2 y}{\cos^2 y}$

51. $\dfrac{\cos^4 u - \sin^4 u}{\cos u - \sin u} = \cos u + \sin u$

52. $\dfrac{\sec^3 y - \tan^3 y}{\sec y - \tan y} = \dfrac{1 + \sin y + \sin^2 y}{\cos^2 y}$

53. $\dfrac{\cos x + \sin x}{\cos x - \sin x} = \dfrac{1 + \tan x}{1 - \tan x}$

54. $\dfrac{1 + \cos x}{1 - \cos x} - \dfrac{1 - \cos x}{1 + \cos x} = 4 \cot x \csc x$

55. $(\sec t + \tan t)(\csc t - 1) = \cot t$

56. $\sec t + \cos t = \sin t \tan t + 2 \cos t$

57. $\dfrac{\cos^3 t + \sin^3 t}{\cos t + \sin t} = 1 - \sin t \cos t$

58. $\dfrac{\tan x}{1 + \tan x} + \dfrac{\cot x}{1 - \cot x} = \dfrac{\tan x + \cot x}{\tan x - \cot x}$

59. $\dfrac{1 - \cos \theta}{1 + \cos \theta} = \left(\dfrac{1 - \cos \theta}{\sin \theta}\right)^2$

60. $\dfrac{(\sec^2 \theta + \tan^2 \theta)^2}{\sec^4 \theta - \tan^4 \theta} = \sec^2 \theta + \tan^2 \theta$

61. $(\csc t - \cot t)^4(\csc t + \cot t)^4 = 1$

62. $(\sec t + \tan t)^5(\sec t - \tan t)^6 = \dfrac{1 - \sin t}{\cos t}$

63. $\sin^6 u + \cos^6 u = 1 - 3 \sin^2 u \cos^2 u$

64. $\dfrac{\cos^2 x - \cos^2 y}{\cot^2 x - \cot^2 y} = \sin^2 x \sin^2 y$

65. In a later section, we will learn that

$$\tan 3x = \frac{3 \tan x - \tan^3 x}{1 - 3 \tan^2 x}$$

Taking this for granted, show that

$$\cot 3x = \frac{3 \cot x - \cot^3 x}{1 - 3 \cot^2 x}$$

Note the similarity in form of these two identities.

66. TEASER Generalize Problem 65 by showing that if $\tan kx = f(\tan x)$ and if k is an odd number, then $\cot kx = f(\cot x)$. *Hint:* Let $x = \pi/2 - y$.

67. Draw the graph of

$$f(x) = \cos^3 x(1 - \tan^4 x + \sec^4 x)$$

From this, conjecture an identity. Prove this identity.

68. Find the (exact) minimum value of $f(x) = (\tan^4 x + 1)/\tan^2 x$. Conjecture an inequality. Prove this inequality.

8-2 ADDITION LAWS

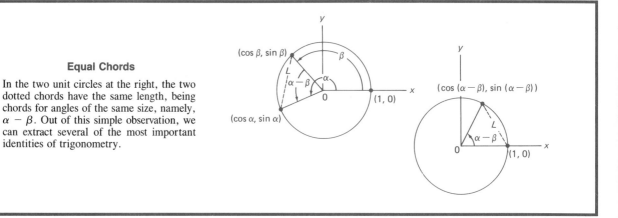

Equal Chords

In the two unit circles at the right, the two dotted chords have the same length, being chords for angles of the same size, namely, $\alpha - \beta$. Out of this simple observation, we can extract several of the most important identities of trigonometry.

When you study calculus, you will meet expressions like $\cos(\alpha + \beta)$ and $\sin(\alpha - \beta)$. It will be very important to rewrite these expressions directly in terms of $\sin \alpha$, $\cos \alpha$, $\sin \beta$, and $\cos \beta$. It might be tempting to replace $\cos(\alpha + \beta)$ by

$\cos \alpha + \cos \beta$ and $\sin(\alpha - \beta)$ by $\sin \alpha - \sin \beta$, but that would be terribly wrong. To see this, let's try $\alpha = \pi/3$ and $\beta = \pi/6$.

$$\cos\left(\frac{\pi}{3} + \frac{\pi}{6}\right) = \cos\frac{\pi}{2} = 0 \qquad \cos\frac{\pi}{3} + \cos\frac{\pi}{6} = \frac{1}{2} + \frac{\sqrt{3}}{2} \approx 1.4$$

$$\sin\left(\frac{\pi}{3} - \frac{\pi}{6}\right) = \sin\frac{\pi}{6} = .5 \qquad \sin\frac{\pi}{3} - \sin\frac{\pi}{6} = \frac{\sqrt{3}}{2} - \frac{1}{2} \approx .4$$

To obtain correct expressions is the goal of this section.

A Key Identity

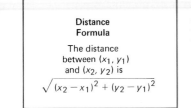

Distance Formula

The distance between (x_1, y_1) and (x_2, y_2) is

$$\sqrt{(x_2 - x_1)^2 + (y_2 - y_1)^2}$$

Figure 1

The opening display shows two chords of equal length L. Using the formula for the distance between two points (Figure 1) and the identity $\sin^2\theta + \cos^2\theta = 1$, we have the following expression for the square of the chord on the right.

$$
\begin{aligned}
L^2 &= [\cos(\alpha - \beta) - 1]^2 + \sin^2(\alpha - \beta) \\
&= \cos^2(\alpha - \beta) - 2\cos(\alpha - \beta) + 1 + \sin^2(\alpha - \beta) \\
&= [\cos^2(\alpha - \beta) + \sin^2(\alpha - \beta)] + 1 - 2\cos(\alpha - \beta) \\
&= 2 - 2\cos(\alpha - \beta)
\end{aligned}
$$

A similar calculation for the square of the chord on the left gives

$$
\begin{aligned}
L^2 &= (\cos\alpha - \cos\beta)^2 + (\sin\alpha - \sin\beta)^2 \\
&= \cos^2\alpha - 2\cos\alpha\cos\beta + \cos^2\beta + \sin^2\alpha - 2\sin\alpha\sin\beta + \sin^2\beta \\
&= 1 - 2\cos\alpha\cos\beta - 2\sin\alpha\sin\beta + 1 \\
&= 2 - 2(\cos\alpha\cos\beta + \sin\alpha\sin\beta)
\end{aligned}
$$

When we equate these two expressions for L^2, we get our key identity

$$\cos(\alpha - \beta) = \cos\alpha\cos\beta + \sin\alpha\sin\beta$$

Our derivation is based on a picture in which α and β are positive angles with $\alpha > \beta$. Minor modifications would establish the identity for arbitrary angles α and β and hence also for their radian measures s and t. Thus for all real numbers s and t,

$$\cos(s - t) = \cos s \cos t + \sin s \sin t$$

In words, this identity says: *The cosine of a difference is the cosine of the first times the cosine of the second plus the sine of the first times the sine of the second.* It is important to memorize this identity in words so you can easily apply it to $\cos(3u - v)$, $\cos[s - (-t)]$, or even $\cos[(\pi/2 - s) - t]$, as we shall have to do soon.

Related Identities

In the boxed identity above, we replace t by $-t$ and use the fundamental identities $\cos(-t) = \cos t$ and $\sin(-t) = -\sin t$ to get

$$\cos[s - (-t)] = \cos s \cos(-t) + \sin s \sin(-t)$$
$$= \cos s \cos t + (\sin s)(-\sin t)$$

This gives us the **addition law for cosines.**

$$\cos(s + t) = \cos s \cos t - \sin s \sin t$$

Example A (Applying the Boxed Identities) Use the boxed identities and knowledge of special angles to obtain the exact values of each of the following.

(a) $\cos \dfrac{\pi}{12}$ (b) $\cos 285°$

Solution

(a) Think of $\dfrac{\pi}{12}$ as $\dfrac{\pi}{3} - \dfrac{\pi}{4}$.

$$\cos \frac{\pi}{12} = \cos\left(\frac{\pi}{3} - \frac{\pi}{4}\right) = \cos \frac{\pi}{3} \cos \frac{\pi}{4} + \sin \frac{\pi}{3} \sin \frac{\pi}{4}$$

$$= \frac{1}{2} \cdot \frac{\sqrt{2}}{2} + \frac{\sqrt{3}}{2} \cdot \frac{\sqrt{2}}{2} = \frac{\sqrt{2} + \sqrt{6}}{4}$$

(b) Since $285° = 225° + 60°$,

$$\cos 285° = \cos 225° \cos 60° - \sin 225° \sin 60°$$

$$= \frac{-\sqrt{2}}{2} \cdot \frac{1}{2} - \frac{-\sqrt{2}}{2} \cdot \frac{\sqrt{3}}{2} = \frac{-\sqrt{2} + \sqrt{6}}{4} \qquad \blacksquare$$

There is also an identity involving $\sin(s + t)$. To derive this identity, we use the cofunction identity $\sin u = \cos(\pi/2 - u)$ to write

$$\sin(s + t) = \cos\left[\frac{\pi}{2} - (s + t)\right] = \cos\left[\left(\frac{\pi}{2} - s\right) - t\right]$$

Then we use our key identity for the cosine of a difference to obtain

$$\cos\left(\frac{\pi}{2} - s\right) \cos t + \sin\left(\frac{\pi}{2} - s\right) \sin t$$

Two applications of cofunction identities give us the result we want, the **addition law for sines.**

$$\sin(s + t) = \sin s \cos t + \cos s \sin t$$

CAUTION

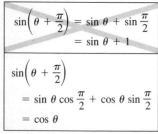

$$\sin\left(\theta + \frac{\pi}{2}\right) = \sin \theta + \sin \frac{\pi}{2}$$
$$= \sin \theta + 1$$

$$\sin\left(\theta + \frac{\pi}{2}\right)$$
$$= \sin \theta \cos \frac{\pi}{2} + \cos \theta \sin \frac{\pi}{2}$$
$$= \cos \theta$$

Finally, replacing t by $-t$ in this last result leads to

$$\sin(s - t) = \sin s \cos t - \cos s \sin t$$

If we let $s = \pi/2$ in the latter, we get another important identity—but it is one we already know.

$$\sin\left(\frac{\pi}{2} - t\right) = \sin\frac{\pi}{2}\cos t - \cos\frac{\pi}{2}\sin t = 1 \cdot \cos t - 0 \cdot \sin t = \cos t$$

Example B (More Identities)   Demonstrate that each of the following equations is an identity.

(a) $\sin\left(t + \dfrac{\pi}{2}\right) = \cos t$

(b) $\sin(\theta - 180°) = -\sin\theta$

(c) $\cos\left(t - \dfrac{3\pi}{2}\right) = -\sin t$

Solution

(a) $\sin\left(t + \dfrac{\pi}{2}\right) = \sin t \cos\dfrac{\pi}{2} + \cos t \sin\dfrac{\pi}{2}$

$\qquad\qquad\quad = (\sin t)(0) + (\cos t)(1) = \cos t$

(b) $\sin(\theta - 180°) = \sin\theta\cos 180° - \cos\theta\sin 180°$

$\qquad\qquad\quad = (\sin\theta)(-1) - (\cos\theta)(0) = -\sin\theta$

(c) $\cos\left(t - \dfrac{3\pi}{2}\right) = \cos t \cos\dfrac{3\pi}{2} + \sin t \sin\dfrac{3\pi}{2}$

$\qquad\qquad\quad = (\cos t)(0) + (\sin t)(-1) = -\sin t$ ∎

Example C (Recognizing Expressions as Single Sines or Cosines)   Write as a single sine or cosine.

(a) $\sin\frac{7}{6}\cos\frac{1}{6} + \cos\frac{7}{6}\sin\frac{1}{6}$

(b) $\cos(x + h)\cos h + \sin(x + h)\sin h$

Solution

(a) By the addition law for sines,

$$\sin\tfrac{7}{6}\cos\tfrac{1}{6} + \cos\tfrac{7}{6}\sin\tfrac{1}{6} = \sin(\tfrac{7}{6} + \tfrac{1}{6}) = \sin\tfrac{4}{3}$$

(b) This we recognize as the cosine of a difference

$$\cos(x + h)\cos h + \sin(x + h)\sin h = \cos[(x + h) - h] = \cos x \qquad ∎$$

Example D (Using the Addition Laws)   Suppose that α is a first quadrant angle with $\cos\alpha = \frac{4}{5}$ and β is a second quadrant angle with $\sin\beta = \frac{12}{13}$. Evaluate $\sin(\alpha + \beta)$ and $\cos(\alpha + \beta)$ and then determine the quadrant for $\alpha + \beta$.

Solution We are going to need $\sin \alpha$ and $\cos \beta$. We can find them by using the identity $\sin^2 \theta + \cos^2 \theta = 1$, but we have to be careful about signs.

$$\sin \alpha = \sqrt{1 - \cos^2 \alpha} = \sqrt{1 - \tfrac{16}{25}} = \tfrac{3}{5}$$

$$\cos \beta = -\sqrt{1 - \sin^2 \beta} = -\sqrt{1 - \tfrac{144}{169}} = -\tfrac{5}{13}$$

We chose the plus sign in the first case because α is a first quadrant angle and the minus sign in the second because β is a second quadrant angle, where the cosine is negative. Then

$$\sin(\alpha + \beta) = \sin \alpha \cos \beta + \cos \alpha \sin \beta$$

$$= \left(\frac{3}{5}\right)\left(\frac{-5}{13}\right) + \left(\frac{4}{5}\right)\left(\frac{12}{13}\right) = \frac{33}{65}$$

$$\cos(\alpha + \beta) = \cos \alpha \cos \beta - \sin \alpha \sin \beta$$

$$= \left(\frac{4}{5}\right)\left(\frac{-5}{13}\right) - \left(\frac{3}{5}\right)\left(\frac{12}{13}\right) = \frac{-56}{65}$$

Since $\sin(\alpha + \beta)$ is positive and $\cos(\alpha + \beta)$ is negative, $\alpha + \beta$ is a second quadrant angle. ∎

The Addition Law for Tangent

The addition laws for the sine and cosine lead naturally to an **addition law for the tangent.**

$$\tan(s + t) = \frac{\tan s + \tan t}{1 - \tan s \tan t}$$

To see that this is an identity, proceed as follows.

$$\tan(s + t) = \frac{\sin(s + t)}{\cos(s + t)} = \frac{\sin s \cos t + \cos s \sin t}{\cos s \cos t - \sin s \sin t}$$

$$= \frac{\dfrac{\sin s \cos t}{\cos s \cos t} + \dfrac{\cos s \sin t}{\cos s \cos t}}{\dfrac{\cos s \cos t}{\cos s \cos t} - \dfrac{\sin s \sin t}{\cos s \cos t}} = \frac{\tan s + \tan t}{1 - \tan s \tan t}$$

The key step was the third one, in which we divided both the numerator and the denominator by $\cos s \cos t$.

Example E (Applying the Tangent Law) Express $\tan(\theta + 60°)$ in terms of $\tan \theta$.

Solution

$$\tan(\theta + 60°) = \frac{\tan \theta + \tan 60°}{1 - \tan \theta \tan 60°} = \frac{\tan \theta + \sqrt{3}}{1 - \sqrt{3} \tan \theta} \qquad ∎$$

PROBLEM SET 8-2

A. Skills and Techniques

Find the value of each expression. Note that in each case, the answers to parts (a) and (b) are different.

1. (a) $\sin\dfrac{\pi}{4} + \sin\dfrac{\pi}{6}$ (b) $\sin\left(\dfrac{\pi}{4} + \dfrac{\pi}{6}\right)$

2. (a) $\cos\dfrac{\pi}{4} + \cos\dfrac{\pi}{6}$ (b) $\cos\left(\dfrac{\pi}{4} + \dfrac{\pi}{6}\right)$

3. (a) $\cos\dfrac{\pi}{4} - \cos\dfrac{\pi}{6}$ (b) $\cos\left(\dfrac{\pi}{4} - \dfrac{\pi}{6}\right)$

4. (a) $\sin\dfrac{\pi}{4} - \sin\dfrac{\pi}{6}$ (b) $\sin\left(\dfrac{\pi}{4} - \dfrac{\pi}{6}\right)$

Use the boxed identities of this section to calculate each value exactly, as in Example A.

5. $\cos\dfrac{13\pi}{12}$ 6. $\sin\dfrac{7\pi}{12}$

7. $\sin 165°$ 8. $\cos 345°$

9. $\tan 75°$ 10. $\sin 105°$

Show as in Example B that each of the following equations is an identity.

11. $\sin(t + \pi) = -\sin t$

12. $\cos(t + \pi) = -\cos t$

13. $\sin\left(t + \dfrac{3\pi}{2}\right) = -\cos t$ 14. $\cos\left(t + \dfrac{3\pi}{2}\right) = \sin t$

15. $\sin\left(t - \dfrac{\pi}{2}\right) = -\cos t$ 16. $\cos\left(t - \dfrac{\pi}{2}\right) = \sin t$

17. $\cos\left(t + \dfrac{\pi}{3}\right) = \dfrac{1}{2}\cos t - \dfrac{\sqrt{3}}{2}\sin t$

18. $\sin\left(t + \dfrac{\pi}{3}\right) = \dfrac{1}{2}\sin t + \dfrac{\sqrt{3}}{2}\cos t$

Write each of the following as a single sine or cosine. See Example C.

19. $\cos\frac{1}{2}\cos\frac{3}{2} - \sin\frac{1}{2}\sin\frac{3}{2}$

20. $\cos 2 \cos 3 + \sin 2 \sin 3$

21. $\sin\dfrac{7\pi}{8}\cos\dfrac{\pi}{8} + \cos\dfrac{7\pi}{8}\sin\dfrac{\pi}{8}$

22. $\sin\dfrac{5\pi}{16}\cos\dfrac{\pi}{16} - \cos\dfrac{5\pi}{16}\sin\dfrac{\pi}{16}$

23. $\cos 33° \cos 27° - \sin 33° \sin 27°$

24. $\sin 49° \cos 41° + \cos 49° \sin 41°$

25. $\sin(\alpha + \beta)\cos\beta - \cos(\alpha + \beta)\sin\beta$

26. $\cos(\alpha + \beta)\cos(\alpha - \beta) - \sin(\alpha + \beta)\sin(\alpha - \beta)$

Reason as in Example D to work Problems 27–30.

27. If α and β are third quadrant angles with $\sin \alpha = -\frac{4}{5}$ and $\cos \beta = -\frac{5}{13}$, find $\sin(\alpha + \beta)$ and $\cos(\alpha + \beta)$. In what quadrant does the terminal side of $\alpha + \beta$ lie?

28. Let α and β be second quadrant angles with $\sin \alpha = \frac{2}{3}$ and $\sin \beta = \frac{3}{4}$. Find $\sin(\alpha + \beta)$ and $\cos(\alpha + \beta)$ and determine the quadrant for $\alpha + \beta$.

29. Let α be a first quadrant angle with $\sin \alpha = 1/\sqrt{10}$ and β be a second quadrant angle with $\cos \beta = -\frac{1}{2}$. Find $\sin(\alpha - \beta)$ and $\cos(\alpha - \beta)$ and determine the quadrant for $\alpha - \beta$.

30. Let α and β be second and third quadrant angles, respectively, with $\cos \alpha = \cos \beta = -\frac{3}{7}$. Find $\sin(\alpha - \beta)$ and $\cos(\alpha - \beta)$ and determine the quadrant for $\alpha - \beta$.

Prove the following tangent identities. See Example E.

31. $\tan(s - t) = \dfrac{\tan s - \tan t}{1 + \tan s \tan t}$

32. $\tan(s + \pi) = \tan s$

33. $\tan\left(t + \dfrac{\pi}{4}\right) = \dfrac{1 + \tan t}{1 - \tan t}$

34. $\tan\left(t - \dfrac{\pi}{3}\right) = \dfrac{\tan t - \sqrt{3}}{1 + \sqrt{3}\tan t}$

B. Applications and Extensions

35. Express in terms of $\sin t$ and $\cos t$.
    (a) $\sin(t - \frac{5}{6}\pi)$ (b) $\cos(\frac{\pi}{6} - t)$

36. Express $\tan(\theta + \frac{3}{4}\pi)$ in terms of $\tan \theta$.

37. Let α and β be first and third quadrant angles, respectively, with $\sin \alpha = \frac{2}{3}$ and $\cos \beta = -\frac{1}{3}$. Evaluate each of the following exactly.
    (a) $\cos \alpha$ (b) $\sin \beta$
    (c) $\cos(\alpha + \beta)$ (d) $\sin(\alpha - \beta)$
    (e) $\tan(\alpha + \beta)$ (f) $\sin(2\beta)$

38. If $0 \le t \le \pi/2$ and $\cos(t + \pi/6) = .8$, find the exact value of $\sin t$ and $\cos t$. *Hint:* $t = (t + \pi/6) - \pi/6$.

39. Evaluate each of the following (the easy way).
    (a) $\sin(t + \pi/3)\cos t - \cos(t + \pi/3)\sin t$
    (b) $\cos 175° \cos 25° + \sin 175° \sin 25°$
    (c) $\sin t \cos(1 - t) + \cos t \sin(1 - t)$

40. Find the exact value of $\cos 85° \cos 40° + \cos 5° \cos 50°$.

41. Show that each of the following is an identity.

(a) $\sin(x + y) \sin(x - y) = \sin^2 x - \sin^2 y$

(b) $\dfrac{\sin(x + y)}{\cos(x - y)} = \dfrac{\tan x + \tan y}{1 + \tan x \tan y}$

(c) $\dfrac{\cos 5t}{\sin t} - \dfrac{\sin 5t}{\cos t} = \dfrac{\cos 6t}{\sin t \cos t}$

42. Show that the following are identities.

(a) $\cot(u + v) = \dfrac{\cot u \cot v - 1}{\cot u + \cot v}$

(b) $\dfrac{\sin(u + v)}{\sin(u - v)} = \dfrac{\tan u + \tan v}{\tan u - \tan v}$

(c) $\dfrac{\cos 2t}{\sin t} + \dfrac{\sin 2t}{\cos t} = \csc t$

43. Let θ be the smallest counterclockwise angle from the line $y = m_1 x + b_1$ to the line $y = m_2 x + b_2$, where $m_1 m_2 \neq -1$. Show that

$$\tan \theta = \frac{m_2 - m_1}{1 + m_1 m_2}$$

44. Find the counterclockwise angle θ from the line $3x - 4y = 1$ to the line $2x + 6y = 3$ (see Problem 43).

45. In Figure 2, $\overline{BC} = 2\,\overline{AB} = 2\,\overline{CD}$. Determine $\alpha + \beta$ exactly.

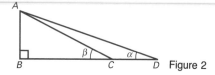

Figure 2

46. Given that $\sin \alpha = \frac{4}{5}$, $\sin \beta = \frac{5}{13}$, and $\sin \gamma = \frac{7}{25}$, where α, β, and γ are first quadrant angles, determine $\sin(\alpha + \beta + \gamma)$ exactly. Without a calculator or tables, estimate $\alpha + \beta + \gamma$.

47. Use addition and subtraction laws (the first four boxed formulas of this section) to prove the following **product identities.**

(a) $\cos s \cos t = \frac{1}{2}[\cos(s + t) + \cos(s - t)]$

(b) $\sin s \sin t = -\frac{1}{2}[\cos(s + t) - \cos(s - t)]$

(c) $\sin s \cos t = \frac{1}{2}[\sin(s + t) + \sin(s - t)]$

(d) $\cos s \sin t = \frac{1}{2}[\sin(s + t) - \sin(s - t)]$

48. Use the identities of Problem 47 to prove the following **factoring identities.** *Hint:* Let $u = s + t$ and $v = s - t$.

(a) $\cos u + \cos v = 2 \cos \dfrac{u + v}{2} \cos \dfrac{u - v}{2}$

(b) $\cos u - \cos v = -2 \sin \dfrac{u + v}{2} \sin \dfrac{u - v}{2}$

(c) $\sin u + \sin v = 2 \sin \dfrac{u + v}{2} \cos \dfrac{u - v}{2}$

(d) $\sin u - \sin v = 2 \cos \dfrac{u + v}{2} \sin \dfrac{u - v}{2}$

49. Evaluate each of the following exactly.

(a) $\cos 105° \cos 45°$ (b) $\sin 15° - \sin 75°$

(c) $\cos 15° + \cos 30° + \cos 45° + \cos 60° + \cos 75°$

50. Show that each of the following is an identity.

(a) $\dfrac{\cos 9t + \cos 3t}{\sin 9t - \sin 3t} = \cot 3t$

(b) $\dfrac{\sin 3u + \sin 7u}{\cos 3u + \cos 7u} = \tan 5u$

(c) $\cos 10\beta + \cos 2\beta + 2 \cos 8\beta \cos 6\beta = 4 \cos^2 6\beta \cos 2\beta$

51. Stack three identical squares and consider angles α, β, and γ as shown in Figure 3. Prove that $\alpha + \beta = \gamma$.

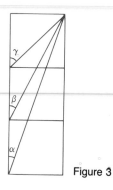

Figure 3

52. **TEASER** Consider an oblique triangle (no right angles) with angles α, β, and γ. Prove that

$$\tan \alpha + \tan \beta + \tan \gamma = \tan \alpha \tan \beta \tan \gamma$$

53. We claim that

$$\sin t + \cos t = A \sin(t - c)$$

is an identity for appropriate choices of A and c. Use the graph of $y = \sin t + \cos t$ to conjecture exact values of A and c. Prove your conjecture.

54. Use the graph of

$$y = \sin^2 t + \sin^2\left(t + \frac{2\pi}{3}\right) + \sin^2\left(t - \frac{2\pi}{3}\right)$$

to make a conjecture. Prove your conjecture.

8-3 DOUBLE-ANGLE AND HALF-ANGLE FORMULAS

I could do this problem if sin (2t) = 2 sin t

Wishful Thinking

"Wishful thinking is imagining good things you don't have . . . . [It] may be bad as too much salt is bad in the soup and even a little garlic is bad in the chocolate pudding. I mean, wishful thinking may be bad if there is too much of it or in the wrong place, but it is good in itself and may be a great help in life and in problem solving."

George Polya
in *Mathematical Discovery*

George Polya would agree that the student in our opening panel is wishing for too much. And there is a better way than wishing to get formulas for sin $2t$ and cos $2t$. All we have to do is to think of $2t$ as $t + t$ and apply the addition laws of the previous section.

$$\sin(t + t) = \sin t \cos t + \cos t \sin t = 2 \sin t \cos t$$

$$\cos(t + t) = \cos t \cos t - \sin t \sin t = \cos^2 t - \sin^2 t$$

Double-Angle Formulas

We have just derived two very important results. They are called *double-angle formulas,* though double-number formulas would perhaps be more appropriate.

$$\sin 2t = 2 \sin t \cos t$$
$$\cos 2t = \cos^2 t - \sin^2 t$$

There are two other forms of the cosine double-angle formula that are often useful. If, in the expression $\cos^2 t - \sin^2 t$, we replace $\cos^2 t$ by $1 - \sin^2 t$, we obtain

$$\cos 2t = 1 - 2 \sin^2 t$$

CAUTION

$\cos 6\theta = 6 \cos \theta$
$\cos 6\theta = 2 \cos^2 3\theta - 1$

and, alternatively, if we replace $\sin^2 t$ by $1 - \cos^2 t$, we have

$$\cos 2t = 2 \cos^2 t - 1$$

Of course, in all that we have done, we may replace the number t by the angle θ; hence the name double-angle formulas.

Example A (Applying Double-Angle Formulas)   Suppose that $\sin t = \frac{2}{5}$ and $\pi/2 < t < \pi$. Calculate $\sin 2t$ and $\cos 2t$ exactly.

Solution   We will need to know $\cos t$. Since $\pi/2 < t < \pi$, the cosine is negative, and therefore

$$\cos t = -\sqrt{1 - \sin^2 t} = -\sqrt{1 - \frac{4}{25}} = -\frac{\sqrt{21}}{5}$$

Applying double-angle formulas, we obtain

$$\sin 2t = 2 \sin t \cos t = 2\left(\frac{2}{5}\right)\left(\frac{-\sqrt{21}}{5}\right) = \frac{-4\sqrt{21}}{25}$$

$$\cos 2t = 1 - 2 \sin^2 t = 1 - 2\left(\frac{2}{5}\right)^2 = \frac{17}{25} \quad \blacksquare$$

Example B (Recognizing Expressions)   Simplify.
(a) $2 \sin 3\theta \cos 3\theta$ (b) $\sin^2 2u - \cos^2 2u$ (c) $2 \cos^2(\frac{3}{2}t) - 1$

Solution
(a) $2 \sin 3\theta \cos 3\theta = \sin(2 \cdot 3\theta) = \sin 6\theta$
(b) $\sin^2 2u - \cos^2 2u = -(\cos^2 2u - \sin^2 2u) = -\cos(2 \cdot 2u) = -\cos 4u$
(c) $2 \cos^2(\frac{3}{2}t) - 1 = \cos(2 \cdot \frac{3}{2}t) = \cos 3t$ $\blacksquare$

Once we grasp the generality of the four boxed formulas, we can write numerous others that follow from them. In particular,

$$\cos t = 1 - 2 \sin^2\left(\frac{t}{2}\right)$$

$$\cos t = 2 \cos^2\left(\frac{t}{2}\right) - 1$$

The last two of these identities lead us directly to the half-angle formulas.

Half-Angle Formulas

In the identity $\cos t = 1 - 2 \sin^2(t/2)$, we solve for $\sin(t/2)$.

$$2 \sin^2\left(\frac{t}{2}\right) = 1 - \cos t$$

$$\sin^2\left(\frac{t}{2}\right) = \frac{1 - \cos t}{2}$$

$$\sin\left(\frac{t}{2}\right) = \pm\sqrt{\frac{1 - \cos t}{2}}$$

Similarly, if we solve $\cos t = 2\cos^2(t/2) - 1$ for $\cos(t/2)$, the result is

$$\cos\left(\frac{t}{2}\right) = \pm\sqrt{\frac{1 + \cos t}{2}}$$

In both of these formulas, the positive or negative sign is determined by the interval in which $t/2$ lies.

Example C (Applying Half-Angle Formulas)   If $\cos\theta = 0.4$ and $270° < \theta < 360°$, find $\sin(\theta/2)$ and $\cos(\theta/2)$.

Solution   Clearly, $\theta/2$ is a second quadrant angle where sine is positive and cosine is negative. Thus

$$\sin\left(\frac{\theta}{2}\right) = \sqrt{\frac{1 - \cos\theta}{2}} = \sqrt{\frac{1 - .4}{2}} \approx .548$$

$$\cos\left(\frac{\theta}{2}\right) = -\sqrt{\frac{1 + \cos\theta}{2}} = -\sqrt{\frac{1 + .4}{2}} \approx -.837 \quad \blacksquare$$

New Identities from Old Ones

Every new identity that we master increases our ability to manipulate trigonometric expressions and prove new identities.

Example D (Using Double-Angle and Half-Angle Formulas to Prove New Identities)   Prove that the following are identities.

(a) $\sin 3t = 3\sin t - 4\sin^3 t$

(b) $\tan\dfrac{t}{2} = \dfrac{\sin t}{1 + \cos t}$

Solution

(a) We think of $3t$ as $2t + t$ and use the addition law for sines and then double-angle formulas.

$$\begin{aligned}
\sin 3t &= \sin(2t + t) \\
&= \sin 2t \cos t + \cos 2t \sin t \\
&= (2\sin t \cos t)\cos t + (1 - 2\sin^2 t)\sin t \\
&= 2\sin t(1 - \sin^2 t) + \sin t - 2\sin^3 t \\
&= 2\sin t - 2\sin^3 t + \sin t - 2\sin^3 t \\
&= 3\sin t - 4\sin^3 t
\end{aligned}$$

(b) This is the unambiguous form for $\tan(t/2)$ (see Problem 35). To prove it, think of t as $2(t/2)$ and apply double-angle formulas to the right side.

$$\frac{\sin t}{1 + \cos t} = \frac{\sin(2(t/2))}{1 + \cos(2(t/2))} = \frac{2\sin(t/2)\cos(t/2)}{1 + 2\cos^2(t/2) - 1} = \frac{\sin(t/2)}{\cos(t/2)} = \tan\frac{t}{2} \quad \blacksquare$$

PROBLEM SET 8-3

A. Skills and Techniques

For Problems 1 and 2, see Example A.

1. If $\cos t = \frac{4}{5}$ with $0 < t < \pi/2$, show that $\sin t = \frac{3}{5}$. Then use formulas from this section to calculate the following.
    (a) $\sin 2t$ **(b)** $\cos 2t$
    (c) $\cos(t/2)$ **(d)** $\sin(t/2)$

2. If $\sin t = -\frac{2}{3}$ with $3\pi/2 < t < 2\pi$, show that $\cos t = \sqrt{5}/3$. Then calculate.
    (a) $\sin 2t$ **(b)** $\cos 2t$
    (c) $\cos(t/2)$ **(d)** $\sin(t/2)$

Use formulas from this section to simplify the expressions in Problems 3–16. See Example B.

3. $2 \sin 5t \cos 5t$ **4.** $2 \sin 3\theta \cos 3\theta$
5. $\cos^2(3t/2) - \sin^2(3t/2)$
6. $\cos^2(7\pi/8) - \sin^2(7\pi/8)$
7. $2 \cos^2(y/4) - 1$ **8.** $2 \cos^2(\alpha/3) - 1$
9. $1 - 2 \sin^2(.6t)$ **10.** $2 \sin^2(\pi/8) - 1$
11. $\sin^2(\pi/8) - \cos^2(\pi/8)$ **12.** $2 \sin(.3) \cos(.3)$
13. $\dfrac{1 + \cos x}{2}$ **14.** $\dfrac{1 - \cos y}{2}$
15. $\dfrac{1 - \cos 4\theta}{2}$ **16.** $\dfrac{1 + \cos 8u}{2}$

Problems 17–22 relate to Example C.

17. Use the half-angle formulas to calculate.
    (a) $\sin(\pi/8)$ **(b)** $\cos(112.5°)$
18. Calculate, using half-angle formulas.
    (a) $\cos 67.5°$ **(b)** $\sin(\pi/12)$
19. If $\cos \theta = -.3$ and $180° < \theta < 270°$, find $\cos(\theta/2)$.
20. If $\cos \theta = -.9$ and $90° < \theta < 180°$, find $\sin(\theta/2)$.
21. If $\sin u = -12/13$ and $3\pi/2 < u < 2\pi$, find $\cos(u/2)$.
22. If $\tan u = \sqrt{5}/2$ and $0 < u < \pi/2$, find $\sin(u/2)$.

Prove that each equation in Problems 23–32 is an identity, as in Example D.

23. $\cos 3t = 4 \cos^3 t - 3 \cos t$
24. $(\sin t + \cos t)^2 = 1 + \sin 2t$
25. $\csc 2t + \cot 2t = \cot t$
26. $\sin^2 t \cos^2 t = \frac{1}{8}(1 - \cos 4t)$
27. $\dfrac{\sin \theta}{1 - \cos \theta} = \cot \dfrac{\theta}{2}$

28. $1 - 2 \sin^2 \theta = 2 \cot 2\theta \sin \theta \cos \theta$
29. $\dfrac{2 \tan \alpha}{1 + \tan^2 \alpha} = \sin 2\alpha$
30. $\dfrac{1 - \tan^2 \alpha}{1 + \tan^2 \alpha} = \cos 2\alpha$
31. $\sin 4\theta = 4 \sin \theta (2 \cos^3 \theta - \cos \theta)$
    Hint: $4\theta = 2(2\theta)$.
32. $\cos 4\theta = 8 \cos^4 \theta - 8 \cos^2 \theta + 1$
33. Use the addition law for tangents (Section 8-2) to show that
$$\tan 2t = \frac{2 \tan t}{1 - \tan^2 t}$$
34. Use the identity of Problem 33 to evaluate $\tan 2t$ given that
    (a) $\tan t = 3$.
    (b) $\cos t = \frac{4}{5}$ and $0 < t < \pi/2$.
35. Use the half-angle formulas for sine and cosine to show that
$$\tan\left(\frac{t}{2}\right) = \pm \sqrt{\frac{1 - \cos t}{1 + \cos t}}$$
36. Use the identity of Problem 35 to evaluate.
    (a) $\tan(\pi/8)$ **(b)** $\tan 112.5°$

B. Applications and Extensions

37. Write a simple expression for each of the following.
    (a) $2 \sin(x/2) \cos(x/2)$ **(b)** $\cos^2 3t - \sin^2 3t$
    (c) $2 \sin^2(y/4) - 1$ **(d)** $(\cos 4t - 1)/2$
    (e) $(1 - \cos 4t)/(1 + \cos 4t)$
    (f) $(\sin 6y)/(1 + \cos 6y)$
38. Find the exact value of each of the following.
    (a) $\sin 15° \cos 15°$
    (b) $\cos^2 105° - \sin^2 105°$
    (c) $\sin 15°$ **(d)** $\cos 105°$
39. If $\pi < t < 3\pi/2$ and $\cos t = -5/13$, find each value.
    (a) $\sin 2t$ **(b)** $\cos(t/2)$
    (c) $\tan(t/2)$
40. If the trigonometric point $P(t)$ on the unit circle has coordinates $(-\frac{3}{5}, \frac{4}{5})$, find the coordinates for each of the following points.
    (a) $P(2t)$ **(b)** $P(t/2)$
    (c) $P(4t)$

In Problems 41–56, prove that each equation is an identity.

41. $\cos^4 z - \sin^4 z = \cos 2z$

42. $(1 - \cos 4x)/\tan^2 2x = 2 \cos^2 2x$

43. $1 + (1 - \cos 8t)/(1 + \cos 8t) = \sec^2 4t$

44. $\sec 2t = (\sec^2 t)/(2 - \sec^2 t)$

45. $\tan(\theta/2) - \sin \theta = (-\sin \theta)/(1 + \sec \theta)$

46. $(2 - \sec^2 2\theta) \tan 4\theta = 2 \tan 2\theta$

47. $3 \cos 2t + 4 \sin 2t =$
$$(3 \cos t - \sin t)(\cos t + 3 \sin t)$$

48. $\csc 2x - \cot 2x = \tan x$

49. $2(\cos 3x \cos x + \sin 3x \sin x)^2 = 1 + \cos 4x$

50. $\dfrac{1 + \sin 2x + \cos 2x}{1 + \sin 2x - \cos 2x} = \cot x$

51. $\tan 3t = \dfrac{3 \tan t - \tan^3 t}{1 - 3 \tan^2 t}$

52. $\cos^4 u = \frac{3}{8} + \frac{1}{2} \cos 2u + \frac{1}{8} \cos 4u$

53. $\sin^4 u + \cos^4 u = \frac{3}{4} + \frac{1}{4} \cos 4u$

54. $\cos^6 u - \sin^6 u = \cos 2u - \frac{1}{4} \sin^2 2u \cos 2u$

55. Prove that $\cos^2 x + \cos^2 2x + \cos^2 3x = 1 + 2 \cos x$ $\cos 2x \cos 3x$ is an identity. *Hint:* Use half-angle formulas and factoring identities.

56. Calculate $(\sin 2t)[3 + (16 \sin^2 t - 16) \sin^2 t] - \sin 6t$ for $t = 1, 2,$ and 3. Guess at an identity and then prove it.

57. If $\alpha, \beta,$ and γ are the three angles of a triangle, prove that
$$\sin 2\alpha + \sin 2\beta + \sin 2\gamma = 4 \sin \alpha \sin \beta \sin \gamma$$

58. Show that $\cos x \cos 2x \cos 4x \cos 8x \cos 16x = (\sin 32x)/(32 \sin x)$

59. Figure 4 shows two abutting circles of radius 1, one centered at the origin, the other at $(2, 0)$. Find the exact coordinates of P, the point where the line through $(-1, 0)$ and tangent to the second circle meets the first circle.

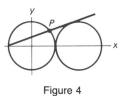

Figure 4

60. TEASER Determine the exact value of
$$\sin 1° \sin 3° \sin 5° \sin 7° \cdots \sin 175° \sin 177° \sin 179°$$

61. Draw the graph of
$$y = 2 \cos x - 4 \sin x \sin 2x$$
and determine the amplitude and period. Conjecture an identity. Prove it.

62. Draw the graph of
$$y = 32 \cos^6 x - 48 \cos^4 x + 18 \cos^2 x - 1$$
and determine its amplitude and period. Conjecture an identity. Prove it.

8-4 INVERSE TRIGONOMETRIC FUNCTIONS

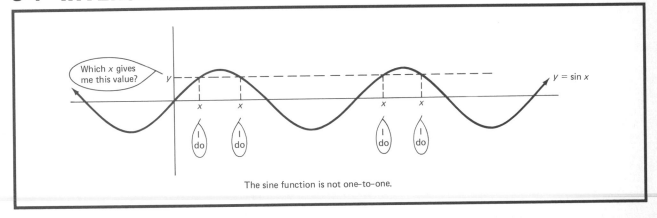

The sine function is not one-to-one.

The diagram above shows why the ordinary sine function does not have an inverse. We learned in Section 5-6 (a section worth reviewing now) that only one-to-one functions have inverses. To be one-to-one means that for each y there is at most one

x that corresponds to it. The sine function is about as far from being one-to-one as possible. For each y between -1 and 1, there are infinitely many x's giving that y-value. To make the sine function have an inverse, we will have to restrict its domain drastically.

The Inverse Sine

Consider the graph of the sine function again (Figure 5). We want to restrict its domain in such a way that the sine assumes its full range of values but takes on each value only once. There are many possible choices, but the one commonly used is $-\pi/2 \le x \le \pi/2$. Notice the corresponding part of the sine graph below. From now on, whenever we need an inverse sine function, we always assume that the domain of the sine has been restricted to $-\pi/2 \le x \le \pi/2$.

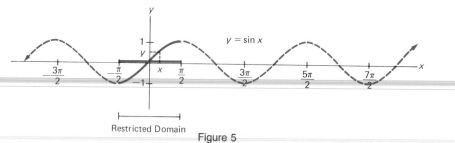

Figure 5

Having done this, we see that each y corresponds to exactly one x. We write $x = \sin^{-1} y$ (x is the inverse sine of y). Thus

$$\sin^{-1}\left(\frac{1}{2}\right) = \frac{\pi}{6} \qquad \sin^{-1}(-1) = -\frac{\pi}{2}$$

Note that $\sin^{-1} y$ does not mean $1/(\sin y)$; you should not think of -1 as an exponent when used as a superscript on a function.

An alternative notation for $x = \sin^{-1} y$ is $x = \arcsin y$ (x is the arcsine of y). This is appropriate notation, since $\pi/6 = \arcsin \frac{1}{2}$ could be interpreted as saying that $\pi/6$ is the arc (on the unit circle) whose sine is $\frac{1}{2}$.

Recall from Section 5-6 that if f is a one-to-one function, then

$$x = f^{-1}(y) \quad \text{if and only if} \quad y = f(x)$$

Here the corresponding statement is

$$x = \sin^{-1} y \quad \text{if and only if} \quad y = \sin x \quad \text{and} \quad -\frac{\pi}{2} \le x \le \frac{\pi}{2}$$

Moreover

$$\sin(\sin^{-1} y) = y \quad \text{for} \quad -1 \le y \le 1$$

$$\sin^{-1}(\sin x) = x \quad \text{for} \quad -\frac{\pi}{2} \le x \le \frac{\pi}{2}$$

The inverse sine function plays a significant role in calculus, where we often want to consider $y = \sin^{-1} x$. You will note that we have interchanged the roles of x and y so that x is now the domain variable for $\sin^{-1}$. On the graph, this corresponds to reflecting (folding) the graph of $y = \sin x$ across the line $y = x$ (Figure 6).

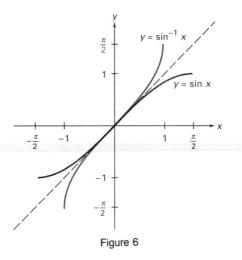

Figure 6

The Inverse Cosine

One look at the graph of $y = \cos x$ should convince you that we cannot restrict the domain of the cosine to the same interval as that for the sine (Figure 7). We choose rather to use the interval $0 \le x \le \pi$, on which the cosine is one-to-one.

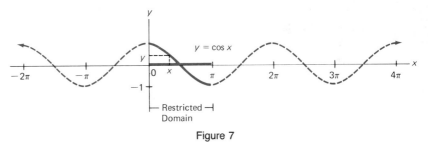

Figure 7

Having made the needed restriction, we may reasonably talk about $\cos^{-1}$. Moreover,

$$x = \cos^{-1} y \quad \text{if and only if} \quad y = \cos x \quad \text{and} \quad 0 \le x \le \pi$$

In particular,

$$\cos^{-1}\left(\frac{\sqrt{3}}{2}\right) = \frac{\pi}{6} \qquad \cos^{-1}(-1) = \pi$$

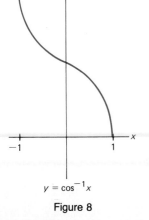

$y = \cos^{-1} x$

Figure 8

The graph of $y = \cos^{-1} x$ is shown in Figure 8. It is the graph of $y = \cos x$ reflected across the line $y = x$.

Example A (Inverse Sine and Cosine Values)   Evaluate.

(a) $\sin^{-1}(1)$ (b) $\sin^{-1}\left(\dfrac{-\sqrt{2}}{2}\right)$ (c) $\cos^{-1}(0)$ (d) $\cos^{-1}\left(\dfrac{-\sqrt{3}}{2}\right)$

Solution

(a) $\sin^{-1}(1) = \dfrac{\pi}{2}$ (b) $\sin^{-1}\left(\dfrac{-\sqrt{2}}{2}\right) = -\dfrac{\pi}{4}$

(c) $\cos^{-1}(0) = \dfrac{\pi}{2}$ (d) $\cos^{-1}\left(\dfrac{-\sqrt{3}}{2}\right) = \dfrac{5\pi}{6}$ ∎

The Inverse Tangent

To make $y = \tan x$ have an inverse, we restrict x to $-\pi/2 < x < \pi/2$. Thus

$$x = \tan^{-1} y \quad \text{if and only if} \quad y = \tan x \quad \text{and} \quad -\frac{\pi}{2} < x < \frac{\pi}{2}$$

The graphs of the tangent function and its inverse are shown in Figure 9.

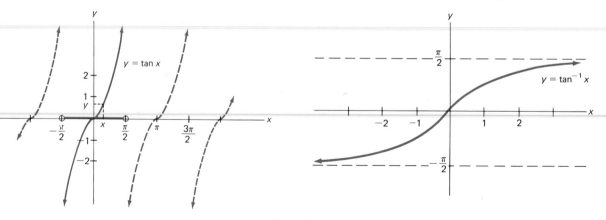

Figure 9

Notice that the graph of $y = \tan^{-1} x$ has horizontal asymptotes at $y = \pi/2$ and $y = -\pi/2$.

The Inverse Secant

The secant function has an inverse, provided we restrict its domain to $0 \le x \le \pi$, excluding $\pi/2$. Thus

$$x = \sec^{-1} y \quad \text{if and only if} \quad y = \sec x \quad \text{and} \quad 0 \le x \le \pi, x \ne \frac{\pi}{2}$$

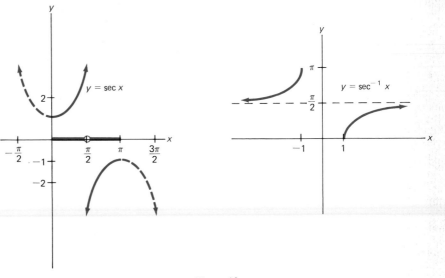

Figure 10

$$\sec^{-1}2 = \frac{1}{\cos 2}$$

$$\sec^{-1}2 = \cos^{-1}\frac{1}{2} = \frac{\pi}{3}$$

Domains

What is the domain of an inverse trigonometric function? It is the range of the corresponding trigonometric function. Here is a list

Function	Domain
$\sin^{-1}x$	$-1 \le x \le 1$
$\cos^{-1}x$	$-1 \le x \le 1$
$\tan^{-1}x$	$-\infty < x < \infty$
$\sec^{-1}x$	$|x| \ge 1$

Thus

$$\sin^{-1}3 \text{ is undefined}$$

whereas

$$\sec^{-1}3 \approx 1.23096$$

(Some authors choose to restrict the domain of the secant to $\{x : 0 \le x < \pi/2$ or $\pi \le x < 3\pi/2\}$. For this reason, check an author's definition before using any stated fact about the inverse secant.) The graphs of $y = \sec x$ and $y = \sec^{-1}x$ are shown in Figure 10.

Since the secant and cosine are reciprocals of each other, it is not surprising that $\sec^{-1}x$ is related to $\cos^{-1}x$. In fact, for every x in the domain of $\sec^{-1}x$, we have

$$\sec^{-1}x = \cos^{-1}\left(\frac{1}{x}\right)$$

This follows from the fact that each side is limited in value to the interval 0 to π and has the same cosine.

$$\cos(\sec^{-1}x) = \frac{1}{\sec(\sec^{-1}x)} = \frac{1}{x}$$

The two other inverse trigonometric functions, $\cot^{-1}x$ and $\csc^{-1}x$. are of less importance. They are introduced in Problem 66.

Example B (Inverse Tangent and Secant Values)   Evaluate.
(a) $\tan^{-1}(-\sqrt{3})$ (b) $\sec^{-1}(-2)$

Solution

(a) $\tan^{-1}(-\sqrt{3}) = -\frac{\pi}{3}$ (b) $\sec^{-1}(-2) = \cos^{-1}\left(-\frac{1}{2}\right) = \frac{5\pi}{6}$ ■

Inverse Trigonometric Functions and Calculators

Most scientific calculators have been programmed to give values of $\sin^{-1}$, $\cos^{-1}$, and $\tan^{-1}$ that are consistent with the definitions we have given. For example, to obtain $\sin^{-1}(.32)$ on many calculators, press the buttons .32 $\boxed{\text{INV}}$ $\boxed{\sin}$; on other calculators, there is a button marked $\sin^{-1}$ and you simply press .32 $\boxed{\sin^{-1}}$. Normally, you cannot get $\sec^{-1}$ directly; instead you must use the identity $\sec^{-1} x = \cos^{-1}(1/x)$. Of course, in every case you must put your calculator in the appropriate mode, depending on whether you want the answer in degrees or radians.

Example C (Using a Calculator) Evaluate.
(a) $\sin^{-1}(.2476)$ (b) $\cos^{-1}(-.8762)$ (c) $\sec^{-1}(3.2345)$

Solution
(a) $\sin^{-1}(.2476) = .25020$ (b) $\cos^{-1}(-.8762) = 2.63872$
(c) $\sec^{-1}(3.2345) = \cos^{-1}(1/3.2345) = 1.25648$ ∎

Here is an example that requires thinking in addition to use of a calculator.

Example D (Solving an Equation) Find all values of t between 0 and 2π for which $\tan t = 2.12345$.

Solution   A calculator immediately gives the solution,

$$t_0 = \tan^{-1}(2.12345) = 1.13067$$

However, there is another solution having t_0 as reference number, namely,

$$t = \pi + t_0 = 4.27227 \quad ∎$$

CAUTION

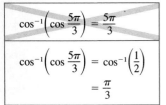
$$\cos^{-1}\left(\cos\frac{5\pi}{3}\right) = \frac{5\pi}{3}$$

$$\cos^{-1}\left(\cos\frac{5\pi}{3}\right) = \cos^{-1}\left(\frac{1}{2}\right)$$
$$= \frac{\pi}{3}$$

Manipulations with Inverse Trigonometric Functions

Some manipulations follow immediately from the definitions. For example,

$$\sin[\sin^{-1}(.3)] = .3 \qquad \tan[\tan^{-1}(4)] = 4$$

Others require insight and a good knowledge of the basic identities of trigonometry.

Example E (Some Calculations with Inverses) Evaluate without using a calculator.
(a) $\sin[\cos^{-1}(.3)]$ (b) $\cos[\tan^{-1}(4)]$

Solution
(a) Let $\theta = \cos^{-1}(.3)$ so that $\cos\theta = .3$ and $0 \le \theta \le \pi$. Then use the identity $\sin^2\theta = 1 - \cos^2\theta$. We obtain

$$\sin\theta = \pm\sqrt{1 - \cos^2\theta} = \pm\sqrt{1 - (.3)^2} = \pm\sqrt{.91}$$

We choose the positive sign since sine is positive on the interval $0 \le \theta \le \pi$. Thus

$$\sin[\cos^{-1}(.3)] = \sqrt{.91}$$

(b) Let $\theta = \tan^{-1}4$ so that $\tan \theta = 4$ and $-\pi/2 < \theta < \pi/2$. Note that $\sec \theta = \sqrt{1 + \tan^2 \theta} = \sqrt{1 + 16} = \sqrt{17}$. Thus

$$\cos[\tan^{-1}(4)] = \cos \theta = \frac{1}{\sec \theta} = \frac{1}{\sqrt{17}} = \frac{\sqrt{17}}{17} \quad \blacksquare$$

Example F (More Calculations with Inverses)   Evaluate without using a calculator.

(a) $\sin[2 \cos^{-1}(\frac{2}{3})]$ (b) $\tan[\tan^{-1}(2) + \sin^{-1}(\frac{4}{5})]$.

Solution

(a) Let $\theta = \cos^{-1}(\frac{2}{3})$ so that $\cos \theta = \frac{2}{3}$ and

$$\sin \theta = \sqrt{1 - \cos^2 \theta} = \sqrt{1 - \frac{4}{9}} = \frac{\sqrt{5}}{3}$$

Then apply the double-angle formula for $\sin 2\theta$ as indicated below.

$$\sin\left(2 \cos^{-1} \frac{2}{3}\right) = \sin 2\theta = 2 \sin \theta \cos \theta = 2\frac{\sqrt{5}}{3} \cdot \frac{2}{3} = \frac{4}{9}\sqrt{5}$$

(b) Let $\alpha = \tan^{-1}(2)$ and $\beta = \sin^{-1}(\frac{4}{5})$ and apply the identity

$$\tan(\alpha + \beta) = \frac{\tan \alpha + \tan \beta}{1 - \tan \alpha \tan \beta}$$

Now $\tan \alpha = 2$ and

$$\tan \beta = \frac{\sin \beta}{\cos \beta} = \frac{\frac{4}{5}}{\sqrt{1 - \left(\frac{4}{5}\right)^2}} = \frac{\frac{4}{5}}{\frac{3}{5}} = \frac{4}{3}$$

Therefore,

$$\tan(\alpha + \beta) = \frac{2 + \frac{4}{3}}{1 - 2 \cdot \frac{4}{3}} = -2 \quad \blacksquare$$

Some Identities for Inverse Trigonometric Functions

Here are three identities connecting sines, cosines, and their inverses.

(i) $$\cos(\sin^{-1}x) = \sqrt{1 - x^2}$$

(ii) $$\sin(\cos^{-1}x) = \sqrt{1 - x^2}$$

(iii) $$\sin^{-1}x + \cos^{-1}x = \frac{\pi}{2}$$

To prove the first identity, we let $\theta = \sin^{-1}x$. Remember that this means that $x = \sin \theta$, with $-\pi/2 \le \theta \le \pi/2$. Then

$$\cos(\sin^{-1}x) = \cos \theta = \pm\sqrt{1 - \sin^2 \theta} = \pm\sqrt{1 - x^2}$$

Finally, choose the plus sign because $\cos \theta$ is positive for $-\pi/2 \le \theta \le \pi/2$. The second identity is proved in a similar manner. We omit the proof of the third identity, choosing rather to offer another example.

Example G (Another Identity)   Show that

$$\cos(2 \tan^{-1}x) = \frac{1 - x^2}{1 + x^2}$$

Solution   We will apply the double-angle formula

$$\cos 2\theta = 2 \cos^2 \theta - 1$$

Here $\theta = \tan^{-1}x$, so that $x = \tan \theta$. Then

$$\cos(2 \tan^{-1}x) = \cos(2\theta) = 2 \cos^2 \theta - 1 = \frac{2}{\sec^2 \theta} - 1$$

$$= \frac{2}{1 + \tan^2 \theta} - 1 = \frac{2}{1 + x^2} - 1 = \frac{1 - x^2}{1 + x^2} \quad \blacksquare$$

PROBLEM SET 8-4

A. Skills and Techniques

Find the exact value of each of the following, as in Examples A and B.

1. $\sin^{-1}(\sqrt{3}/2)$

2. $\cos^{-1}(\frac{1}{2})$

3. $\arcsin(\sqrt{2}/2)$

4. $\arccos(\sqrt{2}/2)$

5. $\tan^{-1}(0)$

6. $\tan^{-1}(1)$

7. $\tan^{-1}(\sqrt{3})$

8. $\tan^{-1}(\sqrt{3}/3)$

9. $\arccos(-\frac{1}{2})$

10. $\arcsin(-\frac{1}{2})$

11. $\sec^{-1}(\sqrt{2})$

12. $\sec^{-1}(-2/\sqrt{3})$

Use a calculator (or Table C) to find each value (in radians) in Problems 13–20. See Example C.

13. $\sin^{-1}(.21823)$

14. $\cos^{-1}(.30582)$

15. $\sin^{-1}(-.21823)$

16. $\cos^{-1}(-.30582)$

17. $\tan^{-1}(.20660)$

18. $\tan^{-1}(1.2602)$

19. Calculate, using $\sec^{-1}x = \cos^{-1}(1/x)$.
  (a) $\sec^{-1}(1.4263)$ **(b)** $\sec^{-1}(-2.6715)$

20. Calculate.
  (a) $\sec^{-1}(\pi + 1)$ **(b)** $\sec^{-1}(-\sqrt{5}/2)$

Solve for t, where $0 \le t < 2\pi$. Use a calculator as in Example D.

21. $\sin t = .3416$

22. $\cos t = .9812$

23. $\tan t = 3.345$

24. $\sec t = 1.342$

Find the following without the use of tables or a calculator. See Example E.

25. $\sin[\sin^{-1}(\frac{2}{3})]$

26. $\cos[\cos^{-1}(-\frac{1}{4})]$

27. $\tan[\tan^{-1}(10)]$

28. $\cos^{-1}[\cos(\pi/2)]$

29. $\sin^{-1}[\sin(\pi/3)]$

30. $\tan^{-1}[\tan(\pi/4)]$

31. $\sin^{-1}[\cos(\pi/4)]$

32. $\cos^{-1}[\sin(-\pi/6)]$

33. $\cos[\sin^{-1}(\frac{4}{5})]$

34. $\sin[\cos^{-1}(\frac{3}{5})]$

35. $\cos[\tan^{-1}(\frac{1}{2})]$

36. $\cos[\tan^{-1}(-\frac{3}{4})]$

37. $\cos[\sec^{-1}(3)]$

38. $\sec[\cos^{-1}(-.4)]$

39. $\sec^{-1}[\sec(2\pi/3)]$

40. $\sec[\sec^{-1}(2.56)]$

Use a calculator to find each value.

41. $\cos[\sin^{-1}(-.2564)]$

42. $\tan^{-1}(\sin 14.1)$

43. $\sin^{-1}(\cos 1.12)$

44. $\cos^{-1}[\cos^{-1}(.91)]$

45. $\tan[\sec^{-1}(2.5)]$

46. $\sec^{-1}(\sin 1.67)$

Evaluate by using the method of Example F, not by using a calculator.

47. $\sin[2 \cos^{-1}(\frac{3}{5})]$

48. $\sin[2 \cos^{-1}(\frac{1}{2})]$

49. $\cos[2 \sin^{-1}(-\frac{3}{5})]$

50. $\tan[2 \tan^{-1}(\frac{1}{3})]$

51. $\sin[\cos^{-1}(\frac{3}{5}) + \cos^{-1}(\frac{5}{13})]$

52. $\tan[\tan^{-1}(\frac{1}{2}) + \tan^{-1}(-3)]$

53. $\cos[\sec^{-1}(\frac{3}{2}) - \sec^{-1}(\frac{4}{3})]$

54. $\sin[\sin^{-1}(\frac{4}{5}) + \sec^{-1}(3)]$

Show that each of the following is an identity, as in Example G.

55. $\tan(\sin^{-1}x) = \dfrac{x}{\sqrt{1 - x^2}}$

56. $\sin(\tan^{-1}x) = \dfrac{x}{\sqrt{1 + x^2}}$

57. $\tan(2 \tan^{-1}x) = \dfrac{2x}{1 - x^2}$

58. $\cos(2 \sin^{-1}x) = 1 - 2x^2$

59. $\cos(2 \sec^{-1}x) = \dfrac{2}{x^2} - 1$

60. $\sec(2 \tan^{-1}x) = \dfrac{1 + x^2}{1 - x^2}$

B. Applications and Extensions

61. Without using tables or a calculator, find each value (in radians).
(a) $\arcsin(-\sqrt{3}/2)$ (b) $\tan^{-1}(-\sqrt{3})$
(c) $\sec^{-1}(-2)$

62. Calculate each of the following (radian mode).
(a) $\dfrac{2 \arccos(.956)}{3 \arcsin(-.846)}$ (b) $.3624 \sec^{-1}(4.193)$
(c) $\cos^{-1}(2 \sin .1234)$
(d) $\sin[\arctan(4.62) - \arccos(-.48)]$

63. Without using tables or a calculator, find each value. Then check using your calcualtor.
(a) $\tan[\tan^{-1}(43)]$ (b) $\cos[\sin^{-1}(\frac{5}{13})]$
(c) $\sin[\frac{\pi}{4} + \sin^{-1}(.8)]$
(d) $\cos[\sin^{-1}(.6) + \sec^{-1}(3)]$

64. Try to calculate each of the following and then explain why your calculator gives you an error message.
(a) $\cos[\sin^{-1}(2)]$ (b) $\cos^{-1}(\tan 2)$
(c) $\tan[\arctan 3 + \arctan(\frac{1}{3})]$

65. Solve for x.
(a) $\cos(\sin^{-1}x) = \frac{3}{4}$
(b) $\sin(\cos^{-1}x) = \sqrt{.19}$
(c) $\sin^{-1}(3x - 5) = \frac{\pi}{6}$
(d) $\tan^{-1}(x^2 - 3x + 3) = \frac{\pi}{4}$

66. To determine inverses for cotangent and cosecant, we restrict their domains to $0 < x < \pi$ and $-\pi/2 \le x \le \pi/2, x \ne 0$, respectively. With these restrictions understood, find each value.
(a) $\cot^{-1}(\sqrt{3})$ (b) $\cot^{-1}(-1/\sqrt{3})$
(c) $\cot^{-1}(0)$ (d) $\csc^{-1}(2)$
(e) $\csc^{-1}(-1)$ (f) $\csc^{-1}(-2/\sqrt{3})$

67. It is always true that $\sin(\sin^{-1}x) = x$, but it is not always true that $\sin^{-1}(\sin x) = x$. For example, $\sin^{-1}(\sin \pi) \ne \pi$. Instead,

$$\sin^{-1}(\sin \pi) = \sin^{-1}(0) = 0$$

Find each value.
(a) $\sin^{-1}[\sin(\pi/2)]$ (b) $\sin^{-1}[\sin(3\pi/4)]$
(c) $\sin^{-1}[\sin(5\pi/4)]$ (d) $\sin^{-1}[\sin(3\pi/2)]$
(e) $\cos^{-1}[\cos(3\pi)]$ (f) $\tan^{-1}[\tan(13\pi/4)]$

68. Sketch the graph of $y = \sin^{-1}(\sin x)$ for $-2\pi \le x \le 4\pi$. *Hint:* See Problem 67.

69. For each of the following right triangles, write θ explicitly in terms of x.

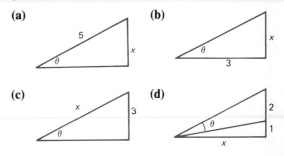

(a) (b) (c) (d)

70. In some computer languages (for example, FORTRAN), the only built-in inverse trigonometric function is $\tan^{-1}$. Establish the following identities, which show why this is sufficient.
(a) $\sin^{-1}x = \tan^{-1}\left(\dfrac{x}{\sqrt{1 - x^2}}\right)$
(b) $\cos^{-1}x = \dfrac{\pi}{2} - \sin^{-1}x = \dfrac{\pi}{2} - \tan^{-1}\left(\dfrac{x}{\sqrt{1 - x^2}}\right)$

71. Assume that your calculator's $\tan^{-1}$ button is working but not the $\sin^{-1}$ or $\cos^{-1}$ buttons. Use the results in Problem 70 to calculate each of the following.
(a) $\sin^{-1}(.6)$ (b) $\sin^{-1}(-.3)$
(c) $\cos^{-1}(.8)$ (d) $\cos^{-1}(-.9)$

72. Show that $\arctan(\frac{1}{4}) + \arctan(\frac{3}{5}) = \pi/4$. *Hint:* Show that both sides have the same tangent, using the formula for $\tan(\alpha + \beta)$.

73. In 1706, John Machin used the following formula to calculate π to 100 decimal places, a tremendous feat for its day. Establish this formula. *Hint:* Apply the addition formula for the tangent to the left side. Think of the right side as $4\theta = 2(2\theta)$ and use the tangent double-angle formula twice.

$$\dfrac{\pi}{4} + \arctan\left(\dfrac{1}{239}\right) = 4 \arctan\left(\dfrac{1}{5}\right)$$

74. Show that

$$\arctan\left(\frac{1}{3}\right) + \arctan\left(\frac{1}{5}\right) + \arctan\left(\frac{1}{7}\right)$$

$$+ \arctan\left(\frac{1}{8}\right) = \frac{\pi}{4}$$

75. A picture 4 feet high is hung on a museum wall so that its bottom is 7 feet above the floor. A viewer whose eye level is 5 feet above the floor stands b feet from the wall.
   (a) Express θ, the vertical angle subtended by the picture at her eye, explicitly in terms of b.
   (b) Calculate θ when $b = 8$.
   (c) Determine b so $\theta = 30°$.

76. TEASER   A goat is tethered to a stake at the edge of a circular pond of radius r by means of a rope of length kr, $0 < k \le 2$. Find an explicit formula for its grazing area in terms of r and k.

77. Draw the graphs of $f(x) = \cos^{-1}(\cos x)$ and $g(x) = \sin^{-1}(\sin x)$. Determine the amplitude and period of these periodic functions.

78. The graph of f in Problem 77 should remind you of the *distance to the nearest integer function*, $\langle \cdot \rangle$ introduced in Problem 53 of Section 5-2. In fact, $\langle x \rangle = A \cos^{-1}(\cos Bx)$ for appropriate choices of A and B. Determine A and B and check your answer by graphing.

8-5 TRIGONOMETRIC EQUATIONS

Two Bugs on a Circle

Two bugs crawl around the unit circle starting together at $(1, 0)$, one moving at one unit per second, the other moving twice as fast. When will one bug be directly above the other bug?

What does the bug problem have to do with trigonometric equations? Well, you should agree that after t seconds the slow bug, having traveled t units along the unit circle, is at $(\cos t, \sin t)$. The fast bug is at $(\cos 2t, \sin 2t)$. One bug will be directly above the other bug when their two x-coordinates are equal. This means we must solve the equation

$$\cos 2t = \cos t$$

Specifically, we must find the first $t > 0$ that makes this equality true. We shall solve this equation in due time, but first we ought to solve some simpler trigonometric equations.

Simple Equations

Suppose we are asked to solve the equation

$$\sin t = \frac{1}{2}$$

for t. The number $t = \pi/6$ occurs to us right away. But that is not the only answer. All numbers that measure angles in the first or second quadrant and have $\pi/6$ as their reference number are solutions. Thus,

$$\ldots, -\frac{11\pi}{6}, -\frac{7\pi}{6}, \frac{\pi}{6}, \frac{5\pi}{6}, \frac{13\pi}{6}, \ldots$$

all work. In fact, one characteristic of trigonometric equations is that, if they have one solution, they have infinitely many solutions.

Let us alter the problem. Suppose we wish to solve $\sin t = \frac{1}{2}$ for $0 \leq t < 2\pi$. Then the answers are $\pi/6$ and $5\pi/6$. In the following pages, we shall assume that, unless otherwise specified, we are to find only those solutions on the interval $0 \leq t < 2\pi$.

Example A (Equations of Linear Form)   Solve for t in the interval $0 \leq t < 2\pi$.
(a) $2 \cos t - \sqrt{2} = 0$ (b) $\sin t = -.5234$

Solution
(a) Rewrite the equation as $\cos t = \sqrt{2}/2$. There are two solutions, namely $\pi/4$ and $7\pi/4$.
(b) Here we will need a calculator (or tables). An unthinking use of this tool would be to calculate $\sin^{-1}(-.5234)$ giving the value $-.55084$, which is not on the required interval. The better way to proceed is to recognize that the given equation has two solutions but that they both have t_0 as their reference number, where

$$t_0 = \sin^{-1}(.5234) = .55084$$

The two solutions we seek are $\pi + t_0$ and $2\pi - t_0$, that is, 3.69243 and 5.73235.   ■

Equations of Quadratic Form

In Section 3-4 we solved quadratic equations by a number of techniques (factoring, taking square roots, and using the quadratic formula). We use the same techniques here. For example,

$$\cos^2 t = \frac{3}{4}$$

is analogous to $x^2 = \frac{3}{4}$. We solve such an equation by taking square roots.

$$\cos t = \pm \frac{\sqrt{3}}{2}$$

The set of solutions on $0 \leq t < 2\pi$ is

$$\left\{ \frac{\pi}{6}, \frac{5\pi}{6}, \frac{7\pi}{6}, \frac{11\pi}{6} \right\}$$

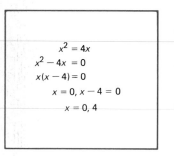

$$x^2 = 4x$$
$$x^2 - 4x = 0$$
$$x(x - 4) = 0$$
$$x = 0, \; x - 4 = 0$$
$$x = 0, \; 4$$

Figure 11

Do you remember how we solved the equation $x^2 = 4x$? We rewrote it with 0 on one side, factored the other side, and then set each factor equal to 0 (see Figure 11). We follow exactly the same procedure with our next trigonometric equation. To solve

$$\cos t \cot t = -\cos t$$

use the following steps:

$$\cos t \cot t + \cos t = 0$$
$$\cos t (\cot t + 1) = 0$$
$$\cos t = 0, \; \cot t + 1 = 0$$
$$\cos t = 0, \qquad \cot t = -1$$

Thus our problem is reduced to solving two simple equations. The first has the two solutions, $\pi/2$ and $3\pi/2$; the second has solutions $3\pi/4$ and $7\pi/4$. Thus the set of all solutions on the interval $0 \leq t < 2\pi$ is

$$\left\{ \frac{\pi}{2}, \frac{3\pi}{4}, \frac{3\pi}{2}, \frac{7\pi}{4} \right\}$$

Example B (Solving by Factoring)   Solve

$$2 \sin^2 t - \sin t - 1 = 0$$

Solution   Think of this equation as being like

$$2x^2 - x - 1 = 0$$

Now

$$2x^2 - x - 1 = (2x + 1)(x - 1)$$

and so

$$2 \sin^2 t - \sin t - 1 = (2 \sin t + 1)(\sin t - 1)$$

When we set each factor equal to zero and solve, we get

$$2 \sin t + 1 = 0 \qquad\qquad \sin t - 1 = 0$$
$$\sin t = -\tfrac{1}{2} \qquad\qquad \sin t = 1$$
$$t = \frac{7\pi}{6}, \frac{11\pi}{6} \qquad\qquad t = \frac{\pi}{2}$$

The set of all solutions on $0 \leq t < 2\pi$ is

$$\left\{ \frac{\pi}{2}, \frac{7\pi}{6}, \frac{11\pi}{6} \right\} \qquad \blacksquare$$

Example C (Using Identities)   Solve

$$\tan^2 x = \sec x + 1$$

Solution   The identity $1 + \tan^2 x = \sec^2 x$ suggests writing everything in terms of $\sec x$.

$$\sec^2 x - 1 = \sec x + 1$$

$$\sec^2 x - \sec x - 2 = 0$$

$$(\sec x + 1)(\sec x - 2) = 0$$

$$\sec x + 1 = 0 \qquad \sec x - 2 = 0$$

$$\sec x = -1 \qquad \sec x = 2$$

$$x = \pi \qquad x = \frac{\pi}{3}, \frac{5\pi}{3}$$

Thus the set of solutions on $0 \le t < 2\pi$ is $\{\pi/3, \pi, 5\pi/3\}$. Unfamiliarity with the secant may hinder you at the last step. If so, use $\sec x = 1/\cos x$ to write the equations in terms of cosines and solve the equations

$$\cos x = -1 \qquad \cos x = \frac{1}{2} \quad \blacksquare$$

Example D (Solving by Squaring Both Sides)   Solve

$$1 - \cos t = \sqrt{3} \sin t$$

Solution   Since the identity relating sines and cosines involves their squares, we begin by squaring both sides. Then we express everything in terms of $\cos t$ and solve.

$$(1 - \cos t)^2 = 3 \sin^2 t$$

$$1 - 2 \cos t + \cos^2 t = 3(1 - \cos^2 t)$$

$$\cos^2 t - 2 \cos t + 1 = 3 - 3 \cos^2 t$$

$$4 \cos^2 t - 2 \cos t - 2 = 0$$

$$2 \cos^2 t - \cos t - 1 = 0$$

$$(2 \cos t + 1)(\cos t - 1) = 0$$

$$\cos t = -\frac{1}{2} \qquad \cos t = 1$$

$$t = \frac{2\pi}{3}, \frac{4\pi}{3} \qquad t = 0$$

Since squaring may introduce extraneous solutions, it is important to check our answers. We find that $4\pi/3$ is extraneous, since substituting $4\pi/3$ for t in the original equation gives us $1 + \frac{1}{2} = -\frac{3}{2}$. However, 0 and $2\pi/3$ are solutions, as you should verify. $\blacksquare$

Solution to the Two-Bug Problem

Our opening display asked when one bug would first be directly above the other. We reduced that problem to solving

$$\cos 2t = \cos t$$

for t. Using a double-angle formula, we may write

$$2 \cos^2 t - 1 = \cos t$$

$$2 \cos^2 t - \cos t - 1 = 0$$

$$(2 \cos t + 1)(\cos t - 1) = 0$$

$$\cos t = -\frac{1}{2} \qquad \qquad \cos t = 1$$

$$t = \frac{2\pi}{3}, \frac{4\pi}{3} \qquad \qquad t = 0$$

The smallest positive solution is $t = 2\pi/3$. After a little over 2 seconds, the slow bug will be directly above the fast bug.

What if we had been asked to find all times when the slow bug and the fast bug will be on the same vertical line? Our next example answers this question.

Example E (Finding All of the Solutions)   Find the entire set of solutions of the equation $\cos 2t = \cos t$.

Solution   We found 0, $2\pi/3$, and $4\pi/3$ to be the solutions for $0 \le t < 2\pi$. Clearly we get new solutions by adding 2π again and again to any of these numbers. The same holds true for subtracting 2π. In fact, the entire solution set consists of all those numbers of the form $2\pi k$, $2\pi/3 + 2\pi k$, or $4\pi/3 + 2\pi k$, where k is any integer.   ∎

Equations Involving Multiple Angles

The two-bug problem involved the double angle $2t$. We used an identity to change the equation to one of quadratic form. If a multiple angle equation is of linear form, we can solve the equation directly, as we now illustrate.

Example F (Multiple-Angle Equations)   Find all solutions of $\cos 4t = \frac{1}{2}$ on the interval $0 \le t < 2\pi$.

Solution   There will be more answers than you think. We know that $\cos 4t$ equals $\frac{1}{2}$ when

$$4t = \frac{\pi}{3}, \frac{5\pi}{3}, \frac{7\pi}{3}, \frac{11\pi}{3}, \frac{13\pi}{3}, \frac{17\pi}{3}, \frac{19\pi}{3}, \frac{23\pi}{3}$$

that is, when

$$t = \frac{\pi}{12}, \frac{5\pi}{12}, \frac{7\pi}{12}, \frac{11\pi}{12}, \frac{13\pi}{12}, \frac{17\pi}{12}, \frac{19\pi}{12}, \frac{23\pi}{12}$$

The reason that there are 8 solutions instead of 2 is that $\cos 4t$ completes 4 cycles on the interval $0 \le t < 2\pi$.   ∎

PROBLEM SET 8-5

A. Skills and Techniques

Solve each of the following, finding all solutions on the interval 0 to 2π, excluding 2π. See Examples A, B, and C.

1. $\sin t = 0$
2. $\cos t = 1$
3. $\sin t = -1$
4. $\tan t = -\sqrt{3}$
5. $\sin t = 2$
6. $\sec t = \frac{1}{2}$
7. $2 \cos x + \sqrt{3} = 0$
8. $2 \sin x + 1 = 0$
9. $\tan^2 x = 1$
10. $4 \sin^2 \theta - 3 = 0$
11. $(2 \cos \theta + 1)(2 \sin \theta - \sqrt{2}) = 0$
12. $(\sin \theta - 1)(\tan \theta + 1) = 0$
13. $\sin^2 x + \sin x = 0$
14. $2 \cos^2 x - \cos x = 0$
15. $\tan^2 \theta = \sqrt{3} \tan \theta$
16. $\cot^2 \theta = -\cot \theta$
17. $2 \sin^2 x = 1 + \cos x$
18. $\sec^2 x = 1 + \tan x$
19. $\tan^2 x - 3 \tan x + 1 = 0$
20. $\cos 2t = \sin t$

Solve each of the following equations on the interval $0 \le t < 2\pi$. See Example D and be sure to check your answer.

21. $\sin t + \cos t = 1$
22. $\sin t - \cos t = 1$
23. $\sqrt{3}(1 - \sin t) = \cos t$
24. $1 + \sin t = \sqrt{3} \cos t$
25. $\sec t + \tan t = 1$
26. $\tan t - \sec t = 1$

Find the entire solution set of each of the following equations, as in Example E.

27. $\sin t = \frac{1}{2}$
28. $\cos t = -\frac{1}{2}$
29. $\tan t = 0$
30. $\tan t = -\sqrt{3}$
31. $\sin^2 t = \frac{1}{4}$
32. $\cos^2 t = 1$

Solve each of the following equations, finding all solutions on the interval $0 \le t < 2\pi$. See Example F.

33. $\sin 2t = 0$
34. $\cos 2t = 0$
35. $\sin 4t = 1$
36. $\cos 4t = 1$
37. $\tan 2t = -1$
38. $\tan 3t = 0$

B. Applications and Extensions

In Problems 38–56, find all solutions to the given equation on $0 \le x < 2\pi$.

39. $2 \sin^2 x = \sin x$
40. $2 \cos x \sin x + \cos x = 0$
41. $\cos^2 x = \frac{1}{3}$
42. $\tan^2 x + 2 \tan x = 0$
43. $2 \tan x - \sec^2 x = 0$
44. $\tan^2 x = 1 + \sec x$
45. $\tan 2x = 3 \tan x$
46. $\cos(x/2) - \cos x = 1$
47. $\sin^2 x + 3 \sin x - 1 = 0$
48. $\tan^2 x - 2 \tan x - 10 = 0$
49. $\sin 2x + \sin x + 4 \cos x = -2$
50. $\cos x + \sin x = \sec x + \sec x \tan x$
51. $\sin x \cos x = -\sqrt{3}/4$
52. $4 \sin x - 4 \sin^3 x + \cos x = 0$
53. $\cos x - 2 \sin x = 2$
54. $\sin x + \cos x = \frac{1}{3}$
55. $\cos^8 x - \sin^8 x = 0$
56. $\cos^6 x + \sin^6 x = \frac{13}{16}$
57. A ray of light from the lamp L in Figure 12 reflects off a mirror to the object O.

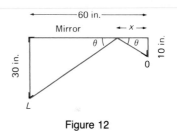

Figure 12

(a) Find the distance x.
(b) Write an equation for θ.
(c) Solve this equation.

58. Tom and John are lost in a desert 1 mile from a highway, at point A in Figure 13. Each strikes out in a different direction to get to the highway. Tom gets to the highway at point B and John arrives at point C, $1 + \sqrt{3}$ miles farther down the road. Write an equation for θ and solve it.

59. Solve the inequality $\sin 2t \ge \cos t$ for $0 \le t \le 2\pi$.

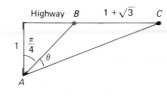

Figure 13

60. In a certain city, the number T of hours of daylight in any given day is approximated by

$$T = 12 + 4 \sin\left[\frac{2\pi}{365}(t - 92)\right]$$

where t is the number of days after midnight on December 31.

(a) How many hours of daylight are there on August 1?

(b) How many days of the year have more than 14 hours of daylight?

61. Mr. Quincy built a slide with a 10-foot rise and 20-foot base (Figure 14). (a) Find the angle α in degrees.

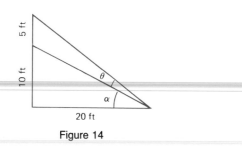

Figure 14

(b) By how much (θ in Figure 14) would the angle of the slide increase if he made the rise 15 feet, keeping the base at 20 feet?

62. Find the angles θ_1, θ_2, and θ_3 shown in Figure 15. Your answers should convince you that the angle ABC is not trisected.

Figure 15

63. Solve the equation

$$\sin 4t + \sin 3t + \sin 2t = 0$$

Hint: Use the identity
$$\sin u + \sin v = 2 \sin[(u + v)/2]\cos[(u - v)/2].$$

64. Solve the equation

$$\cos 5t + \cos 3t - 2 \cos t = 0$$

Hint: Use the identity
$$\cos u + \cos v = 2\cos[(u + v)/2]\cos[(u - v)/2].$$

65. Solve $\cos^8 u + \sin^8 u = \frac{41}{128}$ for $0 \le u \le \pi$. *Hint:* Begin by using half-angle formulas.

66. TEASER Show that $t = \pi/4$ is the only solution on $0 \le t \le \pi$ to the equation

$$\frac{a + b \cos t}{b + a \sin t} = \frac{a + b \sin t}{b + a \cos t}$$

67. Find all solutions to $x = 2 \tan^{-1} x$.

68. Find all solutions to

$$\sin x \sin 2x \sin 3x = \cos x$$

CHAPTER 8 SUMMARY

An **identity** is an equality that is true for all values of the unknown for which both sides of the equality make sense. Our first task was to establish the fundamental identities of trigonometry, here arranged by category.

Basic Identities

1. $\tan t = \dfrac{\sin t}{\cos t}$

2. $\cot t = \dfrac{\cos t}{\sin t} = \dfrac{1}{\tan t}$

3. $\sec t = \dfrac{1}{\cos t}$ 4. $\csc t = \dfrac{1}{\sin t}$

5. $\sin^2 t + \cos^2 t = 1$ 6. $1 + \tan^2 t = \sec^2 t$

7. $1 + \cot^2 t = \csc^2 t$

Cofunction Identities

8. $\sin\left(\dfrac{\pi}{2} - t\right) = \cos t$

9. $\cos\left(\dfrac{\pi}{2} - t\right) = \sin t$

Odd-Even Identities

10. $\sin(-t) = -\sin t$ 11. $\cos(-t) = \cos t$

Addition Formulas

12. $\sin(s + t) = \sin s \cos t + \cos s \sin t$

13. $\sin(s - t) = \sin s \cos t - \cos s \sin t$

14. $\cos(s + t) = \cos s \cos t - \sin s \sin t$

15. $\cos(s - t) = \cos s \cos t + \sin s \sin t$

Double-Angle Formulas

16. $\sin 2t = 2 \sin t \cos t$

17. $\cos 2t = \cos^2 t - \sin^2 t = 1 - 2 \sin^2 t$
$= 2 \cos^2 t - 1$

Half-Angle Formulas

18. $\sin \dfrac{t}{2} = \pm \sqrt{\dfrac{1 - \cos t}{2}}$

19. $\cos \dfrac{t}{2} = \pm \sqrt{\dfrac{1 + \cos t}{2}}$

Once we have memorized the fundamental identities, we can use them to prove thousands of other identities. The suggested technique is to take one side of a proposed identity and show by a chain of equalities that it is equal to the other.

A **trigonometric equation** is an equality involving trigonometric functions that is true only for some values of the unknown (for example, $\sin 2t = \frac{1}{2}$). Here our job is to solve the equation, that is, to find the values of the unknown that make it true.

With their natural domains, the trigonometric functions are not one-to-one and therefore do not have inverses. However, there are standard ways to restrict the domains so that inverses exist. Here are the results.

$x = \sin^{-1} y$ means $y = \sin x$ and $\dfrac{-\pi}{2} \le x \le \dfrac{\pi}{2}$

$x = \cos^{-1} y$ means $y = \cos x$ and $\quad 0 \le x \le \pi$

$x = \tan^{-1} y$ means $y = \tan x$ and $\dfrac{-\pi}{2} < x < \dfrac{\pi}{2}$

$x = \sec^{-1} y$ means $y = \sec x$ and $0 \le x \le \pi, x \ne \dfrac{\pi}{2}$

CHAPTER 8 REVIEW PROBLEM SET

In Problems 1–10, write True or False in the blank. If false, tell why.

_____ **1.** $\sin \theta \cos \theta \tan \theta = 1 - \cos^2 \theta$

_____ **2.** A trigonometric equation can have at most two solutions on the interval 0 to 2π.

_____ **3.** If a trigonometric equation has infinitely many solutions, it is an identity.

_____ **4.** The equation $4 \sec^2 t - 1 = 0$ has no solution.

_____ **5.** $2 \sin 1.5 \cos 1.5 = \sin 3$.

_____ **6.** $\sin(\alpha + \beta) = \sin \alpha + \sin \beta$ is not true for any α and β.

_____ **7.** $\cos 75° \sin 15° - \sin 75° \cos 15° = -\sqrt{3}/2$.

_____ **8.** The domain of $f(x) = 2 \sin^{-1} x$ is the interval $-1 \le x \le 1$.

_____ **9.** $\sin^{-1}(\sin x) = x$ for all real numbers x.

_____ **10.** The function $f(x) = 4 \cos 2x$ has an inverse if its domain is restricted to the interval $-\pi/4 \le x \le \pi/4$.

11. Rewrite each expression in terms of $\cos t$ and simplify.

(a) $2 - 3 \sin^2 t$

(b) $\dfrac{\sin^2 t \sec t}{1 + \sec t}$

(c) $\sin^2 2t \cos^2\left(\dfrac{t}{2}\right)$

12. If $\tan t = -\frac{3}{4}$ and $\pi/2 < t < \pi$, evaluate the following.

(a) $\sin t$ (b) $\cos t$

(c) $\sec t$

Prove that each of the following equations is an identity.

13. $\cot \theta \cos \theta = \csc \theta - \sin \theta$

14. $\sec t - \cos t = \sin t \tan t$

15. $\left(\cos \dfrac{t}{2} + \sin \dfrac{t}{2}\right)^2 = 1 + \sin t$

16. $\sec^4 \theta - \sec^2 \theta = \tan^4 \theta + \tan^2 \theta$

17. $\tan u + \cot u = \sec u \csc u$

18. $\dfrac{1 - \cos x}{\sin x} = \dfrac{\sin x}{1 + \cos x}$

In Problems 19–22, use appropriate identities to simplify each expression and then evaluate.

19. $\cos 153° \cos 33° + \sin 153° \sin 33°$

20. $\sin \dfrac{\pi}{8} \cos \dfrac{3\pi}{8} + \cos \dfrac{\pi}{8} \sin \dfrac{3\pi}{8}$

21. $2 \sin^2 112.5° - 1$

22. $(\tan 20° + \tan 25°)/(1 - \tan 20° \tan 25°)$

23. Express $\tan \theta \tan 2\theta$ in terms of $\sin \theta$.

24. Express $\cos 4t$ in terms of $\sin t$.

25. If $\sin t = -\dfrac{12}{13}$ and $\pi < t < \dfrac{3\pi}{2}$, evaluate the following.

    (a) $\cos t$ **(b)** $\sin 2t$

    (c) $\cos 2t$ **(d)** $\tan(t/2)$

26. If $\sin \alpha = \frac{2}{3}$, $\sin \beta = \frac{4}{5}$, $\cos \alpha < 0$, and $\cos \beta < 0$, evaluate the following.

    (a) $\cos \alpha$ **(b)** $\cos \beta$

    (c) $\sin(\alpha + \beta)$

Prove that each equation is an identity.

27. $\cos\left(u + \dfrac{\pi}{3}\right)\cos(\pi - u)$

    $- \sin\left(u + \dfrac{\pi}{3}\right)\sin(\pi - u) = -\dfrac{1}{2}$

28. $\dfrac{\cos 5t}{\sin t} - \dfrac{\sin 5t}{\cos t} = \dfrac{2 \cos 6t}{\sin 2t}$

29. $\csc 2t + \cot 2t = \cot t$

30. $\sin 3\theta = 3 \sin \theta - 4 \sin^3 \theta$

31. $\dfrac{\sin(\alpha - \beta)}{\cos \alpha \cos \beta} = \tan \alpha - \tan \beta$

32. $\dfrac{1 - \tan^2(u/2)}{1 + \tan^2(u/2)} = \cos u$

In Problems 33–36, calculate each expression without use of a calculator.

33. (a) $\sin^{-1}\left(\dfrac{\sqrt{2}}{2}\right)$ **(b)** $\cos^{-1}\left(-\dfrac{1}{2}\right)$

    (c) $\tan^{-1}(-1)$ **(d)** $\sec^{-1}(\sqrt{2})$

34. (a) $\sec[\sec^{-1}(2.5)]$ **(b)** $\sin^{-1}\left(\sin \dfrac{3\pi}{4}\right)$

    (c) $\csc[\sin^{-1}(.6)]$ **(d)** $\sec[\cos^{-1}(-.2)]$

35. (a) $\sin[2 \cos^{-1}(.8)]$ **(b)** $\cos[2 \cos^{-1}(.7)]$

36. (a) $\sin[\cos^{-1}(.6) + \cos^{-1}(.5)]$

    (b) $\tan[\tan^{-1}(1) + \tan^{-1}(2)]$

Solve the equations in Problems 37–44 for t, $0 \le t < 2\pi$.

37. $\tan t = -\sqrt{3}/3$

38. $2 \cos^2 t - \cos t = 0$

39. $\sin^2 t - 2 \sin t - 3 = 0$

40. $\sec^4 t - 3 \sec^2 t + 2 = 0$

41. $\sin 2t = \frac{1}{2}$

42. $\cos 4t = -\frac{1}{2}$

43. $3 + \cos 2t = 5 \cos t$

44. $\tan^3 t + \tan^2 t - 3 \tan t = 3$

45. State the domain and range of $f(x) = \cos^{-1} x$.

46. Determine an appropriate restricted domain so that $f(x) = \sin 2x$ will have an inverse and then give a formula for $f^{-1}(x)$.

47. Show that $\cot(\sin^{-1} x) = \sqrt{1 - x^2}/x$ for $-1 \le x \le 1$, $x \ne 0$.

48. Show that $\tan^{-1}(2) + \tan^{-1}(3) = 3\pi/4$.

49. Calculate $\tan[\tan^{-1}(2) + \tan^{-1}(3) + \tan^{-1}(5)]$.

50. If $\tan t = .5$, calculate $\tan 3t$ without the help of a calculator or tables.

Applications of Trigonometry

■ Thus one sees in the sciences many brilliant theories which have remained unapplied for a long time suddenly becoming the foundation of most important applications, and likewise applications very simple in appearance giving birth to ideas of the most abstract theories.

Marquis de Condorcet

9-1 OBLIQUE TRIANGLES: LAW OF SINES

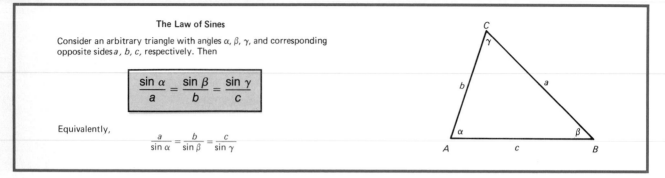

The Law of Sines

Consider an arbitrary triangle with angles α, β, γ, and corresponding opposite sides a, b, c, respectively. Then

$$\frac{\sin \alpha}{a} = \frac{\sin \beta}{b} = \frac{\sin \gamma}{c}$$

Equivalently,

$$\frac{a}{\sin \alpha} = \frac{b}{\sin \beta} = \frac{c}{\sin \gamma}$$

We learned in Section 7-1 how to solve a right triangle. But can we solve an oblique triangle—that is, one without a 90° angle? One valuable tool is the **law of sines,** stated above. It is valid for any triangle, but we initially establish it for the case where all angles are acute.

Proof of the Law of Sines

Figure 1

Consider a triangle with all acute angles, labeled as in Figure 1. Drop a perpendicular of length h from vertex C to the opposite side. Then by right-triangle trigonometry,

$$\sin \alpha = \frac{h}{b} \qquad \sin \beta = \frac{h}{a}$$

If we solve for h in these two equations and equate the results, we obtain

$$b \sin \alpha = a \sin \beta$$

Finally, dividing both sides by ab yields

$$\frac{\sin \alpha}{a} = \frac{\sin \beta}{b}$$

Since the roles of β and γ can be interchanged, the same reasoning gives

$$\frac{\sin \alpha}{a} = \frac{\sin \gamma}{c}$$

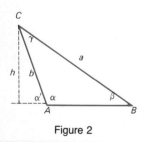

Figure 2

Next consider a triangle with an obtuse angle α ($90° < \alpha < 180°$). Drop a perpendicular of length h from vertex C to the extension of AB (see Figure 2). Notice that angle α' is the reference angle for α and so $\sin \alpha = \sin \alpha'$. It follows from right-triangle trignometry that

$$\sin \alpha = \sin \alpha' = \frac{h}{b} \qquad \sin \beta = \frac{h}{a}$$

just as in the acute case. The rest of the argument is identical with that case.

Solving a Triangle (AAS)

Suppose that we know two angles and any side of a triangle. Then we can solve the triangle, that is, we can find all the remaining parts.

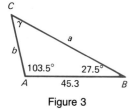

Figure 3

Example A (AAS) Suppose that in triangle ABC, $\alpha = 103.5°$, $\beta = 27.5°$, and $c = 45.3$ (Figure 3). Find γ, a, and b.

Solution

1. Since $\alpha + \beta + \gamma = 180°$, $\gamma = 180° - (103.5° + 27.5°) = 49°$.
2. By the law of sines,

$$\frac{a}{\sin 103.5°} = \frac{45.3}{\sin 49°}$$

$$a = \frac{(45.3)(\sin 103.5°)}{\sin 49°}$$

$$\approx 58.4$$

3. Also by the law of sines,

$$\frac{b}{\sin 27.5°} = \frac{45.3}{\sin 49°}$$

$$b = \frac{(45.3)(\sin 27.5°)}{\sin 49°}$$

$$\approx 27.7 \quad \blacksquare$$

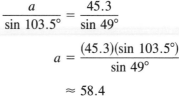
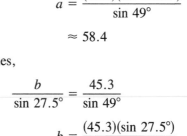

Figure 4

Solving a Triangle (SSA)

Suppose that two sides and the angle opposite one of them are given. This is called the **ambiguous case** because the given information may not determine a unique triangle.

If α, a, and b are given, we consider trying to construct a triangle fitting these data by first drawing angle α, then marking off b on one of its sides thus determining vertex C. Finally, we attempt to locate vertex B by striking off a circular arc of radius a with center at C. If $a \geq b$, this can always be done in a unique way. Figure 4 illustrates this both for α acute and α obtuse. If $a < b$, there are several possibilities (Figure 5).

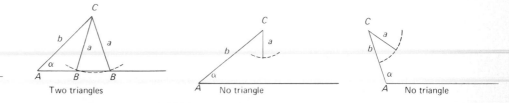

Figure 5

Fortunately, we are able to decide which of these possibilities is the case if we draw an approximate picture and then attempt to apply the law of sines. First, note that if $a \geq b$ there is one triangle corresponding to the data and for it β is an acute angle. Application of the law of sines will give $\sin \beta$, allowing determination of β.

Example B (SSA: One Triangle)   Suppose that in triangle ABC, $\alpha = 142°$, $a = 25.2$, and $b = 19.6$ (Figure 6). Find β, γ, and c.

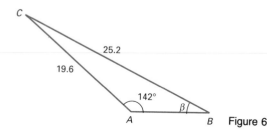

Figure 6

Solution   Since $a > b$, the given information determines a unique triangle. Moreover, β will be an acute angle.

1. $\dfrac{\sin \beta}{19.6} = \dfrac{\sin 142°}{25.2}$

$\sin \beta = \dfrac{19.6 \sin 142°}{25.2} \approx .47885$

$\beta \approx 28.6°$

2. $\gamma \approx 180° - 142° - 28.6° = 9.4°$

3. $\dfrac{c}{\sin 9.4°} = \dfrac{25.2}{\sin 142°}$

$c = \dfrac{25.2 \sin 9.4°}{\sin 142°} \approx 6.69$ ∎

If $a < b$, we may attempt to apply the law of sines. If it yields $\sin \beta = 1$, we have a unique right triangle. If it yields $\sin \beta < 1$, we have two triangles corresponding to the two angles β_1 and β_2 (one acute, the other obtuse) with this sine. If it yields $\sin \beta > 1$, we have an inconsistency in the data; no triangle satisfying the data exists.

Example C (SSA: Two Triangles)   Suppose that in triangle ABC, $\alpha = 36°$, $a = 9.4$, and $b = 13.1$ (Figure 7). Find β, γ, and c.

Solution

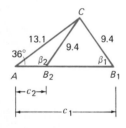

Figure 7

1. $\dfrac{\sin \beta}{13.1} = \dfrac{\sin 36°}{9.4}$

$\sin \beta = \dfrac{(13.1) \sin 36°}{9.4} = \dfrac{(13.1)(.5878)}{9.4} \approx .8191$

Since $\sin \beta < 1$, there are two triangles.

$\beta_1 = 55°$ $\beta_2 = 125°$

2. $\gamma_1 = 180° - (36° + 55°) = 89°$
$\gamma_2 = 180° - (36° + 125°) = 19°$

3. $\dfrac{c_1}{\sin 89°} = \dfrac{9.4}{\sin 36°}$

$c_1 = \dfrac{9.4}{\sin 36°}(\sin 89°) \approx 16.0$

$\dfrac{c_2}{\sin 19°} = \dfrac{9.4}{\sin 36°}$

$c_2 = \dfrac{9.4}{\sin 36°}(\sin 19°) \approx 5.2$ ∎

An Important Area Formula

Consider a triangle with two sides b and c and included angle α. Then the area A of the triangle is given by

$$A = \tfrac{1}{2}bc \sin \alpha$$

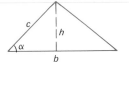

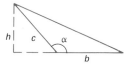

Figure 8

To see that this is so, let h denote the altitude of the triangle as shown in the two diagrams of Figure 8. Whether α is acute or obtuse, we see that $\sin \alpha = h/c$, that is, $h = c \sin \alpha$. We conclude that $A = \tfrac{1}{2}bh = \tfrac{1}{2}bc \sin \alpha$.

Example D (Area of a Triangle) Suppose that in triangle ABC, $b = 143.1$ meters, $c = 421.5$ meters, and $\alpha = 2.312$ radians. Find the area of the triangle.

Solution

$$A = \frac{1}{2}bc \sin \alpha = \frac{1}{2}(143.1)(421.5) \sin 2.312 \approx 22.246$$

The answer is in square meters. ∎

PROBLEM SET 9-1

A. Skills and Techniques

Solve the triangles of Problems 1–10, as in Examples A–C.

1. $\alpha = 42.6°$, $\beta = 81.9°$, $a = 14.3$
2. $\beta = 123°$, $\gamma = 14.2°$, $a = 295$
3. $\alpha = \gamma = 62°$, $b = 50$
4. $\alpha = \beta = 14°$, $c = 30$
5. $\alpha = 115°$, $a = 46$, $b = 34$
6. $\beta = 143°$, $a = 46$, $b = 84$
7. $\alpha = 30°$, $a = 8$, $b = 5$
8. $\beta = 60°$, $a = 11$, $b = 12$
9. $\alpha = 30°$, $a = 5$, $b = 8$

10. $\beta = 60°$, $a = 12$, $b = 11$

11. Two observers stationed 110 meters apart at A and B on the bank of a river are looking at a tower situated at a point C on the opposite bank. They measure angles CAB and CBA to be 43° and 57°, respectively. How far is the first observer from the tower?

12. A telegraph pole leans away from the sun at an angle of 11° to the vertical. The pole casts a shadow 96 feet long on horizontal ground when the angle of elevation of the sun is 23°. Find the length of the pole (see Figure 9 on the next page).

13. A vertical pole 60 feet long is standing by the side of an inclined road. It casts a shadow 138 feet long di-

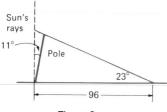

Figure 9

rectly downhill along the road when the angle of elevation of the sun is 58°. Find the angle of inclination θ of the road (see Figure 10).

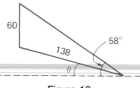

Figure 10

14. Two forest rangers 15 miles apart at points A and B observe a fire at a point C. The ranger at A measures angle CAB as 43.6° and the one at B measures angle CBA as 79.3°. How far is the fire from each ranger? How far is the fire from a straight road that goes from A to B?

For Problems 15–18, refer to Example D.

15. Find the area of the triangle with sides $b = 20$, $c = 30$, and included angle $\alpha = 40°$.

16. Find the area of the triangle with $a = 14.6$, $b = 31.7$, and $\gamma = 130.2°$.

17. Find the area of the triangle with $c = 30.1$, $\alpha = 25.3°$, and $\beta = 112.2°$.

18. Find the area of the triangle with $a = 20$, $\alpha = 29°$, and $\gamma = 46°$.

B. Applications and Extensions

19. The children's slide at the park is 30 feet long and inclines 36° from the horizontal. The ladder to the top is 18 feet long. How steep is the ladder, that is, what angle does it make with the horizontal? Assume the slide is straight and that the bottom end of the slide is at the same level as the bottom end of the ladder.

20. Prevailing winds have caused an old tree to incline 11° eastward from the vertical. The sun in the west is 32° above the horizontal. How long a shadow is cast by

the tree if the tree measures 114 feet from top to bottom?

21. A rectangular room, 16 feet by 30 feet, has an open beam ceiling. The two parts of the ceiling make angles of 65° and 32° with the horizontal (an end view is shown in Figure 11). Find the total area of the ceiling.

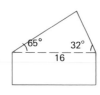

Figure 11

22. Sheila Sather, traveling north on a straight road at a constant rate of 60 miles per hour, sighted flames shooting up into the air at a point 20° west of north. Exactly 1 hour later, the fire was 59° west of south. Determine the shortest distance from the road to the fire.

23. A lighthouse stands at a certain distance out from a straight shoreline. It throws a beam of light that revolves at a constant rate of one revolution per minute. A short time after shining on the nearest point on the shore, the beam reaches a point on the shore that is 2640 feet from the lighthouse, and 3 seconds later it reaches a point 2000 feet farther along the shore. How far is the lighthouse from the shore?

24. In Figure 12, AC is 10 meters longer than CB. Determine the length of CD.

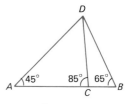

Figure 12

25. Two points A and B are on level ground and in line with the base C of a tower. The angles of elevation of the top of the tower at A and B are 21° and 35°, respectively. How tall is the tower if A and B are 300 feet apart?

26. Tang and Wang left point O together, Tang heading in the direction N58.6°E and Wang in the direction N33.2°E. After 20 seconds, they were 200 feet apart. If Tang ran at 12 feet per second, how fast did Wang run?

27. Spokes OD, OF, and OE of lengths 12, 6, and 10 radiate from a common point O. The angles DOF and FOE are each 20°. Find the area of triangle DEF.

28. The dial of a clock has a radius of 5 inches and the minute and hour hands are 4 and 3 inches long, respectively (Figure 13). D is a fixed point on the rim of the dial at the 12 mark, E is the tip of the minute hand, and F is the tip of the hour hand. The points D, E, and F determine a triangle that changes with time. Let t denote the number of minutes after 12:00 and let $A(t)$ be the area of triangle DEF at time t. Show that

$$A(t) = \frac{1}{2}\left| 20 \sin \frac{\pi t}{30} - 15 \sin \frac{\pi t}{360} - 12 \sin \frac{11\pi t}{360} \right|$$

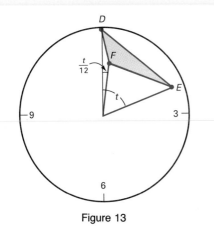

Figure 13

29. Four line segments of lengths 3, 4, 5, and 6 radiate like spokes from a common point. Their outer ends are the vertices of a quadrilateral Q. Determine the maximum possible area of Q.

30. Figure 14 illustrates the Pythagorean Theorem $(a^2 + b^2 = c^2)$. A rubber band is stretched around this figure. Show that the area of the region enclosed by the rubber band is $2(ab + c^2)$.

31. Let 2ϕ denote the angle at a point of the regular six-

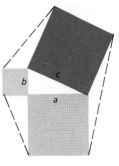

Figure 14

pointed star shown in Figure 15. Express the area A of this star in terms of ϕ and the edge length r.

Figure 15

32. TEASER Figure 16 shows two mirrors intersecting at an angle of 15°. A light ray from S is reflected at P and again at Q and then is absorbed at R. It is given that $ST = RU = 5$, $OT = 50$, and $OU = 20$. Find the length $x + y + z$ of the path of the light ray. As indicated, the angle of incidence equals the angle of reflection.

33. Graph the function $A(t)$ of Problem 28 for $0 \le t \le 60$. Determine the time between 12:00 and 1:00 when the area $A(t)$ is a maximum.

34. Determine the first time after 12:00 when the points D, E, and F of Problem 28 are collinear.

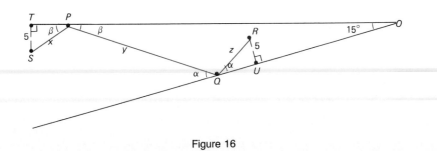

Figure 16

9-2 OBLIQUE TRIANGLES: LAW OF COSINES

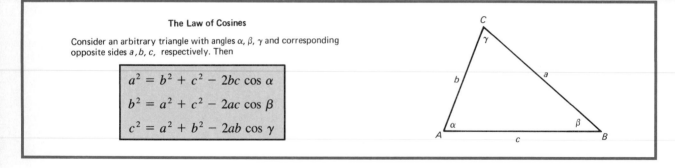

The Law of Cosines

Consider an arbitrary triangle with angles α, β, γ and corresponding opposite sides a, b, c, respectively. Then

$$a^2 = b^2 + c^2 - 2bc \cos \alpha$$
$$b^2 = a^2 + c^2 - 2ac \cos \beta$$
$$c^2 = a^2 + b^2 - 2ab \cos \gamma$$

When two sides and the included angle (SAS) or three sides (SSS) of a triangle are given, we cannot apply the law of sines to solve the triangle. Rather, we need the law of cosines, stated above in symbols. Actually it is wise to learn the law in words.

The square of any side is equal to the sum of the squares of the other two sides minus twice the product of those sides times the cosine of the angle between them.

Notice what happens when $\gamma = 90°$ so that $\cos \gamma = 0$. The law of cosines

$$c^2 = a^2 + b^2 - 2ab \cos \gamma$$

becomes

$$c^2 = a^2 + b^2$$

which is just the Pythagorean Theorem. In fact, you should think of the law of cosines as a generalization of the Pythagorean Theorem, with the term $-2ab \cos \gamma$ acting as a correction term when γ is not 90°.

Proof of the Law of Cosines

Assume first that angle α is acute. Drop a perpendicular CD from vertex C to side AB as shown in Figure 17. Label the lengths of CD, AD, and DB by h, x, and $c - x$, respectively.

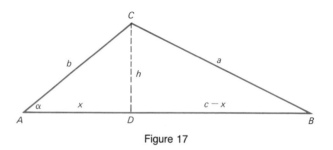

Figure 17

Consider the two right triangles ADC and BDC. By the Pythagorean Theorem,

$$h^2 = b^2 - x^2 \quad \text{and} \quad h^2 = a^2 - (c - x)^2$$

Equating these two expressions for h^2 gives

$$a^2 - (c - x)^2 = b^2 - x^2$$
$$a^2 = b^2 - x^2 + (c - x)^2$$
$$a^2 = b^2 - x^2 + c^2 - 2cx + x^2$$
$$a^2 = b^2 + c^2 - 2cx$$

Now $\cos \alpha = x/b$, and so $x = b \cos \alpha$. Thus

$$a^2 = b^2 + c^2 - 2cb \cos \alpha$$

which is the result we wanted.

Next we give the proof of the law of cosines for the obtuse angle case. Again drop a perpendicular from vertex C to side AB extended and label the resulting diagram as shown in Figure 18.

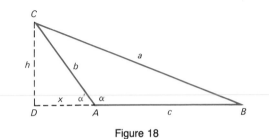

Figure 18

From consideration of triangles ADC and BDC and the Pythagorean Theorem, we obtain

$$h^2 = b^2 - x^2 \quad \text{and} \quad h^2 = a^2 - (c + x)^2$$

Algebra analogous to that used in the acute angle case yields

$$a^2 = b^2 + c^2 + 2cx$$

Now α' is the reference angle for α, and so $\cos \alpha = -\cos \alpha'$. Also $\cos \alpha' = x/b$. Therefore,

$$x = b \cos \alpha' = -b \cos \alpha$$

When we substitute this expression for x in the equation above, we get

$$a^2 = b^2 + c^2 - 2cb \cos \alpha$$

Solving a Triangle (SAS)

A unique triangle is determined by giving the lengths of two sides and the included angle. With this information we can solve the triangle by using the law of cosines to find the third side and then the law of sines to find a second angle.

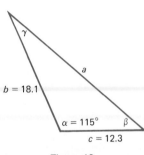

Figure 19

Example A (SAS)   Consider a triangle with $b = 18.1$, $c = 12.3$, and $\alpha = 115°$ (Figure 19). Determine a, β, and γ.

Solution

1. By the law of cosines,

$$a^2 = (18.1)^2 + (12.3)^2 - 2(18.1)(12.3) \cos 115° \approx 667.075$$

$$a \approx 25.8$$

2. Now we can use the law of sines.

$$\frac{\sin \beta}{18.1} = \frac{\sin 115°}{25.8}$$

$$\sin \beta = \frac{(18.1) \sin 115°}{25.8} \approx .63582$$

$$\beta \approx 39.5°$$

3. $\gamma \approx 180° - (115° + 39.5°) = 25.5°$ ∎

Solving a Triangle (SSS)

The law of cosines allows us to solve a triangle when three sides are given.

Example B (SSS) Solve the triangle ABC in which $a = 13.1$, $b = 15.5$, and $c = 17.2$.

Solution

1. By the law of cosines,

$$a^2 = b^2 + c^2 - 2bc \cos \alpha$$

Thus

$$\cos \alpha = \frac{b^2 + c^2 - a^2}{2bc}$$

$$= \frac{(15.5)^2 + (17.2)^2 - (13.1)^2}{2(15.5)(17.2)} = .6836$$

$$\alpha \approx 46.9°$$

2. By the law of sines,

$$\frac{\sin \beta}{15.5} = \frac{\sin 46.9°}{13.1}$$

$$\sin \beta = \frac{(15.5)(\sin 46.9°)}{13.1} = .86393$$

$$\beta \approx 59.8°$$

3. $\gamma = 180° - (46.9° + 59.8°) = 73.3°$. ∎

Heron's Area Formula

Since the three sides of a triangle completely determine the triangle, they must also determine the area of the triangle. About 2000 years ago, Heron of Alexandria found the formula that illustrates this fact. Let a, b, and c denote the three sides of a triangle and let $s = (a + b + c)/2$ be its semiperimeter. Then the area A of the triangle is given by

$$A = \sqrt{s(s - a)(s - b)(s - c)}$$

We will prove that this formula is correct. The proof is subtle, depending on the clever matching of the area formula from the last section with the law of cosines. We begin by writing the law of cosines in the form (see Figure 20)

$$2bc \cos \alpha = b^2 + c^2 - a^2$$

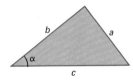

Figure 20

a formula we will use shortly. Next, we take the area formula $A = \frac{1}{2}bc \sin \alpha$ in its squared form and manipulate it very carefully.

$$A^2 = \frac{1}{4}b^2c^2 \sin^2 \alpha = \frac{1}{4}b^2c^2(1 - \cos^2 \alpha)$$

$$= \frac{1}{16}(2bc)(1 + \cos \alpha)(2bc)(1 - \cos \alpha)$$

$$= \frac{1}{16}(2bc + 2bc \cos \alpha)(2bc - 2bc \cos \alpha)$$

$$= \frac{1}{16}(2bc + b^2 + c^2 - a^2)(2bc - b^2 - c^2 + a^2)$$

$$= \frac{1}{16}[(b + c)^2 - a^2][a^2 - (b - c)^2]$$

$$= \frac{(b + c + a)}{2} \frac{(b + c - a)}{2} \frac{(a - b + c)}{2} \frac{(a + b - c)}{2}$$

$$= \left[\frac{a + b + c}{2}\right]\left[\frac{a + b + c}{2} - a\right]\left[\frac{a + b + c}{2} - b\right]\left[\frac{a + b + c}{2} - c\right]$$

$$= s(s - a)(s - b)(s - c)$$

Example C (Area in Terms of Sides) Find the area of the triangle with sides 21.4, 29.6, and 34.2.

Solution   The semiperimeter is

$$s = \frac{21.4 + 29.6 + 34.2}{2} = 42.6$$

Thus, by Heron's formula, the area A of the triangle is

$$A = [(42.6)(42.6 - 21.4)(42.6 - 29.6)(42.6 - 34.2)]^{1/2} \approx 314.04 \quad \blacksquare$$

PROBLEM SET 9-2

A. Skills and Techniques

In Problems 1–8, solve the triangles satisfying the given data, as in Examples A and B.

1. $\alpha = 60°$, $b = 14$, $c = 10$
2. $\beta = 60°$, $a = c = 8$
3. $\gamma = 120°$, $a = 8$, $b = 10$
4. $\alpha = 150°$, $b = 35$, $c = 40$
5. $a = 5$, $b = 6$, $c = 7$
6. $a = 10$, $b = 20$, $c = 25$
7. $a = 12.2$, $b = 19.1$, $c = 23.8$
8. $a = .11$, $b = .21$, $c = .31$
9. At one corner of a triangular field, the angle measures 52.4°. The sides that meet at this corner are 100 meters and 120 meters long. How long is the third side?
10. To approximate the distance between two points A and B on opposite sides of a swamp, a surveyor selects a point C and measures it to be 140 meters from A and 260 meters from B. Then she measures the angle ACB, which turns out to be 49°. What is the calculated distance from A to B?
11. Two runners start from the same point at 12:00 noon, one of them heading north at 6 miles per hour and the other heading 68° east of north at 8 miles per hour (Figure 21). What is the distance between them at 3:00 that afternoon?

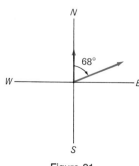

Figure 21

12. A 50-foot pole stands on top of a hill which slants 20° from the horizontal. How long must a rope be to reach from the top of the pole to a point 88 feet directly downhill (that is, on the slant) from the base of the pole?
13. A triangular garden plot has sides of length 35 meters, 40 meters, and 60 meters. Find the largest angle of the triangle.
14. A piece of wire 60 inches long is bent into the shape of a triangle. Find the angles of the triangle if two of the sides have lengths 24 inches and 20 inches.

Problems 15–18 relate to Example C.

15. The area of the right triangle with sides 3, 4, and 5 is 6. Confirm that Heron's formula gives the same answer.
16. Find the area of the triangle with sides 31, 42, and 53.
17. Find the area of the triangle with sides 5.9, 6.7, and 10.3.
18. Use the answer you got to Problem 16 to find the length h of the shortest altitude of the triangle with sides 31, 42, and 53.

B. Applications and Extensions

19. A triangular garden plot has sides measuring 42 meters, 50 meters, and 63 meters. Find the measure of the smallest angle.
20. A diagonal and a side of a parallelogram measure 80 centimeters and 25 centimeters, respectively, and the angle between them measures 47°. Find the length of the other diagonal. Recall that the diagonals of a parallelogram bisect each other.
21. Two cars, starting from the intersection of two straight highways, travel along the highways at speeds of 55 miles per hour and 65 miles per hour, respectively. If the angle of intersection of the highways measures 72°, how far apart are the cars after 36 minutes?
22. Buoys A, B, and C mark the vertices of a triangular racing course on a lake. Buoys A and B are 4200 feet apart, buoys A and C are 3800 feet apart, and angle CAB measures 100°. If the winning boat in a race covered the course in 6.4 minutes, what was its average speed in miles per hour?
23. The sides of a triangular garden plot with area 200 square meters are in the proportion $3:2:2$. Find the length of each side to the nearest hundredth of a meter.
24. Maria, running at 11 feet per second, ran first in the direction N67.9°E for 8 seconds and then in the direc-

tion N29.8°E for the next 12 seconds. How far did this put her from the starting point?

25. Three mutually tangent circles have radii 4, 5, and 6, respectively (Figure 22). Find
    (a) the angles of the triangle with vertices at their centers.
    (b) the area of the white region between the circles.

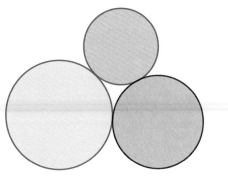

Figure 22

26. Spokes OD, OF, and OE of lengths 12, 6, and 10 radiate from a common point O. The angles DOF and FOE are each 20°. Find the perimeter of triangle DEF (Figure 23).

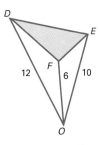

Figure 23

27. A quadrilateral Q has sides of length 1, 2, 3, and 4, respectively. The angle between the first pair of sides is 120°. Find the angle between the other pair of sides and also the exact area of Q.

28. For the triangle ABC in Figure 24, let r be the radius

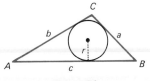

Figure 24

of the inscribed circle and let $s = (a + b + c)/2$ be its semiperimeter.
    (a) Show that the area of the triangle is rs.
    (b) Show that $r = \sqrt{(s - a)(s - b)(s - c)/s}$.
    (c) Find r for a triangle with sides 5, 6, and 7.

29. Consider a triangle with sides of length 4, 5, and 6. Show that one of its angles is twice another. *Hint:* Show that the cosine of twice one angle is equal to the cosine of another angle.

30. In the triangle with sides of length a, b, and c, let a_1, b_1, and c_1 denote the lengths of the corresponding medians to these sides from the opposite vertices. Show that

$$a_1^2 + b_1^2 + c_1^2 = \tfrac{3}{4}(a^2 + b^2 + c^2)$$

31. Determine the length of AB in Figure 25.

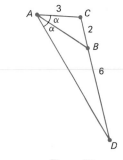

Figure 25

32. **TEASER** The two hands of a clock are 4 and 5 inches long, respectively. At some time between 1:45 and 2, the tips of the hands are 8 inches apart. What time is it then?

33. Refer to the clock illustration in Problem 28 of Section 9-1. Convince yourself that the perimeter $P(t)$ of triangle DEF at time t is given by

$$P(t) = \left(41 - 40 \cos \frac{\pi t}{30}\right)^{1/2}$$
$$+ \left(34 - 30 \cos \frac{\pi t}{360}\right)^{1/2} + \left(25 - 24 \cos \frac{11\pi t}{360}\right)^{1/2}$$

Use the graph of $P(t)$ to determine the time between 12:00 and 1:10 when the perimeter is a maximum.

34. Determine the time between 12:10 and 1:10 when the perimeter of the triangle of Problem 33 is a minimum.

9-3 VECTORS AND SCALARS

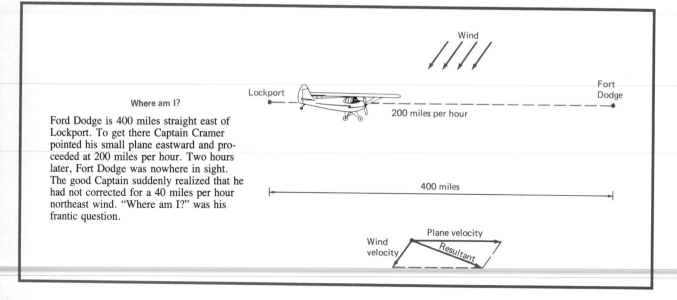

Where am I?

Ford Dodge is 400 miles straight east of Lockport. To get there Captain Cramer pointed his small plane eastward and proceeded at 200 miles per hour. Two hours later, Fort Dodge was nowhere in sight. The good Captain suddenly realized that he had not corrected for a 40 miles per hour northeast wind. "Where am I?" was his frantic question.

Many quantities that occur in science (for example, length, mass, volume, and electric charge) can be specified by giving a single number. We call these quantities *scalar quantities;* the numbers that measure their magnitudes are called **scalars.** Other quantities, such as velocity, force, torque, and displacement, must be specified by giving both a magnitude and a direction. We call such quantities **vectors** and represent them by arrows (directed line segments). The length of the arrow is the **magnitude** of the vector; its **direction** is the direction of the vector. Thus, in our opening diagram, the plane's velocity appears as an arrow 200 units long pointing eastward, while the wind velocity is shown as an arrow 40 units long pointing southwest. But how shall we put these two vectors together, that is, find their resultant? Before we try to answer, we introduce more terminology.

Tail   Head

Figure 26

Arrows that we draw, like those shot from a bow, have two ends. There is the initial or feather end, which we shall call the **tail,** and the pointed or terminal end, which we shall call the **head** (Figure 26). Two vectors are considered to be **equivalent** if they have the same magnitude and direction (Figure 27). We shall symbolize vectors by boldface letters, such as **u** and **v.** (Since this is hard to accomplish in normal writing, you might use $\vec{u}$ and $\vec{v}$.) Also the magnitude (length) of the vector **u** is denoted by $\|\mathbf{u}\|$.

Equivalent
vectors

Figure 27

Addition of Vectors

To find the **sum,** or resultant, of **u** and **v,** move **v** without changing its magnitude or direction until its tail coincides with the head of **u.** Then **u** + **v** is the vector connecting the tail of **u** to the head of **v** (see the left diagram in Figure 28).

As an alternative way to find **u** + **v,** move **v** so that its tail coincides with that of **u.** Then **u** + **v** is the vector with this common tail that is the diagonal of the par-

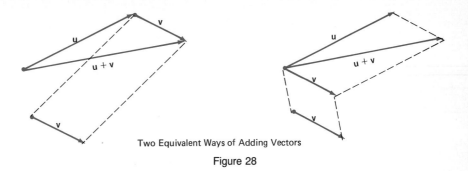

Two Equivalent Ways of Adding Vectors

Figure 28

allelogram with **u** and **v** as sides. This method (called the *parallelogram law*) is illustrated on the right in Figure 28.

You should convince yourself that addition is commutative and associative, that is, **u** + **v** = **v** + **u** and (**u** + **v**) + **w** = **u** + (**v** + **w**).

Scalar Multiplication and Subtraction

If **u** is a vector, then 3**u** is the vector with the same direction as **u** but three times as long; −2**u** is twice as long as **u** and oppositely directed (Figure 29). More generally, c**u** has magnitude $|c|$ times that of **u** and is similarly or oppositely directed, depending on whether c is positive or negative. In particular (−1)**u** (usually written −**u**) has the same length as **u** but the opposite direction. It is called the **negative** of **u** because when we add it to **u**, the result is a vector that has shriveled to a point. This special vector (the only vector without direction) is called the **zero vector** and is denoted by **0**. It is the identity element for addition; that is, **u** + **0** = **0** + **u** = **u**. Finally, subtraction is defined by

$$\mathbf{u} - \mathbf{v} = \mathbf{u} + (-\mathbf{v})$$

Example A (An Algebraic Combination of Two Vectors) For a given **u** and **v**, determine **w** = 3**u** − 2**v**.

Solution   The construction of **w** from **u** and **v** is shown in Figure 30.

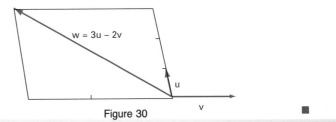

Figure 30

Example B (Expressing a Vector in Terms of Others) *M* is the midpoint of the upper side of the parallelogram in Figure 31. Express **r** and **s** in terms of **u** and **v**.

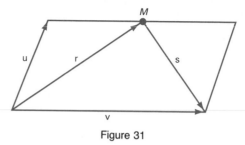

Figure 31

Solution   Clearly, $\mathbf{r} = \mathbf{u} + \frac{1}{2}\mathbf{v}$. Also, $\mathbf{r} + \mathbf{s} = \mathbf{v}$, so

$$\mathbf{s} = \mathbf{v} - \mathbf{r} = \mathbf{v} - (\mathbf{u} + \tfrac{1}{2}\mathbf{v}) = -\mathbf{u} + \tfrac{1}{2}\mathbf{v} \quad \blacksquare$$

Applications

Displacements, velocities, and forces have both magnitude and direction and they behave like vectors under addition and scalar multiplication.

Example C (Displacements Are Vectors)   In navigation, directions are specified by giving an angle, called the **bearing,** with respect to a north-south line. Thus a bearing of N35°E denotes an angle whose initial side points north and whose terminal side is 35° east of north. If a ship sails 70 miles in the direction N35°E and then 90 miles straight east, what is its distance and bearing with respect to its starting point?

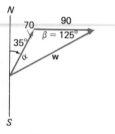

Figure 32

Solution   Our job is to determine the length and bearing of $\mathbf{w}$ (see Figure 32). We first use a little geometry to determine that $\beta = 125°$. Then, by the law of cosines,

$$\|\mathbf{w}\|^2 = (70)^2 + (90)^2 - 2(70)(90) \cos 125° \approx 20{,}227$$

$$\|\mathbf{w}\| \approx 142$$

By the law of sines,

$$\frac{\sin \alpha}{90} = \frac{\sin 125°}{142}$$

$$\sin \alpha = \frac{90 \sin 125°}{142} \approx .5192$$

$$\alpha \approx 31°$$

Thus the bearing of $\mathbf{w}$ is N66°E. $\blacksquare$

Example D (Velocities Are Vectors)   The river is flowing at 6 miles per hour and Jane's boat travels at 20 miles per hour in still water. In what direction should she head her boat if she wants to go straight across the river?

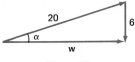

Figure 33

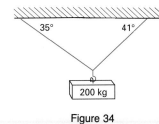

Figure 34

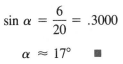

Figure 35

Solution   Our job is simply to determine α in Figure 33.

$$\sin \alpha = \frac{6}{20} = .3000$$

$$\alpha \approx 17° \quad \blacksquare$$

Example E (Forces Are Vectors)   A weight of 200 kilograms is supported by two wires, as shown in Figure 34. Find the magnitude of the tension in each wire.

Solution   The weight and the two tensions are forces which behave as vectors. These three forces must balance; that is, the two forces exerted by the wires must add together and cancel the downward force of the weight. This will happen if their sum is a vector of magnitude 200 pointing upward, as shown in Figure 35. This figure is a parallelogram composed of two congruent triangles. Using the given 35° and 41° angles and the fact that the angles of a triangle have a sum of 180°, we can find all the angles of the figure, as shown. By the law of sines,

$$\frac{\|\mathbf{u}\|}{\sin 49°} = \frac{200}{\sin 76°}$$

$$\|\mathbf{u}\| = \frac{200 \sin 49°}{\sin 76°} \approx 156$$

Similarly,

$$\|\mathbf{v}\| = \frac{200 \sin 55°}{\sin 76°} \approx 169$$

Note that the larger tension is in the shorter wire, as we should expect. $\blacksquare$

Captain Cramer's Question

Consider again the problem posed in the display that opens this section. We know that velocities add as vectors. Consequently, our first problem is to add **u**, which points east and is 200 units long, to a vector **v**, which points southwest and is 40 units long. Specifically, our aim is to find the length of **u** + **v**, denoted by $\|\mathbf{u} + \mathbf{v}\|$, and the angle α that **u** + **v** makes with **u**. The situation is shown in Figure 36. Note that we interpret a northeast wind to mean that $\beta = 45°$.

Now, by the law of cosines,

$$\|\mathbf{u} + \mathbf{v}\|^2 = (200)^2 + (40)^2 - 2(200)(40) \cos 45°$$

$$\approx 30,300$$

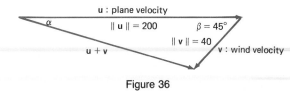

Figure 36

Thus

$$\|\mathbf{u} + \mathbf{v}\| \approx 174$$

Next, by the law of sines,

$$\frac{\sin \alpha}{40} = \frac{\sin 45°}{174}$$

or

$$\sin \alpha = \frac{40 \sin 45°}{174} \approx .1626$$

$$\alpha \approx 9.4°$$

Where is Captain Cramer? Since his true velocity is the vector $\mathbf{u} + \mathbf{v}$ and since he flew at this velocity for 2 hours, he is at a point $2(174) = 348$ miles from Lockport along the line that makes an angle of $9.4°$ with the line between Lockport and Fort Dodge. Of course, he is also 80 miles southwest of Fort Dodge.

PROBLEM SET 9-3

A. Skills and Techniques

In Problems 1–4, draw the vector **w** *so that its tail is at the heavy dot. See Example A.*

1. w = u + v

2. w = u − v

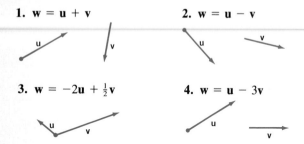

3. w = −2u + ½v

4. w = u − 3v

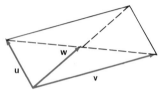

Problems 5–8 relate to Example B.

5. Figure 37 shows a parallelogram. Express **w** in terms of **u** and **v**.

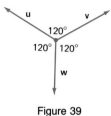

Figure 37

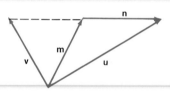

Figure 38

6. In the large triangle of Figure 38, **m** is a median (it bisects the side to which it is drawn). Express **m** and **n** in terms of **u** and **v**.

7. In Figure 39, $\mathbf{w} = -(\mathbf{u} + \mathbf{v})$ and $\|\mathbf{u}\| = \|\mathbf{v}\| = 1$. Find $\|\mathbf{w}\|$.

Figure 39

8. Do Problem 7 if the top angle is $90°$ and the two side angles are $135°$.

For Problems 9–12, see Example C.

9. If I walk 10 miles N45°E and then 10 miles straight north, how far am I from my starting point?

10. In Problem 9, what is the bearing of my final position with respect to my starting point?

11. As airplane flew 100 kilometers in the direction S51°W and then 145 kilometers S 39°W. What was the airplane's distance and bearing with respect to its starting point?

12. A ship sailed 11.2 miles straight north and then 48.3 miles N13.2°W. Find its distance and bearing with respect to the starting point.

13. If the river (see Example D) is $\frac{1}{2}$ mile wide, how long will it take Jane to get across? *Hint:* First determine $\|\mathbf{w}\|$, which is her actual speed with respect to the shore.

14. If Jane (see Example D) had not corrected for the current (that is, if she had pointed her boat straight across), where would she have landed on the opposite shore?

15. A wind with velocity 58 miles per hour is blowing in the direction N20°W. An airplane that flies at 425 miles per hour in still air is supposed to fly straight north. How should the airplane be headed and how fast will it then be flying with respect to the ground?

16. A ship is sailing due south at 20 miles per hour. A man walks west (that is, at right angles to the side of the ship) across the deck at 3 miles per hour. What are the magnitude and direction of his velocity relative to the surface of the water?

17. In Figure 40, $\|\mathbf{u}\| = \|\mathbf{v}\| = 10$. Find the magnitude and direction of a force $\mathbf{w}$ needed to counterbalance $\mathbf{u}$ and $\mathbf{v}$ (see Example E).

18. John pushes on a post from the direction S30°E with a force of 50 pounds. Wayne pushes on the same post from the direction S60°W with a force of 40 pounds. What is the magnitude and direction of the resultant force?

19. A body weighing 237.5 pounds is held in equilibrium by two ropes that make angles of 27.34° and 39.22°, respectively, with the vertical. Find the magnitude of the force exerted on the body by each rope.

20. A 250-kilogram weight rests on a smooth (friction negligible) inclined plane that makes an angle of 30° with the horizontal. What force parallel to the plane will just keep the weight from sliding down the plane? *Hint:* Consider the downward force of 250 kilograms to be the sum of two forces, one parallel to the plane and one perpendicular to it.

B. Applications and Extensions

21. Draw the sum of the three vectors shown in Figure 41.

22. Draw $\mathbf{u} - \mathbf{v} + \frac{1}{2}\mathbf{w}$ for Figure 41.

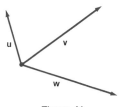

Figure 41

23. Three vectors form the edges of a triangle and are oriented clockwise around it. What is their sum?

24. Four vectors each of length 1 point in the directions N, N30°E, N60°E, and E, respectively. Find the exact length and direction of their sum.

25. Refer to Figure 42. Express each of the following in terms of $\overrightarrow{AD}$ and $\overrightarrow{AB}$.
    (a) $\overrightarrow{BD}$ (b) $\overrightarrow{AF}$ (c) $\overrightarrow{DE}$ (d) $\overrightarrow{AF} - \overrightarrow{DE}$

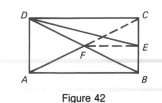

Figure 42

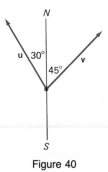

Figure 40

26. Let **u**, **v**, and **w** be the vectors from the vertices of a triangle to the midpoints of the opposite edges (the medians). Show that **u** + **v** + **w** = **0**.

27. Alice and Bette left point P at the same time and met at point Q two hours later. To get there Alice walked a straight path, but Bette first walked 1 mile south and then 2 miles in the direction S60°E. How fast did Alice walk, assuming that she walked at a constant rate?

28. Suppose that **u**, **v**, and **w** of Figure 43 point in the directions N60°E, S45°E, N25°W and have lengths 60, $30\sqrt{2}$, 100, respectively. Find the length of **u** + **v** + **w**.

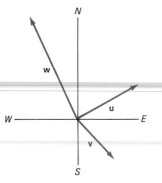

Figure 43

29. Two men are pushing an object along the ground. One is pushing with a force of 50 pounds in the direction N32°W and the other is pushing with a force of 100 pounds in the direction N30°E. In what direction is the object moving?

30. A pilot, flying in a wind which is blowing 80 miles per hour due south, discovers that she is heading due east when she points her plane in the direction N60°E. Find the air speed (speed in still air) of the plane.

31. What heading and air speed are required for a plane to fly 600 miles per hour due north if a wind of 56 miles per hour is blowing in the direction S12°E?

32. A spacecraft designed to softland on the moon has three legs whose feet form the vertices of an equilateral triangle on the ground. Each leg makes an angle of 35° with the vertical. If the impact force of 9000 pounds is evenly distributed, find the compression force on each leg.

33. What is the smallest force needed to keep a car weighing 3625 pounds from rolling down a hill that makes an angle of 10.35° with the horizontal?

34. Work Example E a different way as follows. For equilibrium, the magnitude $\|\mathbf{u}\|\sin 55°$ of the leftward force must equal the magnitude $\|\mathbf{v}\|\sin 49°$ of the rightward force. Similarly, the downward force of 200 must just balance the upward force of $\|\mathbf{u}\|\cos 55° + \|\mathbf{v}\|\cos 49°$. Solve the resulting pair of equations for $\|\mathbf{u}\|$ and $\|\mathbf{v}\|$ and confirm that you get the same answers we got earlier.

35. Suppose as in Example E that a weight w is supported by two wires making angles α and $\beta = 60°$ with the ceiling and creating tensions of 90 pounds in the first wire and 75 pounds in the second wire. Determine α and w by reasoning as in Problem 34.

36. **TEASER** Consider a horizontal triangular table (Figure 44) with each vertex angle being less than 120°. Three strings are knotted together at P and pass over frictionless pulleys at the vertices. Identical weights w are attached to the free ends of the strings. Show that at equilibrium, the angles between the strings at P are equal, that is, show that $\alpha + \beta = \alpha + \gamma = \beta + \gamma$.

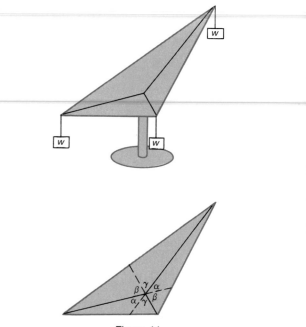

Figure 44

9-4 THE ALGEBRA OF VECTORS

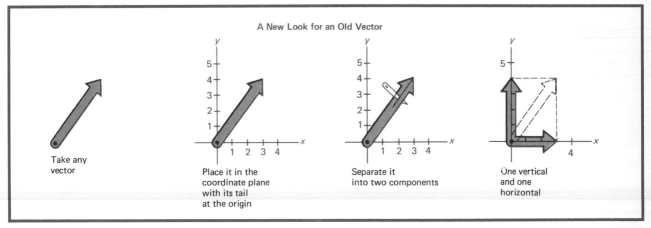

A New Look for an Old Vector

Take any vector

Place it in the coordinate plane with its tail at the origin

Separate it into two components

One vertical and one horizontal

Our treatment of vectors in Section 9-3 was mainly geometric. To give the subject an algebraic appearance, we first suppose that all vectors have been placed in the ordinary cartesian coordinate plane with their tails attached to the origin, (0, 0). In this case, both the magnitude and the direction of a vector are completely determined by the position of its head.

Next we select two vectors to play a permanent and special role. The first, called **i**, is the vector from (0, 0) to (1, 0); the second, called **j**, is the vector from (0, 0) to (0, 1). Then, as Figure 45 makes clear, an arbitrary vector **u** with its head at (a, b) can be expressed uniquely in the form

$$\mathbf{u} = a\mathbf{i} + b\mathbf{j}$$

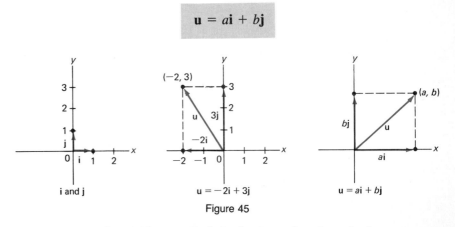

i and j

$u = -2i + 3j$

$u = ai + bj$

Figure 45

The vectors $a\mathbf{i}$ and $b\mathbf{j}$ are called the **horizontal** and **vertical vector components** of **u**, while a and b are called its **scalar components.** Notice that the length of **u** is easily expressed in terms of its scalar components.

$$\|\mathbf{u}\| = \sqrt{a^2 + b^2}$$

Algebraic Operations

To add the vectors $\mathbf{u} = a\mathbf{i} + b\mathbf{j}$ and $\mathbf{v} = c\mathbf{i} + d\mathbf{j}$, simply add the corresponding components; that is

$$\mathbf{u} + \mathbf{v} = (a + c)\mathbf{i} + (b + d)\mathbf{j}$$

Similarly, to multiply $\mathbf{u}$ by the scalar k, multiply each component by k. Thus

$$k\mathbf{u} = (ka)\mathbf{i} + (kb)\mathbf{j}$$

To see that these new algebraic rules are equivalent to the old geometric ones, study Figure 46.

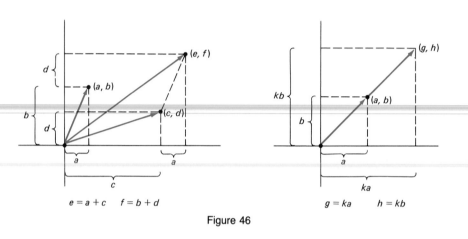

$$e = a + c \qquad f = b + d \qquad\qquad\qquad g = ka \qquad h = kb$$

Figure 46

Once the rules for addition and scalar multiplication are established, the rule for subtraction follows easily.

$$\mathbf{u} - \mathbf{v} = \mathbf{u} + (-1)\mathbf{v} = a\mathbf{i} + b\mathbf{j} + (-1)(c\mathbf{i} + d\mathbf{j}) = (a - c)\mathbf{i} + (b - d)\mathbf{j}$$

Moreover, with vectors written in component form, it is a simple matter to establish all of the following properties. Keep in mind that $\mathbf{0} = 0\mathbf{i} + 0\mathbf{j}$.

Algebraic Properties of Vectors

1. $\mathbf{u} + \mathbf{v} = \mathbf{v} + \mathbf{u}$
2. $\mathbf{u} + (\mathbf{v} + \mathbf{w}) = (\mathbf{u} + \mathbf{v}) + \mathbf{w}$
3. $\mathbf{u} + \mathbf{0} = \mathbf{0} + \mathbf{u} = \mathbf{u}$
4. $\mathbf{u} + (-\mathbf{u}) = -\mathbf{u} + \mathbf{u} = \mathbf{0}$
5. $k(\mathbf{u} + \mathbf{v}) = k\mathbf{u} + k\mathbf{v}$

6. $(k + l)\mathbf{u} = k\mathbf{u} + l\mathbf{u}$
7. $(kl)\mathbf{u} = k(l\mathbf{u}) = l(k\mathbf{u})$
8. $1\mathbf{u} = \mathbf{u}$
9. $0\mathbf{u} = \mathbf{0} = k\mathbf{0}$

Example A (Simple Calculations with Vectors) Let $\mathbf{u} = 3\mathbf{i} - 4\mathbf{j}$ and $\mathbf{v} = 2\mathbf{i} - 5\mathbf{j}$. Calculate the following.
(a) $\|\mathbf{u}\|$ (b) $4\mathbf{u} - 2\mathbf{v}$ (c) $\|4\mathbf{u} - 2\mathbf{v}\|$

Solution

(a) $\|\mathbf{u}\| = \sqrt{3^2 + (-4)^2} = \sqrt{25} = 5$

(b) $4\mathbf{u} - 2\mathbf{v} = 4(3\mathbf{i} - 4\mathbf{j}) - 2(2\mathbf{i} - 5\mathbf{j})$
$= 12\mathbf{i} - 16\mathbf{j} - 4\mathbf{i} + 10\mathbf{j} = 8\mathbf{i} - 6\mathbf{j}$

(c) $\|4\mathbf{u} - 2\mathbf{v}\| = \|8\mathbf{i} - 6\mathbf{j}\| = \sqrt{64 + 36} = 10$ ∎

The Dot Product

Is there a sensible way to multiply two vectors together? Yes; in fact, there are two kinds of products. One, called the vector product, requires three-dimensional space and therefore falls outside of the scope of this course. The other, called the **dot product** or **scalar product,** can be introduced now. If $\mathbf{u} = a\mathbf{i} + b\mathbf{j}$ and $\mathbf{v} = c\mathbf{i} + d\mathbf{j}$, then the dot product of $\mathbf{u}$ and $\mathbf{v}$ is the scalar given by

$$\mathbf{u} \cdot \mathbf{v} = ac + bd$$

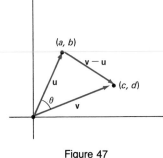

Figure 47

Why would anyone be interested in the dot product? To answer this, we need its geometric interpretation. Suppose that the heads of $\mathbf{u} = a\mathbf{i} + b\mathbf{j}$ and $\mathbf{v} = c\mathbf{i} + d\mathbf{j}$ are at (a, b) and (c, d), as shown in Figure 47. Then we may think of $\mathbf{v} - \mathbf{u}$ as the vector from (a, b) to (c, d). Let θ denote the smallest positive angle between $\mathbf{u}$ and $\mathbf{v}$. By the law of cosines,

$$\|\mathbf{v} - \mathbf{u}\|^2 = \|\mathbf{u}\|^2 + \|\mathbf{v}\|^2 - 2\|\mathbf{u}\| \, \|\mathbf{v}\| \cos \theta$$

$$(a - c)^2 + (b - d)^2 = a^2 + b^2 + c^2 + d^2 - 2\|\mathbf{u}\| \, \|\mathbf{v}\| \cos \theta$$

$$-2ac - 2bd = -2\|\mathbf{u}\| \, \|\mathbf{v}\| \cos \theta$$

$$ac + bd = \|\mathbf{u}\| \, \|\mathbf{v}\| \cos \theta$$

The last equality gives us a geometric formula for the dot product.

$$\mathbf{u} \cdot \mathbf{v} = \|\mathbf{u}\| \, \|\mathbf{v}\| \cos \theta$$

Of what use is this formula? For one thing, it gives us an easy way to tell when two vectors are perpendicular. Since $\cos \theta$ is zero if and only if θ is 90°, we see that:

Two vectors are perpendicular if and only if their dot product is zero.

More generally, we can use the formula to find the angle between any two vectors.

Example B (The Angle Between Two Vectors) Let $\mathbf{u} = 3\mathbf{i} + 4\mathbf{j}$ and $\mathbf{v} = -2\mathbf{i} + 3\mathbf{j}$. Find the (smallest positive) angle between $\mathbf{u}$ and $\mathbf{v}$.

Solution

$$\cos \theta = \frac{\mathbf{u} \cdot \mathbf{v}}{\|\mathbf{u}\| \, \|\mathbf{v}\|} = \frac{(3)(-2) + (4)(3)}{\sqrt{9 + 16}\sqrt{4 + 9}} = \frac{6}{5\sqrt{13}} \approx .3328$$

We conclude that $\theta \approx 70.6°$. ∎

In order to use the formula for the dot product, we must have our vectors in $a\mathbf{i} + b\mathbf{j}$ form.

Example C (Putting Vectors in $a\mathbf{i} + b\mathbf{j}$ Form) Let $\mathbf{u}$ be the vector from the point $P(1, 5)$ to the point $Q(6, 2)$ and let $\mathbf{v}$ be the vector from $P(1, 5)$ to $R(3, -4)$. Write both vectors in $a\mathbf{i} + b\mathbf{j}$ form and then calculate $\mathbf{u} \cdot \mathbf{v}$.

Solution We need the horizontal and vertical components of $\mathbf{u}$. These components are just the coordinates of the head of $\mathbf{u}$ after it has been translated so that its tail is at the origin (Figure 48). These coordinates are obtained by subtracting the coordinates of P from those of Q. Thus

$$\mathbf{u} = (6 - 1)\mathbf{i} + (2 - 5)\mathbf{j} = 5\mathbf{i} - 3\mathbf{j}$$

Similarly,

$$\mathbf{v} = (3 - 1)\mathbf{i} + (-4 - 5)\mathbf{j} = 2\mathbf{i} - 9\mathbf{j}$$

Finally,

$$\mathbf{u} \cdot \mathbf{v} = (5)(2) + (-3)(-9) = 37 \quad \blacksquare$$

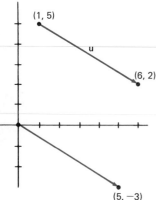

(1, 5)

(6, 2)

(5, −3)

Figure 48

Applications

As a first application, we introduce the concept of the projection of one vector on another. The scalar $\|\mathbf{u}\| \cos \theta$ is called the **scalar projection of u on v,** for reasons that should be apparent from Figure 49. It is positive, zero, or negative, depending on whether θ is acute, right, or obtuse. If we multiply this scalar by the vector $\mathbf{v}/\|\mathbf{v}\|$ of unit length, we get a vector $\mathbf{w}$, called the **vector projection of u on v.** Its geometric interpretation is shown in Figure 50. We can express both of these projections in terms of the dot product. Since

$$\|\mathbf{u}\| \cos \theta = \frac{\|\mathbf{u}\|\|\mathbf{v}\| \cos \theta}{\|\mathbf{v}\|}$$

it follows that

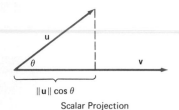

u

θ

v

$\|\mathbf{u}\| \cos \theta$

Scalar Projection

Figure 49

$$\text{scalar proj. } \mathbf{u} \text{ on } \mathbf{v} = \frac{\mathbf{u} \cdot \mathbf{v}}{\|\mathbf{v}\|}$$

$$\text{vector proj. } \mathbf{u} \text{ on } \mathbf{v} = \frac{\mathbf{u} \cdot \mathbf{v}}{\|\mathbf{v}\|^2}\mathbf{v} = \frac{\mathbf{u} \cdot \mathbf{v}}{\mathbf{v} \cdot \mathbf{v}}\mathbf{v}$$

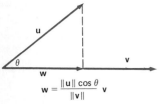

u

θ

w

v

$\mathbf{w} = \dfrac{\|\mathbf{u}\| \cos \theta}{\|\mathbf{v}\|} \mathbf{v}$

Vector Projection

Figure 50

Example D (The Projection of One Vector on Another) Let $\mathbf{u} = 5\mathbf{i} - 2\mathbf{j}$ and $\mathbf{v} = 3\mathbf{i} + 4\mathbf{j}$. Calculate the two projections discussed above.

Solution

$$\text{scalar proj. } \mathbf{u} \text{ on } \mathbf{v} = \frac{\mathbf{u} \cdot \mathbf{v}}{\|\mathbf{v}\|} = \frac{15 - 8}{5} = \frac{7}{5}$$

$$\text{vector proj. } \mathbf{u} \text{ on } \mathbf{v} = \frac{\mathbf{u} \cdot \mathbf{v}}{\mathbf{v} \cdot \mathbf{v}}\mathbf{v} = \frac{15 - 8}{25}(3\mathbf{i} + 4\mathbf{j}) = \frac{21}{25}\mathbf{i} + \frac{28}{25}\mathbf{j} \quad \blacksquare$$

In physics, the **work** done by a force **F** in moving an object from P to Q is defined to be the product of the magnitude of that force times the distance from P to Q. This assumes that the force is in the direction of the motion. In the more general case where the force **F** is at an angle to the motion, we must replace the magnitude of **F** by its scalar projection in the direction of the motion. If both **F** and the displacement **D** are treated as vectors, the work done is

$$\text{(scalar proj. } \mathbf{F} \text{ on } \mathbf{D}) \| \mathbf{D} \| = \frac{\mathbf{F} \cdot \mathbf{D}}{\| \mathbf{D} \|} \| \mathbf{D} \|$$

Thus

$$\text{work} = \mathbf{F} \cdot \mathbf{D}$$

Example E (Work Done by a Force) Find the work done by a force of 80 pounds in the direction N60°E in moving an object from $(1, 0)$ to $(7, -2)$ as in Figure 51.

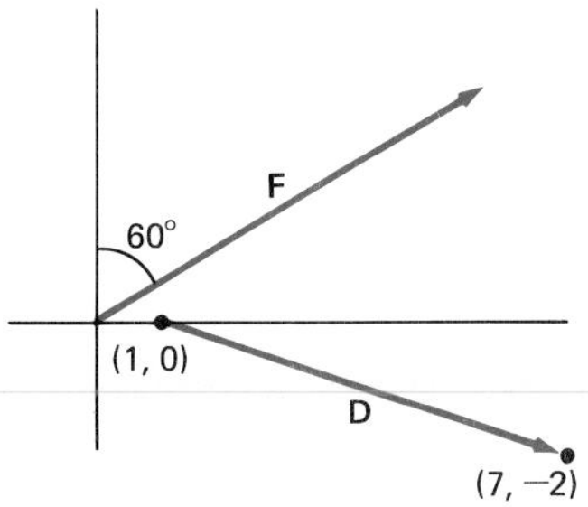

Solution is simply a matter of writing **F** and **D** in the form $a\mathbf{i} + b\mathbf{j}$ and taking their dot product.

$$\mathbf{F} = 80 \cos 30°\mathbf{i} + 80 \sin 30°\mathbf{j}$$
$$= 40\sqrt{3}\mathbf{i} + 40\mathbf{j}$$
$$\mathbf{D} = 6\mathbf{i} - 2\mathbf{j}$$
$$\mathbf{F} \cdot \mathbf{D} = 240\sqrt{3} - 80 \approx 336$$

If the units of distance are feet, the work done is 336 foot-pounds. ∎

Figure 51

PROBLEM SET 9-4

A. Skills and Techniques

In Problems 1–4, find $3\mathbf{u} - \mathbf{v}$, $\mathbf{u} \cdot \mathbf{v}$, and $\cos \theta$ for the given vectors $\mathbf{u}$ and $\mathbf{v}$. See Examples A and B.

1. $\mathbf{u} = 3\mathbf{i} - 4\mathbf{j}$, $\mathbf{v} = 5\mathbf{i} + 12\mathbf{j}$
2. $\mathbf{u} = \mathbf{i} + \sqrt{3}\mathbf{j}$, $\mathbf{v} = 6\mathbf{i} - 8\mathbf{j}$
3. $\mathbf{u} = 2\mathbf{i} - \mathbf{j}$, $\mathbf{v} = 3\mathbf{i} - 4\mathbf{j}$
4. $\mathbf{u} = \mathbf{i} + \mathbf{j}$, $\mathbf{v} = \mathbf{i} - \mathbf{j}$
5. If $\mathbf{u} = 14.1\mathbf{i} + 32.7\mathbf{j}$ and $\mathbf{v} = 19.2\mathbf{i} - 13.3\mathbf{j}$, find θ, the smallest positive angle between $\mathbf{u}$ and $\mathbf{v}$.
6. Determine the length of $2\mathbf{u} - 3\mathbf{v}$, where $\mathbf{u}$ and $\mathbf{v}$ are the vectors in Problem 5.

In Problems 7–10, let $\mathbf{u}$ be the vector from P to Q and $\mathbf{v}$ be the vector from P to R. Write both vectors in the form $a\mathbf{i} + b\mathbf{j}$ and then find $\mathbf{u} \cdot \mathbf{v}$. See Example C.

7. $P(1, 1)$, $Q(6, 3)$, $R(5, -2)$
8. $P(-1, 2)$, $Q(-3, 6)$, $R(0, -5)$

9. $P(1, 1)$, $Q(-3, -4)$, $R(-5, 6)$
10. $P(-1, -1)$, $Q(3, -5)$, $R(2, 4)$
11. If $\mathbf{u}$ is a vector 10 units long pointing in the direction N30°W, write $\mathbf{u}$ in the form $a\mathbf{i} + b\mathbf{j}$.
12. If $\mathbf{u}$ is a vector 9 units long pointing in the direction S21°W, write $\mathbf{u}$ in the form $a\mathbf{i} + b\mathbf{j}$.
13. Determine x so that $x\mathbf{i} + \mathbf{j}$ is perpendicular to $3\mathbf{i} - 4\mathbf{j}$.
14. Determine two vectors that are perpendicular to $2\mathbf{i} + 5\mathbf{j}$. *Hint:* Try $x\mathbf{i} + \mathbf{j}$ and $x\mathbf{i} - \mathbf{j}$.
15. Find a vector of unit length that has the same direction as $\mathbf{u} = 3\mathbf{i} - 4\mathbf{j}$. *Hint:* Try $\mathbf{u}/\|\mathbf{u}\|$.
16. Find two vectors of unit length that are perpendicular to $2\mathbf{i} + 3\mathbf{j}$. *Hint:* See Problems 14 and 15.
17. Find a vector twice as long as $\mathbf{u} = 2\mathbf{i} - 5\mathbf{j}$ and with opposite direction.
18. For any angle θ, show that $\mathbf{u} = (\cos \theta)\mathbf{i} + (\sin \theta)\mathbf{j}$ and $\mathbf{v} = (\sin \theta)\mathbf{i} - (\cos \theta)\mathbf{j}$ are perpendicular unit vectors.

19. Show that $\mathbf{u} \cdot \mathbf{u} = \|\mathbf{u}\|^2$ for any vector $\mathbf{u} = a\mathbf{i} + b\mathbf{j}$.

20. Show that $(k\mathbf{u}) \cdot \mathbf{v} = k(\mathbf{u} \cdot \mathbf{v})$.

In Problems 21–26, let $\mathbf{u} = 2\mathbf{i} + 9\mathbf{j}$, $\mathbf{v} = 4\mathbf{i} + 3\mathbf{j}$, *and* $\mathbf{w} = -5\mathbf{i} - 12\mathbf{j}$. *In each case, sketch the appropriate vectors and then find the indicated quantity, as in Example D.*

21. Scalar projection of $\mathbf{u}$ on $\mathbf{v}$.

22. Vector projection of $\mathbf{u}$ on $\mathbf{v}$.

23. Vector projection of $\mathbf{u}$ on $\mathbf{w}$.

24. Scalar projection of $\mathbf{v}$ on $\mathbf{w}$.

25. Scalar projection of $\mathbf{w}$ on $\mathbf{v}$.

26. Vector projection of $\mathbf{v}$ on $\mathbf{w}$.

Problems 27–30 relate to Example E.

27. Find the work done by the force $\mathbf{F} = 3\mathbf{i} + 10\mathbf{j}$ in moving an object north 10 units.

28. Find the work done by a S70°E force of 100 dynes in moving an object 50 centimeters east.

29. Find the work done by a N45°E force of 50 dynes in moving an object from $(1, 1)$ to $(6, 9)$, with the distance measured in centimeters.

30. Find the work done by $\mathbf{F} = 3\mathbf{i} + 4\mathbf{j}$ in moving an object from $(0, 0)$ to $(-6, 0)$. Interpret the negative answer.

B. Applications and Extensions

31. If $\mathbf{u} = 2\mathbf{i} + 3\mathbf{j}$ and $\mathbf{v} = 3\mathbf{i} + 4\mathbf{j}$, find $\|3\mathbf{u} - \mathbf{v}\|$.

32. For $\mathbf{u}$ and $\mathbf{v}$ in Problem 31, find $\|\mathbf{u}\|$, $\|\mathbf{v}\|$, $\mathbf{u} \cdot \mathbf{v}$, and θ (the angle between $\mathbf{u}$ and $\mathbf{v}$).

33. Find two vectors of length 1 that are perpendicular to $3\mathbf{i} - 4\mathbf{j}$.

34. Find the vector $a\mathbf{i} + b\mathbf{j}$ that is 12 units long with the same direction as $3\mathbf{i} - 4\mathbf{j}$.

35. Find the vector projection of $5\mathbf{i} + 3\mathbf{j}$ on $3\mathbf{i} - 4\mathbf{j}$. Also find the angle between these two vectors.

36. Find the work done by a force of 100 pounds directed N45°E in moving an object along the line from $(1, 1)$ to $(7, 5)$, distances measured in feet.

37. The vectors $\mathbf{u}$, $\mathbf{v}$, $\mathbf{w}$, $\mathbf{p}$, $\mathbf{q}$, have directions N90°E, N60°E, N45°E, N30°W, N0°E and lengths 20, 15, 20, 20, 10, respectively. Write the resultant of these five vectors in the form $a\mathbf{i} + b\mathbf{j}$.

38. An 80-pound weight is suspended from a ceiling by two cables as shown in Figure 52, producing forces $\mathbf{u}$ and $\mathbf{v}$ in the cables. Determine the tensions $\|\mathbf{u}\|$ and $\|\mathbf{v}\|$. *Hint:* $\mathbf{u} + \mathbf{v} = 0\mathbf{i} + 80\mathbf{j}$.

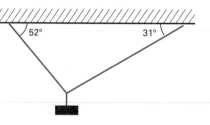

Figure 52

39. The 4-inch dial of a clock is oriented with its center at the origin and the number 12 on the positive x-axis (Figure 53). The minute hand is 3 inches long. Let t denote the number of minutes after 12:00. Show that the minute hand is the vector

$$\left(3 \cos \frac{\pi t}{30}\right)\mathbf{i} - \left(3 \sin \frac{\pi t}{30}\right)\mathbf{j}.$$

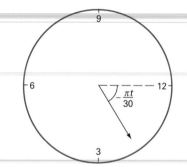

Figure 53

40. Let $d(t)$ denote the distance from the tip of the minute hand of Problem 39 to the point on the rim of the 4-inch dial at 12. Show that

$$d(t) = \left(25 - 24 \cos \frac{\pi t}{30}\right)^{1/2}$$

41. Show that for any two vectors $\mathbf{u}$ and $\mathbf{v}$,

$$|\mathbf{u} \cdot \mathbf{v}| \leq \|\mathbf{u}\| \, \|\mathbf{v}\|$$

When will equality hold?

42. Prove that $\mathbf{u} \cdot \mathbf{v} = \mathbf{v} \cdot \mathbf{u}$ and that $\mathbf{u} \cdot (\mathbf{v} + \mathbf{w}) = \mathbf{u} \cdot \mathbf{v} + \mathbf{u} \cdot \mathbf{w}$.

43. If $\mathbf{u} + \mathbf{v}$ and $\mathbf{u} - \mathbf{v}$ are perpendicular, what can we conclude about $\|\mathbf{u}\|$ and $\|\mathbf{v}\|$?

44. Show that $\|\mathbf{u} + \mathbf{v}\|^2 + \|\mathbf{u} - \mathbf{v}\|^2 = 2(\|\mathbf{u}\|^2 + \|\mathbf{v}\|^2)$.

45. Find the exact value of $\sin 18°$, which may be needed to complete Problem 46. *Hint:* Let $\theta = 18°$ and note that $\cos 3\theta = \sin 2\theta = 2 \sin \theta \cos \theta$ and that $\cos 3\theta = \cos(2\theta + \theta) = (1 - 4 \sin^2 \theta) \cos \theta$.

46. TEASER Let **u** and **v** denote adjacent edges of a regular pentagon that is inscribed in a circle of radius 1 (Figure 54). Let **w** = **u** + **v**.
   (a) Express **u** · **w** and $\|\mathbf{u}\|\,\|\mathbf{w}\|$ in terms of cos 36°.
   (b) Use your calculator to guess the exact value of $(\|\mathbf{u}\|\,\|\mathbf{w}\|)^2$ and then prove that this is the correct value.

47. Draw the graph of $d(t)$ from Problem 40 and determine the amplitude and period of this periodic function.

48. Redo Problem 47 for the 2-inch hour hand. This means you will first have to determine the appropriate formula for $d(t)$.

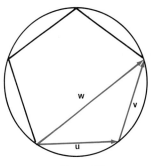

Figure 54

9-5   SIMPLE HARMONIC MOTION

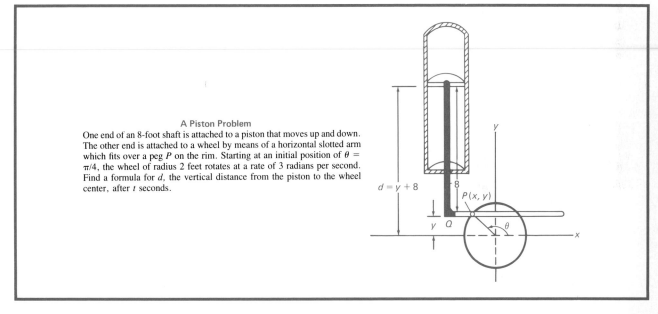

A Piston Problem

One end of an 8-foot shaft is attached to a piston that moves up and down. The other end is attached to a wheel by means of a horizontal slotted arm which fits over a peg P on the rim. Starting at an initial position of $\theta = \pi/4$, the wheel of radius 2 feet rotates at a rate of 3 radians per second. Find a formula for d, the vertical distance from the piston to the wheel center, after t seconds.

The up-and-down motion of the piston above is an example of what is called simple harmonic motion. Notice right away that the motion of the piston is essentially the same as that of the point Q. This means we want to find y; and y is just the y-coordinate of the peg P.

Perhaps the piston-wheel device seems complicated, so let's consider another version of the same problem. Imagine the wheel shown in Figure 55 to be turning at a uniform rate in the counterclockwise direction. Emanating from P, a point attached to the rim, is a horizontal beam of light, which projects a bright spot at Q on a nearby vertical wall. As the wheel turns, the spot at Q moves up and down. Our

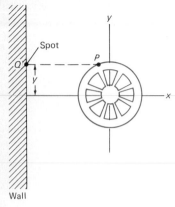

Figure 55

task is to express the y-coordinate of Q (which is also the y-coordinate of P) in terms of the elapsed time t.

The solution to this problem depends on a number of factors (the rate at which the wheel turns, the radius of the wheel, and the location of P at $t = 0$). We think it wise to begin with a simple case and gradually extend to more general situations.

Case 1 Suppose that the wheel has radius 1, that it turns at 1 radian per second, and that it starts at $\theta = 0$. Then at time t, θ will measure t radians and P will have y-coordinate

$$y = \sin t$$

(see Figure 56). Keep in mind that this equation describes the up-and-down motion of Q.

Case 2 Let everything be as in the first case, but now let the wheel turn at 3 radians per second (Figure 57). Then at time t, θ will measure $3t$ radians and both P and Q will have y-coordinate.

$$y = \sin 3t$$

Case 3 Next increase the radius of the wheel to 2 feet, but leave the other information as in Case 2 (Figure 58). Now the coordinates of P are $(2 \cos 3t, 2 \sin 3t)$ and

$$y = 2 \sin 3t$$

Case 4 Finally, let the wheel start at $\theta = \pi/4$ rather than $\theta = 0$. With the help of Figure 59, we see that

$$y = 2 \sin\left(3t + \frac{\pi}{4}\right)$$

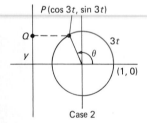

Case 1

Figure 56

Case 2

Figure 57

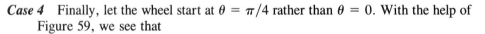

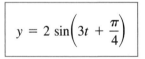

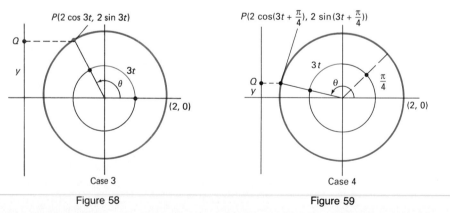

Case 3

Figure 58

Case 4

Figure 59

Case 4 describes the wheel of the original piston-wheel problem. The number y measures the distance between Q and the x-axis, and $d = y + 8$ is the distance from the piston to the x-axis. Thus the answer to the question first posed is

$$d = 8 + 2 \sin\left(3t + \frac{\pi}{4}\right)$$

The number 8 does not interest us; it is the sine expression that is significant. As a matter of fact, equations of the form

$$y = A \sin(Bt + C) \quad \text{and} \quad y = A \cos(Bt + C)$$

with $B > 0$ arise often in physics. Any straight-line motion which can be described by one of these formulas is called **simple harmonic motion.** Cases 1–4 are examples of this motion. Other examples from physics occur in connection with the motion of a weight attached to a vibrating spring (Figure 60) and the motion of a water molecule in an ocean wave. Voltage in an alternating current, although it does not involve motion, is given by the same kind of sine (or cosine) equation.

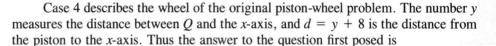

Weight

Simple Harmonic Motion

Figure 60

Graphs

Our first example involves the graphs of the four boxed equations discussed above. Note how these graphs are progressively modified as we move from case 1 to case 4.

Example A (Four Related Graphs)   Sketch the graphs of the four boxed equations.

Solution   The four graphs are shown in Figure 61. Under each graph are listed

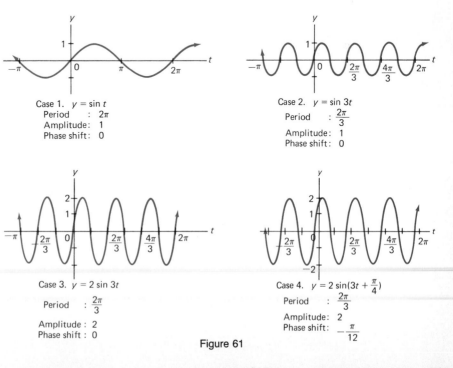

Case 1. $y = \sin t$
Period   : 2π
Amplitude: 1
Phase shift: 0

Case 2. $y = \sin 3t$
Period   : $\dfrac{2\pi}{3}$
Amplitude: 1
Phase shift: 0

Case 3. $y = 2 \sin 3t$
Period : $\dfrac{2\pi}{3}$
Amplitude : 2
Phase shift : 0

Case 4. $y = 2 \sin(3t + \dfrac{\pi}{4})$
Period : $\dfrac{2\pi}{3}$
Amplitude: 2
Phase shift: $-\dfrac{\pi}{12}$

Figure 61

three important numbers, numbers that identify the critical features of the graph. The **period** is the length of the shortest interval after which the graph repeats itself. The **amplitude** is the maximum distance of the graph from its median position (the *t*-axis). The **phase shift** measures the distance the graph is shifted horizontally from its normal position.

You might have expected a phase shift of $-\pi/4$ in Case 4, since the initial angle of the wheel measured $\pi/4$ radians. But, note that factoring 3 from $3t + \pi/4$ gives

$$y = 2 \sin\left(3t + \frac{\pi}{4}\right) = 2 \sin 3\left(t + \frac{\pi}{12}\right)$$

If you recall our discussion of translations (see Section 5-5), you see why the graph is shifted $\pi/12$ units to the left. Note in particular that $y = 0$ when $t = -\pi/12$ instead of when $t = 0$. ■

Graphing in the General Case

If

$$y = A \sin(Bt + C) \quad \text{or} \quad y = A \cos(Bt + C)$$

with $B > 0$, all three concepts (period, amplitude, phase shift) make good sense. We have the following formulas.

Period: $\dfrac{2\pi}{B}$

Amplitude: $|A|$

Phase shift: $\dfrac{-C}{B}$

Knowledge of these three numbers allows us to sketch the graph very quickly.

Example B (Two General Graphs)   Sketch the graph.

(a) $y = 3 \cos\left(4t - \dfrac{\pi}{4}\right)$ (b) $y = 3 \sin\left(\dfrac{1}{2}t + \dfrac{\pi}{8}\right)$

Solution

(a) We recall the graph of $y = \cos t$ and then modify it using the three key numbers.

Period: $\dfrac{2\pi}{B} = \dfrac{2\pi}{4} = \dfrac{\pi}{2}$

Amplitude: $|A| = |3| = 3$

Phase shift: $-\dfrac{C}{B} = \dfrac{\pi/4}{4} = \dfrac{\pi}{16}$

The result is shown in Figure 62.

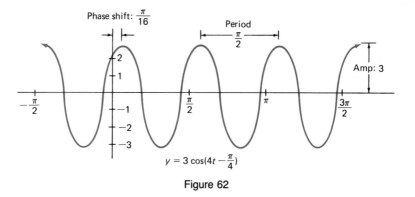

$$y = 3\cos\left(4t - \frac{\pi}{4}\right)$$

Figure 62

(b) We begin by finding the three key numbers.

$$\text{Period:} \qquad \frac{2\pi}{B} = \frac{2\pi}{\frac{1}{2}} = 4\pi$$

$$\text{Amplitude:} \qquad |A| = |3| = 3$$

$$\text{Phase shift:} \qquad \frac{-C}{B} = -\frac{\pi/8}{\frac{1}{2}} = -\frac{\pi}{4}$$

Then we modify the graph of $y = \sin t$ as shown in Figure 63.

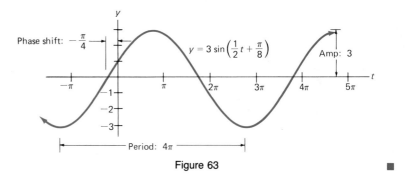

$$y = 3\sin\left(\frac{1}{2}t + \frac{\pi}{8}\right)$$

Figure 63   ■

Example C (Negative A) Sketch the graph of $y = -3\cos 2t$.

Solution   We begin by asking how the graph of $y = -3\cos 2t$ relates to that of $y = 3\cos 2t$. Clearly, every y value has the opposite sign, which has the effect of reflecting the graph about the t-axis. Then we calculate the three crucial numbers.

$$\text{Period:} \qquad \frac{2\pi}{B} = \frac{2\pi}{2} = \pi$$

$$\text{Amplitude:} \qquad |A| = |-3| = 3$$

$$\text{Phase shift:} \qquad \frac{-C}{B} = \frac{0}{2} = 0$$

Finally we sketch the graph (Figure 64).

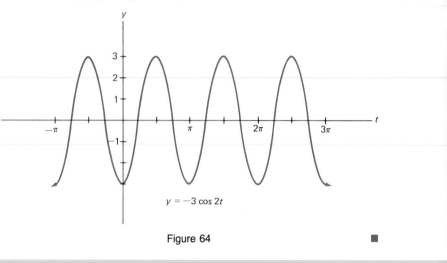

$y = -3 \cos 2t$

Figure 64 ∎

PROBLEM SET 9-5

A. Skills and Techniques

1. Sketch the graphs of the following equations in the order given. Use the interval $-2\pi \le t \le 2\pi$. See Example A.
   (a) $y = \cos t$ **(b)** $y = \cos 2t$
   (c) $y = 4 \cos 2t$
   (d) $y = 4 \cos(2t + \pi/3)$

2. Sketch the graphs of the following on $-2\pi \le t \le 4\pi$.
   (a) $y = \sin t$ **(b)** $y = \sin \frac{1}{2}t$
   (c) $y = 3 \sin \frac{1}{2}t$
   (d) $y = 3 \sin(\frac{1}{2}t + \pi/2)$

In Problem 3–6, find the period, amplitude, and phase shift. Then sketch the graph. See Example B.

3. (a) $y = 4 \sin 2t$ **(b)** $y = 3 \cos\left(t + \dfrac{\pi}{8}\right)$

   (c) $y = \sin\left(4t + \dfrac{\pi}{8}\right)$

   (d) $y = 3 \cos\left(3t - \dfrac{\pi}{2}\right)$

4. (a) $y = \dfrac{1}{2} \cos 3t$ **(b)** $y = 3 \sin\left(t - \dfrac{\pi}{6}\right)$

   (c) $y = 2 \sin\left(\dfrac{1}{2}t + \dfrac{\pi}{8}\right)$ **(d)** $y = \dfrac{1}{2} \sin(2t - 1)$

5. $y = 3 + 2 \cos (\frac{1}{2}t - \pi/16)$. *Hint:* The number 3 lifts the graph of $y = 2 \cos(\frac{1}{2}t - \pi/16)$ up 3 units.

6. $y = 4 + 3 \sin(2t + \pi/16)$

Sketch the graphs of the equations in Problems 7–12. See Example C.

7. $y = -2 \sin 3t$ **8.** $y = -4 \cos \frac{1}{2}t$

9. $y = -\sin(2t + \pi/3)$ **10.** $y = -\cos(3t + \pi)$

11. $y = -2 \cos(t - \frac{1}{6})$ **12.** $y = -3 \sin(3t + 3)$

13. A wheel with center at the origin is rotating counterclockwise at 4 radians per second. There is a small hole in the wheel 5 centimeters from the center. If that hole has initial coordinates (5, 0), what will its coordinates be after t seconds?

14. Answer Problem 13 if the hole is initially at (0, 5).

15. A free-hanging shaft, 8 centimeters long, is attached to the wheel of Problem 13 by putting a bolt through the hole (Figure 65). What are the coordinates of P,

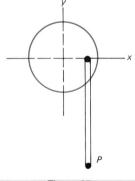

Figure 65

the bottom point of the shaft, at time t (assuming the shaft continues to hang vertically)?

16. Suppose that the wheel of Problem 13 rotates at 3 revolutions per second. What are the coordinates of the hole after t seconds?

B. Applications and Extensions

17. Find the period, amplitude, and phase shift for each graph.
    (a) $y = \sin 5t$ (b) $y = \frac{3}{2} \cos(\frac{1}{2}t)$
    (c) $y = 2 \cos(4t - \pi)$
    (d) $y = -4 \sin(3t + 3\pi/4)$

18. Sketch the graph of the equations in Problem 17 on the interval $-\pi \le t \le 2\pi$.

19. The weight attached to a spring (Figure 66) is bobbing up and down so that

$$y = 8 + 4 \cos\left(\frac{\pi}{2}t + \frac{\pi}{4}\right)$$

where y and t are measured in feet and seconds, respectively. What is the closest the weight gets to the ceiling and when does this first happen for $t > 0$?

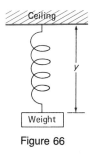

Figure 66

20. The equations $x = 2 + 2 \cos 4t$ and $y = 6 + 2 \sin 4t$ give the coordinates of a point moving along the circumference of a circle. Determine the center and radius of the circle. How long does it take for the point to make a complete revolution?

21. Consider the wheel-piston device shown in Figure 67 (which is analogous to the crankshaft and piston in an automobile engine). The wheel has a radius of 1 foot and rotates counterclockwise at 1 radian per second; the connecting rod is 5 feet long. If the point P is initially at $(1, 0)$, find the y-coordinate of Q after t seconds. Assume the x-coordinate is always zero.

22. Redo Problem 21, but assume the wheel has radius 2 feet and rotates at 60 revolutions per second and that P

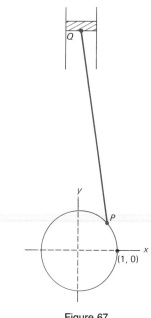

Figure 67

is initially at $(2, 0)$. Is Q executing simple harmonic motion in either of these problems?

23. The voltage drop E across the terminals in a certain alternating current circuit is approximately $E = 156 \sin(110\pi t)$, where t is in seconds. What is the maximum voltage drop and what is the **frequency** (number of cycles per second) for this circuit?

24. The carrier wave for the radio wave of a certain FM station has the form $y = A \sin(2\pi \cdot 10^8 t)$, where t is measured in seconds. What is the frequency for this wave?

25. The AM radio wave for a certain station has the form

$$y = 55[1 + .02 \sin(2400\pi t)] \sin(2 \times 10^5 \pi t)$$

    (a) Find y when $t = 3$.
    (b) Find y when $t = .03216$.
    (c) Find y when $t = .0000321$.

26. In predator-prey systems, the number of predators and the number of prey tend to vary periodically. In a certain region with coyotes as predators and rabbits as prey, the rabbit population R varied according to the formula

$$R = 1000 + 150 \sin 2t$$

where t was measured in years after January 1, 1950.
    (a) What was the maximum rabbit population?
    (b) When was it first reached?
    (c) What was the population on January 1, 1953?

27. The number of coyotes C in Problem 26 satisfied

$$C = 200 + 50 \sin(2t - .7)$$

Sketch the graphs of C and R using the same coordinate system. Attempt to explain the phase shift in C.

28. Sketch the graph of $y = 2^{-t} \cos 2t$ for $0 \le t \le 3\pi$. This is an example of damped harmonic motion, which is typical of harmonic motion where there is friction.

29. Use addition laws to write each of the following in the form $A_1 \sin Bt + A_2 \cos Bt$.

(a) $4 \sin(2t - \frac{\pi}{4})$ **(b)** $3 \cos(3t + \frac{\pi}{3})$

Note: The same idea would work on any expression of the form $A \sin(Bt + C)$ or $A \cos(Bt + C)$.

30. Determine C so that

$$5 \sin 4t + 12 \cos 4t = 13 \sin(4t + C)$$

31. Suppose that A_1 and A_2 are both positive. Show that

$$A_1 \sin Bt + A_2 \cos Bt = A \sin(Bt + C)$$

where $A = \sqrt{A_1^2 + A_2^2}$ and $C = \tan^{-1}(A_2/A_1)$.

32. Generalize Problem 31 by showing that $A_1 \sin Bt + A_2 \cos Bt$ can always be written in the form $A \sin(Bt + C)$. *Hint:* Choose A as in Problem 31 and

let C be the radian measure of an angle that has (A_1, A_2) on its terminal side.

33. Use the result in Problems 31 and 32 to write each of the following in the form $A \sin(Bt + C)$.

(a) $4 \cos 2t + 3 \sin 2t$

(b) $3 \sin 4t - \sqrt{3} \cos 4t$

34. Give an argument to show that

$$A_1 \sin Bt + A_2 \cos Bt, \qquad A_1 A_2 \ne 0, B \ne 0$$

is not a polynomial in t for any choices of A_1, A_2, and B.

35. Find the maximum and minimum values of $\cos t \pm \sin t$.

36. **TEASER** Prove that $\sin(\cos t) < \cos(\sin t)$ for all t. *Hint:* First show that

$$-\frac{\pi}{2} < \cos t < \frac{\pi}{2} - |\sin t|$$

37. Find all solutions to the equation $3 \sin(\frac{1}{2}t + \pi/8) = -5 \cos(\frac{1}{4}t - \frac{1}{10})$. *Hint:* Begin by considering the interval $0 \le t \le 8\pi$.

38. How many intersections do the curves $y = -.3 + e^{\cos x}$ and $y = e^{-.1x} \sin x$ have on $0 \le x < \infty$?

9-6   POLAR REPRESENTATION OF COMPLEX NUMBERS

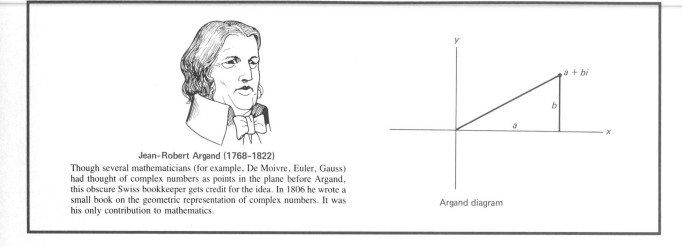

Jean–Robert Argand (1768–1822)

Though several mathematicians (for example, De Moivre, Euler, Gauss) had thought of complex numbers as points in the plane before Argand, this obscure Swiss bookkeeper gets credit for the idea. In 1806 he wrote a small book on the geometric representation of complex numbers. It was his only contribution to mathematics.

Argand diagram

Throughout this book we have used the fact that a real number can be thought of as a point on a line. Now we are going to learn that a complex number can be represented as a point in the plane. This simple idea leads rather quickly to the fruitful notion of the polar form for a complex number. This in turn aids in the multiplication and division of complex numbers and greatly facilitates the calculation of powers and roots of complex numbers. For a discussion of the elementary facts about complex numbers, see Section 1-6.

Complex Numbers as Points in the Plane

Consider a complex number $a + bi$. It is determined by the two real numbers a and b, that is, the ordered pair (a, b). But (a, b), in turn, determines a point in the plane. That point we now label with the complex number $a + bi$. Thus $2 + 4i$, $2 - 4i$, $-3 + 2i$, and all other complex numbers may be used as labels for points in the plane (Figure 68). The plane with points labeled this way is called the **Argand diagram** or **complex plane.** Note that $3i = 0 + 3i$ labels a point on the y-axis, which we now call the **imaginary axis,** while $4 = 4 + 0i$ corresponds to a point on the x-axis (called the **real axis**).

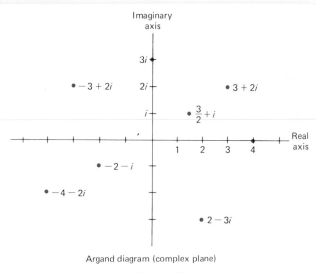

Argand diagram (complex plane)

Figure 68

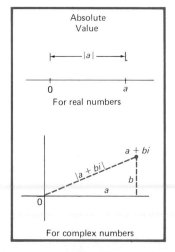

Absolute Value

For real numbers

For complex numbers

Figure 69

Recall that the absolute value of a real number a (written $|a|$) is its distance from the origin on the real line. The concept of absolute value is extended to a complex number $a + bi$ by defining

$$|a + bi| = \sqrt{a^2 + b^2}$$

which is also its distance from the origin (Figure 69). Thus while there are only two real numbers with absolute value of 5, namely -5 and 5, there are infinitely many complex numbers with absolute value 5. They include, 5, -5, $5i$, $3 + 4i$, $3 - 4i$, $-\sqrt{21} + 2i$, and, in fact, all complex numbers on a circle of radius 5 centered at the origin.

Example A (Numbers with Absolute Value 13) Show that $|5 + 12i| = 13$. Then list 11 other numbers with absolute value 13.

Solution

$$|5 + 12i| = \sqrt{5^2 + 12^2} = 13$$

Other numbers with the same absolute value (see Figure 70) are $5 - 12i$, $-5 + 12i$, $-5 - 12i$, $12 + 5i$, $12 - 5i$, $-12 + 5i$, $-12 - 5i$, 13, $13i$, -13, and $-13i$.

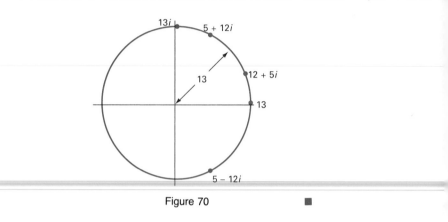

Figure 70 ∎

Polar Form

There is another geometric way to describe complex numbers, a way that will prove very useful to us. For the complex number $a + bi$, which we have already identified with the point (a, b) in the plane, let r denote its distance from the origin and let θ be one of the angles that a ray from the origin through the point makes with the positive x-axis. Then from Figure 71 (or one of its analogues in another quadrant), we see that

$$a = r \cos \theta \qquad b = r \sin \theta$$

Figure 71

This means that we can write

$$a + bi = r \cos \theta + (r \sin \theta)i$$

or

$$a + bi = r(\cos \theta + i \sin \theta)$$

The boxed expression gives the **polar form** of $a + bi$. Notice that r is just the absolute value of $a + bi$; we shall refer to θ as its angle.

To change a complex number from polar form to $a + bi$ form is easy; going in the opposite direction is harder.

Example B (Polar Form to $a + bi$) Change $3(\cos 240° + i \sin 240°)$ to $a + bi$ form.

Solution

$$3(\cos 240° + i \sin 240°) = 3\left(-\frac{1}{2} + i\frac{-\sqrt{3}}{2}\right) = -\frac{3}{2} - \frac{3\sqrt{3}}{2}i \qquad ∎$$

Example C (*a* + *bi* to Polar Form)   Change each of the following to polar form.
(a) −6   (b) 4*i*   (c) $2\sqrt{3} - 2i$

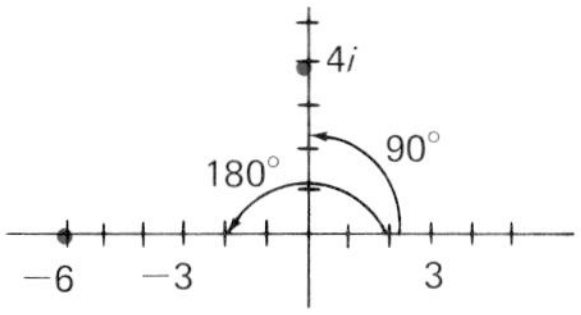

Solution

(a) and (b) Refer to Figure 72 to obtain

$$-6 = 6(\cos 180° + i \sin 180°)$$

$$4i = 4(\cos 90° + i \sin 90°)$$

Figure 72

(c) Use the formulas $r = \sqrt{a^2 + b^2}$ and $\cos \theta = a/r$ to obtain

$$r = \sqrt{(2\sqrt{3})^2 + (-2)^2} = \sqrt{12 + 4} = 4$$

and

$$\cos \theta = \frac{2\sqrt{3}}{4} = \frac{\sqrt{3}}{2}$$

Since $2\sqrt{3} - 2i$ is in quadrant IV and $\cos \theta = \sqrt{3}/2$, θ can be chosen as an angle of $11\pi/6$ radians or 330°. Thus

$$2\sqrt{3} - 2i = 4\left(\cos \frac{11\pi}{6} + i \sin \frac{11\pi}{6}\right)$$

$$= 4(\cos 330° + i \sin 330°) \quad \blacksquare$$

Multiplication and Division

The polar form is ideally suited for multiplying and dividing complex numbers. Let *U* and *V* be complex numbers given in polar form by

$$U = r(\cos \alpha + i \sin \alpha)$$

$$V = s(\cos \beta + i \sin \beta)$$

Then

$$U \cdot V = rs[\cos(\alpha + \beta) + i \sin(\alpha + \beta)]$$

$$\frac{U}{V} = \frac{r}{s}[\cos(\alpha - \beta) + i \sin(\alpha - \beta)]$$

In words, to multiply two complex numbers, we multiply their absolute values and add their angles. To divide two complex numbers, we divide their absolute values and subtract their angles (in the correct order).

To establish the multiplication formula we use a bit of trigonometry.

$$U \cdot V = r(\cos \alpha + i \sin \alpha)s(\cos \beta + i \sin \beta)$$

$$= rs(\cos \alpha \cos \beta + i \cos \alpha \sin \beta + i \sin \alpha \cos \beta + i^2 \sin \alpha \sin \beta)$$

$$= rs[(\cos \alpha \cos \beta - \sin \alpha \sin \beta) + i(\sin \alpha \cos \beta + \cos \alpha \sin \beta)]$$

$$= rs[\cos(\alpha + \beta) + i \sin(\alpha + \beta)]$$

The key step was the last one, where we used the addition laws for the cosine and the sine.

You will be asked to establish the division formula in Problem 56.

Example D (Applying the Formulas)   Let

$$U = 4(\cos 75° + i \sin 75°) \quad \text{and} \quad V = 3(\cos 60° + i \sin 60°)$$

Write $U \cdot V$ and U/V in polar form.

Solution

$$U \cdot V = 12(\cos 135° + i \sin 135°)$$

$$\frac{U}{V} = \frac{4}{3}(\cos 15° + i \sin 15°) \quad \blacksquare$$

Example E (Finding Products in Two Ways)   Find the product

$$(\sqrt{3} + i)(-4 - 4\sqrt{3}i)$$

directly and then by using the polar form.

Solution   *Method 1* We use the definition of multiplication given in Section 1-6 to get

$$(\sqrt{3} + i)(-4 - 4\sqrt{3}i) = (-4\sqrt{3} + 4\sqrt{3}) + (-4 - 12)i$$

$$= -16i$$

Method 2   We change both numbers to polar form, multiply by the method of this section, and finally change to $a + bi$ form.

$$\sqrt{3} + i = 2(\cos 30° + i \sin 30°)$$

$$-4 - 4\sqrt{3}i = 8(\cos 240° + i \sin 240°)$$

$$(\sqrt{3} + i)(-4 - 4\sqrt{3}i) = 16(\cos 270° + i \sin 270°)$$

$$= 16(0 - i)$$

$$= -16i \quad \blacksquare$$

Geometric Addition and Multiplication

Having learned that the complex numbers can be thought of as points in a plane, we should not be surprised that the operations of addition and multiplication have a geometric interpretation. Let U and V be any two complex numbers; that is, let

$$U = a + bi = r(\cos \alpha + i \sin \alpha)$$

$$V = c + di = s(\cos \beta + i \sin \beta)$$

Addition is accomplished algebraically by adding the real parts and imaginary parts separately.

$$U + V = (a + c) + (b + d)i$$

To accomplish the same thing geometrically, we construct the parallelogram that has

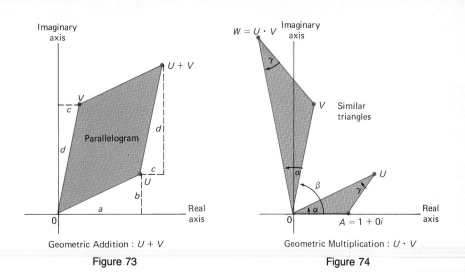

Geometric Addition : $U + V$

Figure 73

Geometric Multiplication : $U \cdot V$

Figure 74

O, U, and V as three of its vertices (see Figure 73). Then $U + V$ corresponds to the vertex opposite the origin, as you should be able to show by finding the coordinates of this vertex.

To multiply algebraically, we use the polar forms of U and V, adding the angles and multiplying the absolute values.

$$U \cdot V = rs[\cos(\alpha + \beta) + i \sin(\alpha + \beta)]$$

To interpret this geometrically (for the case where α and β are between $0°$ and $180°$), first draw triangle OAU, where A is the point $1 + 0i$. Then construct triangle OVW similar to triangle OAU in the manner indicated in Figure 74. We claim that $W = U \cdot V$. Certainly W has the correct angle, namely, $\alpha + \beta$. Moreover, by similarity of triangles,

$$\frac{\overline{OW}}{\overline{OV}} = \frac{\overline{OU}}{\overline{OA}} = \frac{\overline{OU}}{1}$$

(Here we are using $\overline{OW}$ for the length of the line segment from O to W.) Thus

$$|W| = \overline{OW} = \overline{OU} \cdot \overline{OV} = |U| \cdot |V|$$

so W also has the correct absolute value.

PROBLEM SET 9-6

A. Skills and Techniques

In Problems 1–12, plot the given numbers in the complex plane. See Example A.

1. $2 + 3i$

2. $2 - 3i$

3. $-2 - 3i$

4. $-2 + 3i$

5. 5

6. -6

7. $-4i$

8. $6i$

9. $\frac{3}{5} - \frac{4}{5}i$

10. $-\frac{5}{13} + \frac{12}{13}i$

11. $2\left(\cos\frac{\pi}{4} + i \sin\frac{\pi}{4}\right)$

12. $3\left(\cos\frac{7\pi}{6} + i \sin\frac{7\pi}{6}\right)$

13. Find the absolute values of the numbers in Problems 1, 3, 5, 7, 9, and 11.

14. Find the absolute values of the numbers in Problems 2, 4, 6, 8, and 10.

Express each of the following in the form a + bi, as in Example B.

15. $4\left(\cos \dfrac{3\pi}{2} + i \sin \dfrac{3\pi}{2}\right)$ **16.** $5(\cos \pi + i \sin \pi)$

17. $2(\cos 225° + i \sin 225°)$

18. $\frac{3}{2}(\cos 300° + i \sin 300°)$

Express each of the following in polar form. For example, $1 + i = \sqrt{2}(\cos 45° + i \sin 45°)$. See also example C.

19. -4 **20.** 9

21. $-5i$ **22.** $14i$

23. $2 - 2i$ **24.** $-5 - 5i$

25. $2\sqrt{3} + 2i$ **26.** $-4\sqrt{3} + 4i$

27. $5 + 4i$ **28.** $3 + 2i$

For Problems 29–36 let $u = 2(\cos 140° + i \sin 140°)$, $v = 3(\cos 70° + i \sin 70°)$ and $w = \frac{1}{2}(\cos 55° + i \sin 55°)$. Calculate each product or quotient, leaving your answer in polar form, as in Example D. Note also the following calculation.

$$\dfrac{u^2}{w} = \dfrac{u \cdot u}{w} = \dfrac{4(\cos 280° + i \sin 280°)}{\frac{1}{2}(\cos 55° + i \sin 55°)}$$

$$= 8(\cos 225° + i \sin 225°)$$

29. uv **30.** uw

31. vw **32.** uvw

33. u/v **34.** uv/w

35. $1/w$ **36.** $1/v$

Find each of the following products in two ways, giving your final answer in a + bi form. See Example E.

37. $(4 - 4i)(2 + 2i)$

38. $(\sqrt{3} + i)(2 - 2\sqrt{3}i)$

39. $(1 + \sqrt{3}i)(1 + \sqrt{3}i)$

40. $(\sqrt{2} + \sqrt{2}i)(\sqrt{2} + \sqrt{2}i)$

Find the following products and quotients, giving your answers in polar form. Start by changing each of the given complex numbers to polar form.

41. $4i(2\sqrt{3} - 2i)$ **42.** $(-2i)(5 + 5i)$

43. $\dfrac{4i}{2\sqrt{3} - 2i}$ **44.** $\dfrac{-2i}{5 + 5i}$

45. $(2\sqrt{2} - 2\sqrt{2}i)(2\sqrt{2} - 2\sqrt{2}i)$

46. $(1 - \sqrt{3}i)(1 - \sqrt{3}i)$

B. Applications and Extensions

47. Plot the given number in the complex plane and find its absolute value.
   (a) $-5 + 12i$ (b) $-4i$
   (c) $5(\cos 60° + i \sin 60°)$

48. Express in the form $a + bi$.
   (a) $5[\cos(3\pi/2) + i \sin(3\pi/2)]$
   (b) $4(\cos 180° + i \sin 180°)$
   (c) $2(\cos 315° + i \sin 315°)$
   (d) $3[\cos(-2\pi/3) + i \sin(-2\pi/3)]$

49. Express in polar form.
   (a) 12 (b) $-\sqrt{2} + \sqrt{2}i$
   (c) $-3i$ (d) $2 - 2\sqrt{3}i$
   (e) $4\sqrt{3} + 4i$
   (f) $2(\cos 45° - i \sin 45°)$

50. Write in the form $a + bi$.
   (a) $2(\cos 37° + i \sin 37°)8(\cos 113° + i \sin 113°)$
   (b) $6(\cos 123° + i \sin 123°)/[3(\cos 33° + i \sin 33°)]$

51. Perform the indicated operations and write your answer in polar form.
   (a) $1.5(\cos 110° + i \sin 110°)4(\cos 30° + i \sin 30°)$
       $\times 2(\cos 20° + i \sin 20°)$
   (b) $\dfrac{12(\cos 115° + i \sin 115°)}{4(\cos 55° + i \sin 55°)(\cos 20° + i \sin 20°)}$
   (c) $\dfrac{(-\sqrt{2} + \sqrt{2}i)(2 - 2\sqrt{3}i)}{4\sqrt{3} + 4i}$ (See Problem 49.)

52. Calculate and write your answer in the form $a + bi$.
   (a) $\dfrac{(-\sqrt{2} + \sqrt{2}i)^2(2 - 2\sqrt{3}i)^3}{4\sqrt{3} + 4i}$ (See Problem 49.)
   (b) $(-1 + \sqrt{3}i)^5(-1 - \sqrt{3}i)^{-4}$

53. In each case, find two values for $z = a + bi$.
   (a) The imaginary part of z is 5 and $|z| = 13$.
   (b) The number z lies on the line $y = x$ and $|z| = 8$.

54. Find four complex numbers $z = a + bi$ that are located on the hyperbola $y^2 - x^2 = 2$ and satisfy $|z| = \sqrt{10}$.

55. Let $u = r(\cos \theta + i \sin \theta)$. Write each of the following in polar form.
   (a) u^3
   (b) $\bar{u}$ ($\bar{u}$ is the conjugate of u)
   (c) $u\bar{u}$ (d) $1/u$
   (e) u^{-2} (f) $-u$

56. Prove the division formula: If $U = r(\cos \alpha + i \sin \alpha)$ and $V = s(\cos \beta + i \sin \beta)$, then

$$\frac{U}{V} = \frac{r}{s}[\cos(\alpha - \beta) + i \sin(\alpha - \beta)]$$

57. With U and V as in Problem 56, write a formula for U^2/V^3.

58. Note that, by the addition formula for tangent

$$\tan[\tan^{-1}(2) + \tan^{-1}(3)] = \frac{2 + 3}{1 - 2 \cdot 3} = -1$$

(a) Calculate $\tan^{-1}(1) + \tan^{-1}(2) + \tan^{-1}(3)$.
(b) Use part (a) to find the polar form of $z = (1 + i)(1 + 2i)(1 + 3i)$.
(c) Simplify your answer to part (b) and thereby show that $z = -10$.

59. Let U and V be complex numbers. Give a geometric interpretation for (a) $|U - V|$ and (b) the angle of $U - V$.

60. By expanding $(\cos \theta + i \sin \theta)^3$ in two different ways, derive the formulas.
(a) $\cos 3\theta = 4 \cos^3 \theta - 3 \cos \theta$
(b) $\sin 3\theta = -4 \sin^3 \theta + 3 \sin \theta$

61. Let $z_k = 2[\cos(k\pi/4) + i \sin(k\pi/4)]$. Find the exact value of each of the following.
(a) $z_1 z_2 z_3 \cdots z_8$
(b) $z_1 + z_2 + z_3 + \cdots + z_8$ (Think geometrically.)

62. TEASER Let $z_k = \dfrac{k}{k + 1}(\cos k° + i \sin k°)$. Find the exact value of the product

$$z_1 z_2 z_3 \cdots z_{179} z_{180}$$

9-7 POWERS AND ROOTS OF COMPLEX NUMBERS

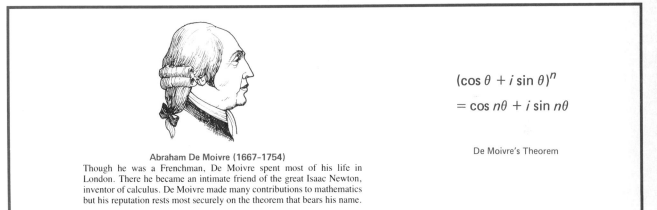

Abraham De Moivre (1667–1754)
Though he was a Frenchman, De Moivre spent most of his life in London. There he became an intimate friend of the great Isaac Newton, inventor of calculus. De Moivre made many contributions to mathematics but his reputation rests most securely on the theorem that bears his name.

$$(\cos \theta + i \sin \theta)^n$$
$$= \cos n\theta + i \sin n\theta$$

De Moivre's Theorem

De Moivre's Theorem tells us how to raise a complex number of absolute value 1 to an integral power. We can easily extend it to cover the case of any complex number, no matter what its absolute value. Then with a little work, we can use it to find roots of complex numbers. Here we are in for a surprise. Take the number $8i$, for example. After some fumbling around, we find that one of its cube roots is $-2i$, because $(-2i)^3 = -8i^3 = 8i$. We shall find that it has two other cube roots (both nonreal numbers). In fact, we shall see that every number has exactly three cube roots, four 4th roots, five 5th roots, and so on. To put it in a spectacular way, we claim that any number, for example $37 + 3.5i$, has 1,000,000 millionth roots.

Powers of Complex Numbers

To raise the complex number $r(\cos \theta + i \sin \theta)$ to the nth power, n a positive integer, we simply find the product of n factors of $r(\cos \theta + i \sin \theta)$. But from Section 9-6, we know that we multiply complex numbers by multiplying their absolute values and adding their angles. Thus

$$[r(\cos \theta + i \sin \theta)]^n$$

$$= \underbrace{r \cdot r \cdots r}_{n \text{ factors}}[\cos(\underbrace{\theta + \theta + \cdots + \theta}_{n \text{ terms}}) + i \sin(\theta + \theta + \cdots + \theta)]$$

In short,

$$[r(\cos \theta + i \sin \theta)]^n = r^n(\cos n\theta + i \sin n\theta)$$

When $r = 1$, this is De Moivre's Theorem.

Example A (Applying the Boxed Formula) Find the sixth power of $2\left(\cos \dfrac{\pi}{6} + i \sin \dfrac{\pi}{6}\right)$

Solution

$$\left[2\left(\cos \frac{\pi}{6} + i \sin \frac{\pi}{6}\right)\right]^6 = 2^6\left[\cos\left(6 \cdot \frac{\pi}{6}\right) + i \sin\left(6 \cdot \frac{\pi}{6}\right)\right]$$

$$= 64(\cos \pi + i \sin \pi)$$

$$= 64(-1 + i \cdot 0)$$

$$= -64 \quad \blacksquare$$

Example B (Raising $a + bi$ to a Power) Find $(1 - \sqrt{3}i)^5$ and write the answer in $a + bi$ form.

Solution   We could use repeated multiplication of $1 - \sqrt{3}i$ by itself. But how much better to change $1 - \sqrt{3}i$ to polar form and use the boxed formula above.

$$1 - \sqrt{3}i = 2(\cos 300° + i \sin 300°)$$

Thus

$$(1 - \sqrt{3}i)^5 = 2^5(\cos 1500° + i \sin 1500°)$$

$$= 32(\cos 60° + i \sin 60°)$$

$$= 32\left(\frac{1}{2} + i\frac{\sqrt{3}}{2}\right)$$

$$= 16 + 16\sqrt{3}i \quad \blacksquare$$

The Three Cube Roots of 8*i*

Because finding roots is tricky, we begin with an example before attempting the general case. We have already noted that $-2i$ is one cube root of $8i$, but now we claim there are two others. How shall we find them? We begin by writing $8i$ in polar form.

$$8i = 8(\cos 90° + i \sin 90°)$$

Finding cube roots is the opposite of cubing. This suggests that we take the real cube root (rather than the cube) of 8 and divide (rather than multiply) the angle $90°$ by 3. This would give us one cube root

$$2(\cos 30° + i \sin 30°)$$

which reduces to

$$2\left(\frac{\sqrt{3}}{2} + \frac{1}{2}i\right) = \sqrt{3} + i$$

Is this really a cube root of $8i$? For fear that you might be suspicious of the polar form, we will cube it the old-fashioned way and check.

$$
\begin{aligned}
(\sqrt{3} + i)^3 &= (\sqrt{3} + i)(\sqrt{3} + i)(\sqrt{3} + i) \\
&= [(3 - 1) + 2\sqrt{3}i](\sqrt{3} + i) \\
&= 2(1 + \sqrt{3}i)(\sqrt{3} + i) \\
&= 2(0 + 4i) \\
&= 8i
\end{aligned}
$$

Of course, the check using polar form is more direct.

$$
\begin{aligned}
[2(\cos 30° + i \sin 30°)]^3 &= 2^3(\cos 90° + i \sin 90°) \\
&= 8(0 + i) \\
&= 8i
\end{aligned}
$$

The process described above yielded one cube root of $8i$ (namely, $\sqrt{3} + i$); there are two others. Let us go back to our representation of $8i$ in polar form. We used the angle $90°$; we could as well have used $90° + 360° = 450°$.

$$8i = 8(\cos 450° + i \sin 450°)$$

Now if we take the real cube root of 8 and divide $450°$ by 3 we get

$$2(\cos 150° + i \sin 150°) = 2\left(-\frac{\sqrt{3}}{2} + \frac{1}{2}i\right) = -\sqrt{3} + i$$

We could again check that this is indeed a cube root of $8i$.

What worked once might work twice. Let us write $8i$ in polar form in a third way, this time adding $2(360°)$ to its angle of $90°$.

$$8i = 8(\cos 810° + i \sin 810°)$$

The corresponding cube root is

$$2(\cos 270° + i \sin 270°) = 2(0 - i) = -2i$$

This does not come as a surprise, since we knew that $-2i$ was one of the cube roots of $8i$.

If we add 3(360°)(that is, 1080°) to 90°, do we get still another cube root of $8i$? No, for if we write

$$8i = 8(\cos 1170° + i \sin 1170°)$$

the corresponding cube root of $8i$ would be

$$2(\cos 390° + i \sin 390°) = 2(\cos 30° + i \sin 30°)$$

But this is the same as the first cube root we found. The truth is that we have found all the cube roots of $8i$, namely, $\sqrt{3} + i$, $-\sqrt{3} + i$, and $-2i$.

Let us summarize. The number $8i$ has three cube roots given by

$$2\left[\cos\left(\frac{90°}{3}\right) + i \sin\left(\frac{90°}{3}\right)\right]$$

$$2\left[\cos\left(\frac{90° + 360°}{3}\right) + i \sin\left(\frac{90° + 360°}{3}\right)\right]$$

$$2\left[\cos\left(\frac{90° + 720°}{3}\right) + i \sin\left(\frac{90° + 720°}{3}\right)\right]$$

We can say the same thing in a shorter way by writing

$$2\left[\cos\left(\frac{90° + k \cdot 360°}{3}\right) + i \sin\left(\frac{90° + k \cdot 360°}{3}\right)\right] \qquad k = 0, 1, 2$$

Roots of Complex Numbers

We are ready to generalize. If $u \neq 0$, then

$$u = r(\cos \theta + i \sin \theta)$$

has n distinct nth roots $u_0, u_1, \ldots, u_{n-1}$ given by

$$u_k = \sqrt[n]{r}\left[\cos\left(\frac{\theta + k \cdot 360°}{n}\right) + i \sin\left(\frac{\theta + k \cdot 360°}{n}\right)\right]$$

$$k = 0, 1, 2, \ldots, n - 1$$

Recall that $\sqrt[n]{r}$ denotes the positive real nth root of $r = |u|$. In our example, it was $\sqrt[3]{|8i|} = \sqrt[3]{8} = 2$. To see that each value of u_k is an nth root, simply raise it to the nth power. In each case, you should get u.

The boxed formula assumes that θ is given in degrees. If θ is in radians, the formula takes the following form.

$$u_k = \sqrt[n]{r}\left[\cos\left(\frac{\theta + 2k\pi}{n}\right) + i \sin\left(\frac{\theta + 2k\pi}{n}\right)\right]$$

$$k = 0, 1, 2, \ldots, n - 1$$

Example C (4th Roots of a Complex Number)   Find the four 4th roots of the complex number $8\sqrt{3} - 8i$.

Solution   First we calculate r and θ. Note that θ must be a fourth quadrant angle since $8\sqrt{3} - 8i$ is in quadrant IV.

$$r = \sqrt{(8\sqrt{3})^2 + (-8)^2} = \sqrt{256} = 16$$

$$\cos \theta = \frac{a}{r} = \frac{8\sqrt{3}}{16} = \frac{\sqrt{3}}{2} \quad \text{so} \quad \theta = 330°$$

Letting $k = 0, 1, 2, 3$ in $\frac{1}{4}(330° + k \cdot 360°)$, we obtain the angles $82.5°$, $172.5°$, $262.5°$, and $352.5°$. The desired roots are

$$u_0 = 2(\cos 82.5° + i \sin 82.5°)$$
$$u_1 = 2(\cos 172.5° + i \sin 172.5°)$$
$$u_2 = 2(\cos 262.5° + i \sin 262.5°)$$
$$u_3 = 2(\cos 352.5° + i \sin 352.5°) \quad \blacksquare$$

Example D (6th Roots of a Real Number)   Find the six 6th roots of 64.

Solution   A real number is a special kind of complex number. In fact, $64 = 64(\cos 0° + i \sin 0°)$. Applying the formula with $r = 64$ and $\theta = 0°$, we obtain

$$u_0 = 2(\cos 0° + i \sin 0°) = 2$$
$$u_1 = 2(\cos 60° + i \sin 60°) = 1 + \sqrt{3}i$$
$$u_2 = 2(\cos 120° + i \sin 120°) = -1 + \sqrt{3}i$$
$$u_3 = 2(\cos 180° + i \sin 180°) = -2$$
$$u_4 = 2(\cos 240° + i \sin 240°) = -1 - \sqrt{3}i$$
$$u_5 = 2(\cos 300° + i \sin 300°) = 1 - \sqrt{3}i$$

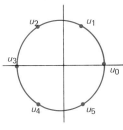

Figure 75

Notice that two of the roots, 2 and -2, are real; the other four are not real.

If you plot these six numbers (Figure 75) you will find that they lie on a circle of radius 2 centered at the origin and that they are equally spaced around the circle. This is typical of what happens in general. $\blacksquare$

Roots of Unity

The nth roots of 1, called the **nth roots of unity,** play an important role in many parts of mathematics.

Example E (5th Roots of Unity)   Find the five 5th roots of 1, plot them, and show that all five of them are powers of one of them.

Solution   First we represent 1 in polar form.

$$1 = 1(\cos 0° + i \sin 0°)$$

The five 5th roots are (according to the formula developed in this section)

$$u_0 = \cos 0° + i \sin 0° = 1$$

$$u_1 = \cos 72° + i \sin 72°$$

$$u_2 = \cos 144° + i \sin 144°$$

$$u_3 = \cos 216° + i \sin 216°$$

$$u_4 = \cos 288° + i \sin 288°$$

These roots are plotted in Figure 76. They lie on the unit circle and are equally spaced around it. Finally notice that

$$u_1 = u_1$$

$$u_2 = u_1^2$$

$$u_3 = u_1^3$$

$$u_4 = u_1^4$$

$$u_0 = u_1^5$$

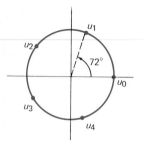

Figure 76

Thus all the roots are powers of u_1. These powers of u_1 repeat in cycles of 5. For example, note that

$$u_1^6 = u_1^5 u_1 = u_1$$

$$u_1^7 = u_1^5 u_1^2 = u_1^2 \quad \blacksquare$$

PROBLEM SET 9-7

A. Skills and Techniques

Find each of the following, leaving your answer in polar form. See Example A.

1. $\left[2\left(\cos \dfrac{\pi}{4} + i \sin \dfrac{\pi}{4}\right)\right]^3$

2. $\left[3\left(\cos \dfrac{5\pi}{6} + i \sin \dfrac{5\pi}{6}\right)\right]^2$

3. $[\sqrt{5}(\cos 11° + i \sin 11°)]^6$

4. $[\frac{1}{3}(\cos 12.5° + i \sin 12.5°)]^4$

5. $(1 + i)^8$

6. $(1 - i)^4$

Find each of the following powers. Write your answer in $a + bi$ form, as in Example B.

7. $(\cos 36° + i \sin 36°)^{10}$

8. $(\cos 27° + i \sin 27°)^{10}$

9. $(\sqrt{3} + i)^5$

10. $(2 - 2\sqrt{3}i)^4$

Find the nth roots of u for the given u and n, leaving your answers in polar form. Plot these roots in the complex plane. See Example C.

11. $u = 125(\cos 45° + i \sin 45°)$; $n = 3$

12. $u = 81(\cos 80° + i \sin 80°)$; $n = 4$

13. $u = 64\left(\cos \dfrac{\pi}{2} + i \sin \dfrac{\pi}{2}\right)$; $n = 6$

14. $u = 3^8\left(\cos \dfrac{2\pi}{3} + i \sin \dfrac{2\pi}{3}\right)$; $n = 8$

15. $u = 4(\cos 112° + i \sin 112°)$; $n = 4$

16. $u = 7(\cos 200° + i \sin 200°)$; $n = 5$

Find the nth roots of u for the given u and n. Write your answers in the $a + bi$ form. See Examples C and D.

17. $u = 16$, $n = 4$

18. $u = -16$, $n = 4$

19. $u = 4i$, $n = 2$

20. $u = -27i$, $n = 3$

21. $u = -4 + 4\sqrt{3}i$, $n = 2$

22. $u = -2 - 2\sqrt{3}i$, $n = 4$

In each of the following, find all the nth roots of unity for the given n and plot them in the complex plane, as in Example E.

23. $n = 4$ **24.** $n = 6$

25. $n = 10$ **26.** $n = 12$

B. Applications and Extensions

27. Calculate each of the following, leaving your answer in polar form.

    (a) $[3(\cos 20° + i \sin 20°)]^4$

    (b) $[2.46(\cos 1.54 + i \sin 1.54)]^5$

    (c) $[2(\cos 50° + i \sin 50°)(\cos 30° + i \sin 30°)]^3$

    (d) $\left(\dfrac{8[\cos(2\pi/3) + i \sin(2\pi/3)]}{4[\cos(\pi/4) + i \sin(\pi/4)]}\right)^4$

28. Change to polar form, calculate, and then change back to $a + bi$ form.

    (a) $(1 - \sqrt{3}i)^5$

    (b) $[(\sqrt{3} + i)(2 - 2i)/(-1 + \sqrt{3}i)]^3$

29. Find the five 5th roots of $32(\cos 255° + i \sin 255°)$, giving your answers in polar form.

30. Find the three cube roots of $-4\sqrt{2} - 4\sqrt{2}i$, giving your answers in polar form.

31. Write the eight 8th roots of 1 in $a + bi$ form and calculate their sum and product.

32. Solve the equation $x^3 - 4 - 4\sqrt{3}i = 0$. You may give your answers in polar form.

33. Find the solution to $x^5 + \sqrt{2} - \sqrt{2}i = 0$ with the largest real part. Write your answer in the form $a + bi$.

34. Solve the equation $x^3 + 27 = 0$ in two ways.

    (a) By finding the three cube roots of -27.

    (b) By writing $x^3 + 27 = (x + 3)(x^2 - 3x + 9)$ and using the quadratic formula.

35. Find the six solutions to $x^6 - 1 = 0$ by two different methods.

36. Show that $\cos(\pi/3) + i \sin(\pi/3)$ is a solution to $2x^4 + x^2 + x + 1 = 0$.

37. Find all six solutions to $x^6 + x^4 + x^2 + 1 = 0$. *Hint:* The left side can be factored as $(x^2 + 1)(x^4 + 1)$.

38. Find all solutions to $x^6 + 2x^3 + 1 = 0$

39. How many of the eight 8th roots of 15 are real? How many of the fifteen 15th roots of -8 are real?

40. Show that De Moivre's Theorem is valid when n is a negative integer.

41. If A is a nonreal number, we agree that $\sqrt{A}$ stands for the one of the two square roots with nonnegative real part. For example, the two square roots of $-4 + 4\sqrt{3}i$ are $\sqrt{2} + \sqrt{6}i$ and $-\sqrt{2} - \sqrt{6}i$, but we agree that

$$\sqrt{-4 + 4\sqrt{3}i} = \sqrt{2} + \sqrt{6}i$$

Evaluate.

    (a) $\sqrt{1 + \sqrt{3}i}$ **(b)** $\sqrt{-1 + \sqrt{3}i}$

42. The quadratic formula is valid even for quadratic equations with nonreal coefficients if we follow the agreement of Problem 41. Solve the following equations.

    (a) $x^2 - 2x + \sqrt{3}i = 0$

    (b) $x^2 - 4ix - 5 + \sqrt{3}i = 0$

43. Let n be an integer that is not divisible by 3. Simplify

$$(-1 + \sqrt{3}i)^n + (-1 - \sqrt{3}i)^n$$

as much as possible.

44. **TEASER** Let $1, u, u^2, u^3, \ldots, u^{15}$ be the sixteen 16th roots of unity. Calculate each of the following. Look for a simple way in each case.

    (a) $1 + u + u^2 + u^3 + \cdots + u^{15}$

    (b) $1 \cdot u \cdot u^2 \cdot u^3 \cdots u^{15}$

    (c) $(1 - u)(1 - u^2)(1 - u^3) \cdots (1 - u^{15})$

    (d) $(1 + u)(1 + u^2)(1 + u^4)(1 + u^8)(1 + u^{16})$

CHAPTER 9 SUMMARY

A triangle like the one in Figure 77 that has no right angle is called an oblique triangle. If any three of the six parts α, β, γ, a, b, and c—including at least one side—are given, we can find the remaining parts by using the **law of sines**

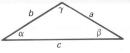

Figure 77

$$\frac{a}{\sin \alpha} = \frac{b}{\sin \beta} = \frac{c}{\sin \gamma}$$

and the **law of cosines**

$$a^2 = b^2 + c^2 - 2bc \cos \alpha \qquad \text{(one of three forms)}$$

There is, however, one case (given two sides and an angle opposite one of them) in which there might be no solution or two solutions. We call it the ambiguous case.

Vectors, represented by arrows, play an important role in science because they have both **magnitude** (length) and **direction.** We call two vectors **equivalent** if they have the same magnitude and direction. Vectors can be added (using the parallelogram law) and multiplied by **scalars,** which are real numbers. The special vectors **i** and **j** allow us to write any vector in the form $a\mathbf{i} + b\mathbf{j}$, where a and b are scalars. If $\mathbf{u} = a\mathbf{i} + b\mathbf{j}$ is a vector, its **length** $\|\mathbf{u}\|$ is given by

$$\|\mathbf{u}\| = \sqrt{a^2 + b^2}$$

If $\mathbf{v} = c\mathbf{i} + d\mathbf{j}$ is another such vector, the **dot product** $\mathbf{u} \cdot \mathbf{v}$ of **u** and **v** is

$$\mathbf{u} \cdot \mathbf{v} = ac + bd = \|\mathbf{u}\|\,\|\mathbf{v}\| \cos \theta$$

where θ is the smallest positive angle between **u** and **v**.

Two vectors are perpendicular if and only if their dot product is zero.

The equations $y = A \sin(Bt + C)$ and $y = A \cos(Bt + C)$ with $B > 0$ describe a common phenomenon known as **simple harmonic motion.** We can quickly draw the graphs of these equations by making use of three key numbers: the **amplitude,** $|A|$, the **period,** $2\pi/B$, and the **phase shift,** $-C/B$.

A complex number $a + bi$ can be represented geometrically as a point (a, b) in a plane called the **complex plane,** or **Argand diagram.** The horizontal and vertical axes are known as the **real axis** and **imaginary axis,** respectively. The distance from the origin to (a, b) is $\sqrt{a^2 + b^2}$; it is also the absolute value of $a + bi$, denoted by $|a + bi|$. If we let $r = \sqrt{a^2 + b^2}$ and θ be the angle that the ray from the origin through (a, b) makes with the positive x-axis, we obtain the **polar form** of $a + bi$, namely,

$$r(\cos \theta + i \sin \theta)$$

This form facilitates multiplication and division and is especially useful in finding powers and roots of complex numbers. A significant result is the formula

$$[r(\cos \theta + i \sin \theta)]^n = r^n(\cos n\theta + i \sin n\theta)$$

CHAPTER 9 REVIEW PROBLEM SET

In Problems 1–10, write True or False in the blank. If false, tell why.

_____ 1. The law of cosines, $c^2 = a^2 + b^2 - 2ab \cos \gamma$, for a right triangle is the Pythagorean Theorem.

_____ 2. For any triangle ABC, $b \sin \alpha = a \sin \beta$.

_____ 3. Addition of vectors is associative but not commutative.

_____ 4. For any vectors **u** and **v**, $\|\mathbf{u} + \mathbf{v}\| \leq \|\mathbf{u}\| + \|\mathbf{v}\|$.

_____ 5. If θ is the smallest nonnegative angle between **u** and **v**, then $\theta = \cos^{-1}[(\mathbf{u} \cdot \mathbf{v})/(\|\mathbf{u}\|\,\|\mathbf{v}\|)]$.

_____ 6. The period of $f(t) = 3 \sin(\frac{2}{3}t + \pi/2)$ is 3π.

_____ 7. The graph of $y = 3 \sin(\frac{3}{2}t + \pi/2)$ is the graph of $y = 3 \sin(\frac{3}{2}t)$ shifted $\pi/2$ units to the left.

_____ 8. If u is a complex number with $|u| = 1$, then $u = \pm 1$.

_____ 9. $-9i = 9(\cos 270° + i \sin 270°)$

_____ 10. Exactly two of the eight 8th roots of a complex number are real.

Solve each of the triangles in Problems 11–14.

11. $\alpha = 30°$, $\beta = 45°$, $c = 10$

12. $a = 2$, $b = 3$, $c = 4$

13. $\beta = 142°$, $b = 94$, $a = 67$

14. $\gamma = 37.6°$, $a = 11.6$, $b = 20.3$

15. Find the area of the triangle in Problem 14.

16. Find the area of a triangle whose sides measure 7, 8, and 9, respectively.

17. At a certain point A along a straight river, my line of sight to a tree B on the opposite bank makes an angle of 30° with the river. After walking 100 yards farther

along the river to the point C, my line of sight to B makes an angle of $45°$ with the river. Determine
(a) the distance $\overline{AB}$.
(b) the width of the river.

18. If $\mathbf{u}$ is a vector 6 units long pointing in the direction N60°W and $\mathbf{v} = -10\mathbf{i} + 15\mathbf{j}$, calculate $\mathbf{u} + \frac{4}{5}\mathbf{v}$ and write it in the form $a\mathbf{i} + b\mathbf{j}$.

19. Let $\mathbf{u} = 5\mathbf{i} - 12\mathbf{j}$ and $\mathbf{v} = 24\mathbf{i} + 7\mathbf{j}$. Calculate each of the following.
(a) $\|\mathbf{u}\|$ (b) $\|\mathbf{v}\|$
(c) $\mathbf{u} \cdot \mathbf{v}$
(d) θ, the angle between $\mathbf{u}$ and $\mathbf{v}$.
(e) A unit length vector with the same direction as $\mathbf{v}$.

20. For $\mathbf{u}$ and $\mathbf{v}$ as in Problem 19, determine
(a) the scalar projection of $\mathbf{u}$ on $\mathbf{v}$.
(b) the vector projection of $\mathbf{u}$ on $\mathbf{v}$.

21. Marge and Fred are pushing an object along the ground. Marge pushes with a force $\mathbf{u}$ of 120 pounds in the direction N arccos $\frac{4}{5}$E; Fred pushes with a force $\mathbf{v}$ of 90 pounds straight north. Calculate the resultant force $\mathbf{w} = \mathbf{u} + \mathbf{v}$ and write it in the form $a\mathbf{i} + b\mathbf{j}$.

22. In Problem 21, find the work done by $\mathbf{w}$ in moving the object
(a) 4 feet in the direction of $\mathbf{w}$.
(b) 4 feet in the direction of $\mathbf{u}$.

Determine the period, amplitude, and phase shift in Problems 23–26.

23. $y = \cos 2t$ 24. $y = 3 \cos 4t$
25. $y = 2 \sin(3t - \pi/2)$ 26. $y = -2 \sin(\frac{1}{2}t + \pi)$
27. Sketch the graph of $y = \cos 2t$ on $-\pi \le t \le \pi$.
28. Sketch the graph of $y = 3 \cos 4t$ on $-\pi \le t \le \pi$.
29. Sketch the graph of $y = 2 \sin(3t - \pi/2)$ on $-\pi \le t \le \pi$.
30. Sketch the graph of $y = -2 \sin(\frac{1}{2}t - \pi)$ on $-\pi \le t \le \pi$.

31. A wheel of radius 3 feet with center at the origin is rotating counterclockwise at $5\pi/6$ radians per second. A paint speck on the rim of the wheel has coordinates (3, 0) initially.
(a) Determine the speck's coordinates after t seconds.
(b) When will the speck first be at $(-3, 0)$?

32. A particle is moving along the x-axis so that its x-coordinate in meters after t seconds is given by $x = 3 \sin(5t + \pi/6)$.
(a) Determine the period of the motion.
(b) Determine the initial position of the particle.

(c) When is the particle at $x = 0$ the first time?

33. Find the distance the particle of Problem 32 travels during the time interval $\pi/6 \le t \le 13\pi/6$.

34. Plot the following numbers in the complex plane.
(a) $3 - 4i$ (b) -6
(c) $5i$
(d) $3\left(\cos \dfrac{3\pi}{4} + i \sin \dfrac{3\pi}{4}\right)$
(e) $4(\cos 300° + i \sin 300°)$

35. Find the absolute value of each number in Problem 34.

36. Express $4\sqrt{2}\left(\cos \dfrac{7\pi}{4} + i \sin \dfrac{7\pi}{4}\right)$ in $a + bi$ form.

Express each number in Problems 37–40 in polar form.

37. -4 38. $9i$
39. $2 + 2i$ 40. $2 - 2\sqrt{3}i$

41. Let $u = r(\cos t + i \sin t)$. Write each of the following in polar form.
(a) u^3 (b) $1/u$
(c) $u\left(\cos \dfrac{\pi}{3} + i \sin \dfrac{\pi}{3}\right)$

42. Let $u = 6(\cos 145° + i \sin 145°)$ and $v = 3(\cos 65° + i \sin 65°)$. Calculate each of the following, leaving your answer in polar form.
(a) uv^2 (b) $2u/v$
(c) u^2/v^3

43. Write the four 4th roots of $2^8(\cos 96° + i \sin 96°)$ in polar form.

44. Find all of the 6th roots of $-i$ and plot them in the complex plane.

45. If u is a square root of $1 - i$ and v is a 4th root of $1 + i$, calculate $u^6 v^{12}$.

46. Let $u = a + bi$ and $v = c + di$. Show that $|uv| = |u||v|$.

47. Find all solutions of $x^3 + 8i = 0$.

48. Let $u = \cos 30° + i \sin 30°$. Then $u, u^2, u^3, \ldots, u^{12}$ are the twelve 12th roots of unity. Which of these are also
(a) 6th roots of unity?
(b) 8th roots of unity?

49. Find two triangles for which $a = 1.5$, $b = 2$, and $\alpha = 40°$.

50. Let O, A, and B denote the corners of the sector of a circle of radius 10 and central angle at O of $36°$. Find the area of the region inside the sector AOB but outside the triangle AOB.

CHAPTER **10**

Theory of Polynomial Equations

■ He who loves practice without theory is like the sailor who boards ship without a rudder and compass and never knows where he may cast.

Leonardo da Vinci

10-1 DIVISION OF POLYNOMIALS

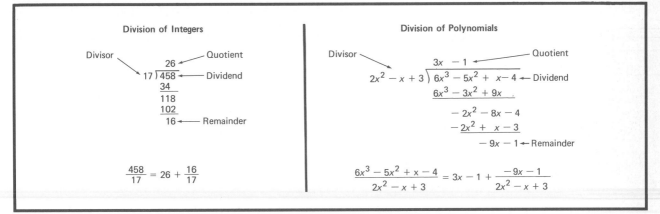

In Section 2-3 we learned that polynomials in x can always be added, subtracted, and multiplied; the result in every case is a polynomial. For example,

$$(3x^2 + x - 2) + (3x - 2) = 3x^2 + 4x - 4$$

$$(3x^2 + x - 2) - (3x - 2) = 3x^2 - 2x$$

$$(3x^2 + x - 2) \cdot (3x - 2) = 9x^3 - 3x^2 - 8x + 4$$

Now we are going to study division of polynomials. Occasionally, the division is exact.

$$(3x^2 + x - 2) \div (3x - 2) = \frac{(3x - 2)(x + 1)}{3x - 2} = x + 1$$

We say $3x - 2$ is an exact divisor, or factor, of $3x^2 + x - 2$. More often than not, the division is inexact and there is a nonzero remainder.

The Division Algorithm

The opening display shows that the process of division for polynomials is much the same as for integers. Both processes (we call them *algorithms*) involve subtraction. The first one shows how many times 17 can be subtracted from 458 before obtaining a remainder less than 17. The answer is 26 times, with a remainder of 16. The second algorithm shows how many times $2x^2 - x + 3$ can be subtracted from $6x^3 - 5x^2 + x - 4$ before obtaining a remainder of lower degree than $2x^2 - x + 3$. The answer is $3x - 1$ times, with a remainder of $-9x - 1$. Thus in these two examples we have

$$458 - 26(17) = 16$$

$$(6x^3 - 5x^2 + x - 4) - (3x - 1)(2x^2 - x + 3) = -9x - 1$$

Of course, we can also write these equalities as

$$458 = (17)(26) + 16$$

$$6x^3 - 5x^2 + x - 4 = (2x^2 - x + 3)(3x - 1) + (-9x - 1)$$

Notice that both of them can be summarized in the words

$$\text{dividend} = (\text{divisor}) \cdot (\text{quotient}) + \text{remainder}$$

This statement is very important and is worth stating again for polynomials in very precise language.

The Division Law for Polynomials

If $P(x)$ and $D(x)$ are any two nonconstant polynomials, then there are unique polynomials $Q(x)$ and $R(x)$ such that

$$P(x) = D(x)Q(x) + R(x)$$

where $R(x)$ is either zero or it is of lower degree than $D(x)$.

Here you should think of $P(x)$ as the dividend, $D(x)$ as the divisor, $Q(x)$ as the quotient, and $R(x)$ as the remainder.

The algorithm we use to find $Q(x)$ and $R(x)$ was illustrated in the opening display. Here we offer another illustration. Note that we arrange both the divisor and the dividend in descending powers of x with blank spaces for missing powers.

Example A (Long Division)   Find the quotient and remainder when $2x^4 - 8x + 5 - 4x^3$ is divided by $x^2 + 3 - 2x$.

Solution

$$
\begin{array}{r}
2x^2 \qquad\quad -6 \\
x^2 - 2x + 3\overline{\smash{\big)}\, 2x^4 - 4x^3 \qquad\quad -8x + 5} \\
\underline{2x^4 - 4x^3 + 6x^2 \qquad\qquad\quad} \\
-6x^2 - 8x + 5 \\
\underline{-6x^2 + 12x - 18} \\
-20x + 23
\end{array}
$$

$$Q(x) = 2x^2 - 6 \qquad R(x) = -20x + 23 \qquad \blacksquare$$

The method works fine even when the coefficients are nonreal numbers.

Example B (Nonreal Coefficients)   Find the quotient and remainder when $3x^2 - 2ix + 7$ is divided by $3x + i$.

Solution

$$
\begin{array}{r}
x - i \\
3x + i\overline{\smash{\big)}\, 3x^2 - 2ix + 7} \\
\underline{3x^2 + ix} \\
-3ix + 7 \\
\underline{-3ix + 1} \\
6
\end{array}
\qquad
\begin{array}{l}
Q(x) = x - i \\
R(x) = 6
\end{array}
$$

$\blacksquare$

Synthetic Division

It is often necessary to divide by polynomials of the form $x - c$. For such a division, there is a shortcut called *synthetic division*. We illustrate how it works for

$$(2x^3 - x^2 + 5) \div (x - 2)$$

Certainly the result depends on the coefficients; the powers of x serve mainly to determine the placement of the coefficients. Below, we show the division in its usual form and then in a skeletal form with the x's omitted. Note that we leave a blank space for the missing first-degree term in the long division but indicate it with a 0 in the skeletal form.

Long Division

$$
\require{enclose}
\begin{array}{r}
2x^2 + 3x\ + 6 \\
x - 2\ \enclose{longdiv}{2x^3 - x^2 + 5} \\
\underline{2x^3 - 4x^2} \\
3x^2 \\
\underline{3x^2 - 6x} \\
6x + 5 \\
\underline{6x - 12} \\
17
\end{array}
$$

First Condensation

$$
\begin{array}{r}
② ③ ⑥ \\
① - 2\ \enclose{longdiv}{2 -1 0 5} \\
2 -4 \\
\underline{} \\
3 ⓪ \\
③ - 6 \\
6 ⑤ \\
⑥ - 12 \\
17
\end{array}
$$

We can condense things still more by discarding all of the circled digits. The coefficients of the quotient, 2, 3, and 6, the remainder, 17, and the numbers from which they were calculated remain. All of the important numbers appear in the diagram below, on the left. On the right, we show the final modification. There we have changed the divisor from -2 to 2 to allow us to do addition rather than subtraction at each stage.

Second Condensation

$$
\begin{array}{r|rrrr}
-2 & 2 & -1 & 0 & 5 \\
& & -4 & -6 & -12 \\
\hline
& 2 & 3 & 6 & 17
\end{array}
$$

Synthetic Division

$$
\begin{array}{r|rrrr}
2 & 2 & -1 & 0 & 5 \\
& & 4 & 6 & 12 \\
\hline
& 2 & 3 & 6 & 17
\end{array}
$$

The process shown in the final format is called **synthetic division.** We can describe it by a series of steps.

CAUTION

In synthetic division, be sure to insert a 0 for each missing term of the dividend.

1. To divide by $x - 2$, use 2 as the synthetic divisor.
2. Write down the coefficients of the dividend according to descending powers. Be sure to write zeros for missing powers.
3. Bring down the first coefficient.
4. Follow the arrows, first multiplying by 2 (the divisor), then adding, multiplying the sum by 2, adding, and so on.
5. The last number in the third row is the remainder and the others are the coefficients of the quotient.

Example C (Synthetic Division by $x + a$) Divide $3x^3 + x^2 - 15x - 5$ by $x + \frac{1}{3}$, that is, by $x - (-\frac{1}{3})$.

Solution

$$
\begin{array}{r|rrrr}
-\frac{1}{3} & 3 & 1 & -15 & -5 \\
 & & -1 & 0 & 5 \\
\hline
 & 3 & 0 & -15 & 0
\end{array}
$$

Since the remainder is 0, the division is exact. We conclude that

$$3x^3 + x^2 - 15x - 5 = (x + \tfrac{1}{3})(3x^2 - 15) \quad \blacksquare$$

Example D (Division by $x - (a + bi)$) Find the quotient and remainder when $x^3 - 2x^2 - 6ix + 18$ is divided by $x - 2 - 3i$.

Solution Synthetic division works just fine even when some or all of the coefficients are nonreal. Since $x - 2 - 3i = x - (2 + 3i)$, we use $2 + 3i$ as the synthetic divisor.

$$
\begin{array}{r|rrrr}
2 + 3i & 1 & -2 & -6i & 18 \\
 & & 2 + 3i & -9 + 6i & -18 - 27i \\
\hline
 & 1 & 3i & -9 & -27i
\end{array}
$$

Quotient: $x^2 + 3ix - 9$; remainder: $-27i$. $\blacksquare$

Proper and Improper Rational Expressions

A rational expression (ratio of two polynomials) is said to be **proper** if the degree of its numerator is smaller than that of its denominator. Thus

$$\frac{x + 1}{x^2 - 3x + 2}$$

is a proper rational expression, but

$$\frac{2x^3}{x^2 - 3}$$

is improper. The division law, $P(x) = D(x)Q(x) + R(x)$, can be written as

$$\frac{P(x)}{D(x)} = Q(x) + \frac{R(x)}{D(x)}$$

It implies that any improper rational expression can be rewritten as the sum of a polynomial and a proper rational expression. For example,

$$\frac{2x^3}{x^2 - 3} = 2x + \frac{6x}{x^2 - 3}$$

a result we obtained by dividing $x^2 - 3$ into $2x^3$.

Example E (Improper Rational Expressions) Write the following improper fraction as a polynomial plus a proper fraction.

$$\frac{2x^4 - 4x^3 - 8x + 5}{x^2 - 2x + 3}$$

Solution   Refer to Example A, where the quotient and remainder were found.

$$\frac{2x^4 - 4x^3 - 8x + 5}{x^2 - 2x + 3} = 2x^2 - 6 + \frac{-20x + 23}{x^2 - 2x + 3} \quad \blacksquare$$

PROBLEM SET 10-1

A. Skills and Techniques

In Problems 1–8, find the quotient and the remainder if the first polynomial is divided by the second. See Example A.

1. $x^3 - x^2 + x + 3; x^2 - 2x + 3$
2. $x^3 - 2x^2 - 7x - 4; x^2 + 2x + 1$
3. $6x^3 + 7x^2 - 18x + 15; 2x^2 + 3x - 5$
4. $10x^3 + 13x^2 + 5x + 12; 5x^2 - x + 4$
5. $4x^4 - x^2 - 6x - 9; 2x^2 - x - 3$
6. $25x^4 - 20x^3 + 4x^2 - 4; 5x^2 - 2x + 2$
7. $2x^5 - 2x^4 + 9x^3 - 12x^2 + 4x - 16;$
   $2x^3 - 2x^2 + x - 4$
8. $3x^5 - x^4 - 8x^3 - x^2 - 3x + 12;$
   $3x^3 - x^2 + x - 4$

In Problems 9–18, use synthetic division to find the quotient and remainder if the first polynomial is divided by the second. See Example C.

9. $2x^3 - x^2 + x - 4; x - 1$
10. $3x^3 + 2x^2 - 4x + 5; x - 2$
11. $3x^3 + 5x^2 + 2x - 10; x - 1$
12. $2x^3 - 5x^2 + 4x - 4; x - 2$
13. $x^4 - 2x^2 - 1; x - 3$
14. $x^4 + 3x^2 - 340; x - 4$
15. $x^3 + 2x^2 - 3x + 2; x + 1$
16. $x^3 - x^2 + 11x - 1; x + 1$
17. $2x^4 + x^3 + 4x^2 + 7x + 4; x + \frac{1}{2}$
18. $2x^4 + x^3 + x^2 + 10x - 8; x - \frac{1}{2}$

Use synthetic division to find the quotient and remainder when the first polynomial is divided by the second, as in Example D.

19. $x^3 - 2x^2 + 5x + 30; x - 2 + 3i$
20. $2x^3 - 11x^2 + 44x + 35; x - 3 - 4i$
21. $x^4 - 17; x - 2i$
22. $x^4 + 18x^2 + 90; x - 3i$

In Problems 23–28, write each rational expression as the sum of a polynomial and a proper rational expression, as in Example E.

23. $\dfrac{x^3 + 2x^2 + 5}{x^2}$

24. $\dfrac{2x^3 - 4x^2 - 3}{x^2 + 1}$

25. $\dfrac{x^3 - 4x + 5}{x^2 + x - 2}$

26. $\dfrac{2x^3 + x - 8}{x^2 - x + 4}$

27. $\dfrac{2x^2 - 4x + 5}{x^2 + 1}$

28. $\dfrac{5x^3 - 6x + 11}{x^3 - x}$

B. Applications and Extensions

29. Express $(2x^4 - x^3 - x^2 - 2)/(x^3 + 1)$ as a polynomial plus a proper rational expression.

30. Find the quotient and the remainder when $x^4 + 6x^3 - 2x^2 + 4x - 15$ is divided by $x^2 - 2x + 3$.

31. Find by inspection the quotient and the remainder when the first polynomial is divided by the second.
    (a) $2x^3 + 3x^2 - 11x + 9; x^2$
    (b) $2(x + 3)^2 + 10(x + 3) - 14; x + 3$
    (c) $(x - 4)^5 + x^2 + x + 1; (x - 4)^3$
    (d) $(x^2 + 3)^3 + 2x(x^2 + 3) + 4x - 1; x^2 + 3$

32. Use synthetic division to find the quotient and the remainder when the first polynomial is divided by the second.
    (a) $x^4 - 4x^3 + 29; x - 3$
    (b) $2x^4 - x^3 + 2x - 4; x + \frac{1}{2}$
    (c) $x^4 + 4x^3 + 4\sqrt{3}x^2 + 3\sqrt{3}x + 3\sqrt{3}; x + \sqrt{3}$
    (d) $x^3 - (3 + 2i)x^2 + 10ix + 20 - 12i; x - 3$

33. Show that the second polynomial is a factor of the first and determine the other factor.
    (a) $x^5 + x^4 - 16x - 16; x - 2$
    (b) $x^5 + 32; x + 2$
    (c) $x^4 - \frac{3}{2}x^3 + 3x^2 + 6x + 2; x + \frac{1}{2}$
    (d) $x^3 - 2ix^2 + x - 2i; x - 2i$

34. Use synthetic division to show the second polynomial is a factor of the first and determine the other factor. *Hint:* In part (a) you will want to use the fact that $x^2 - 1 = (x - 1)(x + 1)$.
    (a) $x^4 + x^3 - x - 1; x^2 - 1$
    (b) $x^4 - x^3 + 2x^2 - 4x - 8; x^2 - x - 2$
    (c) $x^4 + 2x^3 - 4x - 4; x^2 - 2$
    (d) $x^5 + x^4 - 16x - 16; x^2 + 4$

35. Find $h(x)$, given that
$$x^6 + 2x^5 - x^4 - 10x^3 - 16x^2 + 8x + 16$$
$$= (x^2 - 1)(x + 2)h(x)$$

36. Find the quotient and remainder when $(x^2 - 4)^3$ is divided by $x + 3$.

37. Find k so that the second polynomial is a factor of the first.
(a) $x^3 + x^2 - 10x + k$; $x - 4$
(b) $x^4 + kx + 10$; $x + 2$
(c) $k^2x^3 - 4kx + 4$; $x - 1$

38. Determine h and k so that both $x - 3$ and $x + 2$ are factors of $x^4 - x^3 + hx^2 + kx - 6$.

39. Determine a, b, and c so that $(x - 1)^3$ is a factor of $x^4 + ax^3 + bx^2 + cx - 4$.

40. TEASER Let a, b, c, and d be distinct integers and suppose that $x - a$, $x - b$, $x - c$, and $x - d$ are factors of the polynomial $p(x)$, which has integral coefficients. Prove that $p(n)$ is not a prime number for any integer n.

41. Draw the graphs of
$$f(x) = \frac{2x^4 - 4x^3 - 8x + 5}{x^2 - 2x + 3} \quad \text{and} \quad g(x) = 2x^2 - 6$$
in the same plane using the $\boxed{\text{RANGE}}$ values $-8 \le x \le 8$ and $-10 \le y \le 80$. Make a conjecture. Then refer to Example E to make sense of your conjecture.

42. Refer to Problem 41. Find the smallest integer N such that $|x| \ge N$ implies that $|f(x) - g(x)| < .2$.

10-2 FACTORIZATION THEORY FOR POLYNOMIALS

Young Scholar: And does $x^4 + 99x^3 + 21$ have a zero?
Carl Gauss: Yes.
Young Scholar: How about $\pi x^{67} - \sqrt{3x^{19}} + 4i$?
Carl Gauss: It does.
Young Scholar: How can you be sure?
Carl Gauss: When I was young, 22 I think, I proved that every non-constant polynomial, no matter how complicated, has at least one zero. Many people call it the *Fundamental Theorem of Algebra.*

Carl F. Gauss (1777–1855)
"The Prince of Mathematicians"

Our young scholar could have asked a harder question. How do you find the zeros of a polynomial? Even the eminent Gauss would have had trouble with that question. You see, it is one thing to know a polynomial has zeros; it is quite another thing to find them.

Even though it is a difficult task and one at which we will have only limited success, our goal for this and the next two sections is to develop methods for finding zeros of polynomials. Remember that a polynomial is an expression of the form
$$P(x) = a_nx^n + a_{n-1}x^{n-1} + \cdots + a_1x + a_0$$
Unless otherwise specified, the coefficients (the a_i's) are allowed to be *complex* numbers. And by a **zero** of $P(x)$, we mean any complex number c (real or nonreal)

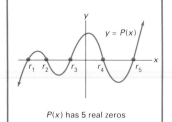
such that $P(c) = 0$. The number c is also called a **solution,** or a **root,** of the equation $P(x) = 0$. Note the use of words: Polynomials have zeros, but polynomial equations have solutions.

The Remainder and Factor Theorems

Recall the division law from Section 10-1, which had as its conclusion

$$P(x) = D(x)Q(x) + R(x)$$

If $D(x)$ has the form $x - c$, this becomes

$$P(x) = (x - c)Q(x) + R$$

where R, which is of lower degree than $x - c$, must be a constant. This last equation is an identity; it is true for all values of x, including $x = c$. Thus

$$P(c) = (c - c)Q(c) + R = 0 + R$$

We have just proved an important result.

Remainder Theorem

If a polynomial $P(x)$ is divided by $x - c$, then the constant remainder R is given by $R = P(c)$.

This theorem means that we may evaluate R or $P(c)$, whichever is easier, and automatically have the value of the other.

Example A [Finding R to Evaluate $P(c)$] Evaluate $P(-3)$ given that $P(x) = 5x^4 + 20x^3 + 14x^2 - 6x + 12$.

Solution We use synthetic division to divide $P(x)$ by $x - (-3)$.

$$
\begin{array}{r|rrrrr}
-3 & 5 & 20 & 14 & -6 & 12 \\
   &   & -15 & -15 & 3 & 9 \\
\hline
   & 5 & 5 & -1 & -3 & 21
\end{array}
$$

we conclude that $P(-3) = R = 21$. ∎

Example B [Evaluating $P(c)$ to Find R] Find the remainder R when $P(x) = x^{100} + x^{22} - 15$ is divided by $x - 1$.

Solution We could do the division—but what a waste of effort. The theorem tells us that

$$R = P(1) = 1^{100} + 1^{22} - 15 = -13$$ ∎

Much more important than the mere calculation of remainders is a consequence called the *Factor Theorem*. Since $R = P(c)$, as we have just seen, we may rewrite the division law as

$$P(x) = (x - c)Q(x) + P(c)$$

It is plain to see that $P(c) = 0$ if and only if the division of $P(x)$ by $x - c$ is exact; that is, if and only if $x - c$ is a factor of $P(x)$.

Factor Theorem

A polynomial $P(x)$ has c as a zero if and only if it has $x - c$ as a factor.

This theorem has many uses.

Example C (Showing That $x - c$ is a Factor)   Show that $x + 1$ is a factor of $P(x) = x^4 - 3x^3 + x^2 + 8x + 3$.

Solution   Since $P(-1) = 1 + 3 + 1 - 8 + 3 = 0$, -1 is a zero of $P(x)$, so $x + 1$ is a factor of $P(x)$.   ■

Sometimes it is easy to spot one zero of a polynomial. If so, the Factor Theorem may help us find the other zeros.

Example D (Factoring Cubics, Given One Zero)   Note that 1 is a zero of $P(x) = 3x^3 - 8x^2 + 3x + 2$. Find the other zeros.

Solution   That 1 is a zero follows from the calculation

$$P(1) = 3 - 8 + 3 + 2 = 0$$

Hence by the Factor Theorem, $x - 1$ is a factor of $P(x)$. We can use synthetic division to find the other factor.

$$
\begin{array}{r|rrrr}
1 & 3 & -8 & 3 & 2 \\
  &   & 3 & -5 & -2 \\
\hline
  & 3 & -5 & -2 & 0
\end{array}
$$

The remainder is 0 as we expected, so

$$P(x) = (x - 1)(3x^2 - 5x - 2)$$

Using the quadratic formula, we find the zeros of $3x^2 - 5x - 2$ to be $(5 \pm \sqrt{49})/6$, which simplify to 2 and $-\frac{1}{3}$. Thus $P(x)$ has 1, 2, and $-\frac{1}{3}$ as its three zeros.   ■

Complete Factorization of Polynomials

In the example above, we did not really need the quadratic formula. If we had been clever, we would have factored $3x^2 - 5x - 2$.

$$3x^2 - 5x - 2 = (3x + 1)(x - 2)$$
$$= 3(x + \tfrac{1}{3})(x - 2)$$

Thus $P(x)$, our original polynomial, may be written as

$$P(x) = 3(x - 1)(x + \tfrac{1}{3})(x - 2)$$

from which all three of the zeros are immediately evident.

But now we make another key observation. Notice that $P(x)$ can be factored as a product of its leading coefficient and three factors of the form $(x - c)$, where the c's are the zeros of $P(x)$. This holds true in general.

Complete Factorization Theorem

If

$$P(x) = a_n x^n + a_{n-1} x^{n-1} + \cdots + a_1 x + a_0$$

is an nth degree polynomial with $n > 0$, then there are n numbers $c_1, c_2, \ldots, c_n$, not necessarily distinct, such that

$$P(x) = a_n(x - c_1)(x - c_2) \cdots (x - c_n)$$

The c's are the zeros of $P(x)$; they may or may not be real numbers.

To prove the Complete Factorization Theorem, we must go back to Carl Gauss and our opening display. In his doctoral dissertation in 1799, Gauss gave a proof of the following important theorem, a proof that unfortunately is beyond the scope of this book.

Fundamental Theorem of Algebra

Every nonconstant polynomial has at least one zero.

Now let $P(x)$ be any polynomial of degree $n > 0$. By the Fundamental Theorem, it has a zero, which we may call c_1. By the Factor Theorem, $x - c_1$ is a factor of $P(x)$; that is,

$$P(x) = (x - c_1)P_1(x)$$

where $P_1(x)$ is a polynomial of degree $n - 1$ and with the same leading coefficient as $P(x)$—namely, a_n.

If $n - 1 > 0$, we may repeat the argument on $P_1(x)$. It has a zero c_2 and hence a factor $x - c_2$; that is,

$$P_1(x) = (x - c_2)P_2(x)$$

where $P_2(x)$ has degree $n - 2$. For our original polynomial $P(x)$ we may now write

$$P(x) = (x - c_1)(x - c_2)P_2(x)$$

Continuing in the pattern now established, we eventually get

$$P(x) = (x - c_1)(x - c_2) \cdots (x - c_n)P_n$$

where P_n has degree zero; that is, P_n is a constant. In fact, $P_n = a_n$, since the leading coefficient stayed the same at each step. This establishes the Complete Factorization Theorem.

About the Number of Zeros

Each of the numbers c_i in

$$P(x) = a_n(x - c_1)(x - c_2) \cdots (x - c_n)$$

is a zero of $P(x)$. Are there any other zeros? No, for if d is any number different from each of the c_i's, then

$$P(d) = a_n(d - c_1)(d - c_2) \cdots (d - c_n) \neq 0$$

All of this tempts us to say that a polynomial of degree n has exactly n zeros. But hold on! The numbers $c_1, c_2, \ldots, c_n$ need not all be different. For example, the sixth-degree polynomial

$$P(x) = 4(x - 2)^3(x + 1)(x - 4)^2$$

has only three distinct zeros, 2, -1, and 4. We have to settle for the following statement.

An nth-degree polynomial has at most n distinct zeros.

There is a way in which we can say that there are exactly n zeros. Call c a **zero of multiplicity k** of $P(x)$ if $x - c$ appears k times in its complete factorization. For example, in

$$P(x) = 4(x - 2)^3(x + 1)(x - 4)^2$$

the zeros 2, -1, and 4 have multiplicities 3, 1, and 2, respectively. A zero of multiplicity 1 is called a **simple zero.** Notice in our example that the multiplicities add to 6, the degree of the polynomial. In general, we may say this.

An nth-degree polynomial has exactly n zeros provided that we count a zero of multiplicity k as k zeros.

Illustrating the Theory

Example E (Finding a Polynomial from Its Zeros)
(a) Find a cubic polynomial having simple zeros 3, $2i$, and $-2i$.
(b) Find a polynomial $P(x)$ with integral coefficients and having $\frac{1}{2}$ and $-\frac{2}{3}$ as simple zeros and 1 as a zero of multiplicity 2.

CAUTION

$$(x - 3i)(x + 3i) = x^2 - 9$$

$$(x - 3i)(x + 3i) = x^2 - 9i^2$$
$$= x^2 + 9$$

Solution
(a) Let us call the required polynomial $P(x)$. Then

$$P(x) = a(x - 3)(x - 2i)(x + 2i)$$

where a can be any nonzero number. Choosing $a = 1$ and multiplying, we have

$$P(x) = (x - 3)(x^2 + 4) = x^3 - 3x^2 + 4x - 12$$

(b) $P(x) = a(x - \frac{1}{2})(x + \frac{2}{3})(x - 1)^2$
We choose $a = 6$ to eliminate fractions.

$$P(x) = 6\left(x - \frac{1}{2}\right)\left(x + \frac{2}{3}\right)(x - 1)^2$$

$$= 2\left(x - \frac{1}{2}\right)3\left(x + \frac{2}{3}\right)(x - 1)^2$$

$$= (2x - 1)(3x + 2)(x^2 - 2x + 1)$$

$$= 6x^4 - 11x^3 + 2x^2 + 5x - 2 \quad \blacksquare$$

Example F (More on Zeros of Polynomials)   Show that 2 is a zero of multiplicity 3 of the polynomial

$$P(x) = 2x^5 - 17x^4 + 51x^3 - 58x^2 + 4x + 24$$

and find the remaining zeros. Then factor $P(x)$.

Solution   We must show that $x - 2$ appears as a factor 3 times in the factored form of $P(x)$. Synthetic division can be used successively, as shown below. The final quo-

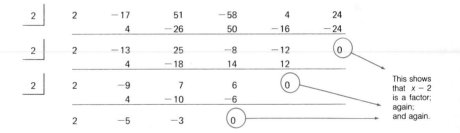

This shows that $x - 2$ is a factor; again; and again.

tient $2x^2 - 5x - 3$ factors as $(2x + 1)(x - 3)$. Therefore, the remaining two zeros are $-\frac{1}{2}$ and 3. The factored form of $P(x)$ is

$$2(x - 2)^3\left(x + \frac{1}{2}\right)(x - 3) \quad \blacksquare$$

Example G (Factoring a Polynomial with a Given Nonreal Zero)   Show that $1 + 2i$ is a zero of $P(x) = x^3 - (1 + 2i)x^2 - 4x + 4 + 8i$. Then factor $P(x)$ into linear factors.

Solution   We start by using synthetic division.

$$
\begin{array}{r|rrrr}
1 + 2i & 1 & -1 - 2i & -4 & 4 + 8i \\
 & & 1 + 2i & 0 & -4 - 8i \\
\hline
 & 1 & 0 & -4 & 0
\end{array}
$$

Therefore $1 + 2i$ is a zero and $x - 1 - 2i$ is a factor of $P(x)$, and

$$P(x) = (x - 1 - 2i)(x^2 - 4)$$

$$= (x - 1 - 2i)(x + 2)(x - 2) \quad \blacksquare$$

PROBLEM SET 10-2

A. Skills and Techniques

Use the remainder theorem to find $P(c)$. Check your answer by substituting c for x. See Example A.

1. $P(x) = 2x^3 - 5x^2 + 3x - 4; c = 2$

2. $P(x) = x^3 + 4x^2 - 11x - 5; c = 3$

3. $P(x) = 8x^4 - 3x^2 - 2; c = \frac{1}{2}$

4. $P(x) = 2x^4 + \frac{3}{4}x + \frac{3}{2}; c = -\frac{1}{2}$

Find the remainder if the first polynomial is divided by the second. Do it without actually dividing, as in Example B.

5. $x^{10} - 15x + 8; x - 1$

6. $2x^{20} + 5; x + 1$

7. $64x^6 + 13; x + \frac{1}{2}$

8. $81x^3 + 9x^2 - 2; x - \frac{1}{3}$

Find all of the zeros of the given polynomial and give their multiplicities.

9. $(x - 1)(x + 2)(x - 3)$

10. $(x + 2)(x + 5)(x - 7)$

11. $(2x - 1)(x - 2)^2 x^3$

12. $(3x + 1)(x + 1)^3 x^2$

13. $3(x - 1 - 2i)(x + \frac{2}{3})$

14. $5(x - 2 + \sqrt{5})(x - \frac{4}{5})$

In Problems 15–18, show that $x - c$ is a factor of $P(x)$. See Example C.

15. $P(x) = 2x^3 - 7x^2 + 9x - 4; c = 1$

16. $P(x) = 3x^3 + 4x^2 - 6x - 1; c = 1$

17. $P(x) = x^3 - 7x^2 + 16x - 12; c = 3$

18. $P(x) = x^3 - 8x^2 + 13x + 10; c = 5$

Problems 19–22 relate to Example D.

19. In Problem 15, you know that $P(x)$ has 1 as a zero. Find the other zeros.

20. Find all the zeros of $P(x)$ in Problem 16. Then factor $P(x)$ completely.

21. Find all of the zeros of $P(x)$ in Problem 17.

22. Find all of the zeros of $P(x)$ in Problem 18.

In Problems 23–26, factor the given polynomial into linear factors. You should be able to do it by inspection.

23. $x^2 - 5x + 6$

24. $2x^2 - 14x + 24$

25. $x^4 - 5x^2 + 4$

26. $x^4 - 13x^2 + 36$

In Problems 27–30, factor $P(x)$ into linear factors given that c is a zero of $P(x)$.

27. $P(x) = x^3 - 3x^2 - 28x + 60; c = 2$

28. $P(x) = x^3 - 2x^2 - 29x - 42; c = -2$

29. $P(x) = x^3 + 3x^2 - 10x - 12; c = -1$

30. $P(x) = x^3 + 11x^2 - 5x - 55; c = -11$

In Problems 31–38, find a polynomial $P(x)$ with integral coefficients having the given zeros. Assume each zero to be simple (multiplicity 1) unless otherwise indicated. See Example E.

31. 2, 1, and -4

32. 3, -2, and 5

33. $\frac{1}{2}, -\frac{5}{6}$

34. $\frac{3}{7}, \frac{3}{4}$

35. 2, $\sqrt{5}, -\sqrt{5}$

36. $-3, \sqrt{7}, -\sqrt{7}$

37. $\frac{1}{2}$ (multiplicity 2), -2 (multiplicity 3)

38. 0, $-2, \frac{3}{4}$ (multiplicity 3)

In Problems 39–42, find a polynomial $P(x)$ having only the indicated simple zeros.

39. 2, -2, i, $-i$

40. $2i, -2i, 3i, -3i$

41. 2, -5, $2 + 3i$, $2 - 3i$

42. $-3, 2, 1 - 4i, 1 + 4i$

Problems 43–46 relate to Example F.

43. Show that 1 is a zero of multiplicity 3 of the polynomial $x^5 + 2x^4 - 6x^3 - 4x^2 + 13x - 6$, find the remaining zeros, and factor completely.

44. Show that the polynomial $x^5 - 11x^4 + 46x^3 - 90x^2 + 81x - 27$ has 3 as a zero of multiplicity 3, find the remaining zeros, and factor completely.

45. Show that the polynomial

$$x^6 - 8x^5 + 7x^4 + 32x^3 + 31x^2 + 40x + 25$$

has -1 and 5 as zeros of multiplicity 2 and find the remaining zeros.

46. Show that the polynomial

$$x^6 + 3x^5 - 9x^4 - 50x^3 - 84x^2 - 72x - 32$$

has 4 as a simple zero and -2 as a zero of multiplicity 3. Find the remaining zeros.

As in Example G, factor $P(x)$ into linear factors given that c is a zero of $P(x)$.

47. $P(x) = x^3 - 2ix^2 - 9x + 18i; c = 2i$

48. $P(x) = x^3 + 3ix^2 - 4x - 12i; c = -3i$

49. $P(x) = x^3 + (1 - i)x^2 - (1 + 2i)x - 1 - i;$
$c = 1 + i$

50. $P(x) = x^3 - 3ix^2 + (3i - 3)x - 2 + 6i;$
$c = -1 + 3i$

B. Applications and Extensions

51. Find the remainder when the first polynomial is divided by the second.
(a) $3x^{44} + 5x^{41} + 4; x + 1$
(b) $1988x^3 - 1989x^2 + 1990x - 1991; x - 1$
(c) $x^9 + 512; x + 2$

52. Prove that the second polynomial is a factor of the first.
(a) $x^n - a^n; x - a$ (n a positive integer)
(b) $x^n + a^n; x + a$ (n an odd positive integer)
(c) $x^5 + 32a^{10}; x + 2a^2$

53. Find all zeros of the following polynomials and give their multiplicities.
(a) $(x^2 - 4)^3$ (b) $(x^2 - 3x + 2)^2$
(c) $(x^2 + 2x - 4)^3(x + 2)^4$

54. Find all zeros of $(x - 1)^4(x^2 + 4)^3(x^3 - 8)^2$ and give their multiplicities.

55. Show that $3x^{31} - 2x^{18} + 4x^3 - x^2 - 4$ has no negative zeros.

56. Show that $2x^{44} + 3x^4 + 5$ has no factor of the form $x - c$, where c is real.

57. Each of the following polynomials has $\frac{1}{2}$ as a zero. Factor each polynomial into linear factors.
(a) $12x^3 + 4x^2 - 3x - 1$
(b) $2x^3 - x^2 - 4x + 2$
(c) $2x^3 - x^2 + 2x - 1$

58. Find a third-degree polynomial $P(x)$ with integral coefficients that has the given numbers as zeros.
(a) $\frac{3}{4}, 2, -\frac{2}{3}$ (b) $3, -2, -2$
(c) $3, \sqrt{2}$

59. Sketch the graph of $y = x^3 + 4x^2 - 2x$. Then find the three values of x for which $y = 8$. *Hint:* One of them is an integer.

60. Refer to the graph of Problem 59 and note that there is only one real x (namely, $x = 2$) for which $y = 20$. What does this tell you about the other two solutions of $x^3 + 4x^2 - 2x = 20$? Find these other two solutions.

61. A tray is to be constructed from a piece of sheet metal 16 inches square by cutting small identical squares of length x from the corners and then folding up the flaps (Figure 1). Determine x if the volume is to be 300 cu-

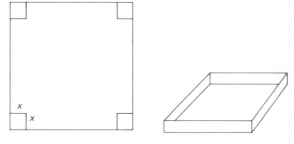

Figure 1

bic inches. *Hint:* There are two possible answers, one of which is an integer.

62. Find a polynomial $P(x) = ax^3 + bx^2 + cx + d$ that has 1, 2, and 3 as zeros and satisfies $P(0) = 36$.

63. Find $P(x) = ax^4 + bx^3 + cx^2 + dx + e$ if $P(x)$ has $1/2$ as a zero of multiplicity 4 and $P(0) = 1$.

64. Write $x^5 + x^4 + x^3 + x^2 + x + 1$ as a product of linear and quadratic factors with integer coefficients.

65. Show that if $P(x) = a_nx^n + a_{n-1}x^{n-1} + a_{n-2}x^{n-2} + \cdots + a_1x + a_0$ has $n + 1$ distinct zeros, then all the coefficients must be 0.

66. Let

$$P_1(x) = x^n + a_{n-1}x^{n-1} + a_{n-2}x^{n-2}$$
$$+ \cdots + a_1x + a_0$$

and

$$P_2(x) = x^n + b_{n-1}x^{n-1} + b_{n-2}x^{n-2}$$
$$+ \cdots + b_1x + b_0$$

Show that if $P_1(x) = P_2(x)$ for n distinct values of x, then the two polynomials are equal for all x. *Hint:* Apply Problem 65 to $P(x) = P_1(x) - P_2(x)$.

67. Show that if c is a zero of $x^6 - 5x^5 + 3x^4 + 7x^3 + 3x^2 - 5x + 1$, then $1/c$ is also a zero.

68. Generalize Problem 67 by showing that if c is a zero of

$$a_nx^n + a_{n-1}x^{n-1} + a_{n-2}x^{n-2} + \cdots + a_1x + a_0$$

where $a_k = a_{n-k}$ for $k = 0, 1, 2, \ldots, n$ and $a_n \neq 0$, then $1/c$ is also a zero.

69. Show that if $c_1, c_2, \ldots, c_n$ are the zeros of

$$a_nx^n + a_{n-1}x^{n-1} + a_{n-2}x^{n-2} + \cdots + a_1x + a_0$$

with $a_n \neq 0$, then
(a) $c_1 + c_2 + \cdots + c_n = -a_{n-1}/a_n$
(b) $c_1c_2 + c_1c_3 + \cdots + c_1c_n + c_2c_3 + \cdots + c_2c_n + \cdots + c_{n-1}c_n = a_{n-2}/a_n$

(c) $c_1 c_2 \cdots c_n = (-1)^n a_0/a_n$

Hint: Use the Complete Factorization Theorem.

70. TEASER Let the polynomial $x^n - kx - 1$, $n > 2$, have zeros $c_1, c_2, \ldots, c_n$. Show that
(a) If $n > 2$, then $c_1^n + c_2^n + \cdots + c_n^n = n$
(b) If $n > 3$, then $c_1^2 + c_2^2 + \cdots + c_n^2 = 0$

71. In Problem 61, the volume V of the box is given by $V = x(16 - 2x)^2$. Determine x that maximizes V; also give the maximum V.

72. Take an ordinary sheet of 8.5- by 11-inch paper and fold a corner over to the opposite side thus determining triangle, *DEF* (Figure 2). Experiment with different folds to see how the area A of the triangle changes.

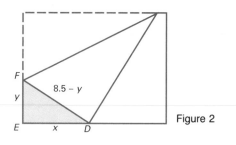

Figure 2

(a) Show that $A = (72.25x - x^3)/34$.
(b) Algebraically determine x so A is minimized.
(c) Graphically determine x so A is maximized.
(d) Determine the largest x that makes $A = 6$.

10-3 POLYNOMIAL EQUATIONS WITH REAL COEFFICIENTS

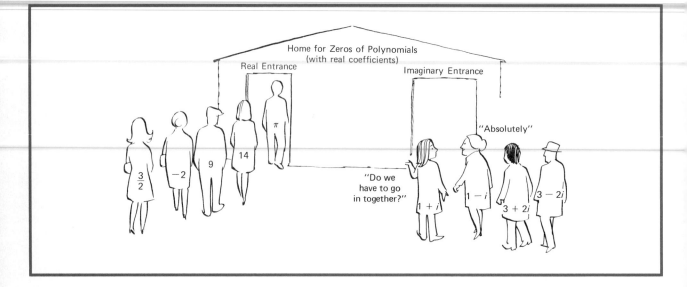

Finding the zeros of a polynomial $P(x)$ is the same as finding the solutions of the equation $P(x) = 0$. We have solved polynomial equations before, especially in Chapter 3. In particular, recall that the two solutions of a quadratic equation with real coefficients are either both real or both nonreal. For example, the equation $x^2 - 7x + 12 = 0$ has the real solutions 3 and 4. On the other hand, $x^2 - 8x + 17 = 0$ has two nonreal solutions $4 + i$ and $4 - i$.

In this section, we shall see that if a polynomial equation has real coefficients, then its nonreal solutions (if any) must occur in pairs—conjugate pairs. In the opening display, $1 + i$ and $1 - i$ must enter together or not at all; so must $3 + 2i$ and $3 - 2i$.

Properties of Conjugates

We indicate the conjugate of a complex number by putting a bar over it. If $u = a + bi$, then $\overline{u} = a - bi$. For example,

$$\overline{1 + i} = 1 - i$$

$$\overline{3 - 2i} = 3 + 2i$$

$$\overline{4} = 4$$

$$\overline{2i} = -2i$$

The operation of taking the conjugate behaves nicely in both addition and multiplication. There are two pertinent properties, stated first in words.

1. The conjugate of a sum is the sum of the conjugates.
2. The conjugate of a product is the product of the conjugates.

In symbols these properties become

1. $\overline{u_1 + u_2 + \cdots + u_n} = \overline{u}_1 + \overline{u}_2 + \overline{u}_3 + \cdots + \overline{u}_n$
2. $\overline{u_1 u_2 u_3 \cdots u_n} = \overline{u}_1 \cdot \overline{u}_2 \cdot \overline{u}_3 \cdots \overline{u}_n$

A third property follows from the second property if we set all u_i's equal to u.

3. $\overline{u^n} = (\overline{u})^n$

Rather than prove these properties, we shall illustrate them.

1. $\overline{(2 + 3i) + (1 - 4i)} = \overline{2 + 3i} + \overline{1 - 4i}$
2. $\overline{(2 + 3i)(1 - 4i)} = \overline{(2 + 3i)}\,\overline{(1 - 4i)}$
3. $\overline{(2 + 3i)^3} = \overline{(2 + 3i)}^3$

Let us check that the second statement is correct. We will do it by computing both sides independently.

$$\overline{(2 + 3i)(1 - 4i)} = \overline{2 + 12 + (-8 + 3)i} = \overline{14 - 5i} = 14 + 5i$$

$$\overline{(2 + 3i)}\,\overline{(1 - 4i)} = (2 - 3i)(1 + 4i) = 2 + 12 + (8 - 3)i = 14 + 5i$$

Example A (Conjugates)   Apply Properties 1–3 to find the following conjugates.

(a) $\overline{2(2 - 3i)(1 + 2i)}$ (b) $\overline{(4 - i)^2 + 3 - 2i}$

Solution

(a) $\overline{2(2 - 3i)(1 + 2i)} = 2(2 + 3i)(1 - 2i) = 2(8 - i) = 16 - 2i$
(b) $\overline{(4 - i)^2 + 3 - 2i} = (4 + i)^2 + 3 + 2i = 15 + 8i + 3 + 2i = 18 + 10i$ ∎

We can use the three properties of conjugates to demonstrate some very important results. For example, we can show that if u is a solution of the equation

$$ax^3 + bx^2 + cx + d = 0$$

where a, b, c, and d are real, then $\bar{u}$ is also a solution. To do this, we show

$$a\bar{u}^3 + b\bar{u}^2 + c\bar{u} + d = 0$$

whenever it is given that

$$au^3 + bu^2 + cu + d = 0$$

In the latter equation, take the conjugate of both sides, using the three properties of conjugates and the fact that a real number is its own conjugate.

$$\overline{au^3 + bu^2 + cu + d} = \bar{0}$$

$$\overline{au^3} + \overline{bu^2} + \overline{cu} + \bar{d} = 0$$

$$\bar{a}\overline{u^3} + \bar{b}\overline{u^2} + \bar{c}\,\bar{u} + \bar{d} = 0$$

$$a\bar{u}^3 + b\bar{u}^2 + c\bar{u} + d = 0$$

Nonreal Solutions Occur in Pairs

We are ready to state the main theorem of this section.

> **Conjugate Pair Theorem**
>
> Let
> $$a_nx^n + a_{n-1}x^{n-1} + \cdots + a_1x + a_0 = 0$$
> be a polynomial equation with real coefficients. If u is a solution, its conjugate $\bar{u}$ is also a solution.

We feel confident that you will be willing to accept the truth of this theorem without further argument. The formal proof would mimic the proof given above for the cubic equation.

Example B (Applying the Theorem)   Given that $3 + 4i$ is a solution to

$$x^3 - 8x^2 + 37x - 50 = 0$$

find the other solutions.

Solution   We know immediately that $3 - 4i$ is also a solution. We can easily find the third solution, which incidentally must be real. (Why?) Here is how we do it.

$$
\begin{array}{r|rrrr}
3 + 4i & 1 & -8 & 37 & -50 \\
       &   & 3 + 4i & -31 - 8i & 50 \\
\hline
3 - 4i & 1 & -5 + 4i & 6 - 8i & 0 \\
       &   & 3 - 4i & -6 + 8i & \\
\hline
       & 1 & -2 & 0 &
\end{array}
$$

From this it follows that

$$x^3 - 8x^2 + 37x - 50 = [x - (3 + 4i)][x - (3 - 4i)][x - 2]$$

and that 2 is the third solution of our equation. ∎

Notice something special about the product of the first two factors just displayed.

$$[x - (3 + 4i)][x - (3 - 4i)] = x^2 - (3 + 4i)x - (3 - 4i)x + (3 + 4i)(3 - 4i)$$
$$= x^2 - 6x + 25$$

The product is a quadratic polynomial with *real* coefficients. This is not an accident. If u is any complex number, then

$$(x - u)(x - \overline{u}) = x^2 - (u + \overline{u})x + u\overline{u}$$

is a real quadratic polynomial, since both $u + \overline{u}$ and $u\overline{u}$ are real (see Problem 41). Thus the Conjugate Pair Theorem, when combined with the Complete Factorization Theorem of Section 10-2, has the following consequence.

Real Factors Theorem

Any polynomial with real coefficients can be factored into a product of linear and quadratic polynomials having real coefficients, where the quadratic polynomials have no real zeros.

Example C (Writing a Polynomial as a Product of Real Factors)   Given that $1 + 2i$ is a solution of the equation

$$P(x) = 2x^4 - 5x^3 + 9x^2 + x - 15 = 0$$

find the other solutions. Then write $P(x)$ as a product of linear and quadratic factors with real coefficients.

Solution   Since the coefficients are real, $1 - 2i$ is also a solution. Apply synthetic division as follows.

$$
\begin{array}{r|ccccc}
1 + 2i & 2 & -5 & 9 & 1 & -15 \\
 & & 2 + 4i & -11 - 2i & 2 - 6i & 15 \\
\hline
1 - 2i & 2 & -3 + 4i & -2 - 2i & 3 - 6i & \circledcirc \\
 & & 2 - 4i & -1 + 2i & -3 + 6i & \\
\hline
 & 2 & -1 & -3 & \circledcirc &
\end{array}
$$

We expected these 0's

The two remaining solutions are the zeros of

$$2x^2 - x - 3 = (x + 1)(2x - 3)$$

We conclude that the four solutions are $1 + 2i$, $1 - 2i$, -1, and $3/2$. Finally,

$$P(x) = [x - (1 + 2i)][x - (1 - 2i)](x + 1)(2x - 3)$$
$$= (x^2 - 2x + 5)(x + 1)(2x - 3) \quad ∎$$

Example D (Obtaining a Polynomial Given Some of Its Zeros)   Find a cubic polynomial with real coefficients that has 3 and $2 + 3i$ as zeros. Make the leading coefficient 1.

Solution   The third zero has to be $2 - 3i$. In factored form, our polynomial is

$$(x - 3)(x - 2 - 3i)(x - 2 + 3i)$$

We multiply this out in stages.

$$(x - 3)[(x - 2)^2 + 9]$$
$$(x - 3)(x^2 - 4x + 13)$$
$$x^3 - 7x^2 + 25x - 39 \quad \blacksquare$$

Rational Solutions

How does one get started on solving an equation of high degree? So far, all we can suggest is to guess. If you are lucky and find a solution, you can use synthetic division to reduce the degree of the equation to be solved. Eventually you may get it down to a quadratic equation, for which we have the quadratic formula.

Guessing would not be so bad if there were not so many possibilities to consider. Is there an intelligent way to guess? There is, but unfortunately, it works only if the coefficients are integers, and then it only helps us find rational solutions. Consider

$$3x^3 + 13x^2 - x - 6 = 0$$

which, as you will note, has integral coefficients. Suppose it has a rational solution c/d which is in reduced form (that is, c and d are integers without common divisors greater than 1 and $d > 0$). Then

$$3 \cdot \frac{c^3}{d^3} + 13 \cdot \frac{c^2}{d^2} - \frac{c}{d} - 6 = 0$$

or, after multiplying by d^3,

$$3c^3 + 13c^2d - cd^2 - 6d^3 = 0$$

We can rewrite this as

$$c(3c^2 + 13cd - d^2) = 6d^3$$

and also as

$$d(13c^2 - cd - 6d^2) = -3c^3$$

The first of these tells us that c divides $6d^3$ and the second that d divides $-3c^3$. But c and d have no common divisors. Therefore, c must divide 6 and d must divide 3.

The only possibilities for c are ± 1, ± 2, ± 3, and ± 6; for d, the only possibilities are 1 and 3. Thus the possible rational solutions must come from the following list.

$$\frac{c}{d}: \quad \pm 1, \pm 2, \pm 3, \pm 6, \pm \tfrac{1}{3}, \pm \tfrac{2}{3}$$

Upon checking all 12 numbers (which takes times, but a bit less time than checking *all* numbers would take!) we find that only $\frac{2}{3}$ works.

$$
\begin{array}{r|rrrr}
\frac{2}{3} & 3 & 13 & -1 & -6 \\
 & & 2 & 10 & 6 \\
\hline
 & 3 & 15 & 9 & 0
\end{array}
$$

We could prove the following theorem by using similar reasoning.

Rational Solution Theorem (Rational Root Theorem)

Let
$$a_n x^n + a_{n-1} x^{n-1} + \cdots + a_1 x + a_0 = 0$$
have integral coefficients. If c/d is a rational solution in reduced form, then c divides a_0 and d divides a_n.

Example E (Finding Rational Solutions)   Find the rational solutions of the equation
$$3x^4 + 2x^3 + 2x^2 + 2x - 1 = 0$$
Then find the remaining solutions.

Solution   The only way that c/d (in reduced form) can be a solution is for c to be 1 or -1 and for d to be 1 or 3. This means that the possibilities for c/d are ± 1 and $\pm \frac{1}{3}$. Synthetic division shows that -1 and $\frac{1}{3}$ work.

$$
\begin{array}{r|rrrrr}
-1 & 3 & 2 & 2 & 2 & -1 \\
 & & -3 & 1 & -3 & 1 \\
\hline
\frac{1}{3} & 3 & -1 & 3 & -1 & 0 \\
 & & 1 & 0 & 1 & \\
\hline
 & 3 & 0 & 3 & 0 &
\end{array}
$$

Setting the final quotient, $3x^2 + 3$, equal to zero and solving, we get
$$3x^2 = -3$$
$$x^2 = -1$$
$$x = \pm i$$

The complete solution set is $\{-1, \frac{1}{3}, i, -i\}$. ■

PROBLEM SET 10-3

A. Skills and Techniques

In Problems 1–10, write the conjugate of the number. See Example A.

1. $2 + 3i$

2. $3 - 5i$

3. $4i$

4. $-6i$

5. $4 + \sqrt{6}$

6. $3 - \sqrt{5}$

7. $(2 - 3i)^8$

8. $(3 + 4i)^{12}$

9. $2(1 + 2i)^3 - 3(1 + 2i)^2 + 5$

10. $4(6 - i)^4 + 11(6 - i) - 23$

11. If $P(x)$ is a cubic polynomial with real coefficients and has -3 and $5 - i$ as zeros, what other zero does it have?

12. If $P(x)$ is a cubic polynomial with real coefficients and has 0 and $\sqrt{2} + 3i$ as zeros, what other zero does it have?

13. Suppose that $P(x)$ has real coefficients and is of the fourth degree. If it has $3 - 2i$ and $5 + 4i$ as two of its zeros, what other zeros does it have?

14. If $P(x)$ is a fourth degree polynomial with real coefficients and has $5 + 6i$ as a zero of multiplicity 2, what are its other zeros?

In Problems 15–18, one or more solutions of the specified equations are given. Find the other solutions. See Examples B and C.

15. $2x^3 - x^2 + 2x - 1 = 0$; i

16. $x^3 - 3x^2 + 4x - 12 = 0$; $2i$

17. $x^4 + x^3 + 6x^2 + 26x + 20 = 0$; $1 + 3i$

18. $x^6 + 2x^5 + 4x^4 + 4x^3 - 4x^2 - 16x - 16 = 0$; $2i, -1 + i$

Find a polynomial with real coefficients that has the indicated degree and the given zero(s). Make the leading coefficient 1. See Example D.

19. Degree 2; zero: $2 + 5i$

20. Degree 2; zero: $\sqrt{6}i$

21. Degree 3; zeros: $-3, 2i$

22. Degree 3; zeros: $5, -3i$

23. Degree 5; zeros: $2, 3i$ (multiplicity 2)

24. Degree 5; zeros: $1, 1 - i$ (multiplicity 2)

In Problems 25–30, find the rational solutions of each equation. If possible, find the other solutions. See Example E.

25. $x^3 - 3x^2 - x + 3 = 0$

26. $x^3 + 3x^2 - 4x - 12 = 0$

27. $2x^3 + 3x^2 - 4x + 1 = 0$

28. $5x^3 - x^2 + 5x - 1 = 0$

29. $\frac{1}{3}x^3 - \frac{1}{2}x^2 - \frac{1}{6}x + \frac{1}{6} = 0$ *Hint:* Clear the equation of fractions.

30. $\frac{2}{3}x^3 - \frac{1}{2}x^2 + \frac{2}{3}x - \frac{1}{2} = 0$

B. Applications and Extensions

31. The number $2 + i$ is one solution to $x^4 - 3x^3 + 2x^2 + x + 5 = 0$. Find the other solutions.

32. Find the fourth degree polynomial with real coefficients and leading coefficient 1 that has $2, -4$, and $2 - 3i$ as three of its zeros.

33. Find all solutions of $x^4 - 3x^3 - 20x^2 - 24x - 8 = 0$.

34. Find all solutions of $2x^4 - x^3 + x^2 - x - 1 = 0$.

35. The equation $2x^5 - 3x^4 + 13x^3 - 22x^2 - 24x + 16 = 0$ has three rational solutions. Find all solutions.

36. The equation $2x^5 - 7x^4 + 34x^3 - 78x^2 + 144x - 135 = 0$ has $1 - 2i$ as one solution. It also has a rational solution. Find all solutions.

37. Solve $x^5 + 6x^4 - 34x^3 + 56x^2 - 39x + 10 = 0$.

38. Solve $x^5 - 2x^4 + x - 2 = 0$. *Hint:* $x^4 + 1$ is easy to factor by adding and subtracting $2x^2$.

39. Show that a polynomial equation with real coefficients and of odd degree has at least one real solution.

40. A cubic equation with real coefficients has either three real solutions (not necessarily distinct), or 1 real solution and a pair of nonreal solutions. State all the possibilities for
   (a) a fourth-degree equation with real coefficients.
   (b) a fifth-degree equation with real coefficients.

41. Show that if u is any complex number, then $u + \bar{u}$ and $u\bar{u}$ are real numbers.

42. Let u and v be complex numbers with $v \neq 0$. Show that $\overline{u/v} = \bar{u}/\bar{v}$. Use this to demonstrate that if $f(x)$ is a real rational function (a quotient of two polynomials with real coefficients) and if $f(a + bi) = c + di$, then $f(a - bi) = c - di$.

43. Write $x^4 + 3x^3 + 3x^2 - 3x - 4$ as a product of linear and quadratic factors with real coefficients as guaranteed by the real factors theorem.

44. Write $x^6 - 9x^5 + 38x^4 - 106x^3 + 181x^2 - 205x + 100$ as a product of linear and quadratic factors with real coefficients. *Hint:* $1 + 2i$ is zero of multiplicity 2.

45. Write $x^8 - 1$ as a product of linear and quadratic factors with real coefficients. *Hint:* First factor as a difference of squares; then see the hint in Problem 38.

46. Let $x^n + a_{n-1}x^{n-1} + \cdots + a_1x + a_0 = 0$ have integral coefficients.
   (a) Show that all real solutions are either integral or irrational.
   (b) From part (a) deduce that if m and n are positive integers and m is not a perfect nth power, then $\sqrt[n]{m}$ is irrational (in particular, $\sqrt[3]{3}, \sqrt[4]{17}$, and $\sqrt[5]{12}$ are irrational).

47. Find the exact value of $x = \sqrt[3]{\sqrt{5} - 2} - \sqrt[3]{\sqrt{5} + 2}$. *Hint:* Begin by showing that $x^3 + 3x + 4 = 0$.

48. **TEASER** Let (u, v, w) satisfy the following nonlinear system of equations.

$$u + v + w = 2$$
$$u^2 + v^2 + w^2 = 8$$
$$u^3 + v^3 + w^3 = 8$$

(a) Determine the cubic equation $x^3 + a_2 x^2 + a_1 x + a_0 = 0$ that has u, v, and w as its solutions. *Hint:* Problem 69 of Section 10-2 should be helpful.

(b) Solve this cubic equation thereby solving the system of equations.

49. Draw the graphs of each of the following.

$$y = (x + 2)^2 \qquad y = (x + 1)^3 \qquad y = (x - 1)^4$$
$$y = (x - 2)^5$$

Make a conjecture about the behavior of the graph of a polynomial at a multiple zero.

50. With the RANGE menu set at $-4 \le x \le 4$ and $-1000 \le y \le 1000$, draw separate graphs of

$$y = (x + 2)^3 x^2 (x - 3)^5$$

and

$$y = (x + 2)^2 x^3 (x - 3)^4$$

Confirm or extend the conjecture you made in Problem 49.

10-4 THE METHOD OF SUCCESSIVE APPROXIMATIONS

Eighteen Months of Genius

The year and a half that began in January 1665 has been called the most fruitful period in the history of scientific thought. In that short time, Isaac Newton discovered the general binomial theorem, the differential and integral calculus, the theory of colors, and the laws of gravity.

"Nature and Nature's laws lay
hid in night:
God said, let Newton be! and
all was light."

Alexander Pope

Isaac Newton
1642–1727

Even with the theory developed so far, we are often unable to get started on solving an equation of high degree. Imagine being given a fifth-degree equation whose true solutions are $\sqrt{2} + 5i$, $\sqrt{2} - 5i$, $\sqrt[3]{19}$, 1.597, and 3π. How would you ever find them? Nothing you have learned until now would be of much help.

There is a general method of solving problems known to all resourceful people. We call it "muddling through" or "trial and error." Given a cup of tea, we add sugar a bit at a time until it tastes just right. Given a stopper too large for a hole, we whittle it down until it fits. We change the solution a step at a time, continually improving the accuracy until we are satisfied. Mathematicians call it **method of successive approximations.**

We explore two such methods in this section. The first is a graphical method; the second is Newton's algebraic method. Both are designed to find the real solutions of polynomial equations with real coefficients. Readers with graphing calculators will find the first method to be slow and clumsy but not unlike the ZOOM procedure on their calculators.

Method of Successive Enlargements

Consider the equation

$$x^3 - 3x - 5 = 0$$

We begin by graphing

$$y = x^3 - 3x - 5$$

looking for the point (or points) where the graph crosses the x-axis (Figure 3). These points correspond to the real solutions.

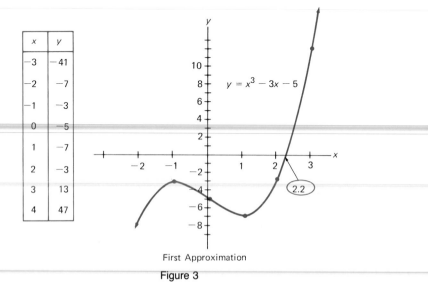

x	y
-3	-41
-2	-7
-1	-3
0	-5
1	-7
2	-3
3	13
4	47

$y = x^3 - 3x - 5$

First Approximation

Figure 3

Clearly there is only one real solution and it is between 2 and 3; a good first guess might be 2.2. Now we calculate y for values of x near 2.2 (for instance, 2.1, 2.2, and 2.3) until we find an interval of length .1 on which y changes sign. The interval is $2.2 \leq x \leq 2.3$. On this interval, we pretend the graph is a straight line. The point at which this line crosses the x-axis gives us our next approximation. It is about 2.28 (Figure 4).

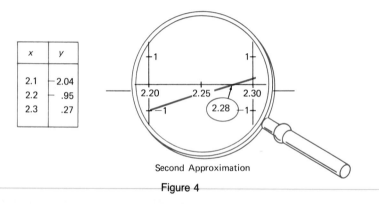

x	y
2.1	-2.04
2.2	$.95$
2.3	$.27$

Second Approximation

Figure 4

x	y
2.27	−.113
2.28	.012
2.29	.139

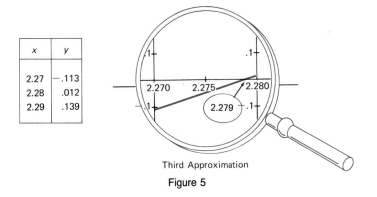

Third Approximation

Figure 5

Now we calculate y for values of x near 2.28 until we find an interval of length .01 on which y changes sign. This occurs on the interval $2.27 \leq x \leq 2.28$. Using a straight-line graph for this interval, we read our next approximation as 2.279 (Figure 5).

In effect, we are enlarging the graph (using a more and more powerful magnifying glass) at each stage, increasing the accuracy by one digit each time. Or to put it in graphing calculator language, we continue to ZOOM in on the root until we locate it to desired accuracy.

Newton's Method

Let $P(x) = 0$ be a polynomial equation with real coefficients. Suppose that by some means (perhaps graphing), we discover that it has a real solution r which we guess to be about x_1. Then, as Figure 6 suggests, a better approximation to r is x_2, the point at which the tangent line to the curve at x_1 crosses the x-axis.

If the slope of the tangent line is m_1, then

$$m_1 = \frac{\text{rise}}{\text{run}} = \frac{P(x_1)}{x_1 - x_2}$$

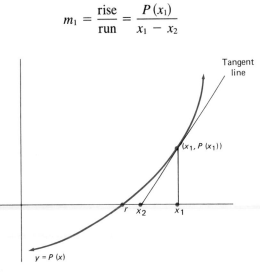

Figure 6

Solving this for x_2 yields

$$x_2 = x_1 - \frac{P(x_1)}{m_1}$$

What has been done once can be repeated. A still better approximation to r would be x_3, obtained from x_2 as follows:

$$x_3 = x_2 - \frac{P(x_2)}{m_2}$$

where m_2 is the slope of the tangent line at $(x_2, P(x_2))$. In general, we find the $(k + 1)$st approximation from the kth by using Newton's formula

$$x_{k+1} = x_k - \frac{P(x_k)}{m_k}$$

where m_k is the slope of the tangent line at $(x_k, P(x_k))$.

The rub, of course, is that we do not know how to calculate m_k. That is precisely where Newton made his biggest contribution. He showed how to find the slope of the tangent line to any curve. This is done in every calculus course; we state one result without proof.

Slope Theorem

If

$$P(x) = a_n x^n + a_{n-1} x^{n-1} + \cdots + a_2 x^2 + a_1 x + a_0$$

is a polynomial with real coefficients, then the slope of the tangent line to the graph of $y = P(x)$ at x is $P'(x)$, where $P'(x)$ is the polynomial

$$P'(x) = n a_n x^{n-1} + (n - 1) a_{n-1} x^{n-2} + \cdots + 2 a_2 x + a_1$$

Example A (The Tangent Line)   Find the slope of the tangent line to the graph of

$$y = P(x) = 2x^4 - 4x^3 - 6x^2 + 2x + 15$$

at x. Then write the equation of the tangent line at $x = 2$.

Solution   Since

$$P'(x) = 8x^3 - 12x^2 - 12x + 2$$

the slope m of the tangent line at $x = 2$ is

$$m = P'(2) = 8 \cdot 2^3 - 12 \cdot 2^2 - 12 \cdot 2 + 2 = -6$$

Also,

$$P(2) = 2 \cdot 2^4 - 4 \cdot 2^3 - 6 \cdot 2^2 + 2 \cdot 2 + 15 = -5$$

The equation of the required tangent line is $y + 5 = -6(x - 2)$.   ∎

Taking the slope theorem for granted, we may write Newton's formula in the useful form

$$x_{k+1} = x_k - \frac{P(x_k)}{P'(x_k)}$$

Using Newton's Method

We considered the equation

$$x^3 - 3x - 5 = 0$$

earlier in the section. Here we look at it again.

Example B (An Equation Revisited)   Use Newton's method to find the real solution of $x^3 - 3x - 5 = 0$ accurate to three decimal places.

Solution   If we let

$$P(x) = x^3 - 3x - 5$$

then, by the slope theorem,

$$P'(x) = 3x^2 - 3$$

and Newton's formula becomes

$$x_{k+1} = x_k - \frac{x_k^3 - 3x_k - 5}{3x_k^2 - 3}$$

If we take $x_1 = 3$ as our initial guess, then

$$x_2 = x_1 - \frac{x_1^3 - 3x_1 - 5}{3x_1^2 - 3}$$

$$= 3 - \frac{3^3 - 3 \cdot 3 - 5}{3 \cdot 3^2 - 3}$$

$$\approx \boxed{2.5}$$

$$x_3 = x_2 - \frac{x_2^3 - 3x_2 - 5}{3x_2^2 - 3}$$

$$= 2.5 - \frac{(2.5)^3 - 3(2.5) - 5}{3(2.5)^2 - 3}$$

$$\approx \boxed{2.30}$$

$$x_4 = x_3 - \frac{x_3^3 - 3x_3 - 5}{3x_3^2 - 3}$$

$$= 2.30 - \frac{(2.30)^3 - 3(2.30) - 5}{3(2.30)^2 - 3}$$

$$\approx \boxed{2.279}$$

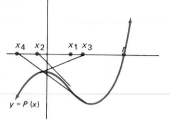

Figure 7

$$x_5 = x_4 - \frac{x_4^3 - 3x_4 - 5}{3x_4^2 - 3}$$

$$= 2.279 - \frac{(2.279)^3 - 3(2.279) - 5}{3(2.279)^2 - 3}$$

$$= 2.2790$$

The real solution to desired accuracy is 2.279. ∎

When you use Newton's method, it is important to make your initial guess reasonably good. Figure 7 shows how badly the method can lead you astray if you choose x_1 too far off the mark.

Example C (Finding All Real Solutions by Newton's Method)   Find all real solutions of the following equation by Newton's method.

$$P(x) = x^4 - 8x^3 + 22x^2 - 24x + 6 = 0$$

Solution   First we sketch the graph of $P(x)$ (Figure 8).

x	y
0	6
1	−3
2	−2
3	−3
4	6

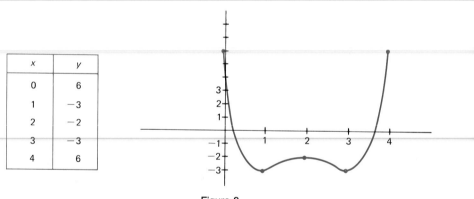

Figure 8

The graph crosses the x-axis at approximately .4 and 3.6. The slope polynomial is

$$P'(x) = 4x^3 - 24x^2 + 44x - 24$$

Take $x_1 = .4$

$$x_2 = .4 - \frac{P(.4)}{P'(.4)}$$

$$\approx .4 - \frac{(-.57)}{(-9.98)} \approx .34$$

$$x_3 = .34 - \frac{P(.34)}{P'(.34)}$$

$$\approx .34 - \frac{.0821}{(-4.657)} \approx .347$$

Take $x_1 = 3.6$

$$x_2 = 3.6 - \frac{P(3.6)}{P'(3.6)}$$

$$\approx 3.6 - \frac{(-.566)}{9.98} \approx 3.66$$

$$x_3 = 3.66 - \frac{P(3.66)}{P'(3.66)}$$

$$\approx 3.66 - \frac{.0821}{11.66} \approx 3.653$$

The two real solutions are approximately .347 and 3.653. ∎

PROBLEM SET 10-4

A. Skills and Techniques

In Problems 1–6, use the Slope Theorem to find the slope polynomial $P'(x)$ if $P(x)$ is the given polynomial. Then find the slope of the tangent line at $x = 1$. See Example A.

1. $2x^2 - 5x + 6$

2. $3x^2 + 4x - 9$

3. $2x^2 + x - 2$

4. $x^3 - 5x + 8$

5. $2x^5 + x^4 - 2x^3 + 8x - 4$

6. $x^6 - 3x^4 + 7x^3 - 4x^2 + 5x - 4$

Each of the equations in Problems 7–10 has exactly one real solution. By means of a graph, make an initial guess at the solution. Then use the method of successive enlargements to find a second and a third approximation.

7. $x^3 + 2x - 5 = 0$

8. $x^3 + x - 32 = 0$

9. $x^3 - 3x - 10 = 0$

10. $2x^3 - 6x - 15 = 0$

Problems 11–14 relate to Example B.

11. Use Newton's method to approximate the real solution of the equation $x^3 + 2x - 5 = 0$. Take as x_1 the initial guess you made in Problem 7, and then find x_2, x_3, and x_4.

12-14. Follow the instructions of Problem 11 for the equations in Problems 8-10.

Each equation in Problems 15–18 has several real solutions. Draw a graph to get your first estimates. Then use Newton's method to find these solutions to three decimal places. See Example C.

15. $x^4 + x^3 - 3x^2 + 4x - 28 = 0$

16. $x^4 + x^3 - 6x^2 - 7x - 7 = 0$

17. $x^3 - 3x + 1 = 0$

18. $x^3 - 12x + 1 = 0$

B. Applications and Extensions

19. The line $y = x + 3$ intersects the curve $y = x^3 - 3x + 4$ at three points. Use Newton's method to find the x-coordinates of these points correct to two decimal places.

20. A spherical shell has a thickness of 1 centimeter. What is the outer radius r of the shell if the volume of the shell is equal to the volume of the space inside? First find an equation for r and then solve it by Newton's method.

21. The problem of the dog running around a marching column of soldiers (Problem 88 of Section 3-4) led to the equation $x^4 - 4x^3 - 2x^2 + 4x + 5 = 0$. Use Newton's method to find the solution of this equation that is near 4, correct to four decimal places. Multiply this answer by 50 to get the answer to the question asked in that earlier problem.

22. The dimensions of a rectangular box are 6, 8, and 10 feet. Suppose that the volume of the box is increased by 300 cubic feet by equal elongations of the three dimensions. Find this elongation correct to two decimal places.

23. What rate of interest compounded annually is implied in an offer to sell a house for $50,000 cash, or in annual installments of $20,000 each payable 1, 2, and 3 years from now? *Hint:* The amount of $50,000 with interest for 3 years should equal the sum of the first payment accumulated for 2 years, the second accumulated for 1 year, and the third payment. Hence, if i is the interest rate, $50,000(1 + i)^3 = 20,000(1 + i)^2 + 20,000(1 + i) + 20,000$. Dividing by 10,000 and writing x for $1 + i$ gives the equation $5x^3 = 2x^2 + 2x + 2$.

24. Find the rate of interest implied if a house is offered for sale at $80,000 cash or in 4 annual installments of $23,000, the first payable now. (See Problem 23.)

25. Find the equation of the tangent line to the graph of the equation $y = 3x^2$ at the point (2, 12). *Hint:* Remember that the line through (2, 12) with slope m has equation $y - 12 = m(x - 2)$. Find m by evaluating the slope polynomial at $x = 2$.

26. Find the equation of the tangent line to the curve
    (a) $y = x^2 + x$ at the point (2, 6).
    (b) $y = 2x^3 - 4x + 5$ at the point $(-1, 7)$.
    (c) $y = \frac{1}{3}x^5$ at the point $(2, \frac{32}{5})$.
    See the hint for Problem 25.

27. The points where the tangent line to a graph is horizontal are of great significance since among them are found the high and low points of the graph. Determine the x-coordinates of such points for the graphs of the following polynomials.
    (a) $P(x) = 2x^3 - 3x^2 - 36x + 10$
    (b) $P(x) = 3x^4 - 8x^3 - 6x^2 + 9$

28. A box is to be made from a rectangular piece of sheet metal 24 inches long and 9 inches wide by cutting out identical squares of side x inches from each of the four corners and turning up the sides. Find the value of x that makes the volume of the box a maximum. What is this maximum volume? *Hint:* See Problem 27.

29. We stated the Slope Theorem without proof, but here we give a hint of its derivation (a subject treated at great length in calculus). Consider the graph of $y = 2x^3$, a part of which is shown in Figure 9. We are interested in finding the slope of the tangent line at the point P with x-coordinate a. Consider P to be a fixed point and let Q be a neighboring movable point with x-coordinate $a + h$. Note that as h tends to 0, the line PQ rotates toward the tangent line.

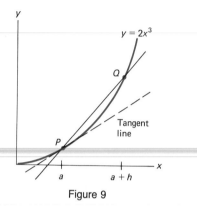

Figure 9

(a) Show that the slope of the line PQ is m_h, where

$$m_h = \frac{2(a + h)^3 - 2a^3}{h}$$

(b) Simplify the expression for m_h above as much as possible.

(c) Find the limiting value of m_h as h tends to 0.

(d) Compare the answer obtained in part (c) with the value of the slope at $x = a$ given by the Slope Theorem.

30. TEASER Let $P(x)$ be a polynomial of degree at least 3. From the division algorithm of Section 10-1, we may write

$$P(x) = (x - a)Q_1(x) + R_1$$

$$Q_1(x) = (x - a)Q_2(x) + R_2$$

from which we conclude that

$$P(x) = (x - a)^2 Q_2(x) + R_2(x - a) + R_1$$

We know that $R_1 = P(a)$, but what is the meaning of R_2? Note that R_2 is the constant remainder we get after dividing $P(x)$ and then its quotient $Q_1(x)$ by $x - a$, a number found easily by a repeated synthetic division.

(a) Carry out this double synthetic division on $P(x) = 2x^3$ to obtain $R_2 = 6a^2$.

(b) Find R_2 for the polynomial $P(x) = x^4 - 2x^3 + x^2 - x + 1$.

(c) Make a conjecture based on parts (a) and (b).

(d) Give an argument to support your conjecture.

31. Graph the polynomial

$$P(x) = x^5 - 4x^4 + 2x^3 + 3x^2 + x + 6$$

(a) Find all its zeros accurate to two decimal places.

(b) How many hills and valleys (local maximum and minimum points) does the graph have?

32. Let M denote the total number of hills and valleys for a graph. Experiment until you find three fifth-degree polynomials: One with $M = 4$, one with $M = 2$, and one with $M = 0$. Can you find one with $M = 3$ or $M = 1$? Think about what you know to be true for second-, third-, and fourth-degree polynomials. Make a conjecture about the value of M for an nth-degree polynomial.

CHAPTER 10 SUMMARY

The **division law for polynomials** asserts that if $P(x)$ and $D(x)$ are any given nonconstant polynomials, then there are unique polynomials $Q(x)$ and $R(x)$ such that

$$P(x) = D(x)Q(x) + R(x)$$

where $R(x)$ is either 0 or of lower degree than $D(x)$. In fact, we can find $Q(x)$ and $R(x)$ by the **division algorithm,** which is just a fancy name for ordinary long division. When $D(x)$ has the form $x - c$, $R(x)$ will have to be a constant R, since it is of lower degree than $D(x)$. The substitution $x = c$ then gives

$$P(c) = R$$

a result known as the **Remainder Theorem.** An immediate consequence is the **Factor Theorem,** which says that c is a zero of $P(x)$ if and only if $x - c$ is a factor of $P(x)$. Division of a polynomial by $x - c$ can be greatly simplified by use of **synthetic division.**

That every nonconstant polynomial has at least one zero is guaranteed by Gauss's **Fundamental Theorem of Algebra.** But we can say much more than that. For any nonconstant polynomial

$$P(x) = a_n x^n + a_{n-1} x^{n-1} + \cdots + a_1 x + a_0$$

there are n numbers $c_1, c_2, \ldots, c_n$ (not necessarily all different) such that

$$P(x) = a_n(x - c_1)(x - c_2) \cdots (x - c_n)$$

We call the latter result the **Complete Factorization Theorem.**

If the polynomial equation

$$P(x) = a_n x^n + a_{n-1} x^{n-1} + \cdots + a_1 x + a_0 = 0$$

has real coefficients, then its nonreal solutions (if

any) must occur in conjugate pairs $a + bi$ and $a - bi$. If the coefficients are integers and if c/d is a rational solution in reduced form, then c divides a_0 and d divides a_n.

To find exact solutions to a polynomial equation may be very difficult; often we are more than happy to find good approximations. Two good methods for doing this are the **method of successive enlargements** and **Newton's method.** Both require plenty of calculating power.

CHAPTER 10 REVIEW PROBLEM SET

In Problems 1–10, write True or False in the blank. If false, tell why.

_____ 1. When $x^4 - 2x^3 + 3x - 4$ is divided by $x - 1$, the remainder is -2.

_____ 2. If $x - a$ is a factor of the polynomial $P(x)$, then $P(-a) = 0$.

_____ 3. An nth degree polynomial has n distinct zeros.

_____ 4. The graph of $y = (x - 2)^3(x + 1)^2$ crosses the x-axis at exactly two places.

_____ 5. If n is a positive even integer, then $x - a$ is a factor of $x^n - a^n$.

_____ 6. The equation $x^{13} - 2x^5 + 4x^2 + 2x - 1 = 0$ has at least one real solution.

_____ 7. If $2 + 3i$ is a zero of a polynomial, then so is $2 - 3i$.

_____ 8. The number 1 is a zero of multiplicity 4 of the polynomial $(x^2 + 2x - 3)^2(x - 1)^4$.

_____ 9. The equation $x^{14} - 7x^3 + 5x^2 - 4x + 1 = 0$ has no rational solutions.

_____ 10. If $P(x)$ is a polynomial with real coefficients such that $P(2) > 0$ and $P(3) < 0$, then $P(x)$ has a real zero between 2 and 3.

Find the quotient Q and the remainder R when the first polynomial is divided by the second.

11. $2x^3 + x^2 - 3x + 4$; $x^2 + x + 1$

12. $2x^4 - 5x^2 + 2x - 3$; $2x^2 + 1$

13. $2x^7 + 8x^6 + 10x^5 + 8x^4 + x^3 + 4x^2 + 5x + 6$; $x^3 + 4x^2 + 5x + 4$

14. $x^3 + 3ix^2 - x + 2i$; $x^2 + ix + 1$

Use synthetic division to find the quotient Q and the remainder R when the first polynomial is divided by the second.

15. $x^5 - x^4 + 2x^3 + 3x^2 + 5x - 1$; $x - 2$

16. $2x^3 - 5x^2 + 5x - 4$; $x - \frac{3}{2}$

17. $x^4 - 2x^2 + 5x - 7$; $x - \sqrt{2} + 1$

18. $2x^3 - 4x^2 + x - 7$; $x - 2i$

19. $x^4 + (-2 + 3i)x^3 - 4x^2 + (16 - 12i)x - 16 + 24i$; $x - 2 + 3i$

Without dividing, find the remainder when the first polynomial is divided by the second.

20. $x^{18} - 2x^5 + 3$; $x + 1$

21. $2x^6 + x^4 - 5x^2 - 7$; $x - \sqrt{2}$

22. $x^4 + 2x^3 + 3x^2 + 4x + 5$; $x - i$

Find the zeros of each polynomial and give their multiplicities.

23. $(x - 2)^2(2x - 3)(x + 4)^3$

24. $(x^2 - 4)^3(x^2 + 1)$

25. $(x^2 - 5x + 6)^2(x^2 - 9)$

26. $x^2(x^2 - 6x + 10)^3(x + 3\sqrt{2})$

Factor each polynomial into linear factors, given that c is a zero.

27. $x^3 - 2x^2 - 4x + 8$; $c = 2$

28. $x^3 - 5x^2 + 2x + 8$; $c = -1$

29. $x^3 - (2 + 3i)x^2 + (-8 + 6i)x + 24i$; $c = 3i$

30. $2x^3 + (1 - 2\pi)x^2 - (\pi + 1)x + \pi$; $c = \pi$

In Problems 31–33, find a polynomial (in expanded form) with integral coefficients that has the given zeros. Zeros are simple unless otherwise specified.

31. -4, 1 (multiplicity 2)

32. $\sqrt{2}$, $-\sqrt{2}$, $\frac{3}{4}$, $-\frac{2}{3}$

33. i, $-i$, $2 - i$, $2 + i$

34. Factor $x^4 - 4x^3 + 5x^2 - 4x + 4$ into real factors, given that 2 is a zero of multiplicity 2.

35. Find a third-degree polynomial $P(x)$ with zeros 1, -4, and $-\frac{2}{3}$ for which $P(-1) = 18$.

36. Determine the value of k so that $\sqrt{3}$ is a zero of $x^3 - 2x^2 - 3x + k$.

37. Find a cubic polynomial with real coefficients that has $5 + 2i$ and -3 as two of its zeros.

38. Solve the equation $2x^4 - 5x^3 + 53x^2 - 24x - 26 = 0$, given that $1 + 5i$ is one of the solutions.

39. Find all solutions of $3x^3 + 4x^2 - 7x + 2 = 0$, given that it has a rational solution.

40. Show that $x^5 - 2x^3 + 4x^2 + 3 = 0$ has one real solution but that it is not rational.

41. Graph $P(x) = x^3 - 8x - 4$ to see that it has one positive zero r. Make an initial guess x_1 for this zero. Use the method of successive enlargements to find a second and a third approximation.

42. Refer to Problem 41. Use x_1 and Newton's Method to find approximations x_2, x_3, and x_4.

43. Find the equation of the tangent line to the curve $y = x^4 - 2x^2 - 3x + 5$ at (1, 1).

44. A 2-centimeter-thick slice is cut from a cube of cheese, leaving a chunk of volume 384 cubic centimeters. Find the length of a side of the original cube.

45. An open box is to be made from a rectangular piece of tin measuring 24 inches by 40 inches by cutting a square x inches by x inches from each corner and then folding up the flaps. The box is to have a volume of 2000 cubic inches. There are two possible values of x. Find the largest of these accurate to three decimal places using Newton's method

46. An object moves along a straight line so that its directed distance s in feet from the origin after t seconds is $s = 3t^3 - 4t^2 + 5t - 6$. Note that initially the object is 6 feet to the left of the origin. When did the object first hit the origin?

47. Let r_1 and r_2 be the zeros of $P(x) = ax^2 + bx + c$ ($a \neq 0$). Show that $r_1 + r_2 = -b/a$ and $r_1 r_2 = c/a$. *Hint:* $P(x) = a(x - r_1)(x - r_2)$.

48. Let r_1, r_2, and r_3 be the zeros of $P(x) = ax^3 + bx^2 + cx + d$ ($a \neq 0$). Show that $r_1 + r_2 + r_3 = -b/a$ and $r_1 r_2 r_3 = -d/a$.

CHAPTER 11

Systems of Equations and Inequalities

■ Geometry may sometimes appear to take the lead over analysis but in fact precedes it only as a servant goes before the master to clear the path and light him on his way.

James Joseph Sylvester

11-1 EQUIVALENT SYSTEMS OF EQUATIONS

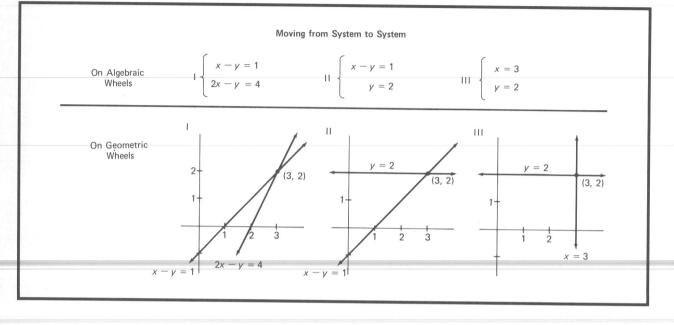

In Section 3-3 you learned how to solve a system of two equations in two unknowns, but you probably did not think of the process as one of replacing the given system by another having the same solutions. It is this point of view that we now want to explore.

In the display above, system I is replaced by system II, which is simpler; system II is in turn replaced by system III, which is simpler yet. To go from system I to system II, we eliminated x from the second equation; we then substituted $y = 2$ in the first equation to get system III. What happened geometrically is shown in the bottom half of our display. Notice that the three pairs of lines have the same point of intersection $(3, 2)$.

Because the notion of changing from one system of equations to another having the same solutions is so important, we make a formal definition. We say that two systems of equations are **equivalent** if they have the same solutions.

Operations that Lead to Equivalent Systems

Now we face a big question. What operations can we perform on a system without changing its solutions?

Operation 1   We can interchange the positions of two equations.
Operation 2   We can multiply an equation by a nonzero constant, that is, we can replace an equation by a nonzero multiple of itself.
Operation 3   We can add a multiple of one equation to another, that is, we can replace an equation by the sum of that equation and a multiple of another.

Operation 3 is the workhorse of the set. We show how it is used in the example of the opening display.

$$\text{I} \quad \begin{cases} x - y = 1 \\ 2x - y = 4 \end{cases}$$

If we add -2 times the first equation to the second, we obtain

$$\text{II} \quad \begin{cases} x - y = 1 \\ \phantom{x - {}} y = 2 \end{cases}$$

We then add the second equation to the first. This gives

$$\text{III} \quad \begin{cases} x = 3 \\ y = 2 \end{cases}$$

This is one way to write the solution. Alternatively, we say that the solution is the ordered pair $(3, 2)$.

The Three Possibilities for a Linear System

We are mainly interested in linear systems—that is, systems of linear equations—and we shall restrict our discussion to the case where the number of equations equals the number of unknowns. There are three possibilities for the set of solutions: The set may be empty, it may have just one point, or it may have infinitely many points. These three case are illustrated in Figure 1.

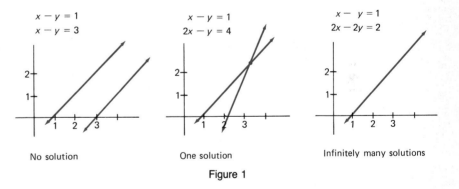

No solution   One solution   Infinitely many solutions

Figure 1

Someone is sure to object and ask if we cannot have a linear system with exactly two solutions or exactly three solutions. The answer is no. If a linear system has two solutions, it has infinitely many. This is obvious in the case of two equations in two unknowns, since two points determine a line, but it is also true for n equations in n unknowns.

Linear Systems in More than Two Unknowns

When we consider large systems, it is a good idea to be very systematic about our method of attack. Our method is to reduce the system to **triangular form** and then use **back substitution**.

Example A (Back Substitution)   Solve the system

$$x - 2y + z = -4$$
$$5y - 3z = 18$$
$$2z = -2$$

Solution   This system is already in triangular form, meaning that the terms within the dashed triangle have zero coefficients. Any system in this form is ready for back substitution. Solve the third equation first, obtaining $z = -1$. Then substitute this value in the second equation,

$$5y - 3(-1) = 18$$

which gives $y = 3$. Finally, substitute these two values in the first equation

$$x - 2(3) + (-1) = -4$$

which gives $x = 3$. Thus the solution of the system is $(3, 3, -1)$. This process, called **back substitution,** works on any linear system that is in triangular form. Just start at the bottom and work your way up.   ∎

If the system is not in triangular form initially, we try to operate on it until it is.

Example B (Changing to Triangular Form)   Change the following system to triangular form and then solve by back substitution.

$$2x - 4y + 2z = -8$$
$$2x + y - z = 10$$
$$3x - y + 2z = 4$$

Solution   Begin by multiplying the first equation by $\frac{1}{2}$ so that its leading coefficient is 1.

$$x - 2y + z = -4$$
$$2x + y - z = 10$$
$$3x - y + 2z = 4$$

Next add -2 times the first equation to the second equation. Also add -3 times the first equation to the third.

$$x - 2y + z = -4$$
$$5y - 3z = 18$$
$$5y - z = 16$$

Finally, add -1 times the second equation to the third.

$$x - 2y + z = -4$$
$$5y - 3z = 18$$
$$2z = -2$$

The system is now in triangular form and can be solved by back substitution. It is, in fact, the triangular system of Example A. The solution is $(3, 3, -1)$. ∎

Systems with Infinitely Many or No Solutions

Our method (change to triangular form, use back substitution) is not restricted to systems with single solutions.

Example C (Infinitely Many Solutions)   Solve

$$x - 2y + 3z = 10$$
$$2x - 3y - z = 8$$
$$4x - 7y + 5z = 28$$

Solution   Using Operation 3, we may eliminate x from the last two equations. First add -2 times the first equation to the second equation. Then add -4 times the first equation to the third equation. We obtain

$$x - 2y + 3z = 10$$
$$y - 7z = -12$$
$$y - 7z = -12$$

Next add -1 times the second equation to the third equation.

$$x - 2y + 3z = 10$$
$$y - 7z = -12$$
$$0z = 0$$

The third equation tells us that z can be any number; we say that z is arbitrary. Solve the second equation for y in terms of z. Finally, substitute this result in the first equation.

$$y = 7z - 12$$
$$x = 2y - 3z + 10 = 2(7z - 12) - 3z + 10 = 11z - 14$$

Notice that there are infinitely many solutions; we can give any value we like to z, calculate the corresponding x and y values, and come up with a solution. Here is the format we use to list all the solutions to the system.

$$x = 11z - 14$$

$$y = 7z - 12$$

$$z \quad \text{arbitrary}$$

We could also say that the set of solutions consists of all ordered triples of the form $(11z - 14, 7z - 12, z)$. Thus if $z = 0$, we get the solution $(-14, -12, 0)$; if $z = 2$, we get $(8, 2, 2)$. Of course, it does not have to be z that is arbitrary; in our example, it could just as well be x or y. For example, if we had arranged things so y is treated as the arbitrary variable, we would have obtained

$$x = \frac{11}{7}y + \frac{34}{7}$$

$$z = \frac{1}{7}y + \frac{12}{7}$$

$$y \quad \text{arbitrary}$$

Note that the solution corresponding to $y = 2$ is $(8, 2, 2)$, which agrees with one found above. ■

Example D (No Solution)   Solve

$$x - 2y + 3z = 10$$

$$2x - 3y - z = 8$$

$$5x - 9y + 8z = 20$$

Solution   Using Operation 3, we eliminate x from the last two equations to obtain

$$x - 2y + 3z = 10$$

$$y - 7z = -12$$

$$y - 7z = -30$$

It is already apparent that the last two equations cannot get along with each other. Let us continue anyway, putting the system in triangular form by adding -1 times the second equation to the third.

$$x - 2y + 3z = 10$$

$$y - 7z = -12$$

$$0z = -18$$

The third equation has no solution, so the system has no solution. We say the system is **inconsistent.** ■

Nonlinear Systems

Models of real-world phenomena often involve nonlinear systems. In general, such systems are very difficult to solve. However, operations 1, 2, and 3 are legal also for these systems and can sometimes lead us to a solution.

Example E (Nonlinear Systems) Solve the following system of equations.

$$x^2 + y^2 = 25$$
$$x^2 + y^2 - 2x - 4y = 5$$

Solution Adding (-1) times the first equation to the second, we get

$$x^2 + y^2 = 25$$
$$-2x - 4y = -20$$

We solve the second equation for x in terms of y, substitute in the first equation, and then solve the resulting quadratic equation in y.

$$x = 10 - 2y$$
$$(10 - 2y)^2 + y^2 = 25$$
$$5y^2 - 40y + 75 = 0$$
$$y^2 - 8y + 15 = 0$$
$$(y - 3)(y - 5) = 0$$

From this we get $y = 3$ or 5. Substituting these values in the equation $x = 10 - 2y$ yields two solutions to the original system, $(4, 3)$ and $(0, 5)$. ■

PROBLEM SET 11-1

A. Skills and Techniques

Solve each of the following systems of equations. See Examples A and B.

1. $2x - 3y = 7$
$y = -1$

2. $5x - 3y = -25$
$y = 5$

3. $x = -2$
$2x + 7y = 24$

4. $x = 5$
$3x + 4y = 3$

5. $x - 3y = 7$
$4x + y = 2$

6. $5x + 6y = 27$
$x - y = 1$

7. $2x - y + 3z = -6$
$2y - z = 2$
$z = -2$

8. $x + 2y - z = -4$
$3y + z = 2$
$z = 5$

9. $3x - 2y + 5z = -10$
$y - 4z = 8$
$2y + z = 7$

10. $4x + 5y - 6z = 31$
$y - 2z = 7$
$5y + z = 2$

11. $x + 2y + z = 8$
$2x - y + 3z = 15$
$-x + 3y - 3z = -11$

12. $x + y + z = 5$
$-4x + 2y - 3z = -9$
$2x - 3y + 2z = 5$

13. $x - 2y + 3z = 0$
$2x - 3y - 4z = 0$
$x + y - 4z = 0$

14. $x + 4y - z = 0$
$-x - 3y + 5z = 0$
$3x + y - 2z = 0$

15. $x + y + z + w = 10$
$y + 3z - w = 7$
$x + y + 2z = 11$
$x - 3y + w = -14$

16. $2x + y + z = 3$
$y + z + w = 5$
$4x + z + w = 0$
$3y - z - 2w = 0$

Solve each of the following systems. Some, but not all, have infinitely many solutions. See Example C.

17. $x - 4y + z = 18$
$2x - 7y - 2z = 4$
$3x - 11y - z = 22$

18. $x + y - 3z = 10$
$2x + 5y + z = 18$
$5x + 8y - 8z = 48$

19. $x - 2y + 3z = -2$
$3x - 6y + 9z = -6$
$-2x + 4y - 6z = 4$

20.
$$-4x + y - z = 5$$
$$4x - y + z = -5$$
$$-24x + 6y - 6z = 30$$

21.
$$2x - y + 4z = 0$$
$$3x + 2y - z = 0$$
$$9x - y + 11z = 0$$

22.
$$x + 3y - 2z = 0$$
$$2x + y + z = 0$$
$$y - z = 0$$

Solve the following systems or show that they are inconsistent. See Example D.

23.
$$x - 4y + z = 18$$
$$2x - 7y - 2z = 4$$
$$3x - 11y - z = 10$$

24.
$$x + y - 3z = 10$$
$$2x + 5y + z = 18$$
$$5x + 8y - 8z = 50$$

25.
$$x + 3y - 2z = 10$$
$$2x + y + z = 4$$
$$5y - 5z = 16$$

26.
$$x - 2y + 3z = -2$$
$$3x - 6y + 9z = -6$$
$$-2x + 4y - 6z = 0$$

Solve each nonlinear system as in Example E.

27.
$$x + 2y = 10$$
$$x^2 + y^2 - 10x = 0$$

28.
$$x + y = 10$$
$$x^2 + y^2 - 10x - 10y = 0$$

29.
$$x^2 + y^2 - 4x + 6y = 12$$
$$x^2 + y^2 + 10x + 4y = 96$$

30.
$$x^2 + y^2 - 16y = 45$$
$$x^2 + y^2 + 4x - 20y = 65$$

31.
$$y = 4x^2 - 2$$
$$y = x^2 + 1$$

32.
$$x = 3y^2 - 5$$
$$x = y^2 + 3$$

B. Applications and Extensions

In Problems 33–40, solve the given system or show that it is inconsistent.

33.
$$2x - 3y = 12$$
$$x + 4y = -5$$

34.
$$2x + 5y = -2$$
$$y = -\frac{2}{3}x + 3$$

35.
$$2x - 3y = 6$$
$$x + 2y = -4$$

36.
$$x - y + 3z = 1$$
$$3x - 2y + 4z = 0$$
$$4x + 2y - z = 3$$

37.
$$x + y + z = 6$$
$$-x + y + z = 18$$
$$4x - y + z = 12$$

38.
$$.43x - .79y + 4.24z = .67$$
$$3.61y - 9.74z = 2$$
$$y + 1.22z = 1.67$$

39.
$$x^2 + y^2 = 4$$
$$x + 2y = 2\sqrt{5}$$

40.
$$x - \log y = 1$$
$$\log y^x = 2$$

41. If the system $x + 2y = 4$ and $ax + 3y = b$ has infinitely many solutions, what are a and b?

42. Helen claims that she has \$4.40 in nickels, dimes, and quarters, that she has four times as many dimes as quarters, and that she has 40 coins in all. Is this possible? If so, determine how many coins of each kind she has.

43. A three-digit number equals 19 times the sum of its digits. If the digits are reversed, the resulting number is greater than the given number by 297. The tens digit exceeds the units digit by 3. Find the number.

44. Find the equation of the parabola $y = ax^2 + bx + c$ that goes through $(-1, 6)$, $(1, 0)$, and $(2, 3)$.

45. Find the equation of the cubic curve $y = ax^3 + bx^2 + cx + d$ that goes through $(-2, -6)$, $(-1, 5)$, $(1, 3)$, and $(2, 14)$.

46. A chemist has three hydrochloric acid solutions with concentrations (by volume) of 20 percent, 35 percent, and 40 percent, respectively. How many liters of each solution should she use if she wishes to obtain 200 liters of solution with a concentration of 34.25 percent and insists on using 30 liters more of the 35 percent solution than of the 20 percent solution?

47. Find the equation and radius of the circle that goes through $(0, 0)$, $(4, 0)$, and $(\frac{72}{25}, \frac{96}{25})$. *Hint:* Writing the equation in the form $(x - h)^2 + (y - k)^2 = r^2$ is not the best way to start. Is there another way to write the equation of a circle?

48. Determine a, b, and c so that
$$\frac{-12x + 6}{(x - 1)(x + 2)(x - 3)} = \frac{a}{x - 1} + \frac{b}{x + 2} + \frac{c}{x - 3}$$

49. Determine a, b, and c so that $(x - 1)^3$ is a factor of $x^4 + ax^2 + bx + c$.

50. Find the dimensions of a rectangle whose diagonal and perimeter measure 25 and 62 meters, respectively.

51. A certain rectangle has an area of 120 square inches. Increasing the width by 4 inches and decreasing the length by 3 inches increases the area by 24 square inches. Find the dimensions of the original rectangle.

52. TEASER The ABC company reported the following statistics about its employees.
(a) Average length of service for all employees: 15.9 years.
(b) Average length of service for male employees: 16.5 years.

(c) Average length of service for female employees: 14.1 years.

(d) Average hourly wage for all employees: $21.40.

(e) Average hourly wage for male employees: $22.50.

(f) Number of male employees: 300.

For reasons not stated, the company did not report the number of female employees nor their average hourly wage, but you can figure them out. Do so.

53. Solve the nonlinear system

$$x^4 + y^4 = 500$$

$$x + y = 1$$

Hint: Rewrite the system as follows.

$$y = (500 - x^4)^{1/4}$$

$$y = -(500 - x^4)^{1/4}$$

$$y = 1 - x$$

54. Solve the nonlinear system

$$x^4 + y^4 = 500$$

$$4x^2 + y^2 = 36$$

11-2 MATRIX METHODS

Arthur Cayley, lawyer, painter, mountaineer, Cambridge professor, but most of all creative mathematician, made his biggest contributions in the field of algebra. To him we owe the idea of replacing a linear system by its matrix.

$$
\begin{array}{l}
2x + 3y - z = 1 \\
x + 4y - z = 4 \\
3x + y + 2z = 5
\end{array}
\qquad
\begin{bmatrix}
2 & 3 & -1 & 1 \\
1 & 4 & -1 & 4 \\
3 & 1 & 2 & 5
\end{bmatrix}
$$

Arthur Cayley (1821-1895)

Contrary to what many people think, mathematicians do not enjoy long, involved calculations. What they do enjoy is looking for shortcuts, for labor-saving devices, and for elegant ways of doing things. Consider the problem of solving a system of linear equations, which as you know can become very complicated. Is there any way to simplify and systematize this process? There is. It is the method of matrices (plural of matrix).

A **matrix** is just a rectangular array of numbers. One example is shown in our opening panel. It has 3 rows and 4 columns and is referred to as a 3×4 matrix. We follow the standard practice of enclosing a matrix in brackets.

An Example with Three Equations

Look at our opening display again. Notice how we obtained the matrix from the system of equations. We just suppressed all the unknowns, the plus signs, and the equal signs, and supplied some 1's. We call this matrix the **matrix of the system.** We are going to solve this system, keeping track of what happens to the matrix as we move from step to step.

$$
\begin{aligned}
2x + 3y - z &= 1 \\
x + 4y - z &= 4 \\
3x + y + 2z &= 5
\end{aligned}
\qquad
\begin{bmatrix}
2 & 3 & -1 & 1 \\
1 & 4 & -1 & 4 \\
3 & 1 & 2 & 5
\end{bmatrix}
$$

Interchange the first and second equations.

$$
\begin{aligned}
x + 4y - z &= 4 \\
2x + 3y - z &= 1 \\
3x + y + 2z &= 5
\end{aligned}
\qquad
\begin{bmatrix}
1 & 4 & -1 & 4 \\
2 & 3 & -1 & 1 \\
3 & 1 & 2 & 5
\end{bmatrix}
$$

Add -2 times the first equation to the second; then add -3 times the first equation to the third.

$$
\begin{aligned}
x + 4y - z &= 4 \\
- 5y + z &= -7 \\
- 11y + 5z &= -7
\end{aligned}
\qquad
\begin{bmatrix}
1 & 4 & -1 & 4 \\
0 & -5 & 1 & -7 \\
0 & -11 & 5 & -7
\end{bmatrix}
$$

Multiply the second equation by $-\frac{1}{5}$.

$$
\begin{aligned}
x + 4y - z &= 4 \\
y - \tfrac{1}{5}z &= \tfrac{7}{5} \\
- 11y + 5z &= -7
\end{aligned}
\qquad
\begin{bmatrix}
1 & 4 & -1 & 4 \\
0 & 1 & -\tfrac{1}{5} & \tfrac{7}{5} \\
0 & -11 & 5 & -7
\end{bmatrix}
$$

Add 11 times the second equation to the third.

$$
\begin{aligned}
x + 4y - z &= 4 \\
y - \tfrac{1}{5}z &= \tfrac{7}{5} \\
\tfrac{14}{5}z &= \tfrac{42}{5}
\end{aligned}
\qquad
\begin{bmatrix}
1 & 4 & -1 & 4 \\
0 & 1 & -\tfrac{1}{5} & \tfrac{7}{5} \\
0 & 0 & \tfrac{14}{5} & \tfrac{42}{5}
\end{bmatrix}
$$

Now the system is in triangular form and can be solved by backward substitution. The result is $z = 3$, $y = 2$, and $x = -1$; we say the solution is $(-1, 2, 3)$.

We make two points about what we have just done. First, the process is not unique. We happen to prefer having a leading coefficient of 1; that was the reason for our first step. One could have started by multiplying the first equation by $-\frac{1}{2}$ and adding to the second, then multiplying the first equation by $-\frac{3}{2}$ and adding to the third. Any process that ultimately puts the system in triangular form is fine.

The second and main point is this. It is unnecessary to carry along all the x's and y's. Why not work with just the numbers? Why not do all the operations on the matrix of the system? Well, why not?

Equivalent Matrices

Guided by our knowledge of systems of equations, we say that matrices **A** and **B** are **equivalent** if **B** can be obtained from **A** by applying the operations below (a finite number of times).

Operation 1 Interchanging two rows.

Operation 2 Multiplying a row by a nonzero number.

Operation 3 Replacing a row by the sum of that row and a multiple of another row.

When **A** and **B** are equivalent, we write **A** ~ **B**. If **A** ~ **B**, then **B** ~ **A**. If **A** ~ **B** and **B** ~ **C**, then **A** ~ **C**.

Unique Solution Examples

We do not know initially whether a system of linear equations has a unique solution, infinitely many solutions or no solution. Our method, carried out systematically, will always determine which case we have.

Example A (Three Equations, Three Unknowns)   Use the matrix method to solve the system

$$
\begin{aligned}
2x - 3y &= 7 \\
3x \quad\quad - 4z &= -10 \\
x + 2y - 3z &= -12
\end{aligned}
$$

Solution   We take the matrix of the system and transform it one step at a time to triangular form using the operations above. Here is one sequence of steps to the desired result.

$$
\begin{bmatrix}
2 & -3 & 0 & 7 \\
3 & 0 & -4 & -10 \\
1 & 2 & -3 & -12
\end{bmatrix}
$$

Interchange the first and third rows to put a 1 in the upper left-hand corner.

$$
\begin{bmatrix}
1 & 2 & -3 & -12 \\
3 & 0 & -4 & -10 \\
2 & -3 & 0 & 7
\end{bmatrix}
$$

Add -3 times the first row to the second row and -2 times the first row to the third row.

$$
\begin{bmatrix}
1 & 2 & -3 & -12 \\
0 & -6 & 5 & 26 \\
0 & -7 & 6 & 31
\end{bmatrix}
$$

Multiply the third row by -6.

$$
\begin{bmatrix}
1 & 2 & -3 & -12 \\
0 & -6 & 5 & 26 \\
0 & 42 & -36 & -186
\end{bmatrix}
$$

Add 7 times the second row to the third row.

$$\begin{bmatrix} 1 & 2 & -3 & -12 \\ 0 & -6 & 5 & 26 \\ 0 & 0 & -1 & -4 \end{bmatrix}$$

The last matrix represents the system

$$
\begin{aligned}
x + 2y - 3z &= -12 \\
-6y + 5z &= 26 \\
-z &= -4
\end{aligned}
$$

Using back substitution, we get $z = 4$, $y = -1$, and $x = 2$. The unique solution to the system is $(2, -1, 4)$. ∎

Example B (Four Equations, Four Unknowns)   Solve the system

$$
\begin{aligned}
x + 3y + z & & &= 1 \\
2x + 7y + z &- w &&= -1 \\
3x - 2y & &+ 4w &= 8 \\
-x + y - 3z &- w &&= -6
\end{aligned}
$$

Solution   We take the matrix of the system and transform it to triangular form. One sequence of steps that does this is as follows.

$$\begin{bmatrix} 1 & 3 & 1 & 0 & 1 \\ 2 & 7 & 1 & -1 & -1 \\ 3 & -2 & 0 & 4 & 8 \\ -1 & 1 & -3 & -1 & -6 \end{bmatrix}$$

Add -2 times the first row to the second; -3 times the first row to the third row; and 1 times the first row to the fourth row.

$$\begin{bmatrix} 1 & 3 & 1 & 0 & 1 \\ 0 & 1 & -1 & -1 & -3 \\ 0 & -11 & -3 & 4 & 5 \\ 0 & 4 & -2 & -1 & -5 \end{bmatrix}$$

Add 11 times the second row to the third and -4 times the second row to the fourth.

$$\begin{bmatrix} 1 & 3 & 1 & 0 & 1 \\ 0 & 1 & -1 & -1 & -3 \\ 0 & 0 & -14 & -7 & -28 \\ 0 & 0 & 2 & 3 & 7 \end{bmatrix}$$

Multiply the third row by $-\frac{1}{14}$.

$$\begin{bmatrix} 1 & 3 & 1 & 0 & 1 \\ 0 & 1 & -1 & -1 & -3 \\ 0 & 0 & 1 & \frac{1}{2} & 2 \\ 0 & 0 & 2 & 3 & 7 \end{bmatrix}$$

Add -2 times the third row to the fourth row.

$$\begin{bmatrix} 1 & 3 & 1 & 0 & 1 \\ 0 & 1 & -1 & -1 & -3 \\ 0 & 0 & 1 & \frac{1}{2} & 2 \\ 0 & 0 & 0 & 2 & 3 \end{bmatrix}$$

This last matrix represents the system

$$\begin{aligned} x + 3y + z \quad\quad &= \quad 1 \\ y - z - \quad w &= -3 \\ z + \tfrac{1}{2}w &= \quad 2 \\ 2w &= \quad 3 \end{aligned}$$

If we use back substitution, we get $w = \frac{3}{2}$, $z = \frac{5}{4}$, $y = -\frac{1}{4}$, and $x = \frac{1}{2}$. The solution is $(\frac{1}{2}, -\frac{1}{4}, \frac{5}{4}, \frac{3}{2})$. ■

Systems with Many Solutions or No Solution

The matrix method works fine even when there is not a unique solution.

Example C (Infinitely Many Solutions)   Solve the system

$$\begin{aligned} x - 2y + 3z &= 10 \\ 2x - 3y - \quad z &= 8 \\ 4x - 7y + 5z &= 28 \end{aligned}$$

Solution   This is the same problem as in Example C of Section 11-1, but now we use the matrix method, abbreviated to its essence.

$$\begin{bmatrix} 1 & -2 & 3 & 10 \\ 2 & -3 & -1 & 8 \\ 4 & -7 & 5 & 28 \end{bmatrix} \sim \begin{bmatrix} 1 & -2 & 3 & 10 \\ 0 & 1 & -7 & -12 \\ 0 & 1 & -7 & -12 \end{bmatrix} \sim \begin{bmatrix} 1 & -2 & 3 & 10 \\ 0 & 1 & -7 & -12 \\ 0 & 0 & 0 & 0 \end{bmatrix}$$

The appearance of a row of zeros tells us that we have infinitely many solutions. The set of solutions is obtained by considering the equations corresponding to the first two rows.

$$\begin{aligned} x - 2y + 3z &= \quad 10 \\ y - 7z &= -12 \end{aligned}$$

When we solve for y in the second equation and substitute in the first, we obtain

$$x = 11z - 14$$
$$y = 7z - 12$$
$$z \quad \text{arbitrary} \quad \blacksquare$$

Example D (No Solution) Solve the system

$$x + 2y - z = 5$$
$$2x - y + 2z = 4$$
$$3x + y + z = -2$$

Solution We considered this system earlier (Example D of Section 11-1). It too can be solved using the matrix method.

$$\begin{bmatrix} 1 & -2 & 3 & 10 \\ 2 & -3 & -1 & 8 \\ 5 & -9 & 8 & 20 \end{bmatrix} \sim \begin{bmatrix} 1 & -2 & 3 & 10 \\ 0 & 1 & -7 & -12 \\ 0 & 1 & -7 & -30 \end{bmatrix} \sim \begin{bmatrix} 1 & -2 & 3 & 10 \\ 0 & 1 & -7 & -12 \\ 0 & 0 & 0 & -18 \end{bmatrix}$$

We are tipped off to the inconsistency of the system by the third row of the matrix. It corresponds to the equation

$$0x + 0y + 0z = -18$$

which has no solution. Consequently, the system as a whole has no solution. $\blacksquare$

We may summarize our discussion as follows. If the process of transforming the matrix of a system of n equations in n unknowns to triangular form leads to a row in which all elements but the last one are zero, then the system is inconsistent; that is, it has no solution. If the above does not occur and we are led to a matrix with one or more rows consisting entirely of zeros, then the system has infinitely many solutions.

PROBLEM SET 11-2

A. Skills and Techniques

Write the matrix of each system in Problems 1–8.

1. $2x - y = 4$
$x - 3y = -2$

2. $x + 2y = 13$
$11x - y = 0$

3. $x - 2y + z = 3$
$2x + y = 5$
$x + y + 3z = -4$

4. $x + 4z = 10$
$2y - z = 0$
$3x - y = 20$

5. $2x = 3y - 4$
$3x + 2 = -y$

6. $x = 4y + 3$
$y = -2x + 5$

7. $x = 5$
$2y + x - z = 4$
$3x - y + 13 = 5z$

8. $z = 2$
$2x - z = -4$
$x + 2y + 4z = -8$

Regard each matrix in Problems 9–18 as a matrix of a linear system of equations. Tell whether the system has a unique solution, infinitely many solutions, or no solution. You need not solve any of the systems. See Examples A–D.

9. $\begin{bmatrix} 1 & -2 & 3 \\ 0 & 1 & -4 \end{bmatrix}$

10. $\begin{bmatrix} 2 & 5 & 0 \\ 0 & -3 & 5 \end{bmatrix}$

11. $\begin{bmatrix} 1 & -3 & 5 \\ 2 & -6 & -10 \end{bmatrix}$

12. $\begin{bmatrix} 2 & 1 & -4 \\ -6 & -3 & 12 \end{bmatrix}$

13. $\begin{bmatrix} 1 & -2 & 4 & -2 \\ 0 & 3 & 1 & 4 \\ 0 & 0 & 1 & -3 \end{bmatrix}$

14. $\begin{bmatrix} 5 & 4 & 0 & -11 \\ 0 & 1 & -4 & 0 \\ 0 & 0 & 2 & -4 \end{bmatrix}$

15. $\begin{bmatrix} 2 & 1 & 5 & 4 \\ 0 & 3 & -2 & 10 \\ 0 & 3 & -2 & 10 \end{bmatrix}$

16. $\begin{bmatrix} 4 & 1 & -3 & 5 \\ 0 & 0 & 1 & -4 \\ 0 & 0 & 1 & -4 \end{bmatrix}$

17. $\begin{bmatrix} 3 & 2 & -1 & 0 \\ 0 & 1 & 0 & -4 \\ 0 & 1 & 0 & 5 \end{bmatrix}$

18. $\begin{bmatrix} -1 & 5 & 6 & -3 \\ 0 & 0 & 0 & 0 \\ 0 & 0 & 0 & 4 \end{bmatrix}$

In Problems 19–30, use matrices to solve each system or to show that it has no solution. See Examples A–D.

19. $x + 2y = 5$
    $2x - 5y = -8$

20. $2x + 4y = 16$
    $3x - y = 10$

21. $3x - 2y = 1$
    $-6x + 4y = -2$

22. $x + 3y = 12$
    $5x + 15y = 12$

23. $3x - 2y + 5z = -10$
    $y - 4z = 8$
    $2y + z = 7$

24. $4x + 5y + 2z = 25$
    $y - 2z = 7$
    $5y + z = 2$

25. $x + y - 3z = 10$
    $2x + 5y + z = 18$
    $5x + 8y - 8z = 48$

26. $x - 4y + z = 18$
    $2x - 7y - 2z = 4$
    $3x - 11y - z = 22$

27. $2x + 5y + 2z = 6$
    $x + 2y - z = 3$
    $3x - y + 2z = 9$

28. $x - 2y + 3z = -2$
    $3x - 6y + 9z = -6$
    $-2x + 4y - 6z = 0$

29. $x + 1.2y - 2.3z = 8.1$
    $1.3x + .7y + .4z = 6.2$
    $.5x + 1.2y + .5z = 3.2$

30. $3x + 2y = 4$
    $3x - 4y + 6z = 16$
    $3x - y + z = 6$

B. Applications and Extensions

Regard each matrix in Problems 31–36 as the matrix of a linear system of equations. Without solving the system, tell whether it has a unique solution, infinitely many solutions, or no solution.

31. $\begin{bmatrix} 3 & -2 & 5 \\ 0 & 1 & -3 \end{bmatrix}$

32. $\begin{bmatrix} 2 & -1 & 5 \\ -4 & 2 & 8 \end{bmatrix}$

33. $\begin{bmatrix} 2 & -1 & 4 & 6 \\ 0 & 4 & -1 & 5 \\ 0 & 0 & 2 & 1 \end{bmatrix}$

34. $\begin{bmatrix} 3 & 3 & 0 & -4 \\ 0 & 1 & -3 & 2 \\ 0 & 0 & 0 & 0 \end{bmatrix}$

35. $\begin{bmatrix} 1 & 2 & 3 & 4 & 5 \\ 0 & 3 & 2 & 1 & 0 \\ 0 & 0 & 0 & 3 & -4 \\ 0 & 0 & 0 & -9 & 15 \end{bmatrix}$

36. $\begin{bmatrix} 0 & 0 & 0 & 2 & 3 \\ 0 & 0 & 3 & 4 & 5 \\ 0 & 4 & 5 & 6 & 7 \\ 5 & 6 & 7 & 8 & 9 \end{bmatrix}$

In Problems 37–40, use matrices to solve each system or to show that it has no solution.

37. $3x - 2y + 4z = 0$
    $x - y + 3z = 1$
    $4x + 2y - z = 3$

38. $-4x + y - z = 5$
    $4x - y + z = -5$
    $-24x + 6y - 6z = 10$

39. $2x + 4y - z = 8$
    $4x + 9y + 3z = 42$
    $8x + 17y + z = 58$

40.
$$2x + y + 2z + 3w = 2$$
$$y - 2z + 5w = 2$$
$$z + 3w = 4$$
$$2z + 7w = 4$$

41. Find a, b, and c so that the parabola $y = ax^2 + bx + c$ passes through the points $(-2, -32)$, $(1, 4)$, and $(3, -12)$.

42. Find the equation and radius of a circle that passes through $(3, -3)$, $(8, 2)$, and $(6, 6)$.

43. Find angles α, β, γ, and δ in Figure 2 given that $\alpha - \beta + \gamma - \delta = 110°$.

Figure 2

44. A chemist plans to mix three different nitric acid solutions with concentrations of 25 percent, 40 percent, and 50 percent to form 100 liters of a 32 percent solution. If she insists on using twice as much of the 25 percent solution as the 40 percent solution, how many liters of each kind should she use?

45. Determine a, b, and c so that

$$\frac{2x^2 - 21x + 44}{(x - 2)^2(x + 3)} = \frac{a}{x - 2} + \frac{b}{(x - 2)^2} + \frac{c}{x + 3}$$

Hint: First clear of fractions.

46. Marie plans to prepare a 100-pound mixture of peanuts, almonds, and cashews that is to sell at $2.88 per pound. Separately, these three kinds of nuts sell for $1.50, $3.50, and $4.50 per pound, respectively. How many pounds of each kind of nut should she use if she is required to use 6 pounds more of peanuts than of almonds?

47. The local garden store stocks three brands of phos-phate-potash-nitrogen fertilizer with compositions indicated in the following table.

Brand	Phosphate	Potash	Nitrogen
A	10%	30%	60%
B	20%	40%	40%
C	20%	30%	50%

Soil analysis shows that Wanda Wiseankle needs fertilizer for her garden that is 19 percent phosphate, 34 percent potash, and 47 percent nitrogen. Can she obtain the right mixture by mixing the three brands? If so, how many pounds of each should she mix together to get 100 pounds of the desired blend?

48. **TEASER** Tom, Dick, and Harry are good friends but have very different work habits. Together, they contracted to paint three identical houses. Tom and Dick painted the first house in $\frac{72}{5}$ hours; Tom and Harry painted the second house in 16 hours; Dick and Harry painted the third house in $\frac{144}{7}$ hours. How long would it have taken each boy to paint a house alone?

49. Some graphing calculators perform matrix operations. The TI-81 will perform the three operations of this section though it requires some study of the instruction book to master the techniques for doing so. Transform the matrix

$$\begin{bmatrix} 1 & 2 & 3 & 4 & 5 \\ 6 & 7 & 8 & 9 & 0 \\ 2 & 1 & 3 & 2 & -1 \\ 4 & 5 & -3 & 6 & 2 \end{bmatrix}$$

to triangular form and then solve the corresponding system of equations.

50. Transform the matrix

$$\begin{bmatrix} 1 & 2 & 3 & 4 & 5 \\ 6 & 7 & 8 & 9 & 0 \\ 2 & 3 & 4 & 5 & 6 \\ 7 & 8 & 9 & 0 & 1 \end{bmatrix}$$

to triangular form and then solve (if possible) the corresponding system of equations.

11-3 THE ALGEBRA OF MATRICES

Cayley's Weapons

$$\begin{bmatrix} a & b \\ c & d \end{bmatrix} + \begin{bmatrix} A & B \\ C & D \end{bmatrix} = \begin{bmatrix} a + A & b + B \\ c + C & d + D \end{bmatrix}$$

$$\begin{bmatrix} a & b \\ c & d \end{bmatrix} \cdot \begin{bmatrix} A & B \\ C & D \end{bmatrix} = \begin{bmatrix} aA + bC & aB + bD \\ cA + dC & cB + dD \end{bmatrix}$$

"Cayley is forging the weapons for future generations of physicists."

P. G. Tait

When Arthur Cayley introduced matrices, he had much more in mind than the application described in the preceding section. There, matrices served as a device to simplify solving systems of equations. Cayley saw that these number boxes could be studied independently of equations, that they could be thought of as a new type of mathematical object. He realized that if he could give appropriate definitions of addition and multiplication, he would create a mathematical system that might stand with the real numbers and the complex numbers as a potential model for many applications. Cayley did all of this in a major paper in 1858. Some of his contemporaries saw little of significance in this new abstraction. But one of them, P. G. Tait, uttered the prophetic words quoted in the opening box. Tait was right. During the 1920s, Werner Heisenberg found that matrices were just the tool he needed to formulate his quantum mechanics. And by 1950, it was generally recognized that matrix theory provides the best model for many problems in economics and the social sciences.

To simplify our discussion, we initially consider only 2×2 matrices, that is, matrices with two rows and two columns. Examples are

$$\begin{bmatrix} -1 & 3 \\ 4 & 0 \end{bmatrix} \qquad \begin{bmatrix} \log .1 & \frac{6}{2} \\ \frac{12}{3} & \log 1 \end{bmatrix} \qquad \begin{bmatrix} a & b \\ c & d \end{bmatrix}$$

The first two of these matrices are said to be equal. In fact, two matrices are **equal** if and only if the entries in corresponding positions are equal. Be sure to distinguish the notion of equality (written $=$) from that of equivalence (written $\sim$) introduced in Section 11-2. For example,

$$\begin{bmatrix} 2 & 1 \\ -3 & 4 \end{bmatrix} \quad \text{and} \quad \begin{bmatrix} -3 & 4 \\ 2 & 1 \end{bmatrix}$$

are equivalent matrices; however, they are not equal.

Addition, Subtraction, and Scalar Multiplication

Cayley's definition of addition is straightforward. To add two matrices, add the entries in corresponding positions. Thus

$$\begin{bmatrix} 1 & 3 \\ -1 & 4 \end{bmatrix} + \begin{bmatrix} 6 & -2 \\ 5 & 1 \end{bmatrix} = \begin{bmatrix} 1+6 & 3+(-2) \\ -1+5 & 4+1 \end{bmatrix} = \begin{bmatrix} 7 & 1 \\ 4 & 5 \end{bmatrix}$$

and in general

$$\begin{bmatrix} a & b \\ c & d \end{bmatrix} + \begin{bmatrix} A & B \\ C & D \end{bmatrix} = \begin{bmatrix} a+A & b+B \\ c+C & d+D \end{bmatrix}$$

It is easy to check that the commutative and associative properties for addition are valid. Let $\mathbf{U}$, $\mathbf{V}$, and $\mathbf{W}$ be any three matrices.

1. **(Commutativity +)** $\mathbf{U} + \mathbf{V} = \mathbf{V} + \mathbf{U}$
2. **(Associativity +)** $\mathbf{U} + (\mathbf{V} + \mathbf{W}) = (\mathbf{U} + \mathbf{V}) + \mathbf{W}$

The matrix

$$\mathbf{O} = \begin{bmatrix} 0 & 0 \\ 0 & 0 \end{bmatrix}$$

behaves as the "zero" for matrices. And the additive inverse of the matrix

$$\mathbf{U} = \begin{bmatrix} a & b \\ c & d \end{bmatrix}$$

is given by

$$-\mathbf{U} = \begin{bmatrix} -a & -b \\ -c & -d \end{bmatrix}$$

We may summarize these statements as follows.

3. **(Neutral element +)**   There is a matrix $\mathbf{O}$ satisfying $\mathbf{O} + \mathbf{U} = \mathbf{U} + \mathbf{O} = \mathbf{U}$.
4. **(Additive inverses)**   For each matrix $\mathbf{U}$, there is a matrix $-\mathbf{U}$ satisfying $\mathbf{U} + (-\mathbf{U}) = (-\mathbf{U}) + \mathbf{U} = \mathbf{O}$

With the existence of an additive inverse settled, we can define subtraction by $\mathbf{U} - \mathbf{V} = \mathbf{U} + (-\mathbf{V})$. This amounts to subtracting the entries of $\mathbf{V}$ from the corresponding entries of $\mathbf{U}$. Thus

$$\begin{bmatrix} 1 & 3 \\ -1 & 4 \end{bmatrix} - \begin{bmatrix} 6 & -2 \\ 5 & 1 \end{bmatrix} = \begin{bmatrix} -5 & 5 \\ -6 & 3 \end{bmatrix}$$

The form of $-\mathbf{U}$ suggests the appropriate definition of **scalar multiplication.** To multiply a matrix by a number, multiply each entry by that number. That is,

$$k\begin{bmatrix} a & b \\ c & d \end{bmatrix} = \begin{bmatrix} ka & kb \\ kc & kd \end{bmatrix}$$

Scalar multiplication satisfies the expected properties.

5. $k(\mathbf{U} + \mathbf{V}) = k\mathbf{U} + k\mathbf{V}$

6. $(k + m)\mathbf{U} = k\mathbf{U} + m\mathbf{U}$

7. $(km)\mathbf{U} = k(m\mathbf{U})$

Example A (Matrix Manipulations)   Given that

$$\mathbf{A} = \begin{bmatrix} 5 & -3 \\ -4 & 2 \end{bmatrix} \quad \text{and} \quad \mathbf{B} = \begin{bmatrix} -2 & -3 \\ 6 & 7 \end{bmatrix}$$

find (a) $-2\mathbf{A}$; (b) $\mathbf{A} - \mathbf{B}$; (c) $-2\mathbf{A} + 3\mathbf{B}$.

Solution

(a) $-2\mathbf{A} = \begin{bmatrix} -2(5) & -2(-3) \\ -2(-4) & -2(2) \end{bmatrix} = \begin{bmatrix} -10 & 6 \\ 8 & -4 \end{bmatrix}$

(b) $\mathbf{A} - \mathbf{B} = \mathbf{A} + (-\mathbf{B}) = \begin{bmatrix} 5 + 2 & -3 + 3 \\ -4 - 6 & 2 - 7 \end{bmatrix} = \begin{bmatrix} 7 & 0 \\ -10 & -5 \end{bmatrix}$

(c) $-2\mathbf{A} + 3\mathbf{B} = \begin{bmatrix} -10 & 6 \\ 8 & -4 \end{bmatrix} + \begin{bmatrix} -6 & -9 \\ 18 & 21 \end{bmatrix} = \begin{bmatrix} -16 & -3 \\ 26 & 17 \end{bmatrix}$ ∎

So far, all has been straightforward and nice. But with multiplication, Cayley hit a snag.

Multiplication

Cayley's definition of multiplication may seem odd at first glance. He was led to it by consideration of a special problem that we do not have time to describe. It is enough to say that Cayley's definition is the one that proves useful in modern applications (as you will see).

Here it is in symbols.

$$\begin{bmatrix} a & b \\ c & d \end{bmatrix} \cdot \begin{bmatrix} A & B \\ C & D \end{bmatrix} = \begin{bmatrix} aA + bC & aB + bD \\ cA + dC & cB + dD \end{bmatrix}$$

Stated in words, we multiply two matrices by multiplying the rows of the left matrix by the columns of the right matrix in pairwise entry fashion, adding the results. For example, the entry in the second row and first column of the product is obtained by multiplying the entries of the second row of the left matrix by the corresponding entries of the first column of the right matrix, adding the results. Until you get used to it, it may help to use your fingers as shown in Figure 3.

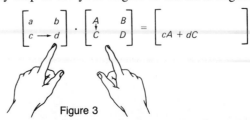

Figure 3

Example B (Multiplying Matrices)   Calculate **AB** and **BA**, given that

$$\mathbf{A} = \begin{bmatrix} 1 & 3 \\ -1 & 4 \end{bmatrix} \quad \text{and} \quad \mathbf{B} = \begin{bmatrix} 6 & -2 \\ 5 & 1 \end{bmatrix}$$

Solution

$$\mathbf{AB} = \begin{bmatrix} 1 & 3 \\ -1 & 4 \end{bmatrix}\begin{bmatrix} 6 & -2 \\ 5 & 1 \end{bmatrix} = \begin{bmatrix} (1)(6) + (3)(5) & (1)(-2) + (3)(1) \\ (-1)(6) + (4)(5) & (-1)(-2) + (4)(1) \end{bmatrix}$$

$$= \begin{bmatrix} 21 & 1 \\ 14 & 6 \end{bmatrix}$$

$$\mathbf{BA} = \begin{bmatrix} 6 & -2 \\ 5 & 1 \end{bmatrix}\begin{bmatrix} 1 & 3 \\ -1 & 4 \end{bmatrix} = \begin{bmatrix} (6)(1) + (-2)(-1) & (6)(3) + (-2)(4) \\ (5)(1) + (1)(-1) & (5)(3) + (1)(4) \end{bmatrix}$$

$$= \begin{bmatrix} 8 & 10 \\ 4 & 19 \end{bmatrix} \quad ■$$

Now you see the snag about which we warned you. The commutative property for multiplication fails. This is troublesome, but not fatal. We manage to get along in a world that is largely noncommutative (try removing your clothes and taking a shower in the opposite order). We just have to remember never to commute matrices under multiplication. Fortunately two other nice properties do hold.

8. **(Associativity ·)** $\mathbf{U} \cdot (\mathbf{V} \cdot \mathbf{W}) = (\mathbf{U} \cdot \mathbf{V}) \cdot \mathbf{W}$
9. **(Distributivity)** $\mathbf{U} \cdot (\mathbf{V} + \mathbf{W}) = \mathbf{U} \cdot \mathbf{V} + \mathbf{U} \cdot \mathbf{W}$
$$(\mathbf{V} + \mathbf{W}) \cdot \mathbf{U} = \mathbf{V} \cdot \mathbf{U} + \mathbf{W} \cdot \mathbf{U}$$

Larger Matrices and Compatibility

So far we have considered only 2×2 matrices. This is an unnecessary restriction; however, to perform operations on arbitrary matrices, we must make sure they are **compatible.** For addition, this simply means that the matrices must be of the same size. Thus

$$\begin{bmatrix} 1 & -1 & 3 \\ 4 & -5 & 2 \end{bmatrix} + \begin{bmatrix} 2 & 6 & 0 \\ -3 & 2 & 4 \end{bmatrix} = \begin{bmatrix} 3 & 5 & 3 \\ 1 & -3 & 6 \end{bmatrix}$$

but

$$\begin{bmatrix} 1 & -1 & 3 \\ 4 & -5 & 2 \end{bmatrix} + \begin{bmatrix} 2 & 6 \\ -3 & 2 \end{bmatrix}$$

makes no sense.

Two matrices are compatible for multiplication if the left matrix has the same number of columns as the right matrix has rows. For example,

$$\begin{bmatrix} 1 & -1 & 3 \\ 4 & -5 & 2 \end{bmatrix}\begin{bmatrix} 2 & 1 & 5 & 0 \\ -1 & 3 & 2 & 1 \\ 1 & -2 & 0 & 2 \end{bmatrix} = \begin{bmatrix} 6 & -8 & 3 & 5 \\ 15 & -15 & 10 & -1 \end{bmatrix}$$

The left matrix is 2×3, the right one is 3×4, and the result is 2×4. In general, we can multiply an $m \times n$ matrix by an $n \times p$ matrix, the result being an $m \times p$ matrix. All of the properties mentioned earlier are valid, provided we work with compatible matrices.

Example C (Compatibility) Let

$$A = \begin{bmatrix} 2 & -1 & 0 \\ 1 & 2 & 4 \end{bmatrix} \quad B = \begin{bmatrix} 5 \\ 6 \end{bmatrix}$$

$$C = \begin{bmatrix} 1 & 0 & 0 \\ -2 & 1 & 0 \\ 0 & 4 & 3 \end{bmatrix} \quad D = \begin{bmatrix} -3 \\ 1 \\ 5 \end{bmatrix}$$

Calculate if possible.

(a) $AD + B$ (b) CD (c) AC (d) CA

Solution

(a) $AD + B = \begin{bmatrix} 2 & -1 & 0 \\ 1 & 2 & 4 \end{bmatrix} \begin{bmatrix} -3 \\ 1 \\ 5 \end{bmatrix} + \begin{bmatrix} 5 \\ 6 \end{bmatrix} = \begin{bmatrix} -7 \\ 19 \end{bmatrix} + \begin{bmatrix} 5 \\ 6 \end{bmatrix} = \begin{bmatrix} -2 \\ 25 \end{bmatrix}$

(b) $CD = \begin{bmatrix} 1 & 0 & 0 \\ -2 & 1 & 0 \\ 0 & 4 & 3 \end{bmatrix} \begin{bmatrix} -3 \\ 1 \\ 5 \end{bmatrix} = \begin{bmatrix} -3 \\ 7 \\ 19 \end{bmatrix}$

(c) $AC = \begin{bmatrix} 2 & -1 & 0 \\ 1 & 2 & 4 \end{bmatrix} \begin{bmatrix} 1 & 0 & 0 \\ -2 & 1 & 0 \\ 0 & 4 & 3 \end{bmatrix} = \begin{bmatrix} 4 & -1 & 0 \\ -3 & 18 & 12 \end{bmatrix}$

(d) Incompatible ∎

A Business Application

The ABC Company sells precut lumber for two types of summer cottages, standard and deluxe. The standard model requires 30,000 board feet of lumber and 100 worker-hours of cutting; the deluxe model takes 40,000 board feet of lumber and 110 worker-hours of cutting. This year, the ABC Company buys its lumber at $.20 per board foot and pays its laborers $9.00 per hour. Next year it expects these costs to be $.25 and $10.00, respectively. This information can be displayed in matrix form as follows.

	Requirements A				Unit Cost B	
	Lumber	Labor			This year	Next year
Standard	30,000	100		Lumber	$.20	$.25
Deluxe	40,000	110		Labor	$9.00	$10.00

Now we ask whether the product matrix **AB** has economic significance. It does: It gives the total dollar cost of standard and deluxe cottages both for this year and next. You can see this from the following calculation.

$$\mathbf{AB} = \begin{bmatrix} (30{,}000)(.20) + (100)(9) & (30{,}000)(.25) + (100)(10) \\ (40{,}000)(.20) + (110)(9) & (40{,}000)(.25) + (110)(10) \end{bmatrix}$$

$$= \begin{bmatrix} \$6900 & \$8500 \\ \$8990 & \$11{,}100 \end{bmatrix} \begin{matrix} \text{Standard} \\ \text{Deluxe} \end{matrix}$$

This year   Next year

PROBLEM SET 11-3

A. Skills and Techniques

Calculate **A** + **B**, **A** − **B**, *and* 3**A** *in Problems 1–4. See Example A.*

1. $\mathbf{A} = \begin{bmatrix} 2 & -1 \\ 3 & 7 \end{bmatrix}$, $\mathbf{B} = \begin{bmatrix} 6 & 5 \\ -2 & 3 \end{bmatrix}$

2. $\mathbf{A} = \begin{bmatrix} -1 & 0 \\ 5 & 4 \end{bmatrix}$, $\mathbf{B} = \begin{bmatrix} 2 & -2 \\ 3 & 7 \end{bmatrix}$

3. $\mathbf{A} = \begin{bmatrix} 3 & -2 & 5 \\ 4 & 0 & -3 \end{bmatrix}$, $\mathbf{B} = \begin{bmatrix} 2 & 6 & -1 \\ 4 & 3 & -3 \end{bmatrix}$

4. $\mathbf{A} = \begin{bmatrix} 1 & 2 & 3 \\ 4 & 5 & 6 \\ 7 & 8 & 9 \end{bmatrix}$, $\mathbf{B} = \begin{bmatrix} -1 & -2 & -2 \\ -4 & -5 & -6 \\ -7 & -8 & -9 \end{bmatrix}$

Calculate **AB** *and* **BA** *if possible in Problems 5–12. See Examples B and C.*

5. $\mathbf{A} = \begin{bmatrix} 2 & -1 \\ 3 & 7 \end{bmatrix}$, $\mathbf{B} = \begin{bmatrix} 6 & 5 \\ -2 & 3 \end{bmatrix}$

6. $\mathbf{A} = \begin{bmatrix} -1 & 0 \\ 5 & 4 \end{bmatrix}$, $\mathbf{B} = \begin{bmatrix} 2 & -2 \\ 3 & 7 \end{bmatrix}$

7. $\mathbf{A} = \begin{bmatrix} 1 & -1 & 2 \\ 3 & 4 & -4 \\ 2 & 1 & 3 \end{bmatrix}$, $\mathbf{B} = \begin{bmatrix} 0 & 2 & -3 \\ 1 & 2 & 3 \\ -1 & -2 & 4 \end{bmatrix}$

8. $\mathbf{A} = \begin{bmatrix} -2 & 5 & 1 \\ 0 & -2 & 3 \\ 1 & 2 & -1 \end{bmatrix}$, $\mathbf{B} = \begin{bmatrix} -3 & 4 & 1 \\ 2 & 5 & 1 \\ 1 & 2 & 3 \end{bmatrix}$

9. $\mathbf{A} = \begin{bmatrix} 1 & -2 & 3 & 4 \\ 3 & 2 & -5 & 1 \end{bmatrix}$, $\mathbf{B} = \begin{bmatrix} 1 & 2 \\ 3 & 4 \end{bmatrix}$

10. $\mathbf{A} = \begin{bmatrix} -1 & 3 \\ 4 & 2 \\ 1 & 5 \end{bmatrix}$, $\mathbf{B} = \begin{bmatrix} -1 & 2 & 3 & 4 \\ 0 & -3 & 2 & 1 \end{bmatrix}$

11. $\mathbf{A} = \begin{bmatrix} 3 & 1 & -1 \\ 2 & 4 & 2 \\ -3 & 2 & -1 \end{bmatrix}$, $\mathbf{B} = \begin{bmatrix} 1 \\ 2 \\ 3 \end{bmatrix}$

12. $\mathbf{A} = \begin{bmatrix} 1 & 2 & -1 \end{bmatrix}$, $\mathbf{B} = \begin{bmatrix} 4 & 3 \\ 0 & 2 \\ -1 & 4 \end{bmatrix}$

13. Calculate **AB** and **BA** for

$$\mathbf{A} = \begin{bmatrix} 0 & 0 \\ 0 & 0 \end{bmatrix} \quad \mathbf{B} = \begin{bmatrix} 2 & -1 \\ 3 & 4 \end{bmatrix}$$

14. State the general property illustrated by Problem 13.

15. Find **X** if

$$\begin{bmatrix} 2 & 1 & -3 \\ 1 & 5 & 0 \end{bmatrix} + \mathbf{X} = 2\begin{bmatrix} -1 & 4 & 3 \\ -2 & 0 & 4 \end{bmatrix}$$

16. Solve for **X**.

$$-3\mathbf{X} + 2\begin{bmatrix} 1 & -2 \\ 5 & 6 \end{bmatrix} = -\begin{bmatrix} 5 & -14 \\ 8 & 15 \end{bmatrix}$$

17. Calculate **A**(**B** + **C**) and **AB** + **AC** for

$$\mathbf{A} = \begin{bmatrix} 2 & -1 \\ 3 & 4 \end{bmatrix} \quad \mathbf{B} = \begin{bmatrix} 2 & 4 \\ 6 & 1 \end{bmatrix} \quad \mathbf{C} = \begin{bmatrix} -1 & -2 \\ 3 & 6 \end{bmatrix}$$

What property does this illustrate?

18. Calculate (**A** + **B**)(**A** − **B**) and **A**² − **B**² for

$$\mathbf{A} = \begin{bmatrix} 3 & -2 \\ 1 & 4 \end{bmatrix} \quad \mathbf{B} = \begin{bmatrix} 6 & -3 \\ 2 & 5 \end{bmatrix}$$

Why are your answers different?

19. Find the entry in the third row and second column of the product

$$\begin{bmatrix} 1.39 & 4.13 & -2.78 \\ 4.72 & -3.69 & 5.41 \\ 8.09 & -6.73 & 5.03 \end{bmatrix} \begin{bmatrix} 5.45 & 6.31 \\ 7.24 & -5.32 \\ 6.06 & 1.34 \end{bmatrix}$$

20. Find the entry in the second row and first column of the product in Problem 19.

B. Applications and Extensions

21. Compute $A - 2B$, AB, and A^2 for

$$A = \begin{bmatrix} 4 & -1 & 3 \\ 2 & 5 & 3 \\ 6 & 2 & 1 \end{bmatrix} \quad B = \begin{bmatrix} 1 & -3 & 2 \\ 5 & 0 & 3 \\ -5 & 2 & 1 \end{bmatrix}$$

22. Let A and B be 3×4 matrices, C a 4×3 matrix, and D a 3×3 matrix. Which of the following do not make sense, that is, do not satisfy the compatibility conditions?

(a) AB (b) AC
(c) $AC - D$ (d) $(A - B)C$
(e) $(AC)D$ (f) $A(CD)$
(g) A^2 (h) $(CA)^2$
(i) $C(A + 2B)$

23. Calculate AB and BA for

$$A = [1 \quad 2 \quad 3 \quad 4] \quad B = \begin{bmatrix} 2 \\ 1 \\ -1 \\ -2 \end{bmatrix}$$

24. Show that if AB and BA both make sense, then AB and BA are both square matrices.

25. If $(A + B)^2 = A^2 + 2AB + B^2$, what conclusions can you draw about A and B?

26. If the ith row of A consists of all zeros, what is true about the ith row of AB (assuming AB makes sense)?

27. Let

$$A = \begin{bmatrix} 0 & a \\ 0 & 0 \end{bmatrix} \quad B = \begin{bmatrix} 0 & a & b \\ 0 & 0 & c \\ 0 & 0 & 0 \end{bmatrix}$$

Calculate A^2 and B^3 and then make a conjecture.

28. A matrix of the form

$$A = \begin{bmatrix} 1 & a & b \\ 0 & 1 & c \\ 0 & 0 & 1 \end{bmatrix}$$

where a, b, and c are any real numbers is called a Heisenberg matrix. What is true about the product of two such matrices?

29. Let

$$A = \begin{bmatrix} 1 & 2 & 0 \\ 0 & 1 & 0 \\ 0 & 0 & 1 \end{bmatrix}$$

Calculate A^2, A^3, A^4, and conjecture the form of A^n.

30. Let

$$B = \begin{bmatrix} 1 & 0 & 3 \\ 0 & 1 & 0 \\ 0 & 0 & 1 \end{bmatrix}$$

Calculate B^2, B^3, B^4, and conjecture the form of B^n.

31. Let

$$A = \begin{bmatrix} 3 & 0 & 0 \\ 0 & -4 & 0 \\ 0 & 0 & 5 \end{bmatrix}$$

If B is any 3×3 matrix, what does multiplication on the left by A do to B? Multiplication on the right by A?

32. Calculate A^2 and A^3 for the matrix A of Problem 31. State a general result about raising a diagonal matrix to a positive integral power.

33. Art, Bob, and Curt work for a company that makes Flukes, Gizmos, and Horks. They are paid for their labor on a piecework basis, receiving \$1 for each Fluke, \$2 for each Gizmo, and \$3 for each Hork. Below are matrices U and V representing their outputs on Monday and Tuesday. Matrix X is the wage/unit matrix.

	Monday's Output U				Tuesday's Output V				Wage/Unit X
	F	G	H		F	G	H		
Art	4	3	2		3	6	1		F [1]
Bob	5	1	2		4	2	2		G [2]
Curt	3	4	1		5	1	3		H [3]

Compute the following matrices and decide what they represent.

(a) UX (b) VX (c) $U + V$ (d) $(U + V)X$

34. Four friends, A, B, C, and D, have unlisted telephone numbers. Whether or not one person knows another's number is indicated by the matrix U below, where 1 indicates knowing and 0 indicates not knowing. For example, the 1 in row 3 and column 1 means that C knows A's number.

$$U = \begin{array}{c} \\ A \\ B \\ C \\ D \end{array} \begin{array}{c} \begin{array}{cccc} A & B & C & D \end{array} \\ \begin{bmatrix} 1 & 0 & 1 & 0 \\ 0 & 1 & 1 & 0 \\ 1 & 0 & 1 & 1 \\ 0 & 1 & 0 & 1 \end{bmatrix} \end{array}$$

(a) Calculate U^2.
(b) Interpret U^2 in terms of the possibility of each person being able to get a telephone message to another.
(c) Can D get a message to A via one other person?
(d) Interpret U^3.

35. Consider the set C of all 2×2 matrices of the form

$$\begin{bmatrix} a & b \\ -b & a \end{bmatrix}$$

where a and b are real numbers.

(a) Let

$$U = \begin{bmatrix} u_1 & u_2 \\ -u_2 & u_1 \end{bmatrix} \quad \text{and} \quad V = \begin{bmatrix} v_1 & v_2 \\ -v_2 & v_1 \end{bmatrix}$$

be two such matrices.
Calculate $U + V$ and UV. Note that both $U + V$ and UV are in C.

(b) Let $I = \begin{bmatrix} 1 & 0 \\ 0 & 1 \end{bmatrix}$ and $J = \begin{bmatrix} 0 & 1 \\ -1 & 0 \end{bmatrix}$. Calculate I^2 and J^2.
(c) Note that $U = u_1 I + u_2 J$ and $V = v_1 I + v_2 J$. Write $U + V$ and UV in terms of I and J.
(d) What does all this have to do with the complex numbers?

36. **TEASER** Find the four square roots of the matrix

$$\begin{bmatrix} 7 & 10 \\ 15 & 22 \end{bmatrix}$$

37. For this problem you will need to discover whether your calculator handles matrices (the TI-81 does), and if so, how it does algebraic operations with them? Let

$$A = \begin{bmatrix} 1 & 2 & 0 & 0 \\ 0 & 1 & 0 & 0 \\ 0 & 0 & 1 & 3 \\ 0 & 0 & 0 & 1 \end{bmatrix} \quad B = \begin{bmatrix} 1 & 0 & 0 & 3 \\ 0 & 1 & 0 & 0 \\ 0 & 0 & 1 & 0 \\ 0 & 0 & 0 & 1 \end{bmatrix}$$

Calculate $A^5 B^2 + BA$.

38. Refer to Problem 37. Calculate various powers of A and B and make a conjecture as to the form of A^n and B^n for n any positive integer.

11-4 MULTIPLICATIVE INVERSES

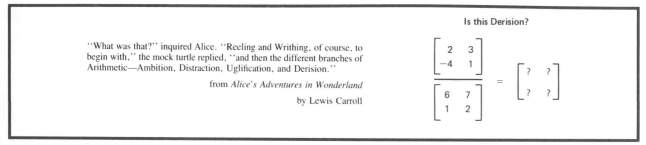

Is this Derision?

"What was that?" inquired Alice. "Reeling and Writhing, of course, to begin with," the mock turtle replied, "and then the different branches of Arithmetic—Ambition, Distraction, Uglification, and Derision."

from *Alice's Adventures in Wonderland*

by Lewis Carroll

$$\frac{\begin{bmatrix} 2 & 3 \\ -4 & 1 \end{bmatrix}}{\begin{bmatrix} 6 & 7 \\ 1 & 2 \end{bmatrix}} = \begin{bmatrix} ? & ? \\ ? & ? \end{bmatrix}$$

Even for ordinary numbers, the notion of division seems more difficult than that of addition, subtraction, and multiplication. Certainly this is true for division of matrices. Look at the example displayed above. It could tempt more than a mock turtle to derision. However, Arthur Cayley saw no need to sneer. He noted that in the case of numbers,

$$\frac{U}{V} = U \cdot \frac{1}{V} = U \cdot V^{-1}$$

What is needed is a concept of "one" for matrices; then we need the concept of multiplicative inverse. The first is easy.

The Multiplicative Identity for Matrices

Let

$$\mathbf{I} = \begin{bmatrix} 1 & 0 \\ 0 & 1 \end{bmatrix}$$

Then for any 2×2 matrix $\mathbf{U}$,

$$\mathbf{UI} = \mathbf{U} = \mathbf{IU}$$

This can be checked by noting that

$$\begin{bmatrix} a & b \\ c & d \end{bmatrix}\begin{bmatrix} 1 & 0 \\ 0 & 1 \end{bmatrix} = \begin{bmatrix} a & b \\ c & d \end{bmatrix} = \begin{bmatrix} 1 & 0 \\ 0 & 1 \end{bmatrix}\begin{bmatrix} a & b \\ c & d \end{bmatrix}$$

The symbol $\mathbf{I}$ is chosen because it is often called the **multiplicative identity.** In accordance with Section 1-4, it is also called the neutral element for multiplication.

For 3×3 matrices, the multiplicative identity has the form

$$\begin{bmatrix} 1 & 0 & 0 \\ 0 & 1 & 0 \\ 0 & 0 & 1 \end{bmatrix}$$

You should be able to guess its form for 4×4 and higher order matrices.

Inverses of 2 × 2 Matrices

Suppose that we want to find the multiplicative inverse of

$$\mathbf{V} = \begin{bmatrix} 6 & 7 \\ 1 & 2 \end{bmatrix}$$

We are looking for a matrix

$$\mathbf{W} = \begin{bmatrix} a & b \\ c & d \end{bmatrix}$$

that satisfies $\mathbf{VW} = \mathbf{I}$ and $\mathbf{WV} = \mathbf{I}$. Taking $\mathbf{VW} = \mathbf{I}$ first, we want

$$\begin{bmatrix} 6 & 7 \\ 1 & 2 \end{bmatrix}\begin{bmatrix} a & b \\ c & d \end{bmatrix} = \begin{bmatrix} 1 & 0 \\ 0 & 1 \end{bmatrix}$$

which means

$$\begin{bmatrix} 6a + 7c & 6b + 7d \\ a + 2c & b + 2d \end{bmatrix} = \begin{bmatrix} 1 & 0 \\ 0 & 1 \end{bmatrix}$$

or

$$6a + 7c = 1 \qquad 6b + 7d = 0$$

$$a + 2c = 0 \qquad b + 2d = 1$$

When these four equations are solved for a, b, c, d, we have

$$\mathbf{W} = \begin{bmatrix} \frac{2}{5} & -\frac{7}{5} \\ -\frac{1}{5} & \frac{6}{5} \end{bmatrix}$$

as a tentative solution to our problem. We say tentative, because so far we know only that $\mathbf{VW} = \mathbf{I}$. Happily, $\mathbf{W}$ works on the other side of $\mathbf{V}$ too, as we can check. (In this exceptional case, we do have commutativity.)

$$\mathbf{WV} = \begin{bmatrix} \frac{2}{5} & -\frac{7}{5} \\ -\frac{1}{5} & \frac{6}{5} \end{bmatrix} \begin{bmatrix} 6 & 7 \\ 1 & 2 \end{bmatrix} = \begin{bmatrix} 1 & 0 \\ 0 & 1 \end{bmatrix}$$

Success! $\mathbf{W}$ is the inverse of $\mathbf{V}$; we denote it by the symbol $\mathbf{V}^{-1}$.

The process just described can be carried out for any specific 2×2 matrix, or better, it can be carried out for a general 2×2 matrix. But before we give the result, we make an important comment. There is no reason to think that every 2×2 matrix has a multiplicative inverse. Remember that the number 0 does not have such an inverse; neither does the matrix $\mathbf{O}$. But here is a mild surprise. Many other 2×2 matrices do not have inverses. The following theorem identifies in a very precise way those that do, and then gives a formula for their inverses.

Theorem (Multiplicative Inverses)

The matrix

$$\mathbf{V} = \begin{bmatrix} a & b \\ c & d \end{bmatrix}$$

has a multiplicative inverse if and only if $D = ad - bc$ is nonzero. If $D \neq 0$, then

$$\mathbf{V}^{-1} = \begin{bmatrix} \dfrac{d}{D} & -\dfrac{b}{D} \\ -\dfrac{c}{D} & \dfrac{a}{D} \end{bmatrix}$$

Thus the number D determines whether a matrix has an inverse. This number, which we shall call a *determinant*, will be studied in detail in the next section. Each 2×2 matrix has such a number associated with it.

Example A (Inverses of 2×2 Matrices)   Determine whether the given matrix has an inverse and if so, find it.

(a) $\mathbf{X} = \begin{bmatrix} 2 & -3 \\ -4 & 6 \end{bmatrix}$ (b) $\mathbf{Y} = \begin{bmatrix} 5 & -3 \\ -4 & 3 \end{bmatrix}$

Solution

(a) $D = (2)(6) - (-3)(-4) = 0$ (b) $D = (5)(3) - (-3)(-4) = 3$

$\mathbf{X}^{-1}$ does not exist $\mathbf{Y}^{-1} = \begin{bmatrix} \frac{3}{3} & \frac{3}{3} \\ \frac{4}{3} & \frac{5}{3} \end{bmatrix}$ ∎

Inverses for Higher-Order Matrices

There is a theorem like the one above for square matrices of any size, which Cayley found in 1858. It is complicated and, rather than try to state it, we are going to illustrate a process which yields the inverse of a matrix whenever it exists. Briefly described, it is this. Take any square matrix $\mathbf{V}$ and write the corresponding identity matrix $\mathbf{I}$ next to it on the right. By using the three row operations of Section 8-2, attempt to reduce $\mathbf{V}$ to the identity matrix while simultaneously performing the same operations on $\mathbf{I}$. If you can reduce $\mathbf{V}$ to $\mathbf{1}$, you will simultaneously turn $\mathbf{I}$ into $\mathbf{V}^{-1}$. If you cannot reduce $\mathbf{V}$ to $\mathbf{I}$, $\mathbf{V}$ has no inverse.

Here is an illustration for the 2×2 matrix $\mathbf{V}$ that we used earlier (page 441)

$$\left[\begin{array}{cc|cc} 6 & 7 & 1 & 0 \\ 1 & 2 & 0 & 1 \end{array}\right] \sim \left[\begin{array}{cc|cc} 1 & 2 & 0 & 1 \\ 6 & 7 & 1 & 0 \end{array}\right] \quad \text{(interchange rows)}$$

$$\sim \left[\begin{array}{cc|cc} 1 & 2 & 0 & 1 \\ 0 & -5 & 1 & -6 \end{array}\right] \quad \begin{array}{l}\text{(add } -6 \text{ times row 1} \\ \text{to row 2)}\end{array}$$

$$\sim \left[\begin{array}{cc|cc} 1 & 2 & 0 & 1 \\ 0 & 1 & -\frac{1}{5} & \frac{6}{5} \end{array}\right] \quad \text{(divide row 2 by } -5)$$

$$\sim \left[\begin{array}{cc|cc} 1 & 0 & \frac{2}{5} & -\frac{7}{5} \\ 0 & 1 & -\frac{1}{5} & \frac{6}{5} \end{array}\right] \quad \begin{array}{l}\text{(add } -2 \text{ times row 2} \\ \text{to row 1)}\end{array}$$

Notice that the matrix on the right is $\mathbf{V}^{-1}$, the inverse of $\mathbf{V}$ that we obtained earlier by another method.

Example B (Inverse of 3×3 Matrices)   Find the multiplicative inverse of

$$\left[\begin{array}{ccc} 2 & 6 & 6 \\ 2 & 7 & 6 \\ 2 & 7 & 7 \end{array}\right]$$

Solution   We use the reduction method described

$$\left[\begin{array}{ccc|ccc} 2 & 6 & 6 & 1 & 0 & 0 \\ 2 & 7 & 6 & 0 & 1 & 0 \\ 2 & 7 & 7 & 0 & 0 & 1 \end{array}\right]$$

$$\sim \left[\begin{array}{ccc|ccc} 1 & 3 & 3 & \frac{1}{2} & 0 & 0 \\ 2 & 7 & 6 & 0 & 1 & 0 \\ 2 & 7 & 7 & 0 & 0 & 1 \end{array}\right] \quad \text{(divide row 1 by 2)}$$

$$\sim \left[\begin{array}{ccc|ccc} 1 & 3 & 3 & \frac{1}{2} & 0 & 0 \\ 0 & 1 & 0 & -1 & 1 & 0 \\ 0 & 1 & 1 & -1 & 0 & 1 \end{array}\right] \quad \text{(add } -2 \text{ times row 1 to row 2 and to row 3)}$$

$$\sim \begin{bmatrix} 1 & 3 & 3 & \frac{1}{2} & 0 & 0 \\ 0 & 1 & 0 & -1 & 1 & 0 \\ 0 & 0 & 1 & 0 & -1 & 1 \end{bmatrix} \quad \text{(add } -1 \text{ times row 2 to row 3)}$$

$$\sim \begin{bmatrix} 1 & 0 & 3 & \frac{7}{2} & -3 & 0 \\ 0 & 1 & 0 & -1 & 1 & 0 \\ 0 & 0 & 1 & 0 & -1 & 1 \end{bmatrix} \quad \text{(add } -3 \text{ times row 2 to row 1)}$$

$$\sim \begin{bmatrix} 1 & 0 & 0 & \frac{7}{2} & 0 & -3 \\ 0 & 1 & 0 & -1 & 1 & 0 \\ 0 & 0 & 1 & 0 & -1 & 1 \end{bmatrix} \quad \text{(add } -3 \text{ times row 3 to row 1)}$$

Thus the desired inverse is

$$\begin{bmatrix} \frac{7}{2} & 0 & -3 \\ -1 & 1 & 0 \\ 0 & -1 & 1 \end{bmatrix} \quad \blacksquare$$

Example C (Matrices Without Inverses) Try to find the multiplicative inverse of

$$U = \begin{bmatrix} 1 & 4 & 2 \\ 0 & 2 & 4 \\ 0 & -3 & -6 \end{bmatrix}$$

Solution

$$\begin{bmatrix} 1 & 4 & 2 & 1 & 0 & 0 \\ 0 & 2 & 4 & 0 & 1 & 0 \\ 0 & -3 & -6 & 0 & 0 & 1 \end{bmatrix} \sim \begin{bmatrix} 1 & 4 & 2 & 1 & 0 & 0 \\ 0 & 1 & 2 & 0 & \frac{1}{2} & 0 \\ 0 & -3 & -6 & 0 & 0 & 1 \end{bmatrix}$$

$$\sim \begin{bmatrix} 1 & 4 & 2 & 1 & 0 & 0 \\ 0 & 1 & 2 & 0 & \frac{1}{2} & 0 \\ 0 & 0 & 0 & 0 & \frac{3}{2} & 1 \end{bmatrix}$$

Since we got a row of zeros in the left half above, we know we can never reduce it to the identity matrix **I**. The matrix **U** does not have an inverse. ■

An Application

Consider the system of equations

$$5x - 3y = 6$$
$$-4x + 2y = 8$$

If we introduce matrices

$$\mathbf{A} = \begin{bmatrix} 5 & -3 \\ -4 & 2 \end{bmatrix} \qquad \mathbf{X} = \begin{bmatrix} x \\ y \end{bmatrix} \qquad \mathbf{B} = \begin{bmatrix} 6 \\ 8 \end{bmatrix}$$

this system can be written in the form

$$\mathbf{AX} = \mathbf{B}$$

CAUTION

$$\mathbf{AX} = \mathbf{B}$$
$$\mathbf{X} = \mathbf{BA}^{-1}$$

$$\mathbf{AX} = \mathbf{B}$$
$$\mathbf{X} = \mathbf{A}^{-1}\mathbf{B}$$

Now divide both sides by $\mathbf{A}$, by which we mean, of course, multiply both sides by $\mathbf{A}^{-1}$. We must be more precise. Multiply both sides on the left by $\mathbf{A}^{-1}$ (do not forget the lack of commutativity).

$$\mathbf{A}^{-1}\mathbf{AX} = \mathbf{A}^{-1}\mathbf{B}$$

$$\mathbf{IX} = \mathbf{A}^{-1}\mathbf{B}$$

$$\mathbf{X} = \mathbf{A}^{-1}\mathbf{B} = \begin{bmatrix} -1 & -\frac{3}{2} \\ -2 & -\frac{5}{2} \end{bmatrix} \begin{bmatrix} 6 \\ 8 \end{bmatrix} = \begin{bmatrix} -18 \\ -32 \end{bmatrix}$$

Thus $(-18, -32)$ is the solution to the system as you may check.

Example D (Solving Systems as Matrix Equations)   Solve

$$2x + 6y + 6z = 8$$
$$2x + 7y + 6z = 10$$
$$2x + 7y + 7z = 9$$

as a matrix equation.

Solution   We introduce the matrices

$$\mathbf{A} = \begin{bmatrix} 2 & 6 & 6 \\ 2 & 7 & 6 \\ 2 & 7 & 7 \end{bmatrix} \qquad \mathbf{X} = \begin{bmatrix} x \\ y \\ z \end{bmatrix} \qquad \mathbf{B} = \begin{bmatrix} 8 \\ 10 \\ 9 \end{bmatrix}$$

and write the system as $\mathbf{AX} = \mathbf{B}$. The inverse of $\mathbf{A}$ was found in Example B.

$$\mathbf{X} = \mathbf{A}^{-1}\mathbf{B} = \begin{bmatrix} \frac{7}{2} & 0 & -3 \\ -1 & 1 & 0 \\ 0 & -1 & 1 \end{bmatrix} \begin{bmatrix} 8 \\ 10 \\ 9 \end{bmatrix} = \begin{bmatrix} 1 \\ 2 \\ -1 \end{bmatrix}$$

and therefore $(1, 2, -1)$ is the solution to our system.   ■

This method of solution is particularly useful when many systems with the same coefficient matrix $\mathbf{A}$ are under consideration. Once we have $\mathbf{A}^{-1}$, we can obtain any solution simply by doing an easy matrix multiplication. If only one system is being studied, the method of Section 11-2 is best.

PROBLEM SET 11-4

A. Skills and Techniques

Find the multiplicative inverse of each matrix. Check by calculating AA^{-1}. See Example A.

1. $\begin{bmatrix} 2 & 3 \\ -1 & -1 \end{bmatrix}$

2. $\begin{bmatrix} 4 & 3 \\ 1 & 2 \end{bmatrix}$

3. $\begin{bmatrix} 6 & -14 \\ 0 & 2 \end{bmatrix}$

4. $\begin{bmatrix} 0 & 3 \\ 2 & 4 \end{bmatrix}$

5. $\begin{bmatrix} 1 & 0 \\ 0 & 1 \end{bmatrix}$

6. $\begin{bmatrix} 4 & 0 \\ 0 & 5 \end{bmatrix}$

7. $\begin{bmatrix} a & 0 \\ 0 & b \end{bmatrix}$

8. $\begin{bmatrix} 3 & 0 & 0 \\ 0 & 4 & 0 \\ 0 & 0 & 5 \end{bmatrix}$

Use the method illustrated in Example B to find the multiplicative inverse of each of the following.

9. $\begin{bmatrix} 1 & 3 \\ 2 & 4 \end{bmatrix}$

10. $\begin{bmatrix} 2 & 6 \\ 3 & 1 \end{bmatrix}$

11. $\begin{bmatrix} 1 & 1 & 1 \\ 1 & -1 & 2 \\ 3 & 2 & 0 \end{bmatrix}$

12. $\begin{bmatrix} 2 & 1 & 1 \\ 1 & 3 & 1 \\ -1 & 4 & 0 \end{bmatrix}$

13. $\begin{bmatrix} 3 & 1 & 2 \\ 4 & 1 & -6 \\ 1 & 0 & 1 \end{bmatrix}$

14. $\begin{bmatrix} 2 & 4 & 6 \\ 3 & 2 & -5 \\ 2 & 3 & 1 \end{bmatrix}$

15. $\begin{bmatrix} 1 & 2 & 1 & 1 \\ 0 & 2 & 3 & 2 \\ 0 & 0 & 1 & 3 \\ 0 & 0 & 0 & 4 \end{bmatrix}$

16. $\begin{bmatrix} 1 & 1 & 1 & 1 \\ 1 & 1 & 1 & -1 \\ 1 & 1 & -1 & 1 \\ 1 & -1 & 1 & 1 \end{bmatrix}$

Show that neither of the following matrices has a multiplicative inverse. See Example C.

17. $\begin{bmatrix} 1 & 3 & 4 \\ 2 & 1 & -1 \\ 4 & 7 & 7 \end{bmatrix}$

18. $\begin{bmatrix} 2 & -2 & 4 \\ 5 & 3 & 2 \\ 3 & 5 & -2 \end{bmatrix}$

Solve the following systems by making use of the inverses
you found in Problems 11–14. Begin by writing the system in
the matrix form $AX = B$. See Example D.

19.
$\begin{aligned}
x + y + z &= 2 \\
x - y + 2z &= -1 \\
3x + 2y &= 5
\end{aligned}$

20.
$\begin{aligned}
2x + y + z &= 4 \\
x + 3y + z &= 5 \\
-x + 4y &= 0
\end{aligned}$

21.
$\begin{aligned}
3x + y + 2z &= 3 \\
4x + y - 6z &= 2 \\
x \quad\quad + z &= 6
\end{aligned}$

22.
$\begin{aligned}
2x + 4y + 6z &= 9 \\
3x + 2y - 5z &= 2 \\
2x + 3y + z &= 4
\end{aligned}$

B. Applications and Extensions

In Problems 23–26, find the multiplicative inverse or indicate that it does not exist.

23. $\begin{bmatrix} 4 & -3 \\ 5 & -\frac{15}{4} \end{bmatrix}$

24. $\begin{bmatrix} 3 & -1 \\ 4 & 2 \end{bmatrix}$

25. $\begin{bmatrix} 1 & -2 & 1 \\ 3 & 0 & 2 \\ 1 & 2 & \frac{1}{2} \end{bmatrix}$

26. $\begin{bmatrix} -2 & 4 & 2 \\ 3 & 5 & 6 \\ 1 & 9 & 8 \end{bmatrix}$

27. Find the multiplicative inverse of

$$\begin{bmatrix} 2 & 0 & 0 \\ 0 & 3 & 0 \\ 0 & 0 & -4 \end{bmatrix}$$

28. Give a formula for U^{-1} if

$$U = \begin{bmatrix} a & 0 & 0 \\ 0 & b & 0 \\ 0 & 0 & c \end{bmatrix}$$

When does the matrix U fail to have an inverse?

29. Use your result from Problem 25 to solve the system

$$x - 2y + z = a$$
$$3x \quad\quad + 2z = b$$
$$x + 2y + \tfrac{1}{2}z = c$$

30. Let A and B be 3×3 matrices with inverses A^{-1} and B^{-1}. Show that AB has an inverse given by $B^{-1}A^{-1}$. *Hint:* The product in either order must be I.

31. Show that

$$\begin{bmatrix} 1 & -1 \\ 3 & -3 \end{bmatrix}\begin{bmatrix} 2 & -4 \\ 2 & -4 \end{bmatrix} = \begin{bmatrix} 0 & 0 \\ 0 & 0 \end{bmatrix}$$

Thus $AB = O$ but neither A nor B is O. This is another way that matrices differ from ordinary numbers.

32. Suppose that $AB = O$ and A has a multiplicative inverse. Show that $B = O$. See Problem 31.

33. Let C be a 3×3 matrix and A and B be as follows:

$$A = \begin{bmatrix} 1 & 0 & 0 \\ 0 & 1 & 2 \\ 0 & 0 & 1 \end{bmatrix} \quad B = \begin{bmatrix} 1 & 0 & 0 \\ 0 & 1 & -2 \\ 0 & 0 & 1 \end{bmatrix}$$

(a) What happens to $\mathbf{C}$ under the multiplication $\mathbf{AC}$? Under the multiplication $\mathbf{BC}$?

(b) What do you conclude about the matrices $\mathbf{A}$ and $\mathbf{B}$?

34. Let

$$\mathbf{A} = \begin{bmatrix} 1 & 0 & 0 \\ 1 & 1 & 0 \\ 1 & 0 & 1 \end{bmatrix}$$

(a) Calculate $\mathbf{A}^{-1}$.

(b) Calculate $\mathbf{A}^2$, $\mathbf{A}^3$, and $\mathbf{A}^{-2} = (\mathbf{A}^{-1})^2$.

(c) Conjecture the form of $\mathbf{A}^n$ for any integer n.

35. Find the inverse of the matrix.

$$\begin{bmatrix} 1 & 1 & 1 & 1 \\ 1 & 2 & 2 & 2 \\ 1 & 2 & 1 & 1 \\ 1 & 2 & 1 & 2 \end{bmatrix}$$

36. Consider the matrices

$$\mathbf{A} = \begin{bmatrix} 0 & 1 & 0 \\ 0 & 0 & 1 \\ 1 & 0 & 0 \end{bmatrix} \quad \text{and} \quad \mathbf{B} = \begin{bmatrix} 0 & 1 & 0 & 0 \\ 0 & 0 & 1 & 0 \\ 0 & 0 & 0 & 1 \\ 1 & 0 & 0 & 0 \end{bmatrix}$$

(a) Show that $\mathbf{A}^3 = \mathbf{I}$ and $\mathbf{B}^4 = \mathbf{I}$.

(b) What does (a) allow you to conclude about $\mathbf{A}^{-1}$ and $\mathbf{B}^{-1}$?

(c) Conjecture a generalization of (a).

37. Show that the inverse of a Heisenberg matrix (see Problem 28 of Section 11-3) is a Heisenberg matrix.

38. TEASER The matrices

$$\mathbf{A} = \begin{bmatrix} 1 & \frac{1}{2} \\ \frac{1}{2} & \frac{1}{3} \end{bmatrix} \quad \text{and} \quad \mathbf{B} = \begin{bmatrix} 1 & \frac{1}{2} & \frac{1}{3} \\ \frac{1}{2} & \frac{1}{3} & \frac{1}{4} \\ \frac{1}{3} & \frac{1}{4} & \frac{1}{5} \end{bmatrix}$$

and their $n \times n$ generalizations are called Hilbert matrices; they play an important role in numerical analysis.

(a) Find $\mathbf{A}^{-1}$ and $\mathbf{B}^{-1}$.

(b) Let $\mathbf{C}$ be the column matrix with entries $(\frac{11}{6}, \frac{13}{12}, \frac{47}{60})$. Solve the equation $\mathbf{BX} = \mathbf{C}$ for $\mathbf{X}$.

39. Let

$$\mathbf{A} = \begin{bmatrix} 1 & 1 & 1 & 1 \\ 1 & 2 & 3 & 4 \\ 1 & 3 & 6 & 10 \\ 1 & 4 & 10 & 20 \end{bmatrix} \qquad \mathbf{X} = \begin{bmatrix} x \\ y \\ z \\ w \end{bmatrix}$$

Solve the equation $\mathbf{AX} = \mathbf{B}$ for $\mathbf{B}$ equal to:

(a) $\begin{bmatrix} 4 \\ 3 \\ 2 \\ 1 \end{bmatrix}$ (b) $\begin{bmatrix} 1 \\ 0 \\ 1 \\ 0 \end{bmatrix}$ (c) $\begin{bmatrix} 1 \\ -2 \\ 3 \\ -4 \end{bmatrix}$

40. Let

$$\mathbf{A} = \begin{bmatrix} 1 & 1 & 1 & 1 & 1 \\ 1 & 2 & 3 & 4 & 5 \\ 1 & 3 & 6 & 10 & 15 \\ 1 & 4 & 10 & 20 & 35 \\ 1 & 5 & 15 & 35 & 70 \end{bmatrix} \qquad \mathbf{X} = \begin{bmatrix} x \\ y \\ z \\ u \\ v \end{bmatrix}$$

Find $\mathbf{A}^{-1}$ and solve $\mathbf{AX} = \mathbf{B}$ for various choices of $\mathbf{B}$.

11-5 SECOND- AND THIRD-ORDER DETERMINANTS

Matrix	Determinant	Value of Determinant
$\begin{bmatrix} a & b \\ c & d \end{bmatrix}$	$\begin{vmatrix} a & b \\ c & d \end{vmatrix}$	$ad - bc$
$\begin{bmatrix} a_1 & b_1 & c_1 \\ a_2 & b_2 & c_2 \\ a_3 & b_3 & c_3 \end{bmatrix}$	$\begin{vmatrix} a_1 & b_1 & c_1 \\ a_2 & b_2 & c_2 \\ a_3 & b_3 & c_3 \end{vmatrix}$	$a_1b_2c_3 + a_2b_3c_1 + a_3b_1c_2$ $- a_1b_3c_2 - a_2b_1c_3 - a_3b_2c_1$

The notion of a determinant is usually attributed to the German mathematician Gottfried Wilhelm Leibniz (1646–1716), but it seems that Seki Kōwa of Japan had the idea somewhat earlier. It grew out of the study of systems of equations.

Second-Order Determinants

Consider the general system of two equations in two unknowns

$$ax + by = r$$

$$cx + dy = s$$

If we multiply the second equation by a and then add $-c$ times the first equation to it, we obtain the equivalent triangular system.

$$ax + \qquad by = r$$

$$(ad - bc)y = as - cr$$

If $ad - bc \neq 0$, we can solve this system by backward substitution.

$$x = \frac{rd - bs}{ad - bc}$$

$$y = \frac{as - rc}{ad - bc}$$

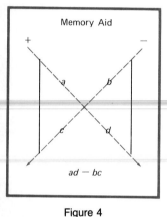

Memory Aid

$ad - bc$

Figure 4

These formulas are hard to remember unless we associate special symbols with the numbers $ad - bc$, $rd - bs$, and $as - rc$. For the first of these, we propose

$$\begin{vmatrix} a & b \\ c & d \end{vmatrix} = ad - bc$$

The symbol on the left is called a **second-order determinant,** and we say that $ad - bc$ is its value. Figure 4 may help you remember how to make this evaluation.

Example A (Values of Second-Order Determinants)   Evaluate.

(a) $\begin{vmatrix} -2 & -1 \\ 5 & 6 \end{vmatrix}$ (b) $\begin{vmatrix} \pi & \sqrt{2} \\ 0 & 2 \end{vmatrix}$

Solution

(a) $\begin{vmatrix} -2 & -1 \\ 5 & 6 \end{vmatrix} = (-2)(6) - (-1)(5) = -7$

(b) $\begin{vmatrix} \pi & \sqrt{2} \\ 0 & 2 \end{vmatrix} = (\pi)(2) - (\sqrt{2})(0) = 2\pi$ ■

With this new symbol, we can write the solution to

$$ax + by = r$$

$$cx + dy = s$$

in a form known as Cramer's Rule.

$$x = \frac{rd - bs}{ad - bc} = \frac{\begin{vmatrix} r & b \\ s & d \end{vmatrix}}{\begin{vmatrix} a & b \\ c & d \end{vmatrix}}$$

$$y = \frac{as - rc}{ad - bc} = \frac{\begin{vmatrix} a & r \\ c & s \end{vmatrix}}{\begin{vmatrix} a & b \\ c & d \end{vmatrix}}$$

These results are easy to remember when we notice that the denominator is the determinant of the coefficient matrix, and that the numerator is the same except that the coefficients of the unknown we are seeking are replaced by the constants from the right side of the system.

Example B (Cramer's Rule: Second-Order Case)   Solve the system

$$3x - 2y = 7$$
$$4x + 5y = 2$$

Solution

$$x = \frac{\begin{vmatrix} 7 & -2 \\ 2 & 5 \end{vmatrix}}{\begin{vmatrix} 3 & -2 \\ 4 & 5 \end{vmatrix}} = \frac{(7)(5) - (-2)(2)}{(3)(5) - (-2)(4)} = \frac{39}{23}$$

$$y = \frac{\begin{vmatrix} 3 & 7 \\ 4 & 2 \end{vmatrix}}{\begin{vmatrix} 3 & -2 \\ 4 & 5 \end{vmatrix}} = \frac{(3)(2) - (7)(4)}{23} = -\frac{22}{23} \quad \blacksquare$$

The choice of the name *determinant* is appropriate, for the determinants of a system completely *determine* its character.

1. If $ad - bc \neq 0$, the system has a unique solution.
2. If $ad - bc = 0$, $as - rc = 0$, and $rd - bs = 0$, then a, b, and r are proportional to c, d, and s and the system has infinitely many solutions. Here is an example.

$$\begin{array}{cc} 3x - 2y = 7 & \frac{3}{6} = \frac{-2}{-4} = \frac{7}{14} \\ 6x - 4y = 14 & \end{array}$$

3. If $ad - bc = 0$ and $as - rc \neq 0$ or $rd - bs \neq 0$, then a and b are proportional to c and d, but this proportionality does not extend to r and s; the system has no solution. This is illustrated by the following.

$$\begin{array}{cc} 3x - 2y = 7 & \frac{3}{6} = \frac{-2}{-4} \neq \frac{7}{10} \\ 6x - 4y = 10 & \end{array}$$

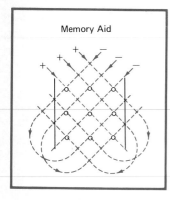

Memory Aid

Figure 5

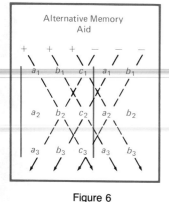

Alternative Memory Aid

Figure 6

Third-Order Determinants

When we consider the general system of three equations in three unknowns

$$a_1x + b_1y + c_1z = d_1$$
$$a_2x + b_2y + c_2z = d_2$$
$$a_3x + b_3y + c_3z = d_3$$

things get more complicated, but the results are similar. The appropriate determinant symbol and its corresponding value are

$$\begin{vmatrix} a_1 & b_1 & c_1 \\ a_2 & b_2 & c_2 \\ a_3 & b_3 & c_3 \end{vmatrix} = a_1b_2c_3 + b_1c_2a_3 + c_1b_3a_2 - c_1b_2a_3 - b_1a_2c_3 - a_1b_3c_2$$

There are six terms in the sum on the right, three with a positive sign and three with a negative sign. The diagrams in Figures 5 and 6 will help you remember the products that enter each term. Just follow the arrows.

Example C (Values of Third-Order Determinants)　Evaluate.

(a) $\begin{vmatrix} 3 & 2 & 4 \\ 4 & -2 & 6 \\ 8 & 3 & 5 \end{vmatrix}$　(b) $\begin{vmatrix} 2 & 3 & -4 \\ 0 & -1 & 5 \\ 0 & 0 & 6 \end{vmatrix}$

Solution

(a) $\begin{vmatrix} 3 & 2 & 4 \\ 4 & -2 & 6 \\ 8 & 3 & 5 \end{vmatrix} = (3)(-2)(5) + (2)(6)(8) + (4)(3)(4)$
$- (4)(-2)(8) - (2)(4)(5) - (3)(3)(6) = 84$

(b) $\begin{vmatrix} 2 & 3 & 4 \\ 0 & -1 & 5 \\ 0 & 0 & 6 \end{vmatrix} = (2)(-1)(6) + (3)(5)(0) + (4)(0)(0)$
$- (4)(-1)(0) - (3)(0)(6) - (2)(0)(5) = -12$

Note that all terms except the main (northwest-southeast) diagonal product were zero. ■

Properties of Determinants

We are interested in how the matrix operations considered in Section 11-2 affect the values of the corresponding determinants.

1. Interchanging two rows changes the sign of the determinant; for example,

$$\begin{vmatrix} a & b \\ c & d \end{vmatrix} = - \begin{vmatrix} c & d \\ a & b \end{vmatrix}$$

2. Multiplying a row by a constant k multiplies the value of the determinant by k; for example,

$$\begin{vmatrix} ka & kb \\ c & d \end{vmatrix} = k \begin{vmatrix} a & b \\ c & d \end{vmatrix}$$

CAUTION

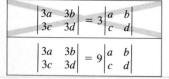

$$\begin{vmatrix} 3a & 3b \\ 3c & 3d \end{vmatrix} = 3 \begin{vmatrix} a & b \\ c & d \end{vmatrix}$$

$$\begin{vmatrix} 3a & 3b \\ 3c & 3d \end{vmatrix} = 9 \begin{vmatrix} a & b \\ c & d \end{vmatrix}$$

3. Adding a multiple of one row to another does not affect the value of the determinant; for example,

$$\begin{vmatrix} a & b \\ c + ka & d + kb \end{vmatrix} = \begin{vmatrix} a & b \\ c & d \end{vmatrix}$$

We mention also the effect of a new operation.

4. Interchanging the rows and corresponding columns does not affect the value of the determinant; for example,

$$\begin{vmatrix} a & c \\ b & d \end{vmatrix} = \begin{vmatrix} a & b \\ c & d \end{vmatrix}$$

Though they are harder to prove in the third-order case, we emphasize that all four properties hold for both second- and third-order determinants. We offer only one proof, a proof of Property 3 in the second-order case.

$$\begin{vmatrix} a & b \\ c + ka & d + kb \end{vmatrix} = a(d + kb) - b(c + ka)$$

$$= ad + akb - bc - bka = ad - bc = \begin{vmatrix} a & b \\ c & d \end{vmatrix}$$

Property 3 can be a great aid in evaluating a determinant. Using it, we can transform a matrix to triangular form without changing the value of its determinant. But *the determinant of a triangular matrix is just the product of the elements on the main diagonal,* since this is the only nonzero term in the determinant formula.

Example D (Using the Properties)   Evaluate.

$$\begin{vmatrix} 1 & 3 & 4 \\ -1 & -2 & 3 \\ 2 & -6 & 11 \end{vmatrix}$$

Solution   We use Property 3 to introduce zeros below the main diagonal.

$$\begin{vmatrix} 1 & 3 & 4 \\ -1 & -2 & 3 \\ 2 & -6 & 11 \end{vmatrix} = \begin{vmatrix} 1 & 3 & 4 \\ 0 & 1 & 7 \\ 0 & -12 & 3 \end{vmatrix} = \begin{vmatrix} 1 & 3 & 4 \\ 0 & 1 & 7 \\ 0 & 0 & 87 \end{vmatrix} = 87 \qquad \blacksquare$$

Cramer's Rule

We saw that the solutions for x and y in a second-order system could be written as the quotients of two determinants. That fact generalizes to the third-order case. We present it without proof. Consider

$$a_1 x + b_1 y + c_1 z = d_1$$

$$a_2 x + b_2 y + c_2 z = d_2$$

$$a_3 x + b_3 y + c_3 z = d_3$$

If

$$D = \begin{vmatrix} a_1 & b_1 & c_1 \\ a_2 & b_2 & c_2 \\ a_3 & b_3 & c_3 \end{vmatrix} \neq 0$$

then the system above has a unique solution given by

$$x = \frac{1}{D}\begin{vmatrix} d_1 & b_1 & c_1 \\ d_2 & b_2 & c_2 \\ d_3 & b_3 & c_3 \end{vmatrix} \qquad y = \frac{1}{D}\begin{vmatrix} a_1 & d_1 & c_1 \\ a_2 & d_2 & c_2 \\ a_3 & d_3 & c_3 \end{vmatrix} \qquad z = \frac{1}{D}\begin{vmatrix} a_1 & b_1 & d_1 \\ a_2 & b_2 & d_2 \\ a_3 & b_3 & d_3 \end{vmatrix}$$

The pattern is the same as in the second-order situation. The denominator D is the determinant of the coefficient matrix. The numerator in each case is obtained from D by replacing the coefficients of the unknown by the constants from the right side of the system.

This method of solving a system of equations is named after one of its discoverers, Gabriel Cramer (1704–1752). Historically, it has been a popular method. However, note that even for a system of three equations in three unknowns, it requires the evaluation of four determinants. The methods of matrices (Section 11-2) is considerably more efficient, both for hand and computer calculation. Consequently, Cramer's Rule is now primarily of theoretical rather than practical interest.

Example E (Cramer's Rule: Third-Order Case)   Solve the following system using Cramer's rule.

$$3x + 2y + 4z = 2$$
$$4x - 2y + 6z = 0$$
$$8x + 3y + 5z = 0$$

Solution   We found that $D = 84$ in Example C.

$$x = \frac{1}{84}\begin{vmatrix} 2 & 2 & 4 \\ 0 & -2 & 6 \\ 0 & 3 & 5 \end{vmatrix} = \frac{1}{84}(-20 - 36) = -\frac{2}{3}$$

$$y = \frac{1}{84}\begin{vmatrix} 3 & 2 & 4 \\ 4 & 0 & 6 \\ 8 & 0 & 5 \end{vmatrix} = \frac{1}{84}(96 - 40) = \frac{2}{3}$$

$$z = \frac{1}{84}\begin{vmatrix} 3 & 2 & 2 \\ 4 & -2 & 0 \\ 8 & 3 & 0 \end{vmatrix} = \frac{1}{84}[24 - (-32)] = \frac{2}{3} \qquad \blacksquare$$

PROBLEM SET 11-5

A. Skills and Techniques

Evaluate each of the determinants in Problems 1–8 by inspection. See Examples A and C.

1. $\begin{vmatrix} 4 & 0 \\ 0 & -2 \end{vmatrix}$

2. $\begin{vmatrix} 8 & 0 \\ 5 & 0 \end{vmatrix}$

3. $\begin{vmatrix} 11 & 4 \\ 0 & 2 \end{vmatrix}$

4. $\begin{vmatrix} 2 & -1 & 5 \\ 0 & 4 & 2 \\ 0 & 0 & -1 \end{vmatrix}$

5. $\begin{vmatrix} -1 & -7 & 9 \\ 0 & 5 & 4 \\ 0 & 0 & 10 \end{vmatrix}$

6. $\begin{vmatrix} 3 & -2 & 1 \\ 0 & 0 & 0 \\ 1 & 5 & -8 \end{vmatrix}$

7. $\begin{vmatrix} 3 & 0 & 8 \\ 10 & 0 & 2 \\ -1 & 0 & -9 \end{vmatrix}$

8. $\begin{vmatrix} 9 & 0 & 0 \\ 0 & 0 & -2 \\ 0 & 4 & 0 \end{vmatrix}$

9. Let

$$\begin{vmatrix} a_1 & b_1 & c_1 \\ a_2 & b_2 & c_2 \\ a_3 & b_3 & c_3 \end{vmatrix} = 12$$

Use the properties of determinants to evaluate:

(a) $\begin{vmatrix} a_1 & a_2 & a_3 \\ b_1 & b_2 & b_3 \\ c_1 & c_2 & c_3 \end{vmatrix}$

(b) $\begin{vmatrix} a_3 & b_3 & c_3 \\ a_2 & b_2 & c_2 \\ a_1 & b_1 & c_1 \end{vmatrix}$

(c) $\begin{vmatrix} a_1 & b_1 & c_1 \\ a_2 & b_2 & c_2 \\ 3a_3 & 3b_3 & 3c_3 \end{vmatrix}$

(d) $\begin{vmatrix} a_1 + 3a_3 & b_1 + 3b_3 & c_1 + 3c_3 \\ a_2 & b_2 & c_2 \\ a_3 & b_3 & c_3 \end{vmatrix}$

Evaluate each of the determinants in Problems 10–17. See Example D.

10. $\begin{vmatrix} 3 & 2 \\ 5 & 6 \end{vmatrix}$

11. $\begin{vmatrix} 5 & 3 \\ 5 & -3 \end{vmatrix}$

12. $\begin{vmatrix} 3 & 0 & 0 \\ -2 & 5 & 4 \\ 1 & 2 & -9 \end{vmatrix}$

13. $\begin{vmatrix} 4 & 8 & -2 \\ 1 & -2 & 0 \\ 2 & 4 & 0 \end{vmatrix}$

14. $\begin{vmatrix} 3 & 2 & -4 \\ 1 & 0 & 5 \\ 4 & -2 & 3 \end{vmatrix}$

15. $\begin{vmatrix} 2 & 4 & 1 \\ 1 & 3 & 6 \\ 2 & 3 & -1 \end{vmatrix}$

16. $\begin{vmatrix} 5.1 & -3.2 & 2.6 \\ 1.3 & 4.5 & 2.3 \\ 3.4 & -2.2 & 1.9 \end{vmatrix}$

17. $\begin{vmatrix} 2.03 & 5.41 & -3.14 \\ 0 & 6.22 & 0 \\ -1.93 & 7.13 & 6.34 \end{vmatrix}$

Use Cramer's rule to solve the system of equations in Problems 18–21. See Examples B and E.

18. $2x - 3y = -11$
$\quad\ x + 2y = -2$

19. $5x + \ \ y = \ \ 7$
$\quad\ 3x - 4y = 18$

20. $2x + 4y + \ \ z = 15$
$\quad\ \ x + 3y + 6z = 15$
$\quad\ 2x + 3y - \ \ z = 11$

21. $5x - 3y + 2z = 18$
$\quad\ \ x + 4y + 2z = -4$
$\quad\ 3x - 2y + \ \ z = 11$

B. Applications and Extensions

22. Establish Property 2 for a general second-order determinant.

23. Evaluate each of the following determinants (the easy way).

(a) $\begin{vmatrix} 1 & 2 & 3 \\ 0 & 0 & 0 \\ 1.9 & 2.9 & 3.9 \end{vmatrix}$

(b) $\begin{vmatrix} 1 & 2 & 3 \\ 1.1 & 2.2 & 3.3 \\ 1.9 & 2.9 & 3.9 \end{vmatrix}$

(c) $\begin{vmatrix} 1.1 & 2.2 & 3.3 \\ 4.4 & 5.5 & 6.6 \\ 5.5 & 7.7 & 9.9 \end{vmatrix}$

(d) $\begin{vmatrix} 1 & 2 & 3 \\ 0 & 2 & 3 \\ 0 & 0 & 3 \end{vmatrix}$

(e) $\begin{vmatrix} 1 & 1 & 1 \\ 1 & 2 & 3 \\ 1 & 3 & 6 \end{vmatrix}$

(f) $\begin{vmatrix} 1 & 1 & 1 \\ 1 & 2 & 4 \\ 1 & 3 & 9 \end{vmatrix}$

24. Show that the value of a third-order determinant is 0 if either of the following is true.
   (a) Two rows are proportional.
   (b) The sum of two rows is the third row.

25. Solve for x.

$$\begin{vmatrix} x^2 & x & 1 \\ 1 & 2 & 3 \\ 4 & 2 & 1 \end{vmatrix} = 0$$

26. Solve for x.

$$\begin{vmatrix} x^2 & x & 1 \\ 1 & 2 & 3 \\ 4 & 5 & 6 \end{vmatrix} = 0$$

27. Consider the system of equations

$$kx + y = k^2$$

$$x + ky = 1$$

For what values of k does this system have (a) unique solution; (b) infinitely many solutions; (c) no solution?

28. If $\mathbf{C} = \mathbf{AB}$, where $\mathbf{A}$ and $\mathbf{B}$ are square matrices of the same size, then their determinants satisfy $|\mathbf{C}| = |\mathbf{A}||\mathbf{B}|$. Prove this fact for 2×2 matrices.

29. Let $\mathbf{A} = \begin{bmatrix} a_1 & b_1 & c_1 \\ a_2 & b_2 & c_2 \\ a_3 & b_3 & c_3 \end{bmatrix}$ with $|\mathbf{A}| = 12$. Evaluate.

(a) $|\mathbf{A}^2|$ **(b)** $|2\mathbf{A}|$

(c) $|\mathbf{A}^{-1}|$

(d) $\begin{vmatrix} 2a_1 & 2b_1 & 2c_1 \\ a_2 & b_2 & c_2 \\ -3a_3 & -3b_3 & -3c_3 \end{vmatrix}$

(e) $\begin{vmatrix} a_1 & b_1 & c_1 \\ a_2 - a_1 & b_2 - b_1 & c_2 - c_1 \\ 2a_2 & 2b_2 & 2c_2 \end{vmatrix}$

30. Show that for all x,

$$\begin{vmatrix} x^3 & x^2 & x \\ 3x^2 & 4x & 5 \\ 2x^2 & 3x & 4 \end{vmatrix} = 0$$

31. Suppose that $\mathbf{A}$ and $\mathbf{B}$ are 3×3 matrices with $|\mathbf{A}| = -2$. Use Problem 28 to evaluate (a) $|\mathbf{A}^5|$; (b) $|\mathbf{A}^{-1}|$; (c) $|\mathbf{B}^{-1}\mathbf{AB}|$; (d) $|3\mathbf{A}^3|$.

32. Show that

$$\begin{vmatrix} 1 & a & a^2 \\ 1 & b & b^2 \\ 1 & c & c^2 \end{vmatrix} = (a - b)(b - c)(c - a)$$

33. Let $(0, 0)$, (a, b), (c, d), and $(a + c, b + d)$ be the vertices of a parallelogram with a, b, c, and d being positive numbers. Show that the area of the parallelogram is the determinant

$$\begin{vmatrix} a & b \\ c & d \end{vmatrix}$$

34. Consider the following determinant equation.

$$\begin{vmatrix} x & y & 1 \\ a & b & 1 \\ c & d & 1 \end{vmatrix} = 0 \qquad \text{where } (a, b) \neq (c, d)$$

(a) Show that the above equation is the equation of a line in the xy-plane.
(b) How can you tell immediately that the points (a, b) and (c, d) are on this line?
(c) Write a determinant equation for the line that passes through the points $(5, -1)$ and $(4, 11)$.

35. Determine the polynomial $P(a, b, c)$ for which

$$\begin{vmatrix} a & b & c \\ b & c & a \\ c & a & b \end{vmatrix} = (a + b + c)P(a, b, c)$$

36. TEASER If $\mathbf{A}$ is a square matrix, then $P(x) = |\mathbf{A} - x\mathbf{I}|$ is called the *characteristic polynomial* of the matrix. Let

$$\mathbf{A} = \begin{bmatrix} 1 & 1 & -1 \\ 0 & 0 & -1 \\ 1 & 1 & 4 \end{bmatrix}$$

(a) Write the polynomial $P(x)$ in the form $ax^3 + bx^2 + cx + d$.
(b) Solve the equation $P(x) = 0$. The solutions are called the *characteristic values* of $\mathbf{A}$.
(c) Show that $P(\mathbf{A}) = 0$, that is, $\mathbf{A}$ satisfies its own characteristic equation.

37. Refer to Problem 36. Find the characteristic values accurate to two decimal places, of

$$\mathbf{A} = \begin{bmatrix} 2 & 5 & 3 \\ -1 & -4 & -1 \\ 1 & 0 & 1 \end{bmatrix}$$

38. By direct calculation, show that

$$\begin{vmatrix} 1.2 & 1 & 1 \\ 1 & 1.3 & 1 \\ 1 & 1 & 1.4 \end{vmatrix} = (.2)(.3)(.4) + (.2)(.3) + (.2)(.4) + (.3)(.4)$$

Do the same with the numerals 2, 3, and 4 replaced by 8, 5, and 9, respectively. Try other replacements. Conjecture a general theorem where the main diagonal elements are replaced by $1 + a$, $1 + b$, and $1 + c$. Prove your conjecture.

11-6 HIGHER-ORDER DETERMINANTS

A Mathematician and a Poet

One of the most consistent workers on the theory of determinants over a period of 50 years was the Englishman, James Joseph Sylvester. Known as a poet, a wit, and a mathematician, he taught in England and in America. During his stay at Johns Hopkins University in Baltimore (1877-1883), he helped establish one of the first graduate programs in mathematics in America. Under his tutelage, mathematics began to flourish in the United States.

James Joseph Sylvester
1814–1897

Having defined determinants for 2×2 and 3×3 matrices, we expect to do it for 4×4 matrices, 5×5 matrices, and so on. Our problem is to do it in such a way that Cramer's Rule and the determinant properties of Section 11-5 still hold. This will take some work.

Minors

We begin by introducing the standard notation for a general $n \times n$ matrix.

$$\begin{bmatrix} a_{11} & a_{12} & a_{13} & \cdots & a_{1n} \\ a_{21} & a_{22} & a_{23} & \cdots & a_{2n} \\ a_{31} & a_{32} & a_{33} & \cdots & a_{3n} \\ \cdot & \cdot & \cdot & & \cdot \\ \cdot & \cdot & \cdot & & \cdot \\ \cdot & \cdot & \cdot & & \cdot \\ a_{n1} & a_{n2} & a_{n3} & \cdots & a_{nn} \end{bmatrix}$$

Note the use of the double subscript on each entry: the first subscript gives the row in which a_{ij} stands and the second gives the column. For example, a_{32} is the entry in the third row and second column.

Associated with each entry a_{ij} in an $n \times n$ is a determinant M_{ij} of order $n - 1$ called the **minor** of a_{ij}. It is obtained by taking the determinant of the submatrix that results when we blot out the row and column in which a_{ij} stands. For example, the minor M_{13} of a_{13} in the 4×4 matrix

$$\begin{bmatrix} a_{11} & a_{12} & a_{13} & a_{14} \\ a_{21} & a_{22} & a_{23} & a_{24} \\ a_{31} & a_{32} & a_{33} & a_{34} \\ a_{41} & a_{42} & a_{43} & a_{44} \end{bmatrix}$$

is the third-order determinant.

$$\begin{vmatrix} a_{21} & a_{22} & a_{24} \\ a_{31} & a_{32} & a_{34} \\ a_{41} & a_{42} & a_{44} \end{vmatrix}$$

The General nth-Order Determinant

Here is the definition to which we have been leading.

$$\begin{vmatrix} a_{11} & a_{12} & \cdots & a_{1n} \\ a_{21} & a_{22} & \cdots & a_{2n} \\ \vdots & \vdots & & \vdots \\ a_{n1} & a_{n2} & \cdots & a_{nn} \end{vmatrix} = a_{11}M_{11} - a_{12}M_{12} + a_{13}M_{13} \cdots + (-1)^{n+1}a_{1n}M_{1n}$$

There are three important questions to answer regarding this definition.

Does this definition really define? Only if the minors M_{ij} can be evaluated. They are themselves determinants, but here is the key point: They are of order $n - 1$, one less than the order of the determinant we started with. They can, in turn, be expressed in terms of determinants of order $n - 2$, and so on, using the same definition. Thus, for example, a fifth-order determinant can be expressed in terms of fourth-order determinants, and these fourth-order determinants can be expressed in terms of third-order determinants. But we know how to evaluate third-order determinants from Section 11-5.

Is this definition consistent with the earlier definition when applied to third-order determinants? Yes, for if we apply it to a general third-order determinant, we get

$$\begin{vmatrix} a_1 & b_1 & c_1 \\ a_2 & b_2 & c_2 \\ a_3 & b_3 & c_3 \end{vmatrix} = a_1 \begin{vmatrix} b_2 & c_2 \\ b_3 & c_3 \end{vmatrix} - b_1 \begin{vmatrix} a_2 & c_2 \\ a_3 & c_3 \end{vmatrix} + c_1 \begin{vmatrix} a_2 & b_2 \\ a_3 & b_3 \end{vmatrix}$$

$$= a_1 b_2 c_3 - a_1 c_2 b_3 - b_1 a_2 c_3 + b_1 c_2 a_3 + c_1 a_2 b_3 - c_1 b_2 a_3$$

This is the same value we gave in Section 11-5.

Example A (Method of Minors)   Use the method just described to evaluate

$$\begin{vmatrix} 3 & -1 & 0 \\ 1 & 4 & 5 \\ 4 & -2 & 1 \end{vmatrix}$$

Solution

$$\begin{vmatrix} 3 & -1 & 0 \\ 1 & 4 & 5 \\ 4 & -2 & 1 \end{vmatrix} = 3 \begin{vmatrix} 4 & 5 \\ -2 & 1 \end{vmatrix} - (-1) \begin{vmatrix} 1 & 5 \\ 4 & 1 \end{vmatrix} = 3(14) + (-19) = 23 \quad \blacksquare$$

Does this definition preserve Cramer's Rule and the properties of Section 11-5? Yes, it does. We shall not prove this because the proofs are lengthy and difficult.

Expansion According to Any Row or Column

Our definition expressed the value of a determinant in terms of the entries and minors of the first row; we call it an expansion according to the first row. It is a remarkable fact that we can expand a determinant according to any row or column (and always get the same answer).

Before we can show what we mean, we must explain a sign convention. We associate a positive or negative sign with every position in a matrix. To the ij-position, we assign a positive sign if $i + j$ is even and a negative sign otherwise. Thus for a 4×4 matrix, we have this pattern of signs.

$$\begin{bmatrix} + & - & + & - \\ - & + & - & + \\ + & - & + & - \\ - & + & - & + \end{bmatrix}$$

There is always a $+$ in the upper left position and then the signs alternate.

With this understanding about signs, we may expand according to any row or column. For example, to evaluate a fourth-order determinant, we can expand according to the second column if we wish. We multiply each entry in that column by its minor, prefixing each product with a positive or negative sign according to the pattern above. Then we add the results.

$$\begin{vmatrix} a_{11} & a_{12} & a_{13} & a_{14} \\ a_{21} & a_{22} & a_{23} & a_{24} \\ a_{31} & a_{32} & a_{33} & a_{34} \\ a_{41} & a_{42} & a_{43} & a_{44} \end{vmatrix} = -a_{12} M_{12} + a_{22} M_{22} - a_{32} M_{32} + a_{42} M_{42}$$

Example B (Expansion by a Column)   Evaluate

$$\begin{vmatrix} 6 & 0 & 4 & -1 \\ 2 & 0 & -1 & 4 \\ -2 & 4 & -2 & 3 \\ 4 & 0 & 5 & -4 \end{vmatrix}$$

Solution   It is clearly best to expand according to the second column, since three of the four resulting terms are zero. The single nonzero term is just $(-1)(4)$ times the minor M_{32}—that is,

$$-4 \begin{vmatrix} 6 & 4 & -1 \\ 2 & -1 & 4 \\ 4 & 5 & -4 \end{vmatrix}$$

We could now evaluate this third-order determinant as in Section 11-5. But having seen the usefulness of zeros, let us take a different tack. It is easy to get two zeros in the first column. Simply add -3 times the second row to the first row and -2 times the second row to the third. We get

$$-4 \begin{vmatrix} 0 & 7 & -13 \\ 2 & -1 & 4 \\ 0 & 7 & -12 \end{vmatrix}$$

Finally, expand according to the first column.

$$(-4)(-1)(2)\begin{vmatrix} 7 & -13 \\ 7 & -12 \end{vmatrix} = 8(-84 + 91) = 56$$

The reason for the factor of -1 is that the entry 2 is in a negative position in the 3×3 pattern of signs. ■

Example C (Introducing Zeros)   Evaluate

$$\begin{vmatrix} 2 & -1 & 3 & 0 \\ 1 & 0 & 4 & -1 \\ 0 & 3 & 6 & 2 \\ 1 & 1 & -1 & -1 \end{vmatrix}$$

Solution   We expand according to the first column, after introducing two more zeros by adding multiples of the fourth row to the first and second rows.

$$\begin{vmatrix} 2 & -1 & 3 & 0 \\ 1 & 0 & 4 & -1 \\ 0 & 3 & 6 & 2 \\ 1 & 1 & -1 & -1 \end{vmatrix} = \begin{vmatrix} 0 & -3 & 5 & 2 \\ 0 & -1 & 5 & 0 \\ 0 & 3 & 6 & 2 \\ 1 & 1 & -1 & -1 \end{vmatrix} = -\begin{vmatrix} -3 & 5 & 2 \\ -1 & 5 & 0 \\ 3 & 6 & 2 \end{vmatrix}$$

$$= -\begin{vmatrix} -6 & -1 & 0 \\ -1 & 5 & 0 \\ 3 & 6 & 2 \end{vmatrix} = -2\begin{vmatrix} -6 & -1 \\ -1 & 5 \end{vmatrix} = 62 \quad ■$$

Example D (Cramer's Rule Again)   Solve for x in the system

$$\begin{aligned} 2x - y + 3z &= 1 \\ x + 4z - w &= -5 \\ 3y + 6z + 2w &= 0 \\ x + y - z - w &= 0 \end{aligned}$$

Solution   Note that the determinant D of the coefficient matrix was evaluated in Example C.

$$x = \frac{1}{62}\begin{vmatrix} 1 & -1 & 3 & 0 \\ -5 & 0 & 4 & -1 \\ 0 & 3 & 6 & 2 \\ 0 & 1 & -1 & -1 \end{vmatrix} = \frac{1}{62}\begin{vmatrix} 1 & -1 & 3 & 0 \\ 0 & -5 & 19 & -1 \\ 0 & 3 & 6 & 2 \\ 0 & 1 & -1 & -1 \end{vmatrix} = \frac{1}{62}\begin{vmatrix} -5 & 19 & -1 \\ 3 & 6 & 2 \\ 1 & -1 & -1 \end{vmatrix}$$

$$= \frac{1}{62}\begin{vmatrix} -5 & 14 & -6 \\ 3 & 9 & 5 \\ 1 & 0 & 0 \end{vmatrix} = \frac{1}{62}\begin{vmatrix} 14 & -6 \\ 9 & 5 \end{vmatrix} = \frac{124}{62} = 2 \quad ■$$

PROBLEM SET 11-6

A. Skills and Techniques

Evaluate each of the determinants in Problems 1–6 according to a row or column of your choice. Make a good choice or suffer the consequences! See Examples A and B.

1. $\begin{vmatrix} 3 & -2 & 4 \\ 1 & 5 & 0 \\ 3 & 10 & 0 \end{vmatrix}$

2. $\begin{vmatrix} 4 & 0 & -6 \\ -2 & 3 & 5 \\ 1 & 0 & 8 \end{vmatrix}$

3. $\begin{vmatrix} 1 & 2 & 3 \\ 0 & 2 & 3 \\ 1 & 3 & 4 \end{vmatrix}$

4. $\begin{vmatrix} 2 & -1 & -1 \\ 3 & 4 & 2 \\ 0 & -1 & -1 \end{vmatrix}$

5. $\begin{vmatrix} 3 & 0 & 0 & 0 \\ -1 & 1 & 4 & 2 \\ 2 & 0 & 2 & -3 \\ -4 & 0 & 1 & 5 \end{vmatrix}$

6. $\begin{vmatrix} 0 & 5 & 0 & 0 \\ 1 & -3 & 0 & 2 \\ 4 & 1 & 2 & 8 \\ -3 & 2 & 0 & 5 \end{vmatrix}$

Evaluate each of the determinants in Problems 7–16 by first getting some zeros in a row or column and then expanding according to that row or column. See Example C.

7. $\begin{vmatrix} 3 & 5 & -10 \\ 2 & 4 & 6 \\ -3 & -5 & 12 \end{vmatrix}$

8. $\begin{vmatrix} 2 & -1 & 2 \\ 4 & 3 & 4 \\ 7 & -5 & 10 \end{vmatrix}$

9. $\begin{vmatrix} 1 & -2 & 1 & 4 \\ -2 & 5 & -3 & 1 \\ 0 & 7 & -4 & 2 \\ 3 & -2 & 2 & 6 \end{vmatrix}$

10. $\begin{vmatrix} 1 & -2 & 0 & -4 \\ 3 & -4 & 3 & -10 \\ 2 & 1 & -2 & 1 \\ 4 & -5 & 1 & 4 \end{vmatrix}$

11. $\begin{vmatrix} 2 & -3 & 2 \\ 1 & 0 & -4 \\ -1 & 0 & 6 \end{vmatrix}$

12. $\begin{vmatrix} 3 & 1 & -5 \\ 2 & -2 & 7 \\ 1 & 0 & -1 \end{vmatrix}$

13. $\begin{vmatrix} 2 & -3 & 4 & 5 \\ 2 & -3 & 4 & 7 \\ 1 & 6 & 4 & 5 \\ 2 & 6 & 4 & -8 \end{vmatrix}$

14. $\begin{vmatrix} 2 & 2 & 3 & 7 \\ 1 & 2 & 3 & -2 \\ 4 & -3 & 9 & 6 \\ 1 & 2 & 3 & -1 \end{vmatrix}$

15. $\begin{vmatrix} 1 & 2 & -3 & 1 & 2 \\ -1 & 0 & 2 & 5 & -3 \\ 5 & 0 & 0 & -2 & 4 \\ 0 & 0 & 0 & 6 & 3 \\ 0 & 0 & 0 & 2 & -7 \end{vmatrix}$

16. $\begin{vmatrix} 1 & 2 & 3 & 4 & 5 \\ 2 & 1 & 1 & 1 & 1 \\ 3 & 1 & 1 & 1 & 1 \\ 4 & 1 & 1 & 1 & 1 \\ 5 & 1 & 1 & 1 & 1 \end{vmatrix}$

17. As in Example D, Solve for x only.

$$x - 2y + z + 4w = 1$$
$$-2x + 5y - 3z + w = -2$$
$$7y - 4z + 2w = 3$$
$$3x - 2y + 2z + 6w = 6$$

(Make use of your answer to Problem 9.)

18. Solve the following system for z only.

$$x - 2y - 4w = -14$$
$$3x - 4y + 3z - 10w = -28$$
$$2x + y - 2z + w = 0$$
$$4x - 5y + z + 4w = 9$$

(Make use of your answer to Problem 10.)

19. Evaluate the following determinant.

$$\begin{vmatrix} a & b & c & d \\ 0 & e & f & g \\ 0 & 0 & h & i \\ 0 & 0 & 0 & j \end{vmatrix}$$

Conjecture a general result about the determinant of a triangular matrix.

20. Use the result of Problem 19 to evaluate

$$\begin{vmatrix} 2.12 & 3.14 & -1.61 & 1.72 \\ 0 & -2.36 & 5.91 & 7.82 \\ 0 & 0 & 1.46 & 3.34 \\ 0 & 0 & 0 & 3.31 \end{vmatrix}$$

21. Evaluate by reducing to triangular form and using Problem 19.

$$\begin{vmatrix} 1 & 2 & 2.6 & 1.5 \\ 2.3 & 5.6 & -1.3 & 9.8 \\ 2.7 & 1.3 & 4.2 & -1.9 \\ 5.5 & 6.2 & 3.0 & 1.4 \end{vmatrix}$$

22. Show that

$$\begin{vmatrix} a_1 + d_1 & b_1 & c_1 \\ a_2 + d_2 & b_2 & c_2 \\ a_3 + d_3 & b_3 & c_3 \end{vmatrix} = \begin{vmatrix} a_1 & b_1 & c_1 \\ a_2 & b_2 & c_2 \\ a_3 & b_3 & c_3 \end{vmatrix} + \begin{vmatrix} d_1 & b_1 & c_1 \\ d_2 & b_2 & c_2 \\ d_3 & b_3 & c_3 \end{vmatrix}$$

B. Applications and Extensions

23. Evaluate each of the following determinants.

(a) $\begin{vmatrix} 1 & 0 & 0 & 2 \\ 2.7 & 5 & 0 & 8.9 \\ 3.4 & 0 & 6 & 9.1 \\ 3 & 0 & 0 & 4 \end{vmatrix}$ (b) $\begin{vmatrix} a & 0 & 0 & b \\ ? & e & 0 & ? \\ ? & 0 & f & ? \\ c & 0 & 0 & d \end{vmatrix}$

24. Solve for x.

$$\begin{vmatrix} 2 - x & 0 & 0 & 5 \\ 2.7 & 3 - x & 0 & 8.9 \\ 3.4 & 0 & 4 - x & 9.1 \\ 5 & 0 & 0 & 2 - x \end{vmatrix} = 0$$

25. Evaluate the determinant below. *Hint:* Subtract the first row from each of the second and third rows.

$$\begin{vmatrix} n + 1 & n + 2 & n + 3 \\ n + 4 & n + 5 & n + 6 \\ n + 7 & n + 8 & n + 9 \end{vmatrix}$$

26. Show that if the entries in a determinant are all integers, then the value of the determinant is an integer. From this and Cramer's Rule, draw a conclusion about the nature of the solution to a system of n linear equations in n unknowns if all the constants in the system are integers and the determinant of coefficients is nonzero.

27. Show that

$$\begin{vmatrix} a_1 & b_1 & 0 & 0 \\ a_2 & b_2 & 0 & 0 \\ 0 & 0 & c_1 & d_1 \\ 0 & 0 & c_2 & d_2 \end{vmatrix} = \begin{vmatrix} a_1 & b_1 \\ a_2 & b_2 \end{vmatrix} \begin{vmatrix} c_1 & d_1 \\ c_2 & d_2 \end{vmatrix}$$

28. Use Problem 27 to help you solve for x.

$$\begin{vmatrix} x - 2 & 3 & 0 & 0 \\ 2 & x + 3 & 0 & 0 \\ 0 & 0 & x - 4 & 2 \\ 0 & 0 & 3 & x + 1 \end{vmatrix} = 0$$

29. Evaluate the given determinants, in which the entries come from Pascal's triangle (see Section 12-7).

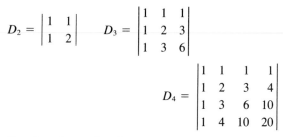

$$D_2 = \begin{vmatrix} 1 & 1 \\ 1 & 2 \end{vmatrix} \qquad D_3 = \begin{vmatrix} 1 & 1 & 1 \\ 1 & 2 & 3 \\ 1 & 3 & 6 \end{vmatrix}$$

$$D_4 = \begin{vmatrix} 1 & 1 & 1 & 1 \\ 1 & 2 & 3 & 4 \\ 1 & 3 & 6 & 10 \\ 1 & 4 & 10 & 20 \end{vmatrix}$$

30. Based on the results of Problem 29, make a conjecture about the value of D_n, the nth-order determinant obtained from Pascal's triangle. Then support your conjecture by describing a systematic way of evaluating D_n.

31. Solve the given system. It should be easy after Problem 29.

$$x + y + z + w = 0$$
$$x + 2y + 3z + 4w = 0$$
$$x + 3y + 6z + 10w = 0$$
$$x + 4y + 10z + 20w = 1$$

32. Evaluate the determinants.

$$E_1 = \begin{vmatrix} 2 & 1 \\ 1 & 2 \end{vmatrix} \qquad E_2 = \begin{vmatrix} 2 & 1 & 0 \\ 1 & 2 & 1 \\ 0 & 1 & 2 \end{vmatrix}$$

$$E_3 = \begin{vmatrix} 2 & 1 & 0 & 0 \\ 1 & 2 & 1 & 0 \\ 0 & 1 & 2 & 1 \\ 0 & 0 & 1 & 2 \end{vmatrix}$$

Now generalize by conjecturing the value of the nth-order determinant E_n that has 2's on the main diagonal, 1's adjacent to this diagonal on either side, and 0's elsewhere. Then prove your conjecture. *Hint:* Expand according to the first row to show that $E_n = 2E_{n-1} - E_{n-2}$ and from this argue that your conjecture must be correct.

33. Let $a_1, a_2, \ldots, a_n$ and $b_1, b_2, \ldots, b_n$ be two sequences of numbers.

    (a) Let $\mathbf{C}_n$ be the $n \times n$ matrix with entries $c_{ij} = a_i b_j$. Evaluate $|\mathbf{C}_n|$ for $n = 2, 3,$ and 4 and make a conjecture.

    (b) Do the same if $c_{ij} = a_i - b_j$.

34. TEASER Generalize Problem 32 of Section 11-5 by evaluating the following determinant and writing your answer as the product of six linear factors.

$$\begin{vmatrix} 1 & a & a^2 & a^3 \\ 1 & b & b^2 & b^3 \\ 1 & c & c^2 & c^3 \\ 1 & d & d^2 & d^3 \end{vmatrix}$$

35. Let $\mathbf{A}_3, \mathbf{A}_4, \mathbf{A}_5$ be the 3×3, 4×4, and 5×5 Hilbert matrices (see Problem 38 of Section 11-4).

Calculate:

(a) $|\mathbf{A}_3|$. **(b)** $|\mathbf{A}_4|$. **(c)** $|\mathbf{A}_5|$.

Conjecture what happens to $|\mathbf{A}_n|$ as n grows larger and larger.

36. Refer to Problem 35 and let

$$\mathbf{A} = \mathbf{A}_5, \quad \mathbf{B} = \begin{bmatrix} 1 \\ 1 \\ 1 \\ 1 \\ 1 \end{bmatrix}, \quad \text{and} \quad \mathbf{C} = \begin{bmatrix} 1 \\ 1 \\ 1.01 \\ 1 \\ 1 \end{bmatrix}$$

Calculate:

(a) $\mathbf{A}^{-1}$. **(b)** $\mathbf{A}^{-1}\mathbf{B}$. **(c)** $\mathbf{A}^{-1}\mathbf{C}$.

What difficulty do you see in trying to solve a system of linear equations in which the coefficient matrix is a Hilbert matrix?

11-7 SYSTEMS OF INEQUALITIES

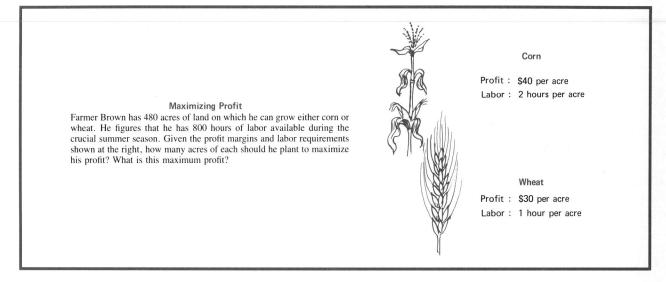

Maximizing Profit

Farmer Brown has 480 acres of land on which he can grow either corn or wheat. He figures that he has 800 hours of labor available during the crucial summer season. Given the profit margins and labor requirements shown at the right, how many acres of each should he plant to maximize his profit? What is this maximum profit?

Corn

Profit : $40 per acre

Labor : 2 hours per acre

Wheat

Profit : $30 per acre

Labor : 1 hour per acre

At first glance you might think that Farmer Brown should put all of his land into corn. Unfortunately, however, that requires 960 hours of labor and he has only 800 available. Well, maybe he should plant 400 acres of corn, using his allocated 800 hours of labor on them, and let the remaining 80 acres lie idle. Or would it be wise to at least plant enough wheat so all his land is in use? This problem is complicated enough so that no one is likely to find the best solution without a lot of work. And would not a method be better than blind experimenting? That is our subject—a method for handling Farmer Brown's problem and others of the same type.

Like all individuals and businesses, Farmer Brown must operate within certain limitations; we call them **constraints.** Suppose that he plants x acres of corn and y acres of wheat. His constraints can be translated into inequalities.

$$\text{Land constraint:} \qquad x + y \le 480$$
$$\text{Labor constraint:} \qquad 2x + y \le 800$$
$$\text{Nonnegativity constraints:} \qquad x \ge 0 \qquad y \ge 0$$

His task is to maximize the profit $P = 40x + 30y$ subject to these constraints. Before we can solve his problem, we will need to know more about inequalities.

The Graph of a Linear Inequality

The best way to visualize an inequality is by means of its graph. Consider, for example,

$$2x + y \le 6$$

which can be rewritten as

$$y \le -2x + 6$$

The complete graph consists of those points which satisfy $y = -2x + 6$ (a line), together with those that satisfy $y < -2x + 6$ (the points below the line). To see that this description is correct, note that for any abscissa x_1, the point $(x_1, -2x_1 + 6)$ is on the line $y = -2x + 6$. The point (x_1, y_1) is directly below that point if and only if $y_1 < -2x_1 + 6$ (Figure 7). Thus the graph of $y \le -2x + 6$ is the **closed half-plane** that we have shaded on the diagram. We refer to it as *closed* because the edge $y = -2x + 6$ is included. Correspondingly, the graph of $y < -2x + 6$ is called an **open half-plane.**

The graph of any linear inequality in x and y is a half-plane, open or closed. To sketch the graph, first draw the corresponding edge. Then determine the |correct half-plane by taking a sample point, not on the edge, and checking to see if it satisfies the inequality.

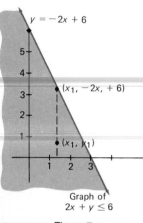

$y = -2x + 6$

$(x_1, -2x, + 6)$

(x_1, y_1)

Graph of
$2x + y \le 6$

Figure 7

Example A (A Single Inequality) Sketch the graph of $2x - 3y < -6$.

Solution The graph does not include the line $2x - 3y = -6$, although that line is crucial in determining the graph. We therefore show it as a dashed line. Since the sample point $(0, 0)$ does not satisfy the inequality, we choose the half-plane on the opposite side of the line from it. The complete graph is the shaded open half-plane shown in Figure 8. ■

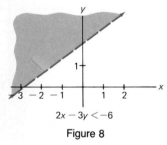

$2x - 3y < -6$

Figure 8

Graphing a System of Linear Inequalities

The graph of a system of inequalities like Farmer Brown's constraints

$$x + y \le 480$$
$$2x + y \le 800$$
$$x \ge 0 \qquad y \ge 0$$

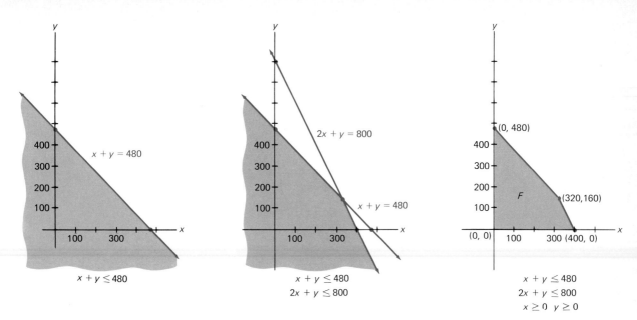

$x + y \leq 480$

$x + y \leq 480$
$2x + y \leq 800$

$x + y \leq 480$
$2x + y \leq 800$
$x \geq 0 \quad y \geq 0$

Figure 9

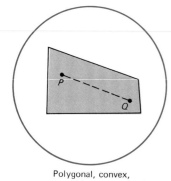

Polygonal, convex,
and bounded

Figure 10

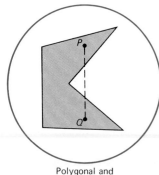

Polygonal and
bounded but
not convex

Figure 11

is simply the intersection of the graphs of the individual inequalities. We can construct the graph in stages as we do in Figure 9, though we are confident that you will quickly learn to do it in one operation.

The diagram on the right of Figure 9 is the one we want. All the points in the shaded region F satisfy the four inequalities simultaneously. The points $(0, 0)$, $(400, 0)$, $(320, 160)$, and $(0, 480)$ are called the **vertices** (or corner points) of F. Incidentally, the point $(320, 160)$ was obtained by solving the two equations $2x + y = 800$ and $x + y = 480$ simultaneously.

The region F has three important properties (Figure 10).

1. It is polygonal (its boundary consists of line segments).
2. It is convex (if points P and Q are in the region, then the line segment PQ lies entirely within the region).
3. It is bounded (it can be enclosed in a circle).

As a matter of fact, every region that arises as the solution set of a system of linear inequalities is polygonal and convex, though it need not be bounded. The shaded region in Figure 11 could not be the solution set for a system of linear inequalities because it is not convex.

Example B (Systems of Inequalities)   Sketch the solution set for the system of inequalities

$$2x - y \geq -1$$
$$x + y \leq 7$$
$$3x + y \leq 15$$
$$x + 5y \geq 5$$

Solution   The solution set, with vertices labeled, is shown in Figure 12.   ■

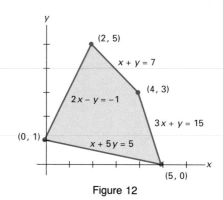

Figure 12

Linear Programming Problems

It is time that we solved Farmer Brown's problem.
Maximize

$$P = 40x + 30y$$

subject to

$$x + y \leq 480$$
$$2x + y \leq 800$$
$$x \geq 0 \qquad y \geq 0$$

Any problem that asks us to find the maximum (or minimum) of a linear function subject to linear inequality constraints is called a **linear programming problem.** Here is a method for solving such problems (in the bounded case).

1. Graph the solution set corresponding to the inequality constraints.
2. Find the coordinates of the vertices of the solution set.
3. Evaluate the linear function that you want to maximize (or minimize) at each of these vertices. The largest of these gives the maximum, while the smallest gives the minimum.

To see why this method works, consider the diagram for Farmer Brown's problem (Figure 13). The dashed lines are profit lines, each with slope $-\frac{4}{3}$; they are the graphs of $40x + 30y = P$ for various values of P. All the points on a dotted line give the same total profit. Imagine a profit line moving from left to right across the shaded region with the slope constant, as indicated in Figure 13. During this motion, the profit increases from zero at $(0, 0)$ to its maximum of 17,600 at $(320, 160)$. It should be clear that such a moving line (no matter what its slope) will always enter the shaded set at a vertex and leave it at another vertex. The particular vertex depends upon the slope of the profit lines. In Farmer Brown's problem, the minimum profit of $0 occurs at $(0, 0)$; the maximum profit of $17,600 occurs at $(320, 160)$. In

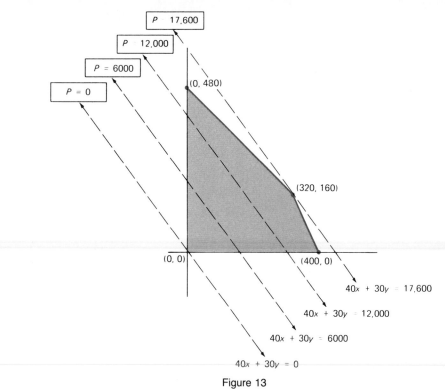

Vertex	$P = 40x + 30y$
(0, 0)	0
(0, 480)	14,400
(320, 160)	17,600
(400, 0)	16,000

Figure 14

Figure 13

Vertex	$P = 80x + 30y$
(0, 0)	0
(0, 480)	14,400
(320, 160)	30,400
(400, 0)	32,000

Figure 15

Vertex	$P = 40x + 40y$
(0, 0)	0
(0, 480)	19,200
(320, 160)	19,200
(400, 0)	16,000

Figure 16

Vertex	$P = 2x + 3y$
(0, 1)	3
(2, 5)	19
(4, 3)	17
(5, 0)	10

Figure 17

Figure 14, we show the total profit for each of the four vertices. Clearly, Farmer Brown should plant 320 acres of corn and 160 acres of wheat.

Now suppose the price of corn goes up so that Farmer Brown can expect a profit of $80 per acre on corn but still only $30 per acre on wheat. Would this change his strategy? In Figure 15, we show his total profit $P = 80x + 30y$ at each of the four vertices. Evidently, he should plant 400 acres of corn and no wheat to achieve maximum profit. Note that this means he should leave 80 acres of his land idle.

Finally, suppose that the profit per acre is $40 both for wheat and for corn. The table for this case (Figure 16) shows the same total profit at the vertices (0, 480) and (320, 160). This means that the moving profit line leaves the shaded region along the side determined by those two vertices, so that every point on that side gives a maximum profit. It is still true, however, that the maximum profit occurs at a vertex.

Example C (A Simple Linear Programming Problem)   Find the maximum and minimum values of $P = 2x + 3y$ subject to the constraints of Example B.

Solution   The coordinates of the vertices of the solution set are shown in Figure 12 and allow us to make the table in Figure 17. The maximum value of P is 19 and the minimum value of P is 3.   ■

The situation with an unbounded constraint set is slightly more complicated, since there may not be a minimum or a maximum value. However, when they exist, they occur at vertices.

Example D (Unbounded Region)   Find the maximum and minimum values of the function $3x + 4y$ subject to the constraints

$$3x + 2y \geq 13$$
$$x + y \geq 5$$
$$x \geq 1 \qquad y \geq 0$$

Solution   We proceed to graph the solution set of our system, noting that the region must lie above the lines $3x + 2y = 13$ and $x + y = 5$ and to the right of the line $x = 1$. It is shown in Figure 18. Notice that the region is unbounded and has (5, 0), (3, 2), and (1, 5) as its vertices. The point (1, 4) at which the lines $x + y = 5$ and $x = 1$ intersect is not a vertex. It should be clear right away that $3x + 4y$ does not assume a maximum value in our region; its values can be made as large as we please by increasing x and y. To find the minimum value, we calculate $3x + 4y$ at the three vertices.

$$(5, 0){:}\quad 3 \cdot 5 + 4 \cdot 0 = 15$$
$$(3, 2){:}\quad 3 \cdot 3 + 4 \cdot 2 = 17$$
$$(1, 5){:}\quad 3 \cdot 1 + 4 \cdot 5 = 23$$

The minimum value of $3x + 4y$ is 15.   ■

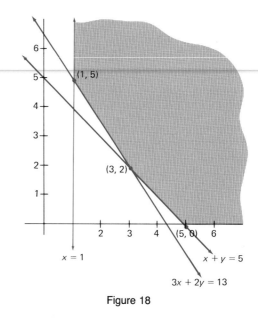

Figure 18

Nonlinear Programming

This title introduces a complicated subject but one of great importance. We offer only a very simple example.

Example E (Systems with Nonlinear Inequalities)   Graph the solution set of the following system of inequalities.

$$y \geq 2x^2$$
$$y \leq 2x + 4$$

then maximize $P = 3x + y$ on this set.

Solution   It helps to find the points at which the parabola $y = 2x^2$ intersects the line $y = 2x + 4$. Eliminating y between the two equations and then solving for x, we get

$$2x^2 = 2x + 4$$
$$x^2 - x - 2 = 0$$
$$(x - 2)(x + 1) = 0$$
$$x = 2 \qquad x = -1$$

The corresponding values of y are 8 and 2, respectively; so the points of intersection are $(-1, 2)$ and $(2, 8)$. Making use of these points, we draw the parabola and the line. Since the point $(0, 2)$ satisfies the inequality $y \geq 2x^2$, the desired region is above and including the parabola. The graph of the linear inequality $y \leq 2x + 4$ is to the right of and including the line. The shaded region in Figure 19 is the graph we want. Now imagine the line $3x + y = P$ (of slope -3) to move across the set from left to right. Clearly, P increases and takes its maximum value of 14 at the vertex $(2, 8)$.   ■

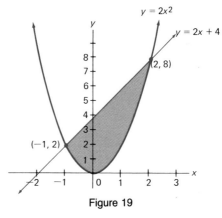

Figure 19

PROBLEM SET 11-7

A. Skills and Techniques

In Problems 1–6, graph the solution set of each inequality in the xy-plane, as in Example A.

1. $4x + y \leq 8$

2. $2x + 5y \leq 20$

3. $x \leq 3$

4. $y \leq -2$

5. $4x - y \geq 8$

6. $2x - 5y \geq -20$

In Problems 7–10, graph the solution set of the given system. On the graph, label the coordinates of the vertices. See Example B.

7. $4x + y \leq 8$
   $2x + 3y \leq 14$
   $x \geq 0 \quad y \geq 0$

8. $2x + 5y \leq 20$
   $4x + y \leq 22$
   $x \geq 0 \quad y \geq 0$

9. $4x + y \leq 8$
$x - y \leq -2$
$x \geq 0$

10. $2x + 5y \leq 20$
$x - 2y \geq 1$
$y \geq 0$

In Problems 11–14, find the maximum and minimum values of the given linear function P subject to the given inequalities, as in Example C.

11. $P = 2x + y$; the inequalities of Problem 7

12. $P = 3x + 2y$; the inequalities of Problem 8

13. $P = 2x - y$; the inequalities of Problem 9

14. $P = 3x - 2y$; the inequalities of Problem 10

Solve each of the following problems. See Example D.

15. Minimize $5x + 2y$ subject to

$$\begin{cases} x + y \geq 4 \\ x \geq 2 \\ y \geq 0 \end{cases}$$

16. Minimize $2x - y$ subject to

$$\begin{cases} x - 2y \geq 2 \\ y \geq 2 \end{cases}$$

17. Minimize $2x + y$ subject to

$$\begin{cases} 4x + y \geq 7 \\ 2x + 3y \geq 6 \\ x \geq 1 \\ y \geq 0 \end{cases}$$

18. Minimize $3x + 2y$ subject to

$$\begin{cases} x - 2y \leq 2 \\ x - 2y \geq -2 \\ 3x - 2y \geq 10 \end{cases}$$

In Problems 19–22, graph the solution set of each system of inequalities. See Example E.

19. $y \leq 4x - x^2$
$y \leq x$
$x \geq 0 \quad y \geq 0$

20. $y \geq 2^x$
$y \leq 8$
$x \geq 0$

21. $y \leq \log_2 x$
$x \leq 8$
$y \geq 0$

22. $x^2 + y^2 \geq 9$
$0 \leq x \leq 3$
$0 \leq y \leq 3$

B. Applications and Extensions

In Problems 23 and 24, graph the solution set and label the coordinates of the vertices.

23. $4x + y \leq 8$
$x - y \geq -2$
$x \geq 0 \quad y \geq 0$

24. $2x + 5y \leq 20$
$x - 2y \leq 1$
$x \geq 0$

25. Find the maximum and minimum values of $P = 2x - y$ subject to the inequalities of Problem 23.

26. Find the maximum and minimum values of $P = 3x - 2y$ subject to the inequalities of Problem 24.

27. Find the maximum and minimum values of $x + y$ (if they exist) subject to the following inequalities.

$$x - y \geq -1$$
$$x - 2y \leq 5$$
$$3x + y \leq 10$$
$$x \geq 0 \quad y \geq 0$$

28. Find the maximum and minimum values of $x - 2y$ (if they exist) subject to the inequalities of Problem 27.

29. Sketch the graph of the solution set for

$$x^2 + y^2 \geq 25$$
$$x^2 + (y - 7)^2 \leq 32$$

Then find the maximum and minimum values of $P = 2x + 2y$ on this set. *Hint:* What happens to P as the line $2x + 2y = P$ moves across the set from left to right? Where does this moving line first touch and last leave the set?

30. Rachel has received a lump sum of $100,000 on her retirement, which she will invest in safe AA bonds paying 12 percent annual interest and risky BBB bonds paying 15 percent annual interest. She plans to put at least $10,000 more in the safe bonds than the risky ones but wants at least two-thirds as much in BBB bonds as in AA bonds. How much should she invest in each type of bond to maximize her annual income while maintaining the constraints? *Suggestion:* To keep the numbers small, measure money in thousands of dollars.

31. A company makes a single product on two production lines, A and B. A labor force of 900 hours per week is available, and weekly running costs shall not exceed $1500. It takes 4 hours to produce one item on production line A and 3 hours on production line B. The cost per item is $5 on line A and $6 on line B. Find the largest number of items that can be produced in one week.

32. An oil refinery has a maximum production of 2000 barrels of oil per day. It produces two types of oil; type A, which is used for gasoline and type B, which is used for heating oil. There is a requirement that at least 300 barrels of type B be produced each day. If the profit is $3 a barrel for type A and $2 a barrel for type B, find the maximum profit per day.

33. A manufacturer of trailers wishes to determine how many camper units and how many house trailers she should produce in order to make the best possible use of her resources. She has 42 units of wood, 56 worker-weeks of labor and 16 units of aluminum. (Assume that all other needed resources are available and have no effect on her decision.) The amount of each resource needed to produce each camper and each trailer is given below.

	Wood	Worker-Weeks	Aluminum
Per camper	3	7	3
Per trailer	6	7	1

If the manufacturer realizes a profit of $600 on a camper and $800 on a trailer, what should be her production in order to maximize her profit?

34. A shoemaker has a supply of 100 square feet of type A leather which is used for soles and 600 square feet of type B leather used for the rest of the shoe. The average shoe uses $\frac{1}{4}$ square feet of type A leather and 1 square foot of type B leather. The average boot uses $\frac{1}{4}$ square feet and 3 square feet of types A and B leather, respectively. If shoes sell at $40 a pair and boots at $60 a pair, find the maximum income.

35. Suppose that the minimum monthly requirements for one person are 60 units of carbohydrates, 40 units of protein, and 35 units of fat. Two foods A and B contain the following numbers of units of the three diet components per pound.

	Carbohydrates	Protein	Fat
A	5	3	5
B	2	2	1

If food A costs $3.00 a pound and food B costs $1.40 a pound, how many pounds of each should a person purchase per month to minimize the cost?

36. A grain farmer has 100 acres available for sowing oats and wheat. The seed oats costs $5 per acre and the seed wheat costs $8 per acre. The labor costs are $20 per acre for oats and $12 per acre for wheat. The farmer expects an income from oats of $220 per acre and from wheat of $250 per acre. How many acres of each crop should he sow to maximize his profit, if he does not wish to spend more than $620 for seed and $1800 for labor?

37. Sketch the polygon with vertices (0, 3), (4, 7), (3, 0), and (2, 4) taken in cyclic order. Then, find the maximum value of $|y - 2x| + y + x$ on this polygon.

38. TEASER If $P = (a, b)$ is a point in the plane, then $tP = (ta, tb)$. Thus, if $0 \le t \le 1$, the set of points of the form $tP + (1 - t)Q$ is just the line segment PQ (see point-of-division formula, page 134). It follows that a set A is convex if whenever P and Q are in A, all points of the form $tP + (1 - t)Q$, $0 \le t \le 1$ are also in A.

(a) Let P, Q, and R be three fixed points in the plane and consider the set H of all points of the form $t_1 P + t_2 Q + t_3 R$, where the t's are nonnegative and $t_1 + t_2 + t_3 = 1$. Show that H is convex.

(b) Let P, Q, R, and S be four fixed points in the plane and consider the set K of all points of the form $t_1 P + t_2 Q + t_3 R + t_4 S$ where the t's are nonnegative and $t_1 + t_2 + t_3 + t_4 = 1$. Show that K is convex.

(c) Describe the sets H and K geometrically.

39. Draw the graph of the set $D = \{(x, y) : x^4 + y^4 \le 6561\}$. Then think of a way to use your calculator to maximize $P = 2x + y$ on this set. Find this maximum value accurate to two decimal places. *Hint:* First, convince yourself that the maximum value of P occurs at a boundary point of D in the first quadrant.

40. Maximize $P = .2x^3 + y^2$ on the set D of Problem 39. Again, you should be able to convince yourself that the maximum value of P occurs at a boundary point of D in the first quadrant.

CHAPTER 11 SUMMARY

Two systems of equations are **equivalent** if they have the same solutions. Three elementary operations (multiplying an equation by a nonzero constant, interchanging two equations, and adding a multiple of one equation to another) lead to equivalent systems. Use of these operations allows us to transform a system of linear equations to **triangular form** and then to solve the system by **back substitution** or to show it is **inconsistent.**

A **matrix** is a rectangular array of numbers. In solving a system of linear equations, it is efficient to work with just the **matrix of the system.** We solve the system by transforming its matrix to triangular form using the three operations mentioned above.

Addition and multiplication are defined for matrices with resulting algebraic rules. Matrices behave much like numbers, with the exception that the commutative law for multiplication fails. Even the notion of **multiplicative inverse** has meaning, though the process for finding such an inverse is lengthy.

Associated with every square matrix is a symbol called its **determinant.** For 2×2 and 3×3 matrices, the value of the determinant can be found by using certain arrow diagrams. For higher-order cases, the value of a determinant is found by expanding it in terms of the elements of a row (or column) and their **minors** (determinants whose order is lower than the given determinant by 1). In doing this, it is helpful to know what happens to the determinant of a matrix when any of the three elementary operations are applied to it. **Cramer's Rule** provides a direct way of solving a system of n equations in n unknowns using determinants.

The graph of a **linear inequality** in x and y is a **half-plane** (closed or open according as the inequality sign does or does not include the equal sign). The graph of a **system of linear inequalities** is the intersection of the half-planes corresponding to the separate inequalities. Such a graph is always **polygonal** and **convex** but may be **bounded** or **unbounded.** A **linear programming problem** asks us to find the maximum (or minimum) of a linear function (such as $2x + 5y$) subject to a system of linear inequalities called **constraints.** The maximum (or minimum) always occurs at a **vertex,** that is, at a corner point of the graph of the inequality constraints. Nonlinear maximization problems are much more complicated.

CHAPTER 11 REVIEW PROBLEM SET

In Problems 1–10, write True or False in the blank. If false, tell why.

_____ **1.** It is possible for a system of three equations in three unknowns to have no solution.

_____ **2.** It is possible for a system of three equations in three unknowns to have exactly three solutions.

_____ **3.** Matrix multiplication of compatible matrices is associative.

_____ **4.** Multiplying each row of a 2×2 matrix by 3 multiplies its determinant by 3.

_____ **5.** If two rows of a 3×3 matrix are multiples of each other, then the determinant of the matrix is zero.

_____ **6.** A square triangular matrix in which the main diagonal elements are nonzero has a nonzero determinant.

_____ **7.** Adding a multiple of one column to another column does not change the value of the determinant of a square matrix.

_____ **8.** Interchanging two rows of a square matrix does not change the value of its determinant.

_____ **9.** A linear function will always achieve a maximum value on a polygonal convex set.

_____ **10.** If $\mathbf{A}$ is an invertible matrix, then $|\mathbf{A}^{-1}| = |\mathbf{A}|^{-1}$.

In Problems 11–20, solve the given system or show that it has no solution.

11. $3x + 2y = 7$
$2x - y = 7$

12. $y = 3x - 2$
$6x - 2y = 5$

13. $\dfrac{3}{x} + \dfrac{1}{y} = 9$
$\dfrac{2}{x} - \dfrac{3}{y} = -5$

14. $x^2 + y^2 = 13$
$2x^2 - 3y^2 = -19$

15. $2x + y - 4z = 3$
$3y + z = 7$
$z = -2$

16. $2x - 3y + z = 4$
$3y - 5z = 6$
$-9y + 15z = 18$

17. $x - 2y + 4z = 16$
$2x - 3y - z = 4$
$x + 3y + 2z = 5$

18. $x - 3y + z = 6$
$2x + y - 2z = 5$
$-4x - 2y + 4z = -10$

19. $x^2 + y^2 = 13$
$2x^2 - 3y^2 = -14$

20. $5 \cdot 2^x - 3^y = 1$
$2^{x+2} + 3^{y+2} = 40$

In Problems 21–28, perform the indicated operations and express as a single matrix.

21. $\begin{bmatrix} 2 & -1 & 4 \\ 0 & 1 & 3 \end{bmatrix} + 2 \begin{bmatrix} 1 & 0 & 3 \\ 4 & -1 & 5 \end{bmatrix}$

22. $\begin{bmatrix} 5 & -4 \\ 3 & 1 \end{bmatrix} \begin{bmatrix} 2 & 5 \\ 3 & -1 \end{bmatrix}$

23. $4 \begin{bmatrix} 2 \\ -1 \\ 0 \end{bmatrix} - 3 \begin{bmatrix} 8 \\ -5 \\ 9 \end{bmatrix}$

24. $\begin{bmatrix} 2 & -3 \\ 4 & 1 \\ 0 & 5 \end{bmatrix} \begin{bmatrix} 4 & 0 & 1 \\ 2 & -3 & 6 \end{bmatrix}$

25. $\begin{bmatrix} 1 & 0 \\ 0 & 1 \end{bmatrix} \begin{bmatrix} 5 & 2 \\ -3 & 4 \end{bmatrix} \begin{bmatrix} b & 0 \\ 0 & b \end{bmatrix}$

26. $\begin{bmatrix} 6 & 11 \\ 0 & 0 \end{bmatrix} \begin{bmatrix} 11 & 0 \\ -6 & 0 \end{bmatrix}$

27. $\begin{bmatrix} 4 & 3 \\ 2 & 3 \end{bmatrix} \left\{ \begin{bmatrix} 4 & 3 \\ 2 & 3 \end{bmatrix}^{-1} + \begin{bmatrix} 1 & 0 \\ 0 & 1 \end{bmatrix} \right\}$

28. $\begin{bmatrix} 1 & -2 & 3 \\ 0 & 2 & 1 \\ 1 & -2 & 4 \end{bmatrix} \begin{bmatrix} 1 & 0 & 0 \\ 0 & \frac{1}{2} & 0 \\ 0 & 0 & \frac{1}{3} \end{bmatrix}^{-1}$

29. Find the inverse of the left matrix in Problem 28.

30. Write the following system in matrix form and then use the result of Problem 29 to solve it.

$$x - 2y + 3z = 2$$
$$2y + z = -4$$
$$x - 2y + 4z = 0$$

Evaluate the determinants in Problems 31–38.

31. $\begin{vmatrix} 3 & -4 \\ -5 & 6 \end{vmatrix}$

32. $\begin{vmatrix} 5 & -2 \\ -10 & 4 \end{vmatrix}^2$

33. $\begin{vmatrix} 3 & 0 & -2 \\ -3 & 1 & 0 \\ 1 & 5 & 4 \end{vmatrix}$

34. $\begin{vmatrix} 5 & 1 & 19 \\ 0 & -\frac{1}{2} & 1 \\ 0 & 0 & 4 \end{vmatrix}$

35. $\begin{vmatrix} 2 & -1 & 4 \\ 1 & 3 & -2 \\ 4 & 5 & 0 \end{vmatrix}$

36. $\begin{vmatrix} 2 & -1 & 4 \\ 1 & 3 & -2 \\ 4 & 5 & 8 \end{vmatrix}$

37. $\begin{vmatrix} 3 & 1 & 0 & 0 \\ 5 & 2 & 0 & 0 \\ 0 & 0 & 4 & -1 \\ 0 & 0 & -6 & 2 \end{vmatrix}$

38. $\begin{vmatrix} 3 & 0 & 2 & 0 & -1 \\ 0 & 2 & 0 & 3 & 2 \\ -3 & 3 & 1 & -3 & 0 \\ 0 & 0 & 3 & 0 & 1 \\ -1 & 1 & 0 & -1 & 0 \end{vmatrix}$

39. Clearly, $(0, 0, 0)$ is a solution to the system

$$3x - 2z = 0$$
$$-3x + y = 0$$
$$x + 5y + 4z = 0$$

Refer to your answer to Problem 33 and then tell how you can know immediately that this is the only solution.

40. Use your answer to Problem 36 and Cramer's Rule to solve for y in the system

$$2x - y + 4z = 0$$
$$x + 3y - 2z = -2$$
$$4x + 5y + 8z = 1$$

In Problems 41–44, graph the solution set for the given system of inequalities and label the coordinates of the vertices.

41. $x + y \leq 7$
$-3x + 4y \geq 0$
$x \geq 0 \qquad y \geq 0$

42. $2x + y \leq 8$
$x + y \leq 6$
$3x + 2y \geq 12$

43. $3x + y \leq 15$
$x + y \leq 7$
$x \geq 0 \qquad y \geq 0$

44. $x - 2y + 4 \geq 0$
$x + y - 11 \geq 0$
$x \geq 0 \qquad y \geq 0$

45. Find the maximum value of $P = 2x + y$ subject to the constraints of Problem 41.

46. Find the maximum value of $P = x + 2y$ subject to the constraints of Problem 42.

47. Find (if possible) the maximum and the minimum values of $P = x + 2y$ subject to the constraints of Problem 43.

48. Find (if possible) the maximum and the minimum values of $P = 2x + 3y$ subject to the constraints of Problem 44.

49. A metallurgist has three alloys containing 30 percent, 40 percent, and 60 percent nickel, respectively. How much of each should be melted and combined to obtain 100 grams of an alloy containing 45 percent nickel, provided that 10 grams more of the 60 percent alloy is used than the 30 percent alloy?

50. A manufacturer of personal computers makes $100 on model A and $120 on model B. Daily production for model A can vary between 50 and 110 and for model B between 75 and 100 but total daily production cannot exceed 180 units. How many units of each model should be produced each day to maximize profit?

CHAPTER **12**

Sequences, Counting Problems, and Probability

■ Method consists entirely in properly ordering and arranging things to which we should pay attention.

René Descartes

12-1 NUMBER SEQUENCES

Try filling in the boxes of our opening display. You will have little trouble with a, b, and c, but d and e may offer quite a challenge. We will give the answers we had in mind later. Right now, we merely point out that each sequence has a pattern; we used a definite rule in writing the first six terms of each of them.

The word *sequence* is commonly used in ordinary language. For example, your history teacher may talk about a sequence of events that led to World War II (for instance, the Versailles Treaty, world depression, Hitler's ascendancy, Munich Agreement). What characterizes this sequence is the notion of one event following another in a definite order. There is a first event, a second event, a third event, and so on. We might even give them labels.

E_1: Versailles Treaty

E_2: World depression

E_3: Hitler's ascendancy

E_4: Munich Agreement

We use a similar notation for number sequences. Thus

$$a_1, a_2, a_3, a_4, \ldots$$

could denote sequence (a) of our opening display. Then

$$a_1 = 1$$
$$a_2 = 4$$
$$a_3 = 7$$
$$a_4 = 10$$
$$\vdots$$

Note that a_3 stands for the 3rd term; a_{10} would represent the tenth term. The subscript indicates the position of the term in the sequence. For the general term, that is, the nth term, we use the symbol a_n. The three dots indicate that the sequence continues indefinitely.

There is another way to describe a number sequence. A **number sequence** is a function whose domain is the set of positive integers. That means it is a rule that associates with each positive integer n a definite number a_n. In conformity with Chapter 5, we could use the notation $a(n)$, but tradition dictates that we use a_n instead. We usually specify functions by giving formulas; this is true of sequences also.

Explicit Formulas

Rather than give the first few terms of a sequence and hope that our readers see the pattern intended (different people sometimes see different patterns in the first few terms of a sequence), it is better to give a formula.

Example A (Using an Explicit Formula)   Find a_5 and a_{19} for the sequence having $a_n = 2^n + n^2$ as its explicit formula.

Solution   We simply replace n by 5 and then by 19.

$$a_5 = 2^5 + 5^2 = 32 + 25 = 57$$

$$a_{19} = 2^{19} + 19^2 = 524{,}288 + 361 = 524{,}649 \quad \blacksquare$$

Seeing how easy it is to use an explicit formula, we ask whether it is possible to find such a formula for each of the sequences in the display at the beginning of this section.

Example B (Finding Explicit Formulas)   Give an explicit formula for sequences (a), (b), and (c) of our opening display.

Solution

(a) Let a_n denote the nth term of sequence (a). Then $a_1 = 1$, $a_2 = 4$, $a_3 = 7$, $a_4 = 10$, and so on. Our job is to relate the value of the nth term to its subscript. After pondering the terms for awhile, we may observe that the value of each term is 2 less than three times its subscript, that is,

$$a_n = 3n - 2$$

As a check, we note that this formula gives $a_5 = 3 \cdot 5 - 2 = 13$ and $a_6 = 3 \cdot 6 - 2 = 16$, as desired.

(b) Let b_n stand for the nth term of sequence (b). Then $b_1 = 2$, $b_2 = 4$, $b_3 = 6$, and it appears that the value of the nth term is just twice its subscript, that is,

$$b_n = 2n$$

(c) Since $c_1 = 1 = 1^2$, $c_2 = 4 = 2^2$, $c_3 = 9 = 3^2$, we infer that

$$c_n = n^2 \quad \blacksquare$$

Things get harder when we consider sequences (d) and (e) of our opening display. Look at sequence (d).

$$1, 4, 9, 16, 27, 40, \ldots$$

The fact that it starts out just like sequence (c) suggests that the pattern is subtle and incidentally warns us that we may have to look at many terms of a sequence before we can discover its rule of construction. Here, as in many sequences, it is a good idea to observe how each term relates to the previous one. Let us write the sequence again, indicating below it the numbers to be added as we progress from term to term.

$$
\begin{array}{ccccccc}
1 & & 4 & & 9 & & 16 & & 27 & & 40 \\
& 3 & & 5 & & 7 & & 11 & & 13 &
\end{array}
$$

You may recognize the second row of numbers as consecutive primes (starting with 3). Thus the next two terms in sequence (d) are

$$40 + 17 = 57$$

$$57 + 19 = 76$$

But observing a pattern does not necessarily mean we can write an explicit formula. Though many have tried, no one has found a simple formula for the nth prime and, thus, no one is likely to find a simple formula for sequence (d).

Sequence (e) is a famous one. It was introduced by Leonardo Fibonacci around A.D. 1200 in connection with rabbit reproduction (see Problem 39). If you are a keen observer, you have noticed that any term (after the second) is the sum of the preceding two. It was not until 1724 that mathematician Daniel Bernoulli found the explicit formula for this sequence. You will agree that it is complicated, but at least you can check it for $n = 1, 2, 3$.

$$e_n = \frac{1}{\sqrt{5}}\left[\left(\frac{1 + \sqrt{5}}{2}\right)^n - \left(\frac{1 - \sqrt{5}}{2}\right)^n\right]$$

If it took 500 years to discover this formula, you should not be surprised when we say that explicit formulas are often difficult to find (the problem set will give more evidence). There is another type of formula that is usually easier to discover.

Recursion Formulas

An explicit formula relates the value of a_n to its subscript n (for example, $a_n = 3n - 2$). Often the pattern we first observe relates a term to the preceding term (or terms). If so, we may be able to describe this pattern by a **recursion formula**. Look at sequence (a) again. To get a term from the preceding one, we always add 3, that is,

$$a_n = a_{n-1} + 3$$

Or look at sequence (b). There we add 2 each time.

$$b_n = b_{n-1} + 2$$

Sequence (e) is more interesting. There we add together the two previous terms, that is,

$$e_n = e_{n-1} + e_{n-2}$$

We summarize our knowledge about the five sequences in Table 2.

Table 2

Sequence	Explicit Formula	Recursion Formula
(a) 1, 4, 7, 10, 13, 16, . . .	$a_n = 3n - 2$	$a_n = a_{n-1} + 3$
(b) 2, 4, 6, 8, 10, 12, . . .	$b_n = 2n$	$b_n = b_{n-1} + 2$
(c) 1, 4, 9, 16, 25, 36, . . .	$c_n = n^2$	$c_n = c_{n-1} + 2n - 1$
(d) 1, 4, 9, 16, 27, 40, . . .	?	$d_n = d_{n-1} + n\text{th prime}$
(e) 1, 1, 2, 3, 5, 8, . . .	$e_n = \dfrac{1}{\sqrt{5}}\left[\left(\dfrac{1 + \sqrt{5}}{2}\right)^n - \left(\dfrac{1 - \sqrt{5}}{2}\right)^n\right]$	$e_n = e_{n-1} + e_{n-2}$

Recursion formulas are themselves not quite enough to determine a sequence. For example, the recursion formula

$$f_n = 3f_{n-1}$$

does not determine a sequence until we specify the first term. But with the additional information that $f_1 = 2$, we can find any term. Thus

$$f_1 = 2$$
$$f_2 = 3f_1 = 3 \cdot 2 = 6$$
$$f_3 = 3f_2 = 3 \cdot 6 = 18$$
$$f_4 = 3f_3 = 3 \cdot 18 = 54$$
$$\vdots$$

The disadvantage of a recursion formula is apparent. To find the 100th term, we must first calculate the 99 previous terms. It may be hard work but it is always possible. Programmable calculators are particularly adept at calculating sequences by means of recursion formulas.

Example C (Using a Recursion Formula) Suppose that $a_n = \frac{1}{2}a_{n-1} + 3$ and $a_1 = -4$. Find a_5.

Solution $a_2 = \frac{1}{2}(-4) + 3 = 1$, $a_3 = \frac{1}{2}(1) + 3 = \frac{7}{2}$, $a_4 = \frac{1}{2}\left(\frac{7}{2}\right) + 3 = \frac{19}{4}$, $a_5 = \frac{1}{2}\left(\frac{19}{4}\right) + 3 = \frac{43}{8}$. ∎

Sum Sequences and Sigma Notation

Corresponding to a sequence $a_1, a_2, a_3, \ldots$, we introduce another sequence A_n, called the **sum sequence,** by

$$A_n = a_1 + a_2 + a_3 + \cdots + a_n$$

Thus

$$A_1 = a_1$$
$$A_2 = a_1 + a_2$$
$$A_3 = a_1 + a_2 + a_3$$
$$\vdots$$

Example D (A Sum Sequence)   Find A_5 for the sequence a_n with explicit formula $a_n = 2^n - 3n$.

Solution   First, we must find $a_1, a_2, \ldots, a_5$.

$$a_1 = 2 - 3 = -1$$
$$a_2 = 2^2 - 3 \cdot 2 = -2$$
$$a_3 = 2^3 - 3 \cdot 3 = -1$$
$$a_4 = 2^4 - 3 \cdot 4 = 4$$
$$a_5 = 2^5 - 3 \cdot 5 = 17$$

Thus

$$A_5 = -1 + (-2) + (-1) + 4 + 17 = 17 \quad \blacksquare$$

There is a convenient shorthand that is frequently employed in connection with sums. The first letter of the word *sum* is *s*; the Greek letter for *S* is Σ (sigma). We use Σ in mathematics to stand for the operation of summation. In particular,

$$\sum_{i=1}^{n} a_i = a_1 + a_2 + a_3 + \cdots + a_n$$

The symbol $i = 1$ underneath the sigma tells where to start adding the terms a_i and then n at the top tells where to stop. Thus,

$$\sum_{i=1}^{4} a_i = a_1 + a_2 + a_3 + a_4$$

$$\sum_{i=3}^{7} b_i = b_3 + b_4 + b_5 + b_6 + b_7$$

$$\sum_{i=1}^{5} i^2 = 1^2 + 2^2 + 3^2 + 4^2 + 5^2$$

Example E (Sigma Notation)   Write each sum in sigma notation.
(a) $3 + 6 + 9 + \cdots + 90$
(b) $2^4 + 3^4 + 4^4 + \cdots + n^4$

Solution
(a) $3 + 6 + 9 + \cdots + 90 = \sum_{i=1}^{30} 3i$
(b) $2^4 + 3^4 + 4^4 + \cdots + n^4 = \sum_{i=2}^{n} i^4$ $\quad \blacksquare$

PROBLEM SET 12-1

A. Skills and Techniques

1. Discover a pattern and use it to fill in the boxes.
   (a) $1, 3, 5, 7, \square, \square, \ldots$
   (b) $17, 14, 11, 8, \square, \square, \ldots$
   (c) $1, \frac{1}{2}, \frac{1}{4}, \frac{1}{8}, \square, \square, \ldots$
   (d) $1, 9, 25, 49, \square, \square, \ldots$

2. Fill in the boxes.
   (a) $1, 3, 9, 27, \square, \square, \ldots$
   (b) $2, 2.5, 3, 3.5, \square, \square, \ldots$
   (c) $1, 8, 27, 64, \square, \square, \ldots$
   (d) $\frac{1}{2}, \frac{2}{3}, \frac{3}{4}, \frac{4}{5}, \square, \square, \ldots$

3. In each case an explicit formula is given. Find the indicated terms (see Example A).
   (a) $a_n = 2n + 3$; $a_4 = \square$; $a_{20} = \square$
   (b) $a_n = \dfrac{n}{n + 1}$; $a_5 = \square$; $a_9 = \square$
   (c) $a_n = (2n - 1)^2$; $a_4 = \square$; $a_5 = \square$
   (d) $a_n = (-3)^n$; $a_3 = \square$; $a_4 = \square$

4. Find the indicated terms.
   (a) $a_n = 2n - 5$; $a_4 = \square$; $a_{20} = \square$
   (b) $a_n = 1/n$; $a_5 = \square$; $a_{50} = \square$
   (c) $a_n = (2n)^2$; $a_5 = \square$; $a_{10} = \square$
   (d) $a_n = 4 - \frac{1}{2}n$; $a_5 = \square$; $a_{10} = \square$

5. Give an explicit formula for each sequence in Problem 1. Recall that you must relate the value of a term to its subscript (see Example B).

6. Give an explicit formula for each sequence in Problem 2.

7. In each case, an initial term and a recursion formula are given. Find a_5. *Hint:* First find $a_2, a_3,$ and a_4, and then find a_5 (see Example C).
   (a) $a_1 = 2$; $a_n = a_{n-1} + 3$
   (b) $a_1 = 2$; $a_n = 3a_{n-1}$
   (c) $a_1 = 8$; $a_n = \frac{1}{2}a_{n-1}$
   (d) $a_1 = 1$; $a_n = a_{n-1} + 8(n - 1)$

8. Find a_4 for each of the following sequences.
   (a) $a_1 = 2$; $a_n = 2a_{n-1} + 1$
   (b) $a_1 = 2$; $a_n = a_{n-1} + 3$
   (c) $a_1 = 1$; $a_n = a_{n-1} + 3n^2 - 3n + 1$
   (d) $a_1 = 3$; $a_n = a_{n-1} + .5$

9. Try to find a recursion formula for each of the sequences in Problem 1.

10. Try to find a recursion formula for each of the sequences in Problem 2.

In each of the following problems, find A_6. See Example D.

11. $a_n = 2n + 1$

12. $a_n = 2^n$

13. $a_n = (-2)^n$

14. $a_n = n^2$

15. $a_n = n^2 - 2$

16. $a_n = 3n - 4$

17. $a_1 = 4$; $a_n = a_{n-1} + 3$

18. $a_1 = 1$; $a_2 = 1$; $a_n = a_{n-1} + 2a_{n-2}$

Problems 19–22 relate to Example E.

19. Write in sigma notation.
   (a) $b_3 + b_4 + \cdots + b_{20}$
   (b) $1^2 + 2^2 + \cdots + 19^2$
   (c) $1 + \dfrac{1}{2} + \dfrac{1}{3} + \cdots + \dfrac{1}{n}$

20. Write in sigma notation.
   (a) $a_6 + a_7 + a_8 + \cdots + a_{70}$
   (b) $2^3 + 3^3 + 4^3 + \cdots + 100^3$
   (c) $1 + 3 + 5 + 7 + \cdots + 99$

21. Calculate each sum.
   (a) $\sum\limits_{i=1}^{5} (i^2 - 1)$
   (b) $\sum\limits_{i=1}^{5} (2^i - i^2)$
   (c) $\sum\limits_{i=2}^{5} \dfrac{1}{i - 1}$
   (d) $\sum\limits_{i=2}^{7} (2i + 5)$

22. Calculate each sum.
   (a) $\sum\limits_{k=1}^{4} (k^3 - 8)$
   (b) $\sum\limits_{k=1}^{4} (k - 1)^3$
   (c) $\sum\limits_{k=1}^{6} k^2$
   (d) $\sum\limits_{k=1}^{10} (3k - 6)$

B. Applications and Extensions

23. Find a_5 and a_{16} for each sequence.
   (a) $a_n = 5n + 2$
   (b) $a_n = n(n - 1)$

24. Find a_5 in each case.
   (a) $a_1 = 9$, $a_n = \frac{2}{3}a_{n-1}$
   (b) $a_1 = 6$, $a_2 = 2$, $a_n = (a_{n-1} + a_{n-2})/2$

25. Let $a_1 = 2$, $a_n = 3a_{n-1} - 1$; $b_n = n^2 - 3n$; $c_n = a_n - b_n$. Evaluate $c_1, c_2, c_3, c_4,$ and c_5.

26. Find a pattern in each of the following sequences and use it to fill in the boxes.
   (a) $2, 6, 18, 54, \square, \square, \ldots$
   (b) $2, 6, 10, 14, \square, \square, \ldots$
   (c) $2, 4, 8, 14, \square, \square, \ldots$
   (d) $2, 4, 6, 10, 16, \square, \square, \ldots$
   (e) $2, 1, \frac{1}{2}, \frac{1}{4}, \square, \square, \ldots$
   (f) $2, 5, 10, 17, \square, \square, \ldots$

27. Find a recursion formula for each sequence in Problem 26.

28. Find (if you can) an explicit formula for each sequence in Problem 26.

29. Let a_n be the nth digit in the decimal expansion of $\frac{1}{7} = .1428 \ldots$. Thus $a_1 = 1$, $a_2 = 4$, $a_3 = 2$, and so on. Find a pattern and use it to determine a_8, a_{27}, and a_{53}.

30. Suppose that January 1 occurs on Wednesday. Let a_n be the day of the week corresponding to the nth day of the year. Thus $a_1 = $ Wednesday, $a_2 = $ Thursday, and so on. Find a_{39}, a_{57}, and a_{84}.

31. Let $a_n = 2n - 1$ and $A_n = a_1 + a_2 + a_3 + \cdots + a_n$.

   (a) Calculate a_1, a_2, a_3, a_4, and a_5.
   (b) Calculate A_1, A_2, A_3, A_4, and A_5.
   (c) Find an explicit formula for A_n.

32. Follow the directions of Problem 31 for

$$a_n = \frac{1}{n(n+1)} = \frac{1}{n} - \frac{1}{n+1}$$

33. Write in sigma notation.

   (a) $c_3 + c_4 + c_5 + \cdots + c_{112}$
   (b) $4^2 + 5^2 + 6^2 + \cdots + 104^2$
   (c) $12 + 18 + 24 + \cdots + 60$

34. Calculate each sum.

   (a) $\displaystyle\sum_{i=1}^{10} (2i + 4)$ (b) $\displaystyle\sum_{i=1}^{100} \left(\frac{i}{i+1} - \frac{i-1}{i} \right)$

35. Oranges are piled in a pyramid with a square base of $15^2 = 225$ oranges rising to a peak consisting of one orange. Write a sum in sigma notation for the total number of oranges in the pile.

36. Repeat Problem 35 but assume that the bottom tier is a rectangle with $12 \times 16 = 192$ oranges and the top tier has $2 \times 6 = 12$ oranges.

37. The Greeks were enchanted with sequences that arose in a geometric way (Figures 1 and 2).

   (a) Find an explicit formula for s_n.
   (b) Find a recursion formula for s_n.
   (c) Find a recursion formula for t_n.
   (d) By considering a square of dots, $n + 1$ on a side, express s_{n+1} in terms of t_n.
   (e) Use (d) to find an explicit formula for t_n.

38. The numbers 1, 5, 12, 22, . . . are called *pentagonal numbers*. See if you can figure out why and then guess at an explicit formula for p_n, the nth pentagonal number. Use diagrams.

39. Following Leonardo Fibonacci, suppose that a pair of rabbits consisting of a male and a female begin to bear young when they are 2 months old, and that thereafter they produce one male-female pair at 1-month intervals forever. Assume that each new male-female pair of rabbits has the same reproductive habits as its parents and that no rabbits die. Beginning with a single pair born on the first day of a month, how many rabbit pairs will there be on the last day of that month? the last day of the second month? Third month? Fourth month? Fifth month? nth month?

40. Let f_n denote the Fibonacci sequence determined by $f_1 = 1$, $f_2 = 1$, and $f_n = f_{n-1} + f_{n-2}$. Also let $F_n = f_1 + f_2 + \cdots + f_n$.

   (a) Calculate F_1, F_2, F_3, F_4, F_5, and F_6.
   (b) Based on (a), find a nice formula connecting F_n to f_{n+2}.

41. By considering adjoining geometric squares as in Figure 3, obtain a nice formula for $f_1^2 + f_2^2 + f_3^2 + \cdots + f_n^2$.

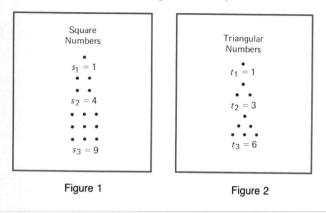

Figure 1   Figure 2

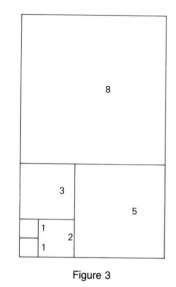

Figure 3

42. TEASER Consider the matrix

$$Q = \begin{bmatrix} 1 & 1 \\ 1 & 0 \end{bmatrix}$$

(a) Find Q^2, Q^3, and Q^4.
(b) Express the matrix Q^n in terms of the Fibonacci numbers f_n.

(c) Use the fact that $|Q^n| = |Q|^n$ to obtain a nice formula for $f_{n+1}f_{n-1} - f_n^2$.
(d) From $Q^{2n} = Q^n Q^n$, obtain a nice formula for f_{2n}.
(e) Use the result in (d) to show that the difference in the squares of alternate Fibonacci numbers is a Fibonacci number.

12-2 ARITHMETIC SEQUENCES

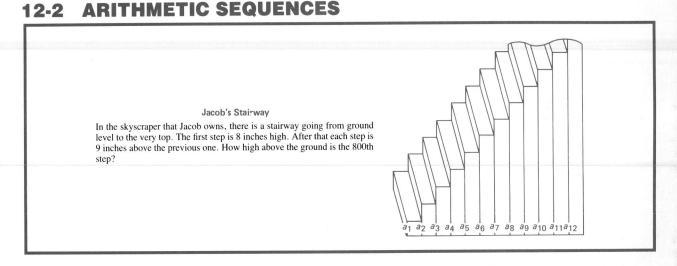

Jacob's Stairway

In the skyscraper that Jacob owns, there is a stairway going from ground level to the very top. The first step is 8 inches high. After that each step is 9 inches above the previous one. How high above the ground is the 800th step?

We are going to answer the question above and others like it. Notice that if a_n denotes the height of the nth step, then

$$a_1 = 8$$
$$a_2 = 8 + 1(9) = 17$$
$$a_3 = 8 + 2(9) = 26$$
$$a_4 = 8 + 3(9) = 35$$
$$\vdots$$
$$a_{800} = 8 + 799(9) = 7199$$

The 800th step of Jacob's stairway is 7199 inches (almost 600 feet) above the ground.

Formulas

Now consider the following number sequences. When you see a pattern, fill in the boxes.

(a) 5, 9, 13, 17, ☐, ☐, . . .

(b) 2, 2.5, 3, 3.5, ☐, ☐, . . .

(c) 8, 5, 2, −1, ☐, ☐, . . .

What is it that these three sequences have in common? Simply this: In each case, you can get a term by adding a fixed number to the preceding term. In (a), you add 4 each time, in (b) you add 0.5, and in (c), you add −3. Such sequences are called **arithmetic sequences.** If we denote such a sequence by $a_1, a_2, a_3, \ldots$, it satisfies the recursion formula

$$a_n = a_{n-1} + d$$

where d is a fixed number called the **common difference.**

Can we also obtain an explicit formula? Yes. Figure 4 should help. Notice that the number of d's to be added to a_1 is one less than the subscript. This means that

$$a_n = a_1 + (n-1)d$$

Now we can give explicit formulas for each of the sequences (a), (b), and (c) given earlier.

$$a_n = 5 + (n - 1)4 = 1 + 4n$$

$$b_n = 2 + (n - 1)(.5) = 1.5 + .5n$$

$$c_n = 8 + (n - 1)(-3) = 11 - 3n$$

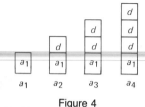

Figure 4

Example A (Two Arithmetic Sequences) Find d and the 41st term for each sequence.

(a) −12, −7, −2, 3, 8, . . .   (b) 4.5, 3.75, 3, 2.25, 1.5, . . .

Solution

(a) $d = 5$ and $a_{41} = -12 + (40)(5) = 188$

(b) $d = -0.75$ and $b_{41} = 4.5 + (40)(-0.75) = -25.5$ ■

Arithmetic Sequences and Linear Functions

We have said that a sequence is a function whose domain is the set of positive integers. Functions are best visualized by drawing their graphs. Consider the sequence b_n discussed above; its explicit formula is

$$b_n = 1.5 + .5n$$

Its graph is shown in Figure 5.

Even a cursory look at this graph suggests that the points lie along a straight line. Consider now the function

$$b(x) = 1.5 + .5x = .5x + 1.5$$

where x is allowed to be any real number. This is a linear function, being of the

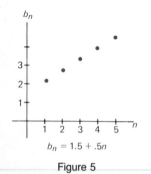

$b_n = 1.5 + .5n$

Figure 5

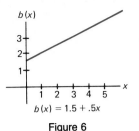

$b(x) = 1.5 + .5x$

Figure 6

form $mx + b$ (see Section 4-3). Its graph is a straight line (Figure 6), and its values at $x = 1, 2, 3, \ldots$ are equal to $b_1, b_2, b_3, \ldots$

The relationship illustrated above between an arithmetic sequence and a linear function holds in general. An arithmetic sequence is just a linear function whose domain has been restricted to the positive integers.

Sums of Arithmetic Sequences

There is an old story about Carl Gauss that aptly illustrates the next idea. We are not sure if the story is true, but if not, it should be.

When Gauss was about 10 years old, he was admitted to an arithmetic class. To keep the class busy, the teacher often assigned long addition problems. One day he asked his students to add the numbers from 1 to 100. Hardly had he made the assignment when young Gauss laid his slate on the teacher's desk with the answer 5050 written on it.

Here is how Gauss probably thought about the problem (Figure 7). Each of the indicated pairs has 101 as its sum and there are 50 such pairs. Thus the answer is $50(101) = 5050$. For a 10-year-old boy, that is good thinking.

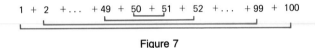

Figure 7

Gauss's trick works perfectly well on any arithmetic sequence where we want to add an even number of terms. And there is a slight modification that works whether the number of terms to be added is even or odd.

Suppose $a_1, a_2, a_3, \ldots$ is an arithmetic sequence and let

$$A_n = a_1 + a_2 + a_3 + \cdots + a_{n-1} + a_n$$

Write this sum twice, once forwards and once backwards, and then add.

$$
\begin{array}{rccccccc}
A_n = & a_1 & + & a_2 & + \cdots + & a_{n-1} & + & a_n \\
A_n = & a_n & + & a_{n-1} & + \cdots + & a_2 & + & a_1 \\
\hline
2A_n = & (a_1 + a_n) & + & (a_2 + a_{n-1}) & + \cdots + & (a_2 + a_{n-1}) & + & (a_1 + a_n)
\end{array}
$$

Each group on the right has the same sum, namely, $a_1 + a_n$. For example,

$$a_2 + a_{n-1} = a_1 + d + a_n - d = a_1 + a_n$$

There are n such groups and so

$$2A_n = n(a_1 + a_n)$$

$$A_n = \frac{n}{2}(a_1 + a_n)$$

We call this the **sum formula** for an arithmetic sequence. You can remember this sum as being n times the average term $(a_1 + a_n)/2$.

Here is how we apply this formula to Gauss's problem. We want the sum of 100 terms of the sequence 1, 2, 3, . . . , that is, we want A_{100}. Here $n = 100$, $a_1 = 1$, and $a_n = 100$. Therefore,

$$A_{100} = \frac{100}{2}(1 + 100) = 50(101) = 5050$$

Example B (Applying the Sum Formula)   Find the sum of the first 350 odd numbers.

Solution   We are to find the sum of the first 350 terms of the arithmetic sequence

$$1, 3, 5, 7, 9, . . .$$

We can calculate the 350th odd number from the formula $a_n = a_1 + (n - 1)d$.

$$a_{350} = 1 + (349)2 = 699$$

Then we use the sum formula with $n = 350$.

$$A_{350} = 1 + 3 + 5 + \cdots + 699$$

$$= \frac{350}{2}(1 + 699) = 122{,}500 \quad \blacksquare$$

Example C (An Arithmetic Sum)   Calculate

$$72 + 78 + 84 + 90 + \cdots + 300$$

Solution   We know that $a_1 = 72$ and $a_n = 300$, but we do not know n. Since $d = 6$, n must satisfy the equation

$$72 + (n - 1)6 = 300$$

Thus

$$n - 1 = \frac{300 - 72}{6} = 38$$

$$n = 39$$

The required sum is

$$A_{39} = \frac{39}{2}(72 + 300) = 7254 \quad \blacksquare$$

We can write the sum formula in sigma notation. If $a_1, a_2, . . . , a_n$ is an arithmetic sequence, then

$$\sum_{i=1}^{n} a_i = \frac{n}{2}(a_1 + a_n)$$

Example D (Sigma Notation)   Calculate

$$\sum_{i=1}^{100} (4i - 5)$$

Solution   The sum written out is

$$-1 + 3 + 7 + 11 + \cdots + 395$$

It has the value

$$\frac{100}{2}(-1 + 395) = 19{,}700 \quad \blacksquare$$

Example E (Total Income)   Maria has just started permanent employment at an annual salary of $40,000 and anticipates a raise of $3000 each year. Estimate her total income over a 30-year career.

Solution   Let s_n denote Maria's salary in the nth year of employment. Then

$$s_{30} = 40{,}000 + (29)(3000) = 127{,}000$$

and

$$S_{30} = \frac{30}{2}(40{,}000 + 127{,}000) = \$2{,}505{,}000 \quad \blacksquare$$

PROBLEM SET 12-2

A. Skills and Techniques

1. Fill in the boxes.
   (a) 1, 4, 7, 10, ☐, ☐, . . .
   (b) 2, 2.3, 2.6, 2.9, ☐, ☐, . . .
   (c) 28, 24, 20, 16, ☐, ☐, . . .

2. Fill in the boxes.
   (a) 4, 6, 8, 10, ☐, ☐, . . .
   (b) 4, 4.2, 4.4, 4.6, ☐, ☐, . . .
   (c) 4, 3.8, 3.6, 3.4, ☐,☐, . . .

3. Determine d and the 30th term of each sequence in Problem 1 (see Example A).

4. Determine d and the 101st term of each sequence in Problem 2.

5. Deterimine $A_{30} = a_1 + a_2 + \cdots + a_{30}$ for the sequence in part (a) of Problem 1. Similarly, find B_{30} and C_{30} for the sequences in parts (b) and (c) (see Example B).

6. Determine A_{100}, B_{100}, and C_{100} for the sequences of Problem 2.

7. If $a_1 = 5$ and $a_{40} = 24.5$ in an arithmetic sequence, determine d.

8. If $b_1 = 6$ and $b_{30} = -52$ in an arithmetic sequence, determine d.

9. Calculate each sum (see Example C).
   (a) $2 + 4 + 6 + \cdots + 200$
   (b) $1 + 3 + 5 + \cdots + 199$
   (c) $3 + 6 + 9 + \cdots + 198$

Hint: Before using the sum formula, you have to determine n. In part (a), n is 100 since we are adding the doubles of the integers from 1 to 100.

10. Calculate each sum.
   (a) $4 + 8 + 12 + \cdots + 100$
   (b) $10 + 15 + 20 + \cdots + 200$
   (c) $6 + 9 + 12 + \cdots + 72$

11. The bottom rung of a tapered ladder (Figure 8) is 30

Figure 8

centimeters long and the top rung is 15 centimeters long. If there are 17 rungs, how many centimeters of rung material are needed to make the ladder, assuming no waste?

12. A clock strikes once at 1:00, twice at 2:00, and so on. How many times does it strike between 10:30 A.M. on Monday and 10:30 P.M. on Tuesday?

13. If 3, a, b, c, d, 7, . . . is an arithmetic sequence, find a, b, c, and d.

14. If 8, a, b, c, 5 is an arithmetic sequence, find a, b, and c.

15. How many multiples of 9 are there between 200 and 300? Find their sum.

16. If Ronnie is paid $10 on January 1, $20 on January 2, $30 on January 3, and so on, how much does he earn during January?

17. Calculate each sum (see Example D).

(a) $\displaystyle\sum_{i=2}^{60} (i - 5)$ (b) $\displaystyle\sum_{i=1}^{40} (4 - 3i)$

(c) $\displaystyle\sum_{i=1}^{100} (3i + 2)$ (d) $\displaystyle\sum_{i=2}^{100} (2i - 3)$

18. Calculate each sum.

(a) $\displaystyle\sum_{i=1}^{60} (2i + 10)$ (b) $\displaystyle\sum_{i=1}^{50} (5 - 2i)$

(c) $\displaystyle\sum_{i=1}^{101} (2i - 6)$ (d) $\displaystyle\sum_{i=3}^{102} (3i + 5)$

19. Write in sigma notation.
(a) $3 \cdot 4 + 4 \cdot 5 + 5 \cdot 6 + \cdots + 99 \cdot 100$
(b) $8 + 12 + 16 + \cdots + 80$
(c) $7 + 10 + 13 + \cdots + 91$

20. Write in sigma notation.
(a) $9 + 16 + 25 + \cdots + 100$
(b) $14 + 16 + 18 + \cdots + 222$
(c) $1 + 3 + 5 + 7 + \cdots + 99$

B. Applications and Extensions

21. Find each of the following for the arithmetic sequence 20, 19.25, 18.5, 17.75, . . . :
(a) The common difference d.
(b) The 51st term.
(c) The sum of the first 51 terms.

22. Let a_n be an arithmetic sequence with $a_{19} = 42$ and $a_{39} = 54$. Find (a) d, (b) a_1, (c) a_{14}, and (d) k, given that $a_k = 64.2$.

23. Calculate the sum
$6 + 6.8 + 7.6 + \cdots + 37.2 + 38$.

24. If 15, a, b, c, d, 24, . . . is an arithmetic sequence, find a, b, c, and d.

25. Let a_n be an arithmetic sequence and, as usual, let $A_n = a_1 + a_2 + \cdots + a_n$. If $A_5 = 50$ and $A_{20} = 650$, find A_{15}.

26. Find the sum of all multiples of 7 between 300 and 450.

27. Calculate each sum.

(a) $\displaystyle\sum_{i=1}^{100} (2i + 1)$ (b) $\displaystyle\sum_{i=1}^{100} (-4i + 2)$

28. Write in sigma notation.
(a) $c_3 + c_4 + c_5 + \cdots + c_{112}$
(b) $9 + 6 + 3 + \cdots + (-90)$
(c) $35 + 40 + 45 + \cdots + 185$

29. At a club meeting with 300 people present, everyone shook hands with every other person exactly once. How many handshakes were there? *Hint:* Person A shook hands with how many people, person B shook hands with how many people not already counted, and so on.

30. Mary learned 20 new French words on January 1, 24 new French words on January 2, 28 new French words on January 3, and so on, through January 31. By how much did she increase her French vocabulary during January?

31. A pile of logs has 70 logs in the bottom layer, 69 logs in the second layer, and so on to the top layer with 10 logs. How many logs are in the pile?

32. Calculate the following sum.

$$-1^2 + 2^2 - 3^2 + 4^2 - 5^2 + 6^2 - \cdots$$
$$- 99^2 + 100^2$$

Hint: Group in a clever way.

33. Jose will invest $1000 today at 9.5 percent (annual) simple interest. How much will this investment be worth 10 years from now? *Note:* Under simple interest, only the original principal of $1000 draws interest.

34. Roberto plans to invest $1000 today and at the beginning of each of the succeeding 9 years. How much will his total investment be worth 10 years from now if his investments draw 9.5 percent simple interest?

35. To pay off a loan of $6000, Ikeda agreed to pay at the end of each month: interest corresponding to 1 percent of the unpaid balance of the principal and then $200 to reduce the principal. What was the total of all his payments?

36. Calculate $\sum_{k=1}^{100} \ln(2^k)$.

37. Calculate the following sum.

$$\frac{1}{2} + \left(\frac{1}{3} + \frac{2}{3}\right) + \left(\frac{1}{4} + \frac{2}{4} + \frac{3}{4}\right) + \cdots$$

$$+ \left(\frac{1}{100} + \frac{2}{100} + \frac{3}{100} + \cdots + \frac{99}{100}\right)$$

38. Show that the sum of n consecutive positive integers plus n^2 is equal to the sum of the next n consecutive integers.

39. Approximately how long is the playing groove in a $33\frac{1}{3}$-rpm record that takes 18 minutes to play if the groove starts 6 inches from the center and ends 3 inches from the center? To approximate, assume that each revolution produces a groove that is circular.

40. **TEASER** Find the sum of all the digits in the integers from 1 to 999,999.

12-3 GEOMETRIC SEQUENCES

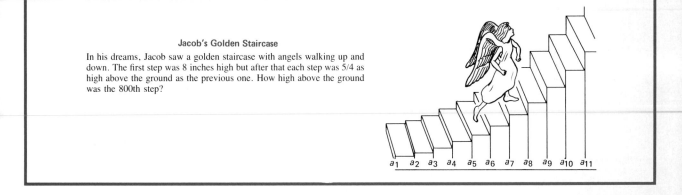

Jacob's Golden Staircase

In his dreams, Jacob saw a golden staircase with angels walking up and down. The first step was 8 inches high but after that each step was 5/4 as high above the ground as the previous one. How high above the ground was the 800th step?

The staircase of Jacob's dream is most certainly one for angels, not for people. The 800th step actually stands 3.4×10^{73} miles high. By way of comparison, it is 9.3×10^7 miles to the sun and 2.5×10^{13} miles to Alpha Centauri, our nearest star beyond the sun. You might say the golden staircase reaches to heaven.

To see how to calculate the height of the 800th step, notice the pattern of heights for the first few steps and then generalize.

$$a_1 = 8$$

$$a_2 = 8\left(\frac{5}{4}\right)$$

$$a_3 = 8\left(\frac{5}{4}\right)^2$$

$$a_4 = 8\left(\frac{5}{4}\right)^3$$

$$\vdots$$

$$a_{800} = 8\left(\frac{5}{4}\right)^{799}$$

With a calculator, it is easy to calculate $8(\frac{5}{4})^{799}$ and then change this number of inches to miles; the result is the figure given above.

Formulas

In the sequence above, each term was $\frac{5}{4}$ times the preceding one. You should be able to find a similar pattern in each of the following sequences. When you do, fill in the boxes.

(a) 3, 6, 12, 24, $\square$, $\square$, . . .

(b) 12, 4, $\frac{4}{3}$, $\frac{4}{9}$, $\square$, $\square$, . . .

(c) .6, 6, 60, 600, $\square$, $\square$, . . .

The common feature of these three sequences is that in each case, you can get a term by multiplying the preceding term by a fixed number. In sequence (a), you multiply by 2; in (b), by $\frac{1}{3}$; and in (c), by 10. We call such sequences **geometric sequences.** Thus a geometric sequence $a_1, a_2, a_3, \ldots$ satisfies the recursion formula

$$a_n = ra_{n-1}$$

where r is a fixed number called the **common ratio.**

To obtain the corresponding explicit formula, note that

$$a_2 = ra_1$$
$$a_3 = ra_2 = r(ra_1) = r^2 a_1$$
$$a_4 = ra_3 = r(r^2 a_1) = r^3 a_1$$

In each case, the exponent on r is one less than the subscript on a. Thus

$$a_n = a_1 r^{n-1}$$

From this, we can get explicit formulas for each of the sequences (a), (b), and (c) above.

$$a_n = 3 \cdot 2^{n-1}$$
$$b_n = 12\left(\frac{1}{3}\right)^{n-1}$$
$$c_n = (.6)(10)^{n-1}$$

Example A (Two Geometric Sequences) Determine r and find the 20th term.
(a) 100, $100(1.04)^2$, $100(1.04)^4$, $100(1.04)^6$, . . .   (b) 81, 54, 36, 24, 16, . . .

Solution
(a) $r = (1.04)^2$ and $a_{20} = 100(1.04)^{38} \approx 443.88135$
(b) $r = \frac{2}{3}$ and $b_{20} = 81(\frac{2}{3})^{19} \approx .0365385$ ∎

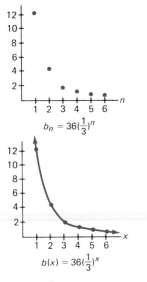

$$b_n = 36(\tfrac{1}{3})^n$$

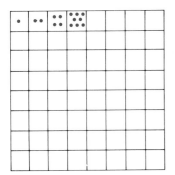

$$b(x) = 36(\tfrac{1}{3})^x$$

Figure 9

Geometric Sequences and Exponential Functions

Let us consider sequence (b) once more; its explicit formula is

$$b_n = 12\left(\frac{1}{3}\right)^{n-1} = 36\left(\frac{1}{3}\right)^n$$

We have graphed this sequence and also the exponential function

$$b(x) = 36\left(\frac{1}{3}\right)^x$$

in Figure 9. (See Section 6-2 for a discussion of exponential functions.) It should be clear that the sequence b_n is the function $b(x)$ with its domain restricted to the positive integers.

What we have observed in this example is true in general. A geometric sequence is simply an exponential function with its domain restricted to the positive integers.

Sums of Geometric Sequences

There is an old legend about geometric sequences and chessboards. When the king of Persia learned to play chess, he was so enchanted with the game that he determined to reward the inventor, a man named Sessa. Calling Sessa to the palace, the king promised to fulfill any request he might make. With an air of modesty, wily Sessa asked for one grain of wheat for the first square of the chessboard, two for the second, four for the third, and so on (Figure 10). The king was amused at such an odd request; nevertheless, he called a servant, told him to get a bag of wheat, and start counting. To the king's surprise, it soon became apparent that Sessa's request could never be fulfilled. Let's see why.

Sessa was really asking for

$$1 + 2 + 2^2 + 2^3 + \cdots + 2^{63}$$

grains of wheat, the sum of the first 64 terms of the geometric sequence 1, 2, 4, 8, ... We are going to develop a formula for this sum and for all others that arise from adding the terms of a geometric sequence.

Let $a_1, a_2, a_3, \ldots$ be a geometric sequence with ratio $r \neq 1$. As usual, let

$$A_n = a_1 + a_2 + a_3 + \cdots + a_n$$

which can be written

$$A_n = a_1 + a_1 r + a_1 r^2 + \cdots + a_1 r^{n-1}$$

Now multiply A_n by r, subtract the result from A_n, and use a little algebra to solve for A_n. We obtain

$$
\begin{array}{l}
A_n = a_1 + a_1 r + a_1 r^2 + \cdots + a_1 r^{n-1} \\
\underline{\quad rA_n = \qquad a_1 r + a_1 r^2 + \cdots + a_1 r^{n-1} + a_1 r^n} \\
A_n - rA_n = a_1 + 0 + 0 + \cdots + 0 - a_1 r^n \\
A_n(1 - r) = a_1(1 - r^n)
\end{array}
$$

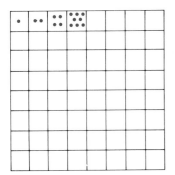

Figure 10

$$A_n = \frac{a_1(1 - r^n)}{1 - r} = \frac{a_1(r^n - 1)}{r - 1} \qquad r \neq 1$$

In sigma notation, this is

$$\sum_{i=1}^{n} a_i = \frac{a_1(1 - r^n)}{1 - r} \qquad r \neq 1$$

In the case where $r = 1$,

$$\sum_{i=1}^{n} a_i = a_1 + a_1 + \cdots + a_1 = na_1$$

Example B (Two Geometric Sums)   Find the sum of the first 20 terms of the geometric sequences of Example A, that is, calculate the following sums.

(a) $100 + 100(1.04)^2 + 100(1.04)^4 + \cdots + 100(1.04)^{38}$

(b) $81 + 54 + 36 + 24 + \cdots + 81(\frac{2}{3})^{19}$

Solution

(a) $A_{20} = \dfrac{a_1(r^{20} - 1)}{r - 1} = \dfrac{100[(1.04)^{40} - 1]}{(1.04)^2 - 1} \approx 4658.1135$

(b) $B_{20} = \dfrac{b_1(1 - r^{20})}{1 - r} = \dfrac{81[1 - (\frac{2}{3})^{20}]}{1 - \frac{2}{3}} \approx 242.92692$ ∎

Applying the sum formula to Sessa's problem (using $n = 64$, $a_1 = 1$, and $r = 2$), we get

$$A_{64} = \frac{1(1 - 2^{64})}{1 - 2} = 2^{64} - 1$$

Ignoring the -1 and using the approximation $2^{10} \approx 1000$ gives

$$A_{64} \approx 2^{64} \approx 2^4(1000)^6 = 1.6 \times 10^{19}$$

If you do this problem on a calculator, you will get $A_{64} \approx 1.845 \times 10^{19}$. Thus, if a bushel of wheat has 1 million grains, A_{64} grains would amount to more than 1.8×10^{13}, or 18 trillion, bushels. That exceeds the world's total production of wheat in one century.

The Sum of the Whole Sequence

Is it possible to add infinitely many numbers? Do the following sums make sense?

$$\frac{1}{2} + \frac{1}{4} + \frac{1}{8} + \frac{1}{16} + \cdots$$

$$1 + 3 + 9 + 27 + \cdots$$

Questions like this have intrigued great thinkers since Zeno first introduced his famous paradoxes of the infinite over 2000 years ago. We now show that we can make sense out of the first of these two sums but not the second.

Consider a string of length 1 kilometer. We may imagine cutting it into infinitely many pieces, as indicated in Figure 11.

$\frac{1}{2}$ $\frac{1}{4}$ $\frac{1}{8}$ $\frac{1}{16}$ $\frac{1}{32}$

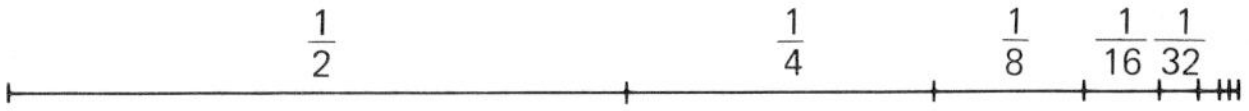

Figure 11

Since these pieces together make a string of length 1, it seems natural to say

$$\frac{1}{2} + \frac{1}{4} + \frac{1}{8} + \frac{1}{16} + \cdots = 1$$

Let us look at it another way. The sum of the first n terms of the geometric sequence $\frac{1}{2}, \frac{1}{4}, \frac{1}{8}, \frac{1}{16}, \ldots$ is given by

$$A_n = \frac{\frac{1}{2}\left[1 - \left(\frac{1}{2}\right)^n\right]}{1 - \frac{1}{2}} = 1 - \left(\frac{1}{2}\right)^n$$

As n gets larger and larger (tends to infinity), $\left(\frac{1}{2}\right)^n$ gets smaller and smaller (approaches 0). Thus A_n tends to 1 as n tends to infinity. We therefore say that 1 is the sum of *all* the terms of this sequence.

Now consider any geometric sequence with ratio r satisfying $|r| < 1$. We claim that when n gets large, r^n approaches 0. [As evidence, try calculating $(.99)^{100}$, $(.99)^{1000}$, and $(.99)^{10,000}$ on your calculator.] Thus, as n gets large,

$$A_n = \frac{a_1(1 - r^n)}{1 - r}$$

approaches $a_1/(1 - r)$. We write

$$\sum_{i=1}^{\infty} a_i = \frac{a_1}{1 - r} \qquad |r| < 1$$

We emphasize that what we have just done is valid if $|r| < 1$. There is no way to make sense out of adding all the terms of a geometric sequence if $|r| \geq 1$.

Example C (Summing Infinite Sequences)   Calculate.

(a) $1 + \dfrac{3}{\pi} + \left(\dfrac{3}{\pi}\right)^2 + \left(\dfrac{3}{\pi}\right)^3 + \cdots$ (b) $\displaystyle\sum_{i=2}^{\infty} \left(\tfrac{3}{4}\right)^i$

Solution

(a) Note that $r = 3/\pi$, so $|r| < 1$. Thus

$$1 + \frac{3}{\pi} + \left(\frac{3}{\pi}\right)^2 + \cdots = \frac{1}{1 - 3/\pi} = \frac{\pi}{\pi - 3} \approx 22.187540$$

(b) The first term is $\left(\tfrac{3}{4}\right)^2 = \tfrac{9}{16}$ and $r = \tfrac{3}{4}$. Thus

$$\sum_{i=2}^{\infty} \left(\frac{3}{4}\right)^i = \frac{\dfrac{9}{16}}{1 - \dfrac{3}{4}} = \frac{9}{4} \qquad\blacksquare$$

Example D (Repeating Decimals)   Show that $.333\overline{3}\ldots$ and $.2323\overline{23}\ldots$ are rational numbers by using the methods of this section.

Solution

$$.333\overline{3} = \frac{3}{10} + \frac{3}{100} + \frac{3}{1000} + \cdots$$

Thus we must add all the terms of an infinite geometric sequence with ratio $\frac{1}{10}$. Using the formula $a_1/(1 - r)$, we get

$$\frac{\dfrac{3}{10}}{1 - \dfrac{1}{10}} = \frac{\dfrac{3}{10}}{\dfrac{9}{10}} = \frac{1}{3}$$

Similarly,

$$.2323\overline{23} = \frac{23}{100} + \frac{23}{10,000} + \frac{23}{1,000,000} + \cdots$$

$$= \frac{\dfrac{23}{100}}{1 - \dfrac{1}{100}} = \frac{\dfrac{23}{100}}{\dfrac{99}{100}} = \frac{23}{99} \quad \blacksquare$$

PROBLEM SET 12-3

A. Skills and Techniques

1. Fill in the boxes.
   (a) $\frac{1}{2}$, 1, 2, 4, ☐, ☐, . . .
   (b) 8, 4, 2, 1, ☐, ☐, . . .
   (c) .3, .03, .003, .0003, ☐, ☐, . . .

2. Fill in the boxes.
   (a) 1, 3, 9, 27, ☐, ☐, . . .
   (b) 27, 9, 3, 1, ☐, ☐, . . .
   (c) .2, .02, .002, .0002, ☐, ☐, . . .

3. Determine r for each of the sequences in Problem 1 and write an explicit formula for the nth term.

4. Write a formula for the nth term of each sequence in Problem 2.

5. Evaluate the 30th term of each sequence in Problem 1 (see Example A).

6. Evaluate the 20th term of each sequence in Problem 2.

7. Use the sum formula to find the sum of the first five terms of each sequence in Problem 1 (see Example B).

8. Use the sum formula to find the sum of the first five terms of each sequence in Problem 2.

9. Find the sum of the first 30 terms of each sequence in Problem 1.

10. Find the sum of the first 20 terms of each sequence in Problem 2.

11. A certain culture of bacteria doubles every week. If there are 100 bacteria now, how many will there be after 10 full weeks?

12. A water lily grows so rapidly that each day it covers twice the area it covered the day before. At the end of 20 days, it completely covers a pond. If we start with two lilies, how long will it take to cover the same pond?

13. Johnny is paid $1 on January 1, $2 on January 2, $4 on January 3, and so on. Approximately how much will he earn during January?

14. If you were offered 1¢ today, 2¢ tomorrow, 4¢ the third day, and so on for 20 days or a lump sum of $10,000, which would you choose? Show why.

15. Calculate, as in Example C.

   (a) $\displaystyle\sum_{i=1}^{\infty} \left(\frac{1}{3}\right)^i$ **(b)** $\displaystyle\sum_{i=2}^{\infty} \left(\frac{2}{5}\right)^i$

16. Calculate.

(a) $\displaystyle\sum_{i=1}^{\infty} \left(\frac{2}{3}\right)^i$ (b) $\displaystyle\sum_{i=3}^{\infty} \left(\frac{1}{6}\right)^i$

17. A ball is dropped from a height of 10 feet. At each bounce, it rises to a height of $\frac{1}{2}$ the previous height. How far will it travel altogether (up and down) by the time it comes to rest? *Hint:* Think of the total distance as being the sum of the "down" distances $(10 + 5 + \frac{5}{2} + \cdots)$ and the "up" distances $(5 + \frac{5}{2} + \frac{5}{4} + \cdots)$.

18. Do Problem 17 assuming the ball rises to $\frac{2}{3}$ its previous height at each bounce.

Use the method of Example D to express each of the following as the ratio of two integers.

19. $.11\overline{1}$ **20.** $.77\overline{7}$

21. $.2525\overline{25}$ **22.** $.99\overline{9}$

23. $1.234\overline{34}$ **24.** $.341\overline{41}$

B. Applications and Extensions

25. If $a_n = 625(0.2)^{n-1}$, find a_1, a_2, a_3, a_4, and a_5.

26. Which of the following sequences are geometric, which are arithmetic, and which are neither?
(a) 130, 65, 32.5, 16.25, . . .
(b) $1, \frac{1}{2}, \frac{1}{3}, \frac{1}{4}, \ldots$
(c) 100(1.05), 100(1.07), 100(1.09), 100(1.11), . . .
(d) 100(1.05), 100(1.05)², 100(1.05)³, 100(1.05)⁴, . . .
(e) 1, 3, 6, 10, . . .
(f) 3, −6, 12, −24, . . .

27. Write an explicit formula for each geometric or arithmetic sequence in Problem 26.

28. Use the formula for the sum of an infinite geometric sequence to express each of the following repeating decimals as a ratio of two integers.
(a) $.499999 \ldots = .4\overline{9}$
(b) $.1234234234 \ldots = .1\overline{234}$

Recall from Section 6-3 that if a sum of P dollars is invested today at a compound rate of i per conversion period, then the accumulated value after n periods is given by $P(1 + i)^n$. The sequence of accumulated values

$$P(1 + i), P(1 + i)^2, P(1 + i)^3, P(1 + i)^4, \ldots$$

is geometric with ratio $1 + i$. In Problems 29–32, write a formula for the answer and then use a calculator to evaluate it.

29. If \$1 is put in the bank at 8 percent interest compounded annually, it will be worth $(1.08)^n$ dollars af-

ter *n* years. How much will \$100 be worth after 10 years? When will the amount first exceed \$250?

30. If \$1 is put in the bank at 8 percent interest compounded quarterly, it will be worth $(1.02)^n$ dollars after *n* quarters. How much will \$100 be worth after 10 years (40 quarters)? When will the amount first exceed \$250?

31. Suppose Karen puts \$100 in the bank today and \$100 at the beginning of each of the following 9 years. If this money earns interest at 8 percent compounded annually, what will it be worth at the end of 10 years?

32. José makes 40 deposits of \$25 each in a bank at intervals of three months, making the first deposit today. If money earns interest at 8 percent compounded quarterly, what will it all be worth at the end of 10 years (40 quarters)?

33. Suppose that the government pumps an extra \$1 billion into the economy. Assume that each business and individual saves 25 percent of its income and spends the rest, so that of the initial \$1 billion, 75 percent is re-spent by individuals and businesses. Of that amount, 75 percent is spent, and so on. What is the total increase in spending due to the government action? (This is called the *multiplier effect* in economics.)

34. Given an arbitrary triangle of perimeter 10, a second triangle is formed by joining the midpoints of the first, a third triangle is formed by joining the midpoints of the second, and so on forever. Find the total length of all line segments in the resulting configuration.

35. Find the area of the painted region in Figure 12, which consists of an infinite sequence of 30°-60°-90° triangles.

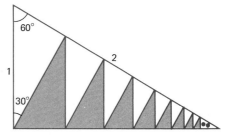

Figure 12

36. If the pattern in Figure 13 is continued indefinitely, what fraction of the area of the original square will be painted?

37. Expand in powers of 3; then evaluate *P*.

$$P = (1 + 3)(1 + 3^2)(1 + 3^4)(1 + 3^8)(1 + 3^{16})$$

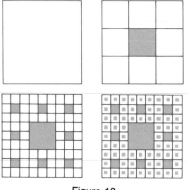

Figure 13

38. By considering $S - rS$, find S.
(a) $S = r + 2r^2 + 3r^3 + 4r^4 + \cdots, |r| < 1$.
(b) $S = \frac{1}{3} + 2(\frac{1}{3})^2 + 3(\frac{1}{3})^3 + 4(\frac{1}{3})^4 + \cdots$

39. In a geometric sequence, the sum of the first 2 terms is 5 and the sum of the first 6 terms is 65. What is the sum of the first 4 terms? *Hint:* Call the first term a and the ratio r. Then $a + ar = 5$, so $a = 5/(1 + r)$. This allows you to write 65 in terms of r alone. Solve for r.

40. Imagine a huge maze with infinitely many adjoining cells, each having a square base 1 meter by 1 meter. The first cell has walls 1 meter high, the second cell has walls $\frac{1}{2}$ meter high, the third cell has walls $\frac{1}{4}$ meter high, and so on.
(a) How much paint would it take to fill the maze with paint?
(b) How much paint would it take to paint the floors of all the cells?
(c) Explain this apparent contradiction.

41. Starting 100 miles apart, Tom and Joel ride toward each other on their bicycles, Tom going at 8 miles per hour and Joel at 12 miles per hour. Tom's dog, Corky, starts with Tom running toward Joel at 25 miles per hour. When Corky meets Joel, he immediately turns tail and heads back to Tom. Reaching Tom, Corky again turns tail and heads toward Joel, and so on. How far did Corky run by the time Tom and Joel met? This can be answered using geometric sequences, but if you are clever, you will find a better way.

42. TEASER Sally walked 4 miles north, then 2 miles east, 1 mile south, $\frac{1}{2}$ mile west, $\frac{1}{4}$ mile north, and so on. If she continued this pattern indefinitely, how far from her initial point did she end?

12-4 MATHEMATICAL INDUCTION

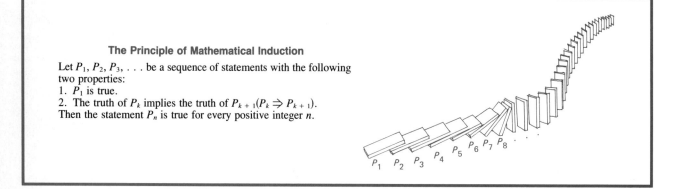

The Principle of Mathematical Induction

Let $P_1, P_2, P_3, \ldots$ be a sequence of statements with the following two properties:
1. P_1 is true.
2. The truth of P_k implies the truth of $P_{k+1}(P_k \Rightarrow P_{k+1})$.
Then the statement P_n is true for every positive integer n.

The principle of mathematical induction deals with a sequence of statements. A **statement** is a sentence that is either true of false. In a sequence of statements, there is a statement corresponding to each positive integer. Here are four examples.

$$P_n: \quad \frac{1}{1 \cdot 2} + \frac{1}{2 \cdot 3} + \frac{1}{3 \cdot 4} + \cdots + \frac{1}{n(n + 1)} = \frac{n}{n + 1}$$

$$Q_n: \quad n^2 - n + 41 \text{ is a prime number}$$

$$R_n: \quad (a + b)^n = a^n + b^n$$

$$S_n: \quad 1 + 2 + 3 + \cdots + n = \frac{n^2 + n - 6}{2}$$

To be sure that we understand the notation, let us write each of these statements for the case $n = 3$.

$$P_3: \quad \frac{1}{1 \cdot 2} + \frac{1}{2 \cdot 3} + \frac{1}{3 \cdot 4} = \frac{3}{4}$$

$$Q_3: \quad 3^2 - 3 + 41 \text{ is a prime number}$$

$$R_3: \quad (a + b)^3 = a^3 + b^3$$

$$S_3: \quad 1 + 2 + 3 = \frac{3^2 + 3 - 6}{2}$$

Of these, P_3 and Q_3 are true, while R_3 and S_3 are false; you should verify this fact. A careful study of these four sequences will indicate the wide range of behavior that sequences of statements can display.

n	$n^2 - n + 41$
1	41
2	43
3	47
4	53
5	61
6	71
7	83
.	.
.	.
.	.
40	1601
41	$1681 = 41^2$

Figure 14

While it certainly is not obvious, we claim that P_n is true for every positive integer n; we are going to prove it soon. Q_n is a well-known sequence. It was thought by some to be true for all n and, in fact, it is true for $n = 1, 2, 3, \ldots, 40$ (see Figure 14). However, it fails for $n = 41$, a fact that allows us to make an important point. Establishing the truth of Q_n for a finite number of cases, no matter how many, does not prove its truth for all n. Sequences R_n and S_n are rather hopeless cases, since R_n is true only for $n = 1$ and S_n is never true.

Proof by Mathematical Induction

How does one prove that something is true for all n? The tool uniquely designed for this purpose is the **principle of mathematical induction;** it was stated in our opening display. Let us use mathematical induction to show that

$$P_n: \quad \frac{1}{1 \cdot 2} + \frac{1}{2 \cdot 3} + \frac{1}{3 \cdot 4} + \cdots + \frac{1}{(n-1)n} + \frac{1}{n(n+1)} = \frac{n}{n+1}$$

is true for every positive integer n. There are two steps to the proof. We must show that

1. P_1 is true;
2. $P_k \Rightarrow P_{k+1}$; that is, the truth of P_k implies the truth of P_{k+1}.

The first step is easy. P_1 is just the statement

$$\frac{1}{1 \cdot 2} = \frac{1}{1 + 1}$$

which is clearly true.

To handle the second step $(P_k \Rightarrow P_{k+1})$, it is a good idea to write down the statements corresponding to P_k and P_{k+1} (at least on scratch paper). We get them by substituting k and $k + 1$ for n in the statement for P_n.

$$P_k: \quad \frac{1}{1 \cdot 2} + \frac{1}{2 \cdot 3} + \cdots + \frac{1}{(k-1)k} + \frac{1}{k(k+1)} = \frac{k}{k+1}$$

$$P_{k+1}: \quad \frac{1}{1 \cdot 2} + \frac{1}{2 \cdot 3} + \cdots + \frac{1}{k(k+1)} + \frac{1}{(k+1)(k+2)} = \frac{k+1}{k+2}$$

Notice that the left side of P_{k+1} is the same as that of P_k except for the addition of one more term, $1/(k+1)(k+2)$.

Suppose for the moment that P_k is true and consider how this assumption allows us to simplify the left side of P_{k+1}.

$$\left[\frac{1}{1 \cdot 2} + \frac{1}{2 \cdot 3} + \cdots + \frac{1}{k(k+1)} \right] + \frac{1}{(k+1)(k+2)} = \frac{k}{k+1} + \frac{1}{(k+1)(k+2)}$$

$$= \frac{k(k+2) + 1}{(k+1)(k+2)}$$

$$= \frac{(k+1)^2}{(k+1)(k+2)}$$

$$= \frac{k+1}{k+2}$$

If you read this chain of equalities from top to bottom, you will see that we have established the truth of P_{k+1}, but under the *assumption that P_k is true*. That is, we have established that the truth of P_k implies the truth of P_{k+1}.

Example A (A Sum of Squares Formula) Prove that

$$P_n: \quad 1^2 + 2^2 + 3^2 + \cdots + n^2 = \frac{n(n+1)(2n+1)}{6}$$

is true for every positive integer n.

Solution Statements P_1, P_k, and P_{k+1} are as follows.

$$P_1: \quad 1^2 = \frac{1(2)(3)}{6}$$

$$P_k: \quad 1^2 + 2^2 + 3^2 + \cdots + k^2 = \frac{k(k+1)(2k+1)}{6}$$

$$P_{k+1}: \quad 1^2 + 2^2 + 3^2 + \cdots + k^2 + (k+1)^2 = \frac{(k+1)(k+2)(2k+3)}{6}$$

Clearly P_1 is true.

Assuming that P_k is true, we can write the left side of P_{k+1} as shown in the following chain of equalities.

$$1^2 + 2^2 + 3^2 + \cdots + k^2 + (k+1)^2 = \frac{k(k+1)(2k+1)}{6} + (k+1)^2$$

$$= \frac{(k+1)[k(2k+1) + 6(k+1)]}{6}$$

$$= \frac{(k + 1)(2k^2 + 7k + 6)}{6}$$

$$= \frac{(k + 1)(k + 2)(2k + 3)}{6}$$

Thus the truth of P_k does imply the truth of P_{k+1}. We conclude by mathematical induction that P_n is true for every positive integer n. Incidentally, the result just proved will be used in calculus. ■

Some Comments about Mathematical Induction

Students never have any trouble with the verification step (showing that P_1 is true). The inductive step (showing that $P_k \Rightarrow P_{k+1}$) is harder and more subtle. In that step, we do *not* prove that P_k or P_{k+1} is true, but rather that the truth of P_k implies the truth of P_{k+1}. For a vivid illustration of the difference, we point out that in the fourth example of our opening paragraph, the truth of S_k does imply the truth of S_{k+1} ($S_k \Rightarrow S_{k+1}$) and yet not a single statement in that sequence is true (see Problem 30). To put it another way, what $S_k \Rightarrow S_{k+1}$ means is that if S_k were true, then S_{k+1} would be true also. It is like saying that if spinach were ice cream, then kids would want two helpings at every meal.

Why They Fall and Why They Don't

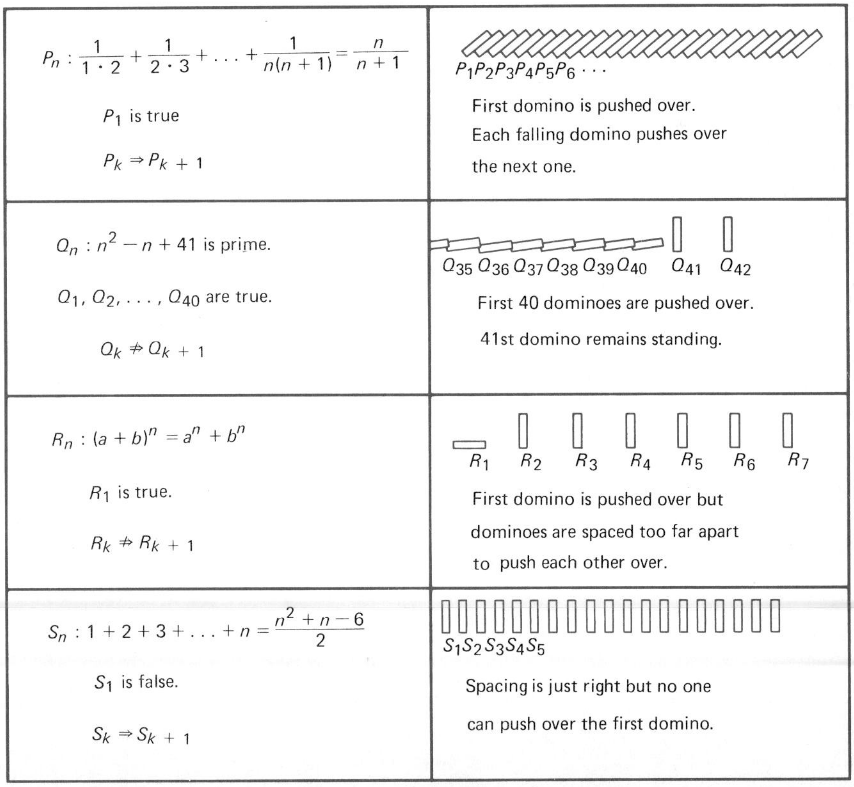

$P_n : \dfrac{1}{1 \cdot 2} + \dfrac{1}{2 \cdot 3} + \ldots + \dfrac{1}{n(n + 1)} = \dfrac{n}{n + 1}$

P_1 is true

$P_k \Rightarrow P_{k+1}$

$P_1 P_2 P_3 P_4 P_5 P_6 \cdots$

First domino is pushed over.

Each falling domino pushes over

the next one.

$Q_n : n^2 - n + 41$ is prime.

$Q_1, Q_2, \ldots, Q_{40}$ are true.

$Q_k \nRightarrow Q_{k+1}$

$Q_{35}\ Q_{36}\ Q_{37}\ Q_{38}\ Q_{39}\ Q_{40} \quad Q_{41} \quad Q_{42}$

First 40 dominoes are pushed over.

41st domino remains standing.

$R_n : (a + b)^n = a^n + b^n$

R_1 is true.

$R_k \nRightarrow R_{k+1}$

$R_1 \quad R_2 \quad R_3 \quad R_4 \quad R_5 \quad R_6 \quad R_7$

First domino is pushed over but

dominoes are spaced too far apart

to push each other over.

$S_n : 1 + 2 + 3 + \ldots + n = \dfrac{n^2 + n - 6}{2}$

S_1 is false.

$S_k \Rightarrow S_{k+1}$

$S_1 S_2 S_3 S_4 S_5$

Spacing is just right but no one

can push over the first domino.

Figure 15

497

Perhaps the dominoes in the opening display can help illuminate the idea. For all the dominoes to fall it is sufficient that

1. the first domino is pushed over.
2. if any domino falls (say the kth one), it pushes over the next one (the $(k + 1)$st one).

Figure 15 (on the previous page) illustrates what happens to the dominoes in the four examples of our opening paragraph. Study them carefully.

The Many Uses of Mathematical Induction

Mathematical induction finds application throughout mathematics. We have already shown that it is a powerful tool for establishing general sum formulas. Our next two examples illustrate quite different uses and the problem set will introduce other applications.

Example B (Mathematical Induction Applied to Inequalities)   Show that the following statement is true for every integer $n \geq 4$.

$$3^n > 2^n + 20$$

Solution   Let P_n represent the given statement. You might check that P_1, P_2, and P_3 are false. However, that does not matter to us. What we need to do is to show that P_4 is true and that $P_k \Rightarrow P_{k+1}$ for any $k \geq 4$.

$$P_4: \quad 3^4 > 2^4 + 20$$

$$P_k: \quad 3^k > 2^k + 20$$

$$P_{k+1}: \quad 3^{k+1} > 2^{k+1} + 20$$

Clearly P_4 is true (81 is greater than 36). Next we assume P_k to be true (for $k \geq 4$) and seek to show that this would force P_{k+1} to be true. Working with the left side of P_{k+1} and using the assumption that $3^k > 2^k + 20$, we get

$$3^{k+1} = 3 \cdot 3^k > 3(2^k + 20) > 2(2^k + 20) = 2^{k+1} + 40 > 2^{k+1} + 20$$

Therefore, P_{k+1} is true, provided P_k is true. We conclude that P_n is true for every integer $n \geq 4$.   ■

Example C (Mathematical Induction and Divisibility)   Prove that the statement

$$P_n: \quad x - y \text{ is a factor of } x^n - y^n$$

is true for every positive integer n.

Solution   Trivially, $x - y$ is a factor of $x - y$ since $x - y = 1(x - y)$; so P_1 is true. Now suppose that P_k is true, that is, that

$$x - y \text{ is a factor of } x^k - y^k$$

This means that there is a polynomial $Q(x, y)$ such that

$$x^k - y^k = Q(x, y)(x - y)$$

Using this assumption, we may write

$$x^{k+1} - y^{k+1} = x^{k+1} - x^k y + x^k y - y^{k+1}$$
$$= x^k(x - y) + y(x^k - y^k)$$
$$= x^k(x - y) + yQ(x, y)(x - y)$$
$$= [x^k + yQ(x, y)](x - y)$$

Thus $x - y$ is a factor of $x^{k+1} - y^{k+1}$. We have shown that $P_k \Rightarrow P_{k+1}$ and that P_1 is true; we therefore conclude that P_n is true for all n. ■

PROBLEM SET 12-4

A. Skills and Techniques

In Problems 1–8, prove by mathematical induction that P_n is true for every positive integer n. See Example A.

1. P_n: $1 + 2 + 3 + \cdots + n = \dfrac{n(n + 1)}{2}$

2. P_n: $1 + 3 + 5 + \cdots + (2n - 1) = n^2$

3. P_n: $3 + 7 + 11 + \cdots + (4n - 1) = n(2n + 1)$

4. P_n: $2 + 9 + 16 + \cdots + (7n - 5) = \dfrac{n(7n - 3)}{2}$

5. P_n: $1 \cdot 2 + 2 \cdot 3 + 3 \cdot 4 + \cdots + n(n + 1)$
$$= \tfrac{1}{3}n(n + 1)(n + 2)$$

6. P_n: $\dfrac{1}{1 \cdot 3} + \dfrac{1}{3 \cdot 5} + \dfrac{1}{5 \cdot 7} + \cdots$

$$+ \dfrac{1}{(2n - 1)(2n + 1)} = \dfrac{n}{2n + 1}$$

7. P_n: $2 + 2^2 + 2^3 + \cdots + 2^n = 2(2^n - 1)$

8. P_n: $1^2 + 3^2 + 5^2 + \cdots + (2n - 1)^2$
$$= \dfrac{n(2n - 1)(2n + 1)}{3}$$

In Problems 9–18, tell what you can conclude from the information given about the sequence of statements. For example, if you are given that P_4 is true and that $P_k \Rightarrow P_{k+1}$ for any k, then you can conclude that P_n is true for every integer $n \geq 4$.

9. P_8 is true and $P_k \Rightarrow P_{k+1}$.

10. P_8 is not true and $P_k \Rightarrow P_{k+1}$.

11. P_1 is true but P_k does not imply P_{k+1}.

12. $P_1, P_2, \ldots, P_{1000}$ are all true.

13. P_1 is true and $P_k \Rightarrow P_{k+2}$.

14. P_{40} is true and $P_k \Rightarrow P_{k-1}$.

15. P_1 and P_2 are true; P_k and P_{k+1} together imply P_{k+2}.

16. P_1 and P_2 are true and $P_k \Rightarrow P_{k+2}$.

17. P_1 is true and $P_k \Rightarrow P_{4k}$.

18. P_1 is true, $P_k \Rightarrow P_{4k}$, and $P_k \Rightarrow P_{k-1}$.

In Problems 19–24, find the smallest positive integer N for which the given statement is true for $n \geq N$. Then prove that the statement is true for all integers greater than or equal to N. See Example B.

19. $n + 5 < 2^n$ **20.** $3n \leq 3^n$

21. $\log_{10} n < n$ *Hint:* $k + 1 < 10k$.

22. $n^2 \leq 2^n$ *Hint:* $k^2 + 2k + 1 = k(k + 2 + 1/k) < k(k + k)$.

23. $(1 + x)^n \geq 1 + nx$, where $x \geq -1$

24. $|\sin nx| \leq |\sin x| \cdot n$ for all x

Use mathematical induction to prove that each of the following is true for every positive integer n. See Example C.

25. $x + y$ is a factor of $x^{2n} - y^{2n}$. *Hint:* $x^{2k+2} - y^{2k+2} = x^{2k+2} - x^{2k}y^2 + x^{2k}y^2 - y^{2k+2}$.

26. $x + y$ is a factor of $x^{2n-1} + y^{2n-1}$.

27. $n^2 - n$ is even (that is, has 2 as a factor).

28. $n^3 - n$ is divisible by 6.

B. Applications and Extensions

29. These four formulas can all be proved by mathematical induction. We proved part (b) in the text; you prove the others.

(a) $1 + 2 + 3 + \cdots + n = \tfrac{1}{2}n(n + 1)$

(b) $1^2 + 2^2 + 3^2 + \cdots + n^2 = \tfrac{1}{6}n(n + 1)(2n + 1)$

(c) $1^3 + 2^3 + 3^3 + \cdots + n^3 = \tfrac{1}{4}n^2(n + 1)^2$

(d) $1^4 + 2^4 + 3^4 + \cdots + n^4 =$
$$\tfrac{1}{30}n(n + 1)(6n^3 + 9n^2 + n - 1)$$

From (a) and (c), another interesting formula follows, namely,

$$1^3 + 2^3 + 3^3 + \cdots + n^3$$
$$= (1 + 2 + 3 + \cdots + n)^2$$

30. Consider the statement

$$S_n: \quad 1 + 2 + 3 + \cdots + n = \frac{n^2 + n - 6}{2}$$

Show that
(a) $S_k \Rightarrow S_{k+1}$ for $k \geq 1$.
(b) S_n is not true for any positive integer n.

31. Use the results of Problem 29 to evaluate each of the following.

(a) $\displaystyle\sum_{k=1}^{100} (3k + 1)$ **(b)** $\displaystyle\sum_{k=1}^{10} (k^2 - 3k)$

(c) $\displaystyle\sum_{k=1}^{10} (k^3 + 3k^2 + 3k + 1)$ **(d)** $\displaystyle\sum_{k=1}^{n} (6k^2 + 2k)$

32. In a popular song titled "The Twelve Days of Christmas," my true love gave me 1 gift on the first day, $(2 + 1)$ gifts on the second day, $(3 + 2 + 1)$ gifts on the third day, and so on.
(a) How many gifts did I get all together during the 12 days?
(b) How many gifts would I get all together in a Christmas that had n days?

33. Prove that for $n \geq 2$,

$$\left(1 - \frac{1}{4}\right)\left(1 - \frac{1}{9}\right)\left(1 - \frac{1}{16}\right) \cdots \left(1 - \frac{1}{n^2}\right) = \frac{n + 1}{2n}$$

34. Prove that for $n \geq 1$,

$$\frac{1}{\sqrt{1}} + \frac{1}{\sqrt{2}} + \frac{1}{\sqrt{3}} + \cdots + \frac{1}{\sqrt{n}} < 2\sqrt{n}$$

35. Prove that for $n \geq 3$,

$$\frac{1}{n + 1} + \frac{1}{n + 2} + \frac{1}{n + 3} + \cdots + \frac{1}{2n} > \frac{3}{5}$$

36. Prove that the number of diagonals in an n-sided convex polygon is $n(n - 3)/2$ for $n \geq 3$. The diagrams in Figure 16 show the situation for $n = 4$ and $n = 5$.

37. Prove that the sum of the measures of the interior angles in an n-sided polygon (without holes or self-intersections) is $(n - 2)180°$. What is the sum of the measures of the exterior angles of such a polygon?

38. Consider n lines in general position (no two parallel, no three meeting in the same point). Prove that I_n, the

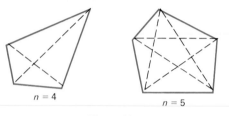

$n = 4$

$n = 5$

Figure 16

number of intersection points, is given by $I_n = (n^2 - n)/2$. *Hint:* $I_{n+1} = I_n + n$.

39. Refer to Problem 38. Prove that R_n, the number of regions, is given by $R_n = (n^2 + n + 2)/2$. *Hint:* First convince yourself that $R_{n+1} = R_n + n + 1$.

40. Let $f_1 = 1$, $f_2 = 1$, and $f_{n+2} = f_{n+1} + f_n$ for $n \geq 1$. Call f_n the Fibonacci sequence (see Section 9-1) and let $F_n = f_1 + f_2 + f_3 + \cdots + f_n$. Prove by mathematical induction that $F_n = f_{n+2} - 1$ for all n.

41. For the Fibonacci sequence of Problem 40, prove that for $n \geq 1$,

$$f_1^2 + f_2^2 + f_3^2 + \cdots + f_n^2 = f_n f_{n+1}$$

42. What is wrong with the following argument?
Theorem. All horses in the world have the same color.
Proof. Let P_n be the statement: All the horses in any set of n horses are identically colored. Certainly, P_1 is true. Suppose that P_k is true, that is, that all the horses in any set of k horses are identically colored. Let W be any set of $k + 1$ horses. Now we may think of W as the union of two overlapping sets X and Y, each with k horses. (The situation for $k = 4$ is shown in Figure 17.) By assumption, the horses in X are identically colored and the horses in Y are identically colored. Since X and Y overlap, all the horses in $X \cup Y$ must be identically colored. We conclude that P_n is true for all n. Thus the set of all horses in the world (some finite number) have the same color.

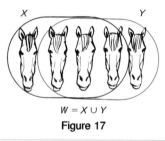

$W = X \cup Y$

Figure 17

43. Let $a_0 = 0$, $a_1 = 1$, and $a_{n+2} = (a_{n+1} + a_n)/2$ for $n \geq 0$. Prove that for $n \geq 0$,

$$a_n = \frac{2}{3}\left[1 - \left(-\frac{1}{2}\right)^n\right]$$

Hint: In the inductive step, show that P_k and P_{k+1} together imply P_{k+2}.

44. TEASER Let f_n be the Fibonacci sequence of Problem 40. Use mathematical induction (as in the hint to Problem 43) to prove that

$$f_n = \frac{1}{\sqrt{5}}\left[\left(\frac{1 + \sqrt{5}}{2}\right)^n - \left(\frac{1 - \sqrt{5}}{2}\right)^n\right]$$

12-5 COUNTING ORDERED ARRANGEMENTS

The Senior Birdwatchers' Club, consisting of 4 women and 2 men, is about to hold its annual meeting. In addition to having a group picture taken, they plan to elect a president, a vice-president, and a secretary. Here are some questions that they (and we) might consider.

1. In how many ways can they line up for their group picture?
2. In how many ways can they elect their three officers if there are no restrictions as to sex?
3. In how many ways can they elect their three officers if the president is required to be female and the vice-president, male?
4. In how many ways can they elect their three officers if the president is to be of one sex and the vice-president and secretary of the other?

In order to answer these questions and others of a similar nature, we need two counting principles. One involves multiplication; the other involves addition.

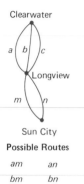

Clearwater

Longview

Sun City

Possible Routes

am	*an*
bm	*bn*
cm	*cn*

Figure 18

Multiplication Principle in Counting

Suppose that there are three roads *a*, *b*, and *c* leading from Clearwater to Longview and two roads *m* and *n* from Longview to Sun City. How many different routes can you choose from Clearwater to Sun City going through Longview? Figure 18 clarifies the situation. For each of the 3 choices from Clearwater to Longview, you have 2 choices from Longview to Sun City. Thus, you have $3 \cdot 2$ routes from Clearwater to Sun City. Here is the general principle.

Multiplication Principle

Suppose that an event *H* can occur in *h* ways and, after it has occurred, event *K* can occur in *k* ways. Then the number of ways in which both *H* **and** *K* can occur is *hk*.

This principle extends in an obvious way to three or more events.

Example A (Applying the Multiplication Principle) Steve has 2 sport coats, 4 pairs of pants, 6 shirts, and 9 ties. How many different outfits can he wear consisting of one item of each kind?

Solution Steve can wear

$$2 \cdot 4 \cdot 6 \cdot 9 = 432$$

different outfits. ∎

Consider now the Birdwatchers' third question. It involves three events.

P: Elect a female president.

V: Elect a male vice-president.

S: Elect a secretary of either sex.

We understand that the election will take place in the order indicated and that no person can fill more than one position. Thus

P can occur in 4 ways (there are 4 women);
V can occur in 2 ways (there are 2 men);
S can occur in 4 ways (after *P* and *V* occur, there are 4 people left from whom to select a secretary).

The entire selection process can be accomplished in $4 \cdot 2 \cdot 4 = 32$ ways.

Permutations

Permutations
of *ART*

ART
ATR
RAT
RTA
TAR
TRA

Figure 19

To permute a set of objects is to rearrange them. Thus a **permutation** of a set of objects is an ordered arrangement of those objects (Figure 19). Take the set of letters in the word *FACTOR* as an example. Imagine that these 6 letters are printed on small cards so they can be arranged at will. Then we may form words like *COTARF*,

TRAFOC, and *FRACTO,* none of which are in a dictionary but all of which are perfectly good words from our point of view. Let us call them code words. How many 6-letter code words can be made from the letters of the word *FACTOR;* that is, how many permutations of 6 objects are there?

Think of this as the problem of filling 6 slots.

We may fill the first slot in 6 ways. Having done that, we may fill the second slot in 5 ways, the third in 4 ways, and so on. By the multiplication principle, we can fill all 6 slots in

$$6 \cdot 5 \cdot 4 \cdot 3 \cdot 2 \cdot 1 = 720$$

ways.

Do you see that this is also the answer to the first question about the Birdwatchers, which asked in how many ways the 6 members could be arranged for a group picture? Let us identify each person by a letter; the letters of *FACTOR* will do just fine. Then to arrange the Birdwatchers is to make a 6-letter code word out of *FACTOR.* It can be done in 720 ways.

What if we want to make 3-letter code words from the letters of the word *FACTOR,* words like *ACT, COF,* and *TAC?* How many such words can be made? This is the problem of filling 3 slots with 6 letters available. We can fill the first slot in 6 ways, then the second in 5 ways, and then the third in 4 ways. Therefore we can make $6 \cdot 5 \cdot 4 = 120$ 3-letter code words from the word *FACTOR.*

The number 120 is also the answer to the second question about the Birdwatchers. If there are no restrictions as to sex, they can elect a president, vice-president, and secretary in $6 \cdot 5 \cdot 4 = 120$ ways (Figure 20).

Figure 20

Example B (Permutations)   Suppose that the 9 letters of LOGARITHM are written on 9 cards.

(a) How many 9-letter code words can be made from them?
(b) How many 4-letter code words can be made from them?

Solution
(a) $9 \cdot 8 \cdot 7 \cdot 6 \cdot 5 \cdot 4 \cdot 3 \cdot 2 \cdot 1 = 362,880$ (b) $9 \cdot 8 \cdot 7 \cdot 6 = 3024$ ∎

Consider the corresponding general problem. Suppose that from n distinguishable objects, we select r of them and arrange them in a row. The resulting arrangement is called a **permutation of n things taken r at a time.** The number of such permutations is denoted by the symbol $_nP_r$. Thus

$$_6P_3 = 6 \cdot 5 \cdot 4 = 120$$

$$_6P_6 = 6 \cdot 5 \cdot 4 \cdot 3 \cdot 2 \cdot 1 = 720$$

$$_8P_2 = 8 \cdot 7 = 56$$

and in general

$$_nP_r = n(n-1)(n-2) \cdots (n-r+2)(n-r+1)$$

Notice that $_nP_r$ is the product of r consecutive positive integers starting with n and going down. In particular, $_nP_n$ is the product of n positive integers starting with n and going all the way down to 1, that is,

$$_nP_n = n(n - 1)(n - 2) \cdots 3 \cdot 2 \cdot 1$$

The symbol $n!$ (read **n factorial**) is also used for this product. Thus

$$_5P_5 = 5! = 5 \cdot 4 \cdot 3 \cdot 2 \cdot 1 = 120$$

$$_4P_4 = 4! = 4 \cdot 3 \cdot 2 \cdot 1 = 24$$

Note also that

$$_nP_r = \frac{n!}{(n - r)!}$$

Addition Principle in Counting

We still have not answered the fourth Birdwatchers' question. In how many ways can they elect their three officers if the president is to be of one sex and the vice-president and secretary of the other? This means that the president should be female and the other two officers male, *or* the president should be male and the other two female. To answer a question like this we need another principle.

Addition Principle

Let H and K be disjoint events, that is, events that cannot happen simultaneously. If H can occur in h ways and K in k ways, then H **or** K can occur in $h + k$ ways.

This principle generalizes to three or more disjoint events.

Applying this principle to the question at hand, we define H and K as follows.

H: Elect a female president, male vice-president, and male secretary.

K: Elect a male president, female vice-president, and female secretary.

Clearly H and K are disjoint. From the multiplication principle,

H can occur in $4 \cdot 2 \cdot 1 = 8$ ways;
K can occur in $2 \cdot 4 \cdot 3 = 24$ ways.

Then by the addition principle, H or K can occur in $8 + 24 = 32$ ways.

Example C (Applying the Addition Principle)   How many code words of any length can be made from the letters FACTOR, assumed written on 6 cards so that letters cannot be repeated?

Solution   We immediately translate the question into asking about 6 disjoint events: make 6-letter words, or 5-letter words, or 4-letter words, or 3-letter words, or 2-letter words, or 1-letter words. We can do this in the following number of ways.

$$_6P_6 + {}_6P_5 + {}_6P_4 + {}_6P_3 + {}_6P_2 + {}_6P_1$$

$$= 6 \cdot 5 \cdot 4 \cdot 3 \cdot 2 \cdot 1 + 6 \cdot 5 \cdot 4 \cdot 3 \cdot 2 + 6 \cdot 5 \cdot 4 \cdot 3 + 6 \cdot 5 \cdot 4 + 6 \cdot 5 + 6$$

$$= \qquad 720 \qquad + \qquad 720 \qquad + \qquad 360 \qquad + \quad 120 \; + \; 30 \; + 6$$

$$= 1956 \quad \blacksquare$$

Students sometimes find it hard to decide whether to multiply or to add in a counting problem. Notice that the words **and** and **or** are in boldface type in the statements of the multiplication principle and of the addition principle. They are the key words; **and** goes with multiplication; **or** goes with addition.

Variations on the Arrangements Problem

All counting problems rely on the two fundamental principles that we have enunciated. However, many variations need to be considered.

Example D (Arrangements with Side Conditions)   Suppose that the letters of the word *COMPLEX* are printed on 7 cards. How many 3-letter code words can be formed from these letters if

(a) the first and last letters must be consonants (that is, C, M, P, L, or X);

(b) all vowels used (if any) must occur in the right-hand portion of a word (that is, a vowel cannot be followed by a consonant)?

Solution

(a) Let c denote consonant, v vowel, and a any letter. We must fill the three slots below.

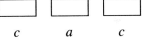

$$\qquad c \qquad\quad a \qquad\quad c$$

We begin by filling the two restricted slots, which can be done in $5 \cdot 4 = 20$ ways. Then we fill the unrestricted slot using one of the 5 remaining letters. It can be done in 5 ways. There are $20 \cdot 5 = 100$ code words of the required type. The following diagram summarizes the procedure.

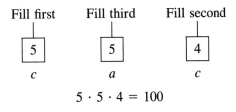

$$5 \cdot 5 \cdot 4 = 100$$

(b) We want to count words of the form cvv, ccv, or ccc. Note the use of the addition principle (as well as the multiplication principle) in the following solution.

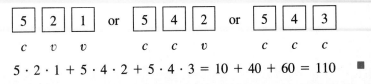

$$5 \cdot 2 \cdot 1 + 5 \cdot 4 \cdot 2 + 5 \cdot 4 \cdot 3 = 10 + 40 + 60 = 110 \quad \blacksquare$$

Example E (Permutations with Some Indistinguishable Objects)   Lucy has 3 identical red flags, 1 white flag, and 1 blue flag. How many different 5-flag signals could she display from the flagpole of her small boat?

Solution   If the 3 red flags were distinguishable, the answer would be $_5P_5 = 5! = 120$. Pretending they are distinguishable leads to counting an arrangement such as *RRBRW* six times, corresponding to the 3! ways of arranging the 3 red flags (Figure 21). For this reason, we must divide by 3!. Thus the number of signals Lucy can make is

$$\frac{5!}{3!} = \frac{5 \cdot 4 \cdot \cancel{3} \cdot \cancel{2} \cdot \cancel{1}}{\cancel{3} \cdot \cancel{2} \cdot \cancel{1}} = 20 \qquad \blacksquare$$

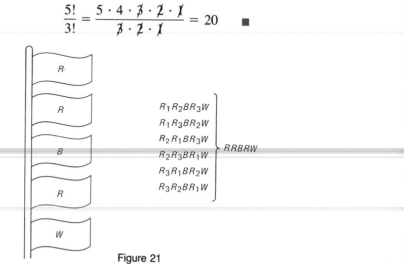

Figure 21

This result can be generalized. For example, given a set of n objects in which j are of one kind, k of a second kind, and m of a third kind, then the number of distinguishable permutations is

$$\frac{n!}{j! \, k! \, m!}$$

PROBLEM SET 12-5

A. Skills and Techniques

1. Calculate.
 (a) 3! (b) (3!)(2!)
 (c) 10!/8!

2. Calculate.
 (a) 7! (b) 7! + 5!
 (c) 12!/9!

3. Calculate.
 (a) $_5P_2$ (b) $_9P_4$
 (c) $_{10}P_3$

4. Calculate.
 (a) $_4P_3$ (b) $_8P_4$
 (c) $_{20}P_3$

Problems 5–18 require use of the multiplication and addition principles. See Examples A–C.

5. In how many ways can a president and a secretary be chosen from a group of 6 people?

6. Suppose that a club consists of 3 women and 2 men. In how many ways can a president and a secretary be chosen if
 (a) the president is to be female and the secretary, male?
 (b) the president is to be male and the secretary, female?
 (c) the president and secretary are to be of opposite sex?

7. A box contains 12 cards numbered 1 through 12. Suppose one card is drawn from the box. Find the number of ways each of the following can occur.
   (a) The number drawn is even.
   (b) The number is greater than 9 or less than 3.

8. Suppose that two cards are drawn in succession from the box in Problem 7. Assume that the first card is not replaced before the second one is drawn. In how many ways can each of the following occur?
   (a) Both numbers are even.
   (b) The two numbers are both even or both odd.
   (c) The first number is greater than 9 and the second number less than 3.

9. Do Problem 8 with the assumption that the first card is replaced before the second one is drawn.

10. In how many ways can a president, a vice-president, and a secretary be chosen from a group of 10 people?

11. How many 4-letter code words can be made from the letters of the word *EQUATION*? (Letters are not to be repeated.)

12. How many 3-letter code words can be made from the letters of the word *PROBLEM* if
   (a) letters cannot be repeated?
   (b) letters can be repeated?

13. Five roads connect Cheer City and Glumville. Starting at Cheer City, how many different ways can Smith drive to Glumville and return, that is, how many different round trips can he make? How many different round trips can he make if he wishes to return by a different road than he took to Glumville?

14. Filipe has 4 ties, 6 shirts, and 3 pairs of trousers. How many different outfits can he wear? Assume that he wears one of each kind of article.

15. Papa's Pizza Place offers 3 choices of salad, 20 kinds of pizza, and 4 different desserts. How many different 3-course meals can one order?

16. Minnesota license plate numbers consist of 3 letters followed by 3 digits (for example, AFF033). How many different plates could be issued? (You need not multiply out your answer.)

17. The letters of the word *CREAM* are printed on 5 cards. How many 3-, 4-, or 5-letter code words can be formed?

18. How many code words of all lengths can be made from the letters of *LOGARITHM*, repeated letters not allowed?

19. Frigid Treats sells 10 flavors of ice cream and makes cones in one- to three-dip sizes. How many different-looking cones can it make?

Problems 20–23 relate to Example D.

20. Using the letters of the word *FACTOR* (without repetition), how many 4-letter code words can be formed
   (a) starting with *R*?
   (b) with vowels in the two middle positions?
   (c) with only consonants?
   (d) with vowels and consonants alternating?
   (e) with all the vowels (if any) in the left-hand portion of a word (that is, a vowel cannot be preceded by a consonant)?

21. Using the letters of the word *EQUATION* (without repetition), how many 4-letter code words can be formed
   (a) starting with *T* and ending with *N*?
   (b) starting and ending with a consonant?
   (c) with vowels only?
   (d) with three consonants?
   (e) with all the vowels (if any) in the right-hand portion of the word?

22. Three brothers and 3 sisters are lining up to be photographed. How many arrangements are there
   (a) altogether?
   (b) with brothers and sisters in alternating positions?
   (c) with the 3 sisters standing together?

23. A baseball team is to be formed from a squad of 12 people. Two teams made up of the same 9 people are different if at least some of the people are assigned different positions. In how many ways can a team be formed if
   (a) there are no restrictions?
   (b) only 2 of the people can pitch and these 2 cannot play any other position?
   (c) only 2 of the people can pitch but they can also play any other position?

Problems 24–31 relate to Example E.

24. How many different signals consisting of 8 flags can be made using 4 white flags, 3 red flags, and 1 blue flag?

25. How many different signals consisting of 7 flags can be made using 3 white, 2 red, and 2 blue flags?

26. How many different 5-letter code words can be made from the 5 letters of the word *MIAMI*?

27. How many different 11-letter code words can be made from the 11 letters of the word *MISSISSIPPI*?

28. In how many different ways can a^4b^6 be written without using exponents? *Hint:* One way is *aaaabbbbbb*.

29. In how many different ways can a^3bc^6 be written without using exponents?

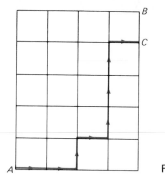

Figure 22

30. Consider the part of a city map shown in Figure 22. How many different shortest routes (no backtracking, no cutting across blocks) are there from A to C? Note that the route shown might be given the designation *EENENNNE*, with E denoting *East* and N denoting *North*.

31. How many different shortest routes are there from A to B in Problem 30?

B. Applications and Extensions

32. Note that $_7P_3 = 7 \cdot 6 \cdot 5 = 7!/4!$. Write each of the following as a quotient of two factorials.
   (a) $_{10}P_5$
   (b) $_{12}P_3$
   (c) $_nP_r$

33. Simplify.
   (a) $\dfrac{11!}{8!}$
   (b) $\dfrac{11!}{8!\,3!}$
   (c) $11! - 8!$
   (d) $\dfrac{8!}{2^6}$
   (e) $\dfrac{(n+1)! - n!}{n!}$
   (f) $\dfrac{(n+1)! + n!}{(n+1)! - n!}$

34. Five chefs enter a pie-baking contest. In how many ways can a blue ribbon, a red ribbon, and a yellow ribbon be awarded for the three best pies?

35. Ten horses run in a race at Canterbury Downs.
   (a) How many different orders of finishing are there?
   (b) How many different possibilities are there for the first three places?

36. The Greek alphabet has 24 letters. How many different 3-letter fraternity names are possible if
   (a) repeated letters are allowed?
   (b) repeated letters are not allowed?

37. The letters of the word *CYCLIC* are written on 6 cards.
   (a) How many 6-letter code words can be obtained?
   (b) How many of these have the three Cs in consecutive positions?

38. I want to arrange my 5 history books, 4 math books, and 3 psychology books on a shelf. In how many ways can I do this if
   (a) there are no restrictions as to arrangement?
   (b) I put the 5 history books on the left, the 4 math books in the middle, and the 3 psychology books on the right?
   (c) I insist only that books on the same subject be together?

39. Obtain a nice formula for
$$\frac{1}{2!} + \frac{2}{3!} + \frac{3}{4!} + \cdots + \frac{n}{(n+1)!}$$
Hint: Show first that
$$\frac{k}{(k+1)!} = \frac{1}{k!} - \frac{1}{(k+1)!}$$

40. Obtain a nice formula for $1 \cdot 1! + 2 \cdot 2! + 3 \cdot 3! + \cdots + n \cdot n!$ *Hint:* $k \cdot k! = (k+1)! - k!$.

41. A telephone number has 10 digits consisting of an area code (three digits, first is not 0 or 1, second is 0 or 1), an exchange (three digits, first is not 0 or 1, second is not 0 or 1), and a line number (four digits, not all are zeros). How many such 10-digit numbers are there?

42. The letters of *ENIGMA* are written on 6 cards. How many code words can be made?

43. How many different numbers are there between 0 and 60,000 that use only the digits 1, 2, 3, 4, or 5?

44. Consider making 6-digit numbers from the digits 1, 2, 3, 4, 5, and 6 without repetition.
   (a) How many such numbers are there?
   (b) Find the sum of these numbers.

45. In how many ways can 6 people be seated at a round table? (We consider two arrangements of people at a round table to be the same if everyone has the same people to the left and right in both arrangements.)

46. A husband and wife plan to invite 4 couples to dinner. The dinner table is rectangular. They decide on a seating arrangement in which the hostess will sit at the end nearest the kitchen, the host at the opposite end, and 4 guests on each side. Furthermore, no man shall sit next to another man, nor shall he sit next to his own wife. In how many ways can this be done?

47. Suppose that n teams enter a tournament in which a team is eliminated as soon as it loses a game. Since $n \geq 2$ is arbitrary, a number of byes may be needed. How many games must be scheduled to determine a winner? *Hint:* There is a clumsy way to do this problem but there is also a very elegant way.

48. TEASER Here is an old problem. Suppose we start with an ordered arrangement $(a_1, a_2, \ldots, a_n)$ of n objects. Let d_n be the number of derangements of this arrangement. By a **derangement,** we mean a permutation that leaves no object fixed. For example, the derangements of *ABC* are *BCA* and *CAB*. Show each of the following.

(a) $d_1 = 0,$ $d_2 = 1,$ and $d_n = (n - 1)d_{n-1} + (n - 1)d_{n-2}$ *Hint:* To derange the *n* objects, we

may either derange the first $n - 1$ objects and then exchange a_n with one of them or we may exchange a_n with a_j and then derange the remaining $n - 2$ objects.

(b) $d_n = n!\left[\dfrac{1}{2!} - \dfrac{1}{3!} + \dfrac{1}{4!} - \dfrac{1}{5!} + \cdots + (-1)^n\dfrac{1}{n!}\right]$

12-6 COUNTING UNORDERED COLLECTIONS

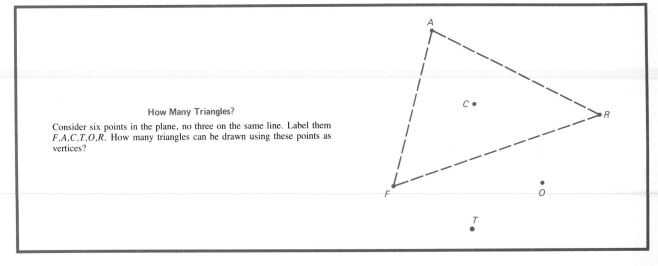

How Many Triangles?

Consider six points in the plane, no three on the same line. Label them *F,A,C,T,O,R*. How many triangles can be drawn using these points as vertices?

For each choice of three points in the opening display, we can draw a triangle. But notice that the order in which we choose the three points does not matter. For example, *FAR*, *FRA*, *AFR*, *ARF*, *RAF*, and *RFA* all determine the same triangle, namely, the one that is shown by dashed lines in the picture. The question about triangles is very different from the question about 3-letter code words raised in the last section; yet there is a connection.

We learned that we can make

$$_6P_3 = 6 \cdot 5 \cdot 4 = 120$$

3-letter code words out of the letters of *FACTOR*. However, every triangle can be labeled with $3! = 6$ different code words. To find the number of triangles, we should therefore divide the number of code words by 3!. We conclude that the number of triangles that can be drawn is

$$\frac{_6P_3}{3!} = \frac{6 \cdot 5 \cdot 4}{3 \cdot 2 \cdot 1} = 20$$

Combinations

An unordered collection of objects is called a **combination** of those objects. If we select *r* objects from a set of *n* distinguishable objects, the resulting subset is called a **combination of *n* things taken *r* at a time.** The number of such combinations is denoted by $_nC_r$ or by $\binom{n}{r}$. Thus $_6C_3$ is the number of combinations of 6 things taken 3 at a time. We calculated it in connection with the triangle problem.

$$_6C_3 = \frac{_6P_3}{3!} = \frac{6 \cdot 5 \cdot 4}{3 \cdot 2 \cdot 1} = 20$$

More generally, if $1 \le r \le n$, the combination symbol $_nC_r$ is given by

$$_nC_r = \frac{_nP_r}{r!} = \frac{n(n-1)(n-2) \cdots (n-r+2)(n-r+1)}{r(r-1)(r-2) \cdots 2 \cdot 1}$$

A good way to remember this is that you want r factors in both numerator and denominator. In the numerator, you start with n and go down; in the denominator you start with r and go down. Incidentally, the answer must be an integer. That means the denominator has to divide the numerator.

Example A (Some Calculations)   Evaluate.
(a) $_5P_2$ (b) $_5C_2$ (c) $_{12}C_4$ (d) $_{10}C_8$ (e) $_{10}C_2$

Solution

(a) $_5P_2 = 5 \cdot 4 = 20$ (b) $_5C_2 = \frac{5 \cdot \overset{2}{\cancel{4}}}{\cancel{2} \cdot 1} = 10$

(c) $_{12}C_4 = \frac{\cancel{12} \cdot 11 \cdot \overset{5}{\cancel{10}} \cdot 9}{\cancel{4} \cdot \cancel{3} \cdot \cancel{2} \cdot 1} = 495$

(d) $_{10}C_8 = \frac{\overset{5}{\cancel{10}} \cdot 9 \cdot \cancel{8} \cdot 7 \cdot \cancel{6} \cdot \cancel{5} \cdot 4 \cdot \cancel{3}}{\cancel{8} \cdot 7 \cdot \cancel{6} \cdot \cancel{5} \cdot 4 \cdot \cancel{3} \cdot \cancel{2} \cdot 1} = 45$ (e) $_{10}C_2 = \frac{\overset{5}{\cancel{10}} \cdot 9}{\cancel{2} \cdot 1} = 45$ ∎

Notice that $_{10}C_8 = {_{10}C_2}$. This is not surprising, since in selecting a subset of 8 objects out of 10, you automatically select 2 to leave behind (you might call it selection by omission). The general fact that follows by the same reasoning is

$$\boxed{_nC_r = {_nC_{n-r}}}$$

Example B (Committees)   In how many ways could the majority leader select a group of 15 senators to serve on an ad hoc committee if 20 are eligible for appointment?

Solution   This is a combination problem; the answer is $_{20}C_{15}$. The boxed formula simplifies the calculation.

$$_{20}C_{15} = {_{20}C_5} = \frac{20 \cdot 19 \cdot 18 \cdot 17 \cdot 16}{5 \cdot 4 \cdot 3 \cdot 2 \cdot 1} = 15{,}504 \quad ∎$$

We can express $_nC_r$ entirely in terms of factorials. Recall that

$$_nP_r = \frac{n!}{(n-r)!}$$

Since $_nC_r = {_nP_r}/r!$, we see that

$$_nC_r = \frac{n!}{r!\,(n-r)!}$$

For this formula to hold when $r = 0$ and $r = n$, it is necessary to define $_nC_0 = 1$ and $0! = 1$. Even in mathematics, there is truth to that old proverb, "Necessity is the mother of invention."

Combinations versus Permutations

Whenever we are faced with the problem of counting the number of ways of selecting r objects from n objects, we are faced with this question. Is the notion of order significant? If the answer is yes, it is a permutation problem; if no, it is a combination problem.

Consider the Birdwatchers' Club of Section 12-5 again. Suppose the club members want to select a president, a vice-president, and a secretary. Is order significant? Yes. The selection can be done in

$$_6P_3 = 6 \cdot 5 \cdot 4 = 120$$

ways.

But suppose they decide simply to choose an executive committee consisting of 3 members. Is order relevant? No. A committee consisting of Filipe, Celia, and Amanda is the same as a committee consisting of Celia, Amanda, and Filipe. A 3-member committee can be chosen from 6 people in

$$_6C_3 = \frac{6 \cdot 5 \cdot 4}{3 \cdot 2 \cdot 1} = 20$$

ways.

The words *arrangement*, *lineup*, and *signal* all suggest order. The words *set*, *committee*, *group*, and *collection* do not.

Example C (Order Not Significant)   From a group of 20 books, Meg will select 4 to read on her vacation. In how many ways can she make this selection?

Solution   Since order is irrelevant, she can do this in $_{20}C_4$ ways

$$_{20}C_4 = \frac{20 \cdot 19 \cdot 18 \cdot 17}{4 \cdot 3 \cdot 2 \cdot 1} = 4845 \quad \blacksquare$$

Example D (Order Significant)   From a group of 20 books, a jury will select four to be awarded first, second, third, and fourth prizes for the quality of their illustrations. In how many ways can this be done?

Solution   Clearly, order is relevant. The jury can make the selection in $_{20}P_4$ ways.

$$_{20}P_4 = 20 \cdot 19 \cdot 18 \cdot 17 = 116,280 \quad \blacksquare$$

Harder Combination Problems

Sometimes we can break a counting task into parts each of which is a combination problem. Then we put the parts together using the multiplication principle and the addition principle of Section 12-5.

Example E (More on Committees)   A committee of 4 is to be selected from a group of 3 seniors, 4 juniors, and 5 sophomores. In how many ways can it be done if
(a) there are no restrictions on the selection?
(b) the committee must have 2 sophomores, 1 junior, and 1 senior?
(c) the committee must have at least 3 sophomores?
(d) the committee must have at least 1 senior?

Solution

(a) $_{12}C_4 = \dfrac{\cancel{12} \cdot 11 \cdot \overset{5}{\cancel{10}} \cdot 9}{\cancel{4} \cdot \cancel{3} \cdot \cancel{2} \cdot 1} = 495$

(b) Two sophomores can be chosen in $_5C_2$ ways, 1 junior in $_4C_1$ ways, and 1 senior in $_3C_1$ ways. By the multiplication principle of counting, the committee can be chosen in

$$_5C_2 \cdot {_4C_1} \cdot {_3C_1} = 10 \cdot 4 \cdot 3 = 120$$

ways. We used the multiplication principle because we choose 2 sophomores *and* 1 junior *and* 1 senior.

(c) At least 3 sophomores means 3 sophomores and 1 nonsophomore *or* 4 sophomores. The word *or* tells us to use the addition principle of counting. We get

$$_5C_3 \cdot {_7C_1} + {_5C_4} = 10 \cdot 7 + 5 = 75$$

(d) Let x be the number of selections with at least one senior and let y be the number of selections with no seniors. Then $x + y$ is the total number of selections, that is, $x + y = 495$ (see part (a)). We calculate y rather than x because it is easier.

$$y = {_9C_4} = \dfrac{9 \cdot 8 \cdot 7 \cdot 6}{4 \cdot 3 \cdot 2 \cdot 1} = 126$$

$$x = 495 - 126 = 369 \quad \blacksquare$$

Example F (Bridge Card Problems)   A standard deck consists of 52 cards. There are 4 suits (spades, clubs, hearts, diamonds), each with 13 cards (2, 3, 4, . . . , 10, jack, queen, king, ace). A bridge hand consists of 13 cards.
(a) How many different possible bridge hands are there?
(b) How many of them have exactly 3 aces?
(c) How many of them have no aces?
(d) How many of them have cards from just 3 suits?

Solution

(a) The order of the cards in a hand is irrelevant; it is a combination problem. We can select 13 cards out of 52 in $_{52}C_{13}$ ways, a number so large we will not bother to calculate it.

(b) The three aces can be selected in $_4C_3$ ways, the 10 remaining cards in $_{48}C_{10}$ ways. The answer (using the multiplication principle) is $_4C_3 \cdot _{48}C_{10}$.

(c) From 48 nonaces, we select 13 cards; the answer is $_{48}C_{13}$.

(d) We think of this as no clubs, or no spades, or no hearts, or no diamonds and use the addition principle.

$$_{39}C_{13} + _{39}C_{13} + _{39}C_{13} + _{39}C_{13} = 4 \cdot _{39}C_{13} \quad \blacksquare$$

PROBLEM SET 12-6

A. Skills and Techniques

1. Calculate each of the following (see Example A).
   (a) $_{10}P_3$ (b) $_{10}C_3$
   (c) $_5P_5$ (d) $_5C_5$
   (e) $_6P_1$ (f) $_6C_1$

2. Calculate each of the following.
   (a) $_{12}P_2$ (b) $_{12}C_2$
   (c) $_4P_4$ (d) $_4C_4$
   (e) $_{10}P_1$ (f) $_{10}C_1$

3. Use the fact that $_nC_r = _nC_{n-r}$ to calculate each of the following.
   (a) $_{20}C_{17}$ (b) $_{100}C_{97}$

4. Calculate each of the following.
   (a) $_{41}C_{39}$ (b) $_{1000}C_{998}$

Problems 5–18 relate to Examples B–D.

5. In how many ways can a committee of 3 be selected from a class of 8 students?

6. In how many ways can a committee of 5 be selected from a class of 8 students?

7. A political science professor must select 4 students from her class of 12 students for a field trip to the state legislature. In how many ways can she do it?

8. The professor of Problem 7 was asked to rank the top 4 students in her class of 12. In how many ways could that be done?

9. A police chief needs to assign officers from the 10 available to control traffic at junctions A, B, and C. In how many ways can he do it?

10. If 12 horses are entered in a race, in how many ways can the first 3 places (win, place, show) be taken?

11. A basket contains 30 apples. In how many ways can a person select
    (a) a sample of 4 apples?
    (b) a sample of 3 apples that excludes the 5 rotten ones?

12. From a group of 9 cards numbered 1, 2, 3, . . . , 9, how many
    (a) sets of 3 cards can be drawn?
    (b) 3-digit numbers can be formed?

13. How many games will be played in a 10-team league if each team plays every other team exactly twice?

14. Determine the maximum number of intersection points for a group of 15 lines.

15. How many 10-member subsets does a set of size 12 have?

16. How many subsets (of all sizes) does a set of size 5 have? The empty set is a subset.

17. From a class of 6 members, in how many ways can a committee of any size be selected (including a committee of one)?

18. From a penny, a nickel, a dime, a quarter, and a half dollar, how many different sums can be made?

For Problems 19–22, see Example E.

19. An investment club has a membership of 4 women and 6 men. A research committee of 3 is to be formed. In how many ways can this be done if
    (a) there are to be 2 women and 1 man on the committee?
    (b) there is to be at least 1 woman on the committee?
    (c) all 3 are to be of the same sex?

20. A senate committee of 4 is to be formed from a group consisting of 5 Republicans and 6 Democrats. In how many ways can this be done if
    (a) there are to be 2 Republicans and 2 Democrats on the committee?
    (b) there are to be no Republicans on the committee?
    (c) there is to be at most one Republican on the committee?

21. Suppose that a bag contains 4 black and 7 white balls. In how many ways can a group of 3 balls be drawn from the bag consisting of

(a) 1 black and 2 white balls?
(b) balls of just one color?
(c) at least 1 black ball?
Note: Assume the balls are distinguishable; for example, they may be numbered.

22. John is going on a vacation trip and wants to take 5 books with him from his personal library, which consists of 6 science books and 10 novels. In how many ways can he make his selection if he wants to take
(a) 2 science books and 3 novels?
(b) at least 1 science book?
(c) 1 book of one kind and 4 books of the other kind?

Problems 23–28 deal with bridge hands. Leave your answers in terms of combination symbols. See Example F.

23. How many of the hands have only red cards? *Note:* Half of the cards are red.
24. How many of the hands have only honor cards (aces, kings, queens, and jacks)?
25. How many of the hands have one card of each kind (1 ace, 1 king, 1 queen, and so on)?
26. How many of the hands have exactly 2 kings?
27. How many of the hands have 2 or more kings?
28. How many of the hands have exactly 2 aces and 2 kings?

Problems 29–32 deal with poker hands, which consist of 5 cards.

29. How many different poker hands are possible?
30. How many of them have exactly 2 hearts and 2 diamonds?
31. How many have 2 pairs of different kinds (for example, 2 aces and 2 fives)?
32. How many are 5-card straights (for example, 7, 8, 9, 10, jack in same suit)? An ace may count either as the highest or the lowest card.

B. Applications and Extensions

33. A quarter, a dime, a nickel, and a penny are tossed. In how many ways can they fall?
34. From a committee of 11, a subcommittee of 4 is to be chosen. In how many ways can this be done?
35. From 5 representatives of labor, 4 representatives of business, and 3 representatives of the general public, how many different mediation committees can be formed with 2 people from each of the three groups?
36. In how many ways can a group of 12 people be split into three nonoverlapping committees of size 5, 4, and 3, respectively?

37. A class of 12 people will select a president, a secretary, a treasurer, and a program committee of 3 with no overlapping of positions. In how many ways can this be done?

38. A committee of 4 is to be formed from a group of 4 freshmen, 3 sophomores, 2 juniors, and 6 seniors. In how many ways can this be done if
(a) each class must be represented?
(b) freshmen are excluded?
(c) the committee must have exactly two seniors?
(d) the committee must have at least one senior?

39. A test consists of 10 true-false items.
(a) How many different sets of answers are possible?
(b) How many of these have exactly 4 right answers?

40. An ice cream parlor has 10 different flavors. How many different double-dip cones can be made if
(a) the two dips must be of different flavors but the order of putting them on the cone does not matter?
(b) the two dips must be different and order does matter?
(c) the two dips need not be different but order does matter?
(d) the two dips need not be different and order does not matter?

41. Mary has a penny, a nickel, a dime, a quarter, a half dollar, and a silver dollar in her purse. How many different possible sums of money (consisting of at least one coin) could she give to her daughter Tosha?

42. In how many ways can 8 presents be split between John and Mary if
(a) each is to get 4 presents?
(b) John is to get 5 presents and Mary 3 presents?
(c) there are no restrictions on how the presents are split?

43. Let 10 fixed points on a circle be given. How many convex polygons can be formed which have vertices chosen from among these points?

44. Calculate the sums in parts (a)–(c).
(a) $_2C_0 + {}_2C_1 + {}_2C_2$
(b) $_3C_0 + {}_3C_1 + {}_3C_2 + {}_3C_3$
(c) $_4C_0 + {}_4C_1 + {}_4C_2 + {}_4C_3 + {}_4C_4$
(d) Conjecture a formula for $_nC_0 + {}_nC_1 + {}_nC_2 + \cdots + {}_nC_n$.

45. Prove your conjecture in Problem 44 by considering two different methods of counting the ways of splitting *n* presents between two people. *Hint:* One method is to let each present choose the person it will go to.

46. Show that

$$_nC_0\,_nC_n + _nC_1\,_nC_{n-1} + _nC_2\,_nC_{n-2}$$
$$+ \cdots + _nC_n\,_nC_0 = _{2n}C_n$$

by counting the number of ways of drawing n balls from an urn that has n red and n black balls.

47. Find a nice formula for $\sum_{j=0}^{n} (_nC_j)^2$. *Hint:* See Problem 46.

48. Use the factorial formula for $_nC_r$ to show that $_nC_{r-1} + _nC_r = _{n+1}C_r$.

49. Obtain a nice formula for

$$S = _{n+1}C_1 + _{n+2}C_2 + _{n+3}C_3 + \cdots + _{n+k}C_k$$

Hint: Add $_{n+1}C_0$ to S on the left; then use the result in Problem 48 repeatedly to collect two terms on the left.

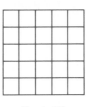

Figure 23

50. Show that if $m \le n$, then $\sum_{j=0}^{m} _nC_j\,_{n-j}C_{m-j} = _nC_m 2^m$.

51. Consider an n-by-n checkerboard (a 5-by-5 checkerboard is shown in Figure 23). How many rectangles of all sizes are determined by this board?

52. TEASER Obtain a formula for the number of squares of all sizes that are determined by the n-by-n checkerboard. Evaluate when $n = 10$.

12-7 THE BINOMIAL FORMULA

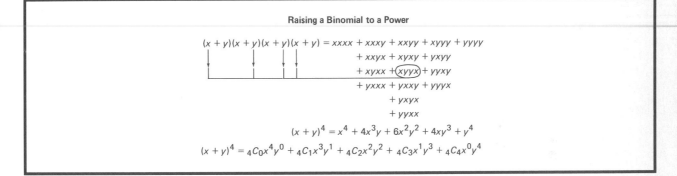

Raising a Binomial to a Power

$$(x + y)(x + y)(x + y)(x + y) = xxxx + xxxy + xxyy + xyyy + yyyy$$
$$+ xxyx + xyxy + yxyy$$
$$+ xyxx + \boxed{xyyx} + yyxy$$
$$+ yxxx + yxxy + yyyx$$
$$+ yxyx$$
$$+ yyxx$$

$$(x + y)^4 = x^4 + 4x^3y + 6x^2y^2 + 4xy^3 + y^4$$
$$(x + y)^4 = _4C_0 x^4 y^0 + _4C_1 x^3 y^1 + _4C_2 x^2 y^2 + _4C_3 x^1 y^3 + _4C_4 x^0 y^4$$

In the opening box, we have shown how to expand $(x + y)^4$. Admittedly, it looks complicated; however, it leads to the remarkable formula at the bottom of the display. That formula generalizes to handle $(x + y)^n$, where n is any positive integer. It is worth a careful investigation.

To produce any given term in the expansion of $(x + y)^4$, each of the four factors $x + y$ contributes either an x or a y. There are $2 \cdot 2 \cdot 2 \cdot 2 = 16$ ways in which they can make this contribution, hence the 16 terms in the long expanded form. But many of these terms are alike; in fact, only five different types occur, namely, x^4, x^3y, x^2y^2, xy^3, and y^4. The number of times each occurs is $_4C_0$, $_4C_1$, $_4C_2$, $_4C_3$, and $_4C_4$, respectively. (Remember we defined $_4C_0$ to be 1.)

Why do the combination symbols arise in this expansion? For example, why is $_4C_2$ the coefficient of x^2y^2? If you follow the arrows in the opening display, you see how the term $xyyx$ comes about. It gets its two y's from the second and third $x + y$ factors (the x's then must come from the first and fourth $x + y$ factors). Thus, the number of terms of the form x^2y^2 is the number of ways of selecting two factors out of four from which to take y's (the x's must come from the remaining two factors).

We can select two objects out of four in $_4C_2$ ways; hence the coefficient of x^2y^2 is $_4C_2$.

The Binomial Formula

What we have just done for $(x + y)^4$ can be carried out for $(x + y)^n$, where n is any positive integer. The result is called the **Binomial Formula.**

$$(x + y)^n = {_nC_0}x^ny^0 + {_nC_1}x^{n-1}y^1 + \cdots + {_nC_{n-1}}x^1y^{n-1} + {_nC_n}x^0y^n$$

$$= \sum_{k=0}^{n} {_nC_k}x^{n-k}y^k$$

Notice that the k in $_nC_k$ is the exponent on y and that the two exponents in each term sum to n.

Let us apply the Binomial Formula with $n = 6$.

$$(x + y)^6 = {_6C_0}x^6 + {_6C_1}x^5y + {_6C_2}x^4y^2 + \cdots + {_6C_6}y^6$$

$$= x^6 + 6x^5y + 15x^4y^2 + 20x^3y^3 + 15x^2y^4 + 6xy^5 + y^6$$

Example A (Complicating the Terms) Write the expansion of $(2a - b^2)^6$ and simplify.

Solution Use the expansion of $(x + y)^6$ given above with x replaced by $2a$ and y by $-b^2$.

$$[2a + (-b^2)]^6 = (2a)^6 + 6(2a)^5(-b^2) + 15(2a)^4(-b^2)^2 + 20(2a)^3(-b^2)^3$$

$$+ 15(2a)^2(-b^2)^4 + 6(2a)(-b^2)^5 + (-b^2)^6$$

$$= 64a^6 - 192a^5b^2 + 240a^4b^4 - 160a^3b^6$$

$$+ 60a^2b^8 - 12ab^{10} + b^{12} \quad \blacksquare$$

Example B (Finding the Initial Terms) Write in simplified form the first three terms in the expansion of $\left(x^3 + \dfrac{3}{x^4}\right)^{10}$.

Solution The first three terms are

$$_{10}C_0(x^3)^{10} + {_{10}C_1}(x^3)^9\left(\frac{3}{x^4}\right) + {_{10}C_2}(x^3)^8\left(\frac{3}{x^4}\right)^2 = x^{30} + (10)(3)\frac{x^{27}}{x^4} + (45)(9)\frac{x^{24}}{x^8}$$

$$= x^{30} + 30x^{23} + 405x^{16} \quad \blacksquare$$

Example C (Finding a Specific Term of a Binomial Expansion) Find the term in the expansion of $(2x + y^2)^{10}$ that involves y^{12}.

Solution This term will arise from raising y^2 to the 6th power. It is, therefore,

$$_{10}C_6(2x)^4(y^2)^6 = 210 \cdot 16x^4y^{12} = 3360x^4y^{12} \quad \blacksquare$$

Example D (An Application to Compound Interest)   If $100 is invested at 12 percent compounded monthly, it will accumulate to $100(1.01)^{12}$ dollars by the end of one year. Use the Binomial Formula to approximate this amount.

Solution

$$100(1.01)^{12} = 100(1 + .01)^{12}$$

$$= 100\left[1 + 12(.01) + \frac{12 \cdot 11}{2}(.01)^2 + \frac{12 \cdot 11 \cdot 10}{6}(.01)^3 + \cdots \right]$$

$$= 100[1 + .12 + .0066 + .00022 + \cdots]$$

$$\approx 100(1.12682) \approx 112.68$$

This answer of $112.68 is accurate to the nearest penny since the last nine terms of the expansion do not add up to as much as a penny.   ∎

The Binomial Coefficients

The combination symbols $_nC_k$ are often called **binomial coefficients,** for reasons that should be obvious. Their remarkable properties have been studied for hundreds of years. Let us see if we can discover some of them. We begin by expanding $(x + y)^k$ for increasing values of k, listing only the coefficients (Figure 24).

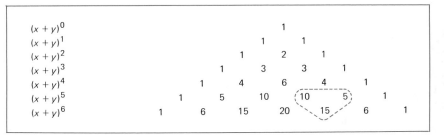

Figure 24

The resulting triangle of numbers composed of the binomial coefficients is called **Pascal's triangle** after the gifted French philosopher and mathematician, Blaise Pascal (1622–1662). Notice its symmetry. If folded across a vertical center line, the numbers match. This corresponds to an algebraic fact you learned earlier.

$$_nC_k = {_nC_{n-k}}$$

Next notice that any number in the body of the triangle is the sum of the two numbers closest to it in the line above the number. For example, $15 = 10 + 5$, as the dashed triangle was meant to suggest. In symbols

$$_{n+1}C_k = {}_nC_{k-1} + {}_nC_k$$

a fact that can be proved rigorously by using the factorial formula for $_nC_k$ (see Problem 48 of Section 12-6).

Now add the numbers in each row of Pascal's triangle.

$$1 = 1 = 2^0$$
$$1 + 1 = 2 = 2^1$$
$$1 + 2 + 1 = 4 = 2^2$$
$$1 + 3 + 3 + 1 = 8 = 2^3$$

This suggests the formula

$$_nC_0 + {}_nC_1 + {}_nC_2 + \cdots + {}_nC_n = 2^n$$

Its truth can be demonstrated by substituting $x = 1$ and $y = 1$ in the Binomial Formula. It has an important interpretation. Take a set with n elements. This set has $_nC_n$ subsets of size n, $_nC_{n-1}$ subsets of size $n - 1$, and so on. The left side of the boxed formula is the total number of subsets (including the empty set) of a set with n elements; remarkably, it is just 2^n. For example, $\{a, b, c\}$ has $2^3 = 8$ subsets.

$$\{a, b, c\} \quad \{a, b\} \quad \{a, c\} \quad \{b, c\} \quad \{a\} \quad \{b\} \quad \{c\} \quad \varnothing$$

A related formula is

$$_nC_0 - {}_nC_1 + {}_nC_2 - \cdots + (-1)^n{}_nC_n = 0$$

which can be obtained by setting $x = 1$ and $y = -1$ in the Binomial Formula. When rewritten as

$$_nC_0 + {}_nC_2 + {}_nC_4 + \cdots = {}_nC_1 + {}_nC_3 + {}_nC_5 + \cdots$$

it says that the number of subsets with an even number of elements is equal to the number of subsets with an odd number of elements.

Example E (Subsets and Committees)   From a group of 11 people, how many of each of the following can be formed?

(a) Subsets
(b) Subsets with an odd number of elements
(c) 8-member committees
(d) Committees with at least 2 members

Solution

(a) The total number of subsets is $2^{11} = 2048$.
(b) Since the number N of subsets with an odd number of elements is equal to the number of subsets with an even number of elements, $N = 2048/2 = 1024$.

(c) $_{11}C_8 = {}_{11}C_3 = \dfrac{11 \cdot 10 \cdot 9}{3 \cdot 2 \cdot 1} = 165.$

(d) The required number of committees is the same as the number of subsets with at least 2 members and this is equal to

$$2^{11} - {}_{11}C_0 - {}_{11}C_1 = 2046 - 1 - 11 = 2036 \quad \blacksquare$$

PROBLEM SET 12-7

A. Skills and Techniques

In Problems 1–8, expand and simplify. See Example A.

1. $(x + y)^3$

2. $(x - y)^3$

3. $(x - 2y)^3$

4. $(3x + b)^3$

5. $(c^2 - 3d^3)^4$

6. $(xy - 2z^2)^4$

7. $(ab^2 - bc)^5$

8. $(ab + 1/a)^5$

Write the first three terms of each expansion in Problems 9–12 in simplified form. See Example B.

9. $(x + y)^{20}$

10. $(x + y)^{30}$

11. $(x + 1/x^5)^{20}$

12. $(xy^2 + 1/y)^{14}$

Problems 13–16 relate to Example C.

13. Find the term in the expansion of $(y^2 - z^3)^{10}$ that involves z^9.

14. Find the term in the expansion of $(3x - y^3)^{10}$ that involves y^{24}.

15. Find the term in the expansion of $(2a - b)^{12}$ that involves a^3.

16. Find the term in the expansion of $(x^2 - 2/x)^5$ that involves x^4.

In Problems 17–20 use the first three terms of a binomial expansion to find an approximate value of the given expression.

17. $20(1.02)^8$

18. $100(1.002)^{20}$

19. $500(1.005)^{20}$

20. $200(1.04)^{10}$

21. Bacteria multiply in a certain medium so that by the end of k hours their number N is $N = 100(1.02)^k$. Approximate the number of bacteria after 20 hours.

22. Do Problem 21 assuming that $N = 1000(1.01)^k$.

Problems 23–28 relate to Example E.

23. Find the number of subsets of each of the following sets.

(a) $\{a, b, c, d\}$

(b) $\{1, 2, 3, 4, 5\}$

(c) $\{x_1, x_2, x_3, x_4, x_5, x_6\}$

24. How many subsets with an even number of elements does a 9-element set have?

25. How many committees with at least 3 members can be formed from a group of 9 people?

26. How many of the committees in Problem 25 have an odd number of members?

27. Let $\mathbf{A}$ and $\mathbf{B}$ be noncommuting 3 by 3 matrices. How many terms are there in the expansions of $(\mathbf{A} + \mathbf{B})^2$? of $(\mathbf{A} + \mathbf{B})^3$?

28. From a group of 10 people, how many committees can be formed whose size is a multiple of 3?

B. Applications and Extensions

29. Expand and simplify each of the following.

(a) $(2x + \frac{1}{2})^8$

(b) $(1 + \sqrt{3})^6$

30. Find and simplify the first three terms of the expansion of $(a^2 + 4b^3)^{12}$.

31. Find and simplify the term in the expansion of $(x - 2z^3)^8$ that involves z^{15}.

32. Expand and simplify each of the following.

(a) $\left(2x^3 - \dfrac{1}{x}\right)^5$

(b) $\dfrac{(x + h)^4 - x^4}{h}$

33. In each of the following, find the term in the expanded and simplified form that does not involve h (a procedure very important in calculus).

(a) $\dfrac{(x + h)^n - x^n}{h}$

(b) $\dfrac{(x + h)^{10} + 2(x + h)^4 - x^{10} - 2x^4}{h}$

34. Given that i is the imaginary unit, calculate the following.

(a) $(1 + i)^5$

(b) $(1 - i)^6$

35. Find the constant term in the expansion of $(3x^2 + 1/3x)^{12}$.

36. Without using a calculator, show that $(1.0003)^{10} > 1.003004$.

37. Without using a calculator, show that $(1.01)^{50} > 1.5$.

38. Without using a calculator, find $(.999)^{10}$ correct to six decimal places.

39. How many committees consisting of 3 or more members can be selected from a group of 12 people?

40. How many subsets with an odd number of elements does a set with 13 members have?

41. In the expansion of the trinomial $(x + y + z)^n$, the coefficient of $x^r y^s z^t$ where $r + s + t = n$ is $n!/r! s! t!$
  (a) Expand $(x + y + z)^3$.
  (b) Find the coefficient of the term $x^2 y^4 z$ in the expansion of $(2x + y + z)^7$.

42. Find the sum of all the coefficients in the expansion of the trinomial $(x + y + z)^n$.

43. Find a simple formula for $\sum_{k=0}^{n} {}_nC_k 2^k$.

44. Find the sum of the coefficients in $(4x^3 - x)^6$ after it is expanded and simplified. *Hint:* This is a simple problem when looked at the right way.

45. Let $P(x)$ be the nth-degree polynomial defined by

$$P(x) = 1 + x + \frac{x(x-1)}{2!} + \frac{x(x-1)(x-2)}{3!}$$

$$+ \cdots + \frac{x(x-1)(x-2)\cdots(x-n+1)}{n!}$$

Find a simple formula for each of the following.
  (a) $P(k)$, $k = 0, 1, 2, \ldots, n$.
  (b) $P(n+1)$
  (c) $P(n+2)$

46. TEASER The integer 3 can be expressed as the sum of one or more positive integers in four ways, namely, as 3, $2 + 1$, $1 + 2$, and $1 + 1 + 1$. In how many ways can the positive integer n be so expressed? It is not enough to give the answer; you must give an argument to show that your answer is correct.

12-8 PROBABILITY

Probability is the very guide of life.

Cicero

If we can be sure of anything, we can be sure that our two college students are not going to study. There may be three outcomes of this experiment, but a coin that balances on edge, when weighed on the scales of justice, is likey to be found wanting. A fair coin landing on a flat surface will show either heads or tails.

More is true. A fair coin is just as likely to show heads as tails. That is, if a fair coin were tossed over and over again, say millions of times, it would show heads approximately one-half of the time and tails about one-half of the time. How do we know that this is true for a particular coin? We do not; but it is an assumption that most of us are willing to make, particularly if the coin appears to be perfectly round and unworn.

Figure 25

There are many situations in which the assumption that certain events are equally likely seems plausible. Toss a standard die (Figure 25). Is there any reason to think that one side has an edge in the battle to show its face? We think not. Draw a card from a well-shuffled deck. Nature is perfectly democratic. It gives each card the same chance of being drawn.

But one should be careful. If a stockbroker tries to convince you that the market can do only one of three things—go up, stay the same, or go down—he or she is, of course, right. But if he or she further insists that these three possibilities are equally likely and therefore that the odds are two to one against losing, hang onto your wallet.

Definition of Probability

Now we are ready to introduce the main notion of this chapter, the probability of an event. Consider the experiment of tossing a die, which can result in six equally likely outcomes. Of these six outcomes, three show an even number of spots. We therefore say that the probability that an even number will show is $\frac{3}{6}$.

Or consider the experiment of picking a card from a well-shuffled deck. Thirteen of the 52 cards are diamonds. Thus we say that the probability of getting a diamond is $\frac{13}{52}$.

These examples serve to illustrate the general situation. If an experiment can result in any one of n equally likely outcomes and if exactly m of them result in the event E, we say that the **probability of E** is m/n. We write this as

$$P(E) = \frac{m}{n}$$

Example A (Simple Probability)   One card is drawn at random from a deck with 99 cards, numbered 1, 2, 3, . . . , 99. Calculate the probability that

(a) the card shows an even number;
(b) the card shows a number greater than 11;
(c) the sum of the digits on the card is greater than 11.

Solution
(a) There are 49 even-numbered cards; thus the required probability is $\frac{49}{99}$.
(b) There are 88 cards numbered above 11; thus the required probability is $\frac{88}{99} = \frac{8}{9}$.
(c) The cards meeting the criterion are those with the numbers 39; 48, 49; 57, 58, 59; and so on. There are $1 + 2 + 3 + \cdots + 7 = 28$ of these cards. The required probability is $\frac{28}{99}$.   ∎

Figure 26

Properties of Probability

We use two dice, one colored and the other white, to illustrate the major properties of probabilities (Figure 26). Since one die can fall in 6 ways, two dice can fall in $6 \times 6 = 36$ ways. The 36 equally likely outcomes are shown in Figure 27. As you read the next several paragraphs, you will want to refer back to this diagram.

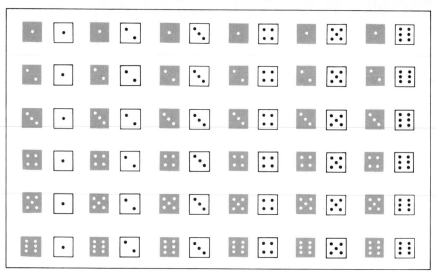

Figure 27

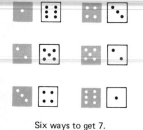

Six ways to get 7.

P (getting 7) $= \frac{6}{36}$

Figure 28

Two ways to get 11.

P (getting 11) $= \frac{2}{36}$

Figure 29

Most questions about a pair of dice have to do with the total number of spots showing after a toss. For example, what is the probability of getting a total of 7? Six of the 36 outcomes result in this event (Figure 28). Therefore, the required probability is $\frac{6}{36} = \frac{1}{6}$.

Example B (Dice Problems)   Let T denote the total number of spots showing after two fair dice are tossed. Calculate the following probabilities.

(a) $P(T = 11)$ (b) $P(T = 12)$ (c) $P(T > 7)$

(d) $P(T < 13)$ (e) $P(T = 13)$

Solution

(a) $P(T = 11) = \frac{2}{36} = \frac{1}{18}$ (see Figure 29)

(b) $P(T = 12) = \frac{1}{36}$

(c) $P(T > 7) = \frac{15}{36} = \frac{5}{12}$

(d) $P(T < 13) = \frac{36}{36} = 1$

(e) $P(T = 13) = \frac{0}{36} = 0$ ■

The last two events in Example B are worthy of comment. The event "getting a total less than 13" is certain to occur; we call it a **sure event.** The probability of a sure event is always 1. However, the event "getting a total of 13" cannot occur; it is called an **impossible event.** The probability of an impossible event is zero.

Next, consider the event "getting 7 or 11," which is important in the dice game called craps. From the diagram, we see that 8 of the 36 outcomes give a total of 7 or 11, so the probability of this event is $\frac{8}{36}$. But note that we could have calculated this probability by adding the probability of getting 7 and the probability of getting 11.

$$P(T = 7 \; or \; T = 11) = P(T = 7) + P(T = 11)$$

$$\frac{8}{36} = \frac{6}{36} + \frac{2}{36}$$

It would appear that we have found a very useful property of probability. When an event is described by using the conjunction *or,* we may find its probability by adding the probabilities of the two parts of the conjunction. However, a different example makes us take second look. Note that

$$P(T \text{ is odd } or \ T > 7) \neq P(T \text{ is odd}) + P(T > 7)$$

$$\frac{27}{36} \neq \frac{18}{36} + \frac{15}{36}$$

Why is it that we can add probabilities in one case but not in the other? The reason is a simple one. The events "getting 7" and "getting 11" are *disjoint* (they cannot both happen). But the events "getting an odd total" and "getting over 7" overlap (a number such as 9 satisfies both conditions).

Considerations like these lead us to the main properties of probability.

Four Properties of Probability

1. $P(\text{impossible event}) = 0;$ $P(\text{sure event}) = 1$
2. $0 \leq P(A) \leq 1$ for any event A.
3. $P(A \text{ or } B) = P(A) + P(B)$, provided that A and B are disjoint (that is, cannot both happen at the same time).
4. $P(A) = 1 - P(\text{not } A)$

Property 4 deserves comment since it follows directly from Properties 1 and 3. The events "A" and "not A" are certainly disjoint. Thus, from Property 3,

$$P(A \text{ or } \text{not } A) = P(A) + P(\text{not } A)$$

The event "A *or* not A" must occur; it is a sure event. Hence from Property 1,

$$1 = P(A) + P(\text{not } A)$$

which is equivalent to Property 4.

Example C (Complementary Events)   Calculate the probability of getting a total T less than 11, when tossing two dice.

Solution

$$P(T < 11) = 1 - P(T \geq 11) = 1 - \frac{3}{36} = \frac{33}{36} = \frac{11}{12} \quad \blacksquare$$

Relation to Set Language

Most American students are introduced to the language of sets in the early grades. Even so, a brief review may be helpful. In everyday language, we talk of a bunch of grapes, a class of students, a herd of cattle, a flock of birds, or perhaps even a team of toads, a passel of possums, or a gaggle of geese. Why are there so many words to express the same idea? We do not know, but we do know that mathematicians prefer to use one word—**set.** The objects that make up a set are called **elements,** or **members,** of the set.

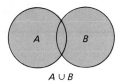

$A \cup B$

Sets can be put together in various ways (Figure 30). We have $A \cup B$ (read "A **union** B"), which consists of the elements in A *or* B. The set $A \cap B$ (read "A **intersection** B") is made up of the elements that are in both A *and* B. We have the notion of an **empty set** $\varnothing$ and of a **universe** S (the set of all elements under discussion). Finally, we have A' (read "A **complement**"), which is composed of all elements in the universe that are not in A.

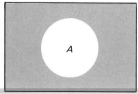

$A \cap B$

The language of sets and the language used in probability are very closely related, as the list below suggests.

Probability Language	Set Language
Outcome	Element
Event	Set
Or	Union
And	Intersection
Not	Complement
Impossible event	Empty set
Sure event	Universe

A'

Figure 30

If we borrow the notation of set theory, we can state the laws of probability in a very succinct form.

1. $P(\varnothing) = 0$; $P(S) = 1$
2. $0 \le P(A) \le 1$
3. $P(A \cup B) = P(A) + P(B)$, provided that $A \cap B = \varnothing$
4. $P(A) = 1 - P(A')$

Gambling and Card Problems

Although gamblers may not realize it, they illustrate the laws of probability.

Example D (Odds) Gamblers often use the language of odds rather than of probability. For example, Las Vegas may report that the odds in favor of the Yankees winning the seventh game of a World Series are 3 to 2. Translate this statement into probability language.

Solution   To say that the odds are 3 to 2 in favor of the Yankees means that if the game were played many times, the Yankees would win 3 times to every 2 times their opponent won. Thus the Yankees would win $\frac{3}{5}$ of the times, which means the probability of the Yankees winning is $\frac{3}{5}$. In general, if the odds in favor of event E are m to n, then the probability of E is $m/(m + n)$. Conversely, if the probability of E is j/k, then the odds in favor of E are j to $k - j$.   ■

Example E (Card Problems)   A standard deck consists of 52 cards. There are 4 suits (spades, clubs, hearts, and diamonds), each with 13 cards (2, 3, . . . , 10, jack, queen, king, ace). A poker hand consists of 5 cards. If a poker hand is to be dealt from a standard deck, what is the probability it will be

(a) a diamond flush (that is, all diamonds)?   (b) a flush?

Solution

(a) Recall our study of combinations from Section 12-6. There are $_{52}C_5$ possible 5-card hands, of which $_{13}C_5$ consists of all diamonds. Thus

$$P(\text{diamond flush}) = \frac{_{13}C_5}{_{52}C_5} = \frac{\dfrac{13 \cdot 12 \cdot 11 \cdot 10 \cdot 9}{5 \cdot 4 \cdot 3 \cdot 2 \cdot 1}}{\dfrac{52 \cdot 51 \cdot 50 \cdot 49 \cdot 48}{5 \cdot 4 \cdot 3 \cdot 2 \cdot 1}} \approx .0005.$$

(b) A flush is a diamond flush or a heart flush or a club flush or a spade flush. Thus

$$P(\text{flush}) = P(\text{diamond flush}) + P(\text{heart flush})$$
$$+ \; P(\text{club flush}) + P(\text{spade flush})$$
$$\approx .0005 + .0005 + .0005 + .0005$$
$$= .002 \quad \blacksquare$$

PROBLEM SET 12-8

A. Skills and Techniques

For Problems 1–14, see Examples A–C.

1. An ordinary die is tossed. What is the probability that the number of spots on the upper face will be
(a) three?   (b) greater than 3?
(c) less than 3?   (d) an even number?
(e) an odd number?

2. Nine balls, numbered 1, 2, . . . , 9, are in a bag. If one is drawn at random, what is the probability that its number is
(a) 9?   (b) greater than 5?
(c) less than 6?   (d) even?
(e) odd?

3. A penny, a nickel, and a dime are tossed. List the 8 possible outcomes of this experiment. What is the probability of
(a) 3 heads?   (b) exactly 2 heads?
(c) more than 1 head?

4. A coin and a die are tossed. Suppose that one side of the coin has a 1 on it and the other a 2. List the 12 possible outcomes of this experiment, for example, (1, 1), (1, 2), (1, 3). What is the probability of
(a) a total of 4?   (b) an even total?
(c) an odd total?

5. Two ordinary dice are tossed. What is the probability of
(a) a double (both showing the same number)?
(b) the number on one of the dice being twice that on the other?

(c) the numbers on the two dice differing by at least 2?

6. A letter is chosen at random from the word *PROBABILITY*. What is the probability that it will be
(a) *P*?   (b) *B*?
(c) *M*?
(d) a vowel? (Treat *Y* as a vowel.)

7. Two regular tetrahedra (tetrahedra have four identical equilateral triangles for faces as shown in Figure 31) have faces numbered 1, 2, 3, and 4. Suppose they are tossed and we keep track of the outcome by listing the numerals on the bottom faces, for example, (1, 1), (1, 2).

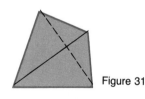

Figure 31

(a) How many outcomes are there?
(b) What is the probability of a sum of 7?
(c) What is the probability of a sum less than 7?

8. Two regular octahedra (polyhedra having eight identical faces) have faces numbered 1, 2, . . . , 8. Suppose they are tossed and we record the outcomes by listing the numerals on the bottom faces.
(a) How many outcomes are there?
(b) What is the probability of a sum of 7?
(c) What is the probability of a sum less than 7?

9. What is wrong with each of the following statements?
   (a) Since there are 50 states, the probability of being born in Wyoming is $\frac{1}{50}$.
   (b) The probability that a person smokes is .45, and the probability that he or she drinks, .54; therefore the probability that he or she smokes or drinks is $.54 + .45 = .99$.
   (c) The probability that a certain candidate for president of the United States will win is $\frac{3}{5}$, and that he or she will lose, $\frac{1}{4}$.
   (d) Two football teams A and B are evenly matched; therefore the probability that A will win is $\frac{1}{2}$.

10. During the past 30 years, Professor Witquick has given only 100 A's and 200 B's in Math 13 to the 1200 students who registered for the class. Based on these data, what is the probability that a student who registers next year
   (a) will get an A or a B?
   (b) will not get either an A or a B?

11. A poll was taken at Podunk University on the question of coeducational dormitories, with the following results.

	Administrators	Faculty	Students	Total
For	4	16	100	120
Against	3	32	100	135
Unsure	3	2	40	45
Total	10	50	240	300

On the basis of this poll, what is the probability that
   (a) a randomly chosen faculty member will favor coed dorms?
   (b) a randomly chosen student will be against coed dorms?
   (c) a person selected at random at Podunk University will favor coed dorms?
   (d) a person selected at random at Podunk University will be a faculty member who is against coed dorms?

12. The well-balanced spinner shown in Figure 32 is spun. What is the probability that the pointer will stop at
   (a) red?   **(b)** green?
   (c) red or green? **(d)** not green?

13. Four balls numbered 1, 2, 3, and 4 are placed in a bag, mixed, and drawn out, one at a time. What is the probability that they will be drawn in the order 1, 2, 3, 4?

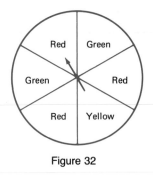

Figure 32

14. A five-volume set of books is placed on a shelf at random. What is the probability they will be in the right order?

For Problems 15–18, see Example D.

15. Suppose that the odds in favor of the Vikings winning the Super Bowl game are given as 2 to 9. What is the probability they will lose the Super Bowl game?

16. What is the probability that David will marry Jane if he says the odds in favor are 1 to 8?

17. What are the odds in favor of getting two heads when tossing two coins?

18. What are the odds in favor of 7 or 11 in tossing two dice?

Problems 19–24 relate to Example E.

19. From a standard deck, one card is drawn. What is the probability that it will be
   (a) red?   **(b)** a spade?
   (c) an ace?
   Note: Two of the suits are red and two are black.

20. Two cards are drawn from a standard deck (there are $_{52}C_2$ ways of doing it). What is the probability both will be
   (a) red?   **(b)** of the same color?
   (c) aces?

21. Three cards are drawn from a standard deck. What is the probability that
   (a) all will be red?
   (b) all will be diamonds?
   (c) exactly one will be a queen?
   (d) all will be queens?

22. A poker hand is drawn from a standard deck. What is the probability of at least one ace? *Hint:* Look at the complementary event consisting of no aces.

23. What is the probability of a poker hand consisting of all kings and queens?

24. What is the probability of a full-house poker hand (two cards of one value, three of another)?

B. Applications and Extensions

25. If three dice are tossed together, what is the probability that the total obtained is
(a) 18?   (b) 16?
(c) greater than 4?

26. A purse contains 3 nickels, 1 dime, 1 quarter, and 1 half-dollar. If three coins are drawn from the purse at random, what is the probability that their value is
(a) 15¢?   (b) 40¢?
(c) $1?   (d) more than 50¢?

27. A single card is drawn from a standard deck. Find the probability of getting the following.
(a) An honor card ($A, J, Q, K, 10$).
(b) A black card or the queen of hearts.

28. A coin is tossed four times. Find the probability of getting
(a) no heads.   (b) at least 1 tail.
(c) exactly 3 heads.

29. If four balls are drawn at random from a bag containing 5 red and 3 black balls, what is the probability of getting each of the following?
(a) One black ball and 3 red balls.
(b) Two balls of each color.
(c) At least 1 black ball.

30. If 4 cards are drawn at random from a standard deck, what is the probability of getting each of the following?
(a) One card from each suit.
(b) Three clubs and 1 diamond.
(c) Two kings and 2 queens.

31. A box contains 15 cards numbered 1 through 15. Three cards are drawn at random from the box. Find the probability of each event.
(a) All 3 numbers are even.
(b) At least 1 number is odd.
(c) The product of the 3 numbers is even.

32. A die has been loaded so that the probabilities of getting 1, 2, 3, 4, 5, and 6 are $\frac{1}{3}, \frac{1}{4}, \frac{1}{6}, \frac{1}{12}, \frac{1}{12}$, and $\frac{1}{12}$, respectively. Assume that the usual properties of probability are still valid in this situation where the outcomes are not equally likely. Find the probability of rolling each of the following.

(a) An even number.
(b) A number less than 5.
(c) An even number or a number less than 5.

33. Three men and 4 women are to be seated in a row at random. Find the probability of each event.
(a) The men and women alternate.
(b) The women are together.
(c) The end positions are occupied by men.

34. Six shoes are to be picked from a pile consisting of 10 identical left shoes and 7 corresponding identical right shoes. What is the probability of getting
(a) three pairs?   (b) exactly two pairs?

35. A single die is to be tossed four times in succession. Find the probability of each event.
(a) The numbers 1, 2, 3, and 4 appear in that order.
(b) The numbers 1, 2, 3, and 4 appear in any order.
(c) At least one 6 appears.
(d) The same number appears each time.

36. If A and B overlap, then the third property of probability takes the form

$$P(A \cup B) = P(A) + P(B) - P(A \cap B)$$

Determine the corresponding formula for

$$P(A \cup B \cup C).$$

37. The letters of *MATHEMATICS* are written on 11 cards and arranged in a row at random. What is the probability that they spell *MATHEMATICS*?

38. Let S be the set of 25 ordered pairs which are the coordinates of the 25 points shown in Figure 33. If an ordered pair (x, y) is chosen at random from S, what is the probability that
(a) $x + y = 4$?   (b) $x + y < 5$?
(c) neither x nor y is 5?   (d) $y > x$?

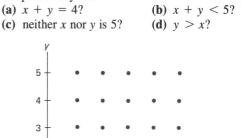

Figure 33

39. Consider the set of triangles that can be formed with vertices from the points (1, 1), (2, 1), (3, 1), (2, 3), (3, 3), and (4, 3). If a triangle is chosen at random from this set, what is the probability that it
(a) is right-angled? (b) has area 2?
Hint: A picture will help.

40. A careless secretary typed 4 letters and 4 envelopes and then inserted the letters at random in the envelopes. Find the probability of each of the following.
(a) No letter went in the correct envelope.
(b) At least 1 letter went in the correct envelope.
(c) Exactly 1 letter went in the correct envelope.
(d) At least 2 letters went in the correct envelopes.
(e) Exactly 3 letters went in the correct envelopes.

41. Three sticks are chosen at random from 8 sticks of lengths 1, 2, . . . , 8, respectively. What is the probability they will form a triangle?

42. TEASER The careless secretary of Problem 40 typed n ($n > 2$) letters and n envelopes and then inserted the letters at random. Find the probabilities of each of the following.
(a) Exactly $n - 1$ letters went in the correct envelopes.
(b) No letter went in the correct envelope (see Problem 48 of Section 12-5).
(c) Exactly 1 letter went in the correct envelope.
Note: In calculus, it is shown that the answer to (b) converges to $1/e \approx .368$ as n increases and that this convergence is very rapid. In fact, .368 is a fine approximation for $n \geq 6$. Thus the probability that 6 letters all go in the wrong envelopes is about the same as the probability that 600 letters all go in the wrong envelopes, a result most people find surprising.

12-9 INDEPENDENT EVENTS

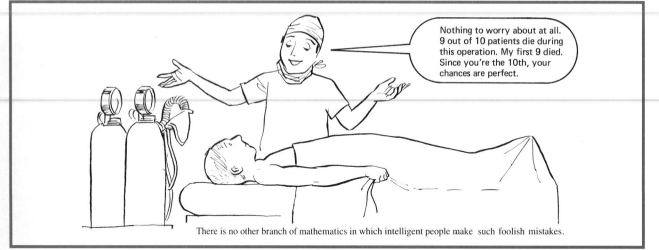

There is no other branch of mathematics in which intelligent people make such foolish mistakes.

A perfectly balanced coin has shown nine tails in a row. What is the probability that it will show a tail on the tenth flip? Some people argue that a mystical law of averages makes the appearance of a head practically certain. It is as if the coin had a memory and a conscience; it must atone for falling on its face nine times in a row. Such thinking is pure, unadulterated nonsense. The probability of showing a tail on the tenth flip is $\frac{1}{2}$, just as it was on each of the previous nine flips. The outcome of any flip is independent of what happened on previous flips.

Here is a different question, not to be confused with the one just answered. If one plans to flip a coin 10 times, what is the probability of getting all tails? To answer, we reason that there are 2 possibilities on the first flip, 2 on the second, and so forth. By the multiplication principle for counting (see Section 12-5), there are $2 \cdot 2 \cdot 2 \cdot 2 \cdot 2 \cdot 2 \cdot 2 \cdot 2 \cdot 2 \cdot 2$ or 1024 possible outcomes of this experiment, only one of which consists of all tails. The probability of 10 tails is $\frac{1}{1024}$, a very unlikely event indeed.

Here are the same questions stated for a game with higher stakes. A couple already has 9 girls. What is the probability that its tenth child will also be a girl? Assuming that boys and girls are equally likely (which is not quite true), the answer is $\frac{1}{2}$. But suppose a newly married couple decides, with great passion and little thought, to have 10 children. Assuming that nature assents to their decision, what is the probability that they will have all girls? It is $\frac{1}{1024}$.

Consider now the worried patient in the opening cartoon. Most of us would not (and should not) find the doctor's logic very comforting. In fact, if we make the assumption that the outcome of the tenth operation is completely independent of the first nine, the probability of a tenth failure is $\frac{9}{10}$. The assumption of independence may be questioned. Perhaps doctors improve with experience, but this particular patient had better hope for a miracle.

Dependence versus Independence

Perhaps we can make the distinction between dependence and independence clear by describing an experiment in which we again toss two dice—one colored and the other white. Let A, B, and C designate the following events.

A: colored die shows 6

B: white die shows 5

C: total on the two dice is greater than 7

Consider first the relationship between B and A. It seems quite clear that the chance of B occurring is not affected by our knowledge that A has occurred. In fact, if we let $P(B \mid A)$ denote the probability of B given that A has occurred, then (see Figure 34)

$$P(B \mid A) = \frac{1}{6}$$

But this is equivalent to the answer we get if we calculate $P(B)$ without any knowledge of A. For then we look at all 36 outcomes for two dice and note that 6 of them have the white die showing 5; that is,

$$P(B) = \frac{6}{36} = \frac{1}{6}$$

We conclude that $P(B \mid A) = P(B)$, just as we expected.

The relation between C and A is very different; C's chances are greatly improved if we know that A has occurred. From the illustration in the margin, we see that

$$P(C \mid A) = \frac{5}{6}$$

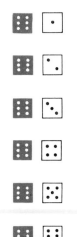

Only 6 outcomes if we know that colored die shows 6

Figure 34

However, if we have no knowledge of A and calculate $P(C)$ by looking at the 36 outcomes (page 522) for two dice, we find

$$P(C) = \frac{15}{36} = \frac{5}{12}$$

Clearly $P(C \mid A) \neq P(C)$.

This discussion has prepared the way for a formal definition. If $P(B \mid A) = P(B)$, we say that A and B are **independent** events. If $P(B \mid A) \neq P(B)$, A and B are dependent events.

And now, recalling that $A \cap B$ means A and B, we can state the **multiplication rule for probabilities.**

$$P(A \cap B) = P(A)P(B \mid A) = P(B)P(A \mid B)$$

In words, the probability of both A and B occurring is equal to the probability that A will occur multiplied by the probability that B will occur given that A has already occurred. In the case of independence, this takes a particularly elegant form.

$$P(A \cap B) = P(A)P(B)$$

Example A (A Two-Card Problem)   Suppose that two cards are to be drawn (one after another) from a well-shuffled deck. Find in two different ways the probability that both will be spades.

Solution   Method 1. Based on previous knowledge (that is, Section 12-8), we respond:

$$P(\text{two spades}) = \frac{_{13}C_2}{_{52}C_2} = \frac{\dfrac{13 \cdot 12}{2 \cdot 1}}{\dfrac{52 \cdot 51}{2 \cdot 1}} = \frac{13 \cdot 12}{52 \cdot 51}$$

Method 2. Consider the events

A: getting a spade on the first draw

B: getting a spade on the second draw

Our interest is in $P(A \cap B)$. According to the rule above, it is given by

$$P(A \cap B) = P(A)P(B \mid A) = \frac{13}{52} \cdot \frac{12}{51}$$

which naturally agrees with our earlier answer.   ∎

Example B (Another Two-Card Problem) Suppose that from a standard deck, we draw one card, place it back in the deck, reshuffle, and then draw a second card. What is the probability that both cards will be spades?

Solution   Let A and B have the meaning ascribed in Example A. Now A and B are independent and

$$P(A \cap B) = P(A)P(B) = \frac{13}{52} \cdot \frac{13}{52} \quad ∎$$

Urns and Balls

For reasons not entirely clear, teachers have always illustrated the central ideas of probability by talking about urns (vases) containing colored balls. Most of us have never seen an urn containing colored balls, but it won't hurt to use a little imagination.

A

6 red and
4 green balls.

Example C (And/Or Problems)   Consider two urns A and B, A containing 6 red balls and 4 green balls, and B containing 18 red balls and 2 green balls (Figure 35).

(a) If a ball is drawn from each urn, what is the probability that both will be red?

(b) If an urn is chosen at random and then a ball drawn, what is the probability that it will be red?

Solution

(a) $P(\text{red from A } and \text{ red from B}) = P(\text{red from A}) \cdot P(\text{red from B})$

$$= \frac{6}{10} \cdot \frac{18}{20} = \frac{6}{10} \cdot \frac{9}{10} = \frac{54}{100} = \frac{27}{50}$$

(b) First describe the desired event as

B

18 red and
2 green balls.

Figure 35

"choose A *and* draw red"   or   "choose B *and* draw red"

The events in quotation marks are disjoint, so their probabilities can be added. We obtain the answer

$$\frac{1}{2} \cdot \frac{6}{10} + \frac{1}{2} \cdot \frac{9}{10} = \frac{15}{20} = \frac{3}{4}$$

You should note the procedure we use. We describe the event using the words *and* and *or*. When we determine probabilities, *and* translates into *times*, and *or* into *plus*.   ∎

Figure 36

Example D (Replacement or Nonreplacement)   An urn contains 2 red, 3 green, and 5 black balls (Figure 36). Two balls are drawn at random. What is the probability that both are green

(a) if we replace the first ball before the second is drawn?

(b) if we do not replace the first ball before the second is drawn?

Solution

(a) Here we have independence.

$$P(\text{green } and \text{ green}) = P(\text{green}) \cdot P(\text{green})$$

$$= \frac{3}{10} \cdot \frac{3}{10} = \frac{9}{100}$$

(b) Now we have dependence. The outcome of the first draw does affect what happens on the second draw.

$$P(\text{green } and \text{ green}) = P(\text{first ball green}) \cdot P(\text{second ball green} \mid \text{first ball green})$$

$$= \frac{3}{10} \cdot \frac{2}{9} = \frac{6}{90} = \frac{1}{15} \quad ∎$$

A Historical Example

Two French mathematicians, Pierre de Fermat (1601–1665) and Blaise Pascal (1623–1662), are usually given credit for originating the theory of probability. This is how it happened: The famous gambler, Chevalier de Méré, was fond of a dice game in which he would bet that a 6 would appear at least once in four throws of a die. He won more often than he lost for, though he probably did not know it,

$$P(\text{at least one } 6) = 1 - P(\text{no } 6\text{'s})$$

$$= 1 - \frac{5}{6} \cdot \frac{5}{6} \cdot \frac{5}{6} \cdot \frac{5}{6} \approx 1 - .48 = .52$$

Growing tired of this game, he introduced a new one played with two dice. Méré then bet that at least one double 6 would appear in 24 throws of two dice. Somehow (perhaps he noted that $\frac{4}{6} = \frac{24}{36}$) he thought he should do just as well as before. But he lost more often than he won. Mystified, he proposed it as a problem to Pascal, who in turn wrote to Fermat. Together they produced the following explanation.

$$P(\text{at least one double } 6) = 1 - P(\text{no double } 6\text{'s})$$

$$= 1 - \left(\frac{35}{36}\right)^{24} \approx 1 - .51 = .49$$

From this humble, slightly disreputable origin grew the science of probability.

Example E (Using the Laws of Probability) Recall two general laws of probability. The first appeared as Problem 36 of Section 12-8; the second came in the present section.

$$P(A \cup B) = P(A) + P(B) - P(A \cap B)$$

$$P(A \cap B) = P(A)P(B \mid A)$$

If $P(A) = .7$, $P(B) = .3$, and $P(B \mid A) = .4$, calculate
(a) $P(A \cap B)$; (b) $P(A \cup B)$; (c) $P(A \mid B)$.

Solution
(a) $P(A \cap B) = P(A)P(B \mid A) = (.7)(.4) = .28$
(b) $P(A \cup B) = P(A) + P(B) - P(A \cap B) = .7 + .3 - .28 = .72$
(c) $P(A \mid B) = P(A \cap B)/P(B) = .28/.3 \approx .93$ ∎

PROBLEM SET 12-9

A. Skills and Techniques

For Problems 1–8, see Examples A and B.

1. Toss a balanced die three times in succession. What is the probability of all 1's?

2. Toss a fair coin four times in succession. What is the probability of all heads?

3. Spin the two spinners pictured in Figure 37. What is the probability that
   (a) both will show red (that is, A shows red *and* B shows red)?
   (b) neither will show red (that is, A shows not red *and* B shows not red)?
   (c) spinner A will show red and spinner B not red?

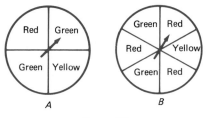

Figure 37

(d) spinner *A* will show red and spinner *B* red or green?
(e) just one of the spinners will show green?

4. The two boxes shown in Figure 38 are shaken thoroughly and a ball is drawn from each. What is the probability that

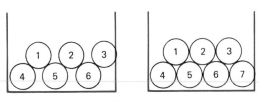

Figure 38

(a) both will be 1's?
(b) exactly one of them will be a 2 (2 from first box and not 2 from second box, or not 2 from first box and 2 from second box)?
(c) both will be even?
(d) exactly one of them will be even?
(e) at least one of them will be even?

5. In each case, indicate whether or not the two events seem independent to you. Explain.
(a) Getting an *A* in math and getting an *A* in physics.
(b) Getting an *A* in math and winning a tennis match.
(c) Getting a new shirt for your birthday and stubbing your toe the next day.
(d) In tossing two dice, getting an odd total and getting a 5 on one of the dice.
(e) Being a woman and being a doctor.
(f) Walking under a ladder and having an accident the next day.

6. In each case indicate whether or not the two events are necessarily disjoint.
(a) Getting an *A* in math and getting an *A* in physics.
(b) Getting an *A* in Math 101 and getting a *B* in Math 101.
(c) In tossing two dice, getting an odd total and getting the same number on both dice.

(d) In tossing two dice, getting an odd total and getting a 5 on one of the dice.
(e) The sun shining on Tuesday and being rainy on Tuesday.
(f) Not losing the college football game and not winning it.

7. A machine produces bolts which are put in boxes. It is known that 1 box in 10 will have at least one defective bolt in it. Assuming that the boxes are independent of each other, what is the probability that a customer who ordered 3 boxes will get all good bolts?

8. Suppose that the probability of being hospitalized during a year is .152. Assuming that family members are hospitalized independently of each other, what is the probability that no one in a family of five will be hospitalized this year? Do you think the assumption of independence is reasonable?

Refer to Example C for Problems 9–12.

9. Consider two urns, one with 3 red balls and 7 white balls, the other with 6 red balls and 6 white balls. If an urn is chosen at random and then a ball drawn, what is the probability it will be red?

10. Suppose that 4 percent of males are colorblind, that 1 percent of females are colorblind, and that males and females each make up 50 percent of the population. If a person is chosen at random, what is the probability that this person will be colorblind?

11. Refer to the urns of Example C. Draw a ball from urn *A*. If it is red replace it and draw 2 balls from urn *A*. If it is green, draw a second ball from urn *B*. Calculate the probability of finishing with 2 balls of the same color.

12. In Figure 38, choose a box at random and draw a ball. If odd, draw a second ball from that box; if even, draw a second ball from the opposite box. Calculate the probability that both balls have the same parity, that is, both are odd or both are even.

Problems 13–18 refer to the urn of Example D.

13. If 2 balls are drawn, what is the probability both are black assuming
(a) replacement between draws?
(b) nonreplacement?

14. If 2 balls are drawn without replacement between draws, what is the probability of getting
(a) red on the first draw and green on the second?
(b) a red and a green ball in either order?
(c) 2 red balls?
(d) 2 nonred balls?

15. If 3 balls are drawn with replacement between draws, what is the probability of getting
    (a) 3 green balls?
    (b) 3 red balls?
    (c) 3 balls of the same color?
    (d) 3 balls of all different colors?

16. Answer the questions of Problem 15 if there is no replacement between draws.

17. Two balls are drawn. What is the probability that the second one drawn is red if
    (a) the first is replaced before the second is drawn?
    (b) the first is not replaced before the second is drawn?

18. If 4 balls are drawn without replacement between draws, what is the probability of getting
    (a) 4 balls of the same color?
    (b) at least two different colors?
    (c) 2 balls of one color and 2 of another?

For Problems 19–22, see Example E.

19. Given that $P(A) = .8$, $P(B) = .5$, and $P(A \cap B) = .4$, find
    (a) $P(A \cup B)$ (b) $P(B|A)$
    (c) $P(A|B)$

20. Given that $P(A) = .8$, $P(B) = .4$, and $P(B|A) = .3$, find
    (a) $P(A \cap B)$ (b) $P(A \cup B)$
    (c) $P(A|B)$

21. In Sudsville, 30 percent of the people are Catholics, 55 percent of the people are Democrats, and 90 percent of the Catholics are Democrats. Find the probability that a person chosen at random from Sudsville is
    (a) both Catholic and Democrat.
    (b) Catholic or Democrat.
    (c) Catholic given that he or she is Democrat.

22. At Podunk U, $\frac{1}{4}$ of the applicants fail the entrance examination in mathematics, $\frac{1}{5}$ fail the one in English, and $\frac{1}{9}$ fail both mathematics and English. What is the probability that an applicant will
    (a) fail mathematics or English?
    (b) fail mathematics, given that he or she failed English?
    (c) fail English, given that he or she failed mathematics?

B. Applications and Extensions

23. Mary and John work independently at deciphering a coded message. If their respective probabilities of deciphering it are $\frac{1}{2}$ and $\frac{2}{3}$, find the probability that

(a) Mary will be the only one of the two to decipher the message.
    (b) the message will be deciphered.

24. Wanda Wiseankle, a candidate for governor, estimates that the probability of winning her party's nomination is $\frac{2}{3}$ and that if she wins the nomination, the probability of her winning the election is $\frac{5}{8}$. Find the probability that
    (a) she will be nominated by her party and then lose the election.
    (b) she will win the election.

25. The probability that Helen will beat Sam in a game of chess is $\frac{2}{3}$. What is the probability that she will beat Sam exactly twice in three games of chess?

26. In the semifinals of a tennis tournament, A will play B and C will play D. The winners will meet in the finals. The probability that A will beat B is $\frac{2}{3}$, that C will beat D is $\frac{5}{6}$, that A will beat C (if they play) is $\frac{1}{4}$, and that A will beat D (if they play) is $\frac{4}{5}$. Find the probability that A will win the tournament.

27. An urn contains 4 red, 5 white, and 7 black balls. Two balls are drawn in succession. What is the probability of drawing 2 white balls if the first ball
    (a) is replaced before the second is drawn?
    (b) is not replaced before the second is drawn?

28. Four balls are drawn from the urn of Problem 27 without replacement between draws. What is the probability that
    (a) all are white? (b) at least one is red?
    (c) the second ball drawn is black?

29. An urn contains 3 red, 2 white, and 5 black balls. If a ball is drawn at random, replaced, and then a second ball drawn, what is the probability that
    (a) both are red?
    (b) one is white and the other is black?
    (c) both balls have the same color?
    (d) the balls have different colors?

30. Consider the urns of Problems 27 and 29. If one of these urns is picked at random and two balls drawn without replacement between draws, what is the probability that
    (a) both are red?
    (b) neither is white?
    (c) one is red and the other is white?

31. In tossing a pair of dice repeatedly, what is the probability of getting a 7 before an 11? *Hint:* Consider the infinite geometric sum whose first term is the probability of getting 7 on the first toss, whose second term is the probability of getting neither 7 nor 11 on the first toss followed by 7 on the second toss, and so on.

32. A coin is tossed. If a head appears, you draw two cards from a standard deck and, if a tail appears, you draw one card. Find the probability that
  (a) no spade is drawn.
  (b) exactly one spade is drawn.
  (c) at least one spade is drawn.

33. A coin, loaded so that the probability of a head is .6, is tossed 8 times. Find the probability of getting each of the following.
  (a) At least 1 head.
  (b) At least 7 tails.

34. If the coin of Problem 33 is tossed repeatedly, what is the probability of getting a string of two heads before getting a string of two tails? *Hint: HH* or *THH* or *HTHH* or *THTHH* or. . . .

35. Amy, Betty, and Candy take turns (in that order) tossing a fair coin until one of them wins by getting a head. Find the probabilities of winning for each of them.

36. (Birthday problem) In a class of 23 students what is the probability that at least two people have the same birthday? Make a guess first. Then make a computation based on the following assumptions: (1) There are

365 days in a year; (2) one day is as likely as another for a birthday; (3) the 23 students were born (chose their birthdays) independently of each other. *Suggestion:* Look at the complementary event. For example, if there were only 3 people in the class, the probability that no two have the same birthday is $1 \cdot \frac{364}{365} \cdot \frac{363}{365}$.

37. Mary's highschool day consists of 6 successive periods. To fill them, she has decided to take history, English, math, social science, physics, and physical education. The times will be scheduled at random by a computer with the one restriction that history must precede English (this because the English course is designed to build on the history course). What is the probability that history is scheduled immediately preceding English?

38. TEASER A stick of length one is divided into three pieces by cutting it at random at two places. What is the probability that the three pieces will form a triangle? *Hint:* Think of the problem this way. Imagine the stick has a scale from 0 to 1 on it. You are to pick two numbers x and y on this interval at which to make the cuts, that is, you are to pick a point (x, y) in the unit square.

CHAPTER 12 SUMMARY

A **number sequence** $a_1, a_2, a_3, \ldots,$ is a function that associates with each positive integer n a number a_n. Such a sequence may be described by an **explicit formula** (for instance, $a_n = 2n + 1$), by a **recursion formula** (for instance, $a_n = 3a_{n-1}$), or by giving enough terms so a pattern is evident (for instance, 1, 11, 21, 31, 41, . . .).

If any term in a sequence can be obtained by adding a fixed number d to the preceding term, we call it an **arithmetic sequence.** There are three key formulas associated with this type of sequence.

Recursion formula: $a_n = a_{n-1} + d$

Explicit formula: $a_n = a_1 + (n - 1)d$

Sum formula: $A_n = \frac{n}{2}(a_1 + a_n)$

In the last formula, A_n represents

$$A_n = a_1 + a_2 + \cdots + a_n = \sum_{i=1}^{n} a_i$$

A **geometric sequence** is one in which any term results from multiplying the previous term by a fixed number r. The corresponding key formulas are

Recursion formula: $a_n = ra_{n-1}$

Explicit formula: $a_n = a_1 r^{n-1}$

Sum formula: $A_n = \frac{a_1(1 - r^n)}{1 - r}, r \neq 1$

In the last formula, we may ask what happens as n grows larger and larger. If $|r| < 1$, the value of A_n gets closer and closer to $a_1/(1 - r)$, which we regard as the sum of *all* the terms of the sequence.

Often in mathematics, we wish to demonstrate that a whole **sequence of statements** P_n is true. For this, a powerful tool is the **principle of mathematical induction,** which asserts that if P_1 is true and if the truth of P_k implies the truth of P_{k+1}, then all the statements of the sequence are true.

Given enough time, anyone can count the number of elements in a set. But if the set consists of arrangements of objects (for example, the letters in a word), the work can be greatly simplified by using two principles, the **multiplication principle** and the **addition principle.** Of special interest are the number of **permutations** (ordered arrangements) of n things taken r at a time and the number of **combinations** (unordered collections) of n things taken r at a time. They can be calculated from the formulas

$$_nP_r = n(n - 1)(n-2) \cdots (n - r + 1)$$

$$_nC_r = \frac{_nP_r}{r!} = \frac{n(n - 1)(n - 2) \cdots (n - r + 1)}{r(r - 1)(r - 2) \cdots 1}$$

One important use of the symbol $_nC_r$ is in the **Binomial Formula**

$$(x + y)^n = {_nC_0}x^ny^0 + {_nC_1}x^{n-1}y^1 + \cdots +$$
$$_nC_{n-1}x^1y^{n-1} + {_nC_n}x^0y^n$$

If an experiment can result in n **equally likely**

outcomes, of which m result in the event E, then the **probability** of E is m/n. There are four main properties of probability. Stated in set language, they are

1. $P(\varnothing) = 0; P(S) = 1;$
2. $0 \le P(A) \le 1;$
3. $P(A \cup B) = P(A) + P(B)$, provided $A \cap B = \varnothing;$
4. $P(A) = 1 - P(A').$

If the probability of event A is unaffected by whether B has occurred or not, we say that A and B are **independent.** In this case, $P(A \cap B) = P(A) \cdot P(B)$. But if A and B are **dependent,** then

$$P(A \cap B) = P(B)P(A \mid B)$$

We call this the **multiplication rule** for probabilities.

CHAPTER 12 REVIEW PROBLEM SET

In Problems 1–10, write True or False in the blank. If false, tell why.

_____ 1. The sequence 1, 1, 1, 1, 1, . . . is the only sequence that is both arithmetic and geometric.

_____ 2. A recursion formula for a sequence relates the value of a term directly to its subscript.

_____ 3. If a sum of money is invested at compound interest, its values at the end of successive years will form a geometric sequence.

_____ 4. If a substance is growing so that the amount S after t years is given by $S = 3t + 14$, then the amounts at the end of successive years will form an arithmetic sequence.

_____ 5. The symbols $.99\overline{9}$ and 1 represent the same number.

_____ 6. If P_1, P_2, and P_3 are true and the truth of P_k implies the truth of P_{k+3}, then P_n is true for all positive integers n.

_____ 7. There are $_{10}P_2$ possible ways to respond to a 10-item true-false quiz.

_____ 8. If m and n are positive integers with $n \ge m$, then $\dfrac{n!}{m! (n - m)!}$ is a positive integer.

_____ 9. If the seventh and eighth terms in $(x + y)^n$ have the same coefficient, then $n = 13$.

_____ 10. In probability theory, the words disjoint and independent have basically the same meaning.

Problems 11–17 refer to the sequences below.

(a) 3, 6, 9, 12, 15, . . .
(b) $-1, \frac{1}{2}, -\frac{1}{4}, \frac{1}{8}, -\frac{1}{16}, \frac{1}{32}, \ldots$
(c) $1, \sqrt{2}, 2, 2\sqrt{2}, 4, \ldots$
(d) $\pi, 3\pi, 5\pi, 7\pi, 9\pi, \ldots$
(e) 1, 1, 1, 3, 5, 9, 17, . . .

11. Which of these sequences are geometric?
12. Which of these sequences are arithmetic?
13. Write recursion formulas for sequences (d) and (e).
14. Find the 51st term of sequence (d).
15. Find the sum of the first 100 terms of sequence (a).
16. Find the sum of the first 10 terms of sequence (c).
17. Find the sum of all the terms of sequence (b).
18. Joe College (after taking college algebra) wants to negotiate a new allowance procedure with his parents. During each month, he would like 1¢ on day one, 2¢ on day two, 4¢ on day three, 8¢ on day four, and so

on. If he is successful, determine Joe's allowance during September.

19. Write $2.22\overline{2}$ as a ratio of two integers.

20. Calculate each sum.
    (a) $\sum_{i=1}^{50} (3i - 2)$
    (b) $\sum_{i=1}^{\infty} 4(\frac{1}{3})^i$

21. Suppose that $a_1 = a_2 = 1$ and $a_n = a_{n-1} + 2a_{n-2}$ for $n \geq 3$. Find the value of a_7.

22. Pat's starting annual salary is \$30,000 and she has been promised an annual raise. Which would give her a better income (and by how much) during her tenth year: a straight \$1800 raise each year or a 5% raise each year?

23. Which of the plans in Problem 22 would give Pat the larger total income during the 10 years and by how much?

24. Evaluate the following.
    (a) $_{13}P_{10}$
    (b) $_{13}C_{10}$

25. There are 10 candidates for a position. In how many ways can you rank your first, second, and third choices?

26. A local ice cream store offers 20 different flavors. How many different double-dip cones can it serve if
    (a) the flavors must be different? *Note:* We consider vanilla on top of chocolate to be the same as chocolate on top of vanilla.
    (b) the flavors may be the same?

27. How many 7-letter code words can be made from
    (a) ALGEBRA?
    (b) SEESAWS?

28. Pizza Heaven offers two types of crust, two types of cheese, five different meat toppings, and three different vegetable toppings. How many different pizzas can one order which have one crust, one cheese, three meats, and two vegetables?

29. King Arthur and 11 of his knights (including Sir Lancelot) wish to sit at the Round Table. In how many ways can they do this if
    (a) there are no special seats?
    (b) King Arthur must sit at the chief chair with Sir Lancelot next to him?

30. A basketball team has 15 members: 6 guards, 4 centers and 5 forwards. How many starting lineups (2 guards, 1 center, and 2 forwards) are possible?

31. A softball team has 12 members. The coach first chooses people to fill the 9 field positions and then

prescribes a batting order. In how many ways can he do this?

32. How many license plates are there that consist of 2, 3, or 4 letters followed by enough digits to make a 7-symbol string?

33. A combination lock has the digits 0, 1, 2, . . . , 9 written around its circular face. How many 3-digit combinations are possible if no combination can have the same or consecutive digits in the final two positions (9-0 is considered consecutive). Thus 6-5-7 and 5-5-4 are legal combinations, but 6-5-5 and 6-9-0 are not.

34. Prove using mathematical induction that a set with n elements has 2^n subsets.

35. Prove using mathematical induction that $n! > 2^n$ for $n \geq 4$.

36. Prove or disprove: $n^2 - n + 11$ is prime for every positive integer n.

37. If P_2 is true and the truth of P_k implies the truth of P_{k+2}, what can you conclude about the sequence P_n of statements?

38. Write out the expansion of $(x + 2y)^6$.

39. Calculate $\sum_{r=0}^{6} {_6}C_r 2^r$. *Hint:* Relate it to Problem 38.

40. Find and simplify the term in the expansion of $(2x - 3y^2)^8$ that involves y^6.

41. A city is laid out in a uniform block pattern with north-south and east-west streets. How many different paths (meaning shortest paths along streets) are there from a corner to another one 5 blocks north and 3 blocks east?

42. Find $(1.00001)^8$ accurate to 15 decimal places.

43. How many subsets with an odd number of elements does the set $\{A, B, C, \ldots , X, Y, Z\}$ have?

44. Suppose that you are to pick a number at random from the first 100 positive integers. What is the probability that you will pick
    (a) an odd number?   (b) a multiple of 5?
    (c) a number whose digits sum to more than 16?
    (d) a number with no ones digits?

45. What is the probability of getting a royal flush (ace, king, queen, jack, 10 in one suit) for a 5-card poker hand?

46. There are 5 red, 6 green, and 7 black balls in an urn. Two balls are to be drawn.
    (a) What is the probability that both will be red?
    (b) What is the probability that both balls will be the same color?

47. Answer the questions of Problem 46 if the first ball is replaced before the second ball is drawn.

48. If the probability that Arnold will marry is .8, that he will graduate from college is .6, and that he will do both is .5, what is the probability that he will do one or the other (or both)?

49. At Westcott College, both English 11 and History 13 are required courses. On their first attempt, 30 percent of the students fail English 11, 20 percent fail History 13, and 8 percent fail both. Find the probability that a student will

(a) fail one or the other of these courses.

(b) fail English 11, given that he or she has already failed History 13

50. What proportion of families with 4 children would you expect to have 2 boys and 2 girls? Assume that boys and girls are equally likely.

Analytic Geometry

■ Appollonius' metric treatment of the conic sections—ellipses, hyperbolas, and parabolas—was one of the great mathematical achievements of antiquity. The importance of conic sections for pure and applied mathematics (for example, the orbits of the planets and of electrons in the hydrogen atom are conic sections) can hardly be overestimated.

Richard Courant and Herbert Robbins

13-1 PARABOLAS

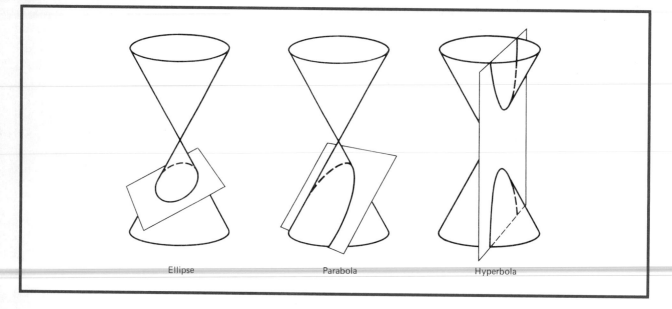

Ellipse   Parabola   Hyperbola

Analytic geometry could well be called algebraic geometry for it is the study of geometric concepts such as curves and surfaces by means of algebra. Among the most important curves are the conic sections, curves that are obtained by intersecting a cone of two nappes with a plane (see our opening display). We are especially interested in the three principal cases of an ellipse, parabola, and hyperbola though we shall also consider certain limiting forms like a circle, two intersecting lines, and so on. We begin with the parabola, a curve already discussed in Section 4-4. Here we give a very general treatment based on the geometric definition given to us by the Greeks.

The Geometric Definition of a Parabola

A parabola is the set of points P that are equidistant from a fixed line l (the directrix) and a fixed point F (the focus). In other words, a parabola is the set of points P in Figure 1 satisfying

$$d(P, L) = d(P, F)$$

Here, L is the point of l closest to P and, as usual, $d(A, B)$ denotes the distance between the points A and B.

A little thought convinces us that a parabola is a two-armed curve opening ever wider and symmetric with respect to the line through the focus perpendicular to the directrix. This line is called the **axis of symmetry** and the point where this line in-

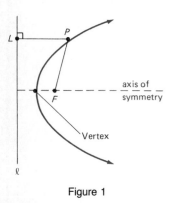

Figure 1

tersects the parabola is called the **vertex.** Note that the vertex is the point of the parabola closest to the directrix.

The Equation of a Parabola

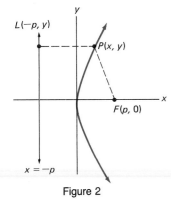

L(-p, y)

P(x, y)

F(p, 0)

x = -p

Figure 2

Place the parabola in the coordinate system so that its axis is the x-axis and its vertex is the origin (Figure 2). Let the focus be to the right of the origin, for example, at $(p, 0)$; then the directrix is the line $x = -p$. If $P(x, y)$ is any point on the curve, it must satisfy

$$d(P, F) = d(P, L)$$

which, because of the distance formula, assumes the form

$$\sqrt{(x - p)^2 + (y - 0)^2} = \sqrt{(x + p)^2 + (y - y)^2}$$

Since both sides are positive, this is equivalent to the result when both sides are squared.

$$x^2 - 2px + p^2 + y^2 = x^2 + 2px + p^2$$

This, in turn, simplifies to

$$y^2 = 4px$$

The final equation is called the **standard equation of the parabola.** It is easy to write, simple to graph, and has a form which is straightforward to interpret. For example, we may replace y by $-y$ without affecting the equation, which means that the graph is symmetric with respect to the x-axis. It crosses the x-axis at the origin which is the vertex. The positive number p measures the distance from the focus to the vertex (also the distance from the vertex to the directrix).

The equation just derived has three other variants. If we interchange the roles of x and y (giving $x^2 = 4py$), we have the equation of a parabola that opens upward with the y-axis as its axis of symmetry. The corresponding parabolas which open to the left and down have equations $y^2 = -4px$ and $x^2 = -4py$, respectively. All of this is summarized in Figure 3 at the top of the next page.

Example A (Finding the Equation)   Determine the equation of the parabola with vertex at the origin and

(a) focus at $(0, -3)$.   (b) directrix $x = \frac{1}{2}$.

Solution

(a) This is a parabola of the fourth type in Figure 3; it turns down with $p = 3$. We conclude that its equation is

$$x^2 = -4(3)y = -12y$$

(b) This is a parabola of the third type, opening left. Since $p = \frac{1}{2}$, the equation is

$$y^2 = -4(\tfrac{1}{2})x = -2x. \quad \blacksquare$$

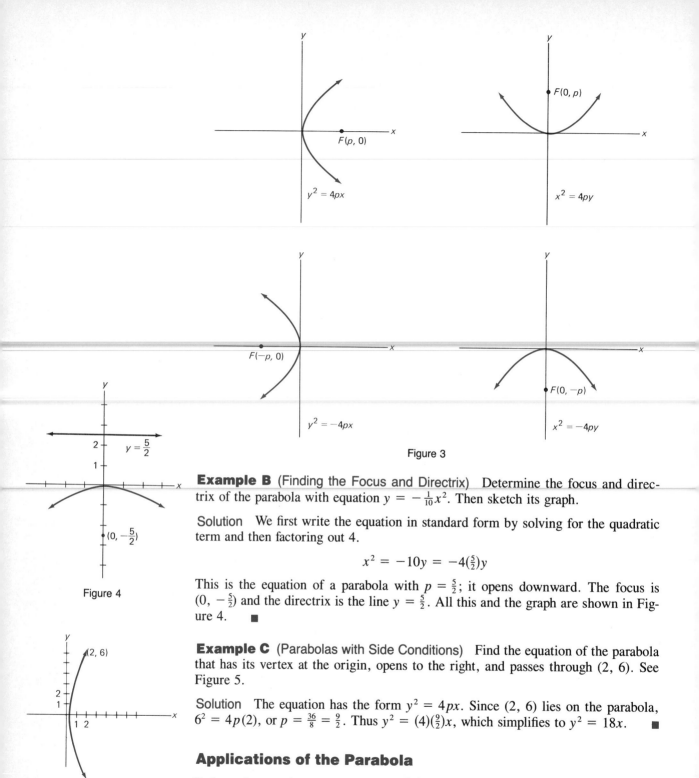

Figure 3

Figure 4

Example B (Finding the Focus and Directrix) Determine the focus and directrix of the parabola with equation $y = -\frac{1}{10}x^2$. Then sketch its graph.

Solution We first write the equation in standard form by solving for the quadratic term and then factoring out 4.

$$x^2 = -10y = -4(\tfrac{5}{2})y$$

This is the equation of a parabola with $p = \frac{5}{2}$; it opens downward. The focus is $(0, -\frac{5}{2})$ and the directrix is the line $y = \frac{5}{2}$. All this and the graph are shown in Figure 4. ∎

Example C (Parabolas with Side Conditions) Find the equation of the parabola that has its vertex at the origin, opens to the right, and passes through $(2, 6)$. See Figure 5.

Solution The equation has the form $y^2 = 4px$. Since $(2, 6)$ lies on the parabola, $6^2 = 4p(2)$, or $p = \frac{36}{8} = \frac{9}{2}$. Thus $y^2 = (4)(\frac{9}{2})x$, which simplifies to $y^2 = 18x$. ∎

Applications of the Parabola

Perhaps the most important property of the parabola is its *optical property*. Consider a paraboloidal mirror, that is, a cup-shaped mirror with a parabolic cross section

Figure 5

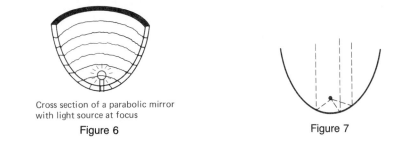

Cross section of a parabolic mirror
with light source at focus

Figure 6

Figure 7

(Figure 6). If a light source is placed at the focus, the resulting rays of light are reflected from the mirror in a beam in which all the rays are parallel to the axis (Figure 7). This fact is used in designing search lights. Conversely, if light rays parallel to the axis (as from a star) hit a paraboloidal mirror, they will be "focused" at the focus. This is the basis for the design of one type of reflecting telescope. The optical property of the parabola is usually demonstrated by means of calculus, but there is a way to do it that involves only geometry and algebra (see Problems 44 and 45). Calculus also allows one to show that the path of a projectile and the shape of the cables for a suspension bridge are approximately parabolic.

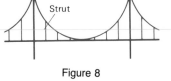

Strut

Figure 8

Example D (A Suspension Bridge Problem)   The cables for a suspension bridge are attached to towers 96 meters apart at points 25 meters above the bridge floor (Figure 8). The cables drop at the center to a point 1 meter above the bridge floor. How long is the supporting strut 12 meters from a tower?

Solution   We suppose that the vertex of the parabola is at the origin so that the equation of the parabola has the form $x^2 = 4py$. The given information implies that the point (48, 24) is on the parabola; thus $48^2 = 4p(24)$ and $p = 24$. We conclude that the equation of the parabola is $x^2 = 96y$. The length L of the required strut is $y + 1$, where $x = 36$. Thus

$$L = \frac{1}{96}(36)^2 + 1 = 14.5 \text{ meters} \quad \blacksquare$$

PROBLEM SET 13-1

A. Skills and Techniques

Each equation below determines a parabola with vertex at the origin. In what direction does the parabola open?

1. $x^2 = 8y$
2. $y^2 = -2x$
3. $y^2 = 6x$
4. $x^2 = -3y$
5. $3y^2 = -5x$
6. $y = -2x^2$

Find the equation of the parabola with vertex at the origin, given the following information. See Example A.

7. The focus is (0, 6).
8. The directrix is $x = -\frac{2}{3}$.

9. The directrix is $x = 3$.
10. The focus is $(-\frac{1}{2}, 0)$.

In Problems 11–18 find p and then sketch the graph, showing the focus and directrix. See Example B.

11. $x^2 = -8y$
12. $y^2 = 3x$
13. $y^2 = \frac{1}{2}x$
14. $y = -3x^2$
15. $y = \frac{1}{2}x^2$
16. $6x = -4y^2$
17. $9x = 4y^2$
18. $x = .125y^2$
19. Determine the coordinates of two points on the parabola $y = 4x^2$ with y-coordinate 1.

20. Determine two points on the parabola $y^2 = 8x$ with the same x-coordinate as the focus.

Find the equation of the parabola with vertex at $(0, 0)$ that satisfies the given conditions. See Example C.

21. Opens up; goes through $(2, 6)$.

22. Opens down; goes through $(-2, -4)$.

23. Opens right; 6 units from focus to directrix.

24. Opens left; goes through $(-1, 4)$.

25. Goes through $(1, 2)$ and $(1, -2)$.

26. Goes through $(1, 4)$ and $(2, 16)$; opens up.

Problems 27–32 involve applications, one of which is discussed in Example D.

27. A paraboloidal reflector is made by revolving the part of the parabola $16y = x^2$ below $y = 2$ about its axis of symmetry. Where should a light source be placed to produce a beam with parallel rays?

28. A paraboloidal mirror, 4 feet in diameter and .5 feet deep, uses the sun's rays to heat objects. Where should an object be placed for most effective heating?

29. The cables for the central span of a suspension bridge are parabolic, as in Figure 8. If the towers are 1000 feet apart and rise 500 feet above the bridge floor and if the cables drop to the bridge floor at the center, how long is the supporting strut 120 feet from the center?

30. A projectile shot from the ground reaches its maximum height of 100 meters after traveling a horizontal distance of M meters. After 40 more meters of horizontal travel, it is at height 80 meters. How far from its starting point will it land?

31. A parabolic door opening is 10 feet high at the center and is 2 feet wide at a height of 8 feet. How wide is it at the base?

32. The focus of a paraboloidal mirror is 10 centimeters above its vertex. A light ray leaving a source at the focus is reflected from the mirror to an absorbing element 10 centimeters to the right of and 30 centimeters above the focus. How far did this ray travel?

B. Applications and Extensions

33. Find the focus and directrix of the parabola with equation $4x = -5y^2$.

34. Find the equation of the parabola with vertex at $(0, 0)$ which in addition satisfies the following condition.
(a) Its focus is at $(\frac{5}{2}, 0)$.
(b) It opens down and passes through $(3, -10)$.

35. The chord of a parabola through the focus and perpendicular to the axis is called the **latus rectum** of the parabola. Find the length of the latus rectum for the parabola $4px = y^2$.

36. The chord of a parabola that is perpendicular to the axis and 1 unit from the vertex has length 3 units. How long is its latus rectum?

37. A door in the shape of a parabolic arch (Figure 9) is 12 feet high at the center and 5 feet wide at the base. A rectangular box 9 feet tall is to be slid through the door. What is the widest the box can be?

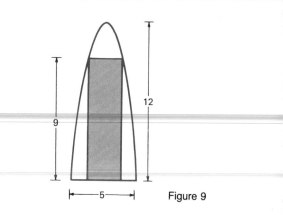

Figure 9

38. The path of a projectile fired from ground level is a parabola opening down. If the greatest height reached by the projectile is 100 meters and if its range (horizontal reach) is 800 meters, what is the horizontal distance from the point of firing to the point where the projectile first reaches a height of 64 meters?

39. The cables for the central span of a suspension bridge take the shape of a parabola. If the towers are 800 meters apart and the cables are attached to them at points 400 meters above the floor of the bridge, how long is the vertical strut that is 100 meters from the tower? Assume the vertex of the parabola is on the floor of the bridge.

40. In Figure 10, AP is parallel to the x-axis and Q is the midpoint of AP. Find the equation of the path of $Q(x_1, y_1)$ as $P(x, y)$ moves along the path $x^2 = 4py$.

41. Suppose that a submarine has been ordered to follow a path that keeps it equidistant from a circular island of radius r and a straight shoreline that is $2p$ units from the edge of the island. Derive the equation of the submarine's path, assuming that the shoreline has equation $x = -p$ and that the center of the island is on the x-axis.

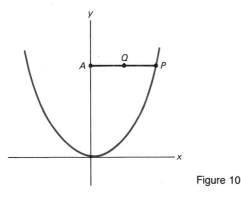

Figure 10

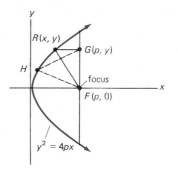

Figure 12

42. Show that there is no point $P(x, y)$ on the parabola $x^2 = 8y$ for which OP is perpendicular to PF, F being the focus of the parabola. *Hint:* Two lines with slopes m_1 and m_2 are perpendicular if and only if $m_1 m_2 = -1$.

43. An equilateral triangle is inscribed in the parabola $y^2 = 4px$ with one vertex at the origin. Find the length of a side of the triangle.

44. Consider a line l, two fixed points P and Q on the same side of l, and a (variable) point R on l. Use Figure 11 to show that the distance $\overline{PR} + \overline{RQ}$ is minimized precisely when $\alpha = \beta$. *Note:* Since a light ray is known to be reflected from a mirror l so that the angle of incidence equals the angle of reflection, we see that a light ray from P to l to Q picks the shortest path.

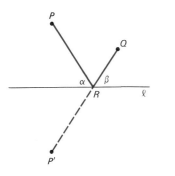

Figure 11

45. (Optical property of the parabola) Imagine the parabola $y^2 = 4px$ of Figure 12 to be a mirror with points F, R, G, and H as indicated and with RG parallel to the x-axis.
(a) Show that $\overline{FR} + \overline{RG} = 2p$.
(b) Show that $\overline{FH} + \overline{HG} > 2p$.
Conclude from Problem 44 that a light ray from the

focus to a parabolic mirror is reflected parallel to the axis of the parabola.

46. TEASER Consider the parabola $y = x^2$ (Figure 13). Let T_1 be the triangle with vertices on this parabola at a, c, and b with c midway between a and b. Let T_2 be the union of the two triangles with vertices on the parabola at a, d, c and c, e, b, respectively, with d midway between a and c and e midway between c and b. In a similar manner, let T_3 be the union of four triangles with vertices on the parabola, and so on.

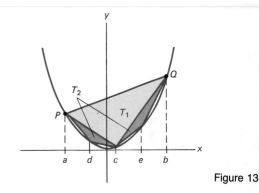

Figure 13

(a) Show that the area of T_1 is given by $A(T_1) = (b - a)^3/8$.
(b) Show that $A(T_2) = A(T_1)/4$.
(c) Find the area of the curved parabolic segment below the line PQ.

47. Draw the graphs of $y = x^2$, $y = 10x - 21$, and $y = -\frac{1}{10}x + 9.3$ using the same axes. Note that all three graphs go through $A = (3, 9)$. Determine the other points B and C where the lines intersect the parabola and then calculate the area of triangle ABC.

48. By experiment, determine k so that the line $y = \frac{5}{3}x + k$ is tangent to the graph of the parabola $y = \frac{2}{7}x^2$. Also find the coordinates (a, b) of the point of tangency accurate to one decimal place.

13-2 ELLIPSES

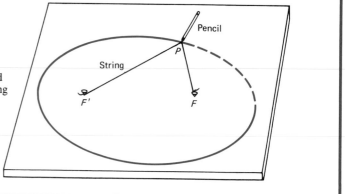

This shows the drawing of an ellipse. A string is tacked down at its ends by thumbtacks. A pencil pulls the string taut.

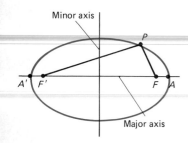

Figure 14

Our opening display suggests the geometric definition of the ellipse. An ellipse is the set of points P for which the sum of the distances from two fixed points F' and F is a constant. In other words, an ellipse is the set of points P in Figure 14 satisfying

$$d(P, F') + d(P, F) = 2a$$

for some positive constant a.

The two fixed points F' and F are called **foci** (plural of focus) and the point midway between the foci is the **center** of the ellipse. We call the line through the two foci the **major axis** of the ellipse; the line through the center and perpendicular to the major axis is its **minor axis.** Note that the ellipse is symmetric with respect to both its major and minor axes. Finally, the intersection of the ellipse with the major axis determines the two points A' and A, which are called **vertices.** All this is shown in Figure 14.

The Equation of an Ellipse

Place the ellipse in the coordinate system so that its center is at the origin with the major axis along the x-axis. We may suppose the two foci F' and F to be located at $(-c, 0)$ and $(c, 0)$, where c is a positive constant (Figure 15). Then $d(P, F') + d(P, F) = 2a$ combined with the distance formula yields

$$\sqrt{(x + c)^2 + (y - 0)^2} + \sqrt{(x - c)^2 + (y - 0)^2} = 2a$$

or, equivalently,

$$\sqrt{(x + c)^2 + y^2} = 2a - \sqrt{(x - c)^2 + y^2}$$

Figure 15

After squaring both sides, we obtain

$$(x + c)^2 + y^2 = 4a^2 - 4a\sqrt{(x - c)^2 + y^2} + (x - c)^2 + y^2$$

and this in turn simplifies to

$$4cx - 4a^2 = -4a\sqrt{(x - c)^2 + y^2}$$

If we now divide both sides by 4 and square again, we get

$$(cx - a^2)^2 = a^2[(x - c)^2 + y^2]$$
$$c^2x^2 - 2a^2cx + a^4 = a^2[x^2 - 2cx + c^2 + y^2]$$
$$a^4 - a^2c^2 = a^2x^2 - c^2x^2 + a^2y^2$$
$$a^2(a^2 - c^2) = (a^2 - c^2)x^2 + a^2y^2$$

Finally, divide both sides by $a^2(a^2 - c^2)$ and interchange the two sides of the equation to obtain

$$\frac{x^2}{a^2} + \frac{y^2}{a^2 - c^2} = 1$$

It is clear from Figure 15 that $a > c$, so we may let $b^2 = a^2 - c^2$. This results in what we shall call the **standard equation of the ellipse,** namely,

$$\frac{x^2}{a^2} + \frac{y^2}{b^2} = 1$$

Interpreting *a*, *b*, and *c*

In this section, we consistently use 2a to denote the major diameter of an ellipse. We do this whether the ellipse is horizontal or vertical. This convention was ignored when we introduced ellipses informally in Section 4-5.

We have used c to denote the distance from the center to a focus. If we apply the defining condition for the ellipse to the vertex A, we obtain

$$d(A, F') + d(A, F) = 2a$$

which implies (see Figure 14) that the distance between the two vertices is $2a$. For this reason, we refer to the number $2a$ as the **major diameter** of the ellipse. Since $b^2 + c^2 = a^2$, b and c are the legs of a right triangle with hypotenuse a and it follows that $2b$ is the **minor diameter** of the ellipse. All this is summarized in Figure 16, where you should note especially the significance of the triangle with sides a, b, and c. We call it the fundamental triangle for the ellipse.

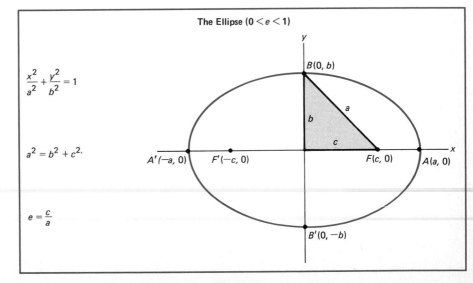

Figure 16

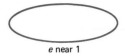

e near 1

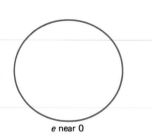

e near 0

Figure 17

$$\frac{x^2}{36} + \frac{y^2}{4} = 1$$

Figure 18

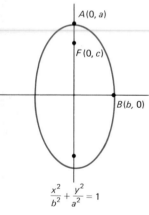

$$\frac{x^2}{b^2} + \frac{y^2}{a^2} = 1$$

Figure 19

The number $e = c/a$, which varies between 0 and 1, measures the **eccentricity** of the ellipse. If e is near 1, the ellipse is very eccentric (long and narrow); if e is near 0, the ellipse is almost circular (Figure 17). Sometimes, a circle is referred to as an ellipse of eccentricity 0; for in this case $c = 0$ and $a = b$ and the standard equation takes the form

$$\frac{x^2}{a^2} + \frac{y^2}{a^2} = 1$$

which is equivalent to the familiar circle equation $x^2 + y^2 = a^2$.

Example A (A Horizontal Ellipse)   Determine the major diameter, minor diameter, and eccentricity of the ellipse with equation

$$\frac{x^2}{36} + \frac{y^2}{4} = 1$$

Solution   Note that $a = 6$ and $b = 2$, so this is the equation of an ellipse with center at the origin, major diameter $2a = 12$ and minor diameter $2b = 4$. Since $c = \sqrt{a^2 - b^2} = \sqrt{32} = 4\sqrt{2}$, the foci are at $(\pm 4\sqrt{2}, 0)$ and the eccentricity is $4\sqrt{2}/6 \approx .94$ (see Figure 18).   ■

Another observation should be made. If we interchange the roles of x and y, then the standard equation takes the form

$$\frac{x^2}{b^2} + \frac{y^2}{a^2} = 1$$

The major axis is now the y-axis; the vertices and foci lie on it (Figure 19). Note that the equation of a vertical ellipse has the larger denominator in the y-term.

Example B (A Vertical Ellipse)   Determine the vertices and the foci of the ellipse with equation

$$\frac{x^2}{16} + \frac{y^2}{25} = 1$$

Solution   Since $a = 5$ and $b = 4$, $c = \sqrt{25 - 16} = 3$. Thus the vertices are $(0, \pm 5)$ and the foci are $(0, \pm 3)$.   ■

Example C (Graphing the Equation of an Ellipse)   Graph the equation $x^2/4 + y^2/9 = 1$, showing all the key features.

Solution   We identify this as the equation of a vertical ellipse (because the larger denominator is in the y term.) Also $a = 3$, $b = 2$, and $c = \sqrt{9 - 4} = \sqrt{5}$, from which we determine the four intercepts and the foci as shown in Figure 20.   ■

Example D (Finding Ellipses with Given Properties)   Write the equations of the three ellipses with vertices at $(0, \pm 8)$ and foci at (a) $(0, \pm 7)$; (b) $(0, \pm 4)$; (c) $(0, \pm 1)$. Determine the eccentricity e in each case. Sketch the graphs.

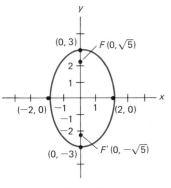

Figure 20

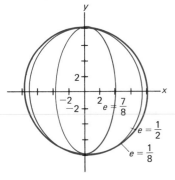

Figure 21

Solution   In each case, the ellipse is vertical. From the formulas $b = \sqrt{a^2 - c^2}$ and $e = c/a$, we determine the following.

(a) $a = 8$, $c = 7$, $b = \sqrt{15}$, $e = \frac{7}{8}$
(b) $a = 8$, $c = 4$, $b = \sqrt{48}$, $e = \frac{1}{2}$
(c) $a = 8$, $c = 1$, $b = \sqrt{63}$, $e = \frac{1}{8}$

The three graphs and the corresponding eccentricities are shown in Figure 21. Note that the smaller e is, the more circular the ellipse.

(a) $\dfrac{x^2}{15} + \dfrac{y^2}{64} = 1$

(b) $\dfrac{x^2}{48} + \dfrac{y^2}{64} = 1$

(c) $\dfrac{x^2}{63} + \dfrac{y^2}{64} = 1$ ∎

Applications

Like the parabola, the ellipse has an important optical property. If we imagine the ellipse to represent a mirror, then a light ray emanating from one focus will be reflected from the ellipse back through the other focus (see Problem 37 for a demonstration of this fact). This is the basis for the whispering gallery effect resulting from the elliptical shaped domes of St. Paul's Cathedral in London and the National Statuary Hall in the United States Capitol.

A much more significant application is the observation by Kepler that the planets move around the sun in elliptical orbits with the sun at one focus. Later, Newton established that this is a consequence of the fact that the gravitational force of attraction between two bodies is inversely proportional to the square of the distance between them. For the same reason, the electrons in the Bohr model of the hydrogen atom travel in elliptical orbits.

Example E (A Planetary Orbit)   The orbit of the planet Mercury is elliptical with $e \approx .21$ and major diameter about 7.2×10^7 miles. Find the equation of the orbit assuming the major and minor axes are the x- and y-axes, respectively. Also determine the distance of closest approach to the sun.

Solution   Clearly, $a = 3.6 \times 10^7$, so

$$c = ea = (.21)(3.6 \times 10^7) \approx 7.56 \times 10^6$$
$$b = \sqrt{a^2 - c^2} = \sqrt{1.2388 \times 10^{15}} \approx 3.52 \times 10^7$$

Thus the equation of the orbit is

$$\frac{x^2}{1.30 \times 10^{15}} + \frac{y^2}{1.24 \times 10^{15}} = 1$$

and the distance of closest approach to the sun is

$$a - c = 3.6 \times 10^7 - 7.6 \times 10^6 \approx 2.8 \times 10^7 \text{ miles}$$ ∎

PROBLEM SET 13-2

A. Skills and Techniques

In Problems 1–8, decide whether the ellipse with the given equation is horizontal or vertical and then determine the major and minor diameters. In Problems 5–8, you will first have to rewrite the equation in standard form. See Examples A and B.

1. $\dfrac{x^2}{7} + \dfrac{y^2}{16} = 1$ **2.** $\dfrac{x^2}{9} + \dfrac{y^2}{8} = 1$

3. $\dfrac{x^2}{36} + \dfrac{y^2}{20} = 1$ **4.** $\dfrac{x^2}{12} + \dfrac{y^2}{25} = 1$

5. $4x^2 + 9y^2 = 4$ **6.** $9x^2 + 8y^2 = 18$

7. $4k^2x^2 + k^2y^2 = 1$ **8.** $k^2x^2 + (k^2 + 1)y^2 = k^2$

In Problems 9–12, decide whether the corresponding ellipse is horizontal or vertical, determine a, b, and c, and sketch the graph. See Example C.

9. $\dfrac{x^2}{25} + \dfrac{y^2}{9} = 1$ **10.** $\dfrac{x^2}{16} + \dfrac{y^2}{9} = 1$

11. $\dfrac{x^2}{1} + \dfrac{y^2}{4} = 1$ **12.** $\dfrac{x^2}{25} + \dfrac{y^2}{169} = 1$

In Problems 13–20, write the equation of the ellipse that satisfies the given conditions, determine its eccentricity, and sketch its graph. See Example D.

13. Vertices at $(0, \pm 5)$, foci at $(0, \pm 3)$.

14. Vertices at $(0, \pm 10)$, foci at $(0, \pm 8)$.

15. Center at $(0, 0)$, a vertex at $(-7, 0)$, a focus at $(3, 0)$.

16. Center at $(0, 0)$, a focus at $(6, 0)$, major diameter 20.

17. Horizontal, center at $(0, 0)$, major diameter 14, minor diameter 4.

18. Foci at $(\pm 10, 0)$, minor diameter 10.

19. Vertices at $(\pm 9, 0)$, curve passes through $(3, \sqrt{8})$.

20. Ends of minor diameter at $(\pm 4, 0)$, curve passes through $(\sqrt{2}, 4\sqrt{3})$.

Problems 21–24 involve applications of which Example E is an illustration.

21. An elliptical pan (obtained by revolving the bottom half of a horizontal ellipse about its minor axis) has diameter 12 and depth 2. Find the eccentricity of the ellipse.

22. An elliptical fireplace arch (a semiellipse) has a base 5 feet long and height at the center of 3 feet. How high is the arch 1 foot from the center?

23. The earth's elliptical orbit has eccentricity $e = .0167$ and major diameter 29,914,000 kilometers. Find the earth's greatest distance from the sun. Assume that the earth and sun are point masses with the sun at a focus.

24. Halley's comet has an elliptical orbit with the sun at a focus. Its major and minor diameters are 36.18 and 9.12 astronomical units (AU), respectively. Determine its closest approach to the sun in AU.

B. Applications and Extensions

25. Determine a, b, c, and e for the ellipse $4x^2 + 25y^2 = 100$.

26. Find the eccentricity of the ellipse $8x^2 + 2y^2 = 8$.

27. Find the equation of the ellipse with eccentricity $\frac{1}{3}$ and foci at $(0, \pm 4)$.

28. Find the equation of the ellipse that goes through $(\frac{1}{4}, \sqrt{3}/2)$ and has vertices at $(0, \pm 1)$.

29. A door has the shape of an elliptical arch (a semiellipse) that is 10 feet wide and 4 feet high at the center (Figure 22). A box 2 feet high is to be pushed through the door. How wide can it be?

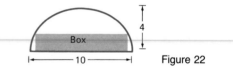

Figure 22

30. How long is the **latus rectum** (chord through the focus perpendicular to the major axis) for the ellipse $x^2/a^2 + y^2/b^2 = 1$?

31. Assume that the center of the earth (a sphere of radius 4000 miles) is at one focus of the elliptical path of a satellite. If the satellite's nearest approach to the surface of the earth is 2000 miles and its farthest distance away is 10,000 miles, what are the major and minor diameters of the elliptical path?

32. ABC is a right triangle with the right angle at B. A and B are the foci of an ellipse and C is on the ellipse. Determine the major and minor diameters of the ellipse given that $\overline{AB} = 8$ and $\overline{BC} = 6$.

33. The area of the ellipse $x^2/a^2 + y^2/b^2 = 1$ is πab. Find the area of the ellipse $11x^2 + 7y^2 = 77$.

34. A square with sides parallel to the coordinate axes is inscribed in the ellipse $b^2x^2 + a^2y^2 = a^2b^2$. Determine the area of the square.

35. A dog's collar is attached by a ring to a loop of rope 32 feet long. The loop of rope is thrown over two stakes 12 feet apart.
(a) How much area can the dog cover?
(b) If the dog should manage to nudge the rope over the top of one of the stakes, how much would this increase the area it can cover?

36. Let P be a point on a 16-foot ladder 7 feet from the top end. As the ladder slides with its top end against a wall (the y-axis) and its bottom end along the ground (the x-axis), P traces a curve. Find the equation of this curve.

37. (Optical property of the ellipse) In Figure 23, let P and Q be the foci of an ellipse, R be a point on the el-

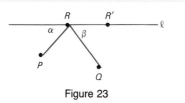

Figure 23

lipse, l be the tangent line at R, and R' be any other point of l.
(a) Show that $\overline{PR'} + \overline{R'Q} > \overline{PR} + \overline{RQ}$.
(b) Show that $\alpha = \beta$. (See Problem 44 of Section 13-1.)
From this we conclude that a light ray from one focus P of an elliptic mirror is reflected back through the other focus.

38. TEASER Two ellipses with the same eccentricity e are such that the major diameter of the smaller ellipse coincides with the minor diameter of the larger ellipse and the area of the smaller ellipse is 19 percent of the area of the larger one. Use this information to determine e.

39. Draw the graph of the ellipse $x^2/64 + y^2/49 = 1$ *Hint:* You will need to solve for y and thereby replace the given equation by two equations.

40. Superimpose the graphs of $3x + y = 10$ and $3x + y = 20$ on the ellipse of Problem 39. Then figure out a way to maximize $P = 3x + y$ on the set consisting of the boundary and interior of this ellipse.

13-3 HYPERBOLAS

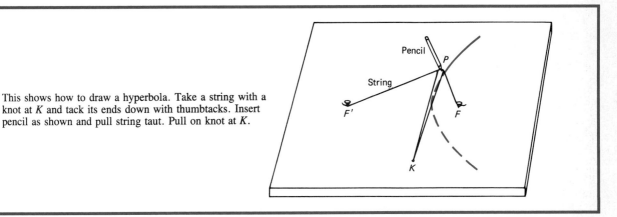

This shows how to draw a hyperbola. Take a string with a knot at K and tack its ends down with thumbtacks. Insert pencil as shown and pull string taut. Pull on knot at K.

Our opening display hints at the geometric definition of a hyperbola. A hyperbola is the set of points P for which the difference of the distances from two fixed points F' and F is a constant. More precisely, a hyperbola is the set of points P in Figure 24 satisfying

$$\left| d(P, F') - d(P, F) \right| = 2a$$

for some constant a.

As with the ellipse, the two fixed points F' and F are called **foci** and the point midway between the foci is the **center** of the hyperbola. The line through the foci is the **major axis** (or transverse axis) of the hyperbola and the line through the center and perpendicular to the major axis is the **minor axis** (or conjugate axis). Also, the points A' and A where the hyperbola intersects the major axis are the **vertices** of the hyperbola. By applying the defining condition with $P = A$, we see that $2a$ is the distance between the vertices. Also, if $2c$ denotes the distance between the foci, then $2c$ must be greater than $2a$, that is, $c > a$.

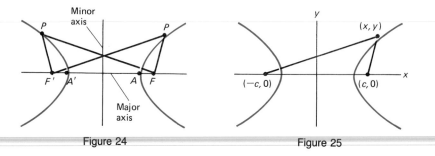

Figure 24   Figure 25

The Equation of a Hyperbola

Place the hyperbola in the coordinate system so that its center is at the origin with the major axis along the x-axis and the foci at $(-c, 0)$ and $(c, 0)$ as in Figure 25. Then the condition $|d(P, F') - d(P, F)| = 2a$ combined with the distance formula yields

$$\left| \sqrt{(x + c)^2 + (y - 0)^2} - \sqrt{(x - c)^2 + (y - 0)^2} \right| = 2a$$

If we now employ the same kind of procedure used in obtaining the equation of the ellipse (square both sides, simplify, square again, and simplify), we get

$$(c^2 - a^2)x^2 - a^2y^2 = a^2(c^2 - a^2)$$

or, equivalently, after dividing both sides by $a^2(c^2 - a^2)$,

$$\frac{x^2}{a^2} - \frac{y^2}{c^2 - a^2} = 1$$

Finally, we let $b^2 = c^2 - a^2$ to obtain what is called the **standard equation of the hyperbola**, namely,

$$\frac{x^2}{a^2} - \frac{y^2}{b^2} = 1$$

Note that both x and y occur to the second power, which corresponds to the fact that the graph of this equation is symmetric with respect to both the x- and y-axes as well as the origin.

Interpreting *a*, *b*, and *c*

We have already noted that $2c$ is the distance between the foci and $2a$ is the distance between the vertices of the hyperbola. To interpret b, observe that if we solve the standard equation for y in terms of x, we get

$$y = \pm \frac{b}{a} \sqrt{x^2 - a^2}$$

For large x, $\sqrt{x^2 - a^2}$ behaves much like x; in fact, as $|x|$ gets larger and larger, $x - \sqrt{x^2 - a^2}$ approaches zero (Problem 26) and hence so does b/a times this quantity. Thus the two branches of the hyperbola $y = \pm \frac{b}{a} \sqrt{x^2 - a^2}$ approach the two lines $y = \pm \frac{b}{a} x$. We say that the hyperbola has these lines as (oblique) asymptotes (see Section 5-4).

Since $c^2 = a^2 + b^2$, the numbers a, b, and c determine a right triangle, which we call the fundamental triangle for the hyperbola. Its role in determining the asymptotes of the hyperbola is clear from Figure 26. This figure summarizes all the key facts for a hyperbola.

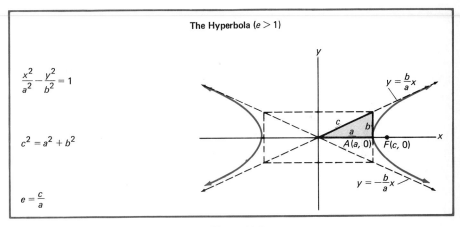

The Hyperbola $(e > 1)$

$$\frac{x^2}{a^2} - \frac{y^2}{b^2} = 1$$

$$c^2 = a^2 + b^2$$

$$e = \frac{c}{a}$$

$$y = \frac{b}{a}x$$

$$y = -\frac{b}{a}x$$

$A(a, 0)$ $F(c, 0)$

Figure 26

The number $e = c/a$, which in this case is greater than 1, is called the **eccentricity** of the hyperbola. If e is near 1, then b is small relative to a and the hyperbola is very thin; if e is large the hyperbola is fat.

Example A (A Horizontal Hyperbola)   Determine the vertices, the asymptotes, and the eccentricity for the hyperbola with equation

$$\frac{x^2}{9} - \frac{y^2}{16} = 1$$

Then sketch its graph.

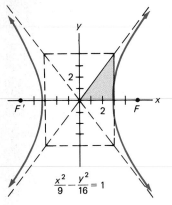

$$\frac{x^2}{9} - \frac{y^2}{16} = 1$$

Figure 27

Solution   We note that $a = 3$, $b = 4$, and $c = \sqrt{a^2 + b^2} = 5$. Thus the vertices are $(\pm 3, 0)$, the asymptotes are $y = \pm \frac{4}{3}x$, and $e = c/a = 5/3$. The graph is shown in Figure 27.   ∎

Again we should consider what happens if we interchange the roles of x and y. The standard equation then takes the form

$$\frac{y^2}{a^2} - \frac{x^2}{b^2} = 1$$

This equation represents a hyperbola with major axis along the y-axis (we call it a *vertical hyperbola*). Its vertices are at $(0, \pm a)$ and its foci are at $(0, \pm c)$.

Example B (A Vertical Hyperbola)   Determine the vertices and sketch the graph of the hyperbola with equation

$$\frac{y^2}{9} - \frac{x^2}{16} = 1$$

Solution   This is the equation of a vertical hyperbola with $a = 3$ and $b = 4$. These numbers allow us to draw the fundamental triangle and the asymptotes which guide the sketching of the graph (Figure 28). The vertices are $(0, \pm 3)$.   ∎

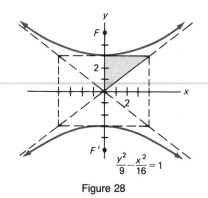

$$\frac{y^2}{9} - \frac{x^2}{16} = 1$$

Figure 28

We make an important observation. It is not the relative sizes of the denominators in the x- and y-terms that determine whether the hyperbola is vertical or horizontal (as it was with the ellipse). Rather, this is determined by whether in the standard form the minus is associated with the x- or the y-term.

Example C (Graphing Another Hyperbola)   Sketch the graph of $x^2/4 - y^2/9 = 1$, showing all the important features.

Solution   The graph is a hyperbola and, since the minus sign is associated with the y-term, the major axis is horizontal. We conclude that $a = 2$, $b = 3$, and

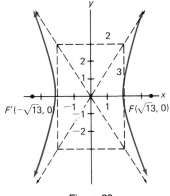

Figure 29

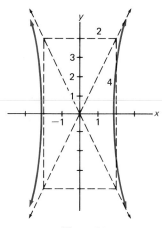

Figure 30

$c = \sqrt{4 + 9} = \sqrt{13}$. The asymptotes are the lines $y = \pm\frac{3}{2}x$. With this information, we may sketch the graph shown in Figure 29. ∎

Example D (Finding Hyperbolas Satisfying Given Conditions)   Find the equation of the hyperbola with vertices at $(\pm 2, 0)$ that passes through $(2\sqrt{2}, 4)$. Sketch its graph.

Solution   Since the vertices are on the x-axis, the hyperbola is horizontal with $a = 2$. The equation has the form

$$\frac{x^2}{4} - \frac{y^2}{b^2} = 1$$

Since the point $(2\sqrt{2}, 4)$ is on the graph,

$$\frac{(2\sqrt{2})^2}{4} - \frac{4^2}{b^2} = 1$$

which gives $b = 4$. Thus the equation is

$$\frac{x^2}{4} - \frac{y^2}{16} = 1$$

The graph is shown in Figure 30. ∎

Applications

The hyperbola, too, has an optical property, as is illustrated in Figure 31. If we imagine one branch of the hyperbola to be a mirror, then a light ray from the opposite focus upon hitting the mirror will be reflected away along the line which passes through the nearby focus. The optical properties of the parabola and the hyperbola are combined in one design for a reflecting telescope (Figure 32). Other applications are treated in the problem set.

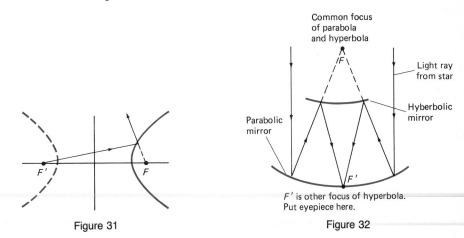

Figure 31

Figure 32

Common focus of parabola and hyperbola

Light ray from star

Hyberbolic mirror

Parabolic mirror

F' is other focus of hyperbola. Put eyepiece here.

PROBLEM SET 13-3

A. Skills and Techniques

In Problems 1–8, decide whether the given equation determines a horizontal or vertical hyperbola. Also find a (the distance from the center to a vertex), b, and c. Be sure the equation is in standard form before you try to give the answers. See Examples A and B.

1. $\dfrac{x^2}{16} - \dfrac{y^2}{36} = 1$

2. $-\dfrac{x^2}{1} + \dfrac{y^2}{8} = 1$

3. $\dfrac{x^2}{16} - \dfrac{y^2}{9} = -1$

4. $\dfrac{x^2}{4} - \dfrac{y^2}{9} = 1$

5. $4x^2 - 16y^2 = 1$

6. $25y^2 - 9x^2 = 1$

7. $4x^2 - y^2 = 16$

8. $k^2y^2 - 4k^2x^2 = 1$

In Problems 9–12, decide whether the corresponding hyperbola is horizontal or vertical, give the values of a, b, and c, and sketch the graph. Be sure to show the asymptotes. See Example C.

9. $\dfrac{x^2}{25} - \dfrac{y^2}{9} = 1$

10. $\dfrac{x^2}{16} - \dfrac{y^2}{9} = 1$

11. $\dfrac{y^2}{64} - \dfrac{x^2}{36} = 1$

12. $\dfrac{x^2}{4} - \dfrac{y^2}{4} = 1$

In Problems 13–18, find the equation of the hyperbola satisfying the given conditions and sketch its graph, displaying the asymptotes. See Example D.

13. Vertices at $(0, \pm 3)$ and going through $(2, 5)$.

14. Vertices at $(\pm 3, 0)$ and going through $(2\sqrt{3}, 9)$.

15. Foci at $(\pm 4, 0)$, vertices at $(\pm 1, 0)$.

16. Vertices at $(\pm 5, 0)$, equations of asymptotes $y = \pm x$.

17. Vertices at $(\pm 3, 0)$, equations of asymptotes $y = \pm 2x$.

18. Vertices at $(0, \pm 3)$, eccentricity $e = \frac{4}{3}$.

B. Applications and Extensions

19. Find the equation of the hyperbola centered at the origin with a focus at $(0, 8)$ and a vertex at $(0, -6)$.

20. Determine the eccentricity of the hyperbola with equation $16x^2 - 20y^2 = 320$.

21. A conic has eccentricity 3 and foci at $(\pm 12, 0)$. Find its equation.

22. Find the equations of the asymptotes of the vertical hyperbola with eccentricity 2 and center at the origin.

23. How long is the **focal chord** (chord through a focus perpendicular to the major axis) of the hyperbola $x^2/9 - y^2/16 = 1$.

24. Generalize Problem 23 by finding the length of the focal chord for the hyperbola $x^2/a^2 - y^2/b^2 = 1$.

25. Find the eccentricity of the hyperbola with asymptotes $y = \pm x$.

26. Show that $x - \sqrt{x^2 - a^2}$ approaches 0 as $|x|$ gets larger and larger. *Hint:* Multiply and divide by $x + \sqrt{x^2 - a^2}$.

27. A ball shot from $(-5, 0)$ hit the right branch of the hyperbolic bangboard $x^2/16 - y^2/9 = 1$ at the point $(8, 3\sqrt{3})$. What was the ball's y-coordinate when its x-coordinate was 10?

28. The rectangle $PQRS$ with sides parallel to the coordinate axes is inscribed in the hyperbola $x^2/4 - y^2/9 = 1$ as shown in Figure 33. Find the coordinates of P if the area of the rectangle is $6\sqrt{5}$.

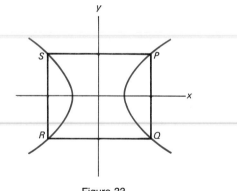

Figure 33

29. Andrew, located at $(0, -2200)$, fired a rifle. The sound echoed off a cliff at $(0, 2200)$ to Brian, located at the point (x, y). Brian heard this echo 6 seconds after he heard the original shot. Find the xy-equation of the curve on which Brian is located. Assume that distances are in feet and that sound travels 1100 feet per second.

30. TEASER Amy, Betty, and Cindy, located at $(-8, 0)$, $(8, 0)$, and $(8, 10)$, respectively, recorded the exact times when they heard an explosion. On comparing notes, they discovered that Betty and Cindy heard the explosion at the same time but that Amy heard it 12 seconds later. Assuming that distances are in kilometers and that sound travels $\frac{1}{3}$ kilometer per second, determine the point of the explosion.

31. Draw the graph of

$$\frac{y^2}{144} - \frac{x^2}{25} = 1$$

32. Superimpose the graph of the line $y = 3.5x - 2$ on the hyperbola of Problem 31 and find the distance between the two intersection points.

13-4 TRANSLATION OF AXES

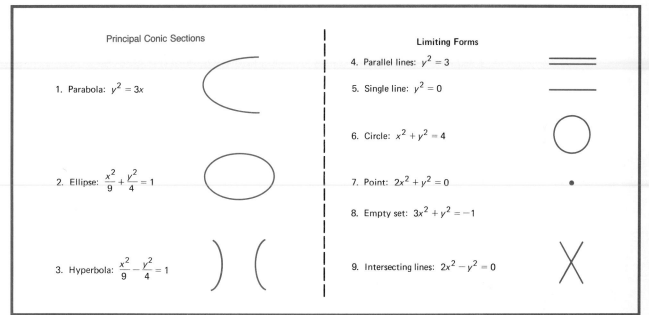

Principal Conic Sections

1. Parabola: $y^2 = 3x$

2. Ellipse: $\dfrac{x^2}{9} + \dfrac{y^2}{4} = 1$

3. Hyperbola: $\dfrac{x^2}{9} - \dfrac{y^2}{4} = 1$

Limiting Forms

4. Parallel lines: $y^2 = 3$

5. Single line: $y^2 = 0$

6. Circle: $x^2 + y^2 = 4$

7. Point: $2x^2 + y^2 = 0$

8. Empty set: $3x^2 + y^2 = -1$

9. Intersecting lines: $2x^2 - y^2 = 0$

Conic Sections

We use the term *principal conic sections* to refer to parabolas, ellipses, and hyperbolas. Actually, circles, points, and single lines can also be called conic sections since they can be obtained as the intersection of a plane with a cone of two nappes. Parallel lines and the empty set cannot be so obtained.

An astute—or perhaps even a casual—observer will note that the standard equations of the three principal conic sections involve the second power of x or y. This observation suggests a question. Suppose we graph a polynomial equation that is of second degree in x and y. Will it always be a conic section? The answer is no; that is, it is no unless we admit the six limiting forms of the conic sections illustrated in the right half of our opening display. But if we do admit them, the answer is yes. In particular, we claim that the graph of any equation of the form

$$Ax^2 + Cy^2 + Dx + Ey + F = 0 \qquad (A, C \text{ not both } 0)$$

is a conic section or one of its limiting forms. We will show you why by moving the coordinate axes in just the right way.

Translations

We begin our task by showing how the substitutions $x = u + h$ and $y = v + k$ can simplify an equation; then we will interpret this change of variables as a translation of axes.

Example A (Change of Variables) In the equation

$$x^2 + y^2 - 4x - 6y - 12 = 0$$

make the substitutions $x = u + 2$ and $y = v + 3$. Identify the resulting equation.

Solution The given substitutions transform the equation to

$$(u + 2)^2 + (v + 3)^2 - 4(u + 2) - 6(v + 3) - 12 = 0$$

or

$$u^2 + 4u + 4 + v^2 + 6v + 9 - 4u - 8 - 6v - 18 - 12 = 0$$

This simplifies to

$$u^2 + v^2 = 25$$

which in the uv-system is the equation of a circle of radius 5. ∎

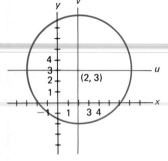

Figure 34

To understand what we have just done, introduce new coordinate axes in the plane (the u- and v-axes) parallel to the old x- and y-axes, but with the new origin at $x = 2$ and $y = 3$ (see Figure 34). Each point now has two sets of coordinates: (x, y) and (u, v). They are related by the equations $x = u + 2$ and $y = v + 3$. In the new coordinate system, which is called a **translation** of the old one, the uv-equation represents a circle of radius 5 centered at the origin. In the old coordinate system, the xy-equation must also have represented a circle of radius 5, but centered at $(2, 3)$.

$$x^2 + y^2 - 4x - 6y - 12 = 0$$
$$\downarrow$$
$$\begin{cases} x = u + 2 \\ y = v + 3 \end{cases}$$
$$\downarrow$$
$$u^2 + v^2 = 25$$

Let us see what happens in general to coordinates of points under a translation of axes. If u- and v-axes are introduced with the same directions as the old x- and y-axes so that the new origin has coordinates (h, k) relative to the old axes (Figure 35), then the two sets of coordinates for a point P are connected by

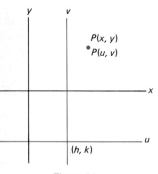

Figure 35

$$u = x - h \qquad v = y - k$$

or, equivalently, by

$$x = u + h \qquad y = v + k$$

The shape of a curve is not changed by a translation of axes since it is the axes, not the curve, that are moved. But as we saw in the example above, the resulting change in the equation may enable us to recognize the curve.

The concept of a translation was discussed from a slightly different perspective in Section 5-5. There we translated the graph; here we are translating the axes.

Completing the Squares

Given an equation, how do we know what translation to make? Here an old algebraic friend, completing the square, comes to our aid. As a typical example, consider

$$x^2 + y^2 - 6x + 8y + 10 = 0$$

We first rewrite the equation and then complete each square.

$$(x^2 - 6x \quad) + (y^2 + 8y \quad) = -10$$

$$(x^2 - 6x + 9) + (y^2 + 8y + 16) = -10 + 9 + 16$$

$$(x - 3)^2 + (y + 4)^2 \quad = 15$$

We can recognize this as the equation of a circle of radius $\sqrt{15}$, centered at $(3, -4)$. In terms of translations (Figure 36), we note that the substitutions $u = x - 3$ and $v = y + 4$ transform the equation into

$$u^2 + v^2 = 15$$

Example B (Complete the Squares: Then Substitute)   Transform the equation

$$4x^2 + 9y^2 - 24x + 18y + 9 = 0$$

to a standard form by means of an appropriate translation of axes (uv-substitution).

Solution   Our first step is to complete the squares by adding the same numbers to both sides of the given equation. This is followed by some simple algebra.

$$4x^2 + 9y^2 - 24x + 18y + 9 = 0$$

$$4(x^2 - 6x \quad) + 9(y^2 + 2y \quad) = -9$$

$$4(x^2 - 6x + 9) + 9(y^2 + 2y + 1) = -9 + 36 + 9$$

$$4(x - 3)^2 + 9(y + 1)^2 \quad = 36$$

$$\frac{(x - 3)^2}{9} + \frac{(y + 1)^2}{4} \quad = 1$$

The substitutions $u = x - 3$ and $v = y + 1$ transform this equation to

$$\frac{u^2}{9} + \frac{v^2}{4} = 1$$

which we recognize as an ellipse with $a = 3$ and $b = 2$ (Figure 37).

$$4x^2 + 9y^2 - 24x + 18y + 9 = 0$$
$$\downarrow$$
$$\begin{cases} x = u + 3 \\ y = v - 1 \end{cases}$$
$$\downarrow$$
$$\frac{u^2}{9} + \frac{v^2}{4} = 1 \quad \blacksquare$$

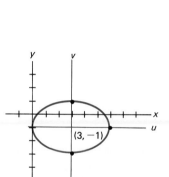

Figure 36

Figure 37

Example C (Graphing Conics)   In the xy-plane, sketch the graphs.
(a) $(x - 2)^2 = 8(y + 3)$ (b) $(x - 2)^2 - (y + 3)^2 = 0$.

Solution
(a) We could formally make the translation of axes corresponding to $u = x - 2$ and $v = y + 3$, which would yield $u^2 = 8v$. But perhaps we can do this mentally and thereby recognize that $(x - 2)^2 = 8(y + 3)$ is the equation of a vertical parabola with vertex at $(2, -3)$ and $p = 2$. With that information, we can make the sketch shown in Figure 38.
(b) The mental substitutions $u = x - 2$ and $v = y + 3$ transform the second equation into $u^2 - v^2 = 0$, which is equivalent to $(u - v)(u + v) = 0$. We recognize this as the equation of two intersecting lines, $v = u$ and $v = -u$. In terms of x and y, this gives $y + 3 = x - 2$ and $y + 3 = -x + 2$ (see Figure 39).

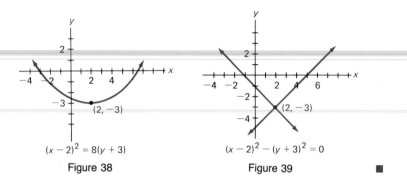

$$(x - 2)^2 = 8(y + 3)$$

Figure 38

$$(x - 2)^2 - (y + 3)^2 = 0$$

Figure 39   ■

Example D (Identifying Conics by Completing Squares)   Identify the conic whose equation is

$$4x^2 - 8x - 2y^2 + 16y = 0$$

Solution   We complete the squares

$$4(x^2 - 2x +) - 2(y^2 - 8y +) = 0$$
$$4(x^2 - 2x + 1) - 2(y^2 - 8y + 16) = 4 - 32$$
$$4(x - 1)^2 - 2(y - 4)^2 = -28$$
$$-\frac{(x - 1)^2}{7} + \frac{(y - 4)^2}{14} = 1$$

In the xy-plane, this is the equation of a vertical hyperbola with center $(1, 4)$.   ■

The General Case

Consider the general equation

$$Ax^2 + Cy^2 + Dx + Ey + F = 0$$

For the moment, we assume that at least one of the coefficients A or C is different from zero. When we apply the process of completing of the square, we transform this equation into one of several forms, the most typical being the following.

1. $(y - k)^2 = 4p(x - h)$

2. $\dfrac{(x - h)^2}{a^2} + \dfrac{(y - k)^2}{b^2} = 1$

3. $\dfrac{(x - h)^2}{a^2} - \dfrac{(y - k)^2}{b^2} = 1$

Perhaps these are already recognizable as the equations of a parabola with vertex at (h, k), an ellipse with center at (h, k), and a hyperbola with center at (h, k). But to remove any doubt, we may translate the axes by the substitutions $u = x - h$ and $v = y - k$, thereby obtaining

1. $v^2 = 4pu$

2. $\dfrac{u^2}{a^2} + \dfrac{v^2}{b^2} = 1$

3. $\dfrac{u^2}{a^2} - \dfrac{v^2}{b^2} = 1$

Our work may also yield these equations with u and v interchanged or we may get one of the six limiting forms illustrated in our opening display. There are no other possibilities.

PROBLEM SET 13-4

A. Skills and Techniques

In Problems 1–6, make the indicated substitutions (a translation of axes) and then name the conic section or limiting form represented by the equation. See Example A.

1. $x^2 + 2y^2 - 4y = 0$; $x = u$, $y = v + 1$
2. $x^2 - 4y^2 - 4x - 5 = 0$; $x = u + 2$, $y = v$
3. $x^2 + y^2 - 4x + 2y = -4$; $x = u + 2$, $y = v - 1$
4. $x^2 + y^2 - 4x + 2y = -5$; $x = u + 2$, $y = v - 1$
5. $x^2 - 6x - 4y + 13 = 0$; $x = u + 3$, $y = v + 1$
6. $x^2 - 4x + 1 = 0$; $x = u + 2$, $y = v$

Change to a standard form by making an appropriate uv-substitution and then name the conic section or limiting form. See Example B.

7. $x^2 + y^2 + 12x - 2y + 33 = 0$
8. $x^2 + y^2 + 12x - 2y + 40 = 0$
9. $x^2 + y^2 + 12x - 2y + 37 = 0$

10. $x^2 + 4y^2 + 12x - 8y + 36 = 0$
11. $x^2 - 4y^2 + 12x - 8y + 28 = 0$
12. $x^2 - 4y^2 + 12x - 8y + 32 = 0$

Sketch the graphs of each of the following equations. See Example C.

13. $\dfrac{(x + 3)^2}{4} + \dfrac{(y + 2)^2}{16} = 1$

14. $(x + 3)^2 + (y - 4)^2 = 25$

15. $\dfrac{(x + 3)^2}{4} - \dfrac{(y + 2)^2}{16} = 1$

16. $4(x + 3) = (y + 2)^2$
17. $(x + 2)^2 = 8(y - 1)$
18. $(x + 2)^2 = 4$
19. $(y - 1)^2 = 16$

20. $\dfrac{(x + 3)^2}{4} + \dfrac{(y - 2)^2}{8} = 0$

Identify the conics determined by the equations in Problems 21–30. See Example D.

21. $4x^2 + 16x + 4y^2 - 8y = 0$

22. $x^2 + 2x + 4y^2 - 8y = 0$

23. $4x^2 - 16x + y^2 - 8y = -6$

24. $4x^2 - 16x - y^2 - 8y = 2$

25. $4x^2 - 16x + y^2 - 8y = -32$

26. $4x^2 - 16x + y^2 - 8y = -40$

27. $4x^2 - 16x + y - 8 = 0$

28. $4x^2 - 16x + 12 = 0$

29. $4x^2 - 16x - 9y^2 + 18y + 7 = 0$

30. $4x^2 - 16x - 9y^2 + 18y + 8 = 0$

31. Sketch the graph of $9x^2 - 18x + 4y^2 + 16y = 11$.

32. Sketch the graph of $4x^2 + 16x - 16y + 32 = 0$.

33. Determine the distance between the vertices of the graph of $-9x^2 + 18x + 4y^2 + 24y = 9$.

34. Find the focus and the directrix of the parabola with equation $x^2 - 4x + 8y = 0$.

35. Find the focus and directrix of the parabola with equation $2y^2 - 4y - 10x = 0$.

36. Find the foci of the ellipse with equation $16(x - 1)^2 + 25(y + 2)^2 = 400$.

B. Applications and Extensions

37. Sketch the graph of each of the following.

(a) $\dfrac{(x + 5)^2}{16} + \dfrac{(y - 3)^2}{9} = 1$

(b) $\dfrac{(x + 5)^2}{16} - \dfrac{(y - 3)^2}{9} = 1$

38. Identify the curve with the given equation.

(a) $4x^2 + 9y^2 - 16x + 54y + 61 = 0$

(b) $x^2 + 8x + 8y = 0$

(c) $x^2 - 4y^2 + 6x + 16y = 16$

(d) $4x^2 + 9y^2 + 16x - 18y + 25 = 0$

39. Name the conic with equation $y^2 + ax^2 = x$ for the various values of a.

40. Find the equation of the parabola with the line $y = 4$ as directrix and the point $(2, -1)$ as focus.

41. Write the equation of the parabola with vertex $(4, 5)$ and focus $(3, 5)$.

42. Write the equation of the ellipse with vertices at $(2, -2)$ and $(2, 10)$ as ends of the major diameter and focus at $(2, 6)$.

43. Write the equation of the ellipse with foci $(\pm 2, 2)$ that goes through the origin.

44. Write the equation of the hyperbola with foci $(0, 0)$ and $(4, 0)$ that passes through $(9, 12)$.

45. Find the equation of the hyperbola with the lines $y = 2x - 10$ and $y = -2x + 2$ as asymptotes and one focus at $(3, 2)$.

46. Transform the equation $xy - 2x + 3y = 18$ by translation of axes so that the new equation has no first degree terms. Sketch the graph showing both sets of axes.

47. A curve C goes through the three points $(-1, 2)$, $(0, 0)$, and $(3, 6)$. Write the equation for C if C is

(a) a vertical parabola.

(b) a horizontal parabola.

(c) a circle.

48. Find the equation of the hyperbola with eccentricity 2 that has the y-axis as one directrix and the corresponding focus at $(6, 0)$.

49. The left and right foci of the ellipse

$$\frac{(x - 3)^2}{25} + \frac{y^2}{16} = 1$$

are A and B, respectively. A light ray from a source at A travels upward along the line $y = x$ to a point C on the ellipse and is reflected back to B. Determine the lengths of the paths ACB and CB.

50. Review the optical property of the hyperbola from Section 13-3. A light ray from the left focus of the hyperbola

$$\frac{(x - 5)^2}{9} - \frac{y^2}{16} = 1$$

travels upward along the line $y = x/3$ through the left branch of the hyperbola and is reflected from the right branch to be absorbed at the point $(a, 10)$. Determine a.

51. Find the equation of the circle that goes through the two foci and the upper y-intercept of the ellipse $36x^2 + 100y^2 = 3600$.

52. TEASER Let C be an arbitrary horizontal ellipse that intersects the parabola $y = x^2$ in four points (x_1, y_1), (x_2, y_2), (x_3, y_3), and (x_4, y_4). Prove that $x_1 + x_2 + x_3 + x_4 = 0$.

53. Verify Problem 52 for the ellipse $(y - 6)^2 + .5(x - 5.5)^2 = 36$ by finding the numbers x_1, x_2, x_3, and x_4 (accurate to three decimal places) and determining their sum. *Hint:* To graph the ellipse, replace its equation by two equations obtained by solving for y.

54. Draw the graph of the ellipse

$$\frac{(x - 8)^2}{25} + \frac{y^2}{16} = 1$$

A light ray from the focus at $(5, 0)$ travels upward along the line $y = x - 5$ and is reflected back through the focus at $(11, 0)$ to a point (a, b) on the bottom semiellipse. Determine a and b. (The TI-81 Draw Line feature may be useful.)

13-5 ROTATION OF AXES

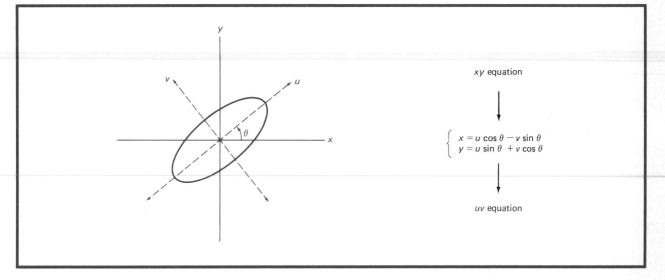

xy equation

$$\begin{cases} x = u \cos \theta - v \sin \theta \\ y = u \sin \theta + v \cos \theta \end{cases}$$

uv equation

We want you to observe two facts about what we have done in the first four sections of this chapter. First, all the conic sections we have considered so far were oriented in the coordinate system with their major axis parallel to either the x-axis or the y-axis. Second, none of the equations of these conics had an xy-term. We would not mention these facts unless there were a connection between them. To see the connection, we need to discuss rotation of axes.

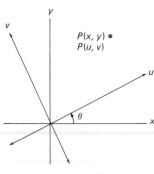

Figure 40

Rotations

Introduce a new pair of axes, called the u- and v-axes, into the xy-plane. These axes have the same origin as the old x- and y-axes, but they are rotated so that the positive u-axis makes an angle θ with the positive x-axis (see the diagram in the opening panel and also the one in Figure 40). A point P then has two sets of coordinates: (x, y) and (u, v). How are they related?

Draw a line segment from the origin O to P, let r denote the length of OP, and let φ denote the angle from the u-axis to OP. Then x, y, u, and v will have the geometric interpretations indicated in Figure 41 at the top of the next page.

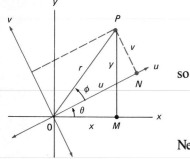

Figure 41

Looking at the right triangle OPM, we see that

$$\cos(\varphi + \theta) = \frac{x}{r}$$

so

$$x = r\cos(\varphi + \theta) = r(\cos\varphi\cos\theta - \sin\varphi\sin\theta)$$
$$= (r\cos\varphi)\cos\theta - (r\sin\varphi)\sin\theta$$

Next, the right triangle OPN tells us that $u = r\cos\varphi$ and $v = r\sin\varphi$. Thus

$$x = u\cos\theta - v\sin\theta$$

Similarly

$$y = r\sin(\varphi + \theta) = r(\sin\varphi\cos\theta + \cos\varphi\sin\theta)$$
$$= (r\sin\varphi)\cos\theta + (r\cos\varphi)\sin\theta$$

so

$$y = u\sin\theta + v\cos\theta$$

We call the boxed results **rotation formulas.**

A Simple Example

When the angle is specified, it is easy to transform an equation by using the rotation formulas.

Example A (Rotation through 45°).   Transform the equation $xy = 1$ by rotating the axes through 45°.

Solution   The required substitutions are

$$x = u\cos 45° - v\sin 45° = \frac{\sqrt{2}}{2}(u - v)$$

$$y = u\sin 45° + v\cos 45° = \frac{\sqrt{2}}{2}(u + v)$$

When we make these substitutions in $xy = 1$, we obtain

$$\frac{\sqrt{2}}{2}(u - v)\frac{\sqrt{2}}{2}(u + v) = 1$$

which simplifies to

$$\frac{u^2}{2} - \frac{v^2}{2} = 1$$

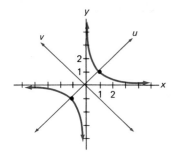

Figure 42

We recognize this as the equation of a hyperbola with $a = b = \sqrt{2}$. Note how the cross-product term xy disappeared as a result of the rotation. The choice of a 45° angle was just right to make this happen (see Figure 42).

$$xy = 1$$
$$\downarrow$$
$$\begin{cases} x = u \cos 45° - v \sin 45° \\ y = u \sin 45° + v \cos 45° \end{cases}$$
$$\downarrow$$
$$\frac{u^2}{2} - \frac{v^2}{2} = 1 \quad \blacksquare$$

The General Second Degree Equation

How do we know what rotation to make? Consider the most general second degree equation in x and y:

$$Ax^2 + Bxy + Cy^2 + Dx + Ey + F = 0$$

If we make the substitutions

$$x = u \cos \theta - v \sin \theta$$
$$y = u \sin \theta + v \cos \theta$$

this equation takes the form

$$au^2 + buv + cv^2 + du + ev + f = 0$$

where a, b, c, d, e and f are numbers which depend upon θ. We could find values for all of them, but we really care only about b. When we do the necessary algebra, we find

$$b = B(\cos^2 \theta - \sin^2 \theta) - 2(A - C) \sin \theta \cos \theta$$
$$= B \cos 2\theta - (A - C) \sin 2\theta$$

We would like to have $b = 0$; that is,

$$B \cos 2\theta = (A - C) \sin 2\theta$$

This will occur if

$$\cot 2\theta = \frac{A - C}{B}$$

This formula is the answer to our question: to eliminate the cross-product (xy) term, choose θ so it satisfies this formula. In the example $xy = 1$, we have $A = 0$, $B = 1$, and $C = 0$ so we choose θ to satisfy

$$\cot 2\theta = \frac{0 - 0}{1} = 0$$

One angle that works is $\theta = 45°$. We could also use $\theta = 135°$ or $\theta = -225°$, but it is customary to choose a first quadrant angle.

More Examples

It is time to apply what we have learned in more complicated situations.

Example B (Rotation through a Special Angle) Make a rotation of axes to remove the cross-product term in

$$4x^2 + 2\sqrt{3}\,xy + 2y^2 + 10\sqrt{3}\,x + 10y = 5$$

Then translate the axes and sketch the graph.

Solution We rotate the axes through an angle θ satisfying

$$\cot 2\theta = \frac{A - C}{B} = \frac{4 - 2}{2\sqrt{3}} = \frac{1}{\sqrt{3}}$$

This means that $2\theta = 60°$ and so $\theta = 30°$. When we use this value of θ in the rotation formulas, we obtain

$$x = u \cdot \frac{\sqrt{3}}{2} - v \cdot \frac{1}{2} = \frac{\sqrt{3}\,u - v}{2}$$

$$y = u \cdot \frac{1}{2} + v \cdot \frac{\sqrt{3}}{2} = \frac{u + \sqrt{3}\,v}{2}$$

Substituting these in the original equation gives

$$4\frac{(\sqrt{3}\,u - v)^2}{4} + 2\sqrt{3}\frac{(\sqrt{3}\,u - v)(u + \sqrt{3}\,v)}{4}$$

$$+ 2\frac{(u + \sqrt{3}\,v)^2}{4} + 10\sqrt{3}\frac{\sqrt{3}\,u - v}{2} + 10\frac{u + \sqrt{3}\,v}{2} = 5$$

After collecting terms and simplifying, we have

$$5u^2 + v^2 + 20u = 5$$

Next we complete the squares.

$$5(u^2 + 4u + 4) + v^2 = 5 + 20$$

$$\frac{(u + 2)^2}{5} + \frac{v^2}{25} = 1$$

As a final step, we make the translation determined by $r = u + 2$ and $s = v$, which gives

$$\frac{r^2}{5} + \frac{s^2}{25} = 1$$

This is the equation of a vertical ellipse in the rs-coordinate system. It has major di-

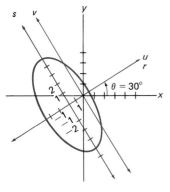

Figure 43

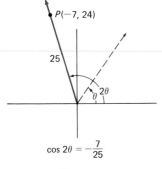

$P(-7, 24)$

25

2θ

θ

$\cos 2\theta = -\dfrac{7}{25}$

Figure 44

ameter of length 10 and minor diameter of length $2\sqrt{5}$. All of this is shown in Figure 43.

$$4x^2 + 2\sqrt{3}\,xy + 2y^2 + 10\sqrt{3}\,x + 10y = 5$$

$$\downarrow$$

$$\begin{cases} x = u \cos 30° - v \sin 30° \\ y = u \sin 30° + v \cos 30° \end{cases}$$

$$\downarrow$$

$$\frac{(u + 2)^2}{5} + \frac{v^2}{25} = 1$$

$$\downarrow$$

$$\begin{cases} r = u + 2 \\ s = v \end{cases}$$

$$\downarrow$$

$$\frac{r^2}{5} + \frac{s^2}{25} = 1 \quad \blacksquare$$

Example C (Rotation through a Nonspecial Angle) By rotation of axes, eliminate the xy-term from

$$x^2 + 24xy + 8y^2 = 136$$

and then sketch the graph.

Solution   Recall that we must choose θ to satisfy

$$\cot 2\theta = \frac{A - C}{B} = \frac{1 - 8}{24} = -\frac{7}{24}$$

Here our problem is complicated by the fact that 2θ is not a special angle. How shall we find the values of $\sin\theta$ and $\cos\theta$ needed for the rotation formulas?

First, we place 2θ in standard position (see Figure 44), noting that $P(-7, 24)$ is on its terminal side. Since P is a distance $r = \sqrt{(-7)^2 + (24)^2} = 25$ from the origin, $\cos 2\theta = -\frac{7}{25}$.

Second, we recall the half-angle formulas.

$$\sin\theta = \pm\sqrt{\frac{1 - \cos 2\theta}{2}} \qquad \cos\theta = \pm\sqrt{\frac{1 + \cos 2\theta}{2}}$$

Since our θ is in the first quadrant, we use the plus sign in both cases. We obtain

$$\sin\theta = \sqrt{\frac{1 + \frac{7}{25}}{2}} = \frac{4}{5} \qquad \cos\theta = \sqrt{\frac{1 - \frac{7}{25}}{2}} = \frac{3}{5}$$

These, in turn, give the rotation formulas

$$x = \frac{3u - 4v}{5} \qquad y = \frac{4u + 3v}{5}$$

All this was preliminary; our main task is to substitute these expressions for x and y in the original equation and simplify.

$$\left(\frac{3u - 4v}{5}\right)^2 + 24\left(\frac{3u - 4v}{5}\right)\left(\frac{4u + 3v}{5}\right) + 8\left(\frac{4u + 3v}{5}\right)^2 = 136$$

After multiplying by 25 and collecting terms, we have

$$425u^2 - 200v^2 = 136 \cdot 25$$

or

$$\frac{u^2}{8} - \frac{v^2}{17} = 1$$

We summarize the process below and show the graph in Figure 45.

$$x^2 + 24xy + 8y^2 = 136$$
$$\downarrow$$
$$\begin{cases} x = \frac{3}{5}u - \frac{4}{5}v \\ y = \frac{4}{5}u + \frac{3}{5}v \end{cases}$$
$$\downarrow$$
$$\frac{u^2}{8} - \frac{v^2}{17} = 1$$

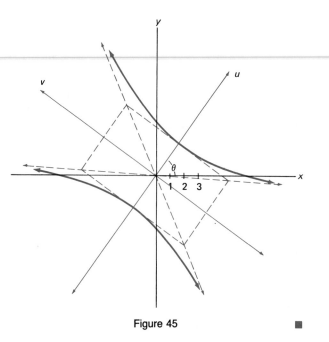

Figure 45

Example D (The Inverse Rotation Formulas)   For a rotation of axes through angle θ, obtain the formulas that express u and v in terms of x and y. Then use the result to obtain the uv-coordinates of the point that has xy-coordinates $(4, 2\sqrt{3})$ if the angle θ is $30°$.

Solution   Consider the rotation formulas at the beginning of this problem set. Multiply the first one by $\cos\theta$ and the second by $\sin\theta$, then add.

$$x\cos\theta = u\cos^2\theta - v\sin\theta\cos\theta$$
$$\underline{y\sin\theta = u\sin^2\theta + v\sin\theta\cos\theta}$$
$$x\cos\theta + y\sin\theta = u(\cos^2\theta + \sin^2\theta) \quad = u$$

Similarly, multiply the first formula by $-\sin\theta$ and the second by $\cos\theta$, and add. The two resulting formulas are

$$u = x\cos\theta + y\sin\theta$$
$$v = -x\sin\theta + y\cos\theta$$

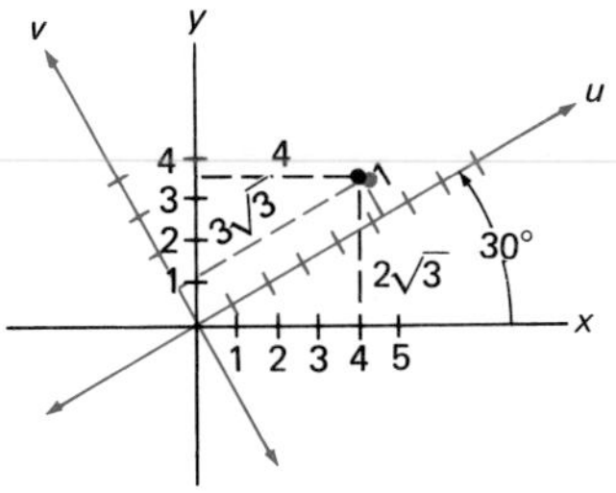

To find the values of u and v, simply substitute $x = 4$, $y = 2\sqrt{3}$, and $\theta = 30°$ in the above formulas. This gives

$$u = 4\cdot\frac{\sqrt{3}}{2} + 2\sqrt{3}\cdot\frac{1}{2} = 3\sqrt{3}$$

$$v = -4\cdot\frac{1}{2} + 2\sqrt{3}\cdot\frac{\sqrt{3}}{2} = 1$$

Figure 46

The geometric interpretation of these numbers is shown in Figure 46.   ■

PROBLEM SET 13-5

A. Skills and Techniques

In the text, we derived the following rotation formulas.

$$x = u\cos\theta - v\sin\theta$$
$$y = u\sin\theta + v\cos\theta$$

In Problems 1–10, transform the given xy-equation to a uv-equation by a rotation through the specified angle θ. See Example A.

1. $y = \sqrt{3}\,x;\quad \theta = 60°$
2. $y = x;\quad \theta = 45°$
3. $x^2 + 4y^2 = 16;\quad \theta = 90°$
4. $4y^2 - x^2 = 4;\quad \theta = 90°$
5. $y^2 = 4\sqrt{2}\,x;\quad \theta = 45°$
6. $x^2 = -\sqrt{2}\,y + 3;\quad \theta = 45°$
7. $x^2 - xy + y^2 = 4;\quad \theta = 45°$
8. $x^2 + 3xy + y^2 = 10;\quad \theta = 45°$
9. $6x^2 - 24xy - y^2 = 30;\quad \theta = \cos^{-1}(\frac{3}{5})$
10. $3x^2 - \sqrt{3}\,xy + 2y^2 = 39;\quad \theta = 60°$

In Problems 11–20, eliminate the xy-term by a suitable rotation of axes and then, if necessary, translate axes (complete the squares) to put the equation in standard form. Finally, graph the equation showing all axes used. (Note: Some problems involve special angles, but several do not.) See Examples B and C.

11. $3x^2 + 10xy + 3y^2 + 8 = 0$

12. $2x^2 + xy + 2y^2 = 90$

13. $4x^2 - 3xy = 18$

14. $4xy - 3y^2 = 64$

15. $x^2 - 2\sqrt{3}\,xy + 3y^2 - 12\sqrt{3}x - 12y = 0$

16. $x^2 + 2\sqrt{3}\,xy + 3y^2 + 8\sqrt{3}x - 8y = 0$

17. $13x^2 + 6\sqrt{3}\,xy + 7y^2 - 32 = 0$

18. $17x^2 + 12xy + 8y^2 + 17 = 0$

19. $9x^2 - 24xy + 16y^2 - 60x + 80y + 75 = 0$

20. $16x^2 + 24xy + 9y^2 - 20x - 15y - 150 = 0$

In Problems 21–26, find u and v for the given values of x, y, and θ. Then make a diagram to check that your answers make sense. See Example D.

21. $(5, -3)$; $60°$

22. $(-2, 5)$; $60°$

23. $(3\sqrt{2}, \sqrt{2})$; $45°$

24. $(5/\sqrt{2}, -5/\sqrt{2})$; $45°$

25. $(3, 4)$; $\tan^{-1}(\frac{4}{3})$

26. $(5, -12)$; $\arctan(\frac{5}{12})$

27. Find the xy-equation that simplifies to $u^2 = 4v$ when the axes are rotated through an angle of $60°$.

28. Find the xy-equation that simplifies to $u^2 - 4v^2 = 4$ when the axes are rotated through an angle of $30°$.

B. Applications and Extensions

29. Transform the equation $2x^2 + \sqrt{3}xy + y^2 = 5$ to a uv-equation by rotating the axes through $30°$. Use the result to identify the corresponding curve.

30. Without any algebra, determine the uv-equation corresponding to the equation $x^2/16 + y^2/9 = 1$ when the axes are rotated through $90°$. Then do the algebra to corroborate your answer.

31. Without any algebra, determine the uv-equation corresponding to $(x - 2\sqrt{2})^2 + (y - 2\sqrt{2})^2 = 16$ when the axes are rotated through $45°$.

32. Find the angle θ through which the axes must be rotated so that the circle $(x - 4)^2 + (y - 3)^2 = 4$ lies above the u-axis and is tangent to it. Write the uv-equation of this circle.

33. By a rotation of axes, remove the cross-product term from $13x^2 + 24xy + 3y^2 = 105$ and identify the corresponding conic section.

34. Transform the equation $(y^2 - x^2)(y + x) = 8\sqrt{2}$ to a uv-equation by rotating the axes through $45°$. Sketch the graph showing both sets of axes.

35. The graph of $x \cos \alpha + y \sin \alpha = d$ is a line. Show that the perpendicular distance from the origin to this line is $|d|$ by making a rotation of axes through the angle α.

36. Use Problem 35 to show that the perpendicular distance from the origin to the line $ax + by = c$ is $|c|/\sqrt{a^2 + b^2}$.

37. Use the result of Problem 36 to find the perpendicular distance from the origin to the line $5x + 12y = 39$.

38. When $Ax^2 + Bxy + Cy^2 = K$ is transformed to $au^2 + buv + cv^2 = K$ by a rotation of axes, it turns out that $A + C = a + c$ and $B^2 - 4AC = b^2 - 4ac$. (Ambitious students will find showing this to be a straightforward but somewhat lengthy algebraic exercise.) Use these results to transform $x^2 - 8xy + 7y^2 = 9$ to $au^2 + cv^2 = 9$ without actually carrying out the rotation.

39. Recall that the area of an ellipse with major diameter $2a$ and minor diameter $2b$ is πab. Use the first sentence of Problem 38 to show that if $A + C$ and $4AC - B^2$ are both positive, then $Ax^2 + Bxy + Cy^2 = 1$ is the equation of an ellipse with area $2\pi/\sqrt{4AC - B^2}$.

40. TEASER The graph of $x^2 - 2xy + 3y^2 = 32$ is an ellipse and therefore can be circumscribed by a rectangle with sides parallel to the x- and y-axes. Find the vertices of this rectangle.

41. Draw the graph of the equation $4x^2 - 2xy + y^2 = 32$ which is equivalent to $y = x \pm \sqrt{32 - 3x^2}$. The graph is an ellipse and can therefore be circumscribed by a rectangle with sides parallel to the x- and y-axes. Find the vertices of this rectangle. *Suggestion:* Use the TRACE feature of your calculator and watch the coordinates of the cursor.

42. Consider the ellipse of Problem 41. Superimpose the line $y = k(x + 6)$ and experiment with various values of k until you find the line through $(-6, 0)$ that is tangent to the lower half of the ellipse. Also find the point of tangency.

13-6 PARAMETRIC EQUATIONS

Every physicist knows that a good way to specify the path of a particle is to give its coordinates as functions of the elapsed time t. Suppose a point $P(x, y)$ moves in the plane so that $x = 3 \cos t$ and $y = 2 \sin t$. What is the shape of its path?

$x = 3 \cos t$

$y = 2 \sin t$

$P(x, y)$

So far in this book, we have described curves by giving their Cartesian equations, that is, their xy-equations. There is another way to describe many curves, including the conic sections. It arises naturally in the study of motion in physics, but its use goes far beyond that particular application.

Imagine that the xy-coordinates of a point on a curve are specified not by giving a relationship between x and y, but rather by telling how x and y are related to a third variable. For example, it may be that as time t advances from $t = a$ to $t = b$, a point $P(x, y)$ traces out a curve in the xy-plane. Then both x and y are functions of t. That is,

$$x = f(t) \quad \text{and} \quad y = g(t) \qquad a \le t \le b$$

We call the boxed equations the **parametric equations** of a curve with t as parameter. A **parameter** is simply an auxiliary variable on which other variables depend.

Graphing, Eliminating Parameters

Let

$$x = 2t \quad \text{and} \quad y = t^2 - 3 \qquad -1 \le t \le 3$$

These equations can be used to make a table of values and then to draw a graph. We illustrate in Figure 47 at the top of the next page.

The curve just drawn looks suspiciously like part of a parabola. We can demonstrate that this is true by **eliminating the parameter** t. Solve the first equation for t, giving $t = x/2$. Then substitute this value of t in the second equation. We obtain

$$y = \left(\frac{x}{2}\right)^2 - 3 = \frac{1}{4}x^2 - 3$$

which we do recognize as the equation of a parabola.

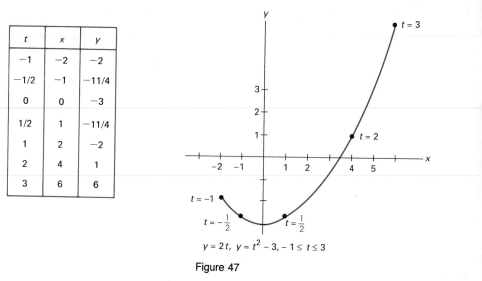

t	x	y
-1	-2	-2
$-1/2$	-1	$-11/4$
0	0	-3
$1/2$	1	$-11/4$
1	2	-2
2	4	1
3	6	6

$$y = 2t, \; y = t^2 - 3, \, -1 \le t \le 3$$

Figure 47

Example A (Eliminating the Parameter) Write the following parametric equations in xy-form.

(a) $x = 3t - 1$, $y = t^3$ (b) $x = 2 \sin t$, $y = 3 \cos t$

Solution

(a) We solve the first equation for t obtaining $t = (x + 1)/3$ and then substitute in the second equation. This gives $y = (x + 1)^3/27$.

(b) Here we play a trick. Solve the first equation for $\sin t$ and the second for $\cos t$. Square the two results and add. We obtain

$$\sin^2 t + \cos^2 t = \frac{x^2}{4} + \frac{y^2}{9}$$

that is,

$$\frac{x^2}{4} + \frac{y^2}{9} = 1 \quad \blacksquare$$

Parametric Equations of a Line

Consider first a line l which passes through the points $(0, 0)$ and (a, b). We claim that

$$x = at \quad \text{and} \quad y = bt$$

are a pair of parametric equations for this line. To see why, note that $t = 0$ and $t = 1$ yield the given points. Moreover, if we eliminate t, we get $y = (b/a)x$, which is the equation of a line. This example shows, incidentally, that a parametric representation is not unique since there are many choices for (a, b). (See Problem 35 for more evidence on this point.)

Next we translate the above line by replacing x by $x - x_1$ and y by $y - y_1$. This gives

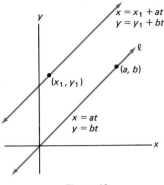

$$x = x_1 + at \quad \text{and} \quad y = y_1 + bt$$

These are the parametric equations of a line through (x_1, y_1) parallel to the line with which we started (Figure 48).

Example B (Parametric Equations for a Line)   Find parametric equations for a line through the points $(3, -1)$ and $(7, 5)$.

Solution   This line is parallel to the line that goes through the origin and the point $(7 - 3, 5 + 1) = (4, 6)$. Thus its parametric equations can be written as $x = 3 + 4t$, $y = -1 + 6t$.   ∎

Finally, consider the line through (x_1, y_1) with slope m. This line is parallel to the line through the points $(0, 0)$ and $(1, m)$ and so has parametric equations

$$x = x_1 + t \quad \text{and} \quad y = y_1 + mt$$

Note that if we eliminate t by solving the first equation for t and substituting it in the second, we get $y - y_1 = m(x - x_1)$, the point-slope form for the equation of a line.

The Circle and the Ellipse

If you think about the definitions of sine and cosine for a moment, you already know one set of parametric equations for a circle of radius a centered at the origin.

$$x = a \cos t \quad \text{and} \quad y = a \sin t \qquad 0 \le t \le 2\pi$$

For other possibilities, see Problem 35.

Consider next an ellipse centered at $(0, 0)$ and passing through $(a, 0)$ and $(0, b)$. We claim that

$$x = a \cos t \quad \text{and} \quad y = b \sin t \qquad 0 \le t \le 2\pi$$

are parametric equations for it. To see that this is correct, we shall eliminate the parameter t. One way to do this is to solve for $\cos t$ and $\sin t$, respectively, square the results, and add. This gives

$$\frac{x^2}{a^2} + \frac{y^2}{b^2} = \cos^2 t + \sin^2 t = 1$$

which we recognize as the xy-equation of an ellipse centered at $(0, 0)$ and passing through $(a, 0)$ and $(0, b)$.

Now we can answer the question asked in the opening display of this section. The parametric equations

$$x = 3 \cos t \quad \text{and} \quad y = 2 \sin t$$

determine an ellipse with center at $(0, 0)$ and passing through $(3, 0)$ and $(0, 2)$.

The Hyperbola

By analogy with the situation for the ellipse, we are led to

$$x = a \sec t \quad \text{and} \quad y = b \tan t \qquad 0 \le t \le 2\pi, t \ne \frac{\pi}{2}, t \ne \frac{3\pi}{2}$$

as parametric equations for a hyperbola. Note that when we solve for $\sec t$ and $\tan t$ in the two equations, square the results, and subtract, we obtain

$$\frac{x^2}{a^2} - \frac{y^2}{b^2} = \sec^2 t - \tan^2 t = 1$$

There is another set of parametric equations for the hyperbola. It involves an important pair of functions called the *hyperbolic sine* and the *hyperbolic cosine*. These functions are discussed in Problem 36.

Example C (Ellipses and Hyperbolas)   Write parametric equations corresponding to each xy-equation.

(a) $\dfrac{x^2}{81} + \dfrac{y^2}{16} = 1$ (b) $\dfrac{x^2}{49} - \dfrac{y^2}{64} = 1$

Solution

(a) Here $a = 9$ and $b = 4$. One appropriate pair of parametric equations is $x = 9 \cos t$, $y = 4 \sin t$, $0 \le t \le 2\pi$.
(b) You may use the identity $\sec^2 t - \tan^2 t = 1$ to show that $x = 7 \sec t$, $y = 8 \tan t$, $0 \le t \le 2\pi$, $t \ne \pi/2$, $t \ne 3\pi/2$ work fine.   ■

The Cycloid

So far, our discussion of parametric equations has concentrated on the conic sections. Actually, there are other important curves where parametric representation is almost essential. The cycloid provides an example where the xy-equation is so complicated it is rarely used.

Consider a wheel of radius a which is free to roll along the x-axis. As the wheel turns, a point P on the rim traces out a curve called the **cycloid** (Figure 49).

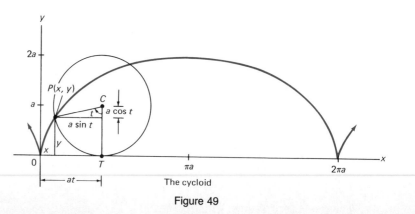

The cycloid

Figure 49

Assume P is initially at the origin and let C and T be as indicated in the diagram, with t denoting the radian measure of angle TCP. Then the arc PT and the segment OT have the same length and so the center C of the rolling circle is at (at, a). Using a little trigonometry, we conclude that

$$x = at - a \sin t = a(t - \sin t)$$

and

$$y = a - a \cos t = a(1 - \cos t)$$

Additional Examples

Example D (More Graphing)   Sketch the graph of the curve with parametric equations $x = 8 \cos^3 t$ and $y = 8 \sin^3 t$, $0 \le t \le 2\pi$.

Solution   A table of values and the graph are shown in Figure 50. This curve is called a *hypocycloid* (see Problem 45).

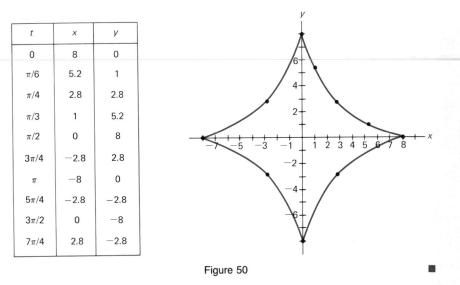

t	x	y
0	8	0
$\pi/6$	5.2	1
$\pi/4$	2.8	2.8
$\pi/3$	1	5.2
$\pi/2$	0	8
$3\pi/4$	−2.8	2.8
π	−8	0
$5\pi/4$	−2.8	−2.8
$3\pi/2$	0	−8
$7\pi/4$	2.8	−2.8

Figure 50   ■

Example E (A Projectile Problem)   The path of a projectile fired at 64 feet per second from ground level at an angle of 60° with the horizontal is given by the parametric equations

$$x = 32t \quad \text{and} \quad y = -16t^2 + 32\sqrt{3}\,t$$

where the origin is at the point of firing and the x-axis is along the (horizontal) ground in the plane of the projectile's flight (Figure 51). Find:

(a) the xy-equation of the path;
(b) the total time of flight;
(c) the range, that is, the value of x where the projectile strikes the ground;
(d) the greatest height reached.

Figure 51

Solution

(a) Since $t = x/32$,

$$y = -16\left(\frac{x}{32}\right)^2 + 32\sqrt{3}\left(\frac{x}{32}\right)$$

$$= -\tfrac{1}{64}x^2 + \sqrt{3}\,x$$

which is the equation of a parabola.

(b) We find the values of t for which $y = 0$.

$$y = -16t^2 + 32\sqrt{3}\,t = -16t\,(t - 2\sqrt{3})$$

Thus $y = 0$ when $t = 0$ (time of firing) and when $t = 2\sqrt{3}$ (time of landing). The time of flight is $2\sqrt{3}$ seconds.

(c) The range is $32(2\sqrt{3}) = 64\sqrt{3}$ feet.

(d) The maximum height occurs when $t = \tfrac{1}{2}(2\sqrt{3}) = \sqrt{3}$. At this time,

$$y = -16(\sqrt{3})^2 + 32\sqrt{3}(\sqrt{3}) = 48 \text{ feet} \quad \blacksquare$$

PROBLEM SET 13-6

A. Skills and Techniques

In Problems 1–10, eliminate the parameter to determine the corresponding xy-equation. See Example A.

1. $x = 3s + 1$, $y = -2s + 5$

2. $x = 3s + 1$, $y = s^2$

3. $x = 2t - 1$, $y = 2t^2 + t$

4. $x = 3t$, $y = t^2 - 3t + 1$

5. $x = 2\cos t$, $y = 2\sin t$

6. $x = 3\sin t$, $y = 3\cos t$

7. $x = 2\cos t$, $y = 3\sin t$

8. $x = 6\sin t$, $y = \cos t$

9. $x = 3t + 1$, $y = t^3$

10. $x = 2\sec t$, $y = 3\tan t$

In Problems 11–14, write parametric equations for the line that passes through the two given points. See Example B.

11. $(0, 0)$, $(2, -3)$ **12.** $(0, 0)$, $(-3, 6)$

13. $(1, 2)$, $(4, -5)$ **14.** $(-2, 3)$, $(3, 7)$

15. Find the slope and y-intercept of the line with parametric equations $x = 3 + 2t$ and $y = -5 - 4t$. *Hint:* Eliminate the parameter t.

16. Write the equation of the line with parametric equations $x = -2 - 3t$ and $y = 4 + 9t$ in the form $Ax + By + C = 0$.

Write the following equations in parametric form. See Example C.

17. $\dfrac{x^2}{25} + \dfrac{y^2}{3} = 1$ **18.** $x^2 + y^2 = 16$

19. $\dfrac{x^2}{25} + \dfrac{y^2}{64} = 1$ **20.** $\dfrac{x^2}{25} - \dfrac{y^2}{64} = 1$

21. $9x^2 - 16y^2 = 144$ **22.** $-9x^2 + 16y^2 = 144$

Sketch the graph of each of the following for the indicated interval. See Example D.

23. $x = 2t - 1$, $y = t^2 + 2$; $-2 \leq t \leq 2$

24. $x = 2t^2$, $y = 3 - 2t$; $-2 \leq t \leq 2$

25. $x = t^3$, $y = t^2$; $-2 \leq t \leq 2$

26. $x = 2^t$, $y = 3t$; $-3 \leq t \leq 2$

27. $x = \dfrac{1 - t^2}{1 + t^2}$, $y = \dfrac{2t}{1 + t^2}$; all t

28. $x = \dfrac{t^2}{1 + t^2}$, $y = \dfrac{t^3}{1 + t^2}$; all t

29. $x = 8t - 4\sin t$, $y = 8 - 4\cos t$; $0 \leq t \leq 4\pi$
Note: This curve is called a *curtate cycloid* (see Problem 43).

30. $x = 6t - 8\sin t$, $y = 6 - 8\cos t$; $0 \leq t \leq 4\pi$
Note: This curve is called a *prolate cycloid* (see Problem 44).

Answer the four questions of Example E for the data below.

31. $x = 64\sqrt{3}\,t,\ y = -16t^2 + 64t$

32. $x = 48\sqrt{2}\,t,\ y = -16t^2 + 48\sqrt{2}\,t$

B. Applications and Extensions

33. Determine the Cartesian equation corresponding to each given pair of parametric equations.
   (a) $x = 3 - 2t,\ y = 4 + 3t$
   (b) $x = 3t,\ y = 4\cos t$
   (c) $x = 2\sec t,\ y = 3\tan t$
   (d) $x = 1 - t^3,\ y = 2t - 1$
   (e) $x = t^2 + 2t,\ y = \sqrt[3]{t} - t^2 - 2t$

34. Write parametric equations for each of the following.
   (a) The line that passes through the points $(2, -1)$ and $(4, 3)$.
   (b) The line through $(4, -2)$ and parallel to the line with parametric equations $x = 3 + 2t$, $y = -2 + t$.
   (c) The ellipse with Cartesian equation $x^2/9 + y^2/25 = 1$.
   (d) The circle $(x - 4)^2 + (y + 2)^2 = 25$.

35. Show that all of the following parametrizations represent the same curve (one quarter of a circle).
   (a) $x = 2\cos t,\ y = 2\sin t;\ 0 \le t \le \pi/2$
   (b) $x = \sqrt{t},\ y = \sqrt{4 - t};\ 0 \le t \le 4$
   (c) $x = t + 1,\ y = \sqrt{3 - 2t - t^2};\ -1 \le t \le 1$
   (d) $x = (2 - 2t)/(1 + t),\ y = 4\sqrt{t}/(1 + t);$
      $0 \le t \le 1$

36. Show that the parametric equations $x = \sqrt{2t + 1}$, $y = \sqrt{8t}$, $t \ge 0$, represent part of a hyperbola. Sketch that part.

37. Sketch the graph of $x = 2 + 3\cos t$, $y = 1 + 4\sin t$, $0 \le t \le 2\pi$, and determine the corresponding Cartesian equation.

38. Sketch the graph of $x = \sin t$, $y = \tan t$, $-\pi/2 < t < \pi/2$, and determine the corresponding Cartesian equation.

39. The path of a projectile fired from level ground with a speed of v_0 feet per second at an angle α with the ground, is given by the parametric equations

$$x = (v_0\cos\alpha)t \qquad y = -16t^2 + (v_0\sin\alpha)t$$

   (a) Show that the path is a parabola.
   (b) Find the time of flight.
   (c) Show that the range is $(v_0^2/32)\sin 2\alpha$.
   (d) For a given v_0, what value of α gives the largest possible range?

40. Sketch the graph of the parametric equations $x = t^{-1/2}\cos t$, $y = t^{-1/2}\sin t$, $t > 0$.

41. Show that the graph of $x = 5\sin^2 t - 4\cos^2 t$, $y = 4\cos^2 t + 5\sin^2 t$, $0 \le t \le \pi/2$, is a line segment and find its endpoints. What curve do you get when t varies from $\pi/2$ to π?

42. Define two (important) functions called the **hyperbolic sine** and **hyperbolic cosine** by

$$\sinh t = \frac{e^t - e^{-t}}{2} \qquad \cosh t = \frac{e^t + e^{-t}}{2}$$

   (a) Show that $\cosh^2 t - \sinh^2 t = 1$
   (b) Show that $x = a\cosh t$, $y = a\sinh t$ give a parameterization for one branch of a hyperbola.

43. Modify the text discussion of the cycloid (and its accompanying diagram) to handle the case where the point P is $b < a$ units from the center of the wheel. You should obtain the parametric equations

$$x = at - b\sin t \qquad y = a - b\cos t$$

   The graph of these equations is called a *curtate cycloid* (see Problem 29).

44. Follow the instructions of Problem 43 for the case $b > a$ (a flanged wheel, as on a train) showing that you get the same parametric equations. The graph of these equations is now called a *prolate cycloid* (see Problem 30).

45. Suppose that a wheel of radius b rolls inside a circle of radius $a = 4b$ (Figure 52). Show that the parametric equations for a point P on the rim of the rolling wheel are

$$x = a\cos^3\theta \qquad y = a\sin^3\theta$$

   provided the fixed wheel is centered at $(0, 0)$ and P is initially at $(a, 0)$. The resulting curve is called a *hypocycloid* (see Example D).

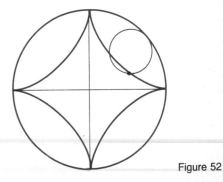

Figure 52

46. TEASER A wheel of radius 3 and centered at the origin is rotating counterclockwise at 2 radians per second so that the point P on its rim is at $(3, 0)$ when time $t = 0$. Another wheel of radius 1 and centered at $(8, 0)$ is also rotating but clockwise at 2 radians per second so that the point Q on its rim is at $(9, 0)$ at $t = 0$. A knot K is tied at the middle of an elastic string and then one end of the string is attached to P and the other to Q. The string is short enough to remain taut at all times.

(a) Obtain the parametric equations for the path of the knot K, using time t as the parameter.

(b) Write the Cartesian equation for the path of K and use it to describe the path in geometric terms.

Most graphing calculators have a parametric mode which allows the graphing of parametric equations. Familiarize yourself with your model by studying the instruction book. Then check your understanding by graphing the parametric equations of Examples C and D.

47. Draw graphs of the following parametric equations. Refer to the discussion of the cycloid and note how small changes in its parametric equations lead to some very different graphs. In each case, allow t to range over the interval $-2\pi \leq t \leq 2\pi$.
(a) $x = 4(t - \sin t)$, $y = 4(1 - \cos t)$
(b) $x = 4(t - \sin 2t)$, $y = 4(1 - \cos t)$
(c) $x = 4(t - \sin 3t)$, $y = 4(1 - \cos t)$
(d) $x = 4(t - \sin 3t)$, $y = 4(1 - \cos 2t)$

48. Draw the graphs of the following parametric equations all related in some way to the hypocloid (the first is actually a hypocycloid). In each case, allow t to range over the interval $0 \leq t \leq 2\pi$.
(a) $x = 6 \cos t + 2 \cos 3t$, $y = 6 \sin t - 2 \cos 3t$
(b) $x = 8 \cos t + \cos 8t$, $y = 8 \sin t - \sin 8t$
(c) $x = 9 \cos t - 3 \cos 3t$, $y = 9 \sin t - 3 \sin 3t$
(d) $x = 8 \cos t - \cos 8t$, $y = 8 \sin t - \sin 8t$

13-7 THE POLAR COORDINATE SYSTEM

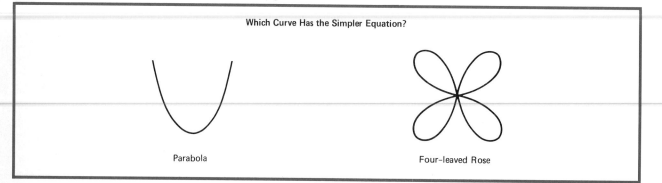

Which Curve Has the Simpler Equation?

Parabola

Four-leaved Rose

The question we have asked above makes no sense unless coordinate axes are present. Most people would then choose the parabola as having the simpler equation; but the question is still more subtle than one might think. You already know that the complexity of the equation of a curve depends on the placement of the coordinate axes. Placed just right, the equation of the parabola might be as simple as $y = x^2$. Placed less wisely, the equation might be as complicated as $x - 3 = -(y + 7)^2$, or even worse. However, the four-leaved rose has a very messy equation no matter where the x- and y-axes are placed.

But there is another aspect to the question, one that Fermat and Descartes did not think about when they gave us Cartesian coordinates. There are many different kinds of coordinate systems, that is, different ways of specifying the position of a point. One of these systems, when placed the best possible way, gives the four-leaved rose a delightfully simple equation (see Example B). This system is called the **polar coordinate system**; it simplifies many problems that arise in calculus.

Polar Coordinates

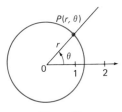

Figure 53

In place of two perpendicular axes as in Cartesian coordinates, we introduce in the plane a single horizontal ray, called the **polar axis,** emanating from a fixed point O, called the **pole.** On the polar axis, we mark off the positive half of a number scale with zero at the pole. Any point P other than the pole is the intersection of a unique circle with center O and a unique ray emanating from O, (Figure 53). If r is the radius of the circle and θ is the angle the ray makes with the polar axis, then (r, θ) are the polar coordinates of P.

Points specified by polar coordinates are easiest to plot if we use polar graph paper. The grid on this paper consists of concentric circles and rays emanating from their common center. We have reproduced such a grid in Figure 54 and plotted a few points.

Of course, we can measure the angle θ in degrees as well as radians. More significantly, notice that while a pair of coordinates (r, θ) determines a unique point $P(r, \theta)$, each point has many different pairs of polar coordinates. For example,

$$\left(2, \frac{3\pi}{2}\right) \qquad \left(2, -\frac{\pi}{2}\right) \qquad \left(2, \frac{7\pi}{2}\right)$$

are all coordinates for the same point.

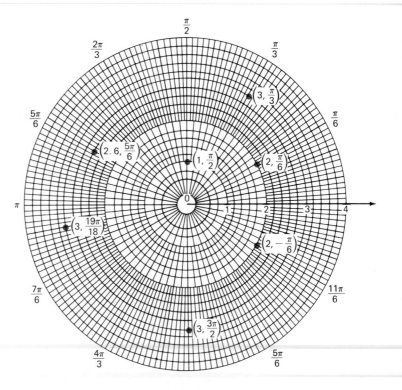

Polar coordinates

Figure 54

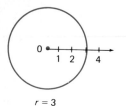

$r = 3$

Polar Graphs

The simplest polar equations are $r = k$ and $\theta = k$, where k is a constant. The graph of the first is a circle; the graph of the second is a ray emanating from the origin. Examples are shown in Figure 55. Equations like

$$r = 4 \sin^2 \theta \quad \text{and} \quad r = 1 + \cos 2\theta$$

$\theta = \dfrac{3\pi}{4}$

Figure 55

are more complicated. To graph such equations, we suggest making a table of values, plotting the corresponding points, and then connecting those points with a smooth curve.

Example A (A Parabola in Polar Form)   Sketch the graph of the polar equation

$$r = \frac{1}{1 - \cos \theta}$$

Solution   In Figure 56 we have constructed a table of values and drawn the corresponding graph. It looks suspiciously like a parabola, and in the next section, we will verify that this suspicion is correct.

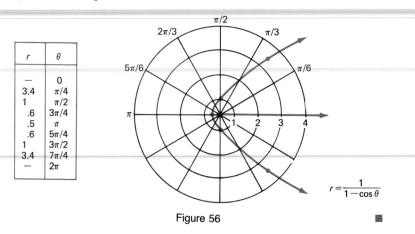

r	θ
—	0
3.4	$\pi/4$
1	$\pi/2$
.6	$3\pi/4$
.5	π
.6	$5\pi/4$
1	$3\pi/2$
3.4	$7\pi/4$
—	2π

$r = \dfrac{1}{1 - \cos \theta}$

Figure 56

Limacons

We consider next equations of the form

$$r = a \pm b \cos \theta \quad \text{and} \quad r = a \pm b \sin \theta$$

with a and b positive. Their graphs are called **limaçons** with the special case $a = b$ giving a curve called a **cardioid** (a heartlike curve). We assert that these graphs have the shapes shown in Figure 57. (In the case $a < b$, we allow r to be negative, a matter discussed in Example C.)

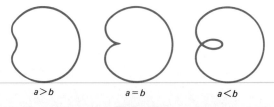

$a > b$ $\qquad$ $a = b$ $\qquad$ $a < b$

Figure 57

Example B (A Cardioid) Sketch the graph of $r = 2(1 + \cos \theta)$.

Solution A table of values and the graph are shown in Figure 58.

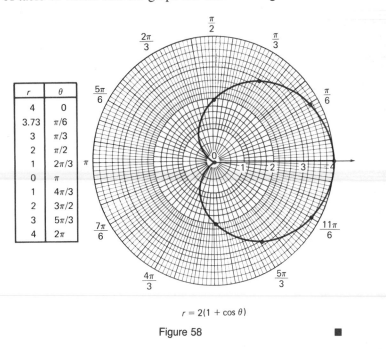

r	θ
4	0
3.73	$\pi/6$
3	$\pi/3$
2	$\pi/2$
1	$2\pi/3$
0	π
1	$4\pi/3$
2	$3\pi/2$
3	$5\pi/3$
4	2π

$r = 2(1 + \cos \theta)$

Figure 58   ■

Example C (Allowing Negative Values for r) It is sometimes useful to allow r to be negative. By the point $(-3, \pi/4)$, we shall mean the point 3 units from the pole on the ray in the opposite direction from the ray for $\theta = \pi/4$ (see Figure 59). Allowing r to be negative, graph

$$r = 2 \sin 2\theta$$

Solution We begin with a table of values (Figure 60), plot the corresponding points, and then sketch the graph (Figure 61 on the next page).

$\pi/4$

$(-3, \pi/4)$

Figure 59

θ	0	$\frac{\pi}{12}$	$\frac{\pi}{6}$	$\frac{\pi}{4}$	$\frac{\pi}{3}$	$\frac{5\pi}{12}$	$\frac{\pi}{2}$	$\frac{7\pi}{12}$	$\frac{3\pi}{4}$	$\frac{11\pi}{12}$	π	$\frac{5\pi}{4}$	$\frac{3\pi}{2}$	$\frac{7\pi}{4}$	2π
2θ	0	$\frac{\pi}{6}$	$\frac{\pi}{3}$	$\frac{\pi}{2}$	$\frac{2\pi}{3}$	$\frac{5\pi}{6}$	π	$\frac{7\pi}{6}$	$\frac{3\pi}{2}$	$\frac{11\pi}{6}$	2π	$\frac{5\pi}{2}$	3π	$\frac{7\pi}{2}$	4π
r	0	1	$\sqrt{3}$	2	$\sqrt{3}$	1	0	-1	-2	-1	0	2	0	-2	0

a b c d

Figure 60

Note: The four leaves correspond to the four parts (a), (b), (c), and (d) of the table of values. For example, leaf (b) results from values of θ between $\pi/2$ and π where r is negative. This graph is the four-leaved rose of our opening display. Its Cartesian equation will be obtained in part (b) of Example E.

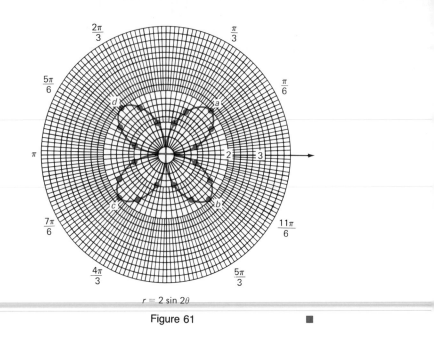

$$r = 2 \sin 2\theta$$

Figure 61 ■

Relation to Cartesian Coordinates

Figure 62

Let the positive x-axis of the Cartesian coordinate system serve also as the polar axis of a polar coordinate system, the origin coinciding with the pole (Figure 62). The Cartesian coordinates and polar coordinates are related by two pairs of simple equations.

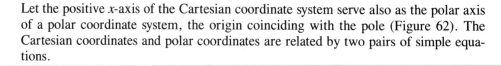

$$x = r \cos \theta \qquad r^2 = x^2 + y^2$$

$$y = r \sin \theta \qquad \tan \theta = \frac{y}{x}$$

Example D (Polar Versus Cartesian Coordinates)
(a) Find the Cartesian coordinates (x, y) for the point with polar coordinates $(4, \pi/6)$.
(b) Find the polar coordinates (r, θ) for the point with Cartesian coordinates $(-3, \sqrt{3})$.

Solution
(a) From the boxed relations, we obtain

$$x = 4 \cos \frac{\pi}{6} = 4 \cdot \frac{\sqrt{3}}{2} = 2\sqrt{3}$$

$$y = 4 \sin \frac{\pi}{6} = 4 \cdot \frac{1}{2} = 2$$

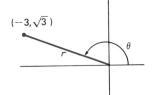

$(-3, \sqrt{3})$

Figure 63

(b) Refer to Figure 63 and the boxed relations.

$$r = \sqrt{(-3)^2 + (\sqrt{3})^2} = \sqrt{12} = 2\sqrt{3}$$

$$\tan \theta = \frac{\sqrt{3}}{-3}$$

Since the point is in the second quadrant, we choose $5\pi/6$ as an appropriate value of θ. Thus one choice of polar coordinates for the point in question is $(2\sqrt{3}, 5\pi/6)$. ∎

Example E (Transforming Equations)

(a) Change the Cartesian equation $(x^2 + y^2)^2 = x^2 - y^2$ to a polar equation.
(b) Change $r = 2 \sin 2\theta$ to a Cartesian equation.

Solution

(a) Replacing $x^2 + y^2$ by r^2, x by $r \cos \theta$, and y by $r \sin \theta$, we get

$$(r^2)^2 = r^2 \cos^2 \theta - r^2 \sin^2 \theta$$

$$r^4 = r^2(\cos^2 \theta - \sin^2 \theta)$$

$$r^2 = \cos 2\theta$$

Dividing by r^2 at the last step did no harm since the graph of the last equation passes through the pole $r = 0$.

(b)
$$r = 2 \sin 2\theta$$

$$r = 2 \cdot 2 \sin \theta \cos \theta$$

Multiplying both sides by r^2 gives

$$r^3 = 4(r \sin \theta)(r \cos \theta)$$

$$(x^2 + y^2)^{3/2} = 4yx \quad ∎$$

PROBLEM SET 13-7

A. Skills and Techniques

Graph each of the following points given in polar coordinates. Polar graph paper will simplify the graphing process.

1. $\left(3, \dfrac{\pi}{4}\right)$

2. $\left(2, \dfrac{\pi}{3}\right)$

3. $\left(\dfrac{3}{2}, \dfrac{5\pi}{6}\right)$

4. $\left(1, \dfrac{5\pi}{3}\right)$

5. $(3, \pi)$

6. $\left(2, \dfrac{\pi}{2}\right)$

7. $(3, -\pi)$

8. $\left(2, -\dfrac{3\pi}{2}\right)$

9. $(4, 70°)$

10. $(3, 190°)$

11. $\left(\dfrac{5}{2}, \dfrac{7\pi}{3}\right)$

12. $\left(\dfrac{7}{2}, \dfrac{11\pi}{4}\right)$

Find the Cartesian coordinates of the point having the given polar coordinates. See Example D.

13. $\left(4, \dfrac{\pi}{4}\right)$

14. $\left(6, \dfrac{\pi}{6}\right)$

15. $(3, \pi)$

16. $\left(2, \dfrac{3\pi}{2}\right)$

17. $\left(10, \dfrac{4\pi}{3}\right)$

18. $\left(8, \dfrac{11\pi}{6}\right)$

19. $\left(2, -\dfrac{\pi}{4}\right)$ **20.** $\left(3, -\dfrac{2\pi}{3}\right)$

Find polar coordinates for the point with the given Cartesian coordinates. See Example D.

21. $(4, 0)$ **22.** $(0, 3)$
23. $(-2, 0)$ **24.** $(0, -5)$
25. $(2, 2)$ **26.** $(2, -2)$
27. $(-2, 2)$ **28.** $(-2, -2)$
29. $(1, -\sqrt{3})$ **30.** $(-2\sqrt{3}, 2)$
31. $(3, -\sqrt{3})$ **32.** $(-\sqrt{3}, -3)$

Graph each of the following equations. Use polar graph paper if it is available. See Examples A and B.

33. $r = 2$ **34.** $r = 5$
35. $\theta = \pi/3$ **36.** $\theta = -2\pi/3$
37. $r = |\theta|$ (with θ in radians)   **38.** $r = \theta^2$
39. $r = 2(1 - \cos \theta)$ **40.** $r = 3(1 + \sin \theta)$
41. $r = 2 + \cos \theta$ **42.** $r = 2 - \sin \theta$

Graph each of the following, allowing r to be negative, as in Example C.

43. $r = 3 \cos 2\theta$ **44.** $r = \cos 3\theta$
45. $r = \sin 3\theta$ **46.** $r = 4 \cos \theta$
47. $r = \sin 4\theta$ **48.** $r = \cos 4\theta$

Transform to a polar equation. See Example E.

49. $x^2 + y^2 = 4$ **50.** $\sqrt{x^2 + y^2} = 6$
51. $y = x^2$ **53.** $x^2 + (y - 1)^2 = 1$

Transform to a Cartesian equation. See Example E.

53. $\tan \theta = 2$ **54.** $r = 3 \cos \theta$
55. $r = \cos 2\theta$ **56.** $r^2 = \cos \theta$

B. Applications and Extensions

57. Transform to a Cartesian equation and identify the corresponding curve.
   (a) $r = 5/(3 \sin \theta - 2 \cos \theta)$
   (b) $r = 4 \cos \theta - 6 \sin \theta$

58. Transform to a polar equation.
   (a) $x^2 = 4y$
   (b) $(x - 5)^2 + (y + 2)^2 = 29$

Graph each of the polar equations in Problems 59–64.

59. $r = 4$ **60.** $\theta = -\pi/3$

61. $r = 2(1 - \sin \theta)$ **62.** $r = 1/\theta,\ \theta > 0$
63. $r^2 = \sin 2\theta$ *Caution: Avoid values of θ that make r^2 negative.*
64. $r = 2^\theta$ *Note:* Use both negative and positive values for θ.
65. Sketch the graphs of each pair of equations and find their points of intersection.
   (a) $r = 4 \cos \theta,\ r \cos \theta = 1$
   (b) $r = 2\sqrt{3} \sin \theta,\ r = 2(1 + \cos \theta)$
66. Find the polar coordinates of the midpoint of the line segment joining the points with polar coordinates $(4, 2\pi/3)$ and $(8, \pi/6)$.
67. Show the distance d between the points with polar coordinates (r_1, θ_1) and (r_2, θ_2) is given by
$$d = \sqrt{r_1^2 + r_2^2 - 2r_1 r_2 \cos(\theta_2 - \theta_1)}$$
and use this result to find the distance between $(4, 2\pi/3)$ and $(8, \pi/6)$.
68. Show that a circle of radius a and center (a, α) has polar equation $r = 2a \cos(\theta - \alpha)$. *Hint:* Law of cosines.
69. Find a formula for the area of the polar rectangle $0 < a < r < b,\ \alpha \le \theta \le \beta,\ \beta - \alpha < \pi$.
70. A point P moves so that its distance from the pole is always equal to its distance from the horizontal line $r \sin \theta = 4$. Show that the equation of the resulting curve (a parabola) is $r = 4/(1 + \sin \theta)$.
71. A line segment L of length 4 has its two endpoints on the x- and y-axes, respectively. The point P is on L and is such that the line OP from the pole to P is perpendicular to L. Show that the set of points P satisfying this condition is a four-leaved rose by finding its polar equation.
72. **TEASER**   Let F and F' be fixed points with polar coordinates $(a, 0)$ and $(-a, 0)$, respectively. A point P moves so that the product of its distances from F and F' is equal to the constant a^2 (that is, $\overline{PF} \cdot \overline{PF'} = a^2$). Find a simple polar equation [of the form $r^2 = f(\theta)$] for the resulting curve and sketch its graph.

To draw the graph of a polar equation $r = f(t)$ using a graphing calculator, change the equation to its parametric form $x = f(t) \cos t,\ y = f(t) \sin t$. To make sure that this works, try reproducing the graphs in Examples A–C.

73. Draw for $0 \le t \le 2\pi$ the graphs of the following limaçons.
   (a) $r = 5 + 4 \sin t$ (b) $r = 5 + 5 \sin t$
   (c) $r = 3 + 5 \sin t$

74. Draw for $0 \le t \le 2\pi$ the graphs of the following roses.
(a) $r = 8 \cos 2t$ (b) $r = 8 \cos 3t$
(c) $r = 8 \cos 4t$
Make a conjecture about the number of petals on the rose $r = 8 \cos nt$.

75. To see some of the beauty and complexity of polar graphs, draw the graph of $r = 6(\cos^4 4t + \sin 3t)$ for $0 \le t \le 2\pi$. To get a good graph, you will need to use a small t-step, say 0.01.

76. Follow the instructions of Problem 75 with $\sin 3t$ replaced by $\sin 4t$.

13-8   POLAR EQUATIONS OF CONICS

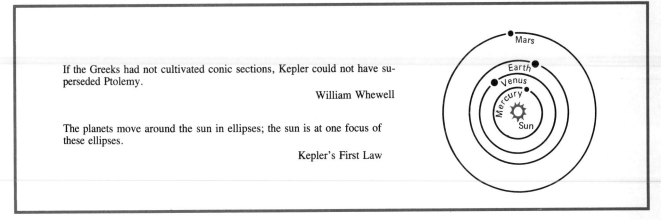

If the Greeks had not cultivated conic sections, Kepler could not have superseded Ptolemy.

William Whewell

The planets move around the sun in ellipses; the sun is at one focus of these ellipses.

Kepler's First Law

Historians have suggested that the Greeks studied the conic sections simply to satisfy their intellectual cravings after the ideal and that they hardly dreamed that these curves would have important physical applications. Today we know that they describe the motions of the moon, of the planets, of comets, of space probes, and even of tiny electrons as they orbit the nucleus of an atom. In the study of such motions, it is not the Cartesian equations of the conics that prove most useful. It is rather their polar equations that play a central role. Before looking at general conics, we consider the simpler cases of lines and circles.

Polar Equations of a Line

If a line passes through the pole, it has the exceedingly simple equation $\theta = \theta_0$, with θ_0 a constant. If a line does not go through the pole, it is some distance d from it. Let θ_0 be the angle from the polar axis to the perpendicular drawn from the pole to the given line (Figure 64). Then if $P(r, \theta)$ is a point on the line,

$$\cos(\theta - \theta_0) = \frac{d}{r}$$

or

$$r = \frac{d}{\cos(\theta - \theta_0)}$$

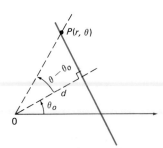

Figure 64

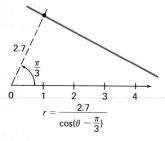

$$r = \frac{2.7}{\cos\left(\theta - \frac{\pi}{3}\right)}$$

Figure 65

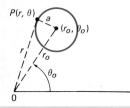

Figure 66

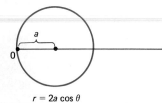

$r = 2a\cos\theta$

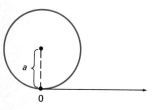

$r = 2a\sin\theta$

Figure 67

Example A (A Line)   Write the polar equation of the line shown in Figure 65.

Solution   Note that $d = 2.7$ and $\theta_0 = \pi/3$. Hence

$$r = \frac{2.7}{\cos(\theta - \pi/3)} \qquad \blacksquare$$

The Polar Equation of a Circle

If the circle is centered at the pole, its polar equation is simply $r = a$, where a is the radius of the circle. If the center is at (r_0, θ_0), then by the law of cosines (see Figure 66)

$$a^2 = r^2 + r_0^2 - 2rr_0\cos(\theta - \theta_0)$$

which is too complicated to be of much use. However, if the circle passes through the pole so $r_0 = a$, the equation simplifies to

$$r^2 = 2ra\cos(\theta - \theta_0)$$

or, after dividing by r,

$$r = 2a\cos(\theta - \theta_0)$$

The cases $\theta_0 = 0$ and $\theta_0 = \pi/2$ are particularly nice. The first gives $r = 2a\cos\theta$; the second gives $r = 2a\cos(\theta - \pi/2)$, which is equivalent to $r = 2a\sin\theta$. The graphs for these two cases are shown in Figure 67.

Example B (A Circle through the Pole)   Write the polar equation of the circle with center $(6, \pi/4)$ of radius 6.

Solution

$$r = 2(6)\cos\left(\theta - \frac{\pi}{4}\right) = 12\cos\left(\theta - \frac{\pi}{4}\right) \qquad \blacksquare$$

A New Approach to the Conics

The definitions we gave in the first part of this chapter for the three conics (parabola, ellipse, and hyperbola) are quite dissimilar in form. However, there is another approach to the three curves that treats them in a uniform way; it is this approach that will eventually lead us to polar equations for these curves.

In the plane, consider a fixed line l (the **directrix**) and a fixed point F (the **focus**). Let a point P move so that its distance $d(P, F)$ from the focus is a constant e times its distance from the directrix $d(P, L)$, that is, so that

$$d(P, F) = ed(P, L)$$

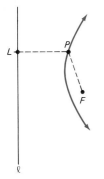

Figure 68

as suggested by Figure 68. Here the constant e (called the **eccentricity**) may be any positive number. Of course, if $e = 1$, we are on familiar ground; the corresponding curve is a parabola. But what if $e \neq 1$?

To get a feeling for this new situation, try graphing the path of P for the two cases $e = \frac{1}{2}$ and $e = 2$. If you do it carefully, you should get curves that look like an ellipse and a hyperbola, respectively. In fact, we claim that the equation $d(P, F) = e\, d(P, L)$ can serve as the definition of an ellipse when $0 < e < 1$ and of a hyperbola when $e > 1$. We demonstrate this now.

The New Definitions Are Equivalent to the Old Ones

Suppose that $e \neq 1$ and consider the curve determined by the defining equation $d(P, F) = e\, d(P, L)$. It is fairly easy to see that this curve must be symmetric with respect to the line through the focus and perpendicular to the directrix and that the curve must cross this line twice say at A' and A. Place the curve in the coordinate system so that A' and A have coordinates $(-a, 0)$ and $(a, 0)$, respectively. Let the directrix be the line $x = k$ and the focus be the point $(c, 0)$ with a, k, and c all positive. There are two possible arrangements (Figure 69) depending on whether $0 < e < 1$ or $e > 1$. (Do not let the appearance of the curves in the figure lead you to the conclusion that we have proved anything yet.)

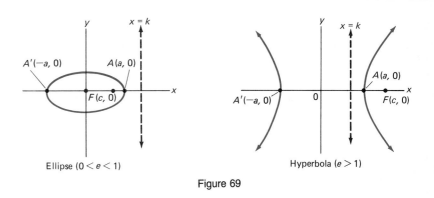

Figure 69

Apply the equation $d(P, F) = e\, d(P, L)$ first with $P = A'$ and then with $P = A$ to obtain the pair of equations

$$a - c = e(k - a) = ek - ea$$
$$a + c = e(k + a) = ek + ea$$

Solve this pair of equations for a and c to get

$$c = ea \quad \text{and} \quad k = \frac{a}{e}$$

and note for later reference that this implies $e = c/a$. Now let $P(x, y)$ be any point

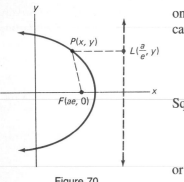

Figure 70

on the curve. Then $L(a/e, y)$ is its projection on the directrix (see Figure 70 for the case $0 < e < 1$) and the condition $d(P, F) = ed(P, L)$ translates to

$$\sqrt{(x - ae)^2 + y^2} = e\sqrt{\left(x - \frac{a}{e}\right)^2}.$$

Squaring both sides and collecting like terms yields

$$x^2 - 2aex + a^2e^2 + y^2 = e^2\left(x^2 - 2\frac{a}{e}x + \frac{a^2}{e^2}\right)$$

or

$$(1 - e^2)x^2 + y^2 = a^2(1 - e^2)$$

or

$$\frac{x^2}{a^2} + \frac{y^2}{(1 - e^2)a^2} = 1$$

If $0 < e < 1$, this is the standard equation of an ellipse; if $e > 1$, it is the standard equation of a hyperbola. Moreover since $e = c/a$, our use of e in this section is consistent with our usage in Sections 13-2 and 13-3.

Polar Equations of the Conics

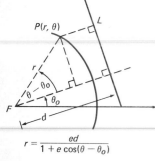

$r = \dfrac{ed}{1 + e\cos(\theta - \theta_o)}$

Figure 71

To simplify matters, we will place the conic in the polar coordinate system so that the focus is at the pole (origin) and the directrix is d units away as in Figure 71. The defining equation $d(P, F) = ed(P, L)$ takes the form

$$r = e(d - r\cos(\theta - \theta_0))$$

which is equivalent to

$$r = \frac{ed}{1 + e\cos(\theta - \theta_0)}$$

As an example, consider a case where $\theta_0 = 0$, namely,

$$r = \frac{2}{1 + \frac{1}{2}\cos\theta} = \frac{\frac{1}{2}\cdot 4}{1 + \frac{1}{2}\cos\theta}$$

Since $e = \frac{1}{2}$, the graph is an ellipse, the one shown in Figure 72. On the other hand,

$$r = \frac{12}{3 + 4\cos\theta} = \frac{4}{1 + \frac{4}{3}\cos\theta} = \frac{\frac{4}{3}\cdot 3}{1 + \frac{4}{3}\cos\theta}$$

is a hyperbola with $e = \frac{4}{3}$ *and* $d = 3$. Examples C and D give a complete discussion of the cases $\theta_0 = 0$, $\pi/2$, π, and $3\pi/2$.

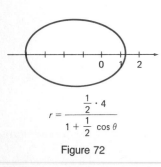

$r = \dfrac{\frac{1}{2}\cdot 4}{1 + \frac{1}{2}\cos\theta}$

Figure 72

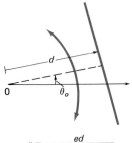

$$r = \frac{ed}{1 + e \cos(\theta - \theta_o)}$$

Figure 73

Example C (Conics with Directrix Perpendicular to the Polar Axis)   Refer to Figure 73. If the directrix l is perpendicular to the polar axis, then $\theta_0 = 0$ or $\theta_0 = \pi$, and the polar equation of the conic takes one of the two forms.

$$\theta_0 = 0 \qquad\qquad\qquad \theta_0 = \pi$$

$$r = \frac{ed}{1 + e \cos \theta} \qquad\qquad r = \frac{ed}{1 - e \cos \theta}$$

In the first case, the directrix is to the right of the focus; in the second, it is to the left.

Identify each of the following conics by name, find its eccentricity, and write the xy-equation of its directrix.

(a) $r = \dfrac{4}{3 + 3 \cos \theta}$ (b) $r = \dfrac{5}{2 - 3 \cos \theta}$

Solution

(a) Divide numerator and denominator by 3, obtaining

$$r = \frac{\frac{4}{3}}{1 + \cos \theta}$$

The conic is a parabola, since the eccentricity $e = 1$. The equation of the directrix is $x = \frac{4}{3}$.

(b) The equation can be rewritten as

$$r = \frac{\frac{5}{2}}{1 - \frac{3}{2}\cos \theta} = \frac{\frac{3}{2} \cdot \frac{5}{3}}{1 - \frac{3}{2}\cos \theta}$$

The conic is a hyperbola with $e = \frac{3}{2}$ and directrix $x = -\frac{5}{3}$.   ∎

Example D (Conics with Directrix Parallel to the Polar Axis)   Refer to the diagram in Figure 73. If the directrix is parallel to the polar axis and above it, then $\theta_0 = \pi/2$; if it is below the polar axis, then $\theta_0 = 3\pi/2$. The corresponding equations can be simplified to

$$\theta_0 = \frac{\pi}{2} \qquad\qquad\qquad \theta_0 = \frac{3\pi}{2}$$

$$r = \frac{ed}{1 + e \sin \theta} \qquad\qquad r = \frac{ed}{1 - e \sin \theta}$$

(a) Derive the first of these equations.
(b) Identify the conic $r = 5/(2 - \sin \theta)$ by name, give its eccentricity, and write the xy-equation of its directrix.

Solution

(a) The equation of the conic with $\theta_0 = \pi/2$ is

$$r = \frac{ed}{1 + e \cos(\theta - \frac{\pi}{2})}$$

Since

$$\cos(\theta - \tfrac{\pi}{2}) = \cos\theta \cos\tfrac{\pi}{2} + \sin\theta \sin\tfrac{\pi}{2} = \sin\theta$$

we get

$$r = \frac{ed}{1 + e\sin\theta}$$

(b) Dividing numerator and denominator by 2, we obtain

$$r = \frac{\tfrac{5}{2}}{1 - \tfrac{1}{2}\sin\theta} = \frac{\tfrac{1}{2}\cdot 5}{1 - \tfrac{1}{2}\sin\theta}$$

The conic is an ellipse with ecentricity $\tfrac{1}{2}$. The directrix is below the polar axis and has xy-equation $y = -5$. ■

Example E (Graphing Conics in Polar Coordinates) Graph the conic whose polar equation is

$$r = \frac{6}{1 + 2\cos\theta}$$

Solution   We recognize this as the polar equation of a hyperbola ($e = 2$) with major axis along the polar axis. Next we make the small table of values shown in Figure 74 and plot the corresponding points (marked with dots). The points marked with a cross are obtained by symmetry ($\cos(-\theta) = \cos\theta$).

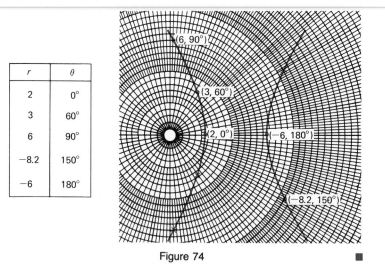

r	θ
2	0°
3	60°
6	90°
−8.2	150°
−6	180°

Figure 74   ■

In Figure 75, we have summarized most of the results of this section.

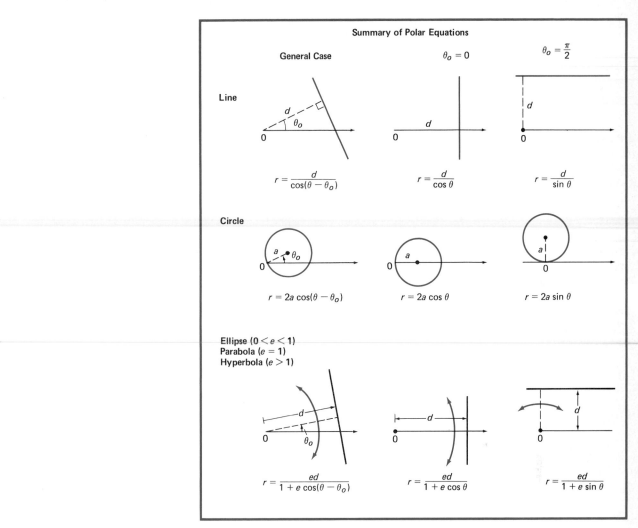

Figure 75

PROBLEM SET 13-8

A. Skills and Techniques

In doing the following problems, it will be helpful to keep three things in mind.

  1. The summary of polar equations in Figure 75.
  2. The addition formulas
     $\cos(\theta \mp \theta_0) = \cos\theta\cos\theta_0 \pm \sin\theta\sin\theta_0.$
  3. The relations $x = r\cos\theta$ and $y = r\sin\theta$.

In Problems 1–4, write a simple polar equation for each equation.

  1. $x = 4$
  2. $y = 3$
  3. $x = -3$
  4. $y = -2$

Each of the polar equations in Problems 5–10 represents a line. Sketch its graph and then write its xy-equation in the form $Ax + By = C$. See Example A.

5. $r = \dfrac{6}{\cos \theta}$

6. $r = \dfrac{3}{\sin \theta}$

7. $r = \dfrac{4}{\cos(\theta - \pi/3)}$

8. $r = \dfrac{10}{\cos(\theta - \pi/4)}$

9. $r = \dfrac{5}{\cos(\theta + \pi/4)}$

10. $r = \dfrac{4}{\cos(\theta + \pi/3)}$

In Problems 11–14, write the polar equation of the circle which passes through the pole and whose center has the given polar coordinates. Then find the corresponding xy-equations. See Example B.

11. $(4, 0)$

12. $(3, 90°)$

13. $(5, \pi/3)$

14. $(8, 150°)$

In Problems 15–22, identify the conic by name, give its eccentricity, and write the equation of its directrix (in xy-coordinates). See Example C.

15. $r = \dfrac{4}{1 + \frac{2}{3}\cos \theta}$

16. $r = \dfrac{\frac{9}{2}}{1 + \frac{3}{4}\cos \theta}$

17. $r = \dfrac{5}{2 + 4 \cos \theta}$

18. $r = \dfrac{3}{1 + \cos \theta}$

19. $r = \dfrac{7}{1 - \cos \theta}$

20. $r = \dfrac{\frac{1}{2}}{1 - \frac{3}{2}\cos \theta}$

21. $r = \dfrac{\frac{1}{2}}{\frac{3}{2} - \cos \theta}$

22. $r = \dfrac{3}{6 - 6 \cos \theta}$

In Problems 23–28, identify the conic by name, give its eccentricity, and write the xy-equation of its directrix. See Example D.

23. $r = \dfrac{5}{1 + \sin \theta}$

24. $r = \dfrac{2}{1 + \frac{2}{3}\sin \theta}$

25. $r = \dfrac{6}{2 - \sin \theta}$

26. $r = \dfrac{5}{2 - 4 \sin \theta}$

27. $r = \dfrac{4}{2 + \frac{5}{2}\sin \theta}$

28. $r = \dfrac{5}{4 - 3 \sin \theta}$

Graph each of the following conics. See Example E.

29. $r = \dfrac{4}{1 + \frac{2}{3}\cos \theta}$

30. $r = \dfrac{6}{1 + \frac{3}{4}\sin \theta}$

31. $r = \dfrac{5}{1 + \sin \theta}$

32. $r = \dfrac{4}{2 + 2 \cos \theta}$

33. $r = \dfrac{18}{2 + 3 \cos \theta}$

34. $r = \dfrac{3}{1 - 2 \cos \theta}$

B. Applications and Extensions

35. Write the polar equation $r = f(\theta)$ that corresponds to each of the following Cartesian equations.
    (a) $y = 3$ **(b)** $x^2 + y^2 = 9$
    (c) $(x + 9)^2 + y^2 = 81$
    (d) $(x - 3)^2 + (y - 3)^2 = 18$

36. Graph each of the following polar equations.
    (a) $r = -4/\cos \theta$ **(b)** $r = -8 \sin \theta$
    (c) $r = 3$
    (d) $r = 6 \cos(\theta - \pi/3)$
    (e) $r = 6/(1 - \cos \theta)$
    (f) $r = 6/(2 + \cos \theta)$

37. Determine the polar equations for the parabolas with Cartesian equations given by (a) $4(x + 1) = y^2$ and (b) $-8(y - 2) = x^2$. *Hint:* In each case, the focus is at the origin.

38. Find the polar equation $r = f(\theta)$ corresponding to the parabola with polar equation $y^2 = 16x$. Why is this polar equation unlike any discussed in this section?

39. Write the Cartesian equation of the curve whose polar equation is

$$\cos^2 \theta + \sin \theta \cos \theta - 6 \sin^2 \theta = 0$$

40. Find the points of intersection (in polar coordinates) of the circles with polar equations $r = 3 \sin \theta$ and $r = \sqrt{3} \cos \theta$.

41. Find e and d for the ellipse with polar equation $r = 8/(2 + \cos \theta)$. Also find the major and minor diameters of this ellipse.

42. Show that the ellipse $r = ed/(1 + e \cos \theta)$ has major diameter $2ed/(1 - e^2)$ and minor diameter $2ed/\sqrt{1 - e^2}$

43. Find the length of the latus rectum (chord through the focus perpendicular to the major axis) for the ellipse $r = 8/(2 + \cos \theta)$.

44. Express the length of the latus rectum of the conic $r = ed/(1 + e \cos \theta)$ in terms of e and d.

45. The graph of $r = 2a + a \cos \theta$ for $a > 0$ is an example of a curve called a *limaçon* (Section 13-7).
    (a) Sketch this graph for the case $a = 3$.
    (b) Show that every chord through the pole has length $4a$ (a nice property it shares with the circle $r = 2a$).

46. TEASER Sketch the graph of the polar equation $r = 1/\theta$ for $\theta \geq \pi/2$ and show that this curve has infinite length.

47. Draw the graphs of each of the following polar equations.
- **(a)** $r = 5/\cos(t - 1)$
- **(b)** $r = 5/\cos(t - 2.5)$
- **(c)** $r = 10\cos(t - 2.5)$
- **(d)** $r = 10\cos(t - 4)$

48. Draw the graphs of each of the following polar equations.
- **(a)** $r = 10/(4 + 3\cos t)$
- **(b)** $r = 10/(4 + 4\cos t)$
- **(c)** $r = 10/(2 + 3\cos t)$
- **(d)** $r = 10/[2 + 3\cos(t - 1)]$

CHAPTER 13 SUMMARY

A **parabola** is the set of points that are equidistant from a fixed line (the **directrix**) and a fixed point (the **focus**). An **ellipse** is the set of points for which the sum of the distances from two fixed points (the **foci**) is a constant. A **hyperbola** is the set of points for which the difference of the distances from two fixed points (the **foci**) is a constant. These three curves (called **conics**) together with their limiting forms (circle, point, empty set, line, two intersecting lines, two parallel lines) are the chief objects of study in this chapter.

When these curves are placed in the Cartesian coordinate plane in an advantageous way, their xy-equations take the **standard forms** below.

Parabola: $\qquad y^2 = 4px$

Ellipse: $\qquad \dfrac{x^2}{a^2} + \dfrac{y^2}{b^2} = 1$

Hyperbola: $\qquad \dfrac{x^2}{a^2} - \dfrac{y^2}{b^2} = 1$

The curves can, of course, be placed in the plane in other ways; in these cases, their equations are more complicated. However, they can be brought to standard form by a **translation** or a **rotation** of axes. Even the complicated equations are always of the form

$$Ax^2 + Bxy + Cy^2 + Dx + Ey + F = 0$$

Conversely, the graph of any equation of this type is one of the conics or the six limiting forms.

Many curves (including the conics) can be described by giving parametric equations $x = f(t)$ and $y = g(t)$ in which both x and y are specified in terms of a **parameter** t.

Cartesian coordinates (x, y) are not the only way to specify the position of a point. **Polar coordinates** (r, θ) also determine points, and **polar equations** determine curves. In fact, some very beautiful curves (**limaçons, cardioids, roses**) have simple polar equations but complicated Cartesian equations. Moreover, for purposes of astronomy, the conics are best described by polar equations, since their equations all take the form

$$r = \frac{ed}{1 + e\cos(\theta - \theta_0)}$$

This equation determines an ellipse if e (called the **eccentricity**) satisfies $0 < e < 1$, a parabola if $e = 1$, and a hyperbola if $e > 1$.

CHAPTER 13 REVIEW PROBLEM SET

In Problems 1–10, write True or False in the blank. If false, tell why.

_____ **1.** It is 6 units from the focus to the directrix of the parabola $x^2 = 6y$.

_____ **2.** The parabola $(y - 2)^2 = 12(x + 3)$ opens to the right.

_____ **3.** The ellipse $\dfrac{(x - 4)^2}{9} + \dfrac{y^2}{16} = 1$ has vertices at $(4, \pm 4)$.

_____ **4.** The hyperbola $\dfrac{x^2}{9} - \dfrac{y^2}{25} = 1$ has eccentricity $\sqrt{34}/5$.

5. The hyperbolas $4x^2 - 9y^2 = 36$ and $-4x^2 + 9y^2 = 36$ have the same asymptotes.

_____ 6. The graph of the equation $x^2 + y^2 + 2x - 4y + 5 = 0$ is a single point.

_____ 7. If $B \neq 0$, the graph of $x^2 + Bxy + y^2 = 1$ is one of these three: parabola, ellipse, hyperbola.

_____ 8. The point with polar coordinates $(6\sqrt{2}, -135°)$ has Cartesian coordinates $(-6, -6)$.

_____ 9. The line with parametric equations $x = 2 + 3t$, $y = -1 - 6t$ has slope 2.

_____ 10. The Cartesian equation of the curve with parametric equations $x = \cos t$, $y = \cos 2t$ is $y = 2x^2 - 1$.

11. Determine the number of the description that best fits the given equation and write it in the blank.

_____ (a) $3x^2 + 3y^2 = 10$
_____ (b) $(x + 2)^2 + 4(y - 1)^2 = 12$
_____ (c) $(x + 2)^2 - 4(y - 1)^2 = -4$
_____ (d) $(x + 2)^2 - 4(y - 1)^2 = 4$
_____ (e) $4(x-1)^2 + y^2 + 20 = 0$
_____ (f) $(y - 2)^2 = 4$
_____ (g) $(y - 2)^2 = 4x^2$
_____ (h) $y = 2x^2 - x - 6$
_____ (i) $(y - 2)^2 = 36 - (x + 4)^2$
_____ (j) $2x - 3y^2 = 6y - 4$
_____ (k) $(3x - 5y)^2 = 0$

  (i) a single point
 (ii) a vertical parabola
(iii) a circle
 (iv) a horizontal hyperbola
  (v) a horizontal ellipse
 (vi) parallel lines
(vii) the empty set
(viii) a vertical hyperbola
 (ix) a horizontal parabola
  (x) a single line
 (xi) intersecting lines

12. Find the vertex and focus of the parabola

$$(x + 1)^2 = -12(y - 2)$$

In Problems 13–15, find the equation of the parabola that satisfies the given conditions.

13. Vertex $(0, 0)$; focus $(5, 0)$.
14. Vertex $(0, 0)$; opens down; passes through $(-2, -4)$.
15. Vertex $(3, 5)$; directrix $y = 1$.

16. Light rays from the sun hit a paraboloidal reflector of diameter 6 feet and depth 1 foot. Where are the rays focused?

17. Determine the vertices and foci of the ellipse with equation $25(x + 2)^2 + 9(y - 1)^2 = 225$.

In Problems 18–20, find the equation of the ellipse that satisfies the given conditions.

18. Vertices $(\pm 4, 0)$; the major diameter is twice the minor diameter.
19. Vertices $(0, \pm 6)$; eccentricity $\frac{1}{2}$.
20. Foci $(\pm 2\sqrt{5}, 0)$; $a = b + 2$.
21. A semielliptical arch is 12 feet wide at its base and 8 feet high at the center. How wide is the arch at a height of 6 feet?
22. Determine the vertices and asymptotes of the hyperbola $25x^2 - 4y^2 = 100$.

Find the equation of the hyperbola that satisfies the given conditions.

23. Vertices $(0, \pm 6)$; eccentricity $\frac{3}{2}$.
24. Vertices $(\pm 4, 0)$; the point $(5, \frac{3}{2})$ is on the hyperbola.
25. Vertices $(\pm 5, 0)$; one asymptote $5y = 2x$.

Sketch the graph of the given equation.

26. $(x + 1)^2 = 12(y - 2)$
27. $4(x - 4)^2 - (y + 1)^2 = 16$
28. $(x + 3)^2 + 4y^2 = 16$

In Problems 29–31, use competing the squares to identify the conic with the given equation.

29. $x^2 + y^2 - 14x + 2y + 25 = 0$
30. $9x^2 + 4y^2 - 90x + 16y + 205 = 0$
31. $9x^2 - 4y^2 - 54x + 16y + 65 = 0$
32. Transform the equation $7x^2 - 4\sqrt{3}xy + 3y^2 = 36$ by rotating the axes through 60°. Then identify the corresponding curve.
33. By rotation of axes, eliminate the cross product term from $3x^2 + 12xy + 8y^2 = 12$. Then find the distance between the foci of this conic.

In Problems 34–36, eliminate the parameter t to obtain the corresponding xy-equation.

34. $x = 4t - 5$, $y = -3t + 2$
35. $x = 4 \cos t$, $y = 3 \sin t$

36. $x = 2t$, $y = 12t^2 + 2t - 4$

37. Sketch the graph of the Witch of Agnesi, which has parametric equations $x = 4 \cot t$, $y = 4 \sin^2 t$.

38. Write the equation of the circle of radius 2 centered at the origin (pole) in
(a) Cartesian coordinates.
(b) polar coordinates.
(c) parametric form using the polar angle θ as parameter.

39. Find the Cartesian coordinates of the point with the given polar coordinates.
(a) $(2, 2\pi/3)$ (b) $(4, -\pi/2)$
(c) $(-10, 240°)$

40. Find polar coordinates of the point with the given Cartesian coordinates.
(a) $(-4, 0)$ (b) $(-4\sqrt{2}, 4\sqrt{2})$
(c) $(3, -3\sqrt{3})$

In Problems 41–46, graph the given polar equation.

41. $r = 4$

42. $r \cos \theta = 4$

43. $r = 4 \sin \theta$

44. $r = 4 \cos 3\theta$

45. $r = 2/(1 + \sin \theta)$

46. $r = 2(1 + \sin \theta)$

47. Transform the equation $x^2 + y^2 + 4x - 2y = 0$ to a polar equation of the form $r = f(\theta)$.

48. Transform the polar equation

$$r = \frac{2 \sin 2\theta}{\cos^3 \theta - \sin^3 \theta}$$

to a Cartesian equation. *Hint:* Clear of fractions and multiply both sides by r^2.

49. Determine the number of the description that best fits the given polar equation and write it in the blank.
_____ (a) $r \cos \theta = -4$
_____ (b) $r = -4 \cos \theta$
_____ (c) $r = 3/(1 + \cos \theta)$
_____ (d) $r = 2/(1 + 4 \cos \theta)$
_____ (e) $r = 5/\cos(\theta - \pi/4)$
_____ (f) $r = 3/(2 + 3 \cos \theta)$
_____ (g) $r = 4/(4 - \cos \theta)$
_____ (h) $r = 2 \cos \theta - 4 \sin \theta$

(i) a parabola
(ii) a hyperbola with $e = 4$
(iii) an ellipse
(iv) a circle
(v) a hyperbola with $e = \frac{3}{2}$
(vi) a nonvertical line
(vii) a vertical line
(viii) none of the above

50. In each case, find the shortest distance from the pole to the curve.
(a) $r = 1/(1 - \cos \theta)$ (b) $r = 3 - 2 \sin \theta$

Appendix

■ Does the pursuit for truth give you as much pleasure as before? Surely it is not the knowing but the learning, not the possessing but the acquiring, not the being-there but the getting-there, that afford the greatest satisfaction. If I have clarified and exhausted something, I leave it in order to go again into the dark. Thus is that insatiable man so strange: when he has completed a structure it is not in order to dwell in it comfortably, but to start another.

Carl Friedrich Gauss

USE OF TABLES

Table A. Natural logarithms (see Section 6-5).

Table B. Trigonometric functions (degrees): See Sections 7-1 and 7-5.

Table C. Trigonometric functions (radians): See Sections 7-2 and 7-5.

To find values between those given in any of these tables, we suggest a process called **linear interpolation.**

If we know $f(a)$ and $f(b)$ and want $f(c)$, where c is between a and b, we may write

$$f(c) \approx f(a) + d$$

where d is obtained by pretending that the graph of $y = f(x)$ is a straight line on the interval $a \le x \le b$ (Figure 1). The following examples illustrate the process.

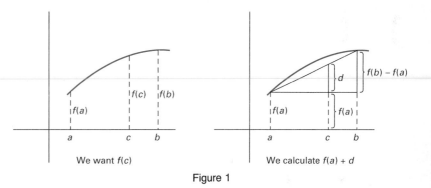

Figure 1

Example A (Natural Logarithms)   Find ln 2.133.

Solution

$$.010\left[.003\begin{bmatrix} \ln 2.130 = .7561 \\ \ln 2.133 = \quad ? \\ \ln 2.140 = .7608 \end{bmatrix} d\right].0047$$

$$\frac{d}{.0047} = \frac{.003}{.010} = .3$$

$$d = .3(.0047) \approx .0014$$

$$\ln 2.133 \approx \ln 2.130 + d = .7561 + .0014 = .7575 \quad \blacksquare$$

Example B (Trigonometric Functions (Degrees))   Find sin 57.44°.

Solution   Note that we must use the degree column on the right and the bottom caption in Table B.

$$.10\left[.04\left[\begin{array}{l}\sin 57.40° = .8425 \\ \sin 57.44° = \quad ? \\ \sin 57.50° = .8434\end{array}\right]d\right].0009$$

$$\frac{d}{.0009} = \frac{.04}{.10} = .4$$

$$d = .4(.0009) \approx .0004$$

$$\sin 57.44° \approx \sin 57.40° + d = .8425 + .0004 = .8429 \quad \blacksquare$$

Example C (Trigonometric Functions (Radians))   Find cos(1.436).

Solution   The cosine is a decreasing function on the interval $0 \le t \le \pi/2$. This causes d to be negative.

$$.010\left[.006\left[\begin{array}{l}\cos 1.430 = .14033 \\ \cos 1.436 = \quad ? \\ \cos 1.440 = .13042\end{array}\right]d\right]-.00991$$

$$\frac{d}{-.00991} = \frac{.006}{.010} = .6$$

$$d = .6(-.00991) \approx -.00595$$

$$\cos(1.436) \approx \cos(1.430) + d = .14033 - .00595 = .13438 \quad \blacksquare$$

Example D (Angle, Given a Trigonometric Function)   Find θ if tan θ = .43600. Give the answer in radians.

Solution   We use Table C and find that θ is between .41 and .42.

$$.01\left[d\left[\begin{array}{l}\tan .41 = .43463 \\ \tan \ \ \theta = .43600\end{array}\right].00137 \\ \tan .42 = .44657\right].01194$$

$$\frac{d}{.01} = \frac{.00137}{.01194} \approx .115$$

$$d \approx (.01)(.115) \approx .001$$

$$\theta \approx .41 + .001 = .411 \quad \blacksquare$$

Table A   Natural Logarithms

	.00	.01	.02	.03	.04	.05	.06	.07	.08	.09
1.0	0.0000	0.0100	0.0198	0.0296	0.0392	0.0488	0.0583	0.0677	0.0770	0.0862
1.1	0.0953	0.1044	0.1133	0.1222	0.1310	0.1398	0.1484	0.1570	0.1655	0.1740
1.2	0.1823	0.1906	0.1989	0.2070	0.2151	0.2231	0.2311	0.2390	0.2469	0.2546
1.3	0.2624	0.2700	0.2776	0.2852	0.2927	0.3001	0.3075	0.3148	0.3221	0.3293
1.4	0.3365	0.3436	0.3507	0.3577	0.3646	0.3716	0.3784	0.3853	0.3920	0.3988
1.5	0.4055	0.4121	0.4187	0.4253	0.4318	0.4383	0.4447	0.4511	0.4574	0.4637
1.6	0.4700	0.4762	0.4824	0.4886	0.4947	0.5008	0.5068	0.5128	0.5188	0.5247
1.7	0.5306	0.5365	0.5423	0.5481	0.5539	0.5596	0.5653	0.5710	0.5766	0.5822
1.8	0.5878	0.5933	0.5988	0.6043	0.6098	0.6152	0.6206	0.6259	0.6313	0.6366
1.9	0.6419	0.6471	0.6523	0.6575	0.6627	0.6678	0.6729	0.6780	0.6831	0.6881
2.0	0.6931	0.6981	0.7031	0.7080	0.7130	0.7178	0.7227	0.7275	0.7324	0.7372
2.1	0.7419	0.7467	0.7514	0.7561	0.7608	0.7655	0.7701	0.7747	0.7793	0.7839
2.2	0.7885	0.7930	0.7975	0.8020	0.8065	0.8109	0.8154	0.8198	0.8242	0.8286
2.3	0.8329	0.8372	0.8416	0.8459	0.8502	0.8544	0.8587	0.8629	0.8671	0.8713
2.4	0.8755	0.8796	0.8838	0.8879	0.8920	0.8961	0.9002	0.9042	0.9083	0.9123
2.5	0.9163	0.9203	0.9243	0.9282	0.9322	0.9361	0.9400	0.9439	0.9478	0.9517
2.6	0.9555	0.9594	0.9632	0.9670	0.9708	0.9746	0.9783	0.9821	0.9858	0.9895
2.7	0.9933	0.9969	1.0006	1.0043	1.0080	1.0116	1.0152	1.0188	1.0225	1.0260
2.8	1.0296	1.0332	1.0367	1.0403	1.0438	1.0473	1.0508	1.0543	1.0578	1.0613
2.9	1.0647	1.0682	1.0716	1.0750	1.0784	1.0818	1.0852	1.0886	1.0919	1.0953
3.0	1.0986	1.1019	1.1053	1.1086	1.1119	1.1151	1.1184	1.1217	1.1249	1.1282
3.1	1.1314	1.1346	1.1378	1.1410	1.1442	1.1474	1.1506	1.1537	1.1569	1.1600
3.2	1.1632	1.1663	1.1694	1.1725	1.1756	1.1787	1.1817	1.1848	1.1878	1.1909
3.3	1.1939	1.1970	1.2000	1.2030	1.2060	1.2090	1.2119	1.2149	1.2179	1.2208
3.4	1.2238	1.2267	1.2296	1.2326	1.2355	1.2384	1.2413	1.2442	1.2470	1.2499
3.5	1.2528	1.2556	1.2585	1.2613	1.2641	1.2669	1.2698	1.2726	1.2754	1.2782
3.6	1.2809	1.2837	1.2865	1.2892	1.2920	1.2947	1.2975	1.3002	1.3029	1.3056
3.7	1.3083	1.3110	1.3137	1.3164	1.3191	1.3218	1.3244	1.3271	1.3297	1.3324
3.8	1.3350	1.3376	1.3403	1.3429	1.3455	1.3481	1.3507	1.3533	1.3558	1.3584
3.9	1.3610	1.3635	1.3661	1.3686	1.3712	1.3737	1.3762	1.3788	1.3813	1.3838
4.0	1.3863	1.3888	1.3913	1.3938	1.3962	1.3987	1.4012	1.4036	1.4061	1.4085
4.1	1.4110	1.4134	1.4159	1.4183	1.4207	1.4231	1.4255	1.4279	1.4303	1.4327
4.2	1.4351	1.4375	1.4398	1.4422	1.4446	1.4469	1.4493	1.4516	1.4540	1.4563
4.3	1.4586	1.4609	1.4633	1.4656	1.4679	1.4702	1.4725	1.4748	1.4770	1.4793
4.4	1.4816	1.4839	1.4861	1.4884	1.4907	1.4929	1.4952	1.4974	1.4996	1.5019
4.5	1.5041	1.5063	1.5085	1.5107	1.5129	1.5151	1.5173	1.5195	1.5217	1.5239
4.6	1.5261	1.5282	1.5304	1.5326	1.5347	1.5369	1.5390	1.5412	1.5433	1.5454
4.7	1.5476	1.5497	1.5518	1.5539	1.5560	1.5581	1.5602	1.5623	1.5644	1.5665
4.8	1.5686	1.5707	1.5728	1.5748	1.5769	1.5790	1.5810	1.5831	1.5851	1.5872
4.9	1.5892	1.5913	1.5933	1.5953	1.5974	1.5994	1.6014	1.6034	1.6054	1.6074
5.0	1.6094	1.6114	1.6134	1.6154	1.6174	1.6194	1.6214	1.6233	1.6253	1.6273
5.1	1.6292	1.6312	1.6332	1.6351	1.6371	1.6390	1.6409	1.6429	1.6448	1.6467
5.2	1.6487	1.6506	1.6525	1.6544	1.6563	1.6582	1.6601	1.6620	1.6639	1.6658
5.3	1.6677	1.6696	1.6715	1.6734	1.6753	1.6771	1.6790	1.6808	1.6827	1.6845
5.4	1.6864	1.6882	1.6901	1.6919	1.6938	1.6956	1.6974	1.6993	1.7011	1.7029

$$\ln(N \cdot 10^m) = \ln N + m \ln 10, \quad \ln 10 = 2.3026$$

Table A Natural Logarithms

	.00	.01	.02	.03	.04	.05	.06	.07	.08	.09
5.5	1.7047	1.7066	1.7084	1.7102	1.7120	1.7138	1.7156	1.7174	1.7192	1.7210
5.6	1.7228	1.7246	1.7263	1.7281	1.7299	1.7317	1.7334	1.7352	1.7370	1.7387
5.7	1.7405	1.7422	1.7440	1.7457	1.7475	1.7492	1.7509	1.7527	1.7544	1.7561
5.8	1.7579	1.7596	1.7613	1.7630	1.7647	1.7664	1.7682	1.7699	1.7716	1.7733
5.9	1.7750	1.7766	1.7783	1.7800	1.7817	1.7834	1.7851	1.7867	1.7884	1.7901
6.0	1.7918	1.7934	1.7951	1.7967	1.7984	1.8001	1.8017	1.8034	1.8050	1.8066
6.1	1.8083	1.8099	1.8116	1.8132	1.8148	1.8165	1.8181	1.8197	1.8213	1.8229
6.2	1.8245	1.8262	1.8278	1.8294	1.8310	1.8326	1.8342	1.8358	1.8374	1.8390
6.3	1.8406	1.8421	1.8437	1.8453	1.8469	1.8485	1.8500	1.8516	1.8532	1.8547
6.4	1.8563	1.8579	1.8594	1.8610	1.8625	1.8641	1.8656	1.8672	1.8687	1.8703
6.5	1.8718	1.8733	1.8749	1.8764	1.8779	1.8795	1.8810	1.8825	1.8840	1.8856
6.6	1.8871	1.8886	1.8901	1.8916	1.8931	1.8946	1.8961	1.8976	1.8991	1.9006
6.7	1.9021	1.9036	1.9051	1.9066	1.9081	1.9095	1.9110	1.9125	1.9140	1.9155
6.8	1.9169	1.9184	1.9199	1.9213	1.9228	1.9242	1.9257	1.9272	1.9286	1.9301
6.9	1.9315	1.9330	1.9344	1.9359	1.9373	1.9387	1.9402	1.9416	1.9430	1.9445
7.0	1.9459	1.9473	1.9488	1.9502	1.9516	1.9530	1.9544	1.9559	1.9573	1.9587
7.1	1.9601	1.9615	1.9629	1.9643	1.9657	1.9671	1.9685	1.9699	1.9713	1.9727
7.2	1.9741	1.9755	1.9769	1.9782	1.9796	1.9810	1.9824	1.9838	1.9851	1.9865
7.3	1.9879	1.9892	1.9906	1.9920	1.9933	1.9947	1.9961	1.9974	1.9988	2.0001
7.4	2.0015	2.0028	2.0042	2.0055	2.0069	2.0082	2.0096	2.0109	2.0122	2.0136
7.5	2.0149	2.0162	2.0176	2.0189	2.0202	2.0215	2.0229	2.0242	2.0255	2.0268
7.6	2.0282	2.0295	2.0308	2.0321	2.0334	2.0347	2.0360	2.0373	2.0386	2.0399
7.7	2.0412	2.0425	2.0438	2.0451	2.0464	2.0477	2.0490	2.0503	2.0516	2.0528
7.8	2.0541	2.0554	2.0567	2.0580	2.0592	2.0605	2.0618	2.0631	2.0643	2.0656
7.9	2.0669	2.0681	2.0694	2.0707	2.0719	2.0732	2.0744	2.0757	2.0769	2.0782
8.0	2.0794	2.0807	2.0819	2.0832	2.0844	2.0857	2.0869	2.0882	2.0894	2.0906
8.1	2.0919	2.0931	2.0943	2.0956	2.0968	2.0980	2.0992	2.1005	2.1017	2.1029
8.2	2.1041	2.1054	2.1066	2.1078	2.1090	2.1102	2.1114	2.1126	2.1138	2.1150
8.3	2.1163	2.1175	2.1187	2.1199	2.1211	2.1223	2.1235	2.1247	2.1258	2.1270
8.4	2.1282	2.1294	2.1306	2.1318	2.1330	2.1342	2.1353	2.1365	2.1377	2.1389
8.5	2.1401	2.1412	2.1424	2.1436	2.1448	2.1459	2.1471	2.1483	2.1494	2.1506
8.6	2.1518	2.1529	2.1541	2.1552	2.1564	2.1576	2.1587	2.1599	2.1610	2.1622
8.7	2.1633	2.1645	2.1656	2.1668	2.1679	2.1691	2.1702	2.1713	2.1725	2.1736
8.8	2.1748	2.1759	2.1770	2.1782	2.1793	2.1804	2.1815	2.1827	2.1838	2.1849
8.9	2.1861	2.1872	2.1883	2.1894	2.1905	2.1917	2.1928	2.1939	2.1950	2.1961
9.0	2.1972	2.1983	2.1994	2.2006	2.2017	2.2028	2.2039	2.2050	2.2061	2.2072
9.1	2.2083	2.2094	2.2105	2.2116	2.2127	2.2138	2.2148	2.2159	2.2170	2.2181
9.2	2.2192	2.2203	2.2214	2.2225	2.2235	2.2246	2.2257	2.2268	2.2279	2.2289
9.3	2.2300	2.2311	2.2322	2.2332	2.2343	2.2354	2.2364	2.2375	2.2386	2.2396
9.4	2.2407	2.2418	2.2428	2.2439	2.2450	2.2460	2.2471	2.2481	2.2492	2.2502
9.5	2.2513	2.2523	2.2534	2.2544	2.2555	2.2565	2.2576	2.2586	2.2597	2.2607
9.6	2.2618	2.2628	2.2638	2.2649	2.2659	2.2670	2.2680	2.2690	2.2701	2.2711
9.7	2.2721	2.2732	2.2742	2.2752	2.2762	2.2773	2.2783	2.2793	2.2803	2.2814
9.8	2.2824	2.2834	2.2844	2.2854	2.2865	2.2875	2.2885	2.2895	2.2905	2.2915
9.9	2.2925	2.2935	2.2946	2.2956	2.2966	2.2976	2.2986	2.2996	2.3006	2.3016

Table B Trigonometric Functions (degrees)

Deg.	Sin	Tan	Cot	Cos		Deg.	Sin	Tan	Cot	Cos	
0.0	0.00000	0.00000	∞	1.0000	**90.0**	**6.0**	0.10453	0.10510	9.514	0.9945	**84.0**
.1	.00175	.00175	573.0	1.0000	89.9	.1	.10626	.10687	9.357	.9943	83.9
.2	.00349	.00349	286.5	1.0000	.8	.2	.10800	.10863	9.205	.9942	.8
.3	.00524	.00524	191.0	1.0000	.7	.3	.10973	.11040	9.058	.9940	.7
.4	.00698	.00698	143.24	1.0000	.6	.4	.11147	.11217	8.915	.9938	.6
.5	.00873	.00873	114.59	1.0000	.5	.5	.11320	.11394	8.777	.9936	.5
.6	.01047	.01047	95.49	0.9999	.4	.6	.11494	.11570	8.643	.9934	.4
.7	.01222	.01222	81.85	.9999	.3	.7	.11667	.11747	8.513	.9932	.3
.8	.01396	.01396	71.62	.9999	.2	.8	.11840	.11924	8.386	.9930	.8
.9	.01571	.01571	63.66	.9999	89.1	.9	.12014	.12101	8.264	.9928	83.1
1.0	0.01745	0.01746	57.29	0.9998	**89.0**	**7.0**	0.12187	0.12278	8.144	0.9925	**83.0**
.1	.01920	.01920	52.08	.9998	88.9	.1	.12360	.12456	8.028	.9923	82.9
.2	.02094	.02095	47.74	.9998	.8	.2	.12533	.12633	7.916	.9921	.8
.3	.02269	.02269	44.07	.9997	.7	.3	.12706	.12810	7.806	.9919	.7
.4	.02443	.02444	40.92	.9997	.6	.4	.12880	.12988	7.700	.9917	.6
.5	.02618	.02619	38.19	.9997	.5	.5	.13053	.13165	7.596	.9914	.5
.6	.02792	.02793	35.80	.9996	.4	.6	.13226	.13343	7.495	.9912	.4
.7	.02967	.02968	33.69	.9996	.3	.7	.13399	.13521	7.396	.9910	.3
.8	.03141	.03143	31.82	.9995	.2	.8	.13572	.13698	7.300	.9907	.2
.9	.03316	.03317	30.14	.9995	88.1	.9	.13744	.13876	7.207	.9905	82.1
2.0	0.03490	0.03492	28.64	0.9994	**88.0**	**8.0**	0.13917	0.14054	7.115	0.9903	**82.0**
.1	.03664	.03667	27.27	.9993	87.9	.1	.14090	.14232	7.026	.9900	81.9
.2	.03839	.03842	26.03	.9993	.8	.2	.14263	.14410	6.940	.9898	.8
.3	.04013	.04016	24.90	.9992	.7	.3	.14436	.14588	6.855	.9895	.7
.4	.04188	.04191	23.86	.9991	.6	.4	.14608	.14767	6.772	.9893	.6
.5	.04362	.04366	22.90	.9990	.5	.5	.14781	.14945	6.691	.9890	.5
.6	.04536	.04541	22.02	.9990	.4	.6	.14954	.15124	6.612	.9888	.4
.7	.04711	.04716	21.20	.9989	.3	.7	.15126	.15302	6.535	.9885	.3
.8	.04885	.04891	20.45	.9988	.2	.8	.15299	.15481	6.460	.9882	.2
.9	.05059	.05066	19.74	.9987	87.1	.9	.15471	.15660	6.386	.9880	81.1
3.0	0.05234	0.05241	19.081	0.9986	**87.0**	**9.0**	0.15643	0.15838	6.314	0.9877	**81.0**
.1	.05408	.05416	18.464	.9985	86.9	.1	.15816	.16017	6.243	.9874	80.9
.2	.05582	.05591	17.886	.9984	.8	.2	.15988	.16196	6.174	.9871	.8
.3	.05756	.05766	17.343	.9983	.7	.3	.16160	.16376	6.107	.9869	.7
.4	.05931	.05941	16.832	.9982	.6	.4	.16333	.16555	6.041	.9866	.6
.5	.06105	.06116	16.350	.9981	.5	.5	.16505	.16734	5.976	.9863	.5
.6	.06279	.06291	15.895	.9980	.4	.6	.16677	.16914	5.912	.9860	.4
.7	.06453	.06467	15.464	.9979	.3	.7	.16849	.17093	5.850	.9857	.3
.8	.06627	.06642	15.056	.9978	.2	.8	.17021	.17273	5.789	.9854	.2
.9	.06802	.06817	14.669	.9977	86.1	.9	.17193	.17453	5.730	.9851	80.1
4.0	0.06976	0.06993	14.301	0.9976	**86.0**	**10.0**	0.1736	0.1763	5.671	0.9848	**80.0**
.1	.07145	.07168	13.951	.9974	85.9	.1	.1754	.1781	5.614	.9845	79.9
.2	.07324	.07344	13.617	.9973	.8	.2	.1771	.1799	5.558	.9842	.8
.3	.07498	.07519	13.300	.9972	.7	.3	.1788	.1817	5.503	.9839	.7
.4	.07672	.07695	12.996	.9971	.6	.4	.1805	.1835	5.449	.9836	.6
.5	.07846	.07870	12.706	.9969	.5	.5	.1822	.1853	5.396	.9833	.5
.6	.08020	.08046	12.429	.9968	.4	.6	.1840	.1871	5.343	.9829	.4
.7	.08194	.08221	12.163	.9966	.3	.7	.1857	.1890	5.292	.9826	.3
.8	.08368	.08397	11.909	.9965	.2	.8	.1874	.1908	5.242	.9823	.2
.9	.08542	.08573	11.664	.9963	85.1	.9	.1891	.1926	5.193	.9820	79.1
5.0	0.08716	0.08749	11.430	0.9962	**85.0**	**11.0**	0.1908	0.1944	5.145	0.9816	**79.0**
.1	.08889	.08925	11.205	.9960	84.9	.1	.1925	.1962	5.079	.9813	78.9
.2	.09063	.09101	10.988	.9959	.8	.2	.1942	.1980	5.050	.9810	.8
.3	.09237	.09277	10.780	.9957	.7	.3	.1959	.1998	5.005	.9806	.7
.4	.09411	.09453	10.579	.9956	.6	.4	.1977	.2016	4.959	.9803	.6
.5	.09585	.09629	10.385	.9954	.5	.5	.1994	.2035	4.915	.9799	.5
.6	.09758	.09805	10.199	.9952	.4	.6	.2011	.2053	4.872	.9796	.4
.7	.09932	.09981	10.019	.9951	.3	.7	.2028	.2071	4.829	.9792	.3
.8	.10106	.10158	9.845	.9949	.2	.8	.2045	.2089	4.787	.9789	.2
.9	.10279	.10334	9.677	.9947	84.1	.9	.2062	.2107	4.745	.9785	78.1
6.0	0.10453	0.10510	9.514	0.9945	**84.0**	**12.0**	0.2079	0.2126	4.705	0.9781	**78.0**
	Cos	Cot	Tan	Sin	Deg.		Cos	Cot	Tan	Sin	Deg.

Table B Trigonometric Functions (degrees)

Deg.	Sin	Tan	Cot	Cos		Deg.	Sin	Tan	Cot	Cos	
12.0	0.2079	0.2126	4.705	0.9781	**78.0**	**18.0**	0.3090	0.3249	3.078	0.9511	**72.0**
.1	.2096	.2144	4.665	.9778	77.9	.1	.3107	.3269	3.060	.9505	71.9
.2	.2113	.2162	4.625	.9774	.8	.2	.3123	.3288	3.042	.9500	.8
.3	.2130	.2180	4.586	.9770	.7	.3	.3140	.3307	3.024	.9494	.7
.4	.2147	.2199	4.548	.9767	.6	.4	.3156	.3327	3.006	.9489	.6
.5	.2164	.2217	4.511	.9763	.5	.5	.3173	.3346	2.989	.9483	.5
.6	.2181	.2235	4.474	.9759	.4	.6	.3190	.3365	2.971	.9478	.4
.7	.2198	.2254	4.437	.9755	.3	.7	.3206	.3385	2.954	.9472	.3
.8	.2215	.2272	4.402	.9751	.2	.8	.3223	.3404	2.937	.9466	.2
.9	.2233	.2290	4.366	.9748	77.1	.9	.3239	.3424	2.921	.9461	71.1
13.0	0.2250	0.2309	4.331	0.9744	**77.0**	**19.0**	0.3256	0.3443	2.904	0.9455	**71.0**
.1	.2267	.2327	4.297	.9740	76.9	.1	.3272	.3463	2.888	.9449	70.9
.2	.2284	.2345	4.264	.9736	.8	.2	.3289	.3482	2.872	.9444	.8
.3	.2300	.2364	4.230	.9732	.7	.3	.3305	.3502	2.856	.9438	.7
.4	.2317	.2382	4.198	.9728	.6	.4	.3322	.3522	2.840	.9432	.6
.5	.2334	.2401	4.165	.9724	.5	.5	.3338	.3541	2.824	.9426	.5
.6	.2351	.2419	4.134	.9720	.4	.6	.3355	.3561	2.808	.9421	.4
.7	.2368	.2438	4.102	.9715	.3	.7	.3371	.3581	2.793	.9415	.3
.8	.2385	.2456	4.071	.9711	.2	.8	.3387	.3600	2.778	.9409	.2
.9	.2402	.2475	4.041	.9707	76.1	.9	.3404	.3620	2.762	.9403	70.1
14.0	0.2419	0.2493	4.011	0.9703	**76.0**	**20.0**	0.3420	0.3640	2.747	0.9397	**70.0**
.1	.2436	.2512	3.981	.9699	75.9	.1	.3437	.3659	2.733	.9391	69.9
.2	.2453	.2530	3.952	.9694	.8	.2	.3453	.3679	2.718	.9385	.8
.3	.2470	.2549	3.923	.9690	.7	.3	.3469	.3699	2.703	.9379	.7
.4	.2487	.2568	3.895	.9686	.6	.4	.3486	.3719	2.689	.9373	.6
.5	.2504	.2586	3.867	.9681	.5	.5	.3502	.3739	2.675	.9367	.5
.6	.2521	.2605	3.839	.9677	.4	.6	.3518	.3759	2.660	.9361	.4
.7	.2538	.2623	3.812	.9673	.3	.7	.3535	.3779	2.646	.9354	.3
.8	.2554	.2642	3.785	.9668	.2	.8	.3551	.3799	2.633	.9348	.2
.9	.2571	.2661	3.758	.9664	75.1	.9	.3567	.3819	2.619	.9342	69.1
15.0	0.2588	0.2679	3.732	0.9659	**75.0**	**21.0**	0.3584	0.3839	2.605	0.9336	**69.0**
.1	.2605	.2698	3.706	.9655	74.9	.1	.3600	.3859	2.592	.9330	68.9
.2	.2622	.2717	3.681	.9650	.8	.2	.3616	.3879	2.578	.9323	.8
.3	.2639	.2736	3.655	.9646	.7	.3	.3633	.3899	2.565	.9317	.7
.4	.2656	.2754	3.630	.9641	.6	.4	.3649	.3919	2.552	.9311	.6
.5	.2672	.2773	3.606	.9636	.5	.5	.3665	.3939	2.539	.9304	.5
.6	.2689	.2792	3.582	.9632	.4	.6	.3681	.3959	2.526	.9298	.4
.7	.2706	.2811	3.558	.9627	.3	.7	.3697	.3979	2.513	.9291	.3
.8	.2723	.2830	3.534	.9622	.2	.8	.3714	.4000	2.500	.9285	.2
.9	.2740	.2849	3.511	.9617	74.1	.9	.3730	.4020	2.488	.9278	68.1
16.0	0.2756	0.2867	3.487	0.9613	**74.0**	**22.0**	0.3746	0.4040	2.475	0.9272	**68.0**
.1	.2773	.2886	3.465	.9608	73.9	.1	.3762	.4061	2.463	.9265	67.9
.2	.2790	.2905	3.442	.9603	.8	.2	.3778	.4081	2.450	.9259	.8
.3	.2807	.2924	3.420	.9598	.7	.3	.3795	.4101	2.438	.9252	.7
.4	.2823	.2943	3.398	.9593	.6	.4	.3811	.4122	2.426	.9245	.6
.5	.2840	.2962	3.376	.9588	.5	.5	.3827	.4142	2.414	.9239	.5
.6	.2857	.2981	3.354	.9583	.4	.6	.3843	.4163	2.402	.9232	.4
.7	.2874	.3000	3.333	.9578	.3	.7	.3859	.4183	2.391	.9225	.3
.8	.2890	.3019	3.312	.9573	.2	.8	.3875	.4204	2.379	.9219	.2
.9	.2907	.3038	3.291	.9568	73.1	.9	.3891	.4224	2.367	.9212	67.1
17.0	0.2924	0.3057	3.271	0.9563	**73.0**	**23.0**	0.3907	0.4245	2.356	0.9205	**67.0**
.1	.2940	.3076	3.251	.9558	72.9	.1	.3923	.4265	2.344	.9198	66.9
.2	.2957	.3096	3.230	.9553	.8	.2	.3939	.4286	2.333	.9191	.8
.3	.2974	.3115	3.211	.9548	.7	.3	.3955	.4307	2.322	.9184	.7
.4	.2990	.3134	3.191	.9542	.6	.4	.3971	.4327	2.311	.9178	.6
.5	.3007	.3153	3.172	.9537	.5	.5	.3987	.4348	2.300	.9171	.5
.6	.3024	.3172	3.152	.9532	.4	.6	.4003	.4369	2.289	.9164	.4
.7	.3040	.3191	3.133	.9527	.3	.7	.4019	.4390	2.278	.9157	.3
.8	.3057	.3211	3.115	.9521	.2	.8	.4035	.4411	2.267	.9150	.2
.9	.3074	.3230	3.096	.9516	72.1	.9	.4051	.4431	2.257	.9143	66.1
18.0	0.3090	0.3249	3.078	0.9511	**72.0**	**24.0**	0.4067	0.4452	2.246	0.9135	**66.0**
	Cos	Cot	Tan	Sin	Deg.		Cos	Cot	Tan	Sin	Deg.

Table B Trigonometric Functions (degrees)

Deg.	Sin	Tan	Cot	Cos	Deg.	Deg.	Sin	Tan	Cot	Cos	Deg.
24.0	0.4067	0.4452	2.246	0.9135	**66.0**	**30.0**	0.5000	0.5774	1.7321	0.8660	**60.0**
.1	.4083	.4473	2.236	.9128	65.9	.1	.5015	.5797	1.7251	.8652	59.9
.2	.4099	.4494	2.225	.9121	.8	.2	.5030	.5820	1.7182	.8643	.8
.3	.4115	.4515	2.215	.9114	.7	.3	.5045	.5844	1.7113	.8634	.7
.4	.4131	.4536	2.204	.9107	.6	.4	.5060	.5867	1.7045	.8625	.6
.5	.4147	.4557	2.194	.9100	.5	.5	.5075	.5890	1.6977	.8616	.5
.6	.4163	.4578	2.184	.9092	.4	.6	.5090	.5914	1.6909	.8607	.4
.7	.4179	.4599	2.174	.9085	.3	.7	.5105	.5938	1.6842	.8599	.3
.8	.4195	.4621	2.164	.9078	.2	.8	.5120	.5961	1.6775	.8590	.2
.9	.4210	.4642	2.154	.9070	65.1	.9	.5135	.5985	1.6709	.8581	59.1
25.0	0.4226	0.4663	2.145	0.9063	**65.0**	**31.0**	0.5150	0.6009	1.6643	0.8572	**59.0**
.1	.4242	.4684	2.135	.9056	64.9	.1	.5165	.6032	1.6577	.8563	58.9
.2	.4258	.4706	2.125	.9048	.8	.2	.5180	.6056	1.6512	.8554	.8
.3	.4274	.4727	2.116	.9041	.7	.3	.5195	.6080	1.6447	.8545	.7
.4	.4289	.4748	2.106	.9033	.6	.4	.5210	.6104	1.6383	.8536	.6
.5	.4305	.4770	2.097	.9026	.5	.5	.5225	.6128	1.6319	.8526	.5
.6	.4321	.4791	2.087	.9018	.4	.6	.5240	.6152	1.6255	.8517	.4
.7	.4337	.4813	2.078	.9011	.3	.7	.5255	.6176	1.6191	.8508	.3
.8	.4352	.4834	2.069	.9003	.2	.8	.5270	.6200	1.6128	.8499	.2
.9	.4368	.4856	2.059	.8996	64.1	.9	.5284	.6224	1.6066	.8490	58.1
26.0	0.4384	0.4887	2.050	0.8988	**64.0**	**32.0**	0.5299	0.6249	1.6003	0.8480	**58.0**
.1	.4399	.4899	2.041	.8980	63.9	.1	.5314	.6273	1.5941	.8471	57.9
.2	.4415	.4921	2.032	.8973	.8	.2	.5329	.6297	1.5880	.8462	.8
.3	.4431	.4942	2.023	.8965	.7	.3	.5344	.6322	1.5818	.8453	.7
.4	.4446	.4964	2.014	.8957	.6	.4	.5358	.6346	1.5757	.8443	.6
.5	.4462	.4986	2.006	.8949	.5	.5	.5373	.6371	1.5697	.8434	.5
.6	.4478	.5008	1.997	.8942	.4	.6	.5388	.6395	1.5637	.8425	.4
.7	.4493	.5029	1.988	.8934	.3	.7	.5402	.6420	1.5577	.8415	.3
.8	.4509	.5051	1.980	.8926	.2	.8	.5417	.6445	1.5517	.8406	.2
.9	.4524	.5073	1.971	.8918	63.1	.9	.5432	.6469	1.5458	.8396	57.1
27.0	0.4540	0.5095	1.963	0.8910	**63.0**	**33.0**	0.5446	0.6494	1.5399	0.8387	**57.0**
.1	.4555	.5117	1.954	.8902	62.9	.1	.5461	.6519	1.5340	.8377	56.9
.2	.4571	.5139	1.946	.8894	.8	.2	.5476	.6544	1.5282	.8368	.8
.3	.4586	.5161	1.937	.8886	.7	.3	.5490	.6569	1.5224	.8358	.7
.4	.4602	.5184	1.929	.8878	.6	.4	.5505	.6594	1.5166	.8348	.6
.5	.4617	.5206	1.921	.8870	.5	.5	.5519	.6619	1.5108	.8339	.5
.6	.4633	.5228	1.913	.8862	.4	.6	.5534	.6644	1.5051	.8329	.4
.7	.4648	.5250	1.905	.8854	.3	.7	.5548	.6669	1.4994	.8320	.3
.8	.4664	.5272	1.897	.8846	.2	.8	.5563	.6694	1.4938	.8310	.2
.9	.4679	.5295	1.889	.8838	62.1	.9	.5577	.6720	1.4882	.8300	56.1
28.0	0.4695	0.5317	1.881	0.8829	**62.0**	**34.0**	0.5592	0.6745	1.4826	0.8290	**56.0**
.1	.4710	.5340	1.873	.8821	61.9	.1	.5606	.6771	1.4770	.8281	55.9
.2	.4726	.5362	1.865	.8813	.8	.2	.5621	.6796	1.4715	.8271	.8
.3	.4741	.5384	1.857	.8805	.7	.3	.5635	.6822	1.4659	.8261	.7
.4	.4756	.5407	1.849	.8796	.6	.4	.5650	.6847	1.4605	.8251	.6
.5	.4772	.5430	1.842	.8788	.5	.5	.5664	.6873	1.4550	.8241	.5
.6	.4787	.5452	1.834	.8780	.4	.6	.5678	.6899	1.4496	.8231	.4
.7	.4802	.5475	1.827	.8771	.3	.7	.5693	.6924	1.4442	.8221	.3
.8	.4818	.5498	1.819	.8763	.2	.8	.5707	.6950	1.4388	.8211	.2
.9	.4833	.5520	1.811	.8755	61.1	.9	.5721	.6976	1.4335	.8202	55.1
29.0	0.4848	0.5543	1.804	0.8746	**61.0**	**35.0**	0.5736	0.7002	1.4281	0.8192	**55.0**
.1	.4863	.5566	1.797	.8738	60.9	.1	.5750	.7028	1.4229	.8181	54.9
.2	.4879	.5589	1.789	.8729	.8	.2	.5764	.7054	1.4176	.8171	.8
.3	.4894	.5612	1.782	.8721	.7	.3	.5779	.7080	1.4124	.8161	.7
.4	.4909	.5635	1.775	.8712	.6	.4	.5793	.7107	1.4071	.8151	.6
.5	.4924	.5658	1.767	.8704	.5	.5	.5807	.7133	1.4019	.8141	.5
.6	.4939	.5681	1.760	.8695	.4	.6	.5821	.7159	1.3968	.8131	.4
.7	.4955	.5704	1.753	.8686	.3	.7	.5835	.7186	1.3916	.8121	.3
.8	.4970	.5727	1.746	.8678	.2	.8	.5850	.7212	1.3865	.8111	.2
.9	.4985	.5750	1.739	.8669	60.1	.9	.5864	.7239	1.3814	.8100	54.1
30.0	0.5000	0.5774	1.732	0.8660	**60.0**	**36.0**	0.5878	0.7265	1.3764	0.8090	**54.0**
	Cos	Cot	Tan	Sin	Deg.		Cos	Cot	Tan	Sin	Deg.

Table B Trigonometric Functions (degrees)

Deg.	Sin	Tan	Cot	Cos		Deg.	Sin	Tan	Cot	Cos	
36.0	0.5878	0.7265	1.3764	0.8090	**54.0**	**40.5**	0.6494	0.8541	1.1708	0.7604	**49.5**
.1	.5892	.7292	1.3713	.8080	53.9	.6	.6508	.8571	1.1667	.7593	.4
.2	.5906	.7319	1.3663	.8070	.8	.7	.6521	.8601	1.1626	.7581	.3
.3	.5920	.7346	1.3613	.8059	.7	.8	.6534	.8632	1.1585	.7570	.2
.4	.5934	.7373	1.3564	.8049	.6	.9	.6547	.8662	1.1544	.7559	49.1
.5	.5948	.7400	1.3514	.8039	.5	**41.0**	0.6561	0.8693	1.1504	0.7547	**49.0**
.6	.5962	.7427	1.3465	.8028	.4	.1	.6574	.8724	1.1463	.7536	48.9
.7	.5976	.7454	1.3416	.8018	.3	.2	.6587	.8754	1.1423	.7524	.8
.8	.5990	.7481	1.3367	.8007	.2	.3	.6600	.8785	1.1383	.7513	.7
.9	.6004	.7508	1.3319	.7997	53.1	.4	.6613	.8816	1.1343	.7501	.6
37.0	0.6018	0.7536	1.3270	0.7986	**53.0**	.5	.6626	.8847	1.1303	.7490	.5
.1	.6032	.7563	1.3222	.7976	52.9	.6	.6639	.8878	1.1263	.7478	.4
.2	.6046	.7590	1.3175	.7965	.8	.7	.6652	.8910	1.1224	.7466	.3
.3	.6060	.7618	1.3127	.7955	.7	.8	.6665	.8941	1.1184	.7455	.2
.4	.6074	.7646	1.3079	.7944	.6	.9	.6678	.8972	1.1145	.7443	48.1
.5	.6088	.7673	1.3032	.7934	.5	**42.0**	0.6691	0.9004	1.1106	0.7431	**48.0**
.6	.6101	.7701	1.2985	.7923	.4	.1	.6704	.9036	1.1067	.7420	47.9
.7	.6115	.7729	1.2938	.7912	.3	.2	.6717	.9067	1.1028	.7408	.8
.8	.6129	.7757	1.2892	.7902	.2	.3	.6730	.9099	1.0990	.7396	.7
.9	.6143	.7785	1.2846	.7891	52.1	.4	.6743	.9131	1.0951	.7385	.6
38.0	0.6157	0.7813	1.2799	0.7880	**52.0**	.5	.6756	.9163	1.0913	.7373	.5
.1	.6170	.7841	1.2753	.7869	51.9	.6	.6769	.9195	1.0875	.7361	.4
.2	.6184	.7869	1.2708	.7859	.8	.7	.6782	.9228	1.0837	.7349	.3
.3	.6198	.7898	1.2662	.7848	.7	.8	.6794	.9260	1.0799	.7337	.2
.4	.6211	.7926	1.2617	.7837	.6	.9	.6807	.9293	1.0761	.7325	47.1
.5	.6225	.7954	1.2572	.7826	.5	**43.0**	0.6820	0.9325	1.0724	0.7314	**47.0**
.6	.6239	.7983	1.2527	.7815	.4	.1	.6833	.9358	1.0686	.7302	46.9
.7	.6252	.8012	1.2482	.7804	.3	.2	.6845	.9391	1.0649	.7290	.8
.8	.6266	.8040	1.2437	.7793	.2	.3	.6858	.9424	1.0612	.7278	.7
.9	.6280	.8069	1.2393	.7782	51.1	.4	.6871	.9457	1.0575	.7266	.6
39.0	0.6293	0.8098	1.2349	0.7771	**51.0**	.5	.6884	.9490	1.0538	.7254	.5
.1	.6307	.8127	1.2305	.7760	50.9	.6	.6896	.9523	1.0501	.7242	.4
.2	.6320	.8156	1.2261	.7749	.8	.7	.6909	.9556	1.0464	.7230	.3
.3	.6334	.8185	1.2218	.7738	.7	.8	.6921	.9590	1.0428	.7218	.2
.4	.6347	.8214	1.2174	.7727	.6	.9	.6934	.9623	1.0392	.7206	46.1
.5	.6361	.8243	1.2131	.7716	.5	**44.0**	0.6947	0.9657	1.0355	0.7193	**46.0**
.6	.6374	.8273	1.2088	.7705	.4	.1	.6959	.9691	1.0319	.7181	45.9
.7	.6388	.8302	1.2045	.7694	.3	.2	.6972	.9725	1.0283	.7169	.8
.8	.6401	.8332	1.2002	.7683	.2	.3	.6984	.9759	1.0247	.7157	.7
.9	.6414	.8361	1.1960	.7672	50.1	.4	.6997	.9793	1.0212	.7145	.6
40.0	0.6428	0.8391	1.1918	0.7660	**50.0**	.5	.7009	.9827	1.0176	.7133	.5
.1	.6441	.8421	1.1875	.7649	49.9	.6	.7022	.9861	1.0141	.7120	.4
.2	.6455	.8451	1.1833	.7638	.8	.7	.7034	.9896	1.0105	.7108	.3
.3	.6468	.8481	1.1792	.7627	.7	.8	.7046	.9930	1.0070	.7096	.2
.4	.6481	.8511	1.1750	.7615	.6	.9	.7059	.9965	1.0035	.7083	45.1
40.5	0.6494	0.8541	1.1708	0.7604	**49.5**	**45.0**	0.7071	1.0000	1.0000	0.7071	**45.0**
	Cos	Cot	Tan	Sin	Deg.		Cos	Cot	Tan	Sin	Deg.

Table C Trigonometric Functions (radians)

Rad.	Sin	Tan	Cot	Cos	Rad.	Sin	Tan	Cot	Cos
.00	.00000	.00000	∞	1.00000	**.50**	.47943	.54630	1.8305	.87758
.01	.01000	.01000	99.997	0.99995	.51	.48818	.55936	1.7878	.87274
.02	.02000	.02000	49.993	.99980	.52	.49688	.57256	1.7465	.86782
.03	.03000	.03001	33.323	.99955	.53	.50553	.58592	1.7067	.86281
.04	.03999	.04002	24.987	.99920	.54	.51414	.59943	1.6683	.85771
.05	.04998	.05004	19.983	.99875	.55	.52269	.61311	1.6310	.85252
.06	.05996	.06007	16.647	.99820	.56	.53119	.62695	1.5950	.84726
.07	.06994	.07011	14.262	.99755	.57	.53963	.64097	1.5601	.84190
.08	.07991	.08017	12.473	.99680	.58	.54802	.65517	1.5263	.83646
.09	.08988	.09024	11.081	.99595	.59	.55636	.66956	1.4935	.83094
.10	.09983	.10033	9.9666	.99500	**.60**	.56464	.68414	1.4617	.82534
.11	.10978	.11045	9.0542	.99396	.61	.57287	.69892	1.4308	.81965
.12	.11971	.12058	8.2933	.99281	.62	.58104	.71391	1.4007	.81388
.13	.12963	.13074	7.6489	.99156	.63	.58914	.72911	1.3715	.80803
.14	.13954	.14092	7.0961	.99022	.64	.59720	.74454	1.3431	.80210
.15	.14944	.15114	6.6166	.98877	.65	.60519	.76020	1.3154	.79608
.16	.15932	.16138	6.1966	.98723	.66	.61312	.77610	1.2885	.78999
.17	.16918	.17166	5.8256	.98558	.67	.62099	.79225	1.2622	.78382
.18	.17903	.18197	5.4954	.98384	.68	.62879	.80866	1.2366	.77757
.19	.18886	.19232	5.1997	.98200	.69	.63654	.82534	1.2116	.77125
.20	.19867	.20271	4.9332	.98007	**.70**	.64422	.84229	1.1872	.76484
.21	.20846	.21314	4.6917	.97803	.71	.65183	.85953	1.1634	.75836
.22	.21823	.22362	4.4719	.97590	.72	.65938	.87707	1.1402	.75181
.23	.22798	.23414	4.2709	.97367	.73	.66687	.89492	1.1174	.74517
.24	.23770	.24472	4.0864	.97134	.74	.67429	.91309	1.0952	.73847
.25	.24740	.25534	3.9163	.96891	.75	.68164	.93160	1.0734	.73169
.26	.25708	.26602	3.7591	.96639	.76	.68892	.95045	1.0521	.72484
.27	.26673	.27676	3.6133	.96377	.77	.69614	.96967	1.0313	.71791
.28	.27636	.28755	3.4776	.96106	.78	.70328	.98926	1.0109	.71091
.29	.28595	.29841	3.3511	.95824	.79	.71035	1.0092	.99084	.70385
.30	.29552	.30934	3.2327	.95534	**.80**	.71736	1.0296	.97121	.69671
.31	.30506	.32033	3.1218	.95233	.81	.72429	1.0505	.95197	.68950
.32	.31457	.33139	3.0176	.94924	.82	.73115	1.0717	.93309	.68222
.33	.32404	.34252	2.9195	.94604	.83	.73793	1.0934	.91455	.67488
.34	.33349	.35374	2.8270	.94275	.84	.74464	1.1156	.89635	.66746
.35	.34290	.36503	2.7395	.93937	.85	.75128	1.1383	.87848	.65998
.36	.35227	.37640	2.6567	.93590	.86	.75784	1.1616	.86091	.65244
.37	.36162	.38786	2.5782	.93233	.87	.76433	1.1853	.84365	.64483
.38	.37092	.39941	2.5037	.92866	.88	.77074	1.2097	.82668	.63715
.39	.38019	.41105	2.4328	.92491	.89	.77707	1.2346	.80998	.62941
.40	.38942	.42279	2.3652	.92106	**.90**	.78333	1.2602	.79355	.62161
.41	.39861	.43463	2.3008	.91712	.91	.78950	1.2864	.77738	.61375
.42	.40776	.44657	2.2393	.91309	.92	.79560	1.3133	.76146	.60582
.43	.41687	.45862	2.1804	.90897	.93	.80162	1.3409	.74578	.59783
.44	.42594	.47078	2.1241	.90475	.94	.80756	1.3692	.73034	.58979
.45	.43497	.48306	2.0702	.90045	.95	.81342	1.3984	.71511	.58168
.46	.44395	.49545	2.0184	.89605	.96	.81919	1.4284	.70010	.57352
.47	.45289	.50797	1.9686	.89157	.97	.82489	1.4592	.68531	.56530
.48	.46178	.52061	1.9208	.88699	.98	.83050	1.4910	.67071	.55702
.49	.47063	.53339	1.8748	.88233	.99	.83603	1.5237	.65631	.54869
.50	.47943	.54630	1.8305	.87758	**1.00**	.84147	1.5574	.64209	.54030
Rad.	Sin	Tan	Cot	Cos	Rad.	Sin	Tan	Cot	Cos

Table C. Trigonometric Functions (radians) **605**

Table C Trigonometric Functions (radians)

Rad.	Sin.	Tan	Cot	Cos	Rad.	Sin	Tan	Cot	Cos
1.00	.84147	1.5574	.64209	.54030	**1.50**	.99749	14.101	.07091	.07074
1.01	.84683	1.5922	.62806	.53186	1.51	.99815	16.428	.06087	.06076
1.02	.85211	1.6281	.61420	.52337	1.52	.99871	19.670	.05084	.05077
1.03	.85730	1.6652	.60051	.51482	1.53	.99917	24.498	.04082	.04079
1.04	.86240	1.7036	.58699	.50622	1.54	.99953	32.461	.03081	.03079
1.05	.86742	1.7433	.57362	.49757	1.55	.99978	48.078	.02080	.02079
1.06	.87236	1.7844	.56040	.48887	1.56	.99994	92.621	.01080	.01080
1.07	.87720	1.8270	.54734	.48012	1.57	1.00000	1255.8	.00080	.00080
1.08	.88196	1.8712	.53441	.47133	1.58	.99996	−108.65	−.00920	−.00920
1.09	.88663	1.9171	.52162	.46249	1.59	.99982	−52.067	−.01921	−.01920
1.10	.89121	1.9648	.50897	.45360	**1.60**	.99957	−34.233	−.02921	−.02920
1.11	.89570	2.0143	.49644	.44466	1.61	.99923	−25.495	−.03922	−.03919
1.12	.90010	2.0660	.48404	.43568	1.62	.99879	−20.307	−.04924	−.04918
1.13	.90441	2.1198	.47175	.42666	1.63	.99825	−16.871	−.05927	−.05917
1.14	.90863	2.1759	.45959	.41759	1.64	.99761	−14.427	−.06931	−.06915
1.15	.91276	2.2345	.44753	.40849	1.65	.99687	−12.599	−.07937	−.07912
1.16	.91680	2.2958	.43558	.39934	1.66	.99602	−11.181	−.08944	−.08909
1.17	.92075	2.3600	.42373	.39015	1.67	.99508	−10.047	−.09953	−.09904
1.18	.92461	2.4273	.41199	.38092	1.68	.99404	−9.1208	−.10964	−.10899
1.19	.92837	2.4979	.40034	.37166	1.69	.99290	−8.3492	−.11977	−.11892
1.20	.93204	2.5722	.38878	.36236	**1.70**	.99166	−7.6966	−.12993	−.12884
1.21	.93562	2.6503	.37731	.35302	1.71	.99033	−7.1373	−.14011	−.13875
1.22	.93910	2.7328	.36593	.34365	1.72	.98889	−6.6524	−.15032	−.14865
1.23	.94249	2.8198	.35463	.33424	1.73	.98735	−6.2281	−.16056	−.15853
1.24	.94578	2.9119	.34341	.32480	1.74	.98572	−5.8535	−.17084	−.16840
1.25	.94898	3.0096	.33227	.31532	1.75	.98399	−5.5204	−.18115	−.17825
1.26	.95209	3.1133	.32121	.30582	1.76	.98215	−5.2221	−.19149	−.18808
1.27	.95510	3.2236	.31021	.29628	1.77	.98022	−4.9534	−.20188	−.19789
1.28	.95802	3.3413	.29928	.28672	1.78	.97820	−4.7101	−.21231	−.20768
1.29	.96084	3.4672	.28842	.27712	1.79	.97607	−4.4887	−.22278	−.21745
1.30	.96356	3.6021	.27762	.26750	**1.80**	.97385	−4.2863	−.23330	−.22720
1.31	.96618	3.7471	.26687	.25785	1.81	.97153	−4.1005	−.24387	−.23693
1.32	.96872	3.9033	.25619	.24818	1.82	.96911	−3.9294	−.25449	−.24663
1.33	.97115	4.0723	.24556	.23848	1.83	.96659	−3.7712	−.26517	−.25631
1.34	.97348	4.2556	.23498	.22875	1.84	.96398	−3.6245	−.27590	−.26596
1.35	.97572	4.4552	.22446	.21901	1.85	.96128	−3.4881	−.28669	−.27559
1.36	.97786	4.6734	.21398	.20924	1.86	.95847	−3.3608	−.29755	−.28519
1.37	.97991	4.9131	.20354	.19945	1.87	.95557	−2.2419	−.30846	−.29476
1.38	.98185	5.1774	.19315	.18964	1.88	.95258	−3.1304	−.31945	−.30430
1.39	.98370	5.4707	.18279	.17981	1.89	.94949	−3.0257	−33.051	−.31381
1.40	.98545	5.7979	.17248	.16997	**1.90**	.94630	−2.9271	−.34164	−.32329
1.41	.98710	6.1654	.16220	.16010	1.91	.94302	−2.8341	−.35284	−.33274
1.42	.98865	6.5811	.15195	.15023	1.92	.93965	−2.7463	−.36413	−.34215
1.43	.99010	7.0555	.14173	.14033	1.93	.93618	−2.6632	−.37549	−.35153
1.44	.99146	7.6018	.13155	.13042	1.94	.93262	−2.5843	−.38695	−.36087
1.45	.99271	8.2381	.12139	.12050	1.95	.92896	−2.5095	−.39849	−.37018
1.46	.99387	8.9886	.11125	.11057	1.96	.92521	−2.4383	−.41012	−.37945
1.47	.99492	9.8874	.10114	.10063	1.97	.92137	−2.3705	−.42185	−.38868
1.48	.99588	10.983	.09105	.09067	1.98	.91744	−2.3058	−.43368	−.39788
1.49	.99674	12.350	.08097	.08071	1.99	.91341	−2.2441	−.44562	−.40703
1.50	.99749	14.101	.07091	.07074	**2.00**	.90930	−2.1850	−.45766	−.41615
Rad.	Sin	Tan	Cot	Cos	Rad.	Sin	Tan	Cot	Cos

Answers
to Odd-Numbered
Problems

PROBLEM SET 1-1 (Page 6)

1. Let x be one number and y the other; $x + \frac{1}{3}y$. **3.** Let x be one number and y the other; $2x/3y$.
5. Let x be the number; $0.10x + x$, or $1.10x$. **7.** Let x be one side and y the other; $x^2 + y^2$. **9.** xy **11.** y/x
13. $(30/x) + (30/y)$ **15.** x^2 **17.** $6x^2$ **19.** $4\pi (x/2)^2 = \pi x^2$ **21.** $10x^2$ **23.** $\frac{4}{3}\pi (x/2)^3 = \pi x^3/6$
25. $x^3 - 4\pi x$ **27.** $A = x^2 + \frac{1}{4}\pi x^2$; $P = 2x + \pi x$ **29.** Let x be the number; $x + \frac{1}{2}x = 45$; $x = 30$.
31. Let x be the smaller odd number; $x + x + 2 = 168$; $x = 83$. **33.** $\frac{9}{2}x = 252$; $x = 56$ **35.** 10.5 meters
37. $800 - 200\pi \approx 171.68$ square feet. **39.** 3 centimeters **41.** \$3.24 **43.** $4\pi r^3$ **45.** 400 miles **47.** 48,000 miles
49. 12π feet

PROBLEM SET 1-2 (Page 12)

1. 12 **3.** -31 **5.** -67 **7.** $-60 + 9x$ **9.** $14t$ **11.** $\frac{8}{9}$ **13.** $-\frac{3}{4}$ **15.** $(1 - 3x)/2$ **17.** $(-2x + 3)/2$
19. $\frac{21}{12} = \frac{7}{4}$ **21.** $\frac{19}{20}$ **23.** $\frac{23}{36}$ **25.** $\frac{106}{108} = \frac{53}{54}$ **27.** $\frac{1}{2}$ **29.** $\frac{3}{4}$ **31.** $\frac{5}{4}$ **33.** $\frac{3}{8}$ **35.** $\frac{54}{7}$ **37.** $\frac{17}{7}$ **39.** -17
41. $\frac{9}{17}$ **43.** $\frac{5}{9}$ **45.** $\frac{1}{18}$ **47.** $\frac{13}{24}$ **49.** $\frac{37}{7}$ **51.** $-\frac{1}{2}$ **53.** 1 **55.** $\frac{33}{5}$ **57.** $\frac{1}{19}$ **59.** 100 **61.** 64
63. 252 meters

PROBLEM SET 1-3 (Page 20)

1. $2 \cdot 5 \cdot 5 \cdot 5$ **3.** $2 \cdot 2 \cdot 2 \cdot 5 \cdot 5$ **5.** $2 \cdot 2 \cdot 3 \cdot 5 \cdot 5 \cdot 7$ **7.** $2 \cdot 2 \cdot 2 \cdot 5 \cdot 5 \cdot 5 = 1000$
9. $2 \cdot 2 \cdot 3 \cdot 5 \cdot 5 \cdot 5 \cdot 7 = 10{,}500$ **11.** $2 \cdot 2 \cdot 2 \cdot 3 \cdot 5 \cdot 5 \cdot 5 \cdot 7 = 21{,}000$ **13.** $97/1000$ **15.** $289/10{,}500$
17. $-1423/21{,}000$ **19.** $.\overline{6}$ **21.** $.62\overline{50}$ **23.** $.\overline{461538}$ **25.** $7/9$ **27.** $235/999$ **29.** $13/40$ **31.** $318/990 = 53/165$
33. $\frac{1}{21}$ **35.** $\frac{50}{99}$
37. Suppose that $\sqrt{3} = m/n$, where m and n are positive integers greater than 1. Then $3n^2 = m^2$. Both n^2 and m^2 must have an even number of 3's as factors. This contradicts $3n^2 = m^2$, since $3n^2$ must have an odd number of 3's.
39. Let $r = a/b$ and $s = c/d$, where a, b, c, and d are integers, $b \neq 0$ and $d \neq 0$. Then $r + s = (ad + bc)/bd$, $rs = ac/bd$, both of which are rational.
41. If $\sqrt{2} + \frac{2}{3} = r$ is rational, then $\sqrt{2} = r + (-\frac{2}{3})$. This is impossible because $\sqrt{2}$ cannot be the sum of two rationals.
43. If $\frac{2}{3}\sqrt{2} = r$ is rational, then $\sqrt{2} = \frac{3}{2}r$. This is impossible because $\sqrt{2}$ cannot be the product of two rationals.
45. (d), (e), (f), and (i)

47.

$a^2 = x^2 + h^2$, $e^2 = (b + x)^2 + h^2$, $d^2 = (b - x)^2 + h^2$,
$d^2 + e^2 = 2h^2 + (b + x)^2 + (b - x)^2 = 2(a^2 - x^2) + 2b^2 + 2x^2 = 2(a^2 + b^2)$

49. $d = \sqrt{(e^2 + f^2 + g^2)/2}$

PROBLEM SET 1-4 (Page 26)

1. 1431   **3.** 1655   **5.** 61   **7.** 3   **9.** 2   **11.** Associative and commutative properties of multiplication.
13. Commutative property of addition.   **15.** Associative property of addition; additive inverse; zero is neutral element for addition.
17. Distributive property.   **19.** Distributive property.
21. True; $a - (b - c) = a + (-1)[b + (-c)] = a + (-1)b + (-1)(-c) = a - b + c$.
23. False; $1 \div (1 + 1) = \frac{1}{2}$, but $(1 \div 1) + (1 \div 1) = 2$.   **25.** True; $ab(a^{-1} + b^{-1}) = aba^{-1} + abb^{-1} = b + a$.
27. False; $(1 + 2)(1^{-1} + 2^{-1}) = 3(1 + \frac{1}{2}) = 3(\frac{3}{2}) = \frac{9}{2}$.   **29.** False; $(1 + 2)(1 + 2) = 3 \cdot 3 = 9$, but $1^2 + 2^2 = 5$.
31. False; $1 \div (2 \div 3) = 1/\frac{2}{3} = \frac{3}{2}$ but $(1 \div 2) \div 3 = \frac{1}{2}/3 = \frac{1}{6}$.
33. (a) 64; 18; 512   (b) No; for example, 4 # 3 = 64 but 3 # 4 = 81.   (c) No; for example, 2 # (3 # 2) = 512 but (2 # 3) # 2 = 64.
35. $(ab)(b^{-1}a^{-1}) = a(bb^{-1})a^{-1} = aa^{-1} = 1$; so $(ab)^{-1} = b^{-1}a^{-1}$. Since $a^{-1} \cdot a = 1$, it follows that a is the multiplicative inverse of a^{-1};
that is, $a = (a^{-1})^{-1}$. Since $\left(\dfrac{a}{b}\right) \cdot \left(\dfrac{b}{a}\right) = (ab^{-1})(ba^{-1}) = a(b^{-1}b)a^{-1} = aa^{-1} = 1$, $\dfrac{b}{a} = \left(\dfrac{a}{b}\right)^{-1}$.
37. Additive inverse; zero is neutral element for addition; distributive property; associative property of addition; additive inverse; zero is neutral element for addition.
39. $(-a)(-b) + -(ab) = (-a)(-b) + (-a)b = -a(-b + b) = -a \cdot 0 = 0$. Thus $(-a)(-b)$ is the additive inverse of $-(ab)$; so $(-a)(-b) = ab$.

PROBLEM SET 1-5 (Page 31)

1. >   **3.** >   **5.** >   **7.** >   **9.** <   **11.** =   **13.** $-\frac{3}{2}\sqrt{2}$; -2; $-\frac{\pi}{2}$, $\frac{3}{4}$; $\sqrt{2}$; $\frac{43}{24}$

15.   **17.**

19.   **21.**

23.   **25.** $-2 \le x \le 3$ **27.** $x \ge 2$ **29.** $-2 < x \le 3$

31. $-1 < x < 2$ **33.** $-4 \le x \le 4$

35. $1 < x < 5$ **37.** $-4 \le x \le 2$

39. $x < 0$ or $x > 10$

41. (a) $x \le 12$ (b) $-11 \le x < 3$ (c) $|x - y| \le 4$ (d) $|x - 7| \ge 3$ (e) $|x - 5| < |x - y|$
43. 1.414, $(\sqrt{2} + 1.414)/2$, $1.4\overline{14}$, $\sqrt{2}$, $1.\overline{414}$, $1.41\overline{4}$ **45.** (a) $\frac{6}{25}$ (b) $\frac{4}{17}$ (c) 17.1/85   (d) They are equal.   **47.** $\dfrac{25}{3} < a < 12$
49. $\dfrac{1}{6} < y < \dfrac{1}{2}$ **51.** $ab \ge 0$ **53.** Suppose that $\sqrt{a^2 + b^2} > |a| + |b|$. Then $a^2 + b^2 > (|a| + |b|)^2 = a^2 + b^2 + 2|a||b|$.
This is impossible. Thus $\sqrt{a^2 + b^2} \le |a| + |b|$.

PROBLEM SET 1-6 (Page 38)

1. $-2 + 8i$ **3.** $-4 - i$ **5.** $0 + 6i$ **7.** $-4 + 7i$ **9.** $11 + 4i$ **11.** $14 + 22i$ **13.** $16 + 30i$ **15.** $61 + 0i$
17. $\frac{3}{2} + \frac{7}{2}i$ **19.** $2 - 5i$ **21.** $\frac{11}{2} + \frac{3}{2}i$ **23.** $\frac{2}{5} + \frac{1}{5}i$ **25.** $\sqrt{3}/4 - i/4$ **27.** $\frac{1}{5} + \frac{8}{5}i$ **29.** $(\sqrt{3} - 1/4) + (-\sqrt{3}/4 - 1)i$
31. -1 **33.** $-i$ **35.** $-729i$ **37.** $-2 + 2i$
39. $i^4 = (i^2)(i^2) = (-1)(-1) = 1$. The four 4th roots of 1 are $1, -1, i,$ and $-i$.
41. $(1 - i)^4 = (1 - i)^2(1 - i)^2 = (-2i)(-2i) = 4i^2 = -4$
43. (a) $-1 - i$ (b) $1 - 2i$ (c) $-8 + 6i$ (d) $-\frac{8}{29} + \frac{20}{29}i$ (e) $\frac{14}{25}$ (f) $2\sqrt{3} - i$ **45.** (a) $a = 2, b = -\frac{4}{5}$ (b) $a = 3, b = 2$
47. (a) 1 (b) 0 (c) 3 **49.** $\pm[(\sqrt{2}/2) + (\sqrt{2}/2)i]$
51. (a) $\overline{x + y} = \overline{a + c + (b + d)i} = a + c - (b + d)i = a - bi + (c - di) = \bar{x} + \bar{y}$
(b) $\overline{xy} = \overline{ac - bd + (bc + ad)i} = ac - bd - (bc + ad)i = (a - bi)(c - di) = \bar{x}\,\bar{y}$
(c) $\overline{x^{-1}} = \overline{[1/(a + bi)]} = \overline{[a/(a^2 + b^2)] - [b/(a^2 + b^2)]i} = a/(a^2 + b^2) + [b/(a^2 + b^2)]i = (a + bi)/(a - bi)(a + bi) = 1/\bar{x} = (\bar{x})^{-1}$
(d) $\overline{x/y} = \overline{xy^{-1}} = \bar{x}\,\overline{y^{-1}} = \bar{x}\,(\bar{y})^{-1} = \bar{x}/\bar{y}$

CHAPTER 1. REVIEW PROBLEM SET (Page 39)

1. T **2.** F (try $x = 0$) **3.** F (try $x = -1$) **4.** T **5.** F (try $a = -2, b = -1$) **6.** T **7.** F (try $a = b = 1$)
8. F (try $x = -1$) **9.** T **10.** F (try $a = b = 1$) **11.** $a \cdot (b \cdot c) = (a \cdot b) \cdot c$ **12.** $a + b = b + a$
13. A number representable as a/b where a and b are integers, $b \neq 0$. **14.** (a) $\frac{29}{24}$ (b) $\frac{1}{5}$ (c) $\frac{1}{18}$ (d) $\frac{17}{4}$
15. (a) $\frac{5}{18}$ (b) $\frac{25}{27}$ (c) $\frac{27}{10}$ (d) $\frac{5}{24}$
16. Suppose that $r = 2 - \sqrt{3}$ is rational. Then $2 - r$ is rational. But $\sqrt{3} = 2 - r$ and so $\sqrt{3}$ is rational, contradicting what was given.
17. $A = \frac{1}{2}x(x - 5)$ **18.** $A = x(10 - x)$ **19.** $V = x(12 - 2x)^2$ **20.** $A = \sqrt{3}x^2/4$ **21.** $S = 2\pi x^2 + 2\pi xy = 2\pi x(x + y)$
22. $A = xy + x^2 + xy = x^2 + 2xy$ **23.** $T = \dfrac{100}{x - y} + \dfrac{100}{x + y}$ **24.** $A = xy + \pi x^2/8$ **25.** $A = xy + x\sqrt{4z^2 - x^2}/4$
26. $500 = 2 \cdot 2 \cdot 5 \cdot 5 \cdot 5, 360 = 2 \cdot 2 \cdot 2 \cdot 3 \cdot 3 \cdot 5$ **27.** $2 \cdot 2 \cdot 2 \cdot 3 \cdot 3 \cdot 5 \cdot 5 \cdot 5 = 9000$ **28.** $-\dfrac{49}{9000}$ **29.** $2^4 3^4 5^2 7$
30. $2^{11}3^{12}5^5 7^4$ **31.** $0.\overline{45}, 0.\overline{384615}$ **32.** $\frac{52}{111}, \frac{357}{110}$ **33.** $\frac{13}{4}$ **34.** $1.4, \sqrt{2}, 1.\overline{4}, \frac{29}{20}, \frac{13}{8}$
35. $a < b \Rightarrow 2a + a < 2a + b \Rightarrow 3a < 2a + b \Rightarrow a < (2a + b)/3$ **36.** $6.5 \le x \le 11.5, -8 < x < 2, x < -4$ or $x > 8$
37. $|x - 8| < 4$ **38.** $a < b$ **39.** $5x + 12y \le 10,000$
40. (a) $-10 + 2i$ (b) $29 + 34i$ (c) $\frac{35}{13} + \frac{7}{13}i$ (d) $-54 - 54i$ (e) $\frac{3}{25} - \frac{4}{25}i$ **41.** $6 - 2i$ **42.** Substitute and simplify.
43. 74 **44.** 320 square meters **45.** 5 feet by 10 feet **46.** 6 inches by 6 inches **47.** The volume is halved. **48.** 22
49. 4.2 miles per hour **50.** Total time allowed: $\frac{2}{60}$ hour; time for first mile: $\frac{1}{30}$ hour; no time left for second mile.

PROBLEM SET 2-1 (Page 48)

1. $3^3 = 27$ **3.** $2^2 = 4$ **5.** 8 **7.** $1/5^2 = 1/25$ **9.** $-1/5^2 = -1/25$ **11.** $1/(-2)^5 = -1/32$ **13.** $-27/8$
15. $27/4$ **17.** $81/16$ **19.** $3/64$ **21.** $1/72$ **23.** $81x^4$ **25.** $x^6 y^{12}$ **27.** $16x^8 y^4/w^{12}$ **29.** $27y^6/(x^3 z^6)$ **31.** $25x^8$
33. $1/(16y^6)$ **35.** $a/(5b^2 x^2)$ **37.** $2z^3/(x^6 y^2)$ **39.** $-x^3 z^2/(2y^7)$ **41.** $a^4 b^3$ **43.** $d^{40}/(32b^{15})$ **45.** $a^3/(a + 1)$
47. 0 **49.** $8x^3/y^6$ **51.** $\frac{1}{5}$ **53.** $2/x^6$ **55.** $4/x^8 y^4$ **57.** xy **59.** (a) 2^{-15} (b) 2^0 **61.** 2^{1000}
63. (a) 15¢, 31¢, 63¢ (b) $2^n - 1$ cents (c) February 7

PROBLEM SET 2-2 (Page 55)

1. 3.41×10^8 **3.** 5.13×10^{-8} **5.** 1.245×10^{-10} **7.** 8.4×10^{-4} **9.** 7.2×10^{-4} **11.** 1.08×10^{10}
13. 4.132×10^4 **15.** 4×10^9 **17.** 9.144×10^2
Note: Answers to Problems 19–43 may vary depending on the calculator used.
19. 48.35 **21.** -2441.7393 **23.** 303.27778 **25.** 2.7721×10^{15} **27.** 1.286×10^{10} **29.** -13.138859
31. 1.7891883 **33.** 1.067068 **35.** .9056964 **37.** .00000081 **39.** About 1.28 seconds **41.** About 4.068×10^{16}

43. About 9.3×10^{32} **45.** (a) 7.04×10^5 (b) 6.72×10^{-5} **47.** 1511.5 square centimeters   **49.** $\bar{x} = 146.17$; $s = 18.85$
51. 3.3923×10^9 **53.** Hans will hear it .1710655 seconds earlier.   **55.** 105.27 square centimeters
57. (a) 6.69×10^{23} (b) 4.10×10^{23} **59.** (a) 1.179×10^{61} (b) x_n grows without bound.   (c) x_n tends toward zero.

PROBLEM SET 2-3   (Page 62)

1. Polynomial of degree 2   **3.** Polynomial of degree 5   **5.** Polynomial of degree 0   **7.** Not a polynomial
9. Not a polynomial   **11.** $-2x + 1$ **13.** $4x^2$ **15.** $-2x^2 + 5x + 1$ **17.** $6x - 15$ **19.** $-10x + 12$
21. $35x^2 - 55x + 19$ **23.** $t^2 + 16t + 55$ **25.** $x^2 - x - 90$ **27.** $2t^2 + 13t - 7$ **29.** $y^2 + 2y - 8$
31. $6x^3 - 19x^2 + 18x - 20$ **33.** $x^2 + 20x + 100$ **35.** $x^2 - 64$ **37.** $4t^2 - 20t + 25$ **39.** $4x^8 - 25x^2$
41. $(t + 2)^2 + 2(t + 2)t^3 + t^6 = t^6 + 2t^4 + 4t^3 + t^2 + 4t + 4$ **43.** $(t + 2)^2 - (t^3)^2 = -t^6 + t^2 + 4t + 4$
45. $5.29x^2 - 6.44x + 1.96$ **47.** $x^3 + 6x^2 + 12x + 8$ **49.** $8t^3 - 36t^2 + 54t - 27$ **51.** $8t^3 + 12t^4 + 6t^5 + t^6$
53. $t^6 + 6t^5 + 15t^4 + 20t^3 + 15t^2 + 6t + 1$ **55.** $x^2 - 6xy + 9y^2$ **57.** $9x^2 - 4y^2$ **59.** $12x^2 + 11xy - 5y^2$
61. $2x^4y^2 - x^2yz - z^2$ **63.** $t^2 + 2t + 1 - s^2$ **65.** $8t^3 - 36t^2s + 54ts^2 - 27s^3$ **67.** $4x^4 - 9y^2$
69. $2s^6 - 5s^3t - 12t^2$ **71.** $8u^3 - 12u^2v^2 + 6uv^4 - v^6$ **73.** $-4x$ **75.** $12s + 18$ **77.** $x^4 + 4x^3 + 4x^2 - 9$
79. $2x^3 + 5x^2 + x - 2$ **81.** $x^5 + x^4y - 4x^3y^2 - 4x^2y^3$ **83.** $x^3 - 8y^3$ **85.** $x^4 + x^2y^2 + y^4$ **87.** -1
89. $(m^2 + 1)^2$ **91.** $2\sqrt{5}$ **93.** 243.5

PROBLEM SET 2-4   (Page 70)

1. $x(x + 5)$ **3.** $(x + 6)(x - 1)$ **5.** $y^3(y - 6)$ **7.** $(y + 6)(y - 2)$ **9.** $(y + 4)^2$ **11.** $(2x - 3y)^2$
13. $(y - 8)(y + 8)$ **15.** $(1 - 5b)(1 + 5b)$ **17.** $(2z - 3)(2z + 1)$ **19.** $(5x + 2y)(4x - y)$ **21.** $(x + 3)(x^2 - 3x + 9)$
23. $(a - 2b)(a^2 + 2ab + 4b^2)$ **25.** $x^3(1 - y^3) = x^3(1 - y)(1 + y + y^2)$ **27.** Does not factor over integers
29. Does not factor over integers   **31.** $(y - \sqrt{5})(y + \sqrt{5})$ **33.** $(\sqrt{5}z - 2)(\sqrt{5}z + 2)$ **35.** $t^2(t - \sqrt{2})(t + \sqrt{2})$
37. $(y - \sqrt{3})^2$ **39.** Does not factor over real numbers   **41.** $(x + 3i)(x - 3i)$ **43.** $(x^3 + 7)(x^3 + 2)$
45. $(2x - 1)(2x + 1)(x - 3)(x + 3)$ **47.** $(x + 4y + 3)^2$ **49.** $(x^2 - 3y^2)(x^2 + 2y^2)$
51. $(x - 2)(x^2 + 2x + 4)(x + 2)(x^2 - 2x + 4)$ **53.** $x^4(x - y)(x + y)(x^2 + y^2)$ **55.** $(x^2 + y^2)(x^4 - x^2y^2 + y^4)$
57. $(x^2 + 1)(x - 4)$ **59.** $(2x - 1 - y)(2x - 1 + y)$ **61.** $(3 - x)(x + y)$ **63.** $(x + 3y)(x + 3y + 2)$
65. $(x + y + 2)(x + y + 1)$ **67.** $(x^2 - 4x + 8)(x^2 + 4x + 8)$ **69.** $(x^2 - x + 1)(x^2 + x + 1)$ **71.** $(2 - 3m)(2 + 3m)$
73. $(3x - 1)(2x - 1)$ **75.** $5x(x - 2)(x + 2)$ **77.** $x(3x - 1)(2x - 1)$ **79.** $(2u^2 - 5)(u - 1)(u + 1)$
81. $(a + 2b - 7)(a + 2b + 4)$ **83.** $(a + 3b - 1)(a + 3b + 1)(a^2 + 6ab + 9b^2 + 1)$ **85.** $(x - 3y)(x - 3y + 4)$
87. $(3x^2 - 4y^2 - y)(3x^2 - 4y^2 + y)$ **89.** $(x^2 - y^2 - xy)(x^2 - y^2 + xy)$ **91.** $(x + 3)(x + 2)^2(x + 7)(x - 2)$
93. $(x^n + 2)(x^n + 1)$ **95.** (a) 94,000   (b) 7   (c) $\frac{15}{29}$
97. $n^3 = [n(n + 1)/2]^2 - [n(n - 1)/2]^2$. It is clear that both $n(n + 1)/2$ and $n(n - 1)/2$ are integers.

PROBLEM SET 2-5   (Page 77)

1. $1/(x - 6)$ **3.** $y/5$ **5.** $(x + 2)^2/(x - 2)$ **7.** $z(x + 2y)/(x + y)$ **9.** $(9x + 2)/(x - 2)(x + 2)$
11. $2(4x - 3)/(x - 2)(x + 2)$ **13.** $(2xy + 3x - 1)/x^2y^2$ **15.** $(x^2 + 10x - 3)/(x - 2)^2(x + 5)$ **17.** $(4 - x)/(2x - 1)$
19. $(6y^2 + 9y + 2)/(3y - 1)(3y + 1)$ **21.** $(3m^2 + m - 1)/3(m - 1)^2$ **23.** $5x/(2x - 1)(x + 1)$ **25.** $1/(x - 3)(x - 2)$
27. $y^2(x^3 - y^3)$ **29.** $x/(x - 4)$ **31.** $5(x + 1/x(2x - 1)$ **33.** $x/(x - 2)$ **35.** $(x - a)(x^2 + a^3)/(x + 2a)(x^2 + ax + a^2)$
37. $(y - 1)/2y$ **39.** $-2/(2x + 2h + 3)(2x + 3)$ **41.** $(-2x - h)/(x + h)^2x^2$ **43.** $(y - 1)(y + 2)/(7y + 19)$
45. $(a^2 + b^2)/ab$ **47.** 1   **49.** $(y - 1)/y$ **51.** $x^2/(x - y)$ **53.** 2   **55.** $(x - 6)/(x + 4)(x + 6)$ **57.** $a^2 - 3$
59. $3x + 2$ **61.** $(x^3 + y^3)/[x^3 + (x - y)^3] = (x + y)(x^2 - xy + y^2)/[x + (x - y)][x^2 - x(x - y) + (x - y)^2] =$
$(x + y)(x^2 - xy + y^2)/[x + (x - y)](x^2 - xy + y^2) = (x + y)/[x + (x - y)]$

63. (a) $2 + \dfrac{1}{1 + \dfrac{1}{1 + \frac{1}{2}}}$ (b) $2 + \dfrac{1}{1 + \dfrac{1}{1 + \dfrac{1}{1 + \frac{1}{3}}}}$ (c) $-2 + \dfrac{1}{1 + \frac{1}{3}}$

CHAPTER 2. REVIEW PROBLEM SET (Page 80)

1. F ($2^m 2^n = 2^{m+n}$) **2.** T **3.** F ($4^{-n} = 1/4^n$) **4.** F (write as 6.2345×10^6) **5.** F (study definition of polynomial)
6. F (pq has degree 7) **7.** T **8.** T **9.** T **10.** F (try $x = 0$) **11.** $\frac{27}{64}$ **12.** $\frac{16}{49}$ **13.** $\frac{36}{169}$ **14.** $24/(a^4 b^2)$
15. $x^9 y^6/64$ **16.** $y/(4x^5)$ **17.** 2.15×10^8 **18.** 1.07×10^{-4} **19.** 8.04×10^{-3} **20.** 3×10^8 **21.** No
22. Yes; 2 **23.** No **24.** No **25.** $-x^2 + x - 7$ **26.** $4x^4 - x^2 - 2x + 15$ **27.** $6y^2 - y - 2$ **28.** $32x^2 + 18$
29. $9x^2 y^2 - 12xyz^2 + 4z^4$ **30.** $10a^6 - 31a^3 b - 14b^2$ **31.** $x^4 - 9x^2 + 12x - 4$ **32.** $x^6 - 27y^3$
33. $4s^4 - 20s^3 + 37s^2 - 30s + 9$ **34.** $25x^4 - 9y^2 z^2$ **35.** $(x + 5)(x - 2)$ **36.** $(x - 10)(x + 3)$ **37.** $(2x + 5)(3x - 1)$
38. $(2x - 1)(x + 11)$ **39.** $(3x^3 - 4)(x^3 + 2)$ **40.** $(2x^3 - 3)(2x^3 + 3)$ **41.** $(2x - 5)(4x^2 + 10x + 25)$ **42.** $(5x - 2y)^2$
43. $(3cd^2 - b)^2$ **44.** $(x - 2)(x + 2)(x^2 + 4)$ **45.** $(2x - y + 3)(2x + y - 3)$ **46.** $(5x - 2y - 3)(5x + 2y)$
47. $(2x - \sqrt{29})(2x + \sqrt{29})$ **48.** $(x + 1 + i)(x + 1 - i)$ **49.** $3x/(x + 3)$ **50.** $(x - 2)^3(x + 2)/(x^2 - 2x + 4)^2$
51. $(5x - 25)/(x^2 - 16)$ **52.** $(x + 3)/(x - 2)$ **53.** $(a^2 - a + 1)/a^2$ **54.** $-x^4 - 2x^2$

PROBLEM SET 3-1 (Page 88)

1. Conditional equation **3.** Identity **5.** Conditional equation **7.** Conditional equation **9.** Identity **11.** 2 **13.** $\frac{2}{5}$
15. $\frac{9}{2}$ **17.** $2/\sqrt{3}$ **19.** 7.57 **21.** -8.71×10^1 **23.** -24 **25.** $-\frac{1}{4}$ **27.** $\frac{22}{5}$ **29.** 3 **31.** 6
33. No solution (2 is extraneous) **35.** $-\frac{3}{4}$ **37.** $-\frac{13}{2}$ **39.** -10 **41.** -21 **43.** $P = A/(1 + rt)$
45. $r = (nE - IR)/nI$ **47.** $h = (A - 2\pi r^2)/2\pi r$ **49.** $R_1 = RR_2/(R_2 - R)$ **51.** $\frac{26}{5}$ **53.** $\frac{5}{14}$ **55.** 2 **57.** $-\frac{1}{4}$
59. No solution **61.** $(1 - 2a)/(2a^2 - 1)$ **63.** (a) 86 (b) -40 (c) 160 **65.** $2700 **67.** (a) $11\frac{1}{9}$ (b) $100p/(100 - p)$

PROBLEM SET 3-2 (Page 95)

1. 9 **3.** 12 **5.** 23 **7.** 31 centimeters **9.** 89 **11.** 7 **13.** 12 **15.** 12:40 A.M. **17.** 4:20 P.M.
19. 9 miles per hour **21.** 15 **23.** 6000 **25.** 142.86 liters **27.** $145 **29.** 9 feet **31.** 7.5 days
33. 4.6 feet from fulcrum **35.** 125 **37.** 44.44 **39.** $17\frac{3}{4}$ feet **41.** 73 years old **43.** 32 **45.** $1511.36
47. 5.25%, 10.5% **49.** 84 years old **51.** (a) $\frac{5}{4}$ hours (b) $\frac{16}{75}$ hours

PROBLEM SET 3-3 (Page 103)

1. $x = -13; y = 13$ **3.** $u = \frac{3}{2}; v = -4$ **5.** $x = -1; y = 3$ **7.** $x = 4; y = 3$ **9.** $x = 6; y = -8$ **11.** $s = 1; t = 4$
13. $x = 3; y = -1$ **15.** $x = 4; y = 7$ **17.** $x = 9; y = -2$ **19.** $x = 16; y = -5$ **21.** $x = \frac{1}{2}; y = \frac{1}{3}$ **23.** $x = 4; y = 9$
25. $x = 1; y = 2; z = 0$ **27.** $x = \frac{4}{3}; y = -\frac{2}{3}; z = -\frac{5}{3}$ **29.** $x = -1; y = -2; z = -3$ **31.** $r = 36; s = 54$
33. $7200 and $5800 **35.** $\frac{5}{3}$ and $-\frac{4}{3}$ **37.** $\frac{11}{19}$ **39.** 18 inches from one end **41.** 36,000 $10 tickets and 9000 $15 tickets
43. 35 hours **45.** $57\frac{1}{7}$ pounds of the first kind and $42\frac{6}{7}$ pounds of the second kind **47.** 8 coats and 100 dresses
49. $x = -10z + 33; y = -7z + 26$

PROBLEM SET 3-4 (Page 111)

1. $5\sqrt{2}$ **3.** $\frac{1}{2}$ **5.** $\frac{3}{2}$ **7.** 22 **9.** $(5 + 6\sqrt{2})/5$ **11.** $(6 + i)/2$ **13.** ± 5 **15.** $-1; 7$ **17.** $-\frac{15}{2}; \frac{5}{2}$ **19.** $\pm 3i$
21. 0; 3 **23.** ± 3 **25.** $\pm .12$ **27.** $-2; 5$ **29.** $-2; \frac{1}{3}$ **31.** $-\frac{4}{3}; \frac{7}{2}$ **33.** $-9; 1$ **35.** $-\frac{1}{2}; \frac{3}{2}$ **37.** $-2 \pm \sqrt{5}i$
39. $-6; -2$ **41.** $(-5 \pm \sqrt{13})/2$ **43.** $(3 + \sqrt{42})/3$ **45.** $(-5 \pm \sqrt{5})/2$ **47.** $(3 \pm \sqrt{13}i)/2$ **49.** $-.2714; 1.8422$
51. $-1.6537; .8302$ **53.** $y = 2 \pm 2x$ **55.** $y = -6x$ or $y = 0$ **57.** $y = -2x + 3$ or $y = -2x + 5$ **59.** $-\frac{1}{4}; \frac{5}{4}$ **61.** 1
63. $-1 \pm \sqrt{5}$ **65.** $(-1 \pm i)/2$ **67.** $\pm\sqrt{6}; \pm i$ **69.** $-1, 4, 2 \pm 2\sqrt{2}$ **71.** 2 **73.** $\frac{1}{4}; \frac{3}{2}$ **75.** 0; 4
77. $x = \frac{5}{2}, y = 8; x = -4, y = -5$ **79.** $x = 2\sqrt{2}, y = \sqrt{2}; x = -2\sqrt{2}, y = -\sqrt{2}$ **81.** 3 by 10 **83.** 12 inches by 12 inches
85. At approximately 1:49 P.M. **87.** 12 miles per hour **89.** $(1 + \sqrt{5})/2$

PROBLEM SET 3-5 (Page 119)

1. Conditional **3.** Unconditional **5.** Conditional **7.** Conditional **9.** Conditional **11.** Unconditional

13. $\{x : x < -6\}$

15. $\{x : x > -24\}$

17. $\{x : x < \frac{30}{7}\}$

19. $\{x : -5 \le x \le 2\}$

21. $\{x : x < -3 \text{ or } > \frac{1}{2}\}$

23. $\{x : x \le 1 \text{ or } x \ge 4\}$

25. $\{x : \frac{1}{2} < x < 3\}$

27. $\{x : -4 \le x \le 0 \text{ or } x \ge 3\}$

29. $\{x : x < 5 \text{ and } x \ne 2\}$

31. $\{x : -2 < x \le 5\}$

33. $\{x : -2 < x < 0 \text{ or } x > 5\}$

35. $\{x : -2 < x < 2 \text{ or } x > 3\}$

37. $-\frac{5}{2} < x < -\frac{1}{2}$ **39.** $-1 \le x \le 0$ **41.** $x \le -\frac{8}{5} \text{ or } x \ge 2$ **43.** $x < -\frac{3}{2} \text{ or } x > \frac{9}{2}$ **45.** $|x - 3| < 3$
47. $|x - 3| \le 4$ **49.** $|x - 6.5| < 4.5$

51. $\{x : -\sqrt{7} < x < \sqrt{7}\}$

53. $\{x : x \le 2 - \sqrt{2} \text{ or } x \ge 2 + \sqrt{2}\}$

55. $\{x : x < -5.71 \text{ or } x > -.61\}$

57. 4   **59.** 100   **61.** $\{x : x < \frac{25}{2}\}$ **63.** $\{x : -3 < x < \frac{1}{2}\}$ **65.** $\{x : x < -1 \text{ or } -1 < x < 1 \text{ or } 4 < x < 8\}$
67. $\{x : -2 < x < 2\}$ **69.** $\{x : x \le \frac{1}{4} \text{ or } x \ge \frac{5}{4}\}$ **71.** $\{x : -1 < x < \frac{5}{2}\}$ **73.** $\{x : x > -\frac{1}{2}\}$
75. $\{x : x < -2 \text{ or } 0 < x < 2 \text{ or } x > 4\}$
77. (a) $\{k : k \le 4\}$ (b) $\{k : k \le -6 \text{ or } k \ge 6\}$ (c) $\{k : k \le 0 \text{ or } k \ge 4\}$ (d) $\{k : k = 0\}$ **79.** $\$25,000 < S < \$53,000$
81. (a) 144 feet   (b) Between $2 - \sqrt{3}$ seconds and $2 + \sqrt{3}$ seconds   (c) After 5 seconds   **83.** $980/\pi^2 < l < 2205/\pi^2$
85. $c^n = c^2 c^{n-2} = (a^2 + b^2)c^{n-2} = a^2 c^{n-2} + b^2 c^{n-2} > a^2 a^{n-2} + b^2 b^{n-2} = a^n + b^n$

PROBLEM SET 3-6 (Page 123)

1. About 18,333 feet   **3.** 525 miles   **5.** About 21.82 minutes   **7.** 4 hours and 48 minutes   **9.** About 4.15 hours after takeoff
11. 12 standard; 20 deluxe   **13.** 60 miles per hour   **15.** 500 meters   **17.** 20 feet   **19.** About 8.17 feet   **21.** $\frac{5}{8}$ kiloliter
23. About 3.29 milligrams   **25.** .298 gram sodium chloride; .202 gram sodium bromide   **27.** 36 shares
29. 14 pounds walnuts; 11 pounds cashews   **31.** 23,500   **33.** \$785,714.29
35. 2340 undergraduate students; 864 graduate students   **37.** 2 feet   **39.** 5; 12; 13   **41.** $a = 8, b = 5, c = 10$
43. About 13.34 feet   **45.** Two solutions: longer piece is 28 inches or $\frac{236}{7}$ inches.

CHAPTER 3. REVIEW PROBLEM SET (Page 127)

1. F ($x = 1$ and $x^2 = x$ are not equivalent since 0 is a solution to the latter) **2.** T **3.** F ($x^2 = 0$ has 0 as its only solution)
4. F [(2, 7) is one of many other solutions] **5.** T **6.** F (its solutions are nonreal) **7.** F ($\sqrt{9} = 3$) **8.** T
9. F ($|x| > 5$ means $x < -5$ or $x > 5$) **10.** F (try $x = -2$)
11. (a) Identity (b) Conditional equation (c) Identity (d) Identity
12. (a) This equation is equivalent to $3 = -2$. (b) This equation is equivalent to $1 = 2$. **13.** (a) $-\frac{5}{6}$ (b) 7 (c) $\frac{1}{9}$ (d) No solution
14. $s = 1/(rt + 3)$ **15.** (a) 80 (b) $F > -40$ **16.** (a) (0, 3) (b) $(\frac{13}{5}, \frac{6}{5})$ **17.** (a) $(2 + \sqrt{2})/4$ (b) $\frac{2}{3}$ (c) 52 (d) $1 - i$
18. None **19.** ± 8 **20.** $-1, 3$ **21.** -2 **22.** 0, 3 **23.** $-5, 7$ **24.** $-\frac{2}{3}, 1$ **25.** $-1, 6$ **26.** $(-1 \pm \sqrt{13})/6$
27. $-1 \pm i$ **28.** $\pm i$ **29.** $(3 \pm \sqrt{5})/2$ **30.** 16 **31.** $-3, 3$ **32.** $-4, 1$ **33.** $y = (3x \pm 3)/2$
34. $y = 4 - 2x^2$ or $y = 1 - 2x^2$ **35.** $x < \frac{7}{5}$ **36.** $x \le -\frac{3}{4}$ **37.** $x < -4$ or $x > 6$ **38.** $x \ne -2$
39. $x \le -\frac{1}{2}$ or $x > 3$ **40.** $-2 < x < 2, x \ne 0$ **41.** $-1 - \sqrt{3} < x < -1 + \sqrt{3}$ **42.** All reals **43.** $-10 < x < 0$
44. $x \le -\frac{11}{3}$ or $x \ge -1$ **45.** $-4, 3$ **46.** 3, 7 **47.** $\frac{3}{2}$ **48.** $\frac{1}{2}$ **49.** We lose the solution $x = 0$.
50. (a) $t = 6$ (b) $1 < t < 5$ **51.** \$9600 at 7 percent, \$15,400 at 8.5 percent **52.** 16 centimeters by 36 centimeters
53. 50 miles per hour **54.** $76.6\overline{6}$ percent **55.** 225 reserved seats, 400 general admission **56.** 36 liters

PROBLEM SET 4-1 (Page 135)

1. 5 **3.** 4 **5.** 1.7 **7.** $2\pi - 5$ **9.** 3; 7 **11.** $-6; 2$ **13.** $\frac{7}{4}, \frac{13}{4}$ **15.** Rectangle **17.** Parallelogram
19. 5; $(\frac{7}{2}, 1)$ **21.** $2\sqrt{2}$; (3, 3) **23.** 3; $(\sqrt{3}/2, \sqrt{6}/2)$ **25.** 14.54; (3.974, 1.605)
27. (a) $\sqrt{10}$; $\sqrt{5}$; $\sqrt{10}$; $\sqrt{5}$ (b) Each midpoint has coordinates (5/2, 5). (c) Opposite sides have the same length; the two diagonals bisect each other.
29. (a) $(-2, 3)$; (4, 0) (b) (2, 7); (8, −1)
31. If the points are labeled A, B, and C, respectively, then $d(A, B) = \sqrt{20}$, $d(B, C) = \sqrt{20}$, and $d(A, C) = \sqrt{40}$. Then note that $(\sqrt{40})^2 = (\sqrt{20})^2 + (\sqrt{20})^2$.
33. (a) The three distances are 5, 10, and 15, and $15 = 5 + 10$. (b) The distances are 5, 10, and 15. **35.** (9, 0) **37.** (17, 35)
39. \$1.47 cheaper by truck **41.** $x \approx 1.9641, y \approx 4.5981$

43. $e^2 = \left(\frac{a}{2}\right)^2 + \left(\frac{b}{2}\right)^2, f^2 = \left(\frac{a}{2}\right)^2 + \left(\frac{b}{2}\right)^2, d^2 = \left(\frac{a}{2}\right)^2 + \left(\frac{b}{2}\right)^2$

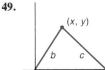

45. $x = 5, y = 7$ **47.** $\sqrt{53} \approx 7.28$ miles

49.

Let the triangle be as shown and suppose that $a^2 + b^2 = c^2$. Now $b^2 = x^2 + y^2$ and $c^2 = (x - a)^2 + y^2$, so $a^2 + b^2 = a^2 + x^2 + y^2 = c^2 = x^2 - 2ax + a^2 + y^2$. This implies that $-2ax = 0$, so $x = 0$, as required.

51. (1, 10), (6, 17), (11, 24), (16, 31) **53.** (a) $\frac{1}{3}(x_1 + x_2 + x_3), \frac{1}{3}(y_1 + y_2 + y_3)$ (b) They intersect at P.

55. $a^2 + c^2 = x^2 + y^2 + (x - e)^2 + (y - f)^2$
$b^2 + d^2 = (x - e)^2 + y^2 + x^2 + (y - f)^2$
Thus $a^2 + c^2 = b^2 + d^2$.

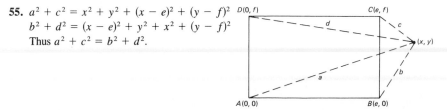

PROBLEM SET 4-2 (Page 143)

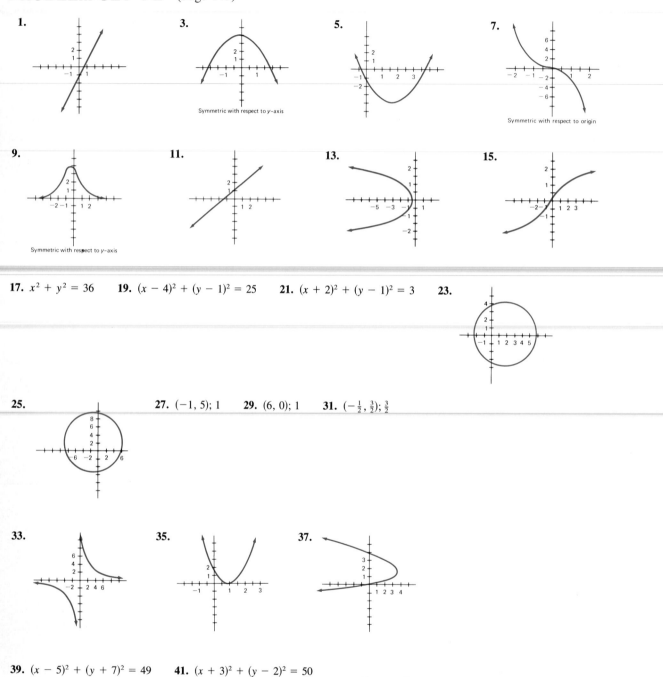

1.

3. Symmetric with respect to y-axis

5.

7. Symmetric with respect to origin

9. Symmetric with respect to y-axis

11.

13.

15.

17. $x^2 + y^2 = 36$ **19.** $(x - 4)^2 + (y - 1)^2 = 25$ **21.** $(x + 2)^2 + (y - 1)^2 = 3$ **23.**

25.   **27.** $(-1, 5); 1$ **29.** $(6, 0); 1$ **31.** $(-\frac{1}{2}, \frac{3}{2}); \frac{3}{2}$

33.   **35.**   **37.**

39. $(x - 5)^2 + (y + 7)^2 = 49$ **41.** $(x + 3)^2 + (y - 2)^2 = 50$

43. (a) Point; $(2, -3)$ (b) Circle: 2; $(\frac{1}{2}, \frac{-3}{2})$ (c) Empty set   (d) Circle: $\sqrt{5}$; $(0, \sqrt{3})$

45. The equation of the given circle may be written as $(x - 3)^2 + (y + 2)^2 = 25$ and $(7 - 3)^2 + (\frac{3}{2} + 2)^2 = \frac{113}{4} > 25$.

47. (a) $(x - 4)^2 + (y - 4)^2 = 25$ (b) $(x - 4)^2 + (y - 4)^2 = \frac{25}{2}$ **49.** $6\sqrt{3} + 6\pi \approx 29.24$

PROBLEM SET 4-3   (Page 150)

1. $\frac{5}{2}$ **3.** $-\frac{2}{7}$ **5.** $-\frac{5}{3}$ **7.** 0.1920   **9.** $4x - y - 5 = 0$ **11.** $2x + y - 2 = 0$ **13.** $2x + y - 4 = 0$
15. $y - 5 = 0$ **17.** $5x - 2y - 4 = 0$ **19.** $5x + 3y - 15 = 0$ **21.** $1.56x + y - 5.35 = 0$ **23.** $x - 2 = 0$ **25.** 3; 5
27. $\frac{2}{3}; -\frac{4}{3}$ **29.** $-\frac{2}{3}; 2$ **31.** $-4; 2$
33. (a) $y + 3 = 2(x - 3)$ (b) $y + 3 = -\frac{1}{2}(x - 3)$ (c) $y + 3 = -\frac{2}{3}(x - 3)$ (d) $y + 3 = \frac{3}{2}(x - 3)$ (e) $y + 3 = -\frac{3}{4}(x - 3)$
(f) $x = 3$ (g) $y = -3$
35. $y + 4 = 2x$ **37.** $(-1, 2); y - 2 = \frac{2}{3}(x + 1)$ **39.** $(3, 1); y - 1 = -\frac{4}{3}(x - 3)$ **41.** $\frac{7}{5}$ **43.** $\frac{18}{13}$ **45.** $\frac{6}{5}$
47. (a) Parallel   (b) Perpendicular   (c) Neither   (d) Perpendicular   **49.** $3x - 2y - 7 = 0$ **51.** $x + 2y - 9 = 0$
53. Note that $(a, 0)$ and $(0, b)$ satisfy the equation $x/a + y/b = 1$, and use the fact that two points determine a line.
55. $2x + y - 8 = 0$ **57.** $P = 4x - 8500; -\$500$ (loss)   **59.** (a) $V = 2600t + 60,000$ (b) 3.5 years after purchase   **61.** $\frac{5}{13}$
63. Draw a picture. We may assume the triangle has vertices $(0, 0)$, $(a, 0)$, and (b, c). Midpoints are $(b/2, c/2)$ and $((a + b)/2, c/2)$. The
line joining midpoints has slope 0 as does the base.
65. $(10 \pm 4\sqrt{2}, 0)$ **67.** $x = (1 + \sqrt{7})/2, y = (1 - \sqrt{7})/2$

PROBLEM SET 4-4   (Page 158)

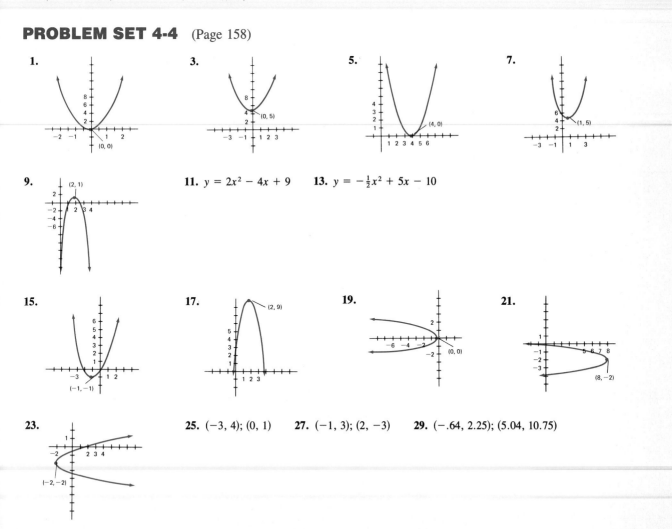

1.   **3.**   **5.**   **7.**

9.   **11.** $y = 2x^2 - 4x + 9$ **13.** $y = -\frac{1}{2}x^2 + 5x - 10$

15.   **17.**   **19.**   **21.**

23.   **25.** $(-3, 4); (0, 1)$ **27.** $(-1, 3); (2, -3)$ **29.** $(-.64, 2.25); (5.04, 10.75)$

31. $y = \frac{1}{12}x^2$ **33.** $y = -\frac{1}{8}(x + 2)^2$ **35.** $y = 2x^2 - 1$ **37.**

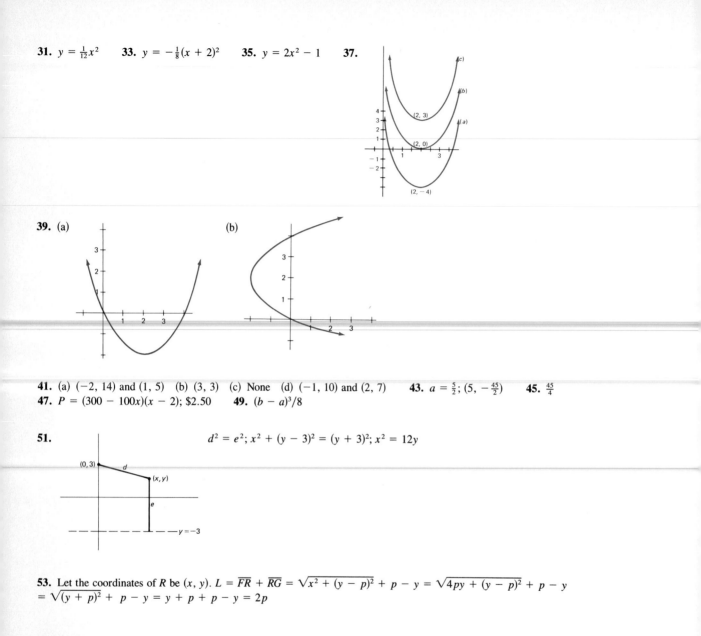

39. (a)   (b)

41. (a) $(-2, 14)$ and $(1, 5)$ (b) $(3, 3)$ (c) None   (d) $(-1, 10)$ and $(2, 7)$ **43.** $a = \frac{5}{2}$; $(5, -\frac{45}{2})$ **45.** $\frac{45}{4}$
47. $P = (300 - 100x)(x - 2)$; $2.50   **49.** $(b - a)^3/8$

51. $d^2 = e^2$; $x^2 + (y - 3)^2 = (y + 3)^2$; $x^2 = 12y$

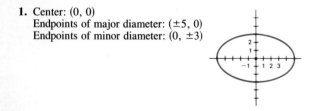

53. Let the coordinates of R be (x, y). $L = \overline{FR} + \overline{RG} = \sqrt{x^2 + (y - p)^2} + p - y = \sqrt{4py + (y - p)^2} + p - y$
$= \sqrt{(y + p)^2} + p - y = y + p + p - y = 2p$

PROBLEM SET 4-5   (Page 166)

1. Center: $(0, 0)$
Endpoints of major diameter: $(\pm 5, 0)$
Endpoints of minor diameter: $(0, \pm 3)$

3. Center: (0, 0)
Endpoints of major diameter: (0, ±5)
Endpoints of minor diameter: (±3, 0)

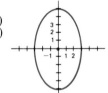

5. Center: (2, −1)
Endpoints of major diameter: (−3, −1), (7, −1)
Endpoints of minor diameter: (2, −4), (2, 2)

7. Center: (−3, 0)
Endpoints of major diameter: (−3, −4), (−3, 4)
Endpoints of minor diameter: (−6, 0), (0, 0)

9. (0, 0); $x^2/36 + y^2/9 = 1$ **11.** (0, 0); $x^2/16 + y^2/36 = 1$ **13.** (2, 3); $(x − 2)^2/36 + (y − 3)^2/4 = 1$

15. Center: (0, 0)
Vertices: (±4, 0)

17. Center: (0, 0)
Vertices: (0, ±3)

19. Center: (3, −2)
Vertices: (0, −2), (6, −2)

21. Center: (0, −3)
Vertices: (0, −5), (0, −1)

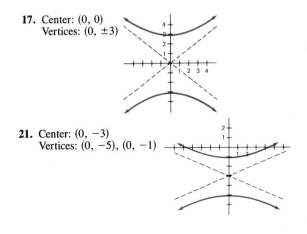

23. $x^2/16 − y^2/25 = 1$ **25.** $(x − 6)^2/4 − (y − 3)^2/4 = 1$ **27.** $(y − 9)^2/36 − (x − 4)^2/16 = 1$
29. $(x + 2)^2/16 + (y − 3)^2/9 = 1$; horizontal ellipse; center: (−2, 3); vertices: (−6, 3), (2, 3)
31. $(x − 2)^2/9 − (y + 1)^2/4 = 1$; horizontal hyperbola; center: (2, −1); vertices: (−1, −1), (5, −1)
33. $x^2/4 + (y − 2)^2/100 = 1$; vertical ellipse; center: (0, 2); vertices: (0, −8), (0, 12)
35. $(y + 2)^2/9 − (x − 4)^2/\frac{9}{4} = 1$; vertical hyperbola; center: (4, −2); vertices: (4, −5), (4, 1)

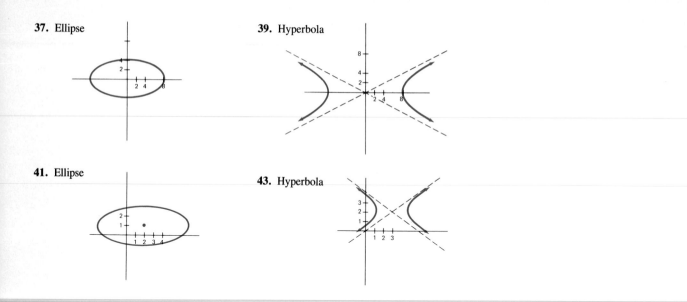

37. Ellipse

39. Hyperbola

41. Ellipse

43. Hyperbola

45. $(x - 3)^2/49 + (y - 2)^2/25 = 1$ **47.** $x^2/36 - y^2/16 = 1$ **49.** ± 3.7486 **51.** (a) $\pi\sqrt{77} \approx 27.5674$
(b) $\pi\sqrt{111} \approx 33.0987$ **53.** $9 - \sqrt{21} \approx 4.42$ centimeters   **55.** $6\sqrt{3} \approx 10.39$ feet   **57.** $y^2/441 + x^2/16 = 1$
59. $x^2/25 + y^2/9 = 1$ **61.** $x^2/9 - y^2/16 = 1$ **63.** $x^2/625 + y^2/400 = 1$

CHAPTER 4. REVIEW PROBLEM SET   (Page 170)

1. T   **2.** T   **3.** F $(d = \sqrt{4\pi^2 + 1})$ **4.** T   **5.** F $(r = 2)$ **6.** F (opens left)   **7.** F (need $a \neq 0$)
8. F (try $D = E = 2, F = -3$)   **9.** T   **10.** T   **11.** Horizontal parabola   **12.** Circle   **13.** Vertical ellipse   **14.** Line
15. Horizontal parabola   **16.** Vertical hyperbola   **17.** Intersecting lines   **18.** Point $(-2, 3)$

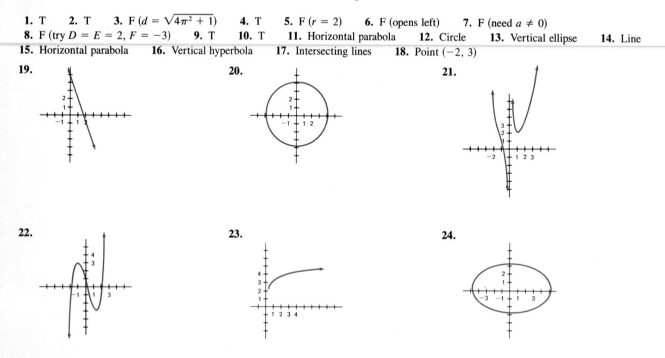

19.

20.

21.

22.

23.

24.

25. (a) $C(-1, 9)$

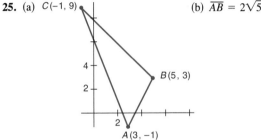

$B(5, 3)$

$A(3, -1)$

(b) $\overline{AB} = 2\sqrt{5}$, $\overline{AC} = 2\sqrt{29}$, $\overline{BC} = 6\sqrt{2}$

26. (a) AB: $m = 2$, AC: $m = -\frac{5}{2}$, BC: $m = -1$ (b) AB: $2x - y - 7 = 0$, AC: $5x + 2y - 13 = 0$, BC: $x + y - 8 = 0$
27. (a) $(4, 1)$, $(1, 4)$ **28.** (a) $x + y - 2 = 0$ (b) $x - y - 4 = 0$ **29.** $x + 0y - 4 = 0$ **30.** $4x + 3y - 17 = 0$
31. $5x + 7y = 0$ **32.** $3x - 2y + 12 = 0$ **33.** $x - 2y - 2 = 0$ **34.** $3x - 4y - 25 = 0$ **35.** $(2, -7)$; opens up
36. $(-2, -3)$; opens right   **37.** $(-1, 3)$; opens down   **38.** $(2, 1)$; opens left   **39.** $-\frac{1}{3}y = x^2$ **40.** $-2(x + 2) = (y - 3)^2$
41. $x^2/4 + y^2/9 = 1$ **42.** $(x - 3)^2/9 + (y - 1)^2/16 = 1$ **43.** $(x - 3)^2 + (y - 1)^2 = 25$
44. $(x + 7)^2/49 + (y - 3)^2/4 = 1$ **45.** $(x - 2)^2/9 - (y + 3)^2/16 = 1$ **46.** $(x - \frac{5}{2})^2 + (y - 3)^2 = \frac{73}{4}$
47. Circle; center: $(3, -1)$; radius: 3   **48.** Parabola; vertex: $(4, -3)$; opens left
49. Horizontal hyperbola; center: $(1, 0)$; asymptotes: $y = \pm\frac{2}{3}(x - 1)$
50. Horizontal ellipse; center: $(-3, 5)$; major diameter: 20; minor diameter: 10
51. $(-3, 1)$, $(2, -4)$ **52.** $(\sqrt{2}, \pm 3\sqrt{2}/2)$, $(-\sqrt{2}, \pm 3\sqrt{2}/2)$ **53.** (a) 5 feet (b) $3\sqrt{5} \approx 6.71$ feet
54. $20 + 6\pi \approx 38.85$ **55.** -14 **56.** (a) 78.125 feet (b) 312.5 feet

PROBLEM SET 5-1   (Page 178)

1. (a) 0 (b) -4 (c) $-15/4$ (d) -3.99 (e) -2 (f) $a^2 - 4$ (g) $(1 - 4x^2)/x^2$ (h) $x^2 + 2x - 3$
3. (a) $\frac{1}{4}$ (b) $-\frac{1}{2}$ (c) 2 (d) -8 (e) Undefined (f) 100 (g) $x/(1 - 4x)$ (h) $1/(x^2 - 4)$ (i) $1/(h - 2)$ (j) $-1/(h + 2)$
5. All real numbers   **7.** $\{x : x \neq \pm 2\}$ **9.** $\{x : x \neq -2 \text{ and } x \neq 3\}$ **11.** All real numbers   **13.** $\{x : x \geq 2\}$
15. $\{x : x \geq 0 \text{ and } x \neq 25\}$ **17.** (a) 9 (b) 5 (c) 3; the positive integers   **19.** $f(x) = \boxed{(}\ x\ \boxed{+}\ 2\ \boxed{)}\ \boxed{x^2}\ \boxed{=}$; $f(2.9) = 24.01$
21. $f(x) = 3\ \boxed{\times}\ \boxed{(}\ x\ \boxed{+}\ 2\ \boxed{)}\ \boxed{x^2}\ \boxed{-}\ 4\ \boxed{=}$; $f(2.9) = 68.03$
23. $f(x) = \boxed{(}\ 3\ \boxed{\times}\ x\ \boxed{+}\ 2\ \boxed{\div}\ x\ \boxed{\sqrt{x}}\ \boxed{)}\ \boxed{y^x}\ 3\ \boxed{=}$; $f(2.9) = 962.80311$
25. $f(x) = \boxed{(}\ x\ \boxed{y^x}\ 5\ \boxed{-}\ 4\ \boxed{)}\ \boxed{\sqrt{x}}\ \boxed{\div}\ \boxed{(}\ 2\ \boxed{+}\ x\ \boxed{1/x}\ \boxed{)}\ \boxed{=}$; $f(2.9) = 6.0479407$ **27.** $x + \sqrt{5}$ **29.** $2x^2 + 3x$
31. $\sqrt{3(x - 2)^2 + 9}$ **33.** 20   **35.** 5   **37.** 8   **39.** Undefined   **41.** $18xy - 10x$ **43.** $3 - 5x$ **45.** $y = 4x$
47. $y = 1/x$ **49.** $I = 324s/d^2$ **51.** (a) $R = 2v^2/45$ (b) About 28,444 feet
53. (a) 3 (b) 3 (c) $-\frac{15}{4}$ (d) $2 - \sqrt{2}$
(e) $4 - 5\sqrt{2}$ (f) -199.9999 (g) $(1 - 2x^3)/x^2$ (h) $(a^6 - 2)/a^z$ (i) $(a^3 + 3a^2b + 3ab^2 + b^3 - 2)/(a + b)$
55. (a) $\{t : t \neq 0, -3\}$ (b) $\{t : t \leq -2 \text{ or } t \geq 2\}$ (c) $\{t : t \geq 0, t \neq \frac{1}{2}\}$ (d) $\{(s, t) : -3 \leq s \leq 3, t \neq \pm 1\}$
57. $d(t) = 6\sqrt{25t^2 + 36t + 36}$ **59.** Domain: $\{x : 0 < x < \frac{1}{\pi}\}$; range: $\{y : 0 < y < \frac{1}{\pi}\}$
61. (a) $\sqrt{3}\ x^2/36$ (b) $3\sqrt{3}\ x^2/2$ (c) $3\pi x^3/64$ (d) $(1300 + 240x)/x$ (e) $F(x) = \begin{cases} 180, & 0 \leq x \leq 100 \\ 180 + .22(x - 100), & x > 100 \end{cases}$
63. $S(x, y, z) = (5000/3)(xy^2/z)$; $333\frac{1}{3}$ pounds   **65.** $f(t) = 1 + t$

PROBLEM SET 5-2 (Page 186)

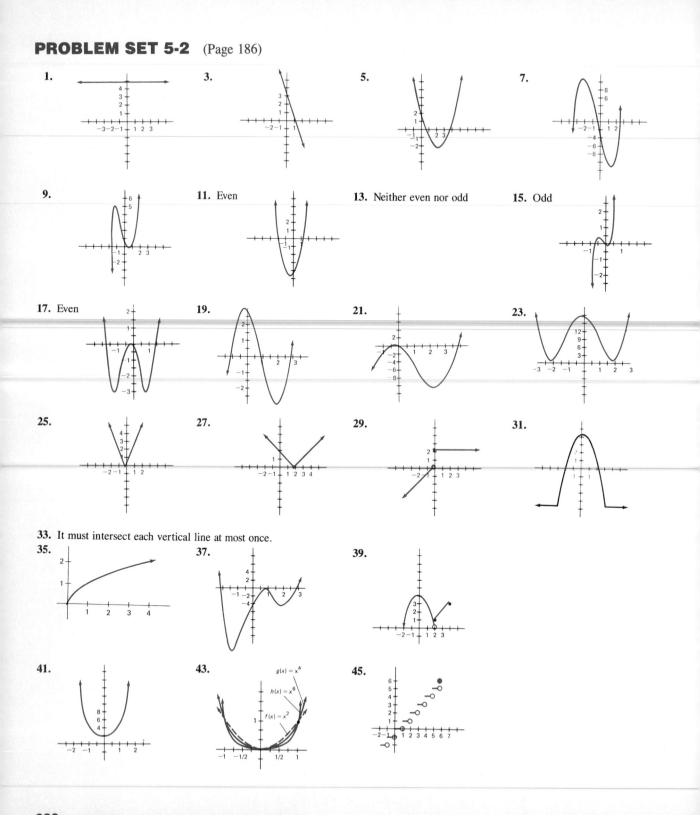

1.

3.

5.

7.

9.

11. Even

13. Neither even nor odd

15. Odd

17. Even

19.

21.

23.

25.

27.

29.

31.

33. It must intersect each vertical line at most once.

35.

37.

39.

41.

43. $g(x) = x^4$ $h(x) = x^6$ $f(x) = x^2$

45.

47. $C(x) = \begin{cases} 15 & \text{if } x < 1 \\ 15 + 10[x] & \text{if } x \geq 1 \end{cases}$

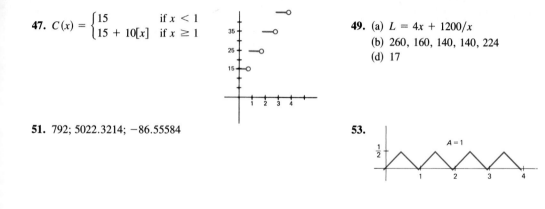

49. (a) $L = 4x + 1200/x$
(b) 260, 160, 140, 140, 224
(d) 17

51. 792; 5022.3214; −86.55584

53.

PROBLEM SET 5-3 (Page 192)

1.

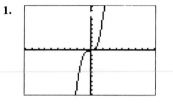

3.

5.

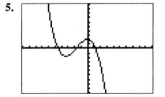

7.

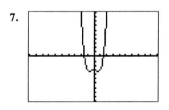

9.

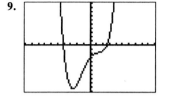

11.

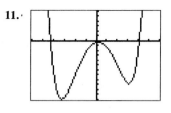

13. (−.393, 1.785)   **15.** (1.488, 3.080)   **17.** (1.940, 7.393)   **19.** (−3.333, −1.704), (10, −298)   **21.** (−3.203, −118.91)
29. 2.279   **31.** (−1, −3), (1, −7)

PROBLEM SET 5-4 (Page 200)

1. **3.** **5.** **7.**

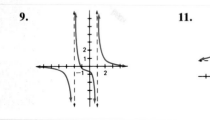 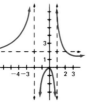 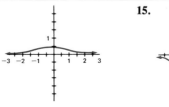 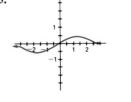

9. **11.** **13.** **15.**

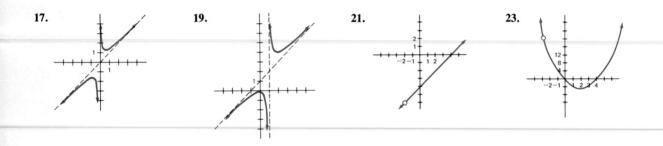

17. **19.** **21.** **23.**

25. **27.** **29.**

31.

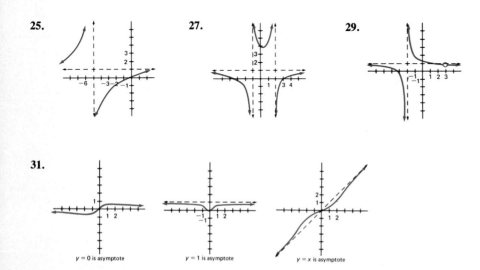

$y = 0$ is asymptote $y = 1$ is asymptote $y = x$ is asymptote

33.

n	Vertical	Horizontal	Oblique
1	$x = 1$	None	None
2	$x = -1, x = 1$	None	None
3	$x = 1$	None	$y = x$
4	$x = -1, x = 1$	$y = 1$	None
5	$x = 1$	$y = 0$	None
6	$x = -1, x = 1$	$y = 0$	None

35.

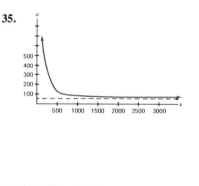

37. (a) $A(t) = 1000t$
(b) $c(t) = 50t/(20 + t)$
(c) $c(t) \to 50$

39. $f(x) = 2x + 3 + 2/(x - 3) = (2x^2 - 3x - 7)/(x - 3)$ **41.** $(2.433, 7.922)$

PROBLEM SET 5-5   (Page 206)

1. (a) 3   (b) Undefined   (c) -1 (d) -10 (e) 2   (f) $\frac{1}{2}$ (g) 10   (h) $\frac{2}{3}$ (i) 3

3. $(f + g)(x) = x^2 + x - 2$, all real numbers; $(f - g)(x) = x^2 - x + 2$, all real numbers; $(f \cdot g)(x) = x^3 - 2x^2$, all real numbers; $(f/g)(x) = x^2/(x - 2)$, $\{x : x \neq 2\}$

5. $(f + g)(x) = x^2 + \sqrt{x}$, $\{x : x \geq 0)\}$ $(f - g)(x) = x^2 - \sqrt{x}$, $\{x : x \geq 0\}$; $(f \cdot g)(x) = x^2\sqrt{x}$, $\{x : x \geq 0\}$; $(f/g)(x) = x^2/\sqrt{x}$, $\{x : x > 0\}$

7. $(f + g)(x) = (x^2 - x - 3)/(x - 2)(x - 3)$, $\{x : x \neq 2$ and $x \neq 3\}$; $(f - g)(x) = (-x^2 + 3x - 3)/(x - 2)(x - 3)$, $\{x : x \neq 2$ and $x \neq 3\}$; $(f \cdot g)(x) = x/(x - 2)(x - 3)$, $\{x : x \neq 2$ and $x \neq 3\}$; $(f/g)(x) = (x - 3)/x(x - 2)$, $\{x : x \neq 0, x \neq 2, x \neq 3\}$

9.

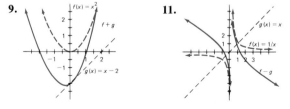

11.

13. $(g \circ f)(x) = x^2 - 2$, all real numbers; $(f \circ g)(x) = (x - 2)^2$, all real numbers

15. $(g \circ f)(x) = (3x + 1)/x$, $\{x : x \neq 0\}$; $(f \circ g)(x) = 1/(x + 3)$, $\{x : x \neq -3\}$

17. $(g \circ f)(x) = x - 4$, $\{x : x \geq 2\}$; $(f \circ g)(x) = \sqrt{x^2 - 4}$, $\{x : |x| \geq 2\}$

19. $(g \circ f)(x) = x$, all real numbers; $(f \circ g)(x) = x$, all real numbers   **21.** $g(x) = x^3; f(x) = x + 4$

23. $g(x) = \sqrt{x}; f(x) = x + 2$ **25.** $g(x) = 1/x^3; f(x) = 2x + 5$ **27.** $g(x) = |x|; f(x) = x^3 - 4$

29.

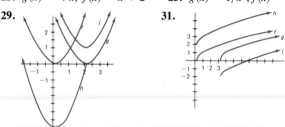

31.

33. (a) $x^3 + 2x + 3$ (b) $x^3 - 2x - 3$ (c) $2x^4 + 3x^3$ (d) $(2x + 3)/x^3$ (e) $2x^3 + 3$ (f) $(2x + 3)^3$ (g) $4x + 9$ (h) x^{27}

35. $1/|x^2 - 4|$; $\{x : x \neq -2$ and $x \neq 2\}$ **37.** (a) $2x + h$ (b) 2   (c) $-1/x(x + h)$ (d) $-2/(x - 2)(x + h - 2)$

39. (a) $x; |x|$ (b) $x; x$ (c) $x^6; x^6$ (d) $1/x^6; 1/x^6$ **41.** $\frac{1}{2}$

43. (a) $x = 3.5t$, $y = 4(t - 1)$ (b) $d = \sqrt{x^2 + y^2}$ **45.**
(c) $d = \sqrt{(3.5t)^2 + 16(t - 1)^2}$ (d) 21.07 miles

47. (a) Odd   (b) Odd   (c) Even   (d) Odd   (e) Neither   (f) Odd   (g) Even   (h) Even   (i) Even

49. (a)   (b) $1; \frac{1}{2}; \frac{1}{4}; \frac{7}{4}$ **51.** (2.481, 1.575)

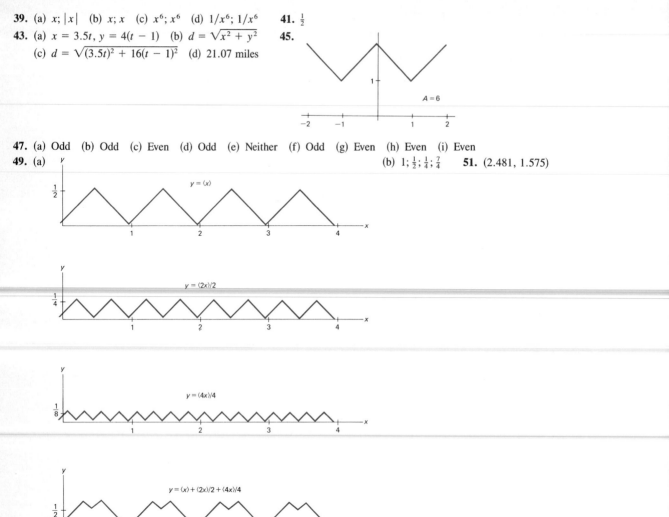

PROBLEM SET 5-6   (Page 215)

1. (a) i, ii; iii; iv; vii; viii   (b) i; ii; viii;   (c) i; ii; viii   **3.** (a) 1   (b) $-\frac{1}{3}$ (c) $\frac{16}{3}$ **5.** $f^{-1}(x) = (\frac{1}{5})x$ **7.** $f^{-1}(x) = (\frac{1}{2})(x + 7)$

9. $f^{-1}(x) = (x - 2)^2$ **11.** $f^{-1}(x) = 3x/(x - 1)$ **13.** $f^{-1}(x) = 2 + \sqrt[3]{(x - 2)}$

15.   **17.** (a)   (b)   (c)

624   ANSWERS TO ODD-NUMBERED PROBLEMS

19. $(f \circ g)(x) = 3\left(\dfrac{2x}{3-x}\right) \Big/ \left(\dfrac{2x}{3-x}+2\right) = \dfrac{6x}{3-x} \Big/ \dfrac{6}{3-x} = x$; $(g \circ f)(x) = 2\left(\dfrac{3x}{x+2}\right) \Big/ \left(3-\dfrac{3x}{x+2}\right) = \dfrac{6x}{x+2} \Big/ \dfrac{6}{x+2} = x$

21. $\{x : x \geq 1\}; f^{-1}(x) = 1 + \sqrt{x}$ **23.** $\{x : x \geq -1\}; f^{-1}(x) = -1 + \sqrt{x+4}$ **25.** $\{x : x \geq -3\}; f^{-1}(x) = -3 + \sqrt{x+2}$

27. $\{x : x \geq -2\}; f^{-1}(x) = x - 2$ **29.** $\{x : x \geq 1\}; f^{-1}(x) = 1 + \sqrt{2x/(1+x)}$

31.

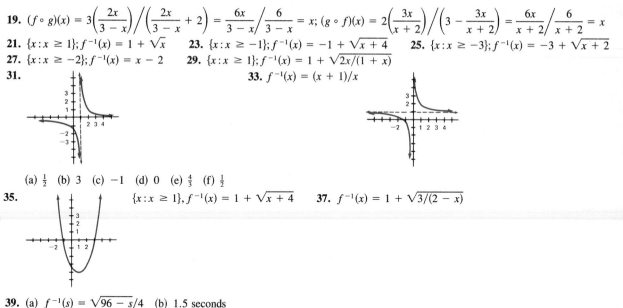

33. $f^{-1}(x) = (x+1)/x$

(a) $\frac{1}{2}$ (b) 3 (c) -1 (d) 0 (e) $\frac{4}{3}$ (f) $\frac{1}{2}$

35.

$\{x : x \geq 1\}, f^{-1}(x) = 1 + \sqrt{x+4}$ **37.** $f^{-1}(x) = 1 + \sqrt{3/(2-x)}$

39. (a) $f^{-1}(s) = \sqrt{96-s}/4$ (b) 1.5 seconds

41. The graph must be symmetric about the line $y = x$. This means the xy-equation determining f is unchanged if x and y are interchanged.

43. (a) $f^{-1}(x) = (b - dx)/(cx - a)$ (b) $(f \circ f^{-1})(x)$ should be x. But $f(f^{-1}(x)) = \left(a\dfrac{b-dx}{cx-a}+b\right) \Big/ \left(c\dfrac{b-dx}{cx-a}+d\right) =$

$(bc - ad)x/(bc - ad) = x$, provided that $bc - ad \neq 0$.   (c) $a = -d$

45. (a) Yes   (b) No   (c) Yes   (d) No

CHAPTER 5. REVIEW PROBLEM SET   (Page 218)

1. T   **2.** F (consider $x = 2$)   **3.** F (Domain $= \{x : x \leq -3 \text{ or } x \geq 4\}$)   **4.** T   **5.** T   **6.** F $[f(f(x)) = x^9]$

7. F ($y = 0$ is a horizontal asymptote)   **8.** T   **9.** T   **10.** F $[f^{-1}(x) = (x+2)^3]$ **11.** No   **12.** Yes   **13.** Yes

14. No   **15.** (a) $\frac{10}{9}$ (b) 3 (c) 96 (d) Undefined (e) 2 (f) -8 **16.** $\{x : x \geq 0, x \neq 1\}$ **17.** $-(2x + h)/[x^2(x + h)^2]$

18. $z = 2x^2/\sqrt[3]{y}$, 180

19.   **20.**   **21.**

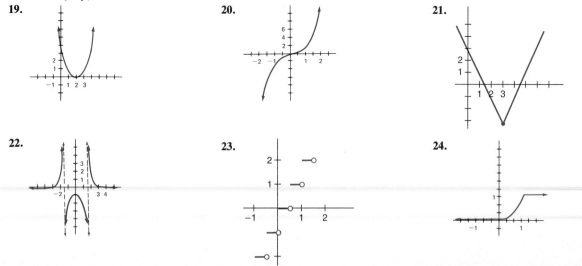

22.   **23.**   **24.**

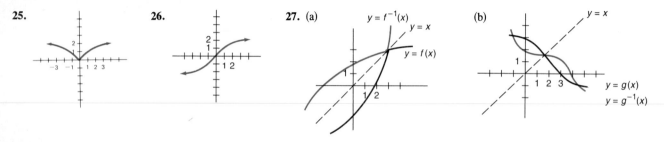

25. **26.** **27.** (a) $y = f^{-1}(x)$ $y = x$ $y = f(x)$ (b) $y = x$ $y = g(x)$ $y = g^{-1}(x)$

28. The graph of f is translated 4 units left. The graph of g is translated 3 units down.

29. The graph is translated 3 units left and 4 units down.

30.

$y = x^2 + \frac{1}{x}$
$y = x^2$
$y = \frac{1}{x}$

31. (a) $f(x + 2) = \sqrt[3]{x} + 3$ (b) $g(f(x)) = x - 2$
     (c) $g^{-1}(x) = \sqrt[3]{x} + 3$ (d) $f^{-1}(x) = (x - 3)^3 + 2$

32. $f^{-1}(x) = 2x - 3$ **33.** $g^{-1}(2) = -4$ **34.** $f^{-1}(x) = 5x/(x - 1)$ **35.** One answer: $f(x) = \sqrt{x}$, $g(x) = x^3 - 7$

36. $(2x + 1)/(x + 2) = 2 \Rightarrow 2x + 1 = 2x + 4 \Rightarrow 1 = 4$ **37.** $\{x : x \ne -2\}$ **38.** $D = \{x : x \ge \frac{5}{2}\}, f^{-1}(x) = (\sqrt{x} + 5)/2$

39. Even   **40.** Even   **41.** Odd   **42.** Neither   **43.** Neither   **44.** Odd

45. $(f \cdot g)(-x) = f(-x) \cdot g(-x) = f(x) \cdot [-g(x)] = -f(x) \cdot g(x) = -(f \cdot g)(x)$
     $(f \circ g)(-x) = f(g(-x)) = f(-g(x)) = f(g(x)) = (f \circ g)(x)$

46. $g(x) = -\sqrt{25 - x^2}$ **47.** (a) $S = \sqrt{9 + x^2} + 12 - x$ (b) $T = \sqrt{9 + x^2}/2.5 + (12 - x)/3.5$

48. (a) $V = \frac{5}{3}\pi x^3$ (b) $A = 5\pi x^2$

PROBLEM SET 6-1 (Page 225)

1. 3   **3.** 2   **5.** 7   **7.** $\frac{3}{2}$ **9.** 25   **11.** 9   **13.** 2   **15.** $\frac{1}{100}$ **17.** $\sqrt{2}/2$ **19.** $\sqrt{5}$ **21.** $3xy\sqrt[3]{2xy^2}$

23. $(x + 2)y\sqrt[4]{y^3}$ **25.** $x\sqrt{1 + y^2}$ **27.** $x\sqrt[3]{x^3 - 9y}$ **29.** $(xz^2/y^2)\sqrt[3]{x}$ **31.** $2(\sqrt{x} - 3)/(x - 9)$ **33.** $2\sqrt{x + 3}/(x + 3)$

35. $\sqrt[4]{2x}/2x$ **37.** $2(\sqrt{3} + \sqrt{2})$ **39.** $(2y/x)\sqrt[3]{x^2}$ **41.** $\sqrt{2}$ **43.** $3\sqrt[3]{3}$ **45.** $-9 + 4\sqrt{5}$ **47.** $\sqrt[4]{4ab}/2b^2$ **49.** 26

51. $\frac{9}{2}$ **53.** $-\frac{32}{15}$ **55.** 0   **57.** 4   **59.** $0, \frac{16}{9}$ **61.** $2[(\sqrt{x} - \sqrt{x + h})/\sqrt{x}\sqrt{x + h}]$ **63.** $(x + 7)/\sqrt{x + 6}$

65. $x/(2\sqrt[3]{x} + 2)$ **67.** $-9/(x^2\sqrt{x^2 + 9})$

69. (a) $2ab^2$ (b) $9b\sqrt{b}$ (c) $3\sqrt{3}$ (d) $5a^2b^3\sqrt{10}$ (e) $-2y^2/x$ (f) $y^2/4x^2$
     (g) $(2a^2 + 3a)\sqrt{2a}$ (h) $4\sqrt[4]{2} - 5\sqrt{2} + 2\sqrt[6]{2}$ (i) $a\sqrt[4]{1 + b^4}$ (j) $\sqrt[3]{49b^2}/(7bc)$ (k) $2(\sqrt{a} + b)/(a - b^2)$ (l) $a + \frac{1}{a} = (a^2 + 1)/a$

71. (a) 13   (b) 0; 2   (c) 6   (d) No solution   (e) 9   (f) -8; 64

73. They are reflections of each other in the line $y = x$.

$f(x) = x^5$
$g(x) = \sqrt[5]{x}$

75. $\overline{AC} = \frac{15}{4}$ **77.** $\sqrt{7}$

79. (a) Both sides are positive. $[(\sqrt{6} + \sqrt{2})/2]^2 = (6 + 2\sqrt{12} + 2)/4 = 2 + \sqrt{3} = (\sqrt{2 + \sqrt{3}})^2$.

(b) Both sides are positive. $(\sqrt{2 + \sqrt{3}} + \sqrt{2 - \sqrt{3}})^2 = 2 + \sqrt{3} + 2\sqrt{1} + 2 - \sqrt{3} = 6$.

(c) $(\sqrt{3} - \sqrt{2})^3 = 3\sqrt{3} - 3 \cdot 3\sqrt{2} + 3\sqrt{3} \cdot 2 - 2\sqrt{2} = 9\sqrt{3} - 11\sqrt{2} = (\sqrt[3]{9\sqrt{3} - 11\sqrt{2}})^3$.

81. 2.167

PROBLEM SET 6-2 (Page 232)

1. $7^{1/3}$ **3.** $7^{2/3}$ **5.** $7^{-1/3}$ **7.** $7^{-2/3}$ **9.** $7^{4/3}$ **11.** $x^{2/3}$ **13.** $x^{5/2}$ **15.** $(x + y)^{3/2}$ **17.** $(x^2 + y^2)^{1/2}$

19. $\sqrt[3]{16}$ **21.** $1/\sqrt{8^3} = \sqrt{2}/32$ **23.** $\sqrt[4]{x^4 + y^4}$ **25.** $y\sqrt[5]{x^4y}$ **27.** $\sqrt{\sqrt{x} + \sqrt{y}}$ **29.** 5 **31.** 4 **33.** $\frac{1}{27}$

35. .04 **37.** .000125 **39.** $\frac{1}{5}$ **41.** $\frac{1}{16}$ **43.** $\frac{1}{16}$ **45.** $-6a^2$ **47.** $8/x^4$ **49.** x^4 **51.** $4y^2/x^4$ **53.** y^9/x^{30}

55. $(2y^3 - 1)/y$ **57.** $x + y + 2\sqrt{xy}$ **59.** $(7x + 2)/3(x + 2)^{1/5}$ **61.** $(1 - x^2)/(x^2 + 1)^{2/3}$ **63.** $\sqrt[6]{32}$ **65.** $\sqrt[12]{8x^2}$

67. $\sqrt{x}$ **69.** 2.53151 **71.** 4.6364 **73.** 1.70777 **75.** .0050463

77. **79.** **81.**

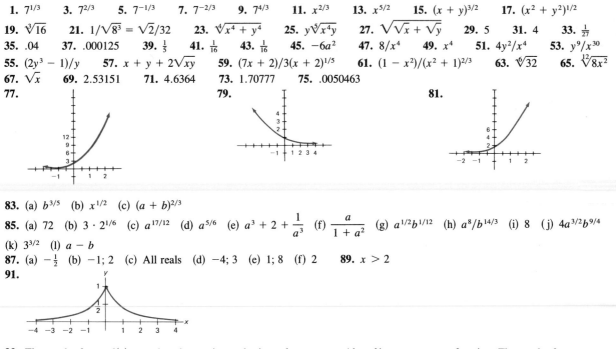

83. (a) $b^{3/5}$ (b) $x^{1/2}$ (c) $(a + b)^{2/3}$

85. (a) 72 (b) $3 \cdot 2^{1/6}$ (c) $a^{17/12}$ (d) $a^{5/6}$ (e) $a^3 + 2 + \dfrac{1}{a^3}$ (f) $\dfrac{a}{1 + a^2}$ (g) $a^{1/2}b^{1/12}$ (h) $a^8/b^{14/3}$ (i) 8 (j) $4a^{3/2}b^{9/4}$

(k) $3^{3/2}$ (l) $a - b$

87. (a) $-\frac{1}{2}$ (b) $-1; 2$ (c) All reals (d) $-4; 3$ (e) $1; 8$ (f) 2 **89.** $x > 2$

91.

93. The graph of $y = f(x) = a^x$ has the x-axis as a horizontal asymptote. Also, f is not a constant function. The graph of a nonconstant polynomial does not have a horizontal asymptote. **95.** 2.382

PROBLEM SET 6-3 (Page 239)

1. (a) Decays (b) Grows (c) Grows (d) Decays **3.** (a) 4.66095714 (b) 17.00006441 (c) 4801.02063 (d) 9750.87832

5. 1480 **7.** (a) 9.751 billion (b) 23.772 billion **9.** $56,000(.98)^{10} \approx 45,760$ **11.** $50(.92)^{15} \approx 14.3$ kg **13.** 8.99 g

15. .298 **17.** 590 years **19.** 15.8 days **21.** (a) $185.09 (b) $247.60 **23.** (a) $76,035.83 (b) $325.678.40

25. $p(1 + r/100)^n$ **27.** $7401.22 **29.** $7102.09 **31.** $7305.57 **33.** $1000(1 + .08/12)^{120} = $2219.64

35. 8100; 5400; 3600; 2400; 1600 **37.** 800 **39.** (a) 8680; 22,497 (b) About 44 years

41. (a) $146.93 (b) $148.59 (c) $148.98 (d) $149.18 **43.** (a) About 9 years. (b) About 11 years

45. (a) $k \approx .0005917$ (b) 10.76 milligrams **47.** 2270 years **49.** $320,057,300 **51.** 80

PROBLEM SET 6-4 (Page 247)

1. $\log_4 64 = 3$ **3.** $\log_{27} 3 = \frac{1}{3}$ **5.** $\log_4 1 = 0$ **7.** $\log_{125}(1/25) = -\frac{2}{3}$ **9.** $\log_{10} a = \sqrt{3}$ **11.** $\log_{10} \sqrt{3} = a$

13. $5^4 = 625$ **15.** $4^{3/2} = 8$ **17.** $10^{-2} = .01$ **19.** $c^1 = c$ **21.** $c^y = Q$ **23.** 2 **25.** -1 **27.** 1/3 **29.** -4

31. 0 **33.** $\frac{4}{3}$ **35.** 2 **37.** $\frac{1}{27}$ **39.** -2.9 **41.** 49 **43.** .778 **45.** 1.204 **47.** $-.602$ **49.** 1.380

51. $-.051$ **53.** .699 **55.** 1.5314789 **57.** -1.9100949 **59.** 3.9878003 **61.** $\log_{10}[(x + 1)^3(4x + 7)]$

63. $\log_2[8x(x + 2)^3/(x + 8)^2]$ **65.** $\log_6(\sqrt{x}\sqrt[3]{x^3 + 3})$ **67.** 47 **69.** $-\frac{11}{4}$ **71.** $\frac{16}{7}$ **73.** 5; 2 is extraneous. **75.** 7

77. 4.0111687 **79.** 2.0446727 **81.** (a) 2 (b) $\frac{1}{2}$ (c) 125 (d) 32 (e) 10 (f) 4 **83.** $\frac{8}{9}$

85. (a) 13 (b) -5 (c) No solution (d) 20 (e) 3 (f) 16 **87.** (a) $y = x/(x - 1)$ (b) $y = \frac{1}{2}(a^x + a^{-x})$

89. (a) $A = C \times 10^M$ (b) $10^{1.8} \approx 63$ times as strong

91. Since $a = b^{\log_b a}$, $\log_a a = \log_b a \cdot \log_a b$, or $1 = \log_b a \cdot \log_a b$. Therefore, $\log_a b = \dfrac{1}{\log_b a}$.

93.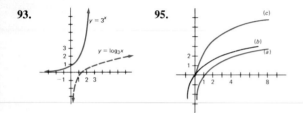

95.

97. (a) They are inverse functions.

(b) $f(g(x)) = \frac{1}{2}[x + \sqrt{x^2 + 1} - 1/(x + \sqrt{x^2 + 1})] = \frac{1}{2}(2x^2 + 2x\sqrt{x^2 + 1})/(x + \sqrt{x^2 + 1}) = x$

PROBLEM SET 6-5 (Page 255)

1. 1 **3.** 0 **5.** $\frac{1}{2}$ **7.** −3 **9.** 25 **11.** 3.5 **13.** −2 **15.** −7.5 **17.** 1.4609379 **19.** −2.0635682
21. −1.8411881 **23.** .50833303 **25.** 11.818915 **27.** 61.146135 **29.** 8.3311375 **31.** .8824969 **33.** 915.98
35. About 2.73 **37.** About −0.737 **39.** About 6.84 **41.** Approximately 6.12 years **43.** Approximately 4.71 years
45. About 55.45 years **47.** (a) 5^{10} (b) 9^{10} (c) 10^{20} (d) 10^{1000} **49.** 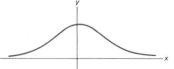 **51.** $y = ba^x$; $a \approx 1.5$, $b \approx 64$

53. $y = bx^a$; $a \approx 4$, $b \approx 12$ **55.** (a) 4.2 (b) 4 (c) $\frac{1}{2}$ **57.** (a) 7 (b) 12.25 (c) $-\frac{1}{2}$ (d) 0 (e) 125 (f) $\frac{1}{3}$
59. (a) −.349 (b) −.823 (c) .633, −3.633 (d) 2.166 (e) 4.560; .219 (f) $e^e \approx 15.154$ (g) $e \approx 2.718$; 1 (h) $e \approx 2.718$
(i) $\pm\sqrt{e + 5} \approx \pm 2.778$; 1
61. $(\ln 2)/3 = .231$ years **63.** $(\ln 2)/240 \approx .00289$
65.

67. $e^{\pi/e-1} > 1 + \pi/e - 1$, $e^{\pi/e}/e > \pi/e$, $e^{\pi/e} > \pi$, $e^\pi > \pi^e$
69. (a) $100(1 + .01)^{120} \approx \$330.04$ (b) $100(1 + .12/365)^{3650} \approx \$331.95$ (c) $100(1 + .12/(365)(24))^{(3650)(24)} \approx \$332.01$
(d) $100e^{(.12)(10)} \approx \$332.01$ **71.** 647.278

CHAPTER 6. REVIEW PROBLEM SET (Page 258)

1. T **2.** F ($\sqrt[4]{16} = 2$) **3.** F ($\sqrt{5}\sqrt[3]{5} = 5^{1/2}5^{1/3} = 5^{5/6} = \sqrt[6]{5^5}$) **4.** T **5.** T **6.** T **7.** T
8. F ($\log_2[(-2)^2 - 2] = \log_2 2 = 1$) **9.** T **10.** F (negative numbers are in the domain of f) **11.** $-3y^2\sqrt[3]{z^2}/z^5$
12. $2x^2yz\sqrt[4]{2yz^4}$ **13.** $\sqrt{5}$ **14.** $(10\sqrt{x} + 15)/(4x - 9)$ **15.** $3\sqrt{9 + x^2}$ **16.** $2\sqrt{13}$ **17.** 43 **18.** 16 **19.** x^2
20. a^2b **21.** $y^2/(xz^3)$ **22.** $6x^{2/3}$ **23.** $27a^{1/4}$ **24.** $2^{5/4}y^{3/4}/x^2$ **25.** They are reflections in the y-axis. **26.** 6,400,000
27. 3200 years **28.** (a) $220 (b) $310.58 (c) $330.04 **29.** $2 = (1 + .085/4)^{4n}$, n is the number of years.
30. $100(1 + .09/365)^{730}$ **31.** 64 **32.** $\frac{1}{125}$ **33.** 3 **34.** π **35.** 2 **36.** $\frac{5}{3}$ **37.** 16 **38.** 1 **39.** ± 3 **40.** 2
41. $\log_4[16(x^2 + 1)^3/x^2]$ **42.** It is translated 1 unit right and 2 units up.

43.

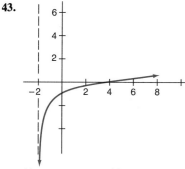

44. $\frac{3}{2}$ **45.** .86342   **46.** About 5.6 years

47. (a) 12.603 years (b) 41.865 years   **48.** .06931   **49.** They are reflections in the line $y = x$.   **50.** 4.277

PROBLEM SET 7-1   (Page 265)

1. .6600   **3.** .6534   **5.** 3.133   **7.** 12.5°   **9.** 66.6°   **11.** 69.3°   **13.** 16.97 ≈ 17   **15.** 41.34 ≈ 41
17. 66.60 ≈ 67   **19.** $\beta = 48°; a = 23.42 \approx 23; b = 26.01 \approx 26$ **21.** $\alpha = 33.8°; a = 50.8; b = 75.9$
23. $\beta = 50.6°; b = 146; c = 189$ **25.** $c = 15; \alpha = 36.9°; \beta = 53.1°$ **27.** $b = 30; \alpha = 53.1°; \beta = 36.9°$
29. $\alpha = 26.7°; \beta = 63.3°; b = 29.0$ **31.** $\alpha = 32.9°; \beta = 57.1°; c = 17.5$ **33.** 14.6°   **35.** 7.0°   **37.** 31.2 feet
39. (a) .25862 (b) 6.6568 (c) 17.445   **41.** 725 feet   **43.** 37.939   **45.** 364 feet   **47.** 448 meters   **49.** 41.77°
51. (a) 76.75 miles (b) 12.03°   **53.** $P = 24, A = 24\sqrt{3}$ **55.** $\sqrt{3}$

PROBLEM SET 7-2   (Page 271)

1. $2\pi/3$ **3.** $4\pi/3$ **5.** $7\pi/6$ **7.** $7\pi/4$ **9.** 3π **11.** $-7\pi/3$ **13.** $8\pi/9$ **15.** $\frac{1}{9}$ **17.** 240°   **19.** $-120°$
21. 540°   **23.** 259.0°   **25.** 18.2°   **27.** (a) 2 (b) 3.14   **29.** (a) 3 centimeters (b) 5.5 inches
31. (a) 12.6 square centimeters (b) 90.75 square inches   **33.** II   **35.** III   **37.** II   **39.** IV
41. $16\pi/3 \approx 16.76$ feet per second   **43.** $320\pi \approx 1005.3$ inches per minute   **45.** (a) -8π (b) 5π (c) $\frac{1}{3}$
47. (a) 25.5 centimeters (b) .0327 centimeter (c) 37.83 centimeters   **49.** $9600\pi \approx 30,159$ centimeters
51. $330\pi \approx 1037$ miles per hour   **53.** $\frac{4}{3}A$ **55.** 8.6×10^5 miles   **57.** $33\pi/20 \approx 5.184$ hours   **59.** $264\pi \approx 829$ miles
61. $7\pi \approx 21.99$ square inches   **63.** 130

PROBLEM SET 7-3   (Page 279)

1. $(\sqrt{3}/2, \frac{1}{2})$ **3.** $(-\sqrt{2}/2, \sqrt{2}/2)$ **5.** $(-\sqrt{2}/2, -\sqrt{2}/2)$ **7.** $(-\sqrt{3}/2, \frac{1}{2})$ **9.** $-\sqrt{2}/2$ **11.** $\sqrt{2}/2$
13. $-\sqrt{2}/2$ **15.** $-\frac{1}{2}$ **17.** 1   **19.** 0   **21.** $-\sqrt{3}/2$ **23.** $\frac{1}{2}$ **25.** $-\sqrt{2}/2$ **27.** $\frac{1}{2}$ **29.** $-\frac{1}{2}$ **31.** $-\sqrt{3}/2$
33. $-.95557; -.29476$ **35.** (a) $(1/\sqrt{5}, 2/\sqrt{5})$ (b) $2/\sqrt{5}, 1/\sqrt{5}$
37. (a) $\sin(\pi + t) = -y = -\sin t$ (b) $\cos(\pi + t) = -x = -\cos t$ **39.** $-\frac{4}{5}$ **41.** $\frac{5}{13}$
43. (a) Negative (b) Negative (c) Positive (d) Positive (e) Positive (f) Negative
45. (a) $\pm\sqrt{3}/2$ (b) $7\pi/6, 11\pi/6$ **47.** 1.98289
49. (a) $\pi/4; 5\pi/4$ (b) $\pi/6 < t < \pi/3, 2\pi/3 < t < 5\pi/6$ (c) $0 \le t \le \pi/3; 2\pi/3 \le t \le 4\pi/3; 5\pi/3 \le t < 2\pi$
(d) $0 \le t < \pi/4; 3\pi/4 < t < 5\pi/4; 7\pi/4 < t < 2\pi$
51. (a) $\frac{3}{5}$ (b) $-\frac{7}{25}$ **53.** (a) $\frac{3}{5}$ (b) $\frac{4}{5}$ (c) $\frac{4}{5}$ (d) $\frac{4}{5}$ (e) $\frac{3}{5}$ (f) $-\frac{4}{5}$
55. (a) Period 1 (b) Period $\frac{1}{3}$ (c) Not periodic (d) Period 1   **57.** 0

PROBLEM SET 7-4   (Page 284)

1. (a) $-\frac{4}{3}$ (b) $-\frac{3}{4}$ (c) $-\frac{5}{3}$ (d) $\frac{5}{4}$ **3.** $-\sqrt{5}/2; \frac{3}{5}\sqrt{5}$ **5.** $\sqrt{3}/3$ **7.** $2\sqrt{3}/3$ **9.** 1   **11.** $2\sqrt{3}/3$ **13.** $-\sqrt{3}/2$
15. $\sqrt{3}$ **17.** 0   **19.** $-\sqrt{3}/3$ **21.** -2

23. (a) $\pi/2$; $3\pi/2$; $5\pi/2$; $7\pi/2$ (b) $\pi/2$; $3\pi/2$; $5\pi/2$; $7\pi/2$ (c) 0; π; 2π; 3π; 4π (d) 0; π; 2π; 3π; 4π
25. (a) $\sqrt{7}$ (b) $-\sqrt{42}/7$ (c) $-\sqrt{6}/6$ **27.** (a) $2\sqrt{2}$ (b) $-\sqrt{2}/4$ (c) $-\frac{1}{3}$ **29.** $-12/13$; $-12/5$; $13/5$
31. $-2\sqrt{5}/5$; 2; $-\sqrt{5}$ **33.** $\frac{3}{5}$; $\frac{5}{4}$ **35.** $-\frac{12}{5}$; $-\frac{12}{5}$ **37.** $(\frac{5}{13}, -\frac{12}{13})$ **39.** $\sqrt{3}x + 3y = 12 - 3\sqrt{3}$ **41.** $111.8°$
43. (a) $-2\sqrt{3}/3$ (b) $\sqrt{3}$ (c) $\sqrt{2}$ (d) -1 (e) -2 (f) 1 **45.** (a) $\frac{24}{25}$ (b) $\frac{-7}{25}$ (c) $\frac{-24}{7}$ (d) $\frac{25}{24}$ (e) $\frac{-24}{7}$ (f) $\frac{-25}{7}$
47. (a) $3\pi/4$; $7\pi/4$ (b) $\pi/4$; $7\pi/4$ (c) $\pi/2$; $3\pi/2$
49. (a) 1 (b) $\sin\theta - 1/\cos\theta$ (c) $1 + 2\sin\theta\cos\theta$ (d) $1/\sin\theta$ (e) $\cos\theta + \sin\theta$ (f) $-(\sin^2\theta + 1)/(\cos^2\theta)$
51. (a) $\tan(t + \pi) = \sin(t + \pi)/\cos(t + \pi) = (-\sin t)/(-\cos t) = \tan t$
(b) $\cot(t + \pi) = \cos(t + \pi)/\sin(t + \pi) = (-\cos t)/(-\sin t) = \cot t$
(c) $\sec(t + \pi) = 1/\cos(t + \pi) = 1/(-\cos t) = -1/(\cos t) = -\sec t$
(d) $\csc(t + \pi) = 1/\sin(t + \pi) = 1/(-\sin t) = -1/(\sin t) = -\csc t$
53. $\frac{119}{169}$ **55.** -10 **57.** 10.47198 **59.** (a) -1.3764 (b) $.3153$ **61.** 428.98 centimeters

PROBLEM SET 7-5 (Page 290)

1. $.98185$ **3.** $.7337$ **5.** $.93309$ **7.** $.9291$ **9.** 1.30 **11.** $.40$ **13.** 1.10 **15.** 1.06 **17.** 1.12 **19.** $.50$
21. $3\pi/8$ **23.** $\pi/3$ **25.** $.24$ **27.** $.24$ **29.** $\pi/2$ **31.** $.15023$ **33.** 5.4707 **35.** $.84147$ **37.** -1.2885
39. $-.82534$ **41.** 1.25; 1.89 **43.** 1.65; 4.63 **45.** 1.37; 4.51 **47.** 1.84; 4.98 **49.** $40.4°$ **51.** $11.3°$ **53.** $80.2°$
55. $.4051$ **57.** $-.1962$ **59.** $.4051$ **61.** $.15126$ **63.** $.9657$ **65.** $21.3°$; $158.7°$ **67.** $26.3°$; $206.3°$
69. $155.3°$; $204.7°$ **71.** (a) $.79608$ (b) $-.79560$ (c) -1.5574 (d) $-.7513$ (e) -1.2349 (f) $-.9877$
73. (a) $.9999997$ (b) $.744399$ (c) 1.2338651 **75.** (a) $.679996$; 2.461597 (b) 1.222007; 5.061178 (c) 1.878966; 5.020558
77. (a) ϕ (b) $90° - \phi$ (c) $90° - \phi$ **79.** $-.514496$ **81.** $\theta \approx 289.47°$ **83.** $\phi \approx 126.9°$

PROBLEM SET 7-6 (Page 298)

1.

t	0	$\frac{\pi}{6}$	$\frac{\pi}{4}$	$\frac{\pi}{3}$	$\frac{\pi}{2}$	$\frac{3\pi}{4}$	π	$\frac{5\pi}{4}$	$\frac{3\pi}{2}$	$\frac{7\pi}{4}$	2π
$\cos t$	1	$\frac{\sqrt{3}}{2}$	$\frac{\sqrt{2}}{2}$	$\frac{1}{2}$	0	$-\frac{\sqrt{2}}{2}$	-1	$-\frac{\sqrt{2}}{2}$	0	$\frac{\sqrt{2}}{2}$	1

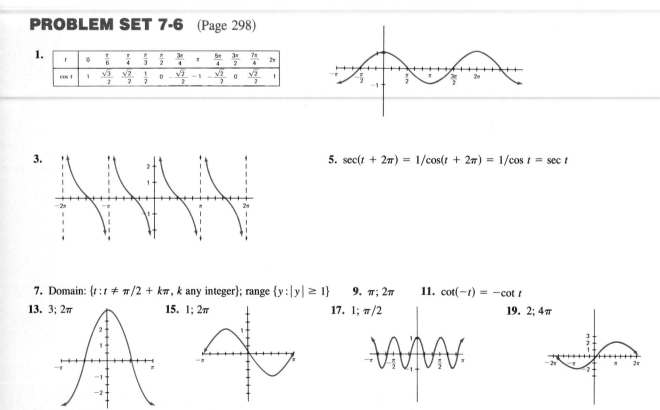

3.

5. $\sec(t + 2\pi) = 1/\cos(t + 2\pi) = 1/\cos t = \sec t$

7. Domain: $\{t : t \neq \pi/2 + k\pi, k \text{ any integer}\}$; range $\{y : |y| \geq 1\}$ **9.** π; 2π **11.** $\cot(-t) = -\cot t$
13. 3; 2π **15.** 1; 2π **17.** 1; $\pi/2$ **19.** 2; 4π

21. 2; $2\pi/3$

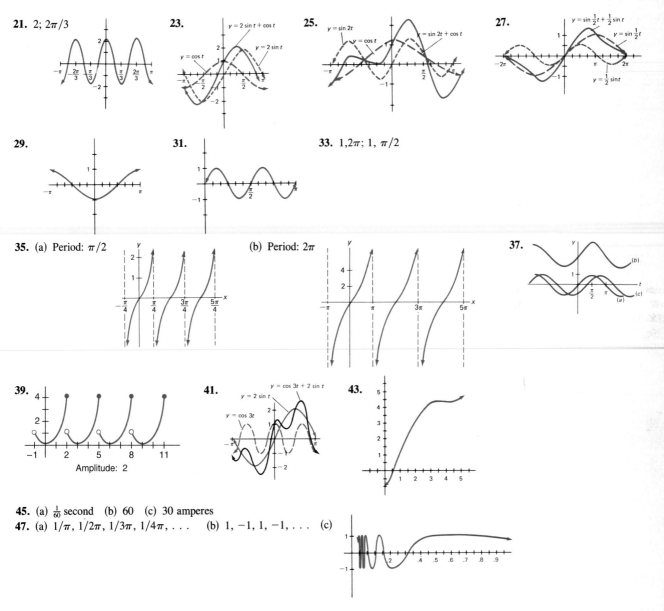

23.

25.

27.

29.

31.

33. $1, 2\pi$; 1, $\pi/2$

35. (a) Period: $\pi/2$ (b) Period: 2π

37.

39.

Amplitude: 2

41.

43.

45. (a) $\frac{1}{60}$ second   (b) 60   (c) 30 amperes
47. (a) $1/\pi$, $1/2\pi$, $1/3\pi$, $1/4\pi$, . . .   (b) 1, -1, 1, -1, . . .   (c)

49. Period: 2π; amplitude: 2.736   **51.** 0, 2.3169

CHAPTER 7. REVIEW PROBLEM SET   (Page 300)

1. T   **2.** T   **3.** F ($\sin 180° = 0$ but $\cos 180° = -1$)   **4.** F [$\cos(-t) = \cos t$]   **5.** F ($\sin \beta = -\sin \alpha$)   **6.** T
7. F (the reference angle is 84°)   **8.** T   **9.** F (the amplitude is 2)   **10.** T   **11.** $b = 12$, $\alpha = 36.87°$, $\beta = 53.13°$
12. $\beta = 17.6°$, $a = 93.3$, $c = 97.9$ **13.** -1 **14.** $\sqrt{3}/3$ **15.** 2   **16.** -1 **17.** $\sqrt{2}/2$ **18.** -1 **19.** -1
20. -1 **21.** $2\pi/3$, $4\pi/3$ **22.** $3\pi/4$, $7\pi/4$ **23.** $\pi/4$, $3\pi/4$ **24.** $\pi/3$, $5\pi/3$ **25.** $-3\sqrt{10}/20$ **26.** $-60°$, $60°$
27. (a) $\frac{3}{5}$ (b) $-\frac{3}{5}$ (c) $-\frac{3}{5}$ (d) $\frac{4}{5}$ **28.** (a) $-\sqrt{1 - \sin^2 \theta}$ (b) $1/\sin \theta$ (c) $-\sin \theta/\sqrt{1 - \sin^2 \theta}$ **29.** (a) $-\frac{12}{5}$ (b) $\frac{13}{12}$

30. (a) $0 < t < \pi$ (b) $0 < t < \pi/2$ and $\pi < t < 3\pi/2$ **31.** $-5 \le y \le 1$
32. $\tan(-t) = \sin(-t)/\cos(-t) = -\sin t/\cos t = -\tan t$ **33.** $a = 2, b = 8 - 2\sqrt{3}$ **34.** (a) $6, -2$ (b) $7\pi/6, 11\pi/6$
35. (a) $\cos t$ (b) $-\cos t$ (c) $1 - \cos^2 t$ (d) $-\cos t$ **36.** $\frac{25}{16} + \frac{27}{125} = 1.7785$ **37.** $A = \frac{1}{2}ab = \frac{1}{2}a \cdot a \cot \alpha = \frac{1}{2}a^2 \cot \alpha$
38. **39.** **40.** $\pi/4, 5\pi/4$

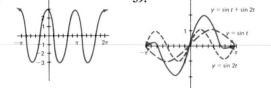

y = sin t + sin 2t
y = sin t
y = sin 2t

41. (a) $50\pi/36 \approx 4.36$ feet per second (b) $2000\pi/12 \approx 523.6$ feet **42.** About 387 **43.** $16 \sin 55° \approx 13.11$ feet
44. .76 feet **45.** $x = 1.10725, y = 2.34386$ **46.** (a) $50\pi \approx 157.08$ meters (b) $1000 \sin 9° \approx 156.43$ meters
47. $\sin(t/2)$ increases from 0 to 1 on $0 \le t \le \pi$, then decreases from 1 to 0 on $\pi \le t \le 2\pi$. **48.** 4
49. $192 \sin 22.5° \approx 73.4752$ centimeters **50.** $A \approx 1,459,380$ square feet

PROBLEM SET 8-1 (Page 307)

1. (a) $1 - \sin^2 t$ (b) $\sin t$ (c) $\sin^2 t$ (d) $(1 - \sin^2 t)/\sin^2 t$ **3.** (a) $1/\tan^2 t$ (b) $1 + \tan^2 t$ (c) $\tan t$ (d) 3
5. $\cos t \sec t = \cos t (1/\cos t) = 1$ **7.** $\tan x \cot x = \tan x (1/\tan x) = 1$ **9.** $\cos y \csc y = \cos y (1/\sin y) = \cot y$
11. $\cot \theta \sin \theta = (\cos \theta/\sin \theta) \sin \theta = \cos \theta$ **13.** $\tan u/\sin u = (\sin u/\cos u)(1/\sin u) = 1/\cos u$
15. $(1 + \sin z)(1 - \sin z) = 1 - \sin^2 z = \cos^2 z = 1/\sec^2 z$ **17.** $(1 - \sin^2 x)(1 + \tan^2 x) = \cos^2 x \sec^2 x = \cos^2 x(1/\cos^2 x) = 1$
19. $\sec t - \sin t \tan t = 1/(\cos t) - (\sin^2 t)/(\cos t) = (1 - \sin^2 t)/(\cos t) = (\cos^2 t)/(\cos t) = \cos t$
21. $(\sec^2 t - 1)/(\sec^2 t) = 1 - 1/(\sec^2 t) = 1 - \cos^2 t = \sin^2 t$
23. $\cos t(\tan t + \cot t) = \sin t + (\cos^2 t)/(\sin t) = (\sin^2 t + \cos^2 t)/(\sin t) = \csc t$
25. $\sec \theta - \cos \theta = 1/(\cos \theta) - \cos \theta = (1 - \cos^2 \theta)/(\cos \theta) = (\sin^2 \theta)/(\cos \theta)$
27. $(\tan \theta - \cot \theta)/(\sin \theta \cos \theta) = (\sin \theta/\cos \theta - \cos \theta/\sin \theta)/(\sin \theta \cos \theta) = (\sin^2 \theta - \cos^2 \theta)/(\sin^2 \theta \cos^2 \theta) = 1/\cos^2 \theta - 1/\sin^2 \theta = \sec^2 \theta - \csc^2 \theta$
29. $\dfrac{\sec t - 1}{\tan t} \cdot \dfrac{\sec t + 1}{\sec t + 1} = \dfrac{\sec^2 t - 1}{\tan t(\sec t + 1)} = \dfrac{\tan^2 t}{\tan t(\sec t + 1)} = \dfrac{\tan t}{\sec t + 1}$
31. $\dfrac{\tan^2 x}{\sec x + 1} = \dfrac{\sec^2 x - 1}{\sec x + 1} = \dfrac{(\sec x - 1)(\sec x + 1)}{\sec x + 1} = \sec x - 1 = \dfrac{1 - \cos x}{\cos x}$
33. $\dfrac{\sin t + \cos t}{\tan^2 t - 1} \cdot \dfrac{\cos^2 t}{\cos^2 t} = \dfrac{(\sin t + \cos t)\cos^2 t}{\sin^2 t - \cos^2 t} = \dfrac{\cos^2 t}{\sin t - \cos t}$
35. $\cot \theta + \tan \theta = \cos \theta/\sin \theta + \sin \theta/\cos \theta = (\cos^2 \theta + \sin^2 \theta)/(\sin \theta \cos \theta) = 1/(\cos \theta \sin \theta) = \sec \theta \csc \theta$
37. $\sin t = (1 - \cos^2 t)^{1/2}$; $\tan t = (1 - \cos^2 t)^{1/2}/\cos t$; $\cot t = \cos t/(1 - \cos^2 t)^{1/2}$; $\sec t = 1/\cos t$; $\csc t = 1/(1 - \cos^2 t)^{1/2}$
39. $\cos t = -3/5$; $\tan t = -4/3$; $\cot t = -3/4$; $\sec t = -5/3$; $\csc t = 5/4$ **41.** (a) $2/\sin^2 x$ (b) $(2 + 2 \tan^2 x)/\tan^2 x$
43. $(1 + \tan^2 t)(\cos t + \sin t) = \sec^2 t \cos t + \sec^2 t \sin t = \sec t + \sec t \tan t = \sec t(1 + \tan t)$
45. $2 \sec^2 y - 1 = \dfrac{2}{\cos^2 y} - 1 = \dfrac{2 - \cos^2 y}{\cos^2 y} = \dfrac{1 + \sin^2 y}{\cos^2 y}$
47. $\dfrac{\sin z}{\sin z + \tan z} = \dfrac{\sin z}{\sin z + \sin z/\cos z} = \dfrac{1}{1 + 1/\cos z} = \dfrac{\cos z}{\cos z + 1}$
49. $(\csc t + \cot t)^2 = \left(\dfrac{1}{\sin t} + \dfrac{\cos t}{\sin t}\right)^2 = \dfrac{(1 + \cos t)^2}{1 - \cos^2 t} = \dfrac{1 + \cos t}{1 - \cos t}$

51. $(\cos^4 u - \sin^4 u)/(\cos u - \sin u) = (\cos^2 u + \sin^2 u)(\cos^2 u - \sin^2 u)/(\cos u - \sin u) =$
$(1)(\cos u - \sin u)(\cos u + \sin u)/(\cos u - \sin u) = \cos u + \sin u$
53. $\dfrac{1 + \tan x}{1 - \tan x} = \dfrac{1 \sin x/\cos x}{1 - \sin x/\cos x} = \dfrac{\cos x + \sin x}{\cos x - \sin x}$ **55.** $(\sec t + \tan t)(\csc t - 1) = \left(\dfrac{1 + \sin t}{\cos t}\right)\left(\dfrac{1 - \sin t}{\sin t}\right) = \dfrac{\cos^2 t}{\cos t \sin t} = \cot t$
57. $\dfrac{\cos^3 t + \sin^3 t}{\cos t + \sin t} = \cos^2 t - \cos t \sin t + \sin^2 t = 1 - \sin t \cos t$ **59.** $\left(\dfrac{1 - \cos \theta}{\sin \theta}\right)^2 = \dfrac{(1 - \cos \theta)^2}{1 - \cos^2 \theta} = \dfrac{1 - \cos \theta}{1 + \cos \theta}$
61. $(\csc t - \cot t)^4(\csc t + \cot t)^4 = (\csc^2 t - \cot^2 t)^4 = 1^4 = 1$

63. $\sin^6 u + \cos^6 u = (1 - \cos^2 u)^3 + \cos^6 u = 1 - 3\cos^2 u + 3\cos^4 u = 1 - 3\cos^2 u(1 - \cos^2 u) = 1 - 3\cos^2 u \sin^2 u$

65. $\cot 3x = \dfrac{1}{\tan 3x} = \dfrac{1 - 3\tan^2 x}{3\tan x - \tan^3 x}\left(\dfrac{\cot^3 x}{\cot^3 x}\right) = \dfrac{\cot^3 x - 3\cot x}{3\cot^2 x - 1} = \dfrac{3\cot x - \cot^3 x}{1 - 3\cot^2 x}$

67. Conjecture: $f(x) = 2\cos x$

$\cos^3 x(1 - \tan^4 x + \sec^4 x) = \cos^3 x[1 + (\sec^2 x + \tan^2 x)(\sec^2 x - \tan^2 x)]$
$\qquad = \cos^3 x[1 + (1 + 2\tan^2 x)(1)] = \cos^3 x[2(1 + \tan^2 x)]$
$\qquad = 2\cos^3 x \sec^2 x = 2\cos x$

PROBLEM SET 8-2   (Page 313)

1. (a) $(\sqrt{2} + 1)/2 \approx 1.21$ (b) $(\sqrt{2}\sqrt{3} + \sqrt{2})/4 \approx .97$ **3.** (a) $(\sqrt{2} - \sqrt{3})/2 \approx -.16$ (b) $(\sqrt{2}\sqrt{3} + \sqrt{2})/4 \approx .97$

5. $\cos(3\pi/4 + \pi/3) = -(\sqrt{2} + \sqrt{6})/4$ **7.** $\sin(120° + 45°) = (\sqrt{6} - \sqrt{2})/4$ **9.** $\tan(45° + 30°) = (3 + \sqrt{3})/(3 - \sqrt{3})$

11. $\sin(t + \pi) = \sin t \cos \pi + \cos t \sin \pi = -\sin t$ **13.** $\sin(t + 3\pi/2) = \sin t \cos(3\pi/2) + \cos t \sin(3\pi/2) = -\cos t$

15. $\sin(t - \pi/2) = \sin t \cos(\pi/2) - \cos t \sin(\pi/2) = -\cos t$

17. $\cos(t + \pi/3) = \cos t \cos(\pi/3) - \sin t \sin(\pi/3) = (1/2)\cos t - (\sqrt{3}/2)\sin t$ **19.** $\cos 2$ **21.** $\sin \pi$ **23.** $\cos 60°$

25. $\sin \alpha$ **27.** $\frac{56}{65}$; $-\frac{33}{65}$; in quadrant II   **29.** $-(1 + 3\sqrt{3})/(2\sqrt{10}) \approx -.9797$; $(-3 + \sqrt{3})/(2\sqrt{10}) \approx -.2005$; quadrant III

31. $\tan(s - t) = \tan(s + (-t)) = (\tan s + \tan(-t))/(1 - \tan s \tan(-t)) = (\tan s - \tan t)/(1 + \tan s \tan t)$

33. $\tan(t + \pi/4) = (\tan t + \tan \pi/4)/(1 - \tan t \tan \pi/4) = (1 + \tan t)/(1 - \tan t)$

35. (a) $-(\cos t + \sqrt{3}\sin t)/2$ (b) $(\sqrt{3}\cos t + \sin t)/2$

37. (a) $\sqrt{5}/3$ (b) $-2\sqrt{2}/3$ (c) $(4\sqrt{2} - \sqrt{5})/9$ (d) $(2\sqrt{10} - 2)/9$ (e) $(-\frac{2}{3})(\sqrt{2} + \sqrt{5})$ (f) $4\sqrt{2}/9$

39. (a) $\sqrt{3}/2$ (b) $-\sqrt{3}/2$ (c) $\sin 1 \approx .84147$

41. (a) $\sin(x + y)\sin(x - y) = (\sin x \cos y + \cos x \sin y)(\sin x \cos y - \cos x \sin y)$
$= \sin^2 x \cos^2 y - \cos^2 x \sin^2 y = \sin^2 x(1 - \sin^2 y) - \cos^2 x \sin^2 y$
$= \sin^2 x - \sin^2 y(\sin^2 x + \cos^2 x) = \sin^2 x - \sin^2 y$

(b) $\dfrac{\sin(x + y)}{\cos(x - y)} = \dfrac{\sin x \cos y + \cos x \sin y}{\cos x \cos y + \sin x \sin y} = \dfrac{\dfrac{\sin x \cos y}{\cos x \cos y} + \dfrac{\cos x \sin y}{\cos x \cos y}}{\dfrac{\cos x \cos y}{\cos x \cos y} + \dfrac{\sin x \sin y}{\cos x \cos y}} = \dfrac{\tan x + \tan y}{1 + \tan x \tan y}$

(c) $\dfrac{\cos 5t}{\sin t} - \dfrac{\sin 5t}{\cos t} = \dfrac{\cos 5t \cos t - \sin 5t \sin t}{\sin t \cos t} = \dfrac{\cos 6t}{\sin t \cos t}$

43.

Since $\theta = \theta_2 - \theta_1$,
$\tan \theta = \dfrac{\tan \theta_2 - \tan \theta_1}{1 + \tan \theta_2 \tan \theta_1} = \dfrac{m_2 - m_1}{1 + m_1 m_2}$

45. $\alpha + \beta = 45°$

47. (a) $\frac{1}{2}[\cos(s + t) + \cos(s - t)] = \frac{1}{2}[\cos s \cos t - \sin s \sin t + \cos s \cos t + \sin s \sin t] = \cos s \cos t$

(b) $-\frac{1}{2}[\cos s \cos t - \sin s \sin t - \cos s \cos t - \sin s \sin t] = \sin s \sin t$

(c) $\frac{1}{2}[\sin s \cos t + \cos s \sin t + \sin s \cos t - \cos s \sin t] = \sin s \cos t$

(d) $\frac{1}{2}[\sin s \cos t + \cos s \sin t - \sin s \cos t + \cos s \sin t] = \cos s \sin t$

49. (a) $(1 - \sqrt{3})/4$ (b) $-\sqrt{2}/2$ (c) $\dfrac{1 + \sqrt{2} + \sqrt{3} + \sqrt{6}}{2}$

51. $\tan(\alpha + \beta) = (\tan \alpha + \tan \beta)/(1 - \tan \alpha \tan \beta) = (\frac{1}{3} + \frac{1}{2})/(1 - \frac{1}{3} \cdot \frac{1}{2}) = 1 = \tan \gamma$. Thus $\alpha + \beta = \gamma$.
53. Conjecture: $\sin t + \cos t = \sqrt{2} \sin(t + \pi/4)$

PROBLEM SET 8-3 (Page 318)

1. (a) $\frac{24}{25}$ (b) $\frac{7}{25}$ (c) $3\sqrt{10}/10$ (d) $\sqrt{10}/10$ **3.** $\sin 10t$ **5.** $\cos 3t$ **7.** $\cos(y/2)$ **9.** $\cos 1.2t$ **11.** $-\cos(\pi/4)$
13. $\cos^2(x/2)$ **15.** $\sin^2 2\theta$ **17.** (a) $\sin(\pi/8) = \sqrt{(1 - \cos \pi/4)/2} \approx .3827$ (b) $\cos 112.5° = -\sqrt{(1 + \cos 225°)/2} \approx -.3827$
19. $-\sqrt{0.35} \approx -.59161$ **21.** $-\sqrt{9/13} \approx -.83205$
23. $\cos 3t = \cos(2t + t) = \cos 2t \cos t - \sin 2t \sin t = (2 \cos^2 t - 1) \cos t - 2 \sin^2 t \cos t =$
$(2 \cos^2 t - 1) \cos t - 2(1 - \cos^2 t) \cos t = 4 \cos^3 t - 3 \cos t$
25. $\csc 2t + \cot 2t = (1 + \cos 2t)/(\sin 2t) = (2 \cos^2 t)/(2 \sin t \cos t) = \cot t$
27. $\sin \theta/(1 - \cos \theta) = 2 \sin(\theta/2)\cos(\theta/2)/2 \sin^2(\theta/2) = \cot(\theta/2)$
29. $2 \tan \alpha/(1 + \tan^2 \alpha) = 2 \tan \alpha/\sec^2 \alpha = 2(\sin \alpha/\cos \alpha)\cos^2 \alpha = 2 \sin \alpha \cos \alpha = \sin 2\alpha$
31. $\sin 4\theta = 2 \sin 2\theta \cos 2\theta = 2(2 \sin \theta \cos \theta)(2 \cos^2 \theta - 1) = 4 \sin \theta (2 \cos^3 \theta - \cos \theta)$
33. $\tan 2t = \tan(t + t) = (\tan t + \tan t)/(1 - \tan^2 t) = 2 \tan t/(1 - \tan^2 t)$
35. $\tan \dfrac{t}{2} = \dfrac{\sin t/2}{\cos t/2} = \dfrac{\pm\sqrt{(1 - \cos t)/2}}{\pm\sqrt{(1 + \cos t)/2}} = \pm\sqrt{\dfrac{1 - \cos t}{1 + \cos t}}$
37. (a) $\sin x$ (b) $\cos 6t$ (c) $-\cos(y/2)$ (d) $-\sin^2 2t$ (e) $\tan^2 2t$ (f) $\tan 3y$
39. (a) $120/169$ (b) $-2\sqrt{13}/13$ (c) $-\frac{3}{2}$ **41.** $\cos^4 z - \sin^4 z = (\cos^2 z + \sin^2 z)(\cos^2 z - \sin^2 z) = 1 \cdot \cos 2z$
43. $1 + (1 - \cos 8t)/(1 + \cos 8t) = 1 + \tan^2 4t = \sec^2 4t$
45. $\tan \frac{\theta}{2} - \sin \theta = (\sin \theta)/(1 + \cos \theta) - \sin \theta = (\sin \theta - \sin \theta - \sin \theta \cos \theta)/(1 + \cos \theta) = (-\sin \theta \cos \theta)/(1 + \cos \theta) =$
$-(\sin \theta)/(\sec \theta + 1)$
47. $(3 \cos t - \sin t)(\cos t + 3 \sin t) = 3 \cos^2 t - 3 \sin^2 t + 8 \sin t \cos t = 3 \cos 2t + 4 \sin 2t$
49. $2(\cos 3x \cos x + \sin 3x \sin x)^2 = 2 \cos^2 2x = 1 + \cos 4x$
51. $\tan 3t = \tan(2t + t) = \dfrac{\tan 2t + \tan t}{1 - \tan 2t \tan t} = \dfrac{\dfrac{2 \tan t}{1 - \tan^2 t} + \tan t}{1 - \dfrac{2 \tan^2 t}{1 - \tan^2 t}} = \dfrac{3 \tan t - \tan^3 t}{1 - 3 \tan^2 t}$
53. $\sin^4 u + \cos^4 u = (\sin^2 u + \cos^2 u)^2 - 2 \sin^2 u \cos^2 u = 1 - \frac{1}{2} \sin^2 2u = 1 - \frac{1}{2} \cdot (1 - \cos 4u)/2 = \frac{3}{4} + \frac{1}{4} \cos 4u$
55. $\cos^2 x + \cos^2 2x + \cos^2 3x = \dfrac{1 + \cos 2x}{2} + \cos^2 2x + \dfrac{1 + \cos 6x}{2} = 1 + \frac{1}{2}(\cos 2x + \cos 6x) + \cos^2 2x =$
$1 + \cos 4x \cos 2x + \cos^2 2x = 1 + \cos 2x(\cos 4x + \cos 2x) = 1 + \cos 2x(2 \cos 3x \cos x) = 1 + 2 \cos x \cos 2x \cos 3x$
57. Since $\alpha + \beta + \gamma = 180°$, $2\gamma = 360° - 2\alpha - 2\beta$. Thus $\sin 2\alpha + \sin 2\beta + \sin 2\gamma = \sin 2\alpha + \sin 2\beta - \sin(2\alpha + 2\beta) =$
$\sin 2\alpha + \sin 2\beta - \sin 2\alpha \cos 2\beta - \cos 2\alpha \sin 2\beta = \sin 2\alpha (1 - \cos 2\beta) + \sin 2\beta (1 - \cos 2\alpha) =$
$2 \sin \alpha \cos \alpha (2 \sin^2 \beta) + 2 \sin \beta \cos \beta (2 \sin^2 \alpha) = 4 \sin \alpha \sin \beta (\cos \alpha \sin \beta + \sin \alpha \cos \beta) = 4 \sin \alpha \sin \beta \sin(\alpha + \beta) =$
$4 \sin \alpha \sin \beta \sin \gamma$.
59. $(\frac{7}{9}, 4\sqrt{2}/9)$ **61.** Amplitude: 2; period: $2\pi/3$; conjecture: $2 \cos x - 4 \sin x \sin 2x = 2 \cos 3x$

PROBLEM SET 8-4 (Page 326)

1. $\pi/3$ **3.** $\pi/4$ **5.** 0 **7.** $\pi/3$ **9.** $2\pi/3$ **11.** $\pi/4$ **13.** $.2200$ **15.** $-.2200$ **17.** $.2037$
19. (a) $.7938$ (b) 1.9545 **21.** $.3486; 2.7930$ **23.** $1.2803; 4.4219$ **25.** $\frac{2}{3}$ **27.** 10 **29.** $\pi/3$ **31.** $\pi/4$ **33.** $\frac{3}{5}$
35. $2/\sqrt{5}$ **37.** $\frac{1}{3}$ **39.** $2\pi/3$ **41.** $.9666$ **43.** $.4508$ **45.** 2.2913 **47.** $\frac{24}{25}$ **49.** $\frac{7}{25}$ **51.** $\frac{56}{65}$
53. $(6 + \sqrt{35})/12 \approx .993$ **55.** $\tan(\sin^{-1}x) = \sin(\sin^{-1}x)/\cos(\sin^{-1}x) = x/\sqrt{1 - x^2}$
57. $\tan(2 \tan^{-1}x) = 2 \tan(\tan^{-1}x)/[1 - \tan^2(\tan^{-1}x)] = 2x/(1 - x^2)$ **59.** $\cos(2 \sec^{-1}x) = \cos[2 \cos^{-1}(1/x)] = 2/x^2 - 1$
61. (a) $-\pi/3$ (b) $-\pi/3$ (c) $2\pi/3$ **63.** (a) 43 (b) $\frac{12}{13}$ (c) $7\sqrt{2}/10$ (d) $(4 - 6\sqrt{2})/15$
65. (a) $\pm\sqrt{7}/4$ (b) $\pm.9$ (c) $\frac{11}{6}$ (d) $1; 2$ **67.** (a) $\pi/2$ (b) $\pi/4$ (c) $-\pi/4$ (d) $-\pi/2$ (e) π (f) $\pi/4$
69. (a) $\sin^{-1}(x/5)$ (b) $\tan^{-1}(x/3)$ (c) $\sin^{-1}(3/x)$ (d) $\tan^{-1}(3/x) - \tan^{-1}(1/x)$

71. (a) .6435011 (b) −.3046927 (c) .6435011 (d) 2.6905658 **73.** Show that the tangent of both sides is 120/119.
75. (a) $\theta = \tan^{-1}(6/b) - \tan^{-1}(2/b)$ (b) 22.83° (c) $2\sqrt{3}$ **77.** Amplitude: $\pi/2$; period: 2π, for both functions

PROBLEM SET 8-5 (Page 333)

1. $\{0, \pi\}$ **3.** $\{3\pi/2\}$ **5.** No solution **7.** $\{5\pi/6, 7\pi/6\}$ **9.** $\{\pi/4, 3\pi/4, 5\pi/4, 7\pi/4\}$ **11.** $\{\pi/4, 2\pi/3, 3\pi/4, 4\pi/3\}$
13. $\{0, \pi, 3\pi/2\}$ **15.** $\{0, \pi/3, \pi, 4\pi/3\}$ **17.** $\{\pi/3, \pi, 5\pi/3\}$ **19.** $\{.3649, 1.2059, 3.5065, 4.3475\}$ **21.** $\{0, \pi/2\}$
23. $\{\pi/6, \pi/2\}$ **25.** $\{0\}$ **27.** $\{\pi/6 + 2k\pi, 5\pi/6 + 2k\pi\colon k$ is an integer$\}$ **29.** $\{k\pi\colon k$ is an integer$\}$
31. $\{\pi/6 + k\pi, 5\pi/6 + k\pi\colon k$ is an integer$\}$ **33.** $\{0, \pi/2, \pi, 3\pi/2\}$ **35.** $\{\pi/8, 5\pi/8, 9\pi/8, 13\pi/8\}$
37. $\{3\pi/8, 7\pi/8, 11\pi/8, 15\pi/8\}$ **39.** $\{0, \pi/6, 5\pi/6, \pi\}$ **41.** $\{.9553, 2.1863, 4.0969, 5.3279\}$ **43.** $\{\pi/4, 5\pi/4\}$
45. $\{0, \pi/6, 5\pi/6, \pi, 7\pi/6, 11\pi/6\}$ **47.** $\{.3076, 2.8340\}$ **49.** $\{2\pi/3, 4\pi/3\}$ **51.** $\{2\pi/3, 5\pi/6, 5\pi/3, 11\pi/6\}$
53. $\{3\pi/2, 5.6397\}$ **55.** $\{\pi/4, 3\pi/4, 5\pi/4, 7\pi/4\}$ **57.** (a) 15 inches (b) $\tan\theta = \frac{2}{3}$ (c) 33.7°
59. $\{x\colon \pi/6 \le t \le \pi/2, 5\pi/6 \le t \le 3\pi/2\}$ **61.** (a) 26.6° (b) 10.3° **63.** $\{k\pi/3\colon k$ is an integer$\}$
65. $\{\pi/6, \pi/3, 2\pi/3, 5\pi/6\}$ **67.** 0, ±2.331

CHAPTER 8 REVIEW PROBLEM SET (Page 335)

1. T **2.** F (consider $\sin 4t = 0$) **3.** F ($\sin t = 0$ has infinitely many solutions) **4.** T **5.** T **6.** F (try $\alpha = \beta = 0$)
7. T **8.** T **9.** F (try $x = \pi$) **10.** F (not one-to-one)
11. (a) $-1 + 3\cos^2 t$ (b) $(1 - \cos^2 t)/(1 + \cos t)$ or $1 - \cos t$ (c) $2(1 - \cos^2 t)(1 + \cos t)\cos^2 t$
12. (a) $\frac{3}{5}$ (b) $-\frac{4}{5}$ (c) $-\frac{5}{4}$ **13.** $\csc\theta - \sin\theta = 1/\sin\theta - \sin\theta = (1 - \sin^2\theta)/\sin\theta = \cos^2\theta/\sin\theta = \cot\theta\cos\theta$
14. $\sec t - \cos t = 1/\cos t - \cos t = (1 - \cos^2 t)/\cos t = \sin^2 t/\cos t = \sin t\tan t$
15. $[\cos(t/2) + \sin(t/2)]^2 = \cos^2(t/2) + 2\sin(t/2)\cos(t/2) + \sin^2(t/2) = 1 + \sin t$
16. $\sec^4\theta - \sec^2\theta = \sec^2\theta(\sec^2\theta - 1) = (1 + \tan^2\theta)\tan^2\theta = \tan^4\theta + \tan^2\theta$
17. $\tan u + \cot u = \dfrac{\sin u}{\cos u} + \dfrac{\cos u}{\sin u} = \dfrac{\sin^2 u + \cos^2 u}{\cos u\sin u} = \dfrac{1}{\cos u\sin u} = \sec u\csc u$
18. $\dfrac{1 - \cos x}{\sin x} = \dfrac{1 - \cos x}{\sin x}\dfrac{1 + \cos x}{1 + \cos x} = \dfrac{1 - \cos^2 x}{\sin x(1 + \cos x)} = \dfrac{\sin^2 x}{\sin x(1 + \cos x)} = \dfrac{\sin x}{1 + \cos x}$ **19.** $-\frac{1}{2}$ **20.** 1 **21.** $\sqrt{2}/2$
22. 1 **23.** $2\sin^2\theta/(1 - 2\sin^2\theta)$ **24.** $1 - 8\sin^2 t + 8\sin^4 t$ **25.** (a) $-\frac{5}{13}$ (b) $\frac{120}{169}$ (c) $-\frac{119}{169}$ (d) $-\frac{3}{2}$
26. (a) $-\sqrt{5}/3$ (b) $-\frac{3}{5}$ (c) $-(6 + 4\sqrt{5})/15$
27. $\cos(u + \pi/3)\cos(\pi - u) - \sin(u + \pi/3)\sin(\pi - u) = \cos(u + \pi/3 + \pi - u) = \cos(4\pi/3) = -\frac{1}{2}$
28. $\dfrac{\cos 5t}{\sin t} - \dfrac{\sin 5t}{\cos t} = \dfrac{\cos 5t\cos t - \sin 5t\sin t}{\sin t\cos t} = \dfrac{\cos(5t + t)}{\sin t\cos t} = \dfrac{2\cos 6t}{\sin 2t}$
29. $\csc 2t + \cot 2t = \dfrac{1}{\sin 2t} + \dfrac{\cos 2t}{\sin 2t} = \dfrac{1 + 2\cos^2 t - 1}{2\sin t\cos t} = \dfrac{\cos^2 t}{\sin t\cos t} = \cot t$
30. $\sin 3\theta = \sin(2\theta + \theta) = \sin 2\theta\cos\theta + \cos 2\theta\sin\theta = 2\sin\theta\cos^2\theta + (1 - 2\sin^2\theta)\sin\theta = 2\sin\theta(1 - \sin^2\theta) + \sin\theta - 2\sin^3\theta = 3\sin\theta - 4\sin^3\theta$
31. $\dfrac{\sin(\alpha - \beta)}{\cos\alpha\cos\beta} = \dfrac{\sin\alpha\cos\beta - \cos\alpha\sin\beta}{\cos\alpha\cos\beta} = \dfrac{\sin\alpha}{\cos\alpha} - \dfrac{\sin\beta}{\cos\beta} = \tan\alpha - \tan\beta$
32. $\dfrac{1 - \tan^2(u/2)}{1 + \tan^2(u/2)} = \dfrac{1 - \sin^2(u/2)/\cos^2(u/2)}{1 + \sin^2(u/2)/\cos^2(u/2)} = \dfrac{\cos^2(u/2) - \sin^2(u/2)}{\cos^2(u/2) + \sin^2(u/2)} = \dfrac{\cos 2(u/2)}{1} = \cos u$
33. (a) $\pi/4$ (b) $2\pi/3$ (c) $-\pi/4$ (d) $\pi/4$ **34.** (a) 2.5 (b) $\pi/4$ (c) $\frac{5}{3}$ (d) -5 **35.** (a) .96 (b) $-.02$
36. (a) $.4 + .3\sqrt{3}$ (b) -3 **37.** $5\pi/6, 11\pi/6$ **38.** $\pi/3, \pi/2, 3\pi/2, 5\pi/3$ **39.** $3\pi/2$
40. $0, \pi/4, 3\pi/4, \pi, 5\pi/4, 7\pi/4$ **41.** $\pi/12, 5\pi/12, 13\pi/12, 17\pi/12$
42. $\pi/6, \pi/3, 2\pi/3, 5\pi/6, 7\pi/6, 4\pi/3, 5\pi/3, 11\pi/6$ **43.** $\pi/3, 5\pi/3$ **44.** $\pi/3, 2\pi/3, 3\pi/4, 4\pi/3, 5\pi/3, 7\pi/4$
45. Domain: $-1 \le x \le 1$, range: $0 \le y \le \pi$ **46.** Domain: $-\pi/4 \le x \le \pi/4, f^{-1}(x) = \frac{1}{2}\sin^{-1}x$
47. Let $\alpha = \sin^{-1}x$. Then $\sin\alpha = x$ and $\cos\alpha = \sqrt{1 - x^2}$. Thus $\cot\alpha = \cos\alpha/\sin\alpha = \sqrt{1 - x^2}/x$.
48. Let $\alpha = \tan^{-1}2$ and $\beta = \tan^{-1}3$ and note that $0 < \alpha + \beta < \pi$. Also, $\tan(\alpha + \beta) = (\tan\alpha + \tan\beta)/(1 - \tan\alpha\tan\beta) = (2 + 3)/(1 - 2\cdot 3) = -1$. Thus $\alpha + \beta = 3\pi/4$
49. $\frac{2}{3}$ **50.** 5.5

PROBLEM SET 9-1 (Page 341)

1. $\gamma = 55.5°; b \approx 20.9; c \approx 17.4$ **3.** $\beta = 56°; a = c \approx 53$ **5.** $\beta \approx 42°; \gamma \approx 23°; c \approx 20$ **7.** $\beta \approx 18°; \gamma \approx 132°; c \approx 12$
9. Two triangles: $\beta_1 \approx 53°, \gamma_1 \approx 97°, c_1 \approx 9.9; \beta_2 \approx 127°, \gamma_2 \approx 23°, c_2 \approx 3.9$ **11.** 93.7 meters **13.** 44.7° **15.** 192.8
17. 265.3 **19.** 78.4° **21.** 694.6 square feet **23.** 1769 feet **25.** About 255 feet **27.** About 16 square units
29. 40 **31.** $6\, r^2 \sin \phi \, (\cos \phi + \sqrt{3} \sin \phi)$ **33.** 12:42.4

PROBLEM SET 9-2 (Page 348)

1. $a \approx 12.5; \beta \approx 76°; \gamma \approx 44°$ **3.** $c \approx 15.6; \alpha \approx 26°; \beta \approx 34°$ **5.** $a \approx 44.4°; \beta \approx 57.1°; \gamma \approx 78.5°$
7. $a \approx 30.6°; \beta \approx 52.9°; \gamma \approx 96.5°$ **9.** 98.8 meters **11.** 24 miles **13.** 106° **15.** $s = 6, A = \sqrt{6 \cdot 3 \cdot 2 \cdot 1} = 6$
17. 18.63 **19.** 41.68° **21.** 42.60 miles **23.** 30.12 meters, 20.08 meters, 20.08 meters
25. (a) 70.53°, 58.99°, 50.48° (b) About 3.85 square units **27.** $\cos^{-1}(3/4) \approx 41.41°; \frac{1}{2}(\sqrt{3} + 3\sqrt{7})$ **31.** $\sqrt{15}$ **33.** 12:31.5

PROBLEM SET 9-3 (Page 354)

1. **3.** **5.** $\mathbf{w} = \frac{1}{2}(\mathbf{u} + \mathbf{v})$ **7.** $\|\mathbf{w}\| = 1$

9. $10\sqrt{2} + \sqrt{2} \approx 18.48$ **11.** 243.7 kilometers; S43.9°W **13.** .026 hour **15.** 479 miles per hour; N2.68°E
17. 15.9; S7.5°W **19.** 163.7, 118.9 **21.** **23. 0**

25. (a) $\overrightarrow{AD} - \overrightarrow{AB}$ (b) $\frac{1}{2}(\overrightarrow{AB} + \overrightarrow{AD})$ (c) $\overrightarrow{AB} - \frac{1}{2}\overrightarrow{AD}$ (d) $\overrightarrow{AD} - \frac{1}{2}\overrightarrow{AB}$ **27.** $\sqrt{7}/2 \approx 1.32$ miles per hour
29. N10.32°E **31.** N1.019°W, 654.88 miles per hour **33.** 651.3 pounds **35.** 65.38°, 146.77 pounds

PROBLEM SET 9-4 (Page 361)

1. $4\mathbf{i} - 24\mathbf{j}; -33; -33/65$ **3.** $3\mathbf{i} + \mathbf{j}; 10; 2/\sqrt{5}$ **5.** 101.385° **7.** $5\mathbf{i} + 2\mathbf{j}; 4\mathbf{i} - 3\mathbf{j}; 14$ **9.** $-4\mathbf{i} - 5\mathbf{j}; -6\mathbf{i} + 5\mathbf{j}; -1$
11. $-5\mathbf{i} + 5\sqrt{3}\mathbf{j}$ **13.** $\frac{4}{3}$ **15.** $\frac{3}{5}\mathbf{i} - \frac{4}{5}\mathbf{j}$ **17.** $-4\mathbf{i} + 10\mathbf{j}$ **19.** $\mathbf{u} \cdot \mathbf{u} = a^2 + b^2 = \|\mathbf{u}\|^2$ **21.** 7 **23.** $(118/169)(5\mathbf{i} + 12\mathbf{j})$
25. $-56/5$ **27.** 100 **29.** $325\sqrt{2}$ dyne-centimeters **31.** $\sqrt{34}$ **33.** $\pm(\frac{4}{5}\mathbf{i} + \frac{3}{5}\mathbf{j})$ **35.** $\frac{9}{25}\mathbf{i} - \frac{12}{25}\mathbf{j}; 84.1°$
37. $37.1325\mathbf{i} + 48.9626\mathbf{j}$ **41.** $|\mathbf{u} \cdot \mathbf{v}| = \|\mathbf{u}\| \|\mathbf{v}\| |\cos \theta| \le \|\mathbf{u}\| \|\mathbf{v}\|$ with equality when $\theta = 0°$ or $180°$ **43.** $\|\mathbf{u}\| = \|\mathbf{v}\|$
45. $(-1 + \sqrt{5})/4$ **47.** Amplitude: 3; period: 60

PROBLEM SET 9-5 (Page 368)

1. (a)   (b)   (c)   (d)

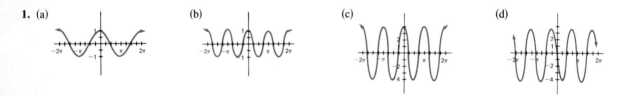

3. (a) π; 4; 0   (b) 2π; 3; $-\pi/8$ (c) $\pi/2$; 1; $-\pi/32$ (d) $2\pi/3$; 3; $\pi/6$

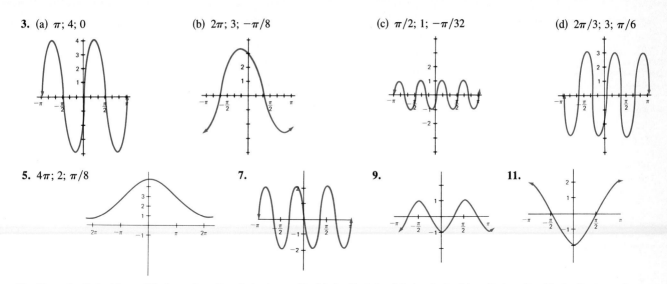

5. 4π; 2; $\pi/8$ **7.**   **9.**   **11.**

13. $(5\cos 4t,\ 5\sin 4t)$ **15.** $5\cos 4t,\ -8 + 5\sin 4t$ **17.** (a) $2\pi/5$; 1; 0   (b) 4π; $\frac{3}{2}$; 0   (c) $\pi/2$; 2; $\pi/4$ (d) $2\pi/3$; 4; $-\pi/4$

19. 4 feet; after 1.5 seconds   **21.** $\sin t + \sqrt{25 - \cos^2 t}$

23. 156; 55   **25.** (a) 0   (b) 0   (c) 53.53   **27.**

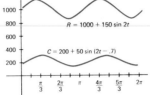

29. (a) $2\sqrt{2}\sin 2t - 2\sqrt{2}\cos 2t$ (b) $-\frac{3}{2}\sqrt{3}\sin 3t + \frac{3}{2}\cos 3t$ **33.** (a) $5\sin\left[2t + \tan^{-1}(4/3)\right]$ (b) $2\sqrt{3}(\sin 4t + 11\pi/6)$

35. $\sqrt{2}$; $-\sqrt{2}$ **37.** $\{6.039 + 8k\pi,\ 21.724 + 8k\pi : k \text{ is an integer}\}$

PROBLEM SET 9-6   (Page 375)

1-11.

13. $\sqrt{13}$; $\sqrt{13}$; 5; 4; 1; 2   **15.** $0 - 4i$ **17.** $-\sqrt{2} - \sqrt{2}i$ **19.** $4(\cos\pi + i\sin\pi)$

21. $5(\cos 270° + i\sin 270°)$ **23.** $2\sqrt{2}(\cos 315° + i\sin 315°)$ **25.** $4(\cos\pi/6 + i\sin\pi/6)$ **27.** $6.403(\cos .6747 + i\sin .6747)$

29. $6(\cos 210° + i\sin 210°)$ **31.** $\frac{3}{2}(\cos 125° + i\sin 125°)$ **33.** $\frac{2}{3}(\cos 70° + i\sin 70°)$ **35.** $2(\cos 305° + i\sin 305°)$

37. $16 + 0i$ **39.** $-2 + 2\sqrt{3}i$ **41.** $16(\cos 60° + i\sin 60°)$ **43.** $1(\cos 120° + i\sin 120°)$ **45.** $16(\cos 270° + i\sin 270°)$

47. (a) $|-5 + 12i| = 13$ (b) $|-4i| = 4$ (c) $|5(\cos 60° + i\sin 60°)| = 5$

49. (a) $12(\cos 0° + i \sin 0°)$ (b) $2(\cos 135° + i \sin 135°)$ (c) $3(\cos 270° + i \sin 270°)$ (d) $4(\cos 300° + i \sin 300°)$
(e) $8(\cos 30° + i \sin 30°)$ (f) $2(\cos 315° + i \sin 315°)$
51. (a) $12(\cos 160° + i \sin 160°)$ (b) $3(\cos 40° + i \sin 40°)$ (c) $\cos 45° + i \sin 45°$
53. (a) $12 + 5i, -12 + 5i$ (b) $\pm(4\sqrt{2} + 4\sqrt{2}i)$
55. (a) $r^3(\cos 3\theta + i \sin 3\theta)$ (b) $r[\cos(-\theta) + i \sin(-\theta)]$ (c) $r^2(\cos 0 + i \sin 0)$ (d) $\frac{1}{r}[\cos(-\theta) + i \sin(-\theta)]$
(e) $r^{-2}[\cos(-2\theta) + i \sin(-2\theta)]$ (f) $r[\cos(\theta + \pi) + i \sin(\theta + \pi)]$ **57.** $(r^2/s^3)[\cos(2\alpha - 3\beta) + i \sin(2\alpha - 3\beta)]$
59. (a) The distance between **U** and **V** in the complex plane   (b) The angle from the positive x-axis to the line joining **U** and **V**
61. (a) -2^8 (b) 0

PROBLEM SET 9-7   (Page 382)

1. $8[\cos(3\pi/4) + i \sin(3\pi/4)]$ **3.** $125(\cos 66° + i \sin 66°)$ **5.** $16(\cos 0° + i \sin 0°)$ **7.** $1 + 0i$ **9.** $-16\sqrt{3} + 16i$
11. $5(\cos 15° + i \sin 15°)$; $5(\cos 135° + i \sin 135°)$; $5(\cos 255° + i \sin 255°)$

13. $2[\cos(\pi/12) + i \sin(\pi/12)]$; $2[\cos(5\pi/12) + i \sin(5\pi/12)]$; $2[\cos(9\pi/12) + i \sin (9\pi/12)]$; $2[\cos13\pi/12) + i \sin13\pi/12)]$;
$2[\cos(17\pi/12) + i \sin(17\pi/12)]$; $2[\cos(21\pi/12) + i \sin(21\pi/12)]$

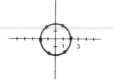

15. $\sqrt{2}(\cos 28° + i \sin 28°)$; $\sqrt{2}(\cos 118° + i \sin 118°)$;   **17.** $\pm 2; \pm 2i$
$\sqrt{2}(\cos 208° + i \sin 208°)$; $\sqrt{2}(\cos 298° + i \sin 298°)$

19. $\pm(\sqrt{2} + \sqrt{2}i)$ **21.** $\pm(\sqrt{2} + \sqrt{6}i)$
23. $\pm 1; \pm i$ **25.** $\cos(k \cdot 36°) + i \sin(k \cdot 36°)$, $k = 0, 1, \ldots , 9$

27. (a) $81(\cos 80° + i \sin 80°)$ (b) $90.09(\cos 7.7 + i \sin 7.7)$ (c) $8(\cos 240° + i \sin 240°)$ (d) $16[\cos(5\pi/3) + i \sin(5\pi/3)]$
29. $2(\cos 51° + i \sin 51°)$, $2(\cos 123° + i \sin 123°)$, $2(\cos 195° + i \sin 195°)$, $2(\cos 267° + i \sin 267°)$, $2(\cos 339° + i \sin 339°)$
31. Sum is 0; product is -1.   **33.** $\sqrt[5]{2}(\cos 27° + i \sin 27°) \approx 1.0235 + .5215i$
35. *Method 1.* Use the formula $\cos(k\pi/3) + i \sin(k\pi/3)$, $k = 0, 1, 2, 3, 4, 5$. *Method 2.* Write $x^6 - 1 =$
$(x - 1)(x^2 + x + 1)(x + 1)(x^2 - x + 1) = 0$ and solve. Both methods give the answers ± 1, $(-1 \pm \sqrt{3}i)/2$, $(1 \pm \sqrt{3}i)/2$.
37. $\pm i, \frac{1}{2}\sqrt{2} \pm \frac{1}{2}\sqrt{2}i, -\frac{1}{2}\sqrt{2} \pm \frac{1}{2}\sqrt{2}i$ **39.** $2, 1$ **41.** (a) $\frac{1}{2}\sqrt{6} + \frac{1}{2}\sqrt{2}i$ (b) $\frac{1}{2}\sqrt{2} + \frac{1}{2}\sqrt{6}i$ **43.** -2^n

CHAPTER 9. REVIEW PROBLEM SET (Page 384)

1. T   **2.** T   **3.** F (it is both associative and commutative)   **4.** T   **5.** T   **6.** T   **7.** F (it is shifted $\pi/3$ units left)
8. F (there are many other such numbers, e.g., $\frac{3}{5} + \frac{4}{5}i$)   **9.** T   **10.** F (i has no real 8th roots)
11. $\gamma = 105°$, $a = 5.18$, $b = 7.32$ **12.** $\alpha = 28.96°$, $\beta = 46.56°$, $\gamma = 104.48°$ **13.** $\alpha = 26.03°$, $\gamma = 11.97°$, $c = 31.67$
14. $\alpha = 32.5°$, $\beta = 109.9°$, $c = 13.2$ **15.** 71.8   **16.** 26.8   **17.** (a) 273.2 yards   (b) 136.6 yards
18. $-(3\sqrt{3} + 8)\mathbf{i} + 15\mathbf{j}$ **19.** (a) 13   (b) 25   (c) 36   (d) 83.64°   (e) $\frac{24}{25}\mathbf{i} + \frac{7}{25}\mathbf{j}$ **20.** (a) $\frac{36}{25}$ (b) $\frac{864}{625}\mathbf{i} + \frac{252}{625}\mathbf{j}$ **21.** $72\mathbf{i} + 186\mathbf{j}$
22. (a) 797.8 foot-pounds   (b) 768 foot-pounds   **23.** π, 1, 0   **24.** $\pi/2$, 3, 0   **25.** $2\pi/3$, 2, $\pi/6$ **26.** 4π, 2, -2π

27.   **28.**   **29.**   **30.**

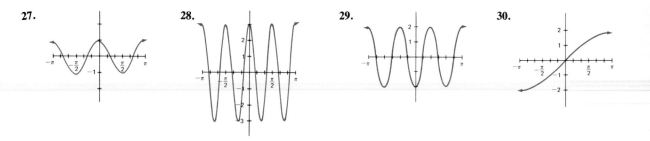

31. (a) $(3 \cos(5\pi t/6), 3 \sin(5\pi t/6))$ (b) $t = \frac{6}{5}$ seconds   **32.** (a) $\frac{2}{5}\pi$ seconds   (b) $x = \frac{3}{2}$ meters   (c) $t = \frac{1}{6}\pi$ seconds
33. 60 meters   **34.**   **35.** (a) 5   (b) 6   (c) 5   (d) 3   (e) 4

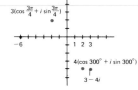

36. $4 - 4i$ **37.** $4(\cos 180° + i \sin 180°)$
38. $9(\cos 90° + i \sin 90°)$ **39.** $2\sqrt{2}(\cos 45° + i \sin 45°)$ **40.** $4(\cos 300° + i \sin 300°)$
41. (a) $r^3(\cos 3t + i \sin 3t)$ (b) $(1/r)[\cos(-t) + i \sin(-t)]$ (c) $r[\cos(t + \pi/3) + i \sin(t + \pi/3)]$
42. (a) $54(\cos 275° + i \sin 275°)$ (b) $4(\cos 80° + i \sin 80°)$ (c) $\frac{4}{3}(\cos 95° + i \sin 95°)$
43. $4(\cos 24° + i \sin 24°)$, $4(\cos 114° + i \sin 114°)$, $4(\cos 204° + i \sin 204°)$, $4(\cos 294° + i \sin 294°)$
44. $\cos 45° + i \sin 45°$, $\cos 105° + i \sin 105°$, $\cos 165° + i \sin 165°$, $\cos 225° + i \sin 225°$, $\cos 285° + i \sin 285°$, $\cos 345° + i \sin 345°$
45. 8   **46.** Calculate each side separately.   **47.** $2i$, $-\sqrt{3} - i$, $\sqrt{3} - i$
48. (a) u^2, u^4, u^6, u^8, u^{10}, u^{12} (b) u^3, u^6, u^9, u^{12}
49. $\beta_1 = 58.99°$, $\gamma_1 = 81.01°$, $c_1 = 2.30$; $\beta_2 = 121.01°$, $\gamma_2 = 18.99°$, $c_2 = 0.76$ **50.** About 2.027

PROBLEM SET 10-1 (Page 391)

1. $x + 1$; 0   **3.** $3x - 1$; 10   **5.** $2x^2 + x + 3$; 0   **7.** $x^2 + 4$; 0   **9.** $2x^2 + x + 2$; -2 **11.** $3x^2 + 8x + 10$; 0
13. $x^3 + 3x^2 + 7x + 21$; 62   **15.** $x^2 + x - 4$; 6   **17.** $2x^3 + 4x + 5$; $\frac{3}{2}$ **19.** $x^2 - 3ix - 4 - 6i$; 4
21. $x^3 + 2ix^2 - 4x - 8i$; -1 **23.** $x + 2 + 5/x^2$ **25.** $x - 1 + (-x + 3)/(x^2 + x - 2)$ **27.** $2 + (3 - 4x)/(x^2 + 1)$
29. $2x - 1 + (-x^2 - 2x - 1)/(x^3 + 1) = 2x - 1 + (-x - 1)/(x^2 - x + 1)$
31. (a) $2x + 3$; $-11x + 9$ (b) $2x + 16$; -14 (c) $x^2 - 8x + 16$; $x^2 + x + 1$ (d) $x^4 + 6x^2 + 2x + 9$; $4x - 1$
33. (a) $x^4 + 3x^3 + 6x^2 + 12x + 8$ (b) $x^4 - 2x^3 + 4x^2 - 8x + 16$ (c) $x^3 - 2x^2 + 4x + 4$ (d) $x^2 + 1$ **35.** $x^3 - 8$
37. (a) -40 (b) 13   (c) 2   **39.** $a = 1$, $b = -9$, $c = 11$ **41.** As $|x| \to \infty$, $|f(x) - g(x)| \to 0$

PROBLEM SET 10-2 (Page 398)

1. -2 **3.** $-\frac{9}{4}$ **5.** -6 **7.** 14   **9.** 1, -2, and 3, each of multiplicity 1
11. $\frac{1}{2}$ (multiplicity 1); 2 (multiplicity 2); 0 (multiplicity 3)   **13.** $1 + 2i$ and $-\frac{2}{3}$, each of multiplicity 1
15. $P(1) = 0$ **17.** $P(3) = 0$ **19.** $(5 \pm \sqrt{7}\,i)/4$ **21.** 3; 2 (multiplicity 2)   **23.** $(x - 2)(x - 3)$
25. $(x - 1)(x + 1)(x - 2)(x + 2)$ **27.** $(x - 6)(x - 2)(x + 5)$ **29.** $(x + 1)(x + 1 - \sqrt{13})(x + 1 + \sqrt{13})$
31. $x^3 + x^2 - 10x + 8$ **33.** $12x^2 + 4x - 5$ **35.** $x^3 - 2x^2 - 5x + 10$
37. $4x^5 + 20x^4 + 25x^3 - 10x^2 - 20x + 8$
39. $x^4 - 3x^2 - 4$ **41.** $x^4 - x^3 - 9x^2 + 79x - 130$ **43.** -3, -2; $(x - 1)^3(x + 3)(x + 2)$ **45.** $\pm i$
47. $(x - 3)(x + 3)(x - 2i)$ **49.** $(x + 1)^2(x - 1 - i)$ **51.** (a) 2   (b) -2 (c) 0
53. (a) ± 2, each of multiplicity 3   (b) 1 and 2, each of multiplicity 2   (c) $-1 \pm \sqrt{5}$, each of multiplicity 3; -2, of multiplicity 4
55. Let $c > 0$ and note that $3(-c)^{31} - 2(-c)^{18} + 4(-c)^3 - (-c)^2 - 4 = -3c^{31} - 2c^{18} - 4c^3 - c^2 - 4 < 0$
57. (a) $(2x - 1)(3x + 1)(2x + 1)$ (b) $(2x - 1)(x - \sqrt{2})(x + \sqrt{2})$ (c) $(2x - 1)(x + i)(x - i)$

59.

61. 3, 2.35   **63.** $16x^4 - 32x^3 + 24x^2 - 8x + 1$
65. An nth-degree polynomial has at most n zeros. Therefore, $P(x)$ must be the zero polynomial (without degree). From this it follows that all coefficients are zero.
67. We are given that $c^6 - 5c^5 + 3c^4 + 7c^3 + 3c^2 - 5c + 1 = 0$. Note that $c \neq 0$ and so we may divide by c^6 to obtain $1 - 5(1/c) + 3(1/c)^2 + 7(1/c)^3 + 3(1/c)^4 - 5(1/c)^5 + (1/c)^6 = 0$, which is the desired conclusion.
69. $a_n x^n + a_{n-1}x^{n-1} + \cdots + a_1 x + a_0 = a_n(x - c_1)(x - c_2) \cdots (x - c_n) =$
$a_n[x^n - (c_1 + c_2 + \cdots + c_n)x^{n-1} + (c_1 c_2 + c_1 c_3 + \cdots + c_{n-1}c_n)x^{n-2} + \cdots + (-1)^n c_1 c_2 \cdots c_n] =$
$a_n x^n - a_n(c_1 + c_2 + \cdots + c_n)x^{n-1} + a_n(c_1 c_2 + c_1 c_3 + \cdots + c_{n-1}c_n)x^{n-2} + \cdots + (-1)^n a_n c_1 c_2 \cdots c_n$.
Thus, $a_{n-1} = -a_n(c_1 + c_2 + \cdots + c_n)$, $a_{n-2} = a_n(c_1 c_2 + \cdots + c_{n-1}c_n)$, $a_0 = (-1)^n a_n c_1 c_2 \cdots c_n$.
71. 2.67, 303.4

PROBLEM SET 10-3 (Page 405)

1. $2 - 3i$ **3.** $-4i$ **5.** $4 + \sqrt{6}$ **7.** $(2 + 3i)^8$ **9.** $2(1 - 2i)^3 - 3(1 - 2i)^2 + 5$ **11.** $5 + i$ **13.** $3 + 2i; 5 - 4i$
15. $-i; \frac{1}{2}$ **17.** $1 - 3i; -2; -1$ **19.** $x^2 - 4x + 29$ **21.** $x^3 + 3x^2 + 4x + 12$
23. $x^5 - 2x^4 + 18x^3 - 36x^2 + 81x - 162$
25. $-1; 1; 3$ **27.** $\frac{1}{2}; -1 \pm \sqrt{2}$ **29.** $\frac{1}{2}; (1 \pm \sqrt{5})/2$ **31.** $2 - i, (-1 \pm \sqrt{3}\,i)/2$ **33.** $-1, -2, 3 \pm \sqrt{13}$
35. $-1, \frac{1}{2}, 2, \pm 2\sqrt{2}\,i$ **37.** 1, 1, 1, 1, -10
39. This follows from the fact that nonreal solutions occur in pairs.
41. If $u = a + bi$, $\bar{u} = a - bi$, then $u + \bar{u} = 2a$, $u\bar{u} = a^2 + b^2$, both of which are real.
43. $(x - 1)(x + 1)(x^2 + 3x + 4)$ **45.** $(x - 1)(x + 1)(x^2 + 1)(x^2 - \sqrt{2}\,x + 1)(x^2 + \sqrt{2}\,x + 1)$
47. Let $r = \sqrt{5} - 2$ and $s = \sqrt{5} + 2$. Then $x = r^{1/3} - s^{1/3}$. When we cube x and simplify, we obtain $x^3 = -4 - 3(rs)^{1/3} = -4 - 3x$. Thus $x^3 + 3x + 4 = 0$, that is, $(x + 1)(x^2 - x + 4) = 0$. Since x is real, x must be -1.
49. The graph touches but does not cross the x-axis at a real zero of even multiplicity; it is tangent to and crosses the x-axis at a real zero of odd multiplicity greater than 1.

PROBLEM SET 10-4 (Page 413)

1. $4x - 5; -1$ **3.** $4x + 1; 5$ **5.** $10x^4 + 4x^3 - 6x^2 + 8; 16$

7.

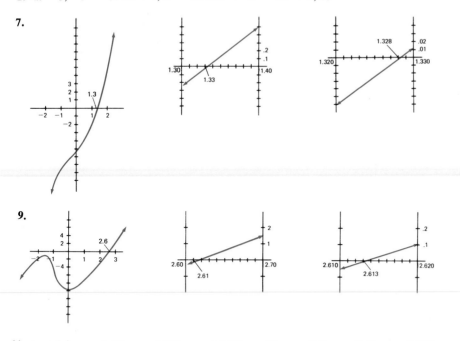

9.

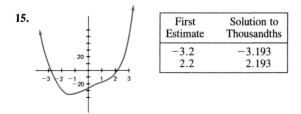

11. $x_1 = 1.3; x_2 = 1.33; x_3 = 1.328; x_4 = 1.3283$ **13.** $x_1 = 2.6; x_2 = 2.61; x_3 = 2.613; x_4 = 2.6129$

15.

First Estimate	Solution to Thousandths
-3.2	-3.193
2.2	2.193

17.

First Estimate	Solution to Thousandths
-1.9	-1.879
$.4$	$.347$
1.5	1.532

19. $-2.11; .25; 1.86$ **21.** $4.1811; 209$ **23.** 9.701 percent   **25.** $y - 12 = 12(x - 2)$ **27.** (a) $-2, 3$ (b) $0, 1 \pm \sqrt{2}$

29. (a) $m_h = [f(a + h) - f(a)]/h = [2(a + h)^3 - 2a^3]/h$ (b) $6a^2 + 6ah + 2h^2$ (c) $6a^2$ (d) They are identical.

31. (a) $-1.04, 2.46, 2.59$ (b) One of each for a total of 2

CHAPTER 10. REVIEW PROBLEM SET (Page 415)

1. T   **2.** F $[P(a) = 0]$ **3.** F $[(x - 1)^3 = 0$ has 1 as its only zero]   **4.** F (the graph crosses the x-axis only at $x = 2$)
5. T   **6.** T   **7.** F (real coefficients are needed for this)   **8.** F (1 is a zero of multiplicity 6)   **9.** T   **10.** T
11. $2x - 1; -4x + 5$ **12.** $x^2 - 3; 2x$ **13.** $2x^4 + 1; 2$ **14.** $x + 2i; 0$ **15.** $x^4 + x^3 + 4x^2 + 11x + 27; 53$
16. $2x^2 - 2x + 2; -1$ **17.** $x^3 + (\sqrt{2} - 1)x^2 + (1 - 2\sqrt{2})x + 3\sqrt{2}; -1 - 3\sqrt{2}$ **18.** $2x^2 + (-4 + 4i)x - 7 - 8i; 9 - 14i$
19. $x^3 - 4x + 8; 0$ **20.** 6   **21.** 3   **22.** $3 + 2i$ **23.** 2 (multiplicity 2); $\frac{3}{2}$; -4 (multiplicity 3)
24. -2 (multiplicity 3); 2 (multiplicity 3), $-i, i$ **25.** -3; 2 (multiplicity 2); 3 (multiplicity 3)
26. $-3\sqrt{2}$; 0 (multiplicity 2); $3 - i$ (multiplicity 3); $3 + i$ (multiplicity 3)
27. $(x - 2)(x - 2)(x + 2)$ **28.** $(x + 1)(x - 2)(x - 4)$ **29.** $(x - 3i)(x - 4)(x + 2)$ **30.** $(x - \pi)(2x - 1)(x + 1)$
31. $x^3 + 2x^2 - 7x + 4$ **32.** $12x^4 - x^3 - 30x^2 + 2x - 6$ **33.** $x^4 - 4x^3 + 6x^2 - 4x + 5$ **34.** $(x - 2)(x - 2)(x^2 + 1)$
35. $9x^3 + 33x^2 - 18x - 24$ **36.** $k = 6$ **37.** $x^3 - 7x^2 - x + 87$ **38.** $-\frac{1}{2}, 1, 1 \pm 5i$ **39.** $\frac{2}{3}, -1 \pm \sqrt{2}$
40. Graph to see that the equation has one real solution. The only possible rational solutions are ± 1 and ± 3 but none of these works.
41. $x_1 = 3, x_2 = 3.0, x_3 = 3.05$ **42.** $x_1 = 3, x_2 = 3.05, x_3 = 3.0514, x_4 = 3.051374$ **43.** $3x + y = 4$ **44.** 8 centimeters
45. $x_1 = 6, x_2 = 6.11, x_3 = 6.107$ **46.** At about $t = 1.265$ seconds
47. $P(x) = ax^2 - a(r_1 + r_2)x + ar_1r_2 = ax^2 + bx + c \Rightarrow -a(r_1 + r_2) = b, ar_1r_2 = c$
48. Follow same procedure as in Problem 47.

PROBLEM SET 11-1 (Page 423)

1. $(2, -1)$ **3.** $(-2, 4)$ **5.** $(1, -2)$ **7.** $(0, 0, -2)$ **9.** $(1, 4, -1)$ **11.** $(2, 1, 4)$ **13.** $(0, 0, 0)$
15. $(5, 6, 0, -1)$ **17.** $(15z - 110, 4z - 32, z)$ **19.** $(2y - 3z - 2, y, z)$ **21.** $(-z, 2z, z)$ **23.** Inconsistent
25. $(-z + \frac{2}{5}, z + \frac{16}{5}, z)$ **27.** $(2, 4); (10, 0)$ **29.** $(5, -7); (6, 0)$ **31.** $(-1, 2); (1, 2)$ **33.** $(3, -2)$
35. $(0, -2)$ **37.** $(-6, -12, 24)$ **39.** $(2\sqrt{5}/5, 4\sqrt{5}/5)$ **41.** $a = \frac{3}{2}, b = 6$ **43.** 285   **45.** $y = 2x^3 - 3x + 4$
47. $x^2 + y^2 - 4x - 3y = 0, r = \frac{5}{2}$ **49.** $a = -6, b = 8, c = -3$ **51.** 15 inches by 8 inches
53. $(-3.383, 4.383), (4.383, -3.383)$

PROBLEM SET 11-2 (Page 430)

1. $\begin{bmatrix} 2 & -1 & 4 \\ 1 & -3 & -2 \end{bmatrix}$ **3.** $\begin{bmatrix} 1 & -2 & 1 & 3 \\ 2 & 1 & 0 & 5 \\ 1 & 1 & 3 & -4 \end{bmatrix}$ **5.** $\begin{bmatrix} 2 & -3 & -4 \\ 3 & 1 & -2 \end{bmatrix}$ **7.** $\begin{bmatrix} 1 & 0 & 0 & 5 \\ 1 & 2 & -1 & 4 \\ 3 & -1 & -5 & -13 \end{bmatrix}$

9. Unique solution   **11.** No solution   **13.** Unique solution   **15.** Infinitely many solutions   **17.** No solution   **19.** $(1, 2)$
21. $(x, \frac{3}{2}x - \frac{1}{2})$ **23.** $(1, 4, -1)$ **25.** $(\frac{16}{3}z + \frac{32}{3}, -\frac{7}{3}z - \frac{2}{3}, z)$ **27.** $(3, 0, 0)$ **29.** $(4.36, 1.26, -.97)$
31. Unique solution   **33.** Unique solution   **35.** No solution   **37.** $(0, 2, 1)$ **39.** $(\frac{21}{2}z - 48, -5z + 26, z)$
41. $a = -4, b = 8, c = 0$ **43.** $\alpha = 80°, \beta = 30°, \gamma = 110°, \delta = 50°$ **45.** $a = -3, b = 2, c = 5$
47. A: 10 pounds; B: 40 pounds; C: 50 pounds   **49.** $(-2.0417, -1.2917, -0.2917, 2.6250)$

PROBLEM SET 11-3 (Page 438)

1. $\begin{bmatrix} 8 & 4 \\ 1 & 10 \end{bmatrix}; \begin{bmatrix} -4 & -6 \\ 5 & 4 \end{bmatrix}; \begin{bmatrix} 6 & -3 \\ 9 & 21 \end{bmatrix}$ **3.** $\begin{bmatrix} 5 & 4 & 4 \\ 8 & 3 & -6 \end{bmatrix}; \begin{bmatrix} 1 & -8 & 6 \\ 0 & -3 & 0 \end{bmatrix}; \begin{bmatrix} 9 & -6 & 15 \\ 12 & 0 & -9 \end{bmatrix}$ **5.** $\begin{bmatrix} 14 & 7 \\ 4 & 36 \end{bmatrix}; \begin{bmatrix} 27 & 29 \\ 5 & 23 \end{bmatrix}$

7. $\begin{bmatrix} -3 & -4 & 2 \\ 8 & 22 & -13 \\ -2 & 0 & 9 \end{bmatrix}; \begin{bmatrix} 0 & 5 & -17 \\ 13 & 10 & 3 \\ 1 & -3 & 18 \end{bmatrix}$ **9.** **AB** not possible; **BA** $= \begin{bmatrix} 7 & 2 & -7 & 6 \\ 15 & 2 & -11 & 16 \end{bmatrix}$

11. $\mathbf{AB} = \begin{bmatrix} 2 \\ 16 \\ -2 \end{bmatrix}$; $\mathbf{BA}$ not possible.   **13.** $\mathbf{AB} = \mathbf{BA} = \begin{bmatrix} 0 & 0 \\ 0 & 0 \end{bmatrix}$ **15.** $\begin{bmatrix} -4 & 7 & 9 \\ -5 & -5 & 8 \end{bmatrix}$

17. $\mathbf{A(B + C)} = \mathbf{AB + AC} = \begin{bmatrix} -7 & -3 \\ 39 & 34 \end{bmatrix}$; the distributive property   **19.** 93.5917

21. $\begin{bmatrix} 2 & 5 & -1 \\ -8 & 5 & -3 \\ 16 & -2 & -1 \end{bmatrix}$; $\begin{bmatrix} -16 & -6 & 8 \\ 12 & 0 & 22 \\ 11 & -16 & 19 \end{bmatrix}$; $\begin{bmatrix} 32 & -3 & 12 \\ 36 & 29 & 24 \\ 34 & 6 & 25 \end{bmatrix}$

23. $\mathbf{AB} = -7$; $\mathbf{BA} = \begin{bmatrix} 2 & 4 & 6 & 8 \\ 1 & 2 & 3 & 4 \\ -1 & -2 & -3 & -4 \\ -2 & -4 & -6 & -8 \end{bmatrix}$ **25.** $\mathbf{A}$ and $\mathbf{B}$ are square, of the same size, and $\mathbf{AB} = \mathbf{BA}$.

27. $\mathbf{A}^2 = \mathbf{0}$, $\mathbf{B}^3 = \mathbf{0}$. The nth power of a strictly upper triangular $n \times n$ matrix is the zero matrix.   **29.** $\mathbf{A}^n = \begin{bmatrix} 1 & 2n & 0 \\ 0 & 1 & 0 \\ 0 & 0 & 1 \end{bmatrix}$

31. Multiplication of $\mathbf{B}$ on the left by $\mathbf{A}$ multiplies the three rows by 3, -4, and 5, respectively. Similarly, multiplication on the right by $\mathbf{A}$ multiplies the columns of $\mathbf{B}$ by 3, -4, and 5, respectively.

33. (a) $\begin{bmatrix} 16 \\ 13 \\ 14 \end{bmatrix}$ → Art's wages on Monday   (b) $\begin{bmatrix} 18 \\ 14 \\ 16 \end{bmatrix}$ Each man's corresponding wages on Tuesday
 → Bob's wages on Monday
 → Curt's wages on Monday

 (c) $\begin{bmatrix} 7 & 9 & 3 \\ 9 & 3 & 4 \\ 8 & 5 & 4 \end{bmatrix}$ The combined output for Monday and Tuesday   (d) $\begin{bmatrix} 34 \\ 27 \\ 30 \end{bmatrix}$ Each man's combined wages for the two days

35. (a) $\mathbf{U + V} = \begin{bmatrix} u_1 + v_1 & u_2 + v_2 \\ -(u_2 + v_2) & u_1 + v_1 \end{bmatrix}$, $\mathbf{UV} = \begin{bmatrix} u_1 v_1 - u_2 v_2 & u_1 v_2 + u_2 v_1 \\ -(u_1 v_2 + u_2 v_1) & u_1 v_1 - u_2 v_2 \end{bmatrix}$
 (b) $\mathbf{I}^2 = \mathbf{I}$, $\mathbf{J}^2 = -\mathbf{I}$ (c) $\mathbf{U + V} = (u_1 + v_1)\mathbf{I} + (u_2 + v_2)\mathbf{J}$, $\mathbf{UV} = (u_1 v_1 - u_2 v_2)\mathbf{I} + (u_1 v_2 + u_2 v_1)\mathbf{J}$
 (d) This system of matrices behaves just like the complex nuumber system provided we identify $\mathbf{I}$ with 1 and $\mathbf{J}$ with i.

37. $\begin{bmatrix} 2 & 12 & 0 & 9 \\ 0 & 2 & 0 & 0 \\ 0 & 0 & 2 & 18 \\ 0 & 0 & 0 & 2 \end{bmatrix}$

PROBLEM SET 11-4   (Page 446)

1. $\begin{bmatrix} -1 & -3 \\ 1 & 2 \end{bmatrix}$ **3.** $\begin{bmatrix} \frac{1}{6} & \frac{7}{6} \\ 0 & \frac{1}{2} \end{bmatrix}$ **5.** $\begin{bmatrix} 1 & 0 \\ 0 & 1 \end{bmatrix}$ **7.** $\begin{bmatrix} 1/a & 0 \\ 0 & 1/b \end{bmatrix}$ **9.** $\begin{bmatrix} -2 & \frac{3}{2} \\ 1 & -\frac{1}{2} \end{bmatrix}$ **11.** $\begin{bmatrix} -\frac{4}{7} & \frac{2}{7} & \frac{3}{7} \\ \frac{6}{7} & -\frac{3}{7} & -\frac{1}{7} \\ \frac{5}{7} & \frac{1}{7} & -\frac{2}{7} \end{bmatrix}$

13. $\begin{bmatrix} -\frac{1}{9} & \frac{1}{9} & \frac{8}{9} \\ \frac{10}{9} & -\frac{1}{9} & -\frac{26}{9} \\ \frac{1}{9} & -\frac{1}{9} & \frac{1}{9} \end{bmatrix}$ **15.** $\begin{bmatrix} 1 & -1 & 2 & -\frac{5}{4} \\ 0 & \frac{1}{2} & -\frac{3}{2} & \frac{7}{8} \\ 0 & 0 & 1 & -\frac{3}{4} \\ 0 & 0 & 0 & \frac{1}{4} \end{bmatrix}$ **19.** $(\frac{5}{7}, \frac{10}{7}, -\frac{1}{7})$ **21.** $(\frac{47}{9}, -\frac{128}{9}, \frac{7}{9})$ **23.** Inverse does not exist.

25. $\begin{bmatrix} -4 & 3 & -4 \\ \frac{1}{2} & -\frac{1}{2} & 1 \\ 6 & -4 & 6 \end{bmatrix}$ **27.** $\begin{bmatrix} \frac{1}{2} & 0 & 0 \\ 0 & \frac{1}{3} & 0 \\ 0 & 0 & -\frac{1}{4} \end{bmatrix}$ **29.** $(-4a + 3b - 4c, \frac{1}{2}a - \frac{1}{2}b + c, 6a - 4b + 6c)$

33. (a) Twice the third row is added to the second row. The negative of twice the third row is added to the second row.
(b) They are multiplicative inverses.

35. $\begin{bmatrix} 2 & -1 & 0 & 0 \\ -1 & 0 & 1 & 0 \\ 0 & 1 & 0 & -1 \\ 0 & 0 & -1 & 1 \end{bmatrix}$ **37.** $\begin{bmatrix} 1 & a & b \\ 0 & 1 & c \\ 0 & 0 & 1 \end{bmatrix}^{-1} = \begin{bmatrix} 1 & -a & ac-b \\ 0 & 1 & -c \\ 0 & 0 & 1 \end{bmatrix}$ **39.** (a) $\begin{bmatrix} 5 \\ -1 \\ 0 \\ 0 \end{bmatrix}$ (b) $\begin{bmatrix} 8 \\ -17 \\ 14 \\ -4 \end{bmatrix}$ (c) $\begin{bmatrix} 32 \\ -79 \\ 68 \\ -20 \end{bmatrix}$

PROBLEM SET 11-5 (Page 453)

1. -8 **3.** 22 **5.** -50 **7.** 0 **9.** (a) 12 (b) -12 (c) 36 (d) 12 **11.** -30 **13.** -16
15. 7 **17.** 42.3582 **19.** $(2, -3)$ **21.** $(2, -2, 1)$ **23.** (a) 0 (b) 0 (c) 0 (d) 6 (e) 1 (f) 2 **25.** $2, \frac{3}{4}$
27. (a) $k \neq \pm 1$ (b) $k = 1$ (c) $k = -1$ **29.** (a) 144 (b) 96 (c) $\frac{1}{12}$ (d) -72 (e) 0
31. (a) -32 (b) $-\frac{1}{2}$ (c) -2 (d) -216

33.
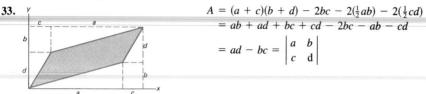

$$A = (a + c)(b + d) - 2bc - 2(\tfrac{1}{2}ab) - 2(\tfrac{1}{2}cd)$$
$$= ab + ad + bc + cd - 2bc - ab - cd$$
$$= ad - bc = \begin{vmatrix} a & b \\ c & d \end{vmatrix}$$

35. $P(a, b, c) = ab + ac + bc - a^2 - b^2 - c^2$ **37.** $-3.14, -0.48, 2.63$

PROBLEM SET 11-6 (Page 459)

1. -20 **3.** -1 **5.** 39 **7.** 4 **9.** 57 **11.** 6 **13.** -72 **15.** -960 **17.** $x = 2$ **19.** $aehj$
21. 156.8659 **23.** (a) -60 (b) $ef(ad - bc)$ **25.** 0 **29.** $D_2 = D_3 = D_4 = 1$ **31.** $x = -1, y = 3, z = -3, w = 1$
33. (a) $|C_2| = |C_3| = |C_4| = 0; |C_n| = 0$ for $n \geq 2$ (b) $|C_2| = (a_2 - a_1)(b_2 - b_1); |C_n| = 0$ for $n \geq 3$
35. (a) $4.629629626 \times 10^{-4}$ (b) $1.653439111 \times 10^{-7}$ (c) $3.749294762 \times 10^{-12}$

PROBLEM SET 11-7 (Page 467)

1. **3.** **5.**
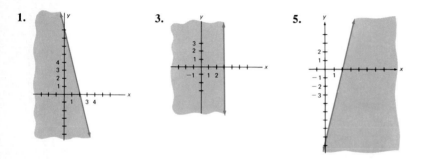

7.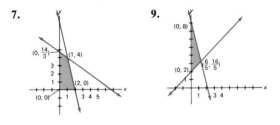

9.

11. Maximum value: 6; minimum value: 0   **13.** Maximum value: $-\frac{4}{5}$; minimum value: -8
15. Minimum value of 14 at (2, 2)   **17.** Minimum value of 4 at $(\frac{3}{2}, 1)$

19.   **21.**   **23.**

25. Maximum value of 4 at (2, 0); minimum value of -2 at (0, 2)   **27.** Maximum value of $\frac{11}{2}$ at $(\frac{9}{4}, \frac{13}{4})$; minimum value of 0 at (0, 0)
29. Maximum: 30; minimum: -2 **31.** 266   **33.** 2 camper units and 6 house trailers
35. 10 pounds of type A and 5 pounds of type B
37.

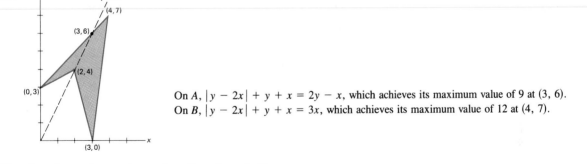

On A, $|y - 2x| + y + x = 2y - x$, which achieves its maximum value of 9 at (3, 6).
On B, $|y - 2x| + y + x = 3x$, which achieves its maximum value of 12 at (4, 7).

Therefore, the maximum value on the entire polygon is 12.
39. 23.13

CHAPTER 11. REVIEW PROBLEM SET   (Page 470)

1. T   **2.** F (if it has more than one solution, it has infinitely many)   **3.** T   **4.** F (the determinant is multiplied by 9)
5. T   **6.** T   **7.** T   **8.** F (this changes the sign of the determinant)   **9.** F (consider $P = 3x + y$ on the first quadrant)
10. T   **11.** (3, -1)   **12.** No solution   **13.** $(\frac{1}{2}, \frac{1}{3})$ **14.** (2, ± 3), (-2, ± 3)   **15.** (-4, 3, -2)   **16.** No solution
17. (2, -1, 3)   **18.** $(\frac{5}{7}z + 3, \frac{4}{7}z - 1, z)$ **19.** $(\sqrt{5}, \pm 2\sqrt{2})$, $(-\sqrt{5}, \pm 2\sqrt{2})$ **20.** $x = 0$, $y = \ln 4/\ln 3 \approx 1.26186$
21. $\begin{bmatrix} 4 & -1 & 10 \\ 8 & -1 & 13 \end{bmatrix}$ **22.** $\begin{bmatrix} -2 & 29 \\ 9 & 14 \end{bmatrix}$ **23.** $\begin{bmatrix} -16 \\ 11 \\ -27 \end{bmatrix}$ **24.** $\begin{bmatrix} 2 & 9 & -16 \\ 18 & -3 & 10 \\ 10 & -15 & 30 \end{bmatrix}$ **25.** $\begin{bmatrix} 5b & 2b \\ -3b & 4b \end{bmatrix}$ **26.** $\begin{bmatrix} 0 & 0 \\ 0 & 0 \end{bmatrix}$

27. $\begin{bmatrix} 5 & 3 \\ 2 & 4 \end{bmatrix}$ **28.** $\begin{bmatrix} 1 & -4 & 9 \\ 0 & 4 & 3 \\ 1 & -4 & 12 \end{bmatrix}$ **29.** $\begin{bmatrix} 5 & 1 & -4 \\ \frac{1}{2} & \frac{1}{2} & -\frac{1}{2} \\ -1 & 0 & 1 \end{bmatrix}$ **30.** (6, -1, -2)   **31.** -2 **32.** 0   **33.** 44

34. -10 **35.** 0   **36.** 56   **37.** 2   **38.** 15

39. Since the determinant of coefficients is nonzero, Cramer's rule implies that the system has a unique solution.   **40.** $\frac{1}{7}$

41.   **42.**

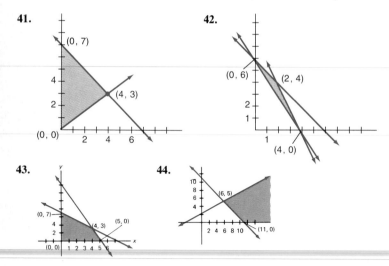

43.   **44.**

45. 11   **46.** 12   **47.** Maximum: 14; minimum: 0   **48.** Maximum: none; minimum: 22   **49.** (30, 30, 40)
50. A: 80; B: 100

PROBLEM SET 12-1 (Page 479)

1. (a) 9; 11   (b) 5; 2   (c) $\frac{1}{16}$; $\frac{1}{32}$ (d) 81; 121   **3.** (a) 11; 43   (b) $\frac{5}{6}$; $\frac{9}{10}$ (c) 49; 81   (d) -27; 81
5. (a) $a_n = 2n - 1$ (b) $b_n = 20 - 3n$ (c) $c_n = (\frac{1}{2})^{n-1}$ (d) $d_n = (2n - 1)^2$ **7.** (a) 14   (b) 162   (c) $\frac{1}{2}$ (d) 81
9. (a) $a_n = a_{n-1} + 2$ (b) $b_n = b_{n-1} - 3$ (c) $c_n = c_{n-1}/2$ (d) $d_n = d_{n-1} + 8(n - 1)$ **11.** 48   **13.** 42   **15.** 79   **17.** 69
19. (a) $\Sigma_{i=3}^{20} b_i$ (b) $\Sigma_{i=1}^{19} i^2$ (c) $\Sigma_{i=1}^{n} \frac{1}{7}$ **21.** (a) 50   (b) 7   (c) $\frac{25}{12}$ (d) 84   **23.** (a) 27; 82   (b) 20; 240
25. 4, 7, 14, 37, 112   **27.** $a_n = 3a_{n-1}$; $b_n = b_{n-1} + 4$; $c_n = c_{n-1} + 2n - 2$; $d_n = d_{n-1} + d_{n-2}$; $e_n = \frac{1}{2}e_{n-1}$; $f_n = f_{n-1} + 2n - 1$
29. 4, 2, 5   **31.** (a) 1, 3, 5, 7, 9   (b) 1, 4, 9, 16, 25   (c) $A_n = n^2$ **33.** (a) $\Sigma_{i=3}^{112} c_i$ (b) $\Sigma_{i=4}^{104} i^2$ (c) $\Sigma_{i=2}^{10} 6i$ **35.** $\Sigma_{i=1}^{15} i^2$
37. (a) $s_n = n^2$ (b) $s_n = s_{n-1} + 2n - 1$ (c) $t_n = t_{n-1} + n$ (d) $s_{n+1} = 2t_n + n + 1$ (e) $t_n = n(n + 1)/2$
39. 1, 1, 2, 3, 5; f_n **41.** $f_n f_{n+1}$

PROBLEM SET 12-2 (Page 485)

1. (a) 13; 16   (b) 3.2; 3.5   (c) 12; 8   **3.** (a) 3; 88   (b) .3; 10.7   (c) -4; -88 **5.** (a) 1335   (b) 190.5   (c) -900
7. .5   **9.** (a) 10,100   (b) 10,000   (c) 6633   **11.** 382.5   **13.** 3.8; 4.6; 5.4; 6.2   **15.** 11; 2772
17. (a) 1534   (b) -2300 (c) 15,350   (d) 9801   **19.** (a) $\Sigma_{i=3}^{99} i(i + 1)$ (b) $\Sigma_{i=2}^{20} 4i$ (c) $\Sigma_{i=2}^{30}(3i + 1)$
21. (a) $-.75$ (b) -17.5 (c) 63.75   **23.** 902   **25.** 375   **27.** (a) 10,200   (b) $-20,000$ **29.** 44,850   **31.** 2440
33. $1950   **35.** $6930   **37.** 2475   **39.** $5400\pi \approx 16,965$ inches

PROBLEM SET 12-3 (Page 492)

1. (a) 8; 16   (b) $\frac{1}{2}$; $\frac{1}{4}$ (c) .00003; .000003   **3.** (a) 2; 2^{n-2} (b) $\frac{1}{2}$; $8(\frac{1}{2})^{n-1} = 1/2^{n-4}$ (c) .1; $.3(.1)^{n-1} = 3 \times 10^{-n}$
5. (a) $2^{28} \approx 2.68 \times 10^8$ (b) $(\frac{1}{2})^{26} \approx 1.49 \times 10^{-8}$ (c) 3×10^{-30} **7.** (a) $\frac{31}{2}$ (b) $\frac{31}{2}$ (c) .33333
9. (a) $\frac{1}{2}(2^{30} - 1) \approx 5.3687 \times 10^8$ (b) $16(1 - 1/2^{30}) = 16 - (\frac{1}{2})^{26}$ (c) $.3[1 - (.1)^{30}]/.9 = \frac{1}{3}(1 - (.1)^{30})$ **11.** $100(2)^{10} = 102,400$

13. \$$(2^{31} - 1)$$, which is over \$2 billion **15.** (a) $\frac{1}{2}$ (b) $\frac{4}{15}$ **17.** 30 feet **19.** $\frac{1}{9}$ **21.** 25/99 **23.** 611/495
25. 625; 125; 25; 5; 1 **27.** $a_n = 130(\frac{1}{2})^{n-1}$, $c_n = 103 + 2n$, $d_n = 100(1.05)^n$, $f_n = 3(-2)^{n-1}$ **29.** \$215.89 **31.** \$1564.55
33. \$3 billion **35.** $3\sqrt{3}/14$ **37.** $(3^{32} - 1)/2 \approx 9.2651 \times 10^{14}$ **39.** $r^2 = 3$. The sum of the first 4 terms is 20.
41. 125 miles

PROBLEM SET 12-4 (Page 499)

Note: In the text, several proofs by mathematical induction are given in complete detail. To save space, we show only the key step here, namely, that P_{k+1} is true if P_k is true.
1. $(1 + 2 + \cdots + k) + (k + 1) = k(k + 1)/2 + k + 1 = [k(k + 1) + 2(k + 1)]/2 = (k + 1)(k + 2)/2$
3. $(3 + 7 + \cdots + (4k - 1)) + (4k + 3) = k(2k + 1) + (4k + 3) = 2k^2 + 5k + 3 = (k + 1)(2k + 3)$
5. $(1 \cdot 2 + 2 \cdot 3 + \cdots + k(k + 1)) + (k + 1)(k + 2) = \frac{1}{3}k(k + 1)(k + 2) + (k + 1)(k + 2) = \frac{1}{3}(k + 1)(k + 2)(k + 3)$
7. $(2 + 2^2 + \cdots + 2^k) + 2^{k+1} = 2(2^k - 1) + 2^{k+1} = 2^{k+1} - 2 + 2^{k+1} = 2(2^{k+1} - 1)$ **9.** P_n is true for $n \geq 8$. **11.** P_1 is true.
13. P_n is true whenever n is odd. **15.** P_n is true for every positive integer n.
17. P_n is true whenever n is a positive integer power of 4. **19.** $n = 4$. If $k + 5 < 2^k$, then $k + 6 < 2^k + 1 < 2^k + 2^k = 2^{k+1}$.
21. $n = 1$. Since $k + 1 < 10k$, $\log(k + 1) < 1 + \log k < 1 + k$.
23. $n = 1$. Multiply both sides of $(1 + x)^k \geq 1 + kx$ by $(1 + x)$: $(1 + x)^{k+1} \geq (1 + x)(1 + kx) = 1 + (k + 1)x + kx^2 > 1 + (k + 1)x$.
25. $x^{2k+2} - y^{2k+2} = x^{2k}(x^2 - y^2) + (x^{2k} - y^{2k})y^2$. Now $(x + y)$ is a factor of both $x^2 - y^2$ and $x^{2k} - y^{2k}$, the latter by assumption.
27. $(k + 1)^2 - (k + 1) = k^2 + 2k + 1 - k - 1 = (k^2 - k) + 2k$. Now 2 divides $k^2 - k$ by assumption and clearly divides $2k$.
29. (a) $(1 + 2 + 3 + \cdots + k) + (k + 1) = \frac{1}{2}k(k + 1) + k + 1 = (k + 1)(\frac{1}{2}k + 1) = \frac{1}{2}(k + 1)(k + 2)$
(c) $(1^3 + 2^3 + \cdots + k^3) + (k + 1)^3 = \frac{1}{4}k^2(k + 1)^2 + (k + 1)^3 = [(k + 1)^2/4][k^2 + 4(k + 1)] = \frac{1}{4}(k + 1)^2(k + 2)^2$
(d) $(1^4 + 2^4 + \cdots + k^4) + (k + 1)^4 = \frac{1}{30}k(k + 1)(6k^3 + 9k^2 + k - 1) + (k + 1)^4 =$
$[(k + 1)/30](6k^4 + 9k^3 + k^2 - k + 30k^3 + 90k^2 + 90k + 30) = [(k + 1)/30](6k^4 + 39k^3 + 91k^2 + 89k + 30) =$
$\frac{1}{30}(k + 1)(k + 2)(6k^3 + 27k^2 + 37k + 15) = \frac{1}{30}(k + 1)(k + 2)[6(k + 1)^3 + 9(k + 1)^2 + (k + 1) - 1]$
31. (a) 15,250 (b) 220 (c) 4355 (d) $2n(n + 1)^2$
33. $(1 - \frac{1}{4})(1 - \frac{1}{9}) \cdots (1 - 1/k^2)(1 - 1/(k + 1)^2) = [(k + 1)/2k][1 - 1/(k + 1)^2] = [(k + 1)/2k][(k^2 + 2k)/(k + 1)^2] = (k + 2)/2(k + 1)$
35. Let $S_k = 1/(k + 1) + 1/(k + 2) + \cdots + 1/2k$ and assume that $S_k > \frac{3}{5}$. Then
$S_{k+1} = 1/(k + 2) + 1/(k + 3) + \cdots + 1/2k + 1/(2k + 1) + 1/(2k + 2) =$
$S_k + 1/(2k + 1) + 1/(2k + 2) - 1/(k + 1) > S_k + 2/(2k + 2) - 1/(k + 1) = S_k > \frac{3}{5}$.
37. The statement is true when $n = 3$ since it asserts that the angles of a triangle have a sum of 180°. Now any $(k + 1)$-sided convex polygon can be dissected into a k-sided polygon and a triangle. Its angles add up to $(k - 2)180° + 180° = (k - 1)180°$.
39. Note that adding a line to n lines creates n new intersection points and thus $n + 1$ additional regions, so $R_{n+1} = R_n + n + 1$. We use induction next.
(i) $R_1 = 2 = (1^2 + 1 + 2)/2$.
(ii) Assume $R_n = (n^2 + n + 2)/2$. Then $R_{n+1} = R_n + n + 1 = \frac{1}{2}(n^2 + n + 2) + n + 1 = \frac{1}{2}(n^2 + 3n + 4) = \frac{1}{2}[(n + 1)^2 + n + 1 + 2]$.
41. $f_1^2 + f_2^2 + \cdots + f_k^2 + f_{k+1}^2 = f_k f_{k+1} + f_{k+1}^2 = f_{k+1}(f_k + f_{k+1}) = f_{k+1}f_{k+2}$
43. Assume the equality holds for a_k and a_{k+1}. Then $a_{k+2} = (a_k + a_{k+1})/2 = \frac{2}{3}[(1 - (-\frac{1}{2})^k + 1 - (-\frac{1}{2})^{k+1})/2] =$
$\frac{2}{3}[1 - \frac{1}{2}(-\frac{1}{2})^k - \frac{1}{2}(-\frac{1}{2})^{k+1}] = \frac{2}{3}[1 - (-\frac{1}{2})^{k+2}]$.

PROBLEM SET 12-5 (Page 506)

1. (a) 6 (b) 12 (c) 90 **3.** (a) 20 (b) 3024 (c) 720 **5.** 30 **7.** (a) 6 (b) 5 **9.** (a) 36 (b) 72 (c) 6
11. 1680 **13.** 25; 20 **15.** 240 **17.** 300 **19.** 1110 **21.** (a) 30 (b) 180 (c) 120 (d) 120 (e) 450
23. (a) $12 \cdot 11 \cdot 10 \cdot 9 \cdot 8 \cdot 7 \cdot 6 \cdot 5 \cdot 4$ (b) $2 \cdot 10 \cdot 9 \cdot 8 \cdot 7 \cdot 6 \cdot 5 \cdot 4 \cdot 3$ (c) $2 \cdot 11 \cdot 10 \cdot 9 \cdot 8 \cdot 7 \cdot 6 \cdot 5 \cdot 4$ **25.** 210 **27.** 34,650
29. 840 **31.** 126 **33.** (a) 990 (b) 165 (c) (8!)(989) (d) 630 (e) n (f) $(n + 2)/n$ **35.** (a) $_{10}P_{10} = 10!$ (b) 720
37. (a) 120 (b) 24 **39.** $1 - 1/(n + 1)!$ **41.** $8 \cdot 2 \cdot 10 \cdot 8 \cdot 8 \cdot 10(10^4 - 1)$ **43.** 3905 **45.** $5! = 120$
47. The number of games = the number of losers = $n - 1$

PROBLEM SET 12-6 (Page 513)

1. (a) 720 (b) 120 (c) 120 (d) 1 (e) 6 (f) 6   **3.** (a) 1140 (b) 161,700   **5.** 56   **7.** 495   **9.** 720
11. (a) $_{30}C_4 = 27{,}405$ (b) $_{25}C_3 = 2300$ **13.** $2 \cdot {_{10}C_2} = 90$ **15.** $_{12}C_{10} = {_{12}C_2} = 66$ **17.** 63
19. (a) 36 (b) 100 (c) 24   **21.** (a) 84 (b) 39 (c) 130   **23.** $_{26}C_{13}$ **25.** 4^{13} **27.** $_4C_2 \cdot {_{48}C_{11}} + {_4C_3} \cdot {_{48}C_{10}} + {_4C_4} \cdot {_{48}C_9}$
29. $_{52}C_5$ **31.** $_{13}C_2 \cdot {_4C_2} \cdot {_4C_2} \cdot 44$ **33.** 16   **35.** $_5C_2 \cdot {_4C_2} \cdot {_3C_2} = 180$ **37.** $12 \cdot 11 \cdot 10 \cdot {_9C_3}$ **39.** (a) 1024 (b) 210
41. 63   **43.** $_{10}C_3 + {_{10}C_4} + \cdots + {_{10}C_{10}} = 2^{10} - {_{10}C_2} - {_{10}C_1} - {_{10}C_0} = 968$
45. One way: $_nC_0 + {_nC_1} + {_nC_2} + \cdots + {_nC_n}$; the other way: $2 \cdot 2 \cdot 2 \cdots 2 = 2^n$
47. $_nC_j \cdot {_nC_j} = {_nC_j} \cdot {_nC_{n-j}}$. So the answer is the same as for Problem 46, namely, $_{2n}C_n$
49. $_{n+1}C_0 + S = \underbrace{_{n+1}C_0 + {_{n+1}C_1}}\; + {_{n+2}C_2} + {_{n+3}C_3} + \cdots + {_{n+k}C_k}$

$$= \underbrace{_{n+2}C_1 + {_{n+2}C_2}}\; + {_{n+3}C_3} + \cdots + {_{n+k}C_k}$$

$$= {_{n+3}C_2} + {_{n+3}C_3} + \cdots + {_{n+k}C_k}$$

$$= \cdots = {_{n+k}C_{k-1}} + {_{n+k}C_k} = {_{n+k+1}C_k}$$

51. The two horizontal lines and the two vertical lines determining the vertices of a rectangle can be chosen in $_{n+1}C_2$ times $_{n+1}C_2$ ways, giving us $[n(n+1)/2]^2$ as the answer.

PROBLEM SET 12-7 (Page 519)

1. $x^3 + 3x^2y + 3xy^2 + y^3$ **3.** $x^3 - 6x^2y + 12xy^2 - 8y^3$ **5.** $c^8 - 12c^6d^3 + 54c^4d^6 - 108c^2d^9 + 81d^{12}$
7. $a^5b^{10} - 5a^4b^9c + 10a^3b^8c^2 - 10a^2b^7c^3 + 5ab^6c^4 - b^5c^5$ **9.** $x^{20} + 20x^{19}y + 190x^{18}y^2$ **11.** $x^{20} + 20x^{14} + 190x^8$
13. $-120y^{14}z^9$ **15.** $-1760a^3b^9$ **17.** 23.424   **19.** 552.375   **21.** 149   **23.** (a) 16 (b) 32 (c) 64   **25.** 466
27. 4, 8   **29.** (a) $256x^8 + 512x^7 + 448x^6 + 224x^5 + 70x^4 + 14x^3 + \frac{7}{4}x^2 + \frac{1}{8}x + \frac{1}{256}$ (b) $208 + 120\sqrt{3}$ **31.** $-1792x^3z^{15}$
33. (a) nx^{n-1} (b) $10x^9 + 8x^3$ **35.** 55/9   **37.** $(1.01)^{50} > 1 + {_{50}C_1}(.01) = 1 + .5$ **39.** $2^{12} - 1 - {_{12}C_1} - {_{12}C_2} = 4017$
41. (a) $x^3 + y^3 + z^3 + 3x^2y + 3xy^2 + 3x^2z + 3xz^2 + 3y^2z + 3yz^2 + 6xyz$ (b) 420
43. $\sum_{k=0}^{n} {_nC_k}\, 2^k = {_nC_0} + {_nC_1} \cdot 2 + {_nC_2} \cdot 2^2 + \cdots + {_nC_n} \cdot 2^n = (1+2)^n = 3^n$ **45.** (a) 2^k (b) $2^{n+1} - 1$ (c) $2^{n+2} - n - 3$

PROBLEM SET 12-8 (Page 525)

1. (a) $\frac{1}{6}$ (b) $\frac{1}{2}$ (c) $\frac{1}{3}$ (d) $\frac{1}{2}$ (e) $\frac{1}{2}$ **3.** (a) $\frac{1}{8}$ (b) $\frac{3}{8}$ (c) $\frac{1}{2}$ **5.** (a) $\frac{1}{6}$ (b) $\frac{1}{6}$ (c) $\frac{5}{9}$ **7.** (a) 16 (b) $\frac{1}{8}$ (c) $\frac{13}{16}$
9. (a) States have different populations and so are not equally likely as a place of birth. (b) The two events are not disjoint; some people smoke and drink. (c) The two events are complementary; their probabilities should add to 1. (d) Ties are possible in football.
11. (a) $\frac{8}{25}$ (b) $\frac{5}{12}$ (c) $\frac{2}{5}$ (d) $\frac{8}{75}$ **13.** $\frac{1}{24}$ **15.** $\frac{9}{11}$ **17.** 1 to 3   **19.** (a) $\frac{1}{2}$ (b) $\frac{1}{4}$ (c) $\frac{1}{13}$
21. (a) $_{26}C_3/{_{52}C_3} \approx .118$ (b) $_{13}C_3/{_{52}C_3} \approx .013$ (c) $_4C_1 \cdot {_{48}C_2}/{_{52}C_3} \approx .204$ (d) $_4C_3/{_{52}C_3} \approx .0002$ **23.** $_8C_5/{_{52}C_5} \approx .00002$
25. (a) $\frac{1}{216}$ (b) $\frac{1}{36}$ (c) $\frac{53}{54}$ **27.** (a) $\frac{5}{13}$ (b) $\frac{27}{52}$ **29.** (a) $\frac{3}{7}$ (b) $\frac{3}{7}$ (c) $\frac{13}{14}$ **31.** (a) $\frac{1}{13}$ (b) $\frac{12}{13}$ (c) $\frac{57}{65}$
33. (a) $\frac{1}{35}$ (b) $\frac{4}{35}$ (c) $\frac{1}{7}$ **35.** (a) $\frac{1}{1296}$ (b) $\frac{1}{54}$ (c) $\frac{671}{1296}$ (d) $\frac{1}{216}$ **37.** 8/11!   **39.** (a) $\frac{4}{9}$ (b) $\frac{1}{3}$ **41.** $\frac{22}{56} = \frac{11}{28}$

PROBLEM SET 12-9 (Page 532)

1. 1/216   **3.** (a) $\frac{1}{8}$ (b) $\frac{3}{8}$ (c) $\frac{1}{8}$ (d) $\frac{5}{24}$ (e) $\frac{1}{2}$
5. (a) No. Most would agree that doing well in physics is heavily dependent on skill in mathematics. (b) Yes. Some might disagree, but we think these two events are quite unrelated. (c) Yes. New shirts and stubbed toes have nothing to do with each other. (d) No. Check the mathematical condition for independence. (e) No. It is more likely that a doctor is a man. (f) Yes. Let old superstitions die.
7. $(\frac{9}{10})^3 = .729$ **9.** $\frac{2}{5}$ **11.** $\frac{8}{25}$ **13.** (a) $\frac{1}{4}$ (b) $\frac{2}{9}$
15. (a) $(\frac{3}{10})^3 = .027$ (b) $(\frac{2}{10})^3 = .008$ (c) $.027 + .008 + .125 = .16$ (d) $6(\frac{2}{10})(\frac{3}{10})(\frac{5}{10}) = .18$ **17.** (a) $\frac{1}{5}$ (b) $\frac{1}{5}$
19. (a) .9 (b) .5 (c) .8   **21.** (a) .27 (b) .58 (c) $.27/.55 \approx .49$ **23.** (a) $\frac{1}{6}$ (b) $\frac{5}{6}$ **25.** $\frac{4}{9}$ **27.** (a) $\frac{25}{256}$ (b) $\frac{1}{12}$
29. (a) $\frac{9}{100}$ (b) $\frac{1}{5}$ (c) $\frac{19}{50}$ (d) $\frac{31}{50}$ **31.** $\frac{3}{4}$ **33.** (a) $1 - (.4)^8 \approx .9993$ (b) $8(.6)(.4)^7 + (.4)^8 \approx .0085$
35. Amy: $\frac{4}{7}$, Betty: $\frac{2}{7}$, Candy: $\frac{1}{7}$ **37.** $\frac{1}{3}$

CHAPTER 12. REVIEW PROBLEM SET (Page 536)

1. F (consider a, a, a, . . .)   **2.** F (this is what an explicit formula does)   **3.** T   **4.** T   **5.** T   **6.** T   **7.** F (2^{10})
8. T   **9.** T   **10.** F (disjoint means not simultaneous)   **11.** (b) and (c)   **12.** (a) and (d)
13. $d_n = d_{n-1} + 2\pi$, $e_n = e_{n-1} + e_{n-2} + e_{n-3}$ **14.** 101π **15.** 15,150   **16.** $31(\sqrt{2} + 1)$ **17.** $-\frac{2}{3}$ **18.** \$10,737,418.23
19. $\frac{20}{9}$ **20.** (a) 3725   (b) 2   **21.** 43   **22.** The second by \$339.85   **23.** The first by \$3663.22
24. (a) 1,037,836,800   (b) 286   **25.** 720   **26.** (a) 190   (b) 210   **27.** (a) 2520   (b) 420   **28.** 120
29. (a) 11! = 39,916,800   (b) $2 \cdot 10! = 7,257,600$ **30.** $_6C_2 \cdot 4 \cdot {}_5C_2 = 600$ **31.** $_{12}P_9 \cdot 9!$ **32.** $26^2 10^5 + 26^3 10^4 + 26^4 10^3$
33. 800   **36.** Not prime; $11^2 - 11 + 11 = 11 \cdot 11$ **37.** P_n is true for even n.
38. $x^6 + 12x^5y + 60x^4y^2 + 160x^3y^3 + 240x^2y^4 + 192xy^5 + 64y^6$ **39.** $3^6 = 279$ **40.** $-48,384x^5y^6$ **41.** 56
42. 1.000080002800056   **43.** 2^{25} **44.** (a) $\frac{1}{2}$ (b) $\frac{1}{5}$ (c) $\frac{3}{100}$ (d) $\frac{4}{5}$ **45.** $4/_{52}C_5$
46. (a) $_5C_2/_{18}C_2$ (b) $(_5C_2 + {}_6C_2 + {}_7C_2)/_{18}C_2$ **47.** (a) $(\frac{5}{18})^2 \approx .077$ (b) $(\frac{5}{18})^2 + (\frac{6}{18})^2 + (\frac{7}{18})^2 \approx .340$ **48.** .9
49. (a) .42   (b) .4   **50.** $\frac{3}{8}$

PROBLEM SET 13-1 (Page 543)

1. Upward   **3.** To the right   **5.** To the left   **7.** $24y = x^2$ **9.** $-12x = y^2$ **11.** $p = 2$

13. $p = \frac{1}{8}$ **15.** $p = \frac{1}{2}$ **17.** $p = \frac{9}{16}$

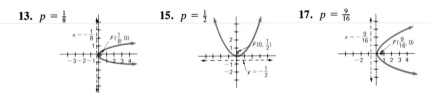

19. $(-\frac{1}{2}, 1)$; $(\frac{1}{2}, 1)$ **21.** $3x^2 = 2y$ **23.** $12x = y^2$ **25.** $y^2 = 4x$ **27.** At $(0, 4)$ **29.** 28.8 feet
31. $2\sqrt{5} \approx 4.47$ feet   **33.** $(-\frac{1}{5}, 0)$; $x = \frac{1}{5}$ **35.** $4p$ **37.** 2.5 feet   **39.** 225 meters   **41.** $y^2 = 4(p + r)x$
43. $8\sqrt{3}\, p$ **47.** $B(-3.10, 9.61)$, $C(7, 49)$, area: 123.22

PROBLEM SET 13-2 (Page 550)

1. Vertical; 8, $2\sqrt{7}$ **3.** Horizontal; 12, $4\sqrt{5}$ **5.** Horizontal; 2, $\frac{4}{3}$ **7.** Vertical; $2/k$, $1/k$

9. Horizontal ellipse; $a = 5$, $b = 3$, $c = 4$ **11.** Vertical ellipse; $a = 2$, $b = 1$, $c = \sqrt{3}$

13. $x^2/16 + y^2/25 = 1$; $e = \frac{3}{5}$ **15.** $x^2/49 + y^2/40 = 1$; $e = \frac{3}{7}$

17. $x^2/49 + y^2/4 = 1$; $e = 3\sqrt{5}/7$ **19.** $x^2/81 + y^2/9 = 1$; $e = 2\sqrt{2}/3$ **21.** $\frac{2}{3}\sqrt{2} \approx 0.943$

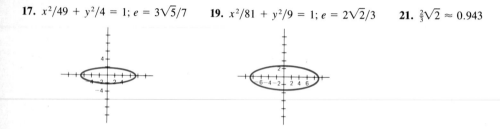

23. About 15,206,800 kilometers   **25.** $a = 5$, $b = 2$, $c = \sqrt{21}$, $e = \sqrt{21}/5$ **27.** $x^2/128 + y^2/144 = 1$ **29.** $5\sqrt{3}$ feet
31. 20,000 miles, $4000\sqrt{21}$ miles   **33.** $\pi\sqrt{77}$ **35.** (a) 80π square feet   (b) 176π square feet
37. (a) $\overline{PR} + \overline{RQ} = 2a$; $\overline{PR'} + \overline{R'Q} > 2a$ since R' is outside the ellipse.
(b) Let Q' be the mirror image of Q about the line l. Show that $\angle Q'RR' = \alpha$.

PROBLEM SET 13-3   (Page 556)

1. Horizontal; $a = 4$, $b = 6$, $c = 2\sqrt{13}$ **3.** Vertical; $a = 3$, $b = 4$, $c = 5$ **5.** Horizontal; $a = \frac{1}{2}$, $b = \frac{1}{4}$, $c = \sqrt{5}/4$
7. Horizontal; $a = 2$, $b = 4$, $c = 2\sqrt{5}$
9. Horizontal hyperbola; $a = 5$, $b = 3$, $c = \sqrt{34}$ **11.** Vertical hyperbola; $a = 8$, $b = 6$, $c = 10$

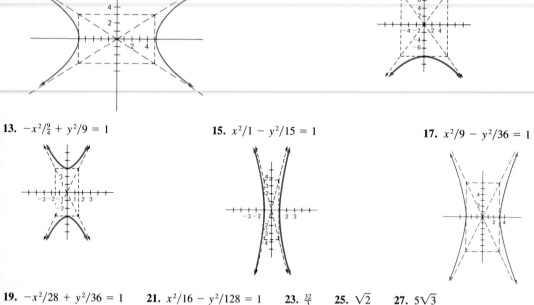

13. $-x^2/\frac{9}{4} + y^2/9 = 1$ **15.** $x^2/1 - y^2/15 = 1$ **17.** $x^2/9 - y^2/36 = 1$

19. $-x^2/28 + y^2/36 = 1$ **21.** $x^2/16 - y^2/128 = 1$ **23.** $\frac{32}{3}$ **25.** $\sqrt{2}$ **27.** $5\sqrt{3}$
29. $-x^2/3,630,000 + y^2/1,210,000 = 1$

PROBLEM SET 13-4   (Page 561)

1. $u^2 + 2v^2 = 2$; ellipse   **3.** $u^2 + v^2 = 1$; circle   **5.** $u^2 = 4v$; parabola   **7.** $u^2 + v^2 = 4$; circle
9. $u^2 + v^2 = 0$; a point   **11.** $u^2/4 - v^2 = 1$; hyperbola

13. 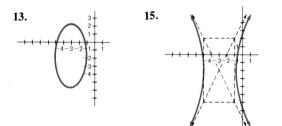 **15.** **17.** 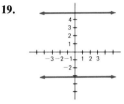 **19.**

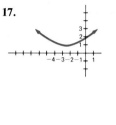

21. $(x + 2)^2 + (y - 1)^2 = 5$; circle: center $(-2, 1)$, radius $\sqrt{5}$ **23.** $(x - 2)^2/\frac{26}{4} + (y - 4)^2/26 = 1$; vertical ellipse: center $(2, 4)$
25. $4(x - 2)^2 + (y - 4)^2 = 0$; point: $(2, 4)$ **27.** $(x - 2)^2 = -\frac{1}{4}(y - 24)$; vertical parabola: vertex $(2, 24)$
29. $4(x - 2)^2 - 9(y - 1)^2 = 0$; two lines intersecting at $(2, 1)$ **31.**

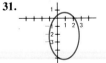

33. $-(x - 1)^2/4 + (y + 3)^2/9 = 1; 6$ **35.** Focus: $(21/20, 1)$; directrix: $x = -29/20$
37. (a)   (b)

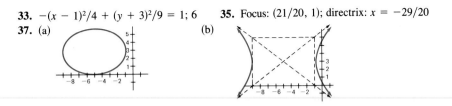

39. $a < 0$ (hyperbola); $a = 0$ (parabola); $a > 0$; $a \neq 1$ (ellipse); $a = 1$ (circle)
41. $-4(x - 4) = (y - 5)^2$ **43.** $x^2/8 + (y - 2)^2/4 = 1$ **45.** $-5(x - 3)^2/36 + 5(y + 4)^2/144 = 1$
47. (a) $y = x^2 - x$ (b) $x = \frac{1}{4}y^2 - y$ (c) $x^2 + y^2 - 5x - 5y = 0$ **49.** $\overline{ACB} = 10, \overline{CB} \approx 4.44$ **51.** $x^2 + (y + \frac{7}{3})^2 = \frac{625}{9}$
53. $x_1 = -2.731, x_2 = -1.671, x_3 = .964, x_4 = 3.438; x_1 + x_2 + x_3 + x_4 = 0$

PROBLEM SET 13-5   (Page 569)

1. $v = 0$ **3.** $4u^2 + v^2 = 16$ **5.** $u^2 + 2uv + v^2 - 8u + 8v = 0$ **7.** $u^2 + 3v^2 = 8$ **9.** $-2u^2 + 3v^2 = 6$
11. $3x^2 + 10xy + 3y^2 + 8 = 0$; $x = (\sqrt{2}/2)(u - v)$ and $y = (\sqrt{2}/2)(u + v)$; $-u^2 + v^2/4 = 1$

13. $4x^2 - 3xy = 18$; $x = (1/\sqrt{10})(u - 3v)$ and $y = (1/\sqrt{10})(3u + v)$; $-u^2/36 + v^2/4 = 1$

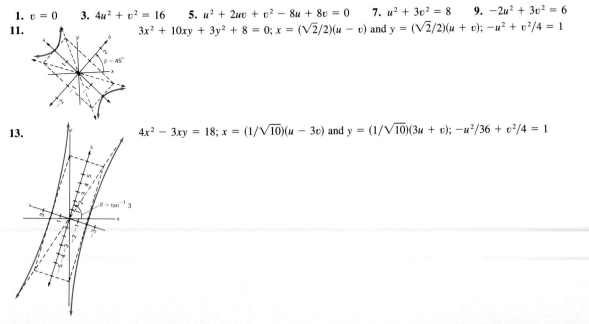

15.

$x^2 - 2\sqrt{3}\,xy + 3y^2 - 12\sqrt{3}\,x - 12y = 0$; $x = \frac{1}{2}(\sqrt{3}\,u - v)$ and $y = \frac{1}{2}(u + \sqrt{3}v)$; $v^2 = 6u$

17.

$13x^2 + 6\sqrt{3}\,xy + 7y^2 - 32 = 0$; $x = \frac{1}{2}(\sqrt{3}\,u - v)$ and $y = \frac{1}{2}(u + \sqrt{3}\,v)$; $u^2/2 + v^2/8 = 1$

19.

$9x^2 - 24xy + 16y^2 - 60x + 80y + 75 = 0$; $x = \frac{1}{5}(4u - 3v)$ and $y = \frac{1}{5}(3u + 4v)$; $v^2 + 4v + 3 = 0$

21. $u = (5 - 3\sqrt{3})/2$; $v = (-5\sqrt{3} - 3)/2$ **23.** $u = 4$; $v = -2$ **25.** $u = 5$; $v = 0$
27. $x^2 + 2\sqrt{3}\,xy + 3y^2 = 8(-\sqrt{3}\,x + y)$ **29.** $u^2/2 + v^2/10 = 1$; ellipse   **31.** $(u - 4)^2 + v^2 = 16$
33. $u^2/5 - v^2/21 = 1$; hyperbola   **35.** Equation transforms to $u = d$, a line $|d|$ units from the origin.   **37.** 3
41. $(-3.26, \pm 6.53)$, $(3.26, \pm 6.53)$

PROBLEM SET 13-6 (Page 576)

1. $2x + 3y = 17$ **3.** $2y = x^2 + 3x + 2$ **5.** $x^2 + y^2 = 4$ **7.** $x^2/4 + y^2/9 = 1$ **9.** $27y = x^3 - 3x^2 + 3x - 1$
11. $x = 2t$; $y = -3t$ **13.** $x = 1 + 3t$; $y = 2 - 7t$ **15.** -2; 1   **17.** $x = 5\cos t$; $y = \sqrt{3}\sin t$
19. $x = 5\cos t$; $y = 8\sin t$ **21.** $x = 4\sec t$; $y = 3\tan t$
23.

25.

27.

29.

31. (a) $y = -x^2/768 + x/\sqrt{3}$ (b) 4 (c) $256\sqrt{3}$ (d) 64
33. (a) $3x + 2y = 17$ (b) $y = 4\cos(x/3)$ (c) $x^2/4 - y^2/9 = 1$ (d) $8x + (y + 1)^3 = 8$ (e) $x = (x + y)^6 + 2(x + y)^3$
35. Show that $x^2 + y^2 = 4$ in each case. Verify that the parameter interval gives the same quarter circle in each case.
37. $(x - 2)^2/9 + (y - 1)^2/16 = 1$

39. (a) $y = (\tan \alpha)x - 16x^2/(v_0^2 \cos^2 \alpha)$ (b) $(v_0 \sin \alpha)/16$ (c) $(v_0 \cos \alpha)(v_0 \sin \alpha)/16 = (v_0^2 \sin 2\alpha)/32$ (d) $\pi/4$

41. Eliminate t to get $-x + 9y = 40$. Endpoints are $(-4, 4)$ and $(5, 5)$. Parameter interval $\pi/2$ to π gives same segment traced in the reverse direction.

47. (a)   (b)

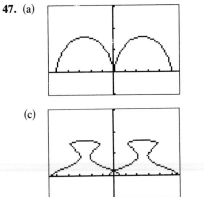 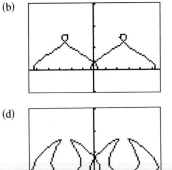

(c)   (d)

PROBLEM SET 13-7   (Page 583)

1–11.   **13.** $(2\sqrt{2}, 2\sqrt{2})$ **15.** $(-3, 0)$ **17.** $(-5, -5\sqrt{3})$ **19.** $(\sqrt{2}, -\sqrt{2})$

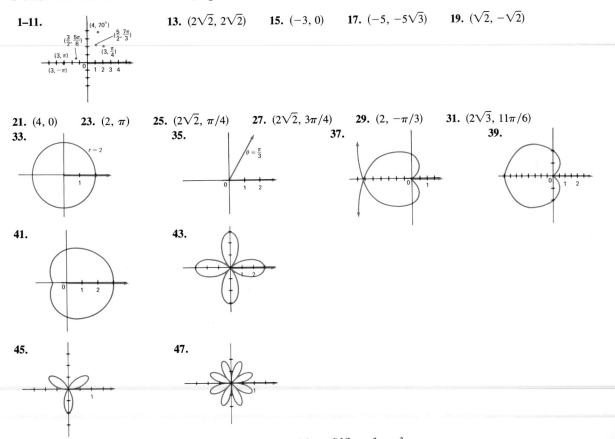

21. $(4, 0)$ **23.** $(2, \pi)$ **25.** $(2\sqrt{2}, \pi/4)$ **27.** $(2\sqrt{2}, 3\pi/4)$ **29.** $(2, -\pi/3)$ **31.** $(2\sqrt{3}, 11\pi/6)$

33.   **35.**   **37.**   **39.**

41.   **43.**

45.   **47.**

49. $r = 2$ **51.** $r = \tan \theta \sec \theta$ **53.** $y = 2x$ **55.** $(x^2 + y^2)^{3/2} = x^2 - y^2$

57. (a) $3y - 2x = 5$; a line.   (b) $(x - 2)^2 + (y + 3)^2 = 13$; a circle.

59.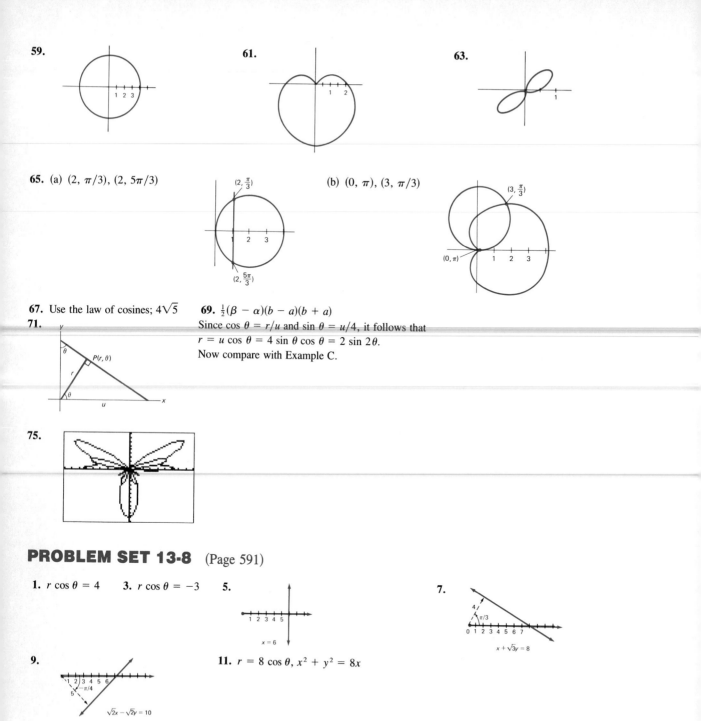

61.

63.

65. (a) $(2, \pi/3)$, $(2, 5\pi/3)$

$(2, \frac{\pi}{3})$

$(2, \frac{5\pi}{3})$

(b) $(0, \pi)$, $(3, \pi/3)$

$(3, \frac{\pi}{3})$

$(0, \pi)$

67. Use the law of cosines; $4\sqrt{5}$

69. $\frac{1}{2}(\beta - \alpha)(b - a)(b + a)$

71.

$P(r, \theta)$

Since $\cos \theta = r/u$ and $\sin \theta = u/4$, it follows that
$r = u \cos \theta = 4 \sin \theta \cos \theta = 2 \sin 2\theta$.
Now compare with Example C.

75.

PROBLEM SET 13-8 (Page 591)

1. $r \cos \theta = 4$ **3.** $r \cos \theta = -3$ **5.**

$x = 6$

7.

$\pi/3$

$x + \sqrt{3}y = 8$

9.

$-\pi/4$

$\sqrt{2}x - \sqrt{2}y = 10$

11. $r = 8 \cos \theta$, $x^2 + y^2 = 8x$

13. $r = 10 \cos (\theta - \pi/3)$; $x^2 + y^2 = 5x + 5\sqrt{3}\, y$ **15.** Ellipse; $e = \frac{2}{3}$; $x = 6$ **17.** Hyperbola; $e = 2$; $x = \frac{5}{4}$
19. Parabola; $e = 1$; $x = -7$ **21.** Ellipse; $e = \frac{2}{3}$; $x = -\frac{1}{2}$ **23.** Parabola; $e = 1$; $y = 5$ **25.** Ellipse; $e = \frac{1}{2}$; $y = -6$
27. Hyperbola; $e = \frac{5}{4}$; $y = \frac{8}{5}$

29.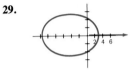

31.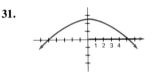

33.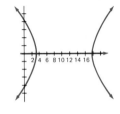

35. (a) $r = 3 \csc \theta$ (b) $r = 3$ (c) $r = -18 \cos \theta$ (d) $r = 6\sqrt{2} \cos (\theta - \pi/4)$
37. (a) $r = 2/(1 - \cos \theta)$ (b) $r = 4/(1 + \sin \theta)$ **39.** $(x + 3y)(x - 2y) = 0$ **41.** $d = 8$, $e = \frac{1}{2}$; $\frac{32}{3}$, $16\sqrt{3}/3$ **43.** 8
45. (a)   (b) $2a + a \cos \theta + 2a + a \cos(\theta + \pi) = 4a$

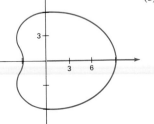

CHAPTER 13. REVIEW PROBLEM SET   (Page 593)

1. F (3 units)   **2.** T   **3.** T   **4.** F ($e = \sqrt{34}/3$)   **5.** T   **6.** T   **7.** F (for example, $B = 2$ gives intersecting lines)
8. T   **9.** F ($m = -2$)   **10.** T   **11.** (a) iii (b) v   (c) viii   (d) iv   (e) vii   (f) vi   (g) xi   (h) ii   (i) iii   (j) ix   (k) x
12. $V(-1, 2)$, $F(-1, -1)$ **13.** $20x = y^2$ **14.** $y = -x^2$ **15.** $16(y - 5) = (x - 3)^2$ **16.** 2.25 feet above the center
17. $V:(-2, -4)$ and $(-2, 6)$, $F:(-2, -3)$ and $(-2, 5)$ **18.** $x^2/16 + y^2/4 = 1$ **19.** $x^2/27 + y^2/36 = 1$
20. $x^2/36 + y^2/16 = 1$ **21.** $3\sqrt{7} \approx 7.94$ feet   **22.** $V(\pm 2, 0)$; $y = \pm\frac{5}{2}x$ **23.** $-x^2/45 + y^2/36 = 1$
24. $x^2/16 - y^2/4 = 1$ **25.** $x^2/25 - y^2/4 = 1$

26.   **27.**   **28.**

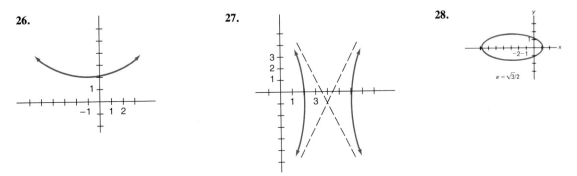

$e = \sqrt{3}/2$

29. Circle; center: $(7, -1)$; radius: 5   **30.** Vertical ellipse; center: $(5, -2)$; $a = 3$, $b = 2$ **31.** Two lines intersecting at $(3, 2)$
32. An ellipse; center: $(0, 0)$; $a = 6$, $b = 2$ **33.** $u^2 - v^2/12 = 1$; $2\sqrt{13}$ **34.** $3x + 4y + 7 = 0$ **35.** $x^2/16 + y^2/9 = 1$
36. $y = 3x^2 + x - 4$

37.

38. (a) $x^2 + y^2 = 4$ (b) $r = 2$ (c) $x = 2\cos\theta$; $y = 2\sin\theta$ **39.** (a) $(-1, \sqrt{3})$ (b) $(0, -4)$ (c) $(5, 5\sqrt{3})$

40. (a) $(4, \pi)$ (b) $(8, 3\pi/4)$ (c) $(6, 5\pi/3)$

41.

42.

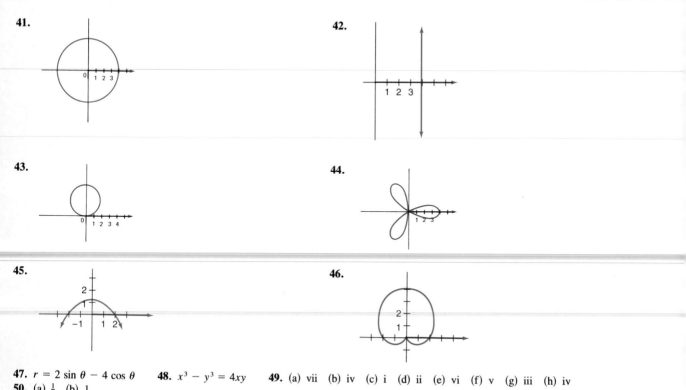

43.

44.

45.

46.

47. $r = 2\sin\theta - 4\cos\theta$ **48.** $x^3 - y^3 = 4xy$ **49.** (a) vii   (b) iv   (c) i   (d) ii   (e) vi   (f) v   (g) iii   (h) iv

50. (a) $\frac{1}{2}$ (b) 1

Credits for quotations in text: Page 1: Philip J. Davis, "Number," in *Scientific American* (Sept. 1964), p. 51. Page 2: Ronald W. Clark, *Einstein: The Life and Times* (New York: World Publishing Co., 1971), p. 12. Page 17: G. H. Hardy: *A Mathematician's Apology* (New York: Cambridge University Press, 1941), Page 49: Harry M. Davis, "Mathematical Machines," in *Scientific American* (April 1949), Page 99: George Polya, *Mathematical Discovery* (New York: John Wiley & Sons, 1962), p. 59. Page 105: Howard Eves, *An Introduction to the History of Mathematics* (New York: Holt, Rinehart & Winston, 1953), p. 32. Page 129: Morris Kline, *Mathematical Thought from Ancient Times* (New York: Oxford University Press, 1972), p. 302. Page 160: I. Bernard Cohen. "Isaac Newton," in *Scientific American* (Dec. 1955), p. 78. Page 180: Morris Kline, "Geometry," in *Scientific American* (Sept. 1964), p. 69. Page 539: Richard Courant and Herbert Robbins, *What is Mathematics?* (New York: Oxford University Press, 1978), p. 198.

Index
of Teaser Problems

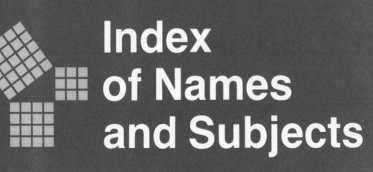

Index
of Names
and Subjects